**Current Topics in
Developmental Biology**

Volume 57

Development of Auditory and Vestibular Systems-3
Molecular Development of the Inner Ear

Series Editor

Gerald P. Schatten
Director, PITTSBURG DEVELOPMENTAL CENTER
Deputy Director, Magee-Women's Research Institute
Professor and Vice-Chair of Ob-Gyn Reproductive Sci. & Cell Biol.- Physiology
University of Pittsburgh School of Medicine
Pittsburg, PA 15213

Editorial Board

Peter Grüss
Max-Planck-Institute of Biophysical Chemistry
Göttingen, Germany

Philip Ingham
University of Sheffield, United Kingdom

Mary Lou King
University of Miami, Florida

Story C. Landis
National Institutes of Health
National Institute of Neurological Disorders and Stroke
Bethesda, Maryland

David R. McClay
Duke University, Durham, North Carolina

Yoshitaka Nagahama
National Institute for Basic Biology, Okazaki, Japan

Susan Strome
Indiana University, Bloomington, Indiana

Virginia Walbot
Stanford University, Palo Alto, California

Founding Editors

A. A. Moscona
Alberto Monroy

Current Topics in Developmental Biology

Volume 57

Development of Auditory and
Vestibular Systems-3
Molecular Development of the Inner Ear

Edited by

Raymond Romand
Institut de Génétique et de Biologie Moléculaire et Cellulaire
Université Louis Pasteur
67404-Illkirch, France

and

Isabel Varela-Nieto
Instituto de Investigaciones Biomédicas "Alberto Sols"
CSIC-UAM
Madrid, Spain

ELSEVIER
ACADEMIC
PRESS

AMSTERDAM • BOSTON • HEIDELBERG • LONDON
NEW YORK • OXFORD • PARIS • SAN DIEGO
SAN FRANCISCO • SINGAPORE • SYDNEY • TOKYO

Academic Press is an imprint of Elsevier

This book is printed on acid-free paper. ∞

Copyright © 2003, Elsevier Inc.

All Rights Reserved.
No part of this publication may be reproduced or transmitted in any form or by any means, electronic or mechanical, including photocopy, recording, or any information storage and retrieval system, without permission in writing from the Publisher.

The appearance of the code at the bottom of the first page of a chapter in this book indicates the Publisher's consent that copies of the chapter may be made for personal or internal use of specific clients. This consent is given on the condition, however, that the copier pay the stated per copy fee through the Copyright Clearance Center, Inc. (222 Rosewood Drive, Danvers, Massachusetts 01923), for copying beyond that permitted by Sections 107 or 108 of the U.S. Copyright Law. This consent does not extend to other kinds of copying, such as copying for general distribution, for advertising or promotional purposes, for creating new collective works, or for resale. Copy fees for pre-2003 chapters are as shown on the title pages. If no fee code appears on the title page, the copy fee is the same as for current chapters.
0070-2153/2003 $35.00

Permissions may be sought directly from Elsevier's Science & Technology Rights Department in Oxford, UK: phone: (+44) 1865 843830, fax: (+44) 1865 853333, e-mail: permissions@elsevier.com.uk. You may also complete your request on-line via the Elsevier homepage (http://elsevier.com), by selecting "Customer Support" and then "Obtaining Permissions."

Elsevier Academic Press.
525 B Street, Suite 1900, San Diego, California 92101-4495, USA
84 Theobald's Road, London WC1X 8RR, UK
http://www.academicpress.com

International Standard Book Number: 0-12-153157-0

PRINTED IN THE UNITED STATES OF AMERICA
03 04 05 06 07 08   9 8 7 6 5 4 3 2 1

Contents

Contributors xiii
Foreword xv
Introduction xvii

1

Molecular Conservation and Novelties in Vertebrate Ear Development
B. Fritzsch and K. W. Beisel

    I. Introduction   2
   II. Overview of Ideas Related to Ear Evolution   4
  III. Making the Ear: Implementing and Expanding
        Genes for Ear Morphogenesis   6
  IV. Evolution of the Ear: Molecular Origin of Mechanosensory
        Cells Predates Formation of the Ear   11
   V. Developmental Molecular Biology of
        Hair Cell and Sensory Neuron Formation   13
  VI. Evolution of Sensory Neurons: Heterochronic
        Alteration of HLH Gene Regulation   22
 VII. Guidance of Afferent Fibers: The Role of Hair
        Cells and Other Mechanisms Revisited   23
VIII. Survival of Afferents: Evolving a Novel Mechanism and
        Expanding it to Fit the Increasing Complexity of
        Ear Development   29
  IX. Splitting Hair Cell and Neuron Populations:
        Coevolving Sensory Epithelia and Their Innervation   33
   X. Summary and Conclusions: Evolving
        Developmental Mechanisms   34
        Acknowledgments   35
        References   35

2

Use of Mouse Genetics for Studying Inner Ear Development
Elizabeth Quint and Karen P. Steel

    I. Introduction: The Genetic Revolution   46
   II. Early Morphogenesis of the Inner Ear   50
  III. Development of the Neuroepithelium   55
  IV. Concluding Remarks   75
       References   75

3

Formation of the Outer and Middle Ear, Molecular Mechanisms
Moisés Mallo

    I. Introduction   85
   II. Basic Anatomical and Embryological Overview   86
  III. Genetic Determinants of Outer and Middle Ear Development   89
  IV. Concluding Remarks   107
       References   108

4

Molecular Basis of Inner Ear Induction
Stephen T. Brown, Kareen Martin, and Andrew K. Groves

     I. Introduction   116
    II. Inner Ear Induction in Fish   118
   III. Inner Ear Induction in Amphibians   127
   IV. Inner Ear Induction in Birds   128
    V. Inner Ear Induction in Mammals   131
   VI. Perspectives on Inner Ear Induction   135
  VII. Conclusions   141
       Acknowledgments   142
       References   142

5

Molecular Basis of Otic Commitment and Morphogenesis: A Role for Homeodomain-Containing Transcription Factors and Signaling Molecules
Eva Bober, Silke Rinkwitz, and Heike Herbrand

    I. Introduction   152
    II. Patterning of the Inner Ear Anlagen   152
    III. Transcription Factors and Diffusible Signals: A Complex Network   162
    IV. Patterning and Morphogenesis   165
    V. Conclusions   170
        Acknowledgments   171
        References   171

6

Growth Factors and Early Development of Otic Neurons: Interactions between Intrinsic and Extrinsic Signals
Berta Alsina, Fernando Giraldez, and Isabel Varela-Nieto

    I. Introduction   178
    II. Early Development of Cochlear (Auditory) and Vestibular Neurons   179
    III. Extrinsic Factors in Inner Ear Neurogenesis: Fibroblast Growth Factor, Nerve Growth Factor, and Insulin-Like Growth Factor-1 Families of Growth Factors   187
    IV. Conclusions   197
        Acknowledgments   197
        References   197

7

Neurotrophic Factors during Inner Ear Development
Ulla Pirvola and Jukka Ylikoski

    I. Neurotrophic Factors   208
    II. The Neurotrophin System   208
    III. The Glial Cell Line–Derived Neurotrophic Factor Family   217
    IV. Conclusions   219
        Acknowledgment   219
        References   219

8

FGF Signaling in Ear Development and Innervation
Tracy J. Wright and Suzanne L. Mansour

    I. Introduction   225
    II. Fibroblast Growth Factor Signaling in Ear Development   232
    III. Conclusions and Future Directions   251
        Acknowledgments   252
        References   252

9

The Roles of Retinoic Acid during Inner Ear Development
Raymond Romand

    I. Introduction   262
    II. Metabolism of Retinoids and Their Receptors   263
    III. Expression of Retinoic Acid Metabolic Enzymes and Receptors in the Developing Inner Ear   265
    IV. Is Retinoic Acid Involved in Otic Placode Induction?   269
    V. Retinoids as Morphogens during Early Embryogenesis   271
    VI. Retinoic Acid and Patterning Genes during Inner Ear Development   278

VII. Retinoic Acid and Hair Cell Differentiation   280
VIII. Concluding Remarks   282
   Acknowledgment   283
   References   283

10

Hair Cell Development in Higher Vertebrates
Wei-Qiang Gao

I. Morphogenesis of the Mammalian Inner Ear   294
II. Control of Hair Cell Differentiation by Specific Genes   294
III. Production and Regeneration of New Hair Cells in Mature Inner Ears   310
IV. Conclusion   312
   Acknowledgments   315
   References   315

11

Cell Adhesion Molecules during Inner Ear and Hair Cell Development, Including Notch and Its Ligands
Matthew W. Kelley

I. Introduction   322
II. Adhesion Molecules   323
III. Adhesion Molecules and Development of the Inner Ear   330
IV. Summary   345
   Acknowledgments   346
   References   346

12

Genes Controlling the Development of the Zebrafish Inner Ear and Hair Cells
Bruce B. Riley

    I. Introduction   358
   II. General Course of Zebrafish Otic Development   359
  III. Otic Induction   359
  IV. Patterning of the Placode and Early Vesicle   367
   V. Development of Sensory Epithelia   373
 VI. Auditory and Vestibular Function   378
VII. Conclusions and Prospects   380
       References   381

13

Functional Development of Hair Cells
Ruth Anne Eatock and Karen M. Hurley

    I. Introduction   390
   II. Maturation of Function   390
  III. Development of Mechanoelectrical Transduction   395
  IV. Developmental Acquisition of Basolateral Ion Channels   406
   V. Outer Hair Cell Electromotility   425
 VI. Interactions between Developing Hair Cells and Innervating Nerve Fibers   427
VII. Summary and Concluding Remarks   433
       Acknowledgments   434
       References   434

14

The Cell Cycle and the Development and Regeneration of Hair Cells
Allen F. Ryan

    I. Introduction   449
   II. The Cell Cycle and the Regulation of Cell Proliferation   450
  III. Hair Cell Development and Regeneration   453
  IV. Cell Cycle Events during Hair Cell Development   455
   V. The Cell Cycle during Hair Cell Regeneration   459
  VI. Conclusions   462
       Acknowledgments   462
       References   463

Index   467
Contents of Previous Volumes   483

Contributors

Numbers in parentheses indicate the pages on which the authors' contributions begin.

Berta Alsina (177), DCEXS-Universitat Pomepu Fabra, Dr Aiguader 80, 08003 Barcelona, Spain

Kirk W. Beisel (1), Creighton University, Department of Biomedical Sciences, Omaha, Nebraska 68178

Eva Bober (151), Institute of Physiological Chemistry, Martin-Luther University Halle-Wittenberg, Holly Strasse 1, D-06097 Halle, Germany

Stephen T. Brown (115), Gonda Department of Cell and Molecular Biology, House Ear Institute, 2100 West Third Street, Los Angeles, California 90057

Ruth Anne Eatock (389), The Bobby R. Alford Department of Otorhinolaryngology and Communicative Sciences, Baylor College of Medicine, Houston, Texas 77030

Bernd Fritzsch (1), Creighton University, Department of Biomedical Sciences, Omaha, Nebraska 68178

Wei-Qiang Gao (293), Department of Molecular Oncology Genetech, Inc., South San Francisco, California 94080

Fernando Giraldez (177), DCEXS-Universitat Pomepu Fabra, Dr Aiguader 80, 08003 Barcelona, Spain

Andrew K. Groves (115), Gonda Department of Cell and Molecular Biology, House Ear Institute, 2100 West Third Street, Los Angeles, California 90057

Heike Herbrand (151), Institute of Immunology, Medical School Hannover, Feodor-Lynen Strasse 21, D-30625 Hannover, Germany

Karen M. Hurley (389), The Bobby R. Alford Department of Otorhinolaryngology and Communicative Sciences, Baylor College of Medicine, Houston, Texas 77030

Matthew W. Kelley (321), Section on Developmental Neuroscience, National Institute on Deafness and Other Communication Disorders, National Institutes of Health, Rockville, Maryland, 20850

Moisés Mallo (85), Instituto Gulbenkian de Ciência, Rua da Quinta Grande 6, 2780–156 Oeiras, Portugal

Suzanne Mansour (225), Department of Human Genetics, University of Utah, Salt Lake City, Utah 84112

Kareen Martin (115), Gonda Department of Cell and Molecular Biology, House Ear Institute, 2100 West Third Street, Los Angeles, California 90057

Ulla Pirvola (207), Institute of Biotechnology and Department of Otaryngology, University of Helsinki, 00014 Helsinki, Finland

Elizabeth Quint (45), MRC Institute of Hearing Research, University Park, Nottingham NG7 2RD, United Kingdom

Bruce B. Riley (357), Biology Department, Texas A & M University, College Station Texas, 77843

Silke Rinkwitz (151), Institute of Biology and Environmental Sciences, Neurogenetics, Carl von Ossietzky University, Carl von Ossietzky Strasse 9–11, D-26129 Oldenberg, Germany

Raymond Romand (261), Institut Clinique de la Souris and Institut de Génétique et de Biologie, Moléculaire et Cellulaire, 67404 Illkirch Cedex, France

Allen F. Ryan (449), Departments of Surgery/Otolaryngology and Neurosciences, University of California San Diego School of Medicine and San Diego Veterans Administration Medical Center, La Jolla, California 92093

Karen P. Steel (45), MRC Institute of Hearing Research, University Park, Nottingham NG7 2RD, United Kingdom

Isabel Varela-Nieto (177), Instituto de Investigaciones Biomédicas "Alberto Sols", CSIC-UAM, Arturo Duperier 4, Madrid 28029, Spain

Tracy, J. Wright (225), Department of Human Genetics, University of Utah, Salt Lake City, Utah 84112

Jukka Ylikoski (207), Institute of Biotechnology and Department of Otaryngology, University of Helsinki, 00014 Helsinki, Finland

Foreword

The development of the inner ear has long been a source of fascination for biologists. The development of a highly sophisticated sensory organ from simple beginnings as a patch of ectoderm parallels the development of the entire embryo from similarly basic origins. Moreover, the accessibility of the developing inner ear, together with its clear visibility from early developmental stages has made it an attractive system for embryological study. The first decades of the twentieth century saw some of the greatest experimental embryologists such as Harrison and Waddington, together with many others, investigate the origins of the inner ear. After the 1960s, however, interest in ear development began to wane – partly perhaps because of the daunting morphological complexity of the inner ear, and also in part because the tools to further dissect the molecular mechanisms underlying its development were not yet available. The last twenty years has seen molecular biology rejuvenate the entire field of development, and this has led to a resurgence of interest in basic developmental questions in the auditory system. The present volume is both a timely summary of the newest and most exciting findings in the field of ear development, as well as an indicator of where the field is likely to go in the coming years.

Where is the field of ear development heading? One of the greatest challenges in the future will be to connect how gene expression controls the highly complex morphogenesis of the inner ear, and the roles played by cell division, cell movements and cell death in this process. As the inner ear has grown progressively more complicated in dierent vertebrate classes, comparative and evolutionary studies will also shed light on how dierent regions of the inner ear form. The question of how the auditory, vestibular and lateral line systems arose in chordates, and to what extent inner ear-like derivatives exist in primitive chordates today will remain a lively focus of investigation.

At present, much attention is focused on the developmental biology of sensory hair cells and neurons, although the mature inner ear also contains a legion of other specialized cell types that play crucial roles in both the auditory and vestibular systems, for example by regulating the composition of endolymph and perilymph. The formation of the endolymphatic duct and sac has clinical relevance to Meniere's disease, and the stria vascularis is perhaps second only to the organ of Corti in its complexity and degree of specialization. To date, very little is known about how these dierent components of the ear form, although the presence of melanocytes in the stria suggests an interesting interaction between the developing inner ear and neural crest.

The diverse variety of cell types in the mature inner ear suggests complicated lineage relationships during the formation of this organ. The use of intravital imaging in birds, fish and amphibians, retroviral lineage analysis, and recombination technology in mice is starting to reveal the lineage relationships between dierent cell types. The ear is somewhat unusual in undergoing a process of neurogenesis, followed by a second phase of sensory hair cell production. Recent evidence suggests that similar sets of molecules are involved in these processes – for example basic helix-loop-helix transcription factors and elements of the Notch pathway – and it will be of great interest to see if these cell types also share common cell lineages. The lineage relationships between dierent classes of hair cells (inner and outer hair cells in the mammalian organ of Corti and type I and II vestibular hair cells) are also unknown at present.

The use of dissociated cell culture systems has greatly advanced our understanding of how the central and peripheral nervous system lineages develop. Until recently, little work had been done to identify progenitor and stem cells from the ear *in vitro*, although the huge interest in stem cells in other parts of the body has led to a flurry of recent activity in the ear field. This important area is likely to be of increasing importance in the future.

The cochlea displays a gradient of frequency sensitivity along its length, such that hair cells at the base respond to high frequencies, whereas cells at the apex are most sensitive to high frequencies. This gradient is reflected in a gradient of hair cell morphology, although the mechanism by which this gradient is set up is completely unknown. Equally fascinating is how neurons project this tonotopic gradient onto the cells of the auditory nuclei in the hindbrain. The huge advances in neuronal mapping and pathfinding made in the invertebrate and vertebrate CNS in the last decade look certain to be applicable to the inner ear.

This book is a testament to the wonderful renaissance in the field of ear development that has taken place in the last decade. Importantly, these chapters highlight the need for combining cell and molecular biology with the more classical field of experimental embryology, and demonstrate the power of genetic, comparative anatomical and evolutionary approaches in a variety of species. While many problems remain to be solved, it is clear from the present volume that progress in ear development will be rapid, diverse and exciting, and something from which developmental biologists in other systems will have much to learn.

Marianne Bronner-Fraser

Introduction

The first edition of Development of Auditory and Vestibular System was published by Academic Press more than 20 years ago and focused on the postnatal development of the central and the peripheral stato-acoustic system. Despite the title, little work was devoted to the developmental processes that biologists would have expected to find. The book reflected rather the research trends and tools of the time including approaches such as electrophysiology, neuroanatomy, and cellular biology. The aim of the first edition was to present cutting edge auditory research, as well as to provide an overview of the past research upon which was built the basic knowledge of the development of the auditory and vestibular systems, with the goal of stimulating and helping young investigators.

The second edition had a similar goal and incorporated new advances, while information on truly developmental aspects of inner ear research which began to reveal itself within a few chapters.

During the past decade, important advances have modified our understanding of inner ear developmental biology. The field has witnessed continuous progress thanks to the understanding of the genes involved in inner ear formation, physiology and hearing disorders. Ear clinical research has taken advantage of genetics, imaging and novel therapies. Alternately, basic research has grown due to molecular and cellular biology techniques, the description of new biological markers and to novel animal models. Advances in basic research has opened new avenues for the understanding of the origins of inner ear pathology and for potential treatments, which should be enhanced by the intelligent integration of basic neurobiology with clinical sciences. At the same time, new, young researchers trained in developmental biology, molecular biology and genetics, joining the field use their experience to generate renewed and more general interest in inner ear developmental biology.

This present book, the third edition, presents a major dierence to the two previous ones since it focuses on inner ear development with a particular emphasis on genes, transcription factors and diusible factors that shape the inner ear. In the first chapter Bernd Fritzsch discusses the current growing interest on inner ear developmental evolutionary biology. This general introduction is followed by a review from Liz Quint and Karen Steel on the use of mouse genetics to generate animal models for the study of the molecular mechanisms underlying inner ear development and deafness. In the third chapter, Moses Mallo reviews dierent aspects of the formation of the outer and middle ear, very often neglected, as a part of sensory processing by developmental biologists. Inner ear development is very similar among vertebrates with the principal dierence being the timing of maturation and the complexity of the final structure. Inner ear development starts with the specification of the otic placode in the ectoderm in a process named induction that is discussed in detail in the chapter by Stephen Brown and his colleagues. Eva Bober and her colleagues contribute a complementary view of ear induction focussed on the participation of homeobox genes. The otic placode invaginates and pinches o the ectoderm to form the otic cup, which continues this morphogenetic process to form a closed, round-shaped structure: the otic vesicle, formed by a single layer of epithelial cells.

The otic vesicle is an autonomous structure that already contains most of the information required to build up the adult inner ear. Considering that the inner ear contains more than a dozen dierent cell types, the otic vesicle epithelial cells are able to generate an amazing diversity of cell types, including the neurones for the cochlear and vestibular ganglia that will innervate the adult organ. Information is coded in intrinsic and extrinsic molecules that co-ordinate a regular inner ear cell proliferation, fate, survival and differentiation. Among the extrinsic molecules with plasma membrane receptors involved in the control of inner ear development, the functions of the insulin-like growth factor-I system are described by Berta Alsina and her colleagues who describe early neurogenesis, the relationship with other diusible factors and the signalling mechanisms that underlie the cellular response and the specification and dierentiation of the cochleo-vestibular ganglia neurones. Ulla Pirvola and Jukka Ylikoski present an overview of the key roles of neurotrophins and glial-derived growth factor for ear development, whereas Tracy Wright with Suzan Mansour review the contribution of the family of fibroblast growth factors to inner ear ontogenesis and innervation. Finally, Raymond Romand delineates the importance of retinoic acid, a diffusible factor with its nuclear receptors, during inner ear ontogenesis.

Recent advances in the understanding of inner ear hair cell dierentiation and function in dierent species are presented over four chapters. The chapters by Wei Quiang Gao and Matthew Kelley review the intrinsic signalling molecules relevant for cell proliferation, dierentiation and survival. Bruce Riley discusses a fish model of hair cell development while Ruth Eatock and Karen Hurley contribute to a comprehensive review of the functional development of vestibular and cochlear hair cells. Deafness is a major health problem with a very high social impact. Up to 6 of the first world population suers from this disease and it will surely increase due to noise and other noxious environmental factors. Therefore, the study of the mechanisms involved in hair-cell formation, degeneration and regeneration together with the clues provided by animal models that, like the avian neuroepithelium, have the ability to regenerate throughout adult life, are fundamental to understand the molecular basis of regeneration and to search for potential treatments.

These aspects are illustrated in the above mentioned chapters and further addressed by Allen Ryan.

By undertaking this project our principal aim has been to serve the current needs of the field experts by reviewing present basic knowledge and pointing aspects in fast development as hair cells regeneration. Our intention is that the present book will serve as a basic reference for future research by encouraging the setting up of further studies of inner ear developmental processes, mechanisms underlying function, causes of disorders and novel therapies. Therefore, this book is aimed at a broad spectrum of ear researchers ranging from molecular and cellular biologists to neurobiologists, geneticists and clinicians.

We would like to thank the generous contribution made to this book by the following reviewers:

Dr. Karen B. Avraham
Dr. Daniel Choo
Dr. Pascal Dollé
Dr. Elisabeth Georges-Labouesse
Dr. Marlies Knipper
Dr. David Kohrman
Dr. Georey Manley
Dr. Felipe Moreno
Dr. Filippo Rijli
Dr. Thomas Schimmang
Dr. Bernard Thisse
Dr. Raphael Yehoash

R. Romand and I. Varela-Nieto

1

Molecular Conservation and Novelties in Vertebrate Ear Development

B. Fritzsch and K. W. Beisel
Creighton University
Department of Biomedical Sciences Omaha, Nebraska 68178

I. Introduction
II. Overview of Ideas Related to Ear Evolution
III. Making the Ear: Implementing and Expanding Genes for Morphogenesis
   A. Forkhead (Winged Helix) Genes
   B. GATA3
   C. The EYA/SIX/DACH Complex
   D. Fibroblast Growth Factors
IV. Evolution of the Ear: Molecular Origin of Mechanosensory Cells Predates Formation of the Ear
V. Developmental Molecular Biology of Hair Cell and Sensory Neuron Formation
   A. Invertebrate Sensory Cell Developmental Factors
   B. Vertebrate Hair Cell Development
VI. Evolution of Sensory Neurons: Heterochronic Alteration of HLH Gene Regulation
VII. Guidance of Afferent Fibers: The Role of Hair Cells and Other Mechanisms Revisited
   A. The Role of Hair Cells in Fiber Guidance and Survival
VIII. Survival of Afferents: Evolving a Novel Mechanism and Expanding it to Fit the Increasing Complexity of Ear Development
IX. Splitting Hair Cell and Neuron Populations: Coevolving Sensory Epithelia and Their Innervation
X. Summary and Conclusions: Evolving Developmental Mechanisms
   References

Evolution shaped the vertebrate ear into a complicated three-dimensional structure and positioned the sensory epithelia so that they can extract specific aspects of mechanical stimuli to govern vestibular and hearing-related responses of the whole organism. This information is conducted from the ear via specific neuronal connections to distinct areas of the hindbrain for proper processing. During development, the otic placode, a simple sheet of epidermal cells, transforms into a complicated system of ducts and recesses. This placode also generates the mechanoelectrical transducers, the hair cells, and sensory neurons of the vestibular and cochlear (spiral) ganglia of the ear. We argue that ear development can be broken down into dynamic processes that use a number of known and unknown genes to govern the formation of the three-dimensional labyrinth in an interactive fashion. Embedded in this process, but in large part independent of it, is an evolutionary conserved process that induces early the development of the neurosensory component of the ear. We present molecular data suggesting that this later process is, in its basic aspects, related to the mechanosensory cell formation across phyla and is extremely conserved at the molecular level. We suggest that sensory neuron development and maintenance are vertebrate or possibly chordate novelties and present the molecular data to support this notion. © 2003, Elsevier Inc.

I. Introduction

The evolution of the vertebrate ear has been an enigma ever since man started to think about it. In part, the problems are similar to the evolution of the vertebrate eye and relate to the fact that it is difficult to understand how a complex three-dimensional system that requires coordinated development of morphology (Fekete and Wu, 2002); nonsensory structures such as cupula and tectorial membranes (Goodyear and Richardson, 2002); and sensory structures, including neurons that connect the ear to the brain (Fritzsch *et al.*, 2002), could evolve in a stepwise fashion, with each step being functional and thus providing selective advantage. Although it is tempting to view the processes of morphogenesis and cell fate assignment as distinctly independent phenomena, in actuality they are intertwined and impact each other.

Formation of two- and three-dimensional patterns during morphogenesis may use two types of major mechanisms: morphostatic and morphodynamic (Salazar-Ciudad *et al.*, 2003). In morphostatic mechanisms cell fate assignment (termed herein "induction") occurs initially, followed by changes in tissue form. In contrast, morphodynamic mechanisms are more complex and simultaneously combine inductive and morphogenetic processes. Generally, morphodynamic mechanisms are likely to appear more often in later development as morphological innovations because they are less likely to disrupt global developmental processes at those stages (Riedl, 1978). Morphostatic mechanisms are usually observed at earlier developmental stages, which would have had more evolutionary time to change from a morphodynamic pattern to a morphostatic mechanism. In later stages of development morphodynamic mechanisms would be more often utilized and would permit already existing complex intermediate phenotypes to produce a wider range of variation and thus respond more easily to selective pressures.

The development of the eye offers an excellent paradigm for developmental mechanisms of sensory organs. In recent years the old notions about multiple parallel evolutions of vertebrate and invertebrate eyes have been reconsidered. In essence, the finding that the transcription factor Pax6 is highly conserved across phyla and is essential for the differentiation of the light-detecting organs in all species studied thus far (Pichaud and Desplan, 2002) has led to a revised perspective of eye evolution. It is now becoming clear that the evolution of developmental transcription factor interactions to generate retina sensors predates the evolution of morphologically distinct eyes and thus represents a highly conserved morphostatic mechanism. The apparent homology at the level of crucial eye transcription factor interactions can now be reconciled with the apparently independent evolution of morphogenetic pathways that generate rather diverse and morphologically distinct eyes in various phyla that are produced using a variety of morphodynamic mechanisms (Pichaud and Desplan, 2002). Basically, the current model of eye evolution suggests two molecularly distinct pathways: one dealing with eye morphogenesis (morphodynamic) and the other with eye histogenesis (morphostatic). Both the morphostatic and morphodynamic processes can be experimentally uncoupled using either teratogens, such as retinoic acid (RA) (Manns and Fritzsch, 1991), or mutations of the Pax6 gene (Pichaud and Desplan, 2002). Thus, eye evolution can provide a useful model for the evolution of ears (Fig. 1).

This chapter builds on this theme and presents the argument that both sensory organs evolved as a combination of two independent processes: an evolutionary conserved morphostatic mechanism involving a set of transcription factors that are essential for neurosensory development and the highly variable morphodynamic processes utilizing sets of transcription factors involved in ear morphogenesis. We base this argument on the evidence suggesting that

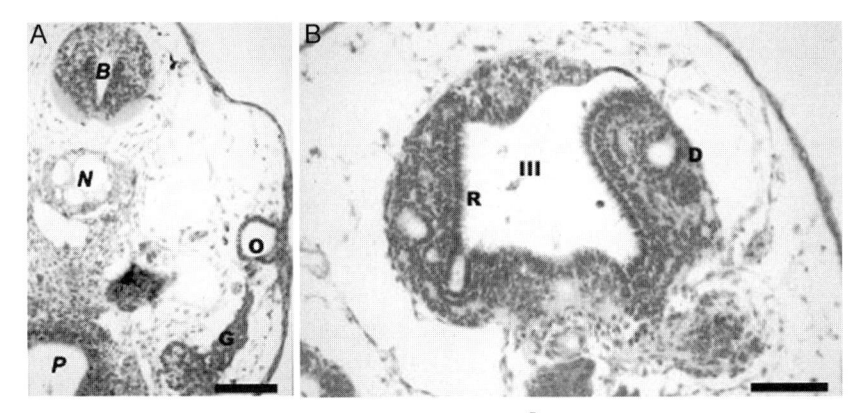

Figure 1 These pictures demonstrate the effect of 5×10^{-7} M retinoic treatment at stage 13 of *Xenopus laevis*. Animals were sacrificed at stage 40, embedded in paraffin, and serially sectioned. (A) Note that the ear is reduced to an otocyst vesicle (O) but nevertheless forms numerous ganglion neurons (G) that delaminate from this reduced otocyst. (B) As with the ear, the morphogenesis of the eyes can be completely blocked by RA treatment, leading to the formation of retina receptors (R) facing into the third ventricle (III) of the diencephalons (D). *N*, notochord; *P*, Pharynx; *B*, midbrain. Bar indicates 100 μm.

in both systems these two processes can be experimentally uncoupled (see Chapter 9 by Romand for additional details). The associated factors involved in cell fate assignment in these epithelia and the interrelationship between these two inner ear lineages are also presented. In addition, a brief discussion is presented on the conserved transcription regulatory factors found in the early stages of ear morphogenesis (i.e., the otic placode and the otocyst). This supports the notion of independent evolution of morphogenetic and neurosensory processes. The vertebrate ear has unique structures related to the mechanoelectrical transduction. These structures—the cupula and the otoconia and tectorial membranes, as well as areas of the ear involved in the unique formation of endolymph—are known to be related to genes expressed in the ear (Cowan *et al.*, 2000; El-Amraoui *et al.*, 2001; Goodyear and Richardson, 2002; Hulander *et al.*, 2003; Simmler *et al.*, 2000). These aspects are not dealt with here.

II. Overview of Ideas Related to Ear Evolution

Herein we integrate the concepts of the involvement of morphostatic mechanism(s) in inner ear cell fate assignment and morphodynamic mechanisms prevailing in morphogenesis in the development and evolution of the ear. There is now uniform agreement that the auditory part of the ear evolved from vestibular organs (Fritzsch, 1999; Wever, 1974). For example, it is likely that the auditory organ of the mammalian ear, the cochlea, evolved through the embryonic transformation of parts of the saccule (Fritzsch, 1992; Fritzsch *et al.*, 2002). In contrast to this universally accepted idea of evolutionary transformation of a vestibular organ into an auditory organ, there is no agreement on the evolution of the vestibular part of the ear. Basically, thinking on the evolution of the ear has revolved around two competing ideas.

One idea suggested a transformation of a preexisting lateral-line–like system through invagination into an ear (the octavolateralis hypothesis [Ayers, 1892]). An improved formulation (van Bergeijk, 1966) suggested that the lateral line and the ear develop from a single placode (the acousticolateral placode), share the same sensory receptors (hair cells), and share a common termination in the same central nuclei. Others have proposed that placodes undergo a stepwise refinement from a general placode system to a more specialized lateral line and ear placode system (Noramly and Grainger, 2002; Streit, 2002), but it is still unclear whether a common developmental program exists for the various sensory placodes (Begbie and Graham, 2001b; Groves and Bronner-Fraser, 2000). Still other work has shown that in many slowly developing vertebrates there are clear spatial and temporal differences in the development of lateral line and inner ear placodes, with the inner ear placode developing typically much earlier (Fritzsch *et al.*, 1998a). Moreover, the uncoupling of lateral line from ear development during evolution has resulted in the complete loss of lateral line placodes in terrestrial vertebrates (Fritzsch, 1999; Schlosser, 2002).

A second idea about ear evolution was based on the apparent morphological conservation of hearing organs across phyla, which implied similarities in function (hearing) that are retained in various forms of statocysts (Wever, 1974). Essentially it was assumed that the formation of an "otocyst" was conserved across phyla and its formation predated the evolution of vertebrates. This concept was revived and further refined by comparison between the atrial chambers of tunicate larvae and the vertebrate ear. Other than the general topology, the tunicate atrial chambers bear sensory organs that function as hydrodynamic sensors (Bone and Ryan, 1978; Mackie and Singla, 2003) and express the homolog of the vertebrate Pax2 gene (Favor *et al.*, 1996, Torres and Giraldez, 1998), the Hr-Pax258 gene (Wada *et al.*, 1998). However, the homologous AmphiPax2/5/8 differs in its expression in the nervous system from mammalian Pax2 (Kozmik *et al.*, 1999), and the presumed homology of the atrial chambers of tunicates with vertebrate ears remains in question.

Basically, these ideas revolve around the issue of whether sensory evolution came first followed by morphological evolution (octavolateralis hypothesis) or were both intertwined from the start (statocyst hypothesis). Our overriding concept is that evolution of the neurosensory system predates the evolution of the morphogenetic system that generates the ear and that evolution progresses from simple, via morphostatic processes, to complex as predicated by a variety of morphodynamic mechanisms. An evolutionarily conserved morphostatic mechanism has been proposed for the ear neurosensory components. This concept builds on the novel finding of a molecularly conserved gene that is essential for sensory cell formation across phyla, mammalian atonal homologue 1 (Aton) (Bermingham *et al.*, 1999; Fritzsch *et al.*, 2000; Wang *et al.*, 2002). The basis of this idea is that evolution of mechanosensory transducers is conserved molecularly to establish a transcription factor link to the still unknown mechanosensory transducer channels (Fritzsch and Biesel, 2001). Comparable to the octavolateralis hypothesis, this hypothesis assumes that the evolution of a mechanosensory transducer predates the evolution of a morphologically distinct vertebrate ear but does not require the prior evolution of a lateral line system. It thus circumvents some of the problems associated with the octavolateralis hypothesis while maintaining the basic idea that the evolution of mechanosensory transducers and their developmental molecular basis predates ear morphogenesis (Fritzsch *et al.*, 1998a; Fritzsch *et al.*, 2000). In line with this idea that ear neurosenory cell lineage formation predates morphogenesis are findings in the established outgroup of craniate vertebrates, the lancet

Amphioxus. This animal lacks an ear and a lateral line but expresses genes involved in chemosensory and mechanosensory cell formation (Holland et al., 2000; Kozmik et al., 1999; Shimeld and Holland, 2000).

III. Making the Ear: Implementing and Expanding Genes for Ear Morphogenesis

Evolution of the ear is to a large extent evolution of ear morphogenesis (Fig. 2). Logically, one would need to show expression of genes clearly involved in ear morphogenesis and their evolutionary changes to unravel the morphostatic or morphodynamic interactions of genes (Salazar-Ciudad et al., 2003). A number of genes essential for ear formation have been identified, as determined by the ear phenotype (lacking parts of the ear) in null mutant mice (see Chapter 2 by Quint and Steel for further discussion). We

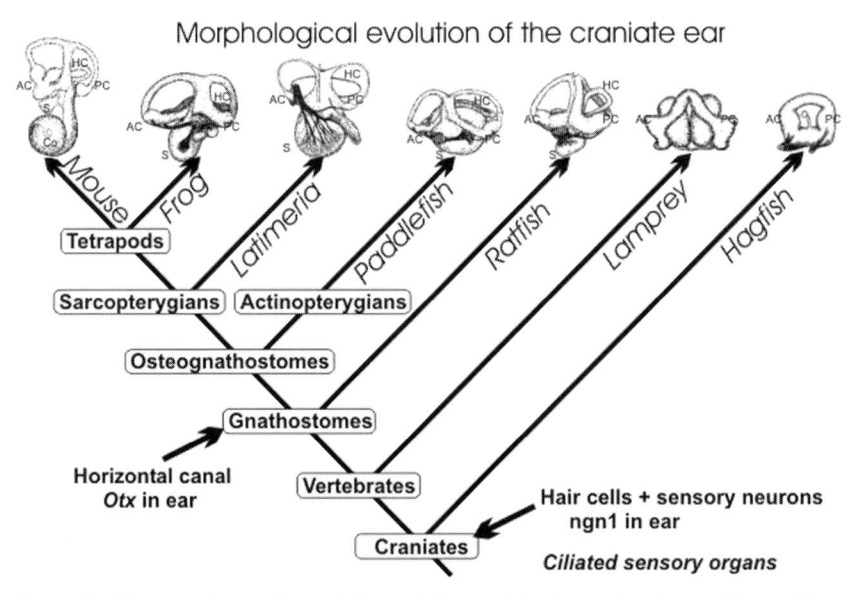

Figure 2 The morphogenetic evolution of the vertebrate ear is shown. How ciliated mechanosensory cells became associated with this morphogenetic process remains unclear. However, the living vertebrates display a morphocline of increasing complexity in the vestibular system that peaks with the nine distinct sensory receptors in some amphibians. Tetrapods, and possibly the coelacanth Latimeria, have evolved a basilar papilla in addition to the vestibular receptors. The major morphogenetic progress of the vestibular system was the split of the single torus of hagfish into two semicircular canals in lampreys and the formation of a horizontal canal in all gnathostomes (jawed vertebrates). It is likely that expression of Otx1 was essential for the evolution of the horizontal canal. *AC*, anterior crista; *Co*, cochlea; *HC*, horizontal canal and crista; *PC*, posterior crista; *S*, sacculus.

concentrate here on the morphostatic mechanisms in ear formation by discussing four genes, forkhead, Gata3, eye absent (Eya), and fibroblast growth factor (Fgf), and their implication for the evolution of a molecular network that governs ear morphogenesis. For more detail on other genes involved in ear morphogenesis and cell fate assignments see the review by Fekete and Wu (2002).

A. Forkhead (Winged Helix) Genes

Overall, forkhead genes are involved in numerous processes, such as patterning, morphogenesis, cell specification, and proliferation (Solomon *et al.*, 2003). In zebrafish, the forkhead/winged helix gene, Foxi1, appears to be the first currently known gene expressed in the otic placode and appears to be essential for otocyst formation (Solomon *et al.*, 2003). A forkhead gene expressed in the mammalian ear is BF1, now designated as Foxg1 (Hatini *et al.*, 1999), which is also found in the developing zebrafish ear (Toresson *et al.*, 1998). As indicated by its expression in sensory neurons and hair cells of the ear, the Foxg1 gene appears to play a role in histogenesis (Hatini *et al.*, 1999). Because it is expressed at a later point in the neurosensory development, Foxg1 may also be part of the morphodynamic mechanism. This suggestion is in line with the lack of Foxg1 expression in the epidermis in *Amphioxus* (Toresson *et al.*, 1998), as well as the lack of any other forkhead genes in the ectoderm (Yu *et al.*, 2002). Certainly Foxi1 may have an initial role in placodal induction and may be a component of the morphostatic cascade (see Chapter 4 by Brown *et al.* and Chapter 12 by Riley for further discussion). Additional data is needed to determine if forkhead genes were co-opted into ear formation as part of the morphostatic mechanism and if they also play a role in the morphodynamic cascade associated with later stages of neurosensory cell development.

Null mutant analyses have shown that the putatively homologous Foxi1 gene in mammals is not essential for ear formation. However, its absence causes ear dysmorphogenesis (Hulander *et al.*, 1998). More recent data suggest that the neonatal dysmorphogenesis is largely due to the lack of pendrin expression in later development (Hulander *et al.*, 2003), which causes endolymphatic hydrops and consequently dysmorphogenesis of the ear. The lack of embryonic dysmorphogenesis suggests either that another forkhead gene is present in the otocyst and provides biological redundancy or that Foxi1 is not part of the evolutionarily conserved morphostatic mechanism. Probably several more of the over 100 forkhead genes will eventually be found in the developing ear. Unlike the Foxi1 nulls, preliminary data on Foxg1 null mutant mice suggest severe phenotypic effects on specific sensory neurons and hair cells. In addition, the horizontal crista and its hair cells are absent (Fritzsch, unpublished data). Collectively, the apparent differences spatiotemporal expression and null mutation phenotypes of these two Fox genes suggest some redundancy of forkhead signaling in ear formation and also indicate that forkhead genes have been co-opted into the morphodynamic mechanisms.

B. GATA3

Like the forkhead genes, Gata3 is expressed at the level of the otic placode (George et al., 1994; Lawoko-Kerali et al., 2002) and has been found in the ear of zebrafish and mammals. Its orthologue, pannier, plays a role in sensory organ formation in insects (Sato and Saigo, 2000). Given the early expression of Gata3, it is not at all surprising to see influences on transcription factors regulating morphogenesis and histogenesis of the ear (Lawoko-Kerali et al., 2002). Targeted disruption of GATA3 leads to an arrest of ear development at the otocyst stage (Karis et al., 2001). GATA3 is the only early marker that identifies delaminating spiral sensory neurons (Karis et al., 2001; Lawoko-Kerali et al., 2002), but its role in spiral ganglion cell fate assignment is unclear. Based on pathfinding errors in inner ear efferents associated with the null phenotype (Karis et al., 2001), it is possible that GATA3 is involved in pathfinding and may have a minor role in morphodynamic processes observed later in development.

How GATA3 interacts with FGFs (Pauley et al., 2003; Pirvola et al., 2000), Nkx (Hadrys et al., 1998), Dlx (Merlo et al., 2002; Solomon and Fritz, 2002), forkheads (Solomon et al., 2003), and other genes involved in early ear histogenesis and morphogenesis (Chang et al., 2002; Fekete and Wu, 2002; Xu et al., 1999) is still uncertain (see Chapter 5 by Bober et al. and Chapter 8 by Wright and Mansour for additional details). However, multiple binding sites for GATA factors in the promoter regions of forkhead genes have been described (David et al., 1999) and are suggestive that GATA3 is part of a morphostatic cascade in the ear.

C. The EYA/SIX/DACH Complex

At early stages in the formation of the embryonic ear, involvement of four genes is observed in the otic vesicle and their presence is conserved across phyla (Noramly and Grainger, 2002). The interaction of these genes has been best documented in *Drosophila* eye development, where they form an evolutionarily conserved gene network consisting of paired-box (Pax)–eye absent (Eya)–sine oculis (SIX)–dachshund (DACH) genes (Hanson, 2001). This gene network is also observed in zebrafish, chicken, and mouse.

A fundamental role may be associated with the process of invagination. For example, this network is observed in involving hypoblast cells at gastrulation and in the forming otic and optic vesicles. It has been suggested that these genes permit cells to migrate without altering their cell fate commitment (Streit, 2002). By altering isoform usage and variation in coexpression patterns, this gene network can be co-opted into a wide variety of different morphogenetic contexts. Thus, these regulatory proteins are usually coexpressed throughout embryogenesis in a wide variety of cell types, tissues, and organs. Their expression in a given tissue does not necessarily imply its homology with other expressing tissues.

In order to understand their individual roles and predict their impact on downstream expression patterns, their protein interactions and functions must be understood. EYA, SIX, and DACH proteins directly interact to form a functional transcription factor. The DNA binding site is contained within SIX proteins, whereas EYA mediates transcriptional transactivation and contains SIX and DACH binding domains. Additional regulatory complexity is provided by DACH, which appears to function as a cofactor by directly interacting with EYA. A conserved expression pattern is found in the otic vesicle where similar isoforms representing these four gene families are found. In the mouse these genes are Pax2, Eya1 and 4, Six1 and 4, and both Dach isoforms (Davis *et al.*, 1999, 2001; Noramly and Grainger, 2002). For ear formation Eya1 and Six1 appear to be individually critical since null mutations in either Eya1 or Six1 do not affect the formation of the otic placode and vesicle, but there is no further morphogenesis (Xu *et al.*, 1997, 1999; Zheng *et al.*, 2003). Biological redundancy is suggested by the formation of the ear in spontaneous and null mutations of the Pax2 gene (Torres and Giraldez, 1998; Xu *et al.*, 1999) and by the absence of any phenotypic changes associated with the loss of the Dach1 gene function (Davis *et al.*, 2001). These data show that this gene network is crucial for ear morphogenesis. Interestingly, the haploid insufficiency observed in human and mouse Eya1 mutations suggests that alterations in gene dosage, protein levels, or function can affect morphogenesis (Abdelhak *et al.*, 1997a,b; Johnson *et al.*, 1999; Vincent *et al.*, 1997).

Similar to other genes implementing ear morphogenesis, these regulatory elements also play a role in histogenesis as indicated by their expression patterns in the developing ear. One example is the Eya4 gene, which is initially expressed in the otic vesicle (Borsani *et al.*, 1999; Wayne *et al.*, 2001). This isoform is present primarily in the upper epithelium of the cochlear duct in a region that develops Reissner's membrane and the stria vascularis. At E18.5 Eya4 is in areas of the cochlear duct destined to become spiral limbus, organ of Corti, and spiral prominence, with the highest level of expression occurring in the basal turn and in the early external auditory meatus. Diminishing levels of expression are found at later stages in these tissues and in the developing cochlear capsule during the period of ossification from birth to P14. In the vestibular system Eya4 is observed in the developing sensory epithelia. Interestingly, mutating Eya4 results in late onset deafness associated with the DFNA10 locus (Pfister *et al.*, 2002; Wayne *et al.*, 2001).

D. Fibroblast Growth Factors

FGFs are well known for their role in branching morphogenesis. However, they play a pivotal, but not necessarily conserved, role in ear formation and are likely important factors in the morphodynamic mechanisms of ear development (see Chapter 4 by Brown *et al.* and Chapter 8 by Wright and Mansour for further discussion). A significant morphogenetic variability can be contributed by the large number of isoforms in the Fgf gene family and their four different receptors. In chicken, *FGF3* and *FGF19* have been suggested for placode induction (Ladher *et al.*, 2000; Vendrell *et al.*, 2000). Other data also suggest a role for FGF2 and FGF8 in chicken ear morphogenesis (Adamska *et al.*, 2001). Zebrafish appears to have an interaction between *Fgf3* and *Fgf8* that is crucial for ear morphogenesis (Liu *et al.*, 2003; Phillips *et al.*, 2001). In mammals, early effects of Fgf8 have not been studied owing to early embryonic lethality, but Fgf8 is later expressed in the ear (Pirvola *et al.*, 2000, 2002). Fgf19 is not known to play any role in mammalian ear formation and an ear forms in Fgf3 null mutants (Mansour, 1994). Biological redundancy is associated with the FGF3, which signals in parallel with FGF10, FGF7, and FGF22 (Satou *et al.*, 2002) through the Fgfr2b receptor. In FGFR2b null mutants there is a loss of almost all ear morphogenesis (Pirvola *et al.*, 2000). Consistent with the signaling redundancy of FGFs through FGFRs, there appears to be a less severe effect of Fgf10 null mutation on ear morphogenesis (Pauley *et al.*, 2003). Most important is the effect on the vestibular canal growth that is consistent with the FGF's role in branching morphogenesis (Pauley *et al.*, 2003; Pirvola *et al.*, 2000). Other FGFs and FGF receptors appear to play a role in cellular differentiation of the mouse ear (Colvin *et al.*, 1996; Pirvola *et al.*, 2002). Modeling their molecular interactions (Davidson *et al.*, 2002) will require a much deeper knowledge of the genes, their qualitative and quantitative expression patterns, and their function in ear development.

In summary, all these genes play a role in both morphogenesis and histogenesis of the inner ear, as well as a survival role in the mature system. Their role in the morphostatic and/or morphodynamic mechanisms of ear development must be understood in terms of their spatiotemporal expression patterns before their complete function is ascertained. Dissection of their role in ear development must be approached by using conditional mutant mouse lines to understand the contextual role these regulatory genes are playing. In addition, it is becoming increasingly obvious that ear formation can be negatively affected by numerous genes, and working out their interaction will be paramount for any further understanding of ear development, evolution, and the underlying morphogenetic and histogenetic mechanisms. If taken at face value, the data also suggest limited conservation and possible changes in the role played by apparently orthologous genes across taxa. This raises the specter that the overall network of gene interactions might be conserved but that the interchangeable use of other members (isoforms) of the same gene family can differ among the vertebrate taxa. If true, this would mean that substitution of one isoform for another will alter the context of all other genes expressed in the ear. Thus, this could cause rapid molecular and morphological evolution of certain aspects of ear morphogenesis while keeping the overall formation and histogenesis of the ear largely unchanged, precisely what has been described in ear evolution (see Fig. 2). The task at hand will be to correlate such gene substitutions with major morphological alterations such as the formation of the mammalian cochlea and gnathostome horizontal canal.

IV. Evolution of the Ear: Molecular Origin of Mechanosensory Cells Predates Formation of the Ear

Hair cells are among the few unique vertebrate features not apparently shared with invertebrates (Jorgensen, 1989). Among the distinguishing features of vertebrate hair cells are the asymmetrical apical specializations, which lead to changes in the resting potential proportional to shearing forces acting on a mechanosensory ion channel (Strassmaier and Gillespie, 2002). In addition, up to 500,000 hair cells (Corwin, 1981) assemble into a large sensory epithelium in a complicated three-dimensional structure, the labyrinth (Lewis *et al.*, 1985). In contrast to most invertebrate sensory systems, vertebrate hair cells are connected to the central nervous system (CNS) via separate neurons grouped into distinct ganglia (Fritzsch, 1988b). Among invertebrates, such sensory cells without axons exist also in some cephalopods (Budelmann, 1992). Among cephalochordates, sensory cells without axons have been described for the lancelet, a likely vertebrate ancestor (Conway Morris, 2000) and ascidians (Burighel *et al.*, 2003). Sensory cells without axons and the generation of a separate set of neurons that connect those sensory cells to the brain might therefore be ancestral features of vertebrates and possibly chordates (Fig. 3) and are shared features of mechanosensory hair cells and taste receptors (Fritzsch *et al.*, 1998a).

A variety of mechanosensory cell types with axons are present in invertebrates, with the touch receptor of *Caenorhabditis* and the mechanosensory bristle of *Drosophila* (Caldwell and Eberl, 2002) representing possible

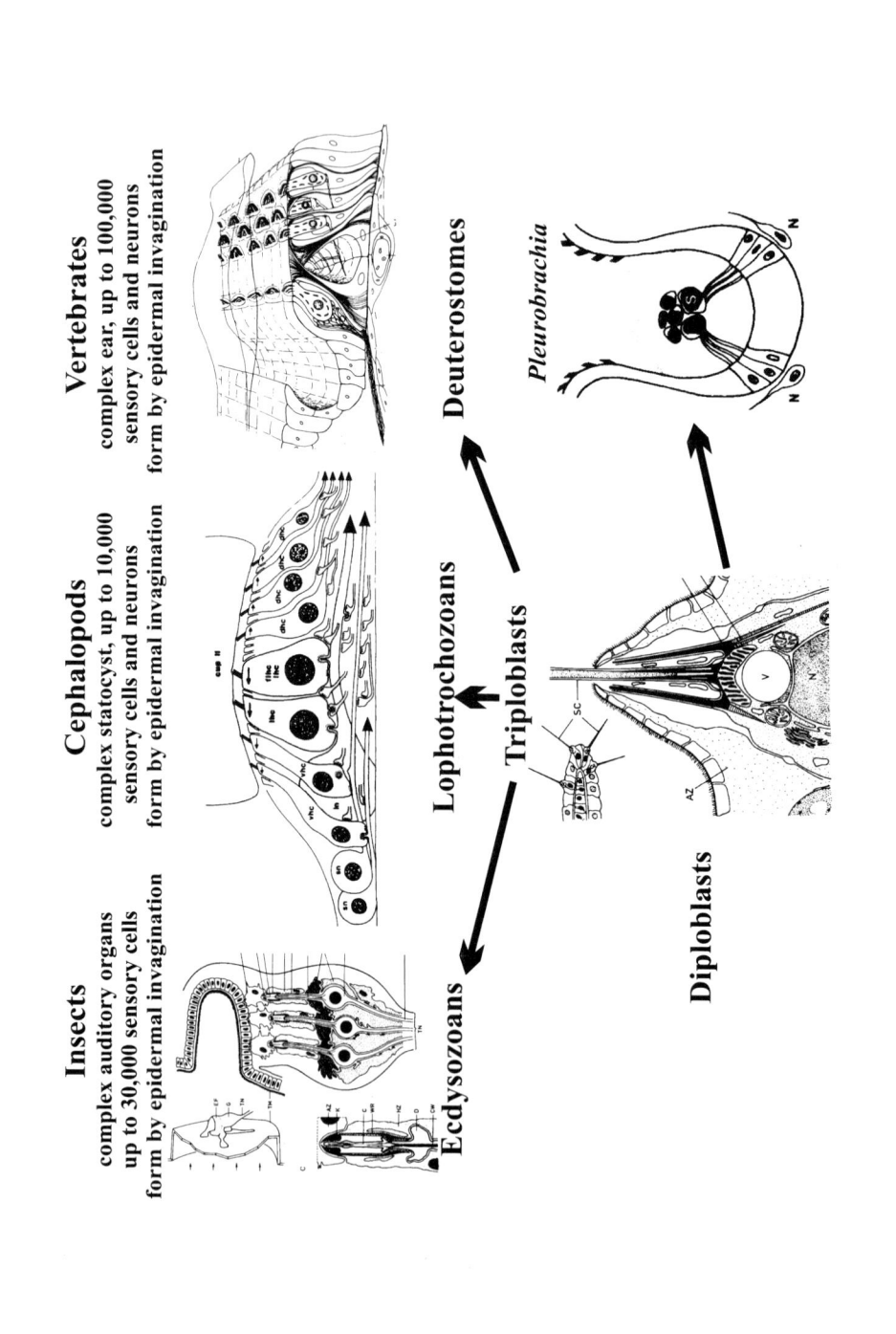

Insects
complex auditory organs up to 30,000 sensory cells form by epidermal invagination

Cephalopods
complex statocyst, up to 10,000 sensory cells and neurons form by epidermal invagination

Vertebrates
complex ear, up to 100,000 sensory cells and neurons form by epidermal invagination

Ecdysozoans

Lophotrochozoans

Deuterostomes

Triploblasts

Diploblasts

Pleurobrachia

prototypical mechanoelectric transducers comparable to those present in the hypothetical ancestor of all bilaterally symmetrical organisms. Moreover, the numbers of mechanosensory cells in some invertebrates (see Fig. 3) can approach or exceed the numbers reported for inner ear hair cells of vertebrates (Fritzsch and Biesel, 2001). Thus, with respect to neither the overall cellular specialization nor the number of individual cells in a given sensory organ are vertebrates unique among triploblastic organisms.

In the past the reconstruction of hair cell evolution has relied almost exclusively on comparison of adult and developmental features of extant invertebrates and chordates with vertebrates. Using the molecularly characterized developmental mechanisms outlined, we want to establish a plausible scenario for the continuity of interacting genetic networks, as well as for their modification, to achieve the unique vertebrate outcome. To obtain this, we need to compare early development of hair cells in vertebrates with invertebrate sensory cell development and analyze whether homologous genes are used to specify the different sets of cells needed to form a mechanosensory transducer organ in divergent phyla.

V. Developmental Molecular Biology of Hair Cell and Sensory Neuron Formation

Before we can discuss the evolutionary link between hair cell and sensory neuron development in the ear, we need to have a detailed understanding of the molecular basis for the formation of these neurosensory components in the vertebrate ear. Recent years have revealed the molecular basis for their formation using mainly the mouse as a model system. These data suggest the role of the following molecules and their interactions in the formation of these cells.

All neurons in vertebrates are derived from ectodermal cells that are transformed by a cascade of genes into neuronal precursor cells. Several genes have been identified that appear crucial for this switch in fate. These genes are referred to as "proneural genes" because of their apparent capacity

---

Figure 3 The evolution of sensory cells from simple single cells in the epidermis of diploblastic animals (bottom center) to complex multicellular organs in all three major radiations of triploblasts and in certain diploblasts is shown. Note that among ecdysozoans, insects have complex auditory receptors that consist of sensory cells with axons projecting to the central nervous system. Among lophotrochozoans, cephalopods have equally large statocysts with sensory cells with an axon, as well as sensory cells innervated by specialized sensory neurons. Vertebrates also have complex sensory organs that consist exclusively of sensory hair cells that are innervated by sensory neurons that derive from the same otocyst. bHLH genes that regulate sensory cell development in insects (atonal) and hair cell development in vertebrates (Aton1) can be experimentally interchanged. It remains to be seen, however, whether diploblasts have a conserved atonal-like bHLH gene, which may be the basis of the cell fate commitment toward a mechanosensory cell.

to transform ectodermal cells into neurons (Lee, 1997). They all belong to the growing family of basic helix-loop-helix (bHLH) genes that encode an ancient protein family with a highly conserved DNA binding domain (Bertrand et al., 2002). Proneural bHLH proteins form heterodimers with the ubiquitous E-proteins (i.e., the insect daughterless proteins) that enable them to bind to DNA and exert their function. These proteins have not only the unique capacity to turn ectodermal cells into neurons in gain-of-function experiments (Lee, 1997; Ma et al., 1996) but can also determine cell fate in rather unrelated tissues such as pancreas (Liu et al., 2000) or gut (Yang et al., 2001). They can also generate neuron-like cells such as Merkel cells (Bermingham et al., 2001).

Given this unique capacity of the proneural genes, it is obvious that these genes are tightly regulated in their spatiotemporal expression through interaction with a number of other transcript regulatory factors. Some of these factors interfere with the heterodimerization of bHLH proteins by binding to the E-proteins. These genes encode proteins that inhibit differentiation and are thus referred to as "inhibitors of differentiation" or Id genes. Others, such as the vertebrate hairy and enhancer of split paralogs, Hes/Her/Esr, act as classical DNA binding repressors of proneural gene transcription. The activation of the latter family appears to be regulated by the ubiquitously present delta–notch system, which negatively regulates the proneural commitment among neighboring cells. However, this so-called lateral inhibition requires the upregulation of bHLH genes in a limited number of cells to prompt the start of the delta–notch system. Most of the factors that drive this initial upregulation of proneural bHLH genes are still unknown, but the Zic genes are good candidates.

In general, the bHLH genes can be divided into three functional groups: true proneural bHLH genes that generate a neural lineage, bHLH genes that drive neural differentiation, and bHLH genes that drive the switch from neural to glial cell lineage (Bertrand et al., 2002, Zhou and Anderson, 2002).

Loss-of-function (targeted null mutations of the respective genes) experiments have clarified some of the proneural genes crucial for inner ear primary sensory neuron development (Fig. 4). The work of Ma and colleagues (Ma et al., 1998) showed that inner ear primary sensory neuron formation requires the vertebrate bHLH gene neurogenin 1 (Ngn1). Indeed, a follow-up study showed that no primary sensory neurons ever form in these mutants (Ma et al., 2000). Nevertheless, such ears develop fairly normally in their overall histology, suggesting that ear formation and even development of many hair cells is largely autonomous of innervation. Although those hair cells that do form are morphologically normal (except for some minor disorientation), hair cell numbers are reduced to various degrees in Ngn1 null mutant mice. Most interestingly, the cochlea is shortened and the saccule is almost completely lost. In addition, extra rows

1. Ear Development and Evolution

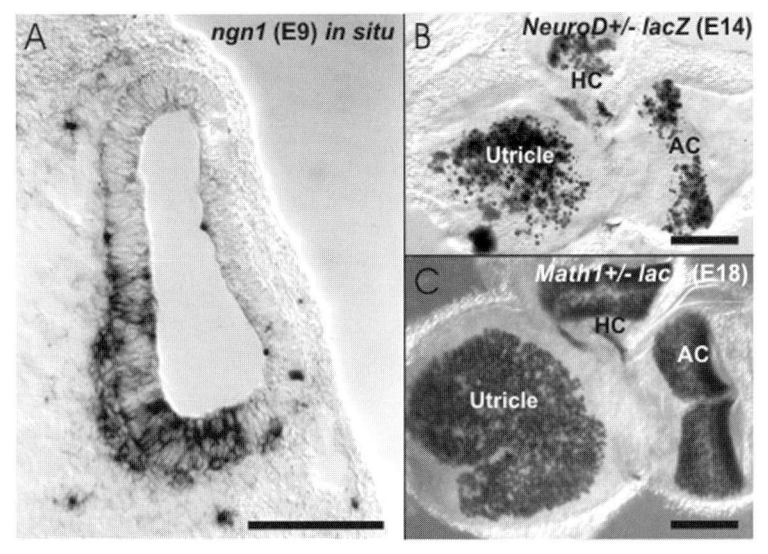

Figure 4 The three known bHLH genes that are essential for neurosensory development in the ear are shown. (A) The cross-sectioned otocyst of a E9 day-old mouse embryo shows the *in situ* expression of Ngn1 in the ventromedial aspect of the otocyst as well as in delaminating cells. (B) NeuroD is revealed with a LacZ reporter and shows a patchy distribution in many hair cells of the vestibular epithelia. (C) In contrast, Math1 is found in every hair cell of the vestibular and cochlear sensory epithelium at this developmental stage. Bar indicates 100 μm.

of hair cells form in the shortened cochlea (Fig. 5). These data suggest a significant interaction between progenitor cells that form primary neurons and progenitor cells that give rise to hair cells, supporting cells, and other inner ear epithelial cells. The simplest explanation would be a clonal relationship between primary sensory clones and hair cells and supporting cells (Fekete and Wu, 2002; Fritzsch *et al.*, 2001). However, other possible interactions cannot be excluded (Fritzsch *et al.*, 2002).

Other data have shown that hair cells require the expression of the bHLH gene Math1(mouse Aton1) for their formation (Bermingham *et al.*, 1999). However, Math1, the bHLH gene responsible for hair cell differentiation, does not function as a true proneural gene to establish a neural lineage in the mammalian brain (Bertrand *et al.*, 2002). In contrast to how it functions in most vertebrates, in mammals Aton functions only to select progenitor cells from a pool of already specified neuroepithelial stem cells (Bertrand *et al.*, 2002). In this context data suggest that the gene that establishes the hair cell lineage is still unknown (Chen *et al.*, 2002). Confirming and extending this conclusion are data on organ of Corti tissue culture showing that Math1 is not required for the initial expression of MyoVIIa, MyoVI, α9AchR, fimbrin, Brn3c, and other hair cell markers (Rivolta *et al.*, 2002).

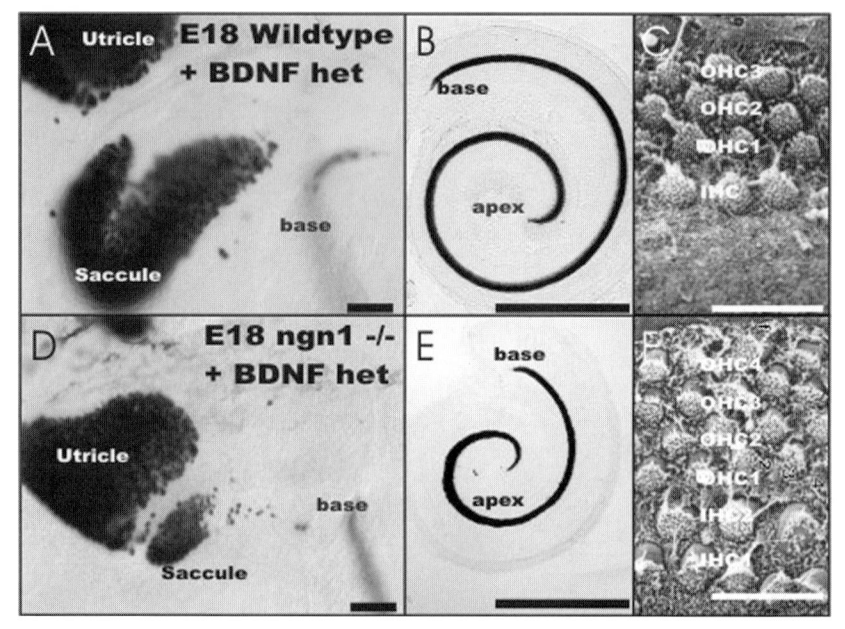

Figure 5 The effects of a null mutation of Ngn1 on the development of hair cells is shown. Using a BDNF LacZ reporter to label hair cells (A, B, D, E) obvious differences appear. The saccule (A, D) is almost completely lost in Ngn1 null mutant mice. The cochlea (B, E) is shortened and shows only a single turn. In addition to being shorter, the upper middle turn is wider and shows up to six rather than the usual four rows of hair cells (C, F). Bar indicates 100 μm (A, D), 1 mm (B, E), and 10 μm (C, F).

This conclusion is also supported by the precocious expression of the neurotrophin BDNF in the cochlea, prior to the upregulation of Math1, in what appears to be hair cell progenitors (Chen et al., 2002; Farinas et al., 2001). Preliminary data of our group suggest that BDNF is even expressed in hair cell progenitors in Math1 null mutants. Interestingly, some hair cells apparently require Math1 to drive BDNF expression, whereas other hair cells express BDNF even in the undifferentiated hair cell precursors that form in Math1 null mutants. However, expression of BDNF is in some but not all undifferentiated precursors of Math1 null mutants. This suggests that other genes also regulate BDNF expression in hair cell precursors and supports the conclusion of the Math1–LacZ analysis that hair cell precursor formation does not require Math1.

Either Ngn1 and/or an as yet unspecified proneural gene could play the role of inducing neuroepithelial commitment in the parts of the otocyst that give rise to Math1-mediated hair cell differentiation. As a consequence of the absence of primary sensory neuron formation in Ngn1 null mutants, the ear develops completely isolated from direct brainstem connections because

1. Ear Development and Evolution

afferents do not form and neither efferents nor autonomic fibers appear to reach the ear in these animals (Ma et al., 2000).

Absence of proneural genes in insects leads to the complete collapse of sensory organ formation (Caldwell and Eberl, 2002). We therefore analyzed double null mutants for both Ngn1 and Math1. If these were the only proneural bHLH genes that were independently expressed in the ear, one would expect severe consequences for ear morphology. However, our preliminary results on a single double-null mutant that also expresses BDNF–LacZ show that absence of both Ngn1 and Math1 is compatible with ear formation (Fig. 6). Either a third as yet undescribed bHLH gene is present in the mammalian ear or the morphogenesis of the ear is fully independent of the neurogenesis of sensory neurons and the formation of hair cells. The fact that RA treatment can block entirely the morphogenesis but still allow neurogenesis of the ear placode to proceed (Fritzsch et al., 1998a) would be compatible with this latter suggestion. Further studies on this subject are clearly warranted.

The next step in primary sensory neuron differentiation is apparently mediated by the bHLH genes of the NeuroD family (Kim et al., 2001; Lee, 1997). As is the case in other true proneural genes, Ngn1 is only transiently upregulated in primary neuron precursors. As primary sensory neuron precursors delaminate from the otocyst wall, the primary sensory neuroblasts downregulate Ngn1 and upregulate NeuroD (Kim et al., 2001; Liu et al., 2000; Ma et al., 2000). Based on data in the CNS, the presence of either Ngn1 or NeuroD (alternate symbol for Neurod1) seems to be involved in

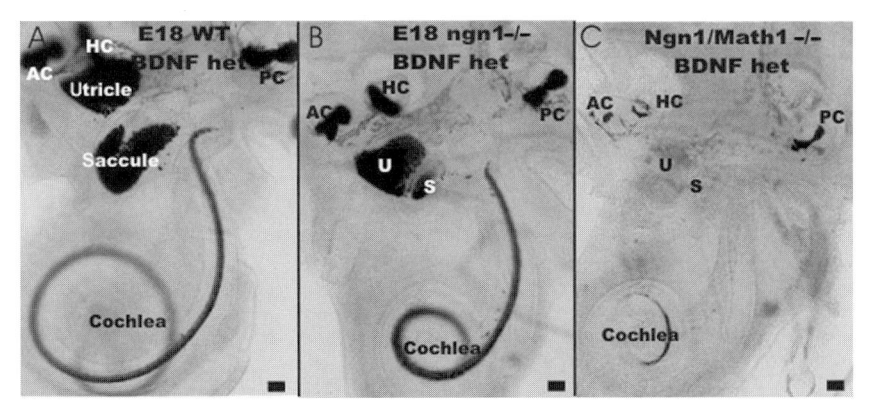

Figure 6 The effect of loss of both Ngn1 and Math1 on ear development and development of hair cell precursors is shown in three littermates. Note the absence of all sensory neurons and the reduction in size of the saccule and the cochlea in Ngn1 null mice (A, B). Combining Ngn1 null with Math1 null (C) is a simple combination of the Ngn1 phenotype (reduction in cochlea and saccule) with the Math1 null phenotype (absence of BDNF–LacZ-positive cells in the utricle, saccule, and basal turn of the cochlea). In contrast, the apex and the three canal cristae show BDNF–LacZ-positive cells even in the double-null mutants. Bar indicates 100 μm.

the continued proliferation of neuroblasts through interference with the cell cycle (Bertrand et al., 2002). Indeed, factors involved in regulating proliferation are only now becoming apparent in other developing systems such as muscle and olfactory epithelium (Wu et al., 2003). How many of these regulatory elements are utilized in the ear remains speculative at the moment.

It appears that NeuroD affects delamination of neuroblasts from the otocyst (Liu et al., 2000) and causes aberrant migration of surviving vestibular neurons in NeuroD null mutants (Kim et al., 2001). In NeuroD null mice the surviving neurons form disorganized projections to the cochlea and the utricle (Fig. 7). These resemble the vestibular projection defects known from mutation in the Pou domain factor, Pou4f1 (also designated as Brn3a) (Huang et al., 2001). It remains to be seen whether all the migration and projection defects in the NeuroD mutant can be attributed to the fact that NeuroD is upstream of Brn3a.

NeuroD null mutation differential affects the cochlear primary neurons (almost completely lost) more than the vestibular primary neurons (many survive, but they are dislocated and project aberrantly). This may relate to the fact that the acquisition of specific neuronal fates requires interaction with homeobox genes such as Pax genes (Lawoko-Kerali et al., 2002). In contrast to the vestibular primary sensory neurons, cochlea (spiral) primary sensory neurons express Gata3 (Karis et al., 2001; Lawoko-Kerali et al., 2002). GATA factors, including pannier and presumably its vertebrate homologue GATA3 (Bertrand et al., 2002; Karis et al., 2001), can interact with bHLH dimers for transcriptional regulation and may be directly involved in specific aspects of cochlear and vestibular fate determination via regulation of bHLH gene transcription. Unfortunately, most Gata3 null mutants result in embryonic lethality before hair cell differentiation is achieved. Nevertheless, it is important to note that a human *GATA3* mutation exists and causes deafness (Van Esch et al., 2000).

In summary, the genes that regulate formation of the neurosensory aspects of ear development are beginning to emerge and their function has been tested in specific null mutant mice. Although the beginning of this line of research has provided dramatic breakthroughs in our understanding of this process, numerous open questions remain before this research can be extended to guide aspects of neurosensory regeneration that would benefit humans with neurosensory hearing loss.

A. Invertebrate Sensory Cell Developmental Factors

Drosophila sensory organ development requires a set of genes that have been shown to be conserved in vertebrates. The early specification of position is influenced by a number of genes such as zinc finger and TGFβ-related genes

1. Ear Development and Evolution   19

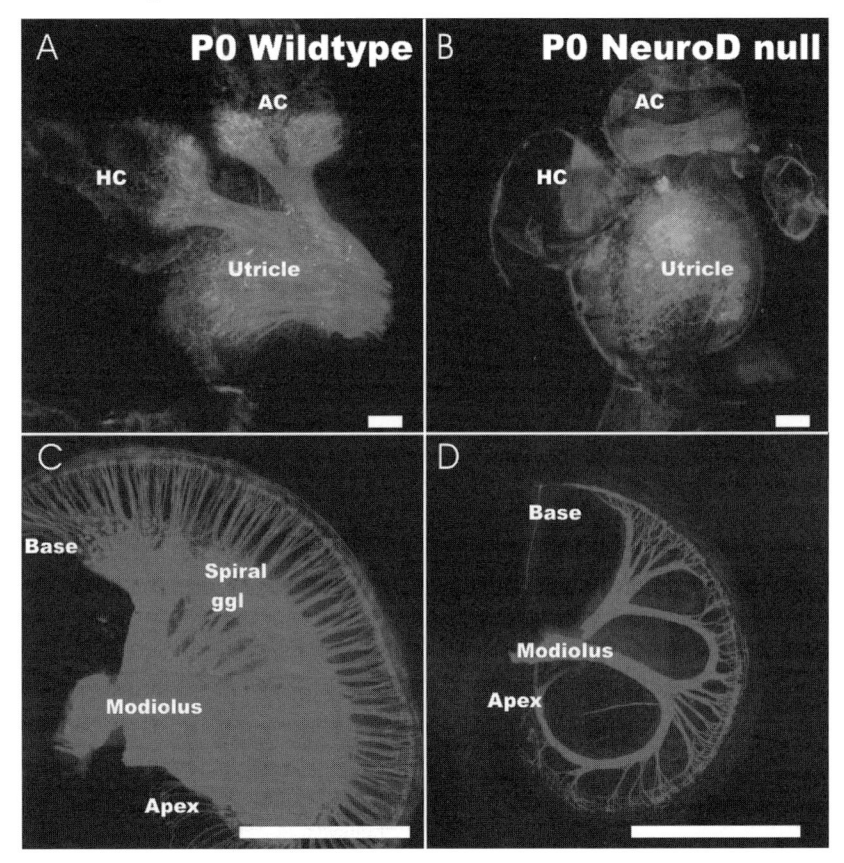

Figure 7 The effect of a NeuroD null mutation on the pattern of innervation of the cochlea and vestibular system is revealed with DiI tracing and confocal microscopy. All vestibular and cochlear epithelia show hair cells, but some hair cells have no innervation. In the vestibular system, the utricle shows a somewhat disorganized innervation in the NeuroD mutant (B) but may lack all innervation of canal cristae (A, B). The cochlea shows a reduction in size and a loss of most spiral sensory neurons (C, D). The remaining neurons expand their peripheral processes to innervate most inner hair cells and a few outer hair cells. Bar indicates 1 mm (C, D) and 100 μm (A, B). (See Color Insert.)

(Caldwell and Eberl, 2002). Homologues of these genes exist in vertebrates and their role in ear development is considerable (Brigande *et al.*, 2000; Cantos *et al.*, 2000; Fekete and Wu, 2002; Karis *et al.*, 2001). These genes may regulate expression of various bHLH genes such as achaete/scute in external sensilla and atonal in chordotonal organs. In addition, in vertebrates Bmp4 and its antagonists play a role in patterning of prepatterning gene expression (Bally-Cuif and Hammerschmidt, 2003). Such genes are crucial for overall development of sensory elements in both vertebrates and invertebrates by setting up proneuronal clusters and by controlling their cell cycle.

Further development of these proneuronal clusters requires a set of intrinsic signaling genes that apparently act in a stereotyped fashion, are highly conserved across evolution, but have increased in the number of genes between unicellular and multicellular organisms (Venter et al., 2001). One cell, the sensory organ precursor cell, strongly expresses a bHLH gene. This bHLH gene expression upregulates Delta gene expression. DELTA downregulates prosensory bHLH gene expression in adjacent cells through its receptor, NOTCH. In fact, Delta/Notch expression can cause irreversible loss of neurogenic capacity (Morrison et al., 2000), thus being able to fine-tune the pattern of proneuronal clusters. This process leads to the selection of clusters or single proneuronal cells in a rich variety of patterns (Chan and Jan, 1999). Acting in parallel, other factors, such as NUMB and PROSPERO, help to reinforce cellular commitment (Sen et al., 2003).

Most interesting is the high degree of conservation of the cellular function of the atonal/Math1 genes across phyla. Whereas minor differences such as the loss of proneural function in the vertebrate Math1 gene are intriguing, the common function in mechanosensory transducing cells appears to be conserved. Indeed, the conservation goes so far that the Math1 transgenes can rescue the atonal mutant phenotype in the fly (Ben-Arie et al., 2000). In contrast to many other transcription factors, atonal transgenes in Math1 null mutant mice rescue hair cell differentiation (Wang et al., 2002), thus suggesting an extraordinary conservation of function, if embedded in the right context. However, Math1 is a powerful differentiation factor but not a true proneural gene in mammals. It remains to be shown whether other bHLH genes can rescue hair cell differentiation if expressed under Math1 promoter control. Such partial rescue of function has been demonstrated in Mash1/Ngn2 transgenic mice (Cau et al., 2002) and need to be tested for Math1.

Beyond the similarity in developmental regulatory genes, other unique hair cell–specific markers such as the unconventional myosin VIIa are also present in the fly scolopidial attachment (Caldwell et al., 2002). It has been suggested that this high degree of conservation of transcriptional factors across phyla relates to the evolution of the mechanoelectrical transducer system (Fritzsch et al., 2000) that might be common in great detail across phyla (Walker et al., 2000). However, the simple fact that this mechanoelectrical transducer channel has not yet been identified leaves this argument open to future scrutiny.

B. Vertebrate Hair Cell Development

Relying in part on the distribution of Delta and Notch, several investigators (Adam et al., 1998; Eddison et al., 2000) have proposed homology between cells in the insect sensilla and those of the vertebrate ear sensory epithelium. However, this idea does not take the full complement of gene complexity of vertebrates and insects into account (Fritzsch et al., 2000). Briefly, it appears that the mouse atonal homologue 1 gene Math1, a proneuronal bHLH gene, is necessary for hair cell formation (Bermingham et al., 1999). Yet a mammalian atonal paralogue gene, Ngn1, is essential for sensory neuron development (Ma et al., 1998, 2000). Neurogenin-like genes have been identified in insects and the lancelet (Holland et al., 2000). These data could simply indicate that gene multiplication has already in urbilaterians generated various bHLH genes (Dehal et al., 2002). Those genes are now assigned to different functions in the various existing metazoan lines (see Fig. 3). Also, continued further increase in bHLH genes exists in various metazoans, thus complicating an evolutionary analysis.

Unfortunately, current analysis of the bHLH genes and their regulators does not indicate expression outside the CNS in the lancelet, except in presumed chemosensory cells (Holland et al., 2000). Interestingly, Ngn1 is also essential for olfactory sensory neuron formation (Wu et al., 2003). Assuming conservation of bHLH function, this suggests that the many sensory organs in the skin of the lancelet may be directed in their development by an as yet undetected proneuronal bHLH gene(s). Finding this gene(s) is essential to establish molecular continuity between chordate and vertebrates in sensory cell development. Alternatively, in light of atonal involvement in internal proprioreceptors (Wang et al., 2002), it is possible that the ear is related to an as yet unidentified proprioreceptor in the common bilaterian ancestor, lost in the lancelet.

The sequencing of the second potential vertebrate outgroup among chordates, the tunicate *Ciona* (sea squirt), indicates the presence of a number of bHLH genes (Dehal et al., 2002). However, in the absence of more detailed information about their expression, in particular in the atrium (Imai et al., 2002), no clear conclusion about conservation across chordates can be drawn for these genes. The presence of sensory cells in the atrium, if combined with *in situ* hybridization for the Math1 and Ngn1 orthologues in tunicates, could potentially clarify the cellular and molecular origin of the vertebrate ear. However, the forkhead Eya and Fgf genes, crucial for ear formation and morphogenesis, either are not expressed outside the CNS in tunicates (Imai et al., 2002) or their expression in the atrium is not known. It is entirely possible that both tunicates and lancelet have diverged so far from the common ancestral pattern through regressive evolution, an event well known in the mechanosensory system of the ear and the lateral line (Fritzsch, 1988a; Fritzsch and Wake, 1988), that they will not reveal the sensory neuronal origin of the vertebrate ear.

Similar to data for other developmental genes that are highly conserved across phyla, these data suggest that essential components of a common genetic network play major roles in both invertebrate and vertebrate sensory cell formation. In fact, null mutants of Atonal (ato) or Math1 cause the absence of specific mechanosensory cells in insects and vertebrates, respectively, and thus show that these bHLH genes constitute a crucial node for the overall cellular development network, hence its conservation across phyla. Although we have at least one important universal developmental gene for a subset of insect and vertebrate mechanoreceptors, a number of issues still need to be resolved. For example, which other bHLH genes (if any) are also able to substitute for atonal (assuming that many bHLH genes can be interchanged to varying degrees) as a consequence of the properties of their conserved DNA binding domain.

VI. Evolution of Sensory Neurons: Heterochronic Alteration of HLH Gene Regulation

One novel feature of the ear, compared to insect and worm mechanosensory cells, is the existence of sensory neurons connecting the hair cells to the brain. It is likely that sensory neurons formed after an ancestral bHLH gene duplication that allowed separate assignment of fate to Ngn1 (sensory neurons) and Math1 (hair cell) expressing cells by retaining Math1 for the sensory (hair) cell and recruiting neurogenin for the sensory neuron. It appears that neurogenin and atonal coexisted in invertebrates. In the lancelet neurogenins are not recruited for peripheral nervous system development (Holland *et al.*, 2000). The role of neurogenins and atonal in salp development is still unknown (Dehal *et al.*, 2002), but paralogues of bHLH genes exist. In mammals, neurogenin expression exists in the olfactory system formation, where it is downstream to achaete-scute complex homolog-like 1 (Ascl1) (Calof *et al.*, 2002). It is thus possible to have a sequential expression of several bHLH genes in a developing cluster of prosensory cells. We presume that such sequential expression in mechanosensory precursors was transformed in evolution into separate expression in distinct clones. This separate expression will ultimately lead to the formation of different cell types in the vertebrate mechanosensory precursors, sensory neurons, hair cells, and supporting cells.

Transforming proneuronal clusters, which give rise to ciliated sensory neurons in the bilaterian ancestor of insects and vertebrates, into cell

1. Ear Development and Evolution

clusters that give rise to both hair cells and sensory neurons in vertebrates could have been accomplished by recruiting an existing neurogenin gene to govern the divergent sensory neuron development in the vertebrate lineage. We propose that another round of division of sensory neuron precursor cells gave rise to the hair cell (retaining Aton) and sensory neuron (recruiting Ngn1). Such a scenario suggests that both hair cells and sensory neurons together are homologous to ciliated sensory neurons of insects. It also suggests a clonal relationship between vertebrate hair cells and sensory neurons.

Interestingly, sensory neurons emigrate from the area of the future sensory epithelia (Farinas *et al.*, 2001, Fritzsch *et al.*, 2002) much like the equally Ngn1-dependent dorsal root ganglion cell precursors delaminate from the spinal cord and hindbrain (Ma *et al.*, 1999). It is possible that this happens for a similar reason, escaping the antineuronal effect of BMP4 expressed in the dorsal part of the spinal cord (Gowan *et al.*, 2001) and the developing sensory epithelia (Morsli *et al.*, 1998). It seems reasonable to assume, by analogy to insects, that some of the pathfinding properties for these sensory neurons are coextensive with the program that leads to the formation of sensory organs (Ghysen and Dambly-Chandiere, 2000). A set of novel downstream genes seems to have evolved, in part related to further specifying neuronal connectivity (Gu *et al.*, 2003; Pauley *et al.*, 2003; Xiang *et al.*, 2003). It is conceivable that the function of Ngn1, which is expressed in cells of neural crest descent, will lead to different sensory projections than when the same gene is expressed in cells of otocyst descent.

The early formation of sensory neurons in inner ear development suggests a significant reorganization of the developmental pathways to implement this evolutionary novelty (Ma *et al.*, 1998). In fact, existing data suggest that Ngn1 is expressed at least one day prior to Math1 (Ma *et al.*, 1998; Zine *et al.*, 2001). Moreover, the behavior of sensory neurons (being formed early and migrating away) is reminiscent of the formation of a glial cell in insect sensilla development (Reddy and Rodrigues, 1999). It is possible that the insect sensilla development was modified in ancestral vertebrates by the early upregulation of the newly recruited Ngn1 in the common precursor of sensory neurons, hair cells, and supporting cells.

VII. Guidance of Afferent Fibers: The Role of Hair Cells and Other Mechanisms Revisited

Primary neuron primordia can be identified either as delaminating cells (Carney and Silver, 1983; Farinas *et al.*, 2001; Fritzsch *et al.*, 2002) emigrating through the basal lamina surrounding the otocyst or as cells that express specific markers such as Ngn1 (Ma *et al.*, 1998), NeuroD (Kim *et al.*, 2001;

Liu et al., 2000), neurotrophins (Farinas et al., 2001), or other genes (Fekete and Wu, 2002). Such delaminating cells are first apparent shortly after the otocyst closes in mouse (E9.5) or even before the otocyst is completely formed in chicken (Adam et al., 1998). Later, primary neuron primordia express many other genes such as Gata3 (Karis et al., 2001; Lawoko-Kerali et al., 2002); Brn3a (Huang et al., 2001); and others such as neurotrophin receptors (Farinas et al., 2001), Shh (Riccomagno et al., 2002), and Fgfs (Pauley et al., 2003).

The possibility of unique identities for early primary sensory neuron precursors is underscored by differential expression of several genes, even in early delaminating cells (Lawoko-Kerali et al., 2002), and the fact that in Neurod1 null mutants most cochlear neurons die in contrast to the loss of many fewer vestibular neurons (see Fig. 7) (Kim et al., 2001). Overall, the already known diversity of gene expression indicates that the various areas of the otocyst could provide unique identities to delaminating precursors based on overlapping and discrete regions of transcription factor expression. Despite this interesting start, it remains to be seen how differential areas of origin in the otocyst relate to differential gene expression and, ultimately, differential projection of primary neurons to specific sensory epithelia of the ear and specific areas of the brain (Maklad and Fritzsch, 2002). Ultimately it is possible, given sufficient nested expression patterns of various transcription factors within the ear, that primary neuron precursors acquire a unique cell fate assignment in the ear by analogy to that of neural-crest–derived primary sensory neurons and motoneurons in the CNS (Brunet and Pattyn, 2002; Gowan et al., 2001; Qian et al., 2001). If these initial data in the ear can be confirmed and extended by future work, development of distinct peripheral and central projections would be a consequence of such molecularly acquired cell fates already predetermined in the otocyst.

In this context it is important to realize that proliferation, delamination, migration to the final position, and development of central and peripheral projections is a prolonged phase in mammals and birds that lasts for several days (Rubel and Fritzsch, 2002; Ruben, 1967). As previously pointed out by Carney and Silver (1983) and recently confirmed (Farinas et al., 2001; Fritzsch et al., 2002), delaminating cells (which are likely neuronal precursors based on Neurod1 expression) apparently migrate away from the otocyst along fibers of more differentiated neurons that project toward the future sensory epithelia. Indeed, it appears that spatiotemporally distinct populations of primary sensory neuron precursors specifically extend along the existing neuronal fibers that reach toward the future primary sensory epithelium. Thus, it is possible that fate acquisition, as specified through the gene expression mosaic in the otocyst, results in restricted areas of primary sensory neuron delamination with specific and predetermined fates.

Primary sensory neurons with acquired identities may subsequently project back to the area from which they delaminated using other delaminating cells as substrate to extend their peripheral processes. Such a scenario would allow primary sensory neurons to be randomly distributed in the ganglia and to project nevertheless specifically to the ear and the brain, using distinct and different guidance cues to navigate to their various targets. In fact, recent data clearly show that most primary neurons projecting to distinct sensory epithelia are mixed in their distribution rather than completely sorted within the ganglion (Maklad and Fritzsch, 1999). Among the ear sensory epithelia, the cochlea of mammals is an exception with its highly organized peripheral and central projection (Lorente de No, 1933). But even here, topologically mismatched primary sensory neurons can comingle (Fritzsch, 2003). In contrast, the distribution of primary sensory neurons in vestibular ganglion is more random and the peripheral, exclusive projection to distinct endorgans contrasts with the highly overlapping topology of the central auditory nuclear representation of individual sensory epithelia (Maklad and Fritzsch, 1999, 2002). Indeed, tracing of early primary neurons shows that some may already have extended an axon toward the brain before they delaminate from the otic epithelium.

The literature is filled with suggestions that cell fate commitment occurs after either central or peripheral contacts are established (Gompel *et al.*, 2001; Northcutt *et al.*, 1995). Alternatively, it seems possible that primary sensory neuron cell fate determination starts in the otocyst prior to delamination. Such cell fate determination may, through unknown mechanisms, determine first where those primary sensory neuron precursors migrate to and, second, where their peripheral and central projections extend. Clearly, NeuroD and Brn3a are important players for this pathfinding (Huang *et al.*, 2001; Kim *et al.*, 2001), but next to nothing is known about their downstream molecular partners outside the primary sensory neurons. In the simplest scenario, primary sensory neurons project their dendrites back to the area from which the cell bodies originated but must subsequently navigate to target endorgan-specific hair cells. However, this is unlikely to be the only way peripheral fiber projections are established as evidenced by numerous anastomoses between sensory epithelia. Given the multitude of pathfinding molecules that have been characterized (Huang *et al.*, 2002; Tessier-Lavigne and Goodman, 1996), it seems likely that multiple mechanisms come into play in the ear as well. Such mechanisms may be modifications of a common theme specific for a given primary sensory modality (i.e., cochlea and gravistatic and angular acceleration endorgans).

Based on insights gained from other systems, it seems logical to explore the function of the ephrin ligands and receptors, known to be expressed in the developing ear (Bianchi and Liu, 1999). However, no serious defect on fiber outgrowth has been reported thus far in the relevant null mutant mouse lines (Cowan et al., 2000). Other factors associated with guidance are the members of cell adhesion molecules, some of which are expressed in intricate patterns in the developing ear (Davies and Holley, 2002). Again, no experimental studies exist that support their function in fiber guidance *in vivo*. The semaphorins and their receptors, the neuropilins and plexins, are well known for their roles in neuronal pathfinding and regeneration (Cloutier et al., 2002; Pasterkamp and Verhaagen, 2001). Plexins and semaphorins have been described in the developing ear (Miyazaki et al., 1999; Murakami et al., 2001), but their potential function has not yet been explored using existing null mutant mice, in particular the plexin null line (Cloutier et al., 2002). Recent work on semaphorin receptor mutants indicates that these receptors may play a crucial role in guiding afferents (Gu et al., 2003). Given that many semaphorins signal via just two receptors, a large degree of redundancy can be expected and thus understanding the functions of these receptors and their ligands in the ear may take some time (Suto et al., 2003). As with the FGFs, both qualitative and quantitive expression patterns may further complicate this issue.

Last but not least, data suggest that glia–axon interactions may play a role in proper pathfinding in the lateral line system (Gilmour et al., 2002). By logical extension, such a role could also be played by glial cells in the developing ear, in particular for the outgrowth of fibers toward the brain (Begbie and Graham, 2001a). Such issues could be studied in erbB receptor mutant mice, which are known to eliminate glial cells (Morris et al., 1999). Indeed, preliminary data on Erbb2 null mutant mice strongly support this notion (Morris and Fritzsch, unpublished data).

A. The Role of Hair Cells in Fiber Guidance and Survival

Numerous observations suggest that hair cells attract fibers to innervate them (Bianchi and Cohan, 1991, 1993), and some neurotrophic secretion might in part mediate these attractions (Cajal, 1919). One way of exploring neurotrophin effects in the ear would be to eliminate neurotrophins in the target of the inner ear afferents, the hair cells, by eliminating hair cells or preventing their differentiation. This is expected to eliminate expression of BDNF because in older embryos BDNF is expressed exclusively in hair cells within the ear (see Figs. 5 and 6). In addition, if no hair cells form, it is likely that supporting cells may not form normally because both are apparently linked in their differentiation via reciprocal interactions mediated through the delta–notch regulatory system (Zine et al., 2001). Recent investigations into two mutations that result in undifferentiated hair cells provide no indication for losses of sensory neurons resulting from neurotrophin deficiency. However, there are examples of sensory neuron loss that appear to be caused by abnormally low expression of the trk receptors in other mutants (Huang et al., 2001; Kim et al., 2001; Liu et al., 2000).

Mice lacking the Pou-domain–containing transcription factor, Brn3c, develop only a limited complement of morphologically undifferentiated hair cells, which can be identified as hair cells by hair-cell–specific molecular marker expression (Xiang et al., 1998). Closer examination of afferent innervation (Fig. 8) showed no correlation of fiber loss with failure to form differentiated hair cells (Xiang et al., 2003). Specifically, a robust sensory innervation persists through embryogenesis into early neonatal life. The survival throughout embryogenesis is apparently mediated by the limited expression of two neurotrophins, BDNF and NT-3, in undifferentiated sensory epithelia as revealed with *in situ* hybridization studies. Even several-month-old animals had a considerable innervation of the apical turn of the cochlea. However, this long-term retention of cochlear innervation is likely not mediated by neurotrophins because these are downregulated in neonatal animals (Wheeler et al., 1994).

It also seemed possible that examination of mutants lacking the bHLH transcription factor Math1 would make it possible to test the possible functions of guidance and survival factors released from hair cells on the guidance and survival of sensory afferent neurons. Analysis shows that Math1 is required for hair cell differentiation and probably acts upstream of Brn3c (Bermingham et al., 1999; Fritzsch et al., 2000). Surprisingly, there is very little effect of this mutation on the initial fiber growth of sensory neurons (Fritzsch et al., 2003). However, older embryos show a severe reduction of afferents that does not correspond to the pattern of loss observed in neurotrophin null mutations. Closer examination of the expression of BDNF using the BDNF–LacZ reporter showed that even in Math1 null mutants some undifferentiated hair cell precursors form and express BDNF (see Fig. 6). Thus, at least in some hair cell precursors BDNF expression does not require Math1-mediated hair cell differentiation. Even complete elimination of hair cells as in the posterior crista of FGF10 null mutant mice (Pauley et al., 2003) is compatible with directed fiber growth (see Fig. 8).

Together the data show that at the time of this publication none of the attempts to eliminate neurotrophin expression in the ear through mutation of essential transcription factors has made it possible to test the proposal of Ramon and Cajal (Cajal, 1919) that hair cells secrete neurotrophic substance(s) that attract sensory afferents. Unfortunately, the finding of neurotrophin expression in delaminating sensory neurons has made the interpretation of neurotrophin effects even more complex. Clearly, the limited expression of BDNF in the undifferentiated hair cell precursors of Math1 null mutant mice is apparently enough to support many afferents throughout embryonic life. Consequently, we cannot exclude a biologically significant effect of the limited expression of neurotrophins within delaminating sensory neurons.

28   Fritzsch and Beisel

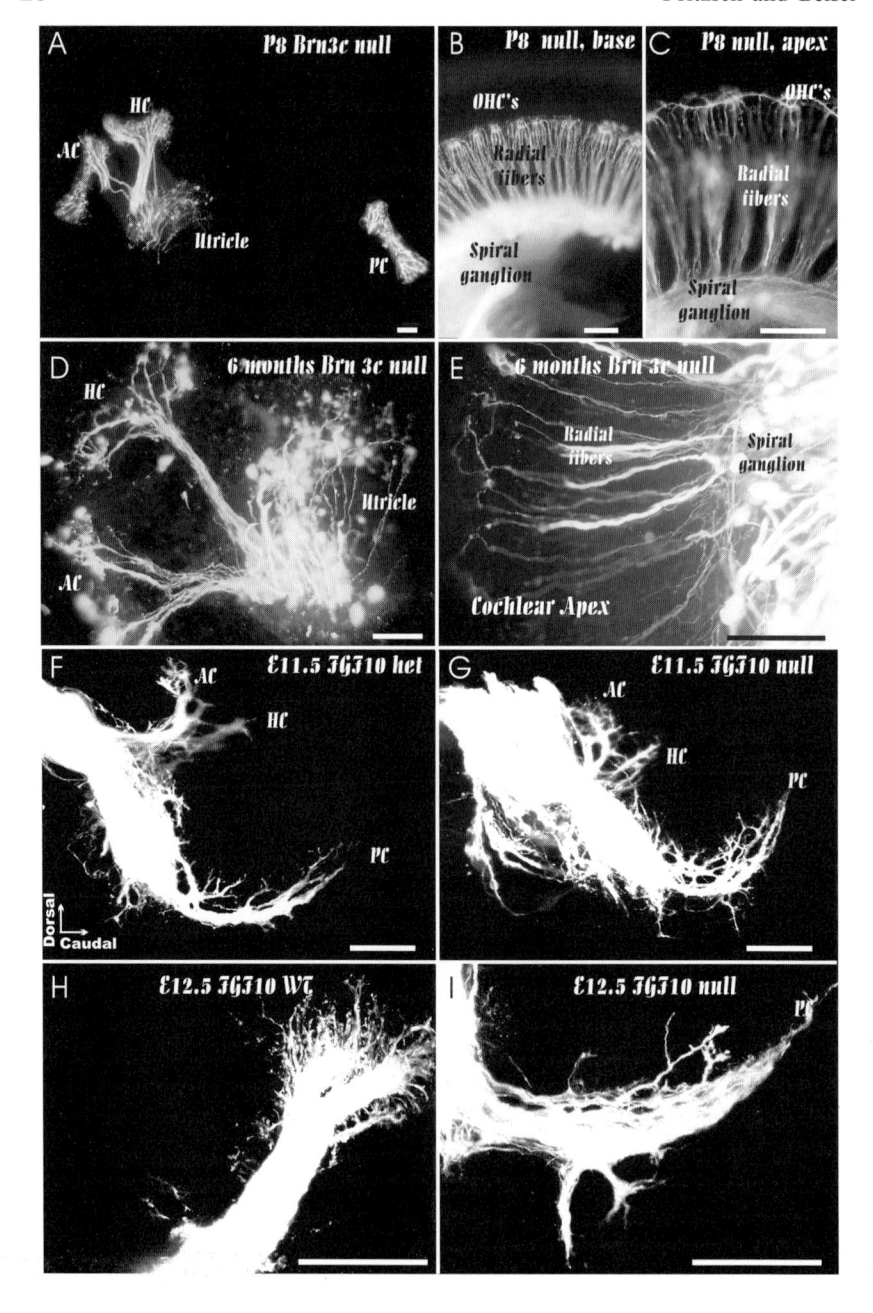

Thus none of the mutations described has critically tested an exclusive role for hair cells in attraction and maintenance of inner ear sensory neurons.

VIII. Survival of Afferents: Evolving a Novel Mechanism and Expanding it to Fit the Increasing Complexity of Ear Development

One of the major novelties in vertebrate development, the formation of a neurotrophin-mediated cell survival loop, appears to be unique to vertebrates, with perhaps some genes of unclear function present in invertebrates (Hallbook, 1999; Hallbook *et al.*, 1998). Interestingly, some Mollusca, but not insects, have separate sensory cells and sensory neurons (Budelmann, 1992). Thus, neurotrophin-mediated cell survival may have evolved in conjunction with the bHLH gene recruitment to distinct sensory neuron formation. It is also possible that the neurotrophins and their receptors may have evolved during the ancestral vertebrate gene duplication from primordial proteins (Hallbook, 1999, Jaaro *et al.*, 2001).

Two neurotrophic factors, BDNF and NT-3, and their high-affinity receptors, trkB and trkC, have been identified by *in situ* hybridization and other molecular techniques in the ear (Farinas *et al.*, 2001; Fritzsch *et al.*, 1999; Pirvola *et al.*, 1992; Wheeler *et al.*, 1994). Targeted mutations of each of these neurotrophins and receptors have clarified their relative contributions to the survival of different sensory neurons in the ear (Fritzsch *et al.*, 1999). These data have shown that there is a dramatic loss of 85% of cochlear sensory (spiral ganglion) neurons in Ntf3 null mutants (Farinas *et al.*, 1994, 2001; Fritzsch *et al.*, 1997) and 80–85% of vestibular neurons in the Bdnf null mutant (Bianchi *et al.*, 1996; Ernfors *et al.*, 1995; Jones *et al.*, 1994; Schimmang *et al.*, 1995). Somewhat similar effects have been described in neurotrophin receptor null mutants (Fritzsch *et al.*, 1995, 2001b). Detailed counting has shown that neuronal loss happens within 2–3 days after the fibers have first extended toward the sensory epithelia (Bianchi *et al.*, 1996; Farinas *et al.*, 2001). Together these data suggest that vestibular and cochlear sensory neurons have distinct but complementary neurotrophin requirements (Ernfors *et al.*, 1995).

Figure 8 These images show that directed fiber growth and even long term retention of fibers is possible without differentiated hair cells. In Brn3c null mutants, fibers grow to all sensory epithelia (A–C) and some are retained even 6 month old mice in the absence of hair cells (D, E). In FGF10 null mutant mice, there is no formation of a posterior crista (PC) but fibers grow toward this part of the ear at embryonic day 11 (F, G). However, in the absence of a posterior crista, fibers can not expand to innervate a sensory epithelium (H, I) and eventually disappear. AC, anterior crista; HC, horizontal crista; OHC's, outer hair cells; PC, posterior crista. Bar indicates 100 μm.

The relative distribution of neurotrophins with more prominent expression of BDNF in the vestibular system and of NT-3 in the cochlea does support the evidence for complementary roles of these neurotrophins in the vestibular and cochlear sensory epithelia, respectively (Farinas et al., 2001; Pirvola et al., 1992). In the context of the hypothesized role of neurotrophin function in numerical matching of pre- and postsynaptic targets, it needs to be pointed out that in the ear there is no uniform relationship between afferents and hair cells, which can vary from 30–1 (convergence on a single inner hair cell) to 1–30 (divergence on outer hair cells and some vestibular fibers). It remains questionable how a single neurotrophin, such as BDNF, distributed fairly uniform in all hair cells, should be able to mediate these differences. Clearly, quantitative data on specific amounts of BDNF expressed in different types of hair cells are needed to evaluate this aspect of ear innervation.

Whereas specific null mutations have shown significant effects on sensory neuron survival projecting to distinct endorgans of the ear, it remains unclear whether more neurotrophins or other neurotrophic factors might add to the survival of inner ear sensory neurons. However, double mutant mice, which lack both the neurotrophin receptors Ntrk2 and Ntrk3 or both neurotrophins Bdnf and Ntf3, have no surviving sensory neurons in the inner ear at birth (Ernfors et al., 1995; Liebl et al., 1997; Silos-Santiago et al., 1997). This dramatic effect of double mutations on ear innervation therefore puts to rest any speculations about additional neurotrophins and neurotrophin receptor requirements. These data on double mutants also show that even if other ligands and receptors are present their function for the development of the inner ear sensory neurons is not critical compared with BDNF/NT-3 and their receptors trkB/trkC, similar to the ubiquitous p75 neurotrophin receptor.

Specific losses of distinct cochlear and vestibular afferents occur in single neurotrophin and neurotrophin receptor null mutant mice, and these losses are related to a highly dynamic pattern of expression of neurotrophins but not of neurotrophin receptors in the ear (Farinas et al., 2001; Pirvola et al., 1992). Both during development and in the adult sensory neurons there appears to be a uniform expression of both trkB and trkC on all of the sensory neurons in the ear (Farinas et al., 2001; Fritzsch et al., 1999).

In addition, primary sensory neurons express neurotrophins soon after delamination is initiated from the placodal epithelium (Farinas et al., 2001). Furthermore, a given primary sensory neuron expresses the same neurotrophin, which is present in the area of the otocyst from which it delaminated (Fritzsch et al., 2002). Comparison of BDNF and NT-3–LacZ-positive cells with delaminating neurons marked by NeuroD–LacZ also suggests that the delaminating BDNF or NT-3-positive precursors are in fact NeuroD-expressing neuronal precursors (Fritzsch et al., 2002; Kim *et al.*, 2001; Liu *et al.*, 2000), but proof will require colabeling for NeuroD and each of the neurotrophins.

These data indicate that initial fiber growth occurs normally in the absence of neurotrophins. Indeed, initial fiber growth may use the same molecular cues recognized by the delaminating primary sensory neuron precursors. However, there is subsequently a critical period of neurotrophin dependency that will result in elimination of all connections that reach areas normally or experimentally deprived of specific neurotrophins. Such aberrantly growing fibers have been reported in the developing, but not in the adult, ear (Lorente de No, 1926). The partially overlapping expression of neurotrophins reported in the developing mammalian ear (Farinas *et al.*, 2001; Pirvola *et al.*, 1992) seems to translate into a spatiotemporal loss of primary sensory neurons in specific mutants (Farinas *et al.*, 2001). Particularly, in the cochlea there is a delayed upregulation of BDNF expression in the basal turn, leaving all basal turn neurons solely dependent on NT-3 during a brief but critical period of embryogenesis. Thus, if NT-3 is absent, there is a progressive loss of spiral neurons, especially in the basal turn, where BDNF is not present to compensate for the absence of NT-3. Overall, sensory neuron loss will occur in an embryo with a targeted mutation in Bdnf or Ntf3 only where the other is not present to compensate for its absence.

This suggestion has led to the prediction that in the ear, NT-3 and BDNF are functionally equivalent and can be substituted for each other without compromising the survival and development of sensory neurons. Consistent with this prediction, data show that the topological loss of sensory neurons in the basal turn in the Ntf3 mutant can be rescued by transgenic expression of Bdnf under the control of Ntf3 gene regulatory elements (Coppola *et al.*, 2001). Likewise, the corresponding transgenic animal in which the Ntf3 coding region is inserted into the Bdnf gene is equally effective in rescuing the BDNF phenotype in the cochlea, but not in the vestibular system (Agerman *et al.*, 2003).

Overall, these data support a role for a very early onset of elimination of exuberant or unconnected afferents and the primary sensory neurons that generate these axons. This verification of proper connections occurs immediately after the fibers have reached and started to invade their target organs (as early as E11 in the canal epithelia and E13 in the basal turn of the cochlea). The most interesting and striking effect of the single neurotrophin null mutation is the dependence of the basal turn cochlear neurons on NT-3 (Fig. 9). This is because BDNF shows a delayed expression in the basal turn. Clearly, BDNF can not only compensate for NT-3 and rescue the basal turn neurons (Coppola *et al.*, 2001) but can also attract vestibular fibers from the nearby nerve to the posterior crista to innervate the basal turn instead. It is conceivable that evolutionary pressures have resulted in the delayed expression of BDNF in the basal turn because of the need to avoid

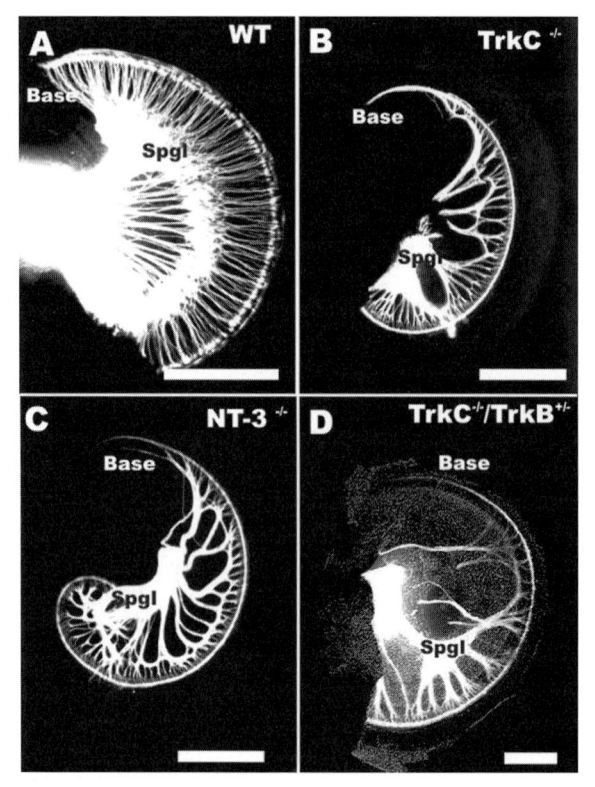

Figure 9 The effects of neurotrophin and neurotrophin receptor null mutations on the pattern of cochlear innervation is shown. Note the severe reduction in density of cochlear innervation in a pair of neurotrophin ligand (NT-3) and neurotrophin receptor(trkC) null mutants (B, C). In addition, there is a complete loss of all sensory neurons in the basal tip (Base) of either neurotrophin or neurotrophin receptor null mutant (compare A with B, C). A Ntf3 null mutation combined with Ntrk3 heterozygosity (designated as TrkC$^{-/-}$/TrkB$^{+/-}$) results in further loss of innervation and a patchy distribution of the remaining sensory neurons (D). Note that afferents from middle turn spiral sensory neurons (Spgl) expand along inner hair cells (shown as a DIC image) to reach the base. Bar indicates 1 mm.

misrouting of vestibular afferents. Misrouting of vestibular fibers to cochlear hair cells, assuming they maintain their normal connections in the CNS, will result in auditory information interfering with perceptions of position and motion.

Interestingly, neurotrophins are apparently downregulated in neonates (Fritzsch *et al.*, 1999; Wheeler *et al.*, 1994), and they appear to be largely lost in adults, despite the fact that their trk receptors are still expressed in the sensory neurons (Fritzsch *et al.*, 1999). This has led to the suggestion that other neurotrophic factors may play a role in the neonatal death of sensory neurons (Echteler and Nofsinger, 2000; Hashino *et al.*, 1999), a suggestion

1. Ear Development and Evolution

that requires further experimental verification in mutants with conditional targeting of neurotrophin genes.

IX. Splitting Hair Cell and Neuron Populations: Coevolving Sensory Epithelia and Their Innervation

In general, neurosensory evolution of the ear is based on the multiplication of existing sensory patches and their innervation (Fritzsch and Wake, 1988; Fritzsch et al., 2002). These modifications are likely followed by functional diversification through creation of modified, unique acellular covering structures (Goodyear and Richardson, 2002) that allow transduction of a previously unexplored property of the mechanical energy that reaches the ear. Clearly, formation of separate epithelia can occur within the same compartment of the ear and such compartments, like the lagenar recess, can be retained in the absence of sensory patch formation (Fritzsch and Wake, 1988). Nevertheless, it appears reasonable to assume that position within a given otic recess will influence hair cell differentiation. It seems possible that the forming cochlear duct is providing a morphodynamic influence on the differentiation of cochlear hair cells. Such tests require transplantation of hair cells from one sensory patch (i.e., the horizontal crista) into another patch (i.e., the cochlea), an experiment that has never been carried out.

The simplest ear to be found in extant vertebrates is the hagfish ear. This ear has only three sensory epithelia: one common macula and two crista organs in a single canal (Fritzsch, 2001a, b; Lewis et al., 1985). The largest number of sensory patches in the ear is found in certain species of limbless amphibians (see Fig. 2), which have nine different sensory patches: three canal cristae, utricle, saccule, lagena, neglected papilla, basilar papilla, and amphibian papilla (Fritzsch and Wake, 1988; Sarasin and Sarasin, 1892). Descriptive developmental evidence has long suggested that the evolution of multiple sensory epithelia comes about through developmental splitting of a single sensory anlage (Fritzsch and Wake, 1998; Norris, 1892). Moreover, it appears that organs such as the lagena may have evolved independently three times in vertebrates, each time by splitting off from the saccule (Fritzsch, 1992). In contrast, other organs such as the basilar papilla may have evolved only once, namely during the segregation of the lagena from the saccule in the tetrapod ancestors (Fritzsch, 1987).

Such segregation and functional diversification are best documented for the neglected/amphibian papilla system. It appears that location of each of the two sensory patches to either the utricle or the saccule and association with the perilymphatic sound conduction system may determine the future function as a sound pressure receiver or as an additional vestibular receptor

(Brichta and Goldberg, 1998; Fritzsch and Wake, 1988). Another well-documented example is the utricle in some bony fishes (e.g., herring) that forms three distinct sensory patches. Although two of these patches retain their ancestral function as gravistatic receptors, one patch has acquired a novel function as a sound pressure receiver by association with the perilymphatic sound conduction system (Fritzsch, 2001a; Lewis *et al.*, 1985).

Clearly, forming a sensory epithelium that can access a novel sensory stimulus requires transformation of existing mechanoelectrical transducers through morphological alterations to tap into this novel mechanical energy source. Sound pressure reception is not accomplished by the mere formation of a novel sensory epithelium that can be dedicated to perceive this energy. However, it is reasonable to assume that once such an uncommitted receptor is available, changes in the ear morphology using some of the genes outlined may achieve changes upon which further refinement can act in the slow process of selection of appropriate function-based modification. It is conceivable that such alterations will take place as soon as a new receptor forms simply because of the invariable alteration of FGF and BMP interaction that comes with that change of the formation of a new sensory epithelium. Indeed, comparing just BMP4 expression in chicken and mice (Morsli *et al.*, 1998; Wu and Oh, 1996) shows differences in the expression patterns that need to be further explored by comparing the expression of BMP4 with those of FGF ligands and receptors (Pirvola *et al.*, 2002).

It is noteworthy that incomplete separation of normally segregated sensory patches has been reported in mutations with altered morphologies (Fritzsch *et al.*, 2001a; Pauley *et al.*, 2003). This implies that morphogenesis and segregation of sensory patches are linked, but not necessary causal. One alternative explanation is simply that reduced morphogenesis limits the capacity of the sensory patches to segregate.

X. Summary and Conclusions: Evolving Developmental Mechanisms

We have seen the beginning of a process that will untangle the evolution of developmental genetic networks that combines mechanoelectrical cellular conservation via evolutionary "preserved" morphostatic pathways with novel morphogenetic features upon which specific aspects of mechanical stimulation are maintained. This apparent contradiction of extreme conservation in the molecular governance of mechanoelectrical transducer cell formation across phyla (Wang *et al.*, 2002) contrasts sharply with the apparent evolution of unique genes that are expressed exclusively in the vertebrate ear to govern development of vertebrate-specific acellular structures of the ear (El-Amraoui *et al.*, 2001). Between these two extremes the genes for ear morphogenesis are represented mostly by developmental modules specifically modified to suit aspects of ear development. Within the context of these modified morphogenic modules are the "ancient" genetic pathways for the ear-specific neuronal elements. Overall, ear evolution resembles the networks of interactions and their alterations known from complex ecosystems. Any alteration in frequency of expression of a single aspect can have significant effects on the system as a whole. Keeping the mechanotransducer sensory element constant allows evolution to optimize aspects of ear morphogenesis to maximize acquisition of specific signals. Evolution of the vertebrate ear can thus be best understood as an iterative developmental transformation toward optimized signal transmission.

Such aspects of ear evolution are particularly clear when comparing the morphocline of hagfish, lamprey, and jawed vertebrate ears (see Fig. 2). Hagfish have a single canal with two epithelia that bear no cupula. Thus, hagfish will see similar rotational movements in the plane of the single torus with opposing activities in either sensory epithelium. Lampreys have evolved a two-canal system that allows extraction of two distinct angular movements at an angle of about 90° to each other in each ear. Jawed vertebrates have evolved a third canal that allows them to extract selectively horizontal angular movements, a stimulus that needs computational work to be extracted from the single or two-canal labyrinth of jawless vertebrates. Clearly, optimizing the extraction of angular stimuli in three orthogonal planes is not a matter of optimizing the mechanoelectrical transducer but rather of optimizing morphogenesis of the ear. It is therefore expected that diversity in morphogenetically important transcription factors will be equivalent to that of ear morphogenesis, whereas transcription factors involved in hair cell formation should be much less variable among vertebrates.

Acknowledgments

This work was supported by grants from NIDCD (R01 DC005590; BF; R01 DC04279; KWB) and NASA (NRA 01-OBPR-06). We wish to express our gratitude to Drs. D. Nichols, H. Zoghbi, Q. Ma, and J. Lee for providing material and intellectual stimuli.

References

Abdelhak, S., Kalatzis, V., Heilig, R., Compain, S., Samson, D., *et al.* (1997a). Clustering of mutations responsible for branchio-oto-renal (BOR) syndrome in the eyes absent homologous region (eyaHR) of EYA1. *Hum. Mol. Genet.* **6,** 2247–2255.

Abdelhak, S., Kalatzis, V., Heilig, R., Compain, S., Samson, D., *et al.* (1997b). A human homologue of the *Drosophila* eyes absent gene underlies branchio-oto-renal (BOR) syndrome and identifies a novel gene family. *Nat. Genet.* **15,** 157–164.

Adam, J., Myat, A., Le Roux, I., Eddison, M., Henrique, D., et al. (1998). Cell fate choices and the expression of Notch, Delta and Serrate homologues in the chick inner ear: Parallels with *Drosophila* sense-organ development. *Development* **125**, 4645–4654.

Adamska, M., Herbrand, H., Adamski, M., Kruger, M., Braun, T., and Bober, E. (2001). FGFs control the patterning of the inner ear but are not able to induce the full ear program. *Mech. Dev.* **109**, 303–313.

Agerman, K., Hjerling-Leffler, J., Blanchard, M. P., Scarfone, E., Canlon, B., et al. (2003). BDNF gene replacement reveals multiple mechanisms for establishing neurotrophin specificity during sensory nervous system development. *Development* **130**, 1479–1491.

Ayers, H. (1892). Vertebrate cephalogenesis. *J. Morphol.* **6**, 1–360.

Bally-Cuif, L., and Hammerschmidt, M. (2003). Induction and patterning of neuronal development, and its connection to cell cycle control. *Curr. Opin. Neurobiol.* **13**, 16–25.

Begbie, J., and Graham, A. (2001a). Integration between the epibranchial placodes and the hindbrain. *Science* **294**, 595–598.

Begbie, J., and Graham, A. (2001b). The ectodermal placodes: A dysfunctional family. *Philos. Trans. R Soc. Lond. B Biol. Sci.* **356**, 1655–1660.

Ben-Arie, N., Hassan, B. A., Bermingham, N. A., Malicki, D. M., Armstrong, D., et al. (2000). Functional conservation of atonal and Math1 in the CNS and PNS. *Development* **127**, 1039–1048.

Bermingham, N. A., Hassan, B. A., Price, S. D., Vollrath, M. A., Ben-Arie, N., et al. (1999). Math1: An essential gene for the generation of inner ear hair cells. *Science* **284**, 1837–1841.

Bermingham, N. A., Hassan, B. A., Wang, V. Y., Fernandez, M., Banfi, S., et al. (2001). Proprioceptor pathway development is dependent on Math1. *Neuron* **30**, 411–422.

Bertrand, N., Castro, D. S., and Guillemot, F. (2002). Proneural genes and the specification of neural cell types. *Nat. Rev. NeuroSci.* **3**, 517–530.

Bianchi, L. M., and Cohan, C. S. (1991). Developmental regulation of a neurite-promoting factor influencing statoacoustic neurons. *Brain. Res. Dev. Brain Res.* **64**, 167–174.

Bianchi, L. M., and Cohan, C. S. (1993). Effects of the neurotrophins and CNTF on developing statoacoustic neurons: Comparison with an otocyst-derived factor. *Dev. Biol.* **159**, 353–365.

Bianchi, L. M., Conover, J. C., Fritzsch, B., DeChiara, T., Lindsay, R. M., and Yancopoulos, G. D. (1996). Degeneration of vestibular neurons in late embryogenesis of both heterozygous and homozygous BDNF null mutant mice. *Development* **122**, 1965–1973.

Bianchi, L. M., and Liu, H. (1999). Comparison of ephrin-A ligand and EphA receptor distribution in the developing inner ear. *Anat. Rec.* **254**, 127–134.

Bone, Q., and Ryan, K. P. (1978). Cupular sense organs in Ciona (Tunicata: Ascidiacea). *J. Zool. (London)* **186**, 417–429.

Borsani, G., DeGrandi, A., Ballabio, A., Bulfone, A., Bernard, L., et al. (1999). EYA4, a novel vertebrate gene related to *Drosophila* eyes absent. *Hum. Mol. Genet.* **8**, 11–23.

Brichta, A. M., and Goldberg, J. M. (1998). The papilla neglecta of turtles: A detector of head rotations with unique sensory coding properties. *J. Neurosci.* **18**, 4314–4324.

Brigande, J. V., Kiernan, A. E., Gao, X., Iten, L. E., and Fekete, D. M. (2000). Molecular genetics of pattern formation in the inner ear: Do compartment boundaries play a role? *Proc. Natl. Acad. Sci. USA* **97**, 11700–11706.

Brunet, J. F., and Pattyn, A. (2002). Phox2 genes—from patterning to connectivity. *Curr. Opin. Genet. Dev.* **12**, 435–440.

Budelmann, B. U. (1992). Hearing in nonarthropod invertebrates. *In* "The Evolutionary Biology of Hearing" (A. N. Popper, Ed.), pp. 141–156. Springer Verlag, New York.

Burighel, P., Lane, N. J, Fabio, G., Stefano, T., Zaniolo, G., Carnevali, M. D., and Manni, L. (2003). Novel, secondary sensory cell organ in ascidians: In search of the ancestor of the vertebrate lateral line. *J. Comp. Neurol.* **461**, 236–249.

Cajal, S. R. (1919). Accion neurotropica de los epitelios. *Trab. del Lab. de Invest. Biol.* **17**, 1–153.
Caldwell, J. C., and Eberl, D. F. (2002). Towards a molecular understanding of *Drosophila* hearing. *J. Neurobiol.* **53**, 172–189.
Calof, A. L., Bonnin, A., Crocker, C., Kawauchi, S., Murray, R. C., *et al.* (2002). Progenitor cells of the olfactory receptor neuron lineage. *Microsc. Res. Tech.* **58**, 176–188.
Cantos, R., Cole, L. K., Acampora, D., Simeone, A., and Wu, D. K. (2000). Patterning of the mammalian cochlea. *Proc. Natl. Acad. Sci. USA* **97**, 11707–11713.
Carney, P. R., and Silver, J. (1983). Studies on cell migration and axon guidance in the developing distal auditory system of the mouse. *J. Comp. Neurol.* **215**, 359–369.
Cau, E., Casarosa, S., and Guillemot, F. (2002). Mash1 and Ngn1 control distinct steps of determination and differentiation in the olfactory sensory neuron lineage. *Development* **129**, 1871–1880.
Chan, Y. M., and Jan, Y. N. (1999). Conservation of neurogenic genes and mechanisms. *Curr. Opin. Neurobiol.* **9**, 582–588.
Chang, W., ten Dijke, P., and Wu, D. K. (2002). BMP pathways are involved in otic capsule formation and epithelial-mesenchymal signaling in the developing chicken inner ear. *Dev. Biol.* **251**, 380–394.
Chen, P., Johnson, J. E., Zoghbi, H. Y., and Segil, N. (2002). The role of Math1 in inner ear development: Uncoupling the establishment of the sensory primordium from hair cell fate determination. *Development* **129**, 2495–2505.
Cloutier, J. F., Giger, R. J., Koentges, G., Dulac, C., Kolodkin, A. L., and Ginty, D. D. (2002). Neuropilin-2 mediates axonal fasciculation, zonal segregation, but not axonal convergence, of primary accessory olfactory neurons. *Neuron* **33**, 877–892.
Colvin, J. S., Bohne, B. A., Harding, G. W., McEwen, D. G., and Ornitz, D. M. (1996). Skeletal overgrowth and deafness in mice lacking fibroblast growth factor receptor 3. *Nat. Genet.* **12**, 390–397.
Conway Morris, S. (2000). The Cambrian "explosion": Slow-fuse or megatonnage? *Proc. Natl. Acad. Sci. USA* **97**, 4426–4429.
Coppola, V., Kucera, J., Palko, M. E., Martinez-De Velasco, J., Lyons, W. E., *et al.* (2001). Dissection of NT3 functions in vivo by gene replacement strategy. *Development* **128**, 4315–4327.
Corwin, J. T. (1981). Postembryonic production and aging of inner ear hair cells in sharks. *J. Comp. Neurol.* **201**, 541–553.
Cowan, C. A., Yokoyama, N., Bianchi, L. M., Henkemeyer, M., and Fritzsch, B. (2000). EphB2 guides axons at the midline and is necessary for normal vestibular function. *Neuron* **26**, 417–430.
David, E. S., Luke, N. H., and Livingston, B. T. (1999). Characterization of a gene encoding a developmentally regulated winged helix transcription factor of the sea urchin *Strongylocentrotus purpuratus*. *Gene* **236**, 97–105.
Davidson, E. H., Rast, J. P., Oliveri, P., Ransick, A., Calestani, C., *et al.* (2002). A genomic regulatory network for development. *Science* **295**, 1669–1678.
Davies, D., and Holley, M. C. (2002). Differential expression of alpha3 and alpha6 integrins in the developing mouse inner ear. *J. Comp. Neurol.* **445**, 122–132.
Davis, R. J., Shen, W., Heanue, T. A., and Mardon, G. (1999). Mouse Dach, a homologue of *Drosophila dachshund*, is expressed in the developing retina, brain and limbs. *Dev. Genes Evol.* **209**, 526–536.
Davis, R. J., Shen, W., Sandler, Y. I., Heanue, T. A., and Mardon, G. (2001). Characterization of mouse Dach2, a homologue of *Drosophila dachshund*. *Mech. Dev.* **102**, 169–179.
Dehal, P., Satou, Y., Campbell, R. K., Chapman, J., Degnan, B., *et al.* (2002). The draft genome of *Ciona intestinalis*: Insights into chordate and vertebrate origins. *Science* **298**, 2157–2167.

Echteler, S. M., and Nofsinger, Y. C. (2000). Development of ganglion cell topography in the postnatal cochlea. *J. Comp. Neurol.* **425**, 436–446.

Eddison, M., Le Roux, I., and Lewis, J. (2000). Notch signaling in the development of the inner ear: Lessons from *Drosophila*. *Proc. Natl. Acad. Sci. USA* **97**, 11692–11699.

El-Amraoui, A., Cohen-Salmon, M., Petit, C., and Simmler, M. C. (2001). Spatiotemporal expression of otogelin in the developing and adult mouse inner ear. *Hear. Res.* **158**, 151–159.

Ernfors, P., Van De Water, T., Loring, J., and Jaenisch, R. (1995). Complementary roles of BDNF and NT-3 in vestibular and auditory development. *Neuron* **14**, 1153–1164.

Farinas, I., Jones, K. R., Backus, C., Wang, X. Y., and Reichardt, L. F. (1994). Severe sensory and sympathetic deficits in mice lacking neurotrophin-3. *Nature* **369**, 658–661.

Farinas, I., Jones, K. R., Tessarollo, L., Vigers, A. J., Huang, E., *et al.* (2001). Spatial shaping of cochlear innervation by temporally regulated neurotrophin expression. *J. NeuroSci.* **21**, 6170–6180.

Favor, J., Sandulache, R., Neuhauser-Klaus, A., Pretsch, W., Chatterjee, B., *et al.* (1996). The mouse Pax2(1Neu) mutation is identical to a human PAX2 mutation in a family with renal-coloboma syndrome and results in developmental defects of the brain, ear, eye, and kidney. *Proc. Natl. Acad. Sci. USA* **93**, 13870–13875.

Fekete, D. M., and Wu, D. K. (2002). Revisiting cell fate specification in the inner ear. *Curr. Opin. Neurobiol.* **12**, 35–42.

Fritzsch, B. (1987). The inner ear of the coelacanth fish *Latimeria* has tetrapod affinities. *Nature* **327**, 153–154.

Fritzsch, B. (1988a). The amphibian octavo-lateralis system and its regressive and progressive evolution. *Acta. Biol. Hung.* **39**, 305–322.

Fritzsch, B. (1988b). The lateral-line and inner-ear afferents in larval and adult urodeles. *Brain Behav. Evol.* **31**, 325–348.

Fritzsch, B. (1992). The water-to-land transition: Evolution of the tetrapod basilar papilla, middle ear and auditory nuclei. *In* "The Evolutionary Biology of Hearing" (A. N. Popper, Ed.), pp. 351–375. Springer Verlag, New York.

Fritzsch, B. (1999). Hearing in two worlds: Theoretical and realistic adaptive changes of the aquatic and terrestrial ear for sound reception. *In* "Springer Handbook of Auditory Research Comparative Hearing: Fish and Amphibians" (R. R. Fay and A. N. Popper, Eds.), pp. 15–42. Springer-Verlag, New York.

Fritzsch, B. (2001a). The morphology and function of fish ears. *In* "The Laboratory Fish" (G. Ostrander, Ed.), pp. 250–259. Academic Press, Exeter.

Fritzsch, B. (2001b). The cellular organization of the fish ear. *In* "The Laboratory Fish." (G. Ostrander, Ed.), pp. 480–487. Academic Press, Exeter.

Fritzsch, B. (2003). Development of inner ear afferent connections: Forming primary neurons and connecting them to the developing sensory epithelia. *Brain Res. Bull.* **60**, 423–433.

Fritzsch, B., Barald, K., and Lomax, M. (1998a). Early embryology of the vertebrate ear. *In* "Springer Handbook of Auditory Research. Vol XII. Development of the Auditory System" (R. R. Fay, Ed.), pp. 80–145. Springer Verlag, New York.

Fritzsch, B., Barbacid, M., and Silos-Santiago, I. (1998b). Nerve dependency of developing and mature sensory receptor cells. *Ann. NY Acad. Sci.* **855**, 14–27.

Fritzsch, B., and Beisel, K. (1998). Development and maintenance of ear innervation and function: Lessons from mutations in mouse and man. *Am. J. Hum. Genet.* **63**, 1263–1270.

Fritzsch, B., and Beisel, K. W. (2001). Evolution and development of the vertebrate ear. *Brain Res. Bull.* **55**, 711–721.

Fritzsch, B., Beisel, K. W., and Bermingham, N. A. (2000). Developmental evolutionary biology of the vertebrate ear: Conserving mechanoelectric transduction and developmental pathways in diverging morphologies. *Neuroreport* **11**, R35–44.

Fritzsch, B., Beisel, K. W., Jones, K., Farinas, I., Maklad, A., *et al.* (2002). Development and evolution of inner ear sensory epithelia and their innervation. *J. Neurobiol.* **53,** 143–156.

Fritzsch, B., Farinas, I., and Reichardt, L. F. (1997). Lack of neurotrophin 3 causes losses of both classes of spiral ganglion neurons in the cochlea in a region-specific fashion. *J. Neurosci.* **17,** 6213–6225.

Fritzsch, B., Nichols, D. H., Bermingham, N. A., Zoghbi, H. Y., Wang, V. Y., *et al.* (2003). Afferent projections in Math1 null mutants follow the limited expression of the neurotrophin BDNF in undifferentiated hair cell precursors. *Development*(submitted for publication).

Fritzsch, B., Pirvola, U., and Ylikoski, J. (1999). Making and breaking the innervation of the ear: Neurotrophic support during ear development and its clinical implications. *Cell Tissue Res.* **295,** 369–382.

Fritzsch, B., Signore, M., and Simeone, A. (2001a). Otx1 null mutant mice show partial segregation of sensory epithelia comparable to lamprey ears. *Dev. Genes Evol.* **211,** 388–396.

Fritzsch, B., Silos-Santiago, I., Farinas, I., and Jones, K. (2001b). Neurotrophins and neurotrophin receptors involved in supporting afferent inner ear innervation. *In* "The Neurotrophins" (I. Mocchetti, Ed.). Salzburger and Graham.

Fritzsch, B., Silos-Santiago, I., Smeyne, R., Fagan, A. M., and Barbacid, M. (1995). Reduction and loss of inner ear innervation in trkB and trkC receptor knockout mice: A whole mount DiI and scanning electron microscopic analysis. *Audit. Neurosci.* **1,** 401–417.

Fritzsch, B., and Wake, M. H. (1988). The inner ear of gymnophione amphibians and its nerve supply: A comparative study of regressive events in a complex sensory system. *Zoomorphol.* **108,** 210–217.

George, K. M., Leonard, M. W., Roth, M. E., Lieuw, K. H., Kioussis, D., *et al.* (1994). Embryonic expression and cloning of the murine GATA-3 gene. *Development* **120,** 2673–2686.

Ghysen, A., and Dambly-Chaudiere, C. (2000). A genetic programme for neuronal connectivity. *Trends Genet.* **16,** 221–226.

Gilmour, D. T., Maischein, H. M., and Nusslein-Volhard, C. (2002). Migration and function of a glial subtype in the vertebrate peripheral nervous system. *Neuron* **34,** 577–588.

Gompel, N., Dambly-Chaudiere, C., and Ghysen, A. (2001). Neuronal differences prefigure somatotopy in the zebrafish lateral line. *Development* **128,** 387–393.

Goodyear, R. J., and Richardson, G. P. (2002). Extracellular matrices associated with the apical surfaces of sensory epithelia in the inner ear: Molecular and structural diversity. *J. Neurobiol.* **53,** 212–227.

Gowan, K., Helms, A. W., Hunsaker, T. L., Collisson, T., Ebert, P. J., *et al.* (2001). Cross-inhibitory activities of Ngn1 and Math1 allow specification of distinct dorsal interneurons. *Neuron* **31,** 219–232.

Groves, A. K., and Bronner-Fraser, M. (2000). Competence, specification and commitment in otic placode induction. *Development* **127,** 3489–3499.

Gu, C., Reimert, D. C., Shu T., Fritzsch, B., Richards, L. J., *et al.,* (2003) Neuropilin-1 integrates *Semaphorin* and VEGF signaling during neural and cardiovascular development. *Developmental Cell* **5,** 45–67.

Hadrys, T., Braun, T., Rinkwitz-Brandt, S., Arnold, H. H., and Bober, E. (1998). Nkx5-1 controls semicircular canal formation in the mouse inner ear. *Development* **125,** 33–39.

Hallbook, F. (1999). Evolution of the vertebrate neurotrophin and Trk receptor gene families. *Curr. Opin. Neurobiol.* **9,** 616–621.

Hallbook, F., Lundin, L. G., and Kullander, K. (1998). *Lampetra fluviatilis* neurotrophin homolog, descendant of a neurotrophin ancestor, discloses the early molecular evolution of neurotrophins in the vertebrate subphylum. *J. NeuroSci.* **18,** 8700–8711.

Hanson, I. M. (2001). Mammalian homologues of the *Drosophila* eye specification genes. *Semin. Cell Dev. Biol.* **12,** 475–484.

Hashino, E., Dolnick, R. Y., and Cohan, C. S. (1999). Developing vestibular ganglion neurons switch trophic sensitivity from BDNF to GDNF after target innervation. *J. Neurobiol.* **38,** 414–427.

Hatini, V., Ye, X., Balas, G., and Lai, E. (1999). Dynamics of placodal lineage development revealed by targeted transgene expression. *Dev. Dyn.* **215,** 332–343.

Holland, L. Z., Schubert, M., Holland, N. D., and Neuman, T. (2000). Evolutionary conservation of the presumptive neural plate markers AmphiSox1/2/3 and AmphiNeurogenin in the invertebrate chordate amphioxus. *Dev. Biol.* **226,** 18–33.

Huang, E. J., Liu, W., Fritzsch, B., Bianchi, L. M., Reichardt, L. F., and Xiang, M. (2001). Brn3a is a transcriptional regulator of soma size, target field innervation and axon pathfinding of inner ear sensory neurons. *Development* **128,** 2421–2432.

Huang, X., Cheng, H. J., Tessier-Lavigne, M., and Jin, Y. (2002). MAX-1, a novel PH/MyTH4/FERM domain cytoplasmic protein implicated in netrin-mediated axon repulsion. *Neuron* **34,** 563–576.

Hulander, M., Kiernan, A. E., Blomqvist, S. R., Carlsson, P., Samuelsson, E. J., et al. (2003). Lack of pendrin expression leads to deafness and expansion of the endolymphatic compartment in inner ears of Foxi1 null mutant mice. *Development* **130,** 2013–2025.

Hulander, M., Wurst, W., Carlsson, P., and Enerback, S. (1998). The winged helix transcription factor Fkh10 is required for normal development of the inner ear. *Nat. Genet.* **20,** 374–376.

Imai, K. S., Satoh, N., and Satou, Y. (2002). Region specific gene expressions in the central nervous system of the ascidian embryo. *Gene Expr. Patterns* **2,** 319–321.

Jaaro, H., Beck, G., Conticello, S. G., and Fainzilber, M. (2001). Evolving better brains: A need for neurotrophins? *Trends NeuroSci.* **24,** 79–85.

Johnson, K. R., Cook, S. A., Erway, L. C., Matthews, A. N., Sanford, L. P., et al. (1999). Inner ear and kidney anomalies caused by IAP insertion in an intron of the Eya1 gene in a mouse model of BOR syndrome. *Hum. Mol. Genet.* **8,** 645–653.

Jones, K. R., Farinas, I., Backus, C., and Reichardt, L. F. (1994). Targeted disruption of the BDNF gene perturbs brain and sensory neuron development but not motor neuron development. *Cell* **76,** 989–999.

Jorgensen, J. M. (1989). Evolution of octavolateralis sensory cells. *In* "The Mechanosensory Lateral Line. Neurobiology and Evolution" (Muenz, Ed.), pp. 99–115. Springer Verlag, New York.

Karis, A., Pata, I., van Doorninck, J. H., Grosveld, F., de Zeeuw, C. I., et al. (2001). Transcription factor GATA-3 alters pathway selection of olivocochlear neurons and affects morphogenesis of the ear. *J. Comp. Neurol* **429,** 615–630.

Kim, W. Y., Fritzsch, B., Serls, A., Bakel, L. A., Huang, E. J., et al. (2001). NeuroD-null mice are deaf due to a severe loss of the inner ear sensory neurons during development. *Development* **128,** 417–426.

Kozmik, Z., Holland, N. D., Kalousova, A., Paces, J., Schubert, M., and Holland, L. Z. (1999). Characterization of an amphioxus paired box gene, AmphiPax2/5/8: Developmental expression patterns in optic support cells, nephridium, thyroid-like structures and pharyngeal gill slits, but not in the midbrain-hindbrain boundary region. *Development* **126,** 1295–1304.

Ladher, R. K., Anakwe, K. U., Gurney, A. L., Schoenwolf, G. C., and Francis-West, P. H. (2000). Identification of synergistic signals initiating inner ear development. *Science* **290,** 1965–1967.

Lawoko-Kerali, G., Rivolta, M. N., and Holley, M. (2002). Expression of the transcription factors GATA3 and Pax2 during development of the mammalian inner ear. *J. Comp. Neurol.* **442,** 378–391.

Lee, J. E. (1997). Basic helix-loop-helix genes in neural development. *Curr. Opin. Neurobiol.* **7**, 13–20.

Lewis, E. R., Leverenz, E. L., and Bialek, W. S. (1985). "The Vertebrate Inner Ear." CRC Press, Boca Raton.

Liebl, D. J., Tessarollo, L., Palko, M. E., and Parada, L. F. (1997). Absence of sensory neurons before target innervation in brain-derived neurotrophic factor-, neurotrophin3-, and trkC-deficient embryonic mice. *J. NeuroSci.* **17**, 9113–9127.

Liu, D., Chu, H., Maves, L., Yan, Y. L., Morcos, P. A., *et al.* (2003). Fgf3 and Fgf8 dependent and independent transcription factors are required for otic placode specification. *Development* **130**, 2213–2224.

Liu, M., Pereira, F. A., Price, S. D., Chu, M. J., Shope, C., *et al.* (2000). Essential role of BETA2/NeuroD1 in development of the vestibular and auditory systems. *Genes Dev.* **14**, 2839–2854.

Lorente de No, R. (1926). Etudes sur l'anatomie et la physiologie du labyrinthe de l'oreille et du VIII nerf. II. Quelques donnees au sujet de l'anatomie des organes sensoriels du labyrinthe. *Trav. Lab. Rech. Biol. Univ. Madrid* **24**, 53–153.

Lorente de No, R. (1933). Anatomy of the eighth nerve: The central projections of the nerve endings of the internal ear. *Laryngoscope* **43**, 1–38.

Ma, Q., Anderson, D. J., and Fritzsch, B. (2000). Neurogenin 1 null mutant ears develop fewer, morphologically normal hair cells in smaller sensory epithelia devoid of innervation. *J. Assoc. Res. Otolaryngol.* **1**, 129–143.

Ma, Q., Chen, Z., del Barco Barrantes, I., de la Pompa, J. L., and Anderson, D. J. (1998). Neurogenin1 is essential for the determination of neuronal precursors for proximal cranial sensory ganglia. *Neuron* **20**, 469–482.

Ma, Q., Fode, C., Guillemot, F., and Anderson, D. J. (1999). Neurogenin1 and neurogenin2 control two distinct waves of neurogenesis in developing dorsal root ganglia. *Genes Dev.* **13**, 1717–1728.

Ma, Q., Kintner, C., and Anderson, D. J. (1996). Identification of neurogenin, a vertebrate neuronal determination gene. *Cell* **87**, 43–52.

Mackie, G. O., and Singla, C. L. (2003). The capsular organ of *Chelyosoma productum* (Ascidiacea: Corellidae): A new tunicate hydrodynamic sense organ. *Brain Behav. Evol.* **61**, 45–58.

Maklad, A., and Fritzsch, B. (1999). Incomplete segregation of endorgan-specific vestibular ganglion cells in mice and rats. *J. Vestib. Res.* **9**, 387–399.

Maklad, A., and Fritzsch, B. (2002). The developmental segregation of posterior crista and saccular vestibular fibers in mice: A carbocyanine tracer study using confocal microscopy. *Dev. Brain Res.* **135**, 1–17.

Manns, M., and Fritzsch, B. (1991). The eye in the brain: Retinoic acid effects morphogenesis of the eye and pathway selection of axons but not the differentiation of the retina in *Xenopus laevis*. *Neurosci. Lett.* **127**, 150–154.

Mansour, S. L. (1994) Targeted disruption of int-2 (fgf-3) causes developmental defects in the tail and inner ear. *Mol. Reprod. Dev.* **39**, 62–67; discussion 67–68.

Merlo, G. R., Paleari, L., Mantero, S., Zerega, B., Adamska, M., *et al.* (2002). The Dlx5 homeobox gene is essential for vestibular morphogenesis in the mouse embryo through a BMP4-mediated pathway. *Dev. Biol.* **248**, 157–169.

Miyazaki, N., Furuyama, T., Takeda, N., Inoue, T., Kubo, T., and Inagaki, S. (1999). Expression of mouse semaphorin H mRNA in the inner ear of mouse fetuses. *Neurosci. Lett.* **261**, 127–129.

Morris, J. K., Lin, W., Hauser, C., Marchuk, Y., Getman, D., and Lee, K. F. (1999). Rescue of the cardiac defect in ErbB2 mutant mice reveals essential roles of ErbB2 in peripheral nervous system development. *Neuron* **23**, 273–283.

Morrison, S. J., Perez, S. E., Qiao, Z., Verdi, J. M., Hicks, C., et al. (2000). Transient notch activation initiates an irreversible switch from neurogenesis to gliogenesis by neural crest stem cells. *Cell* **101**, 499–510.

Morsli, H., Choo, D., Ryan, A., Johnson, R., and Wu, D. K. (1998). Development of the mouse inner ear and origin of its sensory organs. *J. NeuroSci.* **18**, 3327–3335.

Murakami, Y., Suto, F., Shimizu, M., Shinoda, T., Kameyama, T., and Fujisawa, H. (2001). Differential expression of plexin-A subfamily members in the mouse nervous system. *Dev. Dyn.* **220**, 246–258.

Noramly, S., and Grainger, R. M. (2002). Determination of the embryonic inner ear. *J. Neurobiol.* **53**, 100–128.

Norris, H. W. (1892). Studies on the development of the ear in *Amblystoma*. I. Development of the auditory vesicle. *J. Morphol.* **7**, 23–34.

Northcutt, R. G., Brandle, K., and Fritzsch, B. (1995). Electroreceptors and mechanosensory lateral line organs arise from single placodes in axolotls. *Dev. Biol.* **168**, 358–373.

Pasterkamp, R. J., and Verhaagen, J. (2001). Emerging roles for semaphorins in neural regeneration. *Brain Res. Brain Res. Rev.* **35**, 36–54.

Pauley, S., Wright, T. J., Pirvola, U., Ornitz, D., Beisel, K., and Fritzsch, B. (2003). Expression and function of FGF10 in mammalian inner ear development. *Dev. Dyn.* **227**, 203–215.

Pfister, M., Toth, T., Thiele, H., Haack, B., Blin, N., et al. (2002). A 4-bp insertion in the eya-homologous region (eyaHR) of EYA4 causes hearing impairment in a Hungarian family linked to DFNA10. *Mol. Med.* **8**, 607–611.

Phillips, B. T., Bolding, K., and Riley, B. B. (2001). Zebrafish fgf3 and fgf8 encode redundant functions required for otic placode induction. *Dev. Biol.* **235**, 351–365.

Pichaud, F., and Desplan, C. (2002). Pax genes and eye organogenesis. *Curr. Opin. Genet. Dev.* **12**, 430–434.

Pirvola, U., Spencer-Dene, B., Xing-Qun, L., Kettunen, P., Thesleff, I., et al. (2000). FGF/FGFR-2(IIIb) signaling is essential for inner ear morphogenesis. *J. NeuroSci.* **20**, 6125–6134.

Pirvola, U., Ylikoski, J., Palgi, J., Lehtonen, E., Arumae, U., and Saarma, M. (1992). Brain-derived neurotrophic factor and neurotrophin 3 mRNAs in the peripheral target fields of developing inner ear ganglia. *Proc. Natl. Acad. Sci. USA* **89**, 9915–9919.

Pirvola, U., Ylikoski, J., Trokovic, R., Hebert, J., McConnell, S., and Partanen, J. (2002). FGFR1 is required for the development of the auditory sensory epithelium. *Neuron* **35**, 671.

Qian, Y., Fritzsch, B., Shirasawa, S., Chen, C. L., Choi, Y., and Ma, Q. (2001). Formation of brainstem (nor)adrenergic centers and first-order relay visceral sensory neurons is dependent on homeodomain protein Rnx/Tlx3. *Genes Dev.* **15**, 2533–2545.

Reddy, G. V., and Rodrigues, V. (1999). A glial cell arises from an additional division within the mechanosensory lineage during development of the microchaete on the *Drosophila* notum. *Development* **126**, 4617–4622.

Riccomagno, M. M., Martinu, L., Mulheisen, M., Wu, D. K., and Epstein, D. J. (2002). Specification of the mammalian cochlea is dependent on sonic hedgehog. *Genes. Dev.* **16**, 2365–2378.

Riedl, R. (1978). "Order in Living Systems: A Systems Analysis of Evolution." Wiley, New York.

Rivolta, M. N., Halsall, A., Johnson, C. M., Tones, M. A., and Holley, M. C. (2002). Transcript profiling of functionally related groups of genes during conditional differentiation of a mammalian cochlear hair cell line. *Genome Res.* **12**, 1091–1099.

Rubel, E. W., and Fritzsch, B. (2002). Auditory system development: Primary auditory neurons and their targets. *Annu. Rev. NeuroSci.* **25**, 51–101.

Ruben, R. J. (1967). Development of the inner ear of the mouse: A radioautographic study of terminal mitoses. *Acta Otolaryngol. Suppl.* **220**, 221–244.

Salazar-Ciudad, I., Jernvall, J., and Newman, S. A. (2003). Mechanisms of pattern formation in development and evolution. *Development* **130**, 2027–2037.

Sarasin, P., and Sarasin, F. (1892). Über das Gehörorgan der Caeciliiden. *Anat Anz.* **7**, 812–815.

Sato, M., and Saigo, K. (2000). Involvement of pannier and u-shaped in regulation of decapentaplegic-dependent wingless expression in developing *Drosophila notum*. *Mech. Dev.* **93**, 127–138.

Satou, Y., Imai, K. S., and Satoh, N. (2002). Fgf genes in the basal chordate *Ciona intestinalis*. *Dev. Genes Evol.* **212**, 432–438.

Schimmang, T., Minichiello, L., Vazquez, E., San Jose, I., Giraldez, F., et al. (1995). Developing inner ear sensory neurons require TrkB and TrkC receptors for innervation of their peripheral targets. *Development* **121**, 3381–3389.

Schlosser, G. (2002). Development and evolution of lateral line placodes in amphibians. I. Development. *Zoology* **105**, 119–146.

Sen, A., Reddy, G. V., and Rodrigues, V. (2003). Combinatorial expression of Prospero, Seven-up, and Elav identifies progenitor cell types during sense-organ differentiation in the *Drosophila* antenna. *Dev. Biol.* **254**, 79–92.

Shimeld, S. M., and Holland, P. W. (2000). Vertebrate innovations. *Proc. Natl. Acad. Sci. USA* **97**, 4449–4452.

Silos-Santiago, I., Fagan, A. M., Garber, M., Fritzsch, B., and Barbacid, M. (1997). Severe sensory deficits but normal CNS development in newborn mice lacking TrkB and TrkC tyrosine protein kinase receptors. *Eur. J. NeuroSci.* **9**, 2045–2056.

Simmler, M. C., Cohen-Salmon, M., El-Amraoui, A., Guillaud, L., Benichou, J. C., et al. (2000). Targeted disruption of otog results in deafness and severe imbalance. *Nat. Genet.* **24**, 139–143.

Solomon, K. S., and Fritz, A. (2002). Concerted action of two dlx paralogs in sensory placode formation. *Development* **129**, 3127–3136.

Solomon, K. S., Kudoh, T., Dawid, I. B., and Fritz, A. (2003). Zebrafish foxi1 mediates otic placode formation and jaw development. *Development* **130**, 929–940.

Strassmaier, M., and Gillespie, P. G. (2002). The hair cell's transduction channel. *Curr. Opin. Neurobiol.* **12**, 380–386.

Streit, A. (2002). Extensive cell movements accompany formation of the otic placode. *Dev. Biol.* **249**, 237–254.

Suto, F., Murakami, Y., Nakamura, F., Goshima, Y., and Fujisawa, H. (2003). Identification and characterization of a novel mouse plexin, plexin-A4. *Mech. Dev.* **120**, 385–396.

Tessier-Lavigne, M., and Goodman, C. S. (1996). The molecular biology of axon guidance. *Science* **274**, 1123–1133.

Toresson, H., Martinez-Barbera, J. P., Bardsley, A., Caubit, X., and Krauss, S. (1998). Conservation of BF-1 expression in amphioxus and zebrafish suggests evolutionary ancestry of anterior cell types that contribute to the vertebrate telencephalon. *Dev. Genes Evol.* **208**, 431–439.

Torres, M., and Giraldez, F. (1998). The development of the vertebrate inner ear. *Mech. Dev.* **71**, 5–21.

van Bergeijk, W. A. (1966). Evolution of the sense of hearing in vertebrates. *Am. Zoologist* **6**, 371–377.

Van Esch, H., Groenen, P., Nesbit, M. A., Schuffenhauer, S., Lichtner, P., et al. (2000). GATA3 haplo-insufficiency causes human HDR syndrome. *Nature* **406**, 419–422.

Vendrell, V., Carnicero, E., Giraldez, F., Alonso, M. T., and Schimmang, T. (2000). Induction of inner ear fate by FGF3. *Development* **127**, 2011–2019.

Venter, J. C., Adams, M. D., Myers, E. W., Li, P. W., Mural, R. J., et al. (2001). The sequence of the human genome. *Science* **291,** 1304–1351.

Vincent, C., Kalatzis, V., Abdelhak, S., Chaib, H., Compain, S., et al. (1997). BOR and BO syndromes are allelic defects of EYA1. *Eur. J. Hum. Genet.* **5,** 242–246.

Wada, H., Saiga, H., Satoh, N., and Holland, P. W. (1998). Tripartite organization of the ancestral chordate brain and the antiquity of placodes: Insights from ascidian Pax-2/5/8, Hox and Otx genes. *Development* **125,** 1113–1122.

Walker, R. G., Willingham, A. T., and Zuker, C. S. (2000). A *Drosophila* mechanosensory transduction channel. *Science* **287,** 2229–2234.

Wang, V. Y., Hassan, B. A., Bellen, H. J., and Zoghbi, H. Y. (2002). *Drosophila atonal* fully rescues the phenotype of Math1 null mice: New functions evolve in new cellular contexts. *Curr. Biol.* **12,** 1611–1616.

Wayne, S., Robertson, N. G., DeClau, F., Chen, N., Verhoeven, K., et al. (2001). Mutations in the transcriptional activator EYA4 cause late-onset deafness at the DFNA10 locus. *Hum. Mol. Genet.* **10,** 195–200.

Wever, E. G. (1974). The evolution of vertebrate hearing. *In* "Auditory System, Vol. V/1, Handbook of Sensory Physiology" (W. D. Neff, Ed.), pp. 423–454. Springer Verlag, Berlin.

Wheeler, E. F., Bothwell, M., Schecterson, L. C., and von Bartheld, C. S. (1994). Expression of BDNF and NT-3 mRNA in hair cells of the organ of Corti: Quantitative analysis in developing rats. *Hear. Res.* **73,** 46–56.

Wu, D. K., and Oh, S. H. (1996). Sensory organ generation in the chick inner ear. *J. NeuroSci.* **16,** 6454–6462.

Wu, H. H., Ivkovic, S., Murray, R. C., Jaramillo, S., Lyons, K. M., et al. (2003). Autoregulation of neurogenesis by GDF11. *Neuron* **37,** 197–207.

Xiang, M., Gao, W. Q., Hasson, T., and Shin, J. J. (1998). Requirement for Brn-3c in maturation and survival, but not in fate determination of inner ear hair cells. *Development* **125,** 3935–3946.

Xiang, M., Maklad, A., Pirvola, U., and Fritzsch, B. (2003). Brn3c null mutant mice show long-term, incomplete retention of some afferent inner ear innervation. *BMC Neurosci*, Jan 30 (e-pub).

Xu, P. X., Adams, J., Peters, H., Brown, M. C., Heaney, S., and Maas, R. (1999). Eya1-deficient mice lack ears and kidneys and show abnormal apoptosis of organ primordia. *Nat. Genet.* **23,** 113–117.

Xu, P. X., Woo, I., Her, H., Beier, D. R., and Maas, R. L. (1997). Mouse Eya homologues of the *Drosophila* eyes absent gene require Pax6 for expression in lens and nasal placode. *Development* **124,** 219–231.

Yang, Q., Bermingham, N. A., Finegold, M. J., and Zoghbi, H. Y. (2001). Requirement of Math1 for secretory cell lineage commitment in the mouse intestine. *Science* **294,** 2155–2158.

Yu, J. K., Holland, N. D., and Holland, L. Z. (2002). An amphioxus winged helix/forkhead gene, AmphiFoxD: Insights into vertebrate neural crest evolution. *Dev. Dyn.* **225,** 289–297.

Zheng, W. L., Huang, L., Wei, Z. B., Silvius, D., Tang, B., and Xu, P. X. (2003). The role of six1 in mammalian auditory system development. *Development* **130,** 3989–4000.

Zhou, Q., and Anderson, D. J. (2002). The bHLH transcription factors OLIG2 and OLIG1 couple neuronal and glial subtype specification. *Cell* **109,** 61–73.

Zine, A., Aubert, A., Qiu, J., Therianos, S., Guillemot, F., Kageyama, R., and de Ribaupierre, F. (2001). Hes1 and Hes5 activities are required for the normal development of the hair cells in the mammalian inner ear. *J. NeuroSci.* **21,** 4712–4720.

2
Use of Mouse Genetics for Studying Inner Ear Development

Elizabeth Quint and Karen P. Steel
MRC Institute of Hearing Research
University Park, Nottingham NG7 2RD, United Kingdom

I. Introduction: The Genetic Revolution
   A. What Is "Mouse Genetics"?
II. Early Morphogenesis of the Inner Ear
   A. Mutations in Genes Expressed in the Early Otocyst
   B. Mutations that Disrupt Mesenchymal/Epithelial Interactions
   C. Mutations that Disrupt Endolymph Homeostasis
   D. Mutations that Affect Sensory Patch Specification
III. Development of the Neuroepithelium
   A. Formation of the Sensory Patches
   B. Hair Cell Patterning in the Sensory Patches: Evidence from the Cochlea
   C. Hair Cell Differentiation: Specializations of the Cytoskeleton
   D. Endolymph Homeostasis
   E. Survival of the Hair Cell
   F. Neurogenesis and Neural Defects
   G. Extracellular Membranes
IV. Concluding Remarks
   References

The analysis of mouse mutants has become an almost indispensable tool in modern biology. The understanding of inner ear development has been facilitated by the use of mice from chemical- and radiation-induced mutagenesis screens, as well as gene-targeting strategies such as knockout models. In particular, our understanding of hereditary gross malformations and neuroepithelial defects that underlie human deafness has been aided by the analysis of comparable gene defects in mice. Mice with mutations in transcription factor genes such as *Pax2* and *Hmx3* have illustrated the early compartmentalization of the otocyst, whereas defects in endolymph homeostasis, such as those found in Pendred syndrome, show how malformations of the inner ear can arise through fluid changes within it. In addition, a surprisingly large number of mice with hearing and balance problems exhibit defects in sensory hair cell differentiation. From these animals we have been able to identify key gene pathways, such as the *Notch* signaling pathway, that are involved in cell fate determination and maturation in the inner ear. We have also been able to identify some of the major structural, cytoskeletal components of the mechanotransducing hair cells, such as unconventional myosins and other motor proteins such as prestin. Analysis of mutant mice has been incredibly useful in the identification of functionally, and therefore clinically, important genes and together with recent molecular advances, such as the sequencing of the mouse genome, have taken mouse genetics to a new level that will surely make it indispensable in the future analysis of mammalian development. © 2003, Elsevier Inc.

I. Introduction: The Genetic Revolution

These are exciting times for mouse genetics. In the twentieth century we saw the introduction of defined mouse strains into the laboratory setting and shortly after, the establishment of The Jackson Laboratory in Maine, probably the largest archive of mouse strains in the world (*http://www.jax.org/*). Subsequently, as part of the "genetic revolution," the first transgenic mouse was produced (Gordon *et al.*, 1980; Palmiter *et al.*, 1982a, b), followed by the refinement of gene targeting and homologous recombination to generate knock-in and knockout mice (e.g., Doetschman *et al.*, 1987; Hooper *et al.*, 1987; Thomas and Capecchi, 1987) and more recently the introduction of cloning technology (Wakayama *et al.*, 1998). Now, in the twenty-first century and barely 100 years since the introduction of the laboratory mouse, we have the publication of both the human (Lander *et al.*, 2001; Venter *et al.*, 2001) and mouse (Waterston *et al.*, 2002) genome sequences, and several large-scale random mutagenesis programs have been established (e.g., Hrabe de Angelis *et al.*, 2000; Nolan *et al.*, 2000). As a consequence of these advances in molecular biology and recombinant DNA technology, our understanding of the genetic mechanisms that regulate embryonic development, including mammalian inner ear development, has also been enhanced.

From a clinical point of view, studying the genetic mechanisms that regulate the development of the mouse inner ear is very worthwhile. Hearing loss is the most common form of sensory loss known to man. In the United Kingdom alone, 1 in 1000 newborn babies have a significant (greater than 40 dB HL) hearing impairment, and this rises to at least 1 in 500 by the age of 9 (Fortnum *et al.*, 2001). This has a tangible impact on society. Children with undiagnosed hearing impairment have greater problems associated with social integration and communication, educational attainment, and quality of life. It is estimated that around half of all congenital deafness is due to single-gene mutations (Marazita *et al.*, 1993; Morton, 1991) manifested in altered development of the auditory system. Of the known cases of human hereditary deafness to date, where the gene has been identified, almost all affect the periphery and most of these affect the sensorineural components of the cochlea and vestibular apparatus (reviewed in Ahituv and Avraham, 2002; Kiernan and Steel, 2000). Understanding the biological processes involved in inner ear development is essential for the future development of clinical therapies for early-onset hereditary deafness and for later-onset deafness, whether this be of genetic or environmental etiology.

But what will the analysis of the mouse inner ear tell us about human deafness? First, the mouse inner ear is almost identical in every respect except size to the human inner ear. This similarity is also likely to extend to the genetic processes that control development of the cochlear and vestibular portions and their peripheral innervation. The similarities between the mouse and human genomes should not be underestimated. Despite the 75 or so million years of divergent evolution, according to the draft sequence of the mouse genome, 99% of the 30,000 predicted mouse genes appear to have direct counterparts in human. Around 40% of the 2.5 Gigabases (Gb) of mouse genome can be *directly* aligned with the human genome, and conserved synteny (i.e., the same order of orthologous genes) is apparent in approximately 90% of the mouse and human sequences (*http://www.nature.com/nature/mousegenome/*). The high degree of similarity between the mouse and human genomic sequences, in addition to the anatomical and physiological similarities between man and mouse, provide the perfect forum in which to interpret mutant mice in the context of human disease. If we can identify genes that affect the development of the mouse inner ear and lead to deafness, then we can screen human populations for mutations in the equivalent genes. In doing this we will identify causative genes underlying human genetic deafness. This is not a pipe dream. Half of the 50 or so genes known to cause hereditary deafness in humans have an equivalent mouse model. Furthermore, approximately half of these models were identified as potential human "deafness genes" in the mouse first. There are also a growing number of mouse deafness mutants from random mutagenesis programs for which the causative gene has yet to be identified. These mutants are sure to provide more models of human deafness and, in particular, may help to resolve the current paucity of candidate genes for human nonsyndromic deafness. For more comprehensive reviews of this subject, see Ahituv and Avraham, 2002; Kiernan and Steel, 2000; and Steel and Kros, 2001.

A. What Is Mouse Genetics?

Mouse genetics can be defined as the study of the inheritance of different traits within the mouse with the purpose of understanding the molecular and cellular basis of the trait. However, within this definition lies a host of molecular tools with which gene expression can be manipulated. These vary from the generation of mice with targeted gene insertions (knock-ins) and gene deletions (knockouts) to mice with chemical- and radiation-induced random mutations derived from phenotype-driven screens. Each type of mutagenesis approach is associated with its own set of advantages and limitations.

Gene-driven approaches use site-directed or targeted mutagenesis (knockout and knock-in strategies) to provide a direct and efficient vehicle with which to study gene function. Although this approach requires that a given gene is already known and sequenced, it is nevertheless a logical progression from discovering that this gene is expressed in the organ of interest (e.g., the inner ear) to understanding its function there. A useful aspect of this approach is that the mouse can be used in a retrospective way to investigate disease; if gene "X" is mapped to a particular human disease locus and subsequently shown to be the causative gene, a mouse knockout model of this gene can then be generated and used to understand the disease process further. As an example, *EYA1* was identified as the gene responsible for branchio-oto-renal syndrome in humans (Abdelhak et al., 1997). The mouse orthologue *Eya1* was subsequently inactivated in mice and shown to give a comparable phenotype to the human condition (Xu et al., 1999).

Another use of transgenic mouse technology is the production of reporter constructs to investigate the factors that regulate gene expression. Reporter constructs are typically segments of DNA that contain several elements. These include a general or tissue-specific promotor and sometimes also a regulatory element that is specific to the tissue of interest (see the example that follows). The promotor and/or elements drive the expression of a reporter gene that is also contained within the construct. The reporter gene encodes a protein product that can be visualized and is typically either the *LacZ* or *green fluorescent protein* (*GFP* or *enhanced GFP*) gene. The LacZ protein is visualized indirectly using an enzymatic reaction postmortem whereas (e)GFP has the advantage of being visible directly, under fluorescent lighting, in live tissue (Okabe et al., 1997; Pratt et al., 2000). This construct is then used to generate a transgenic mouse. As an example, the *Dlx* homeobox gene family is thought to be important in the development of the forebrain, branchial arches, and sensory placodes (Stock et al., 1996). In the mouse, the gene pair *Dlx5* and *Dlx6* are expressed in overlapping regions in the forebrain, but it is not clear whether they share common regulatory elements driving their expression. They are arranged in a tail-to-tail configuration with a small enhancer sequence contained within the intergenic region. This enhancer sequence is thought to have been highly conserved through evolution and when inserted into a reporter construct, this enhancer element, whether from mouse or zebrafish, can direct expression of the reporter gene to the same regions as endogeneous *Dlx5/6* in the mouse, indicating that in fact they do share regulatory mechanisms (Quint et al., 2000; Zerucha et al., 2000).

In gene-driven strategies as described, the gene of interest is already known and the aim is to understand how that gene functions. In contrast, phenotype-driven strategies rely on the behavior, anatomy, or physiology of an animal to lead to the identification of new genes involved in the process of interest, such as the development of the ear. The identification of the gene is usually carried out via positional cloning and candidate gene approaches (e.g. Kiernan *et al.*, 2001; Steel and Kimberling, 1996). Phenotype-driven strategies include the analysis of randomly generated and spontaneously occurring mutants and are incredibly useful in the hunt for *novel* genes involved in embryogenesis and disease (reviewed in Justice *et al.*, 1999; Schimenti and Bucan, 1998). By definition, these phenotype-driven analyses identify functionally important genes that are often unsuspected candidates for a particular disease process. As an example, the role of unconventional myosins in the normal development of the sensory hair cells within the inner ear was not understood until the gene affected in the shaker1 mouse was identified as *Myo7a* (Gibson *et al.*, 1995). The role of this gene is described in detail in later sections. Identification of the shaker1 gene subsequently led to the identification of *MYO7A* as the causative gene in Usher syndrome type 1B (USH1B) and two forms of nonsyndromic deafness, DFNB2 and DFNA11 (Liu *et al.*, 1997a,b; Weil *et al.*, 1995).

Subsequently other unconventional myosin mutations have been implicated in deafness both in the human and mouse, such as in *MYO6* (DFNA22; Snell's waltzer mouse), *MYO3A* (DFNB30), and *MYO15* (DFNB3; shaker2 mouse) (Avraham *et al.*, 1995; Probst *et al.*, 1998; Wang *et al.*, 1998). Random mutagenesis also provides clues about the regulation of genes, through the identification of mutations in noncoding regions of the genome, and the importance of different regions within the encoded protein for DNA–protein and protein–protein interactions and other aspects of protein function.

One drawback of the random mutagenesis approach is the sheer scale it needs to be carried out on. However, two large-scale European phenotype-driven mutagenesis programs have been conducted using the chemical mutagen *N*-ethyl-*N*-nitrosourea (ENU) (Hrabe de Angelis *et al.*, 2000; Nolan *et al.*, 2000) and animals from these screens are available to the scientific community. Indeed, many of the mutants described in this chapter are derived from these programs because both included screens for hearing and balance defects. In each case, they assessed vestibular dysfunction by screening for the presence of head bobbing, circling, and hyperactivity and screened for deafness by the use of the Preyer's reflex test (a pinna flick in response to a click). It may seem incongruous to screen animals for vestibular defects in the search for "deafness genes", but it turns out that many of the mutants with vestibular defects also exhibit cochlear phenotypes at the ultrastructural level. An example of this is the ENU-derived mutant, tailchaser,

Table I Useful Websites

Mouse genome browsers and bioinformatic resources
http://www.ensembl.org/
http://www.genome.ucsc.edu/
http://www.ncbi.nih.gov/genome/guide/mouse/
http://www.informatics.jax.org/

European mutagenesis program homepages
http://www.mgu.har.mrc.ac.uk/mut.html
http://www.gsf.de/ieg/groups/enu-mouse.html

Gene expression databases and hereditary deafness sites
http://www.ihr.mrc.ac.uk/hereditary/mousemutants.shtml
http://www.uia.ac.be/dnalab/hhh/
http://www.jax.org/hmr/index.html
http://www.ncbi.nlm.nih.gov/entrez/query.fcgi?db=OMIM

which shows a clear behavioral phenotype suggestive of vestibular dysfunction but has a normal Preyer reflex. However, analysis of the cochlea clearly identified a cochlear hair cell phenotype, and physiological measures of cochlear function show that thresholds are significantly raised (Kiernan et al., 1999). The overlap of genes involved in both vestibular and cochlear morphogenesis is not only useful as a method of expanding the screening process for deafness genes but illustrates how fundamental developmental mechanisms can be used by different structures during embryogenesis.

All of these methods of genome disruption, whether gene-or phenotype-driven, have provided valuable clues about the molecular mechanisms of development. As a result, mouse mutational analysis is now a fundamental and integral part of biology. A comprehensive description of every mouse mutant exhibiting inner ear defects is beyond the scope of this chapter. Instead, this chapter concentrates on recent insights into several key aspects of inner ear development that have been provided by mutant mouse models. A list of useful websites of mouse genetic databases and mutagenesis program homepages is provided in Table I.

II. Early Morphogenesis of the Inner Ear

Although the development of the inner ear is clearly regulated by complex interactions between different genes and groups of genes, the analysis of existing mutant mice with hearing and balance defects suggests that this process can be conveniently (albeit oversimplistically) described in terms of a series of ontogenetic events. Mutations in genes that act early in inductive and patterning processes tend to result in gross morphogenetic malformations of the inner ear. On the other hand, genes expressed later tend to affect discrete regions of the inner ear such as the neuroepithelium or accessory structures such as the tectorial membrane. The resulting phenotype of a particular gene mutation can therefore be used to predict broadly the stage of development at which that gene is involved. There is inevitably some overlap, however, between these broad classifications, particularly because some genes may have separate early and late developmental roles, such as the *Notch* ligand *Jagged1* (Kiernan et al., 2001) (see Sections II.D and III.B). Based on phenotypic analysis of mouse mutants, it seems that there are four general mechanisms by which early gross malformations of the inner ear can occur. These are outlined as follows with examples.

A. Mutations in Genes Expressed in the Early Otocyst

In mouse, the otic placode is first visible at around E8.5 (Kaufman, 1992) as an ectodermal thickening of either side of the neural plate in the prospective hindbrain region. With the exception of neural-crest–derived melanocytes and Schwann cells, the otic placode gives rise to all of the cell types that form the inner ear, including the vestibuloacoustic ganglion of the VIII cranial nerve that innervates the sensory hair cells (Carney and Silver, 1983; Li et al., 1978). The otic placode forms on the surface of the embryo and then invaginates to form an otic cup, which in turn closes to form the otic vesicle (Fig. 1A) from which the cochlea and vestibular components arise (Fig. 1B).

Genes that are expressed within the early otocyst are thought to specify or pattern the future compartments of the ear (reviewed in Fekete and Wu, 2002) and are typically transcription factors. Examples include *Pax2* and *Hmx3* that are expressed in the ventromedial and dorsolateral walls of the ototcyst, respectively. Disruption of *Pax2* results in agenesis of the cochlea (Favor et al., 1996; Torres et al., 1996), whereas mutations in *Hmx3* result in severe disruption of the vestibular system (Hadrys et al., 1998; Wang et al., 1998). For more comprehensive reviews of this area, see Cantos et al., (2000) and Fekete and Wu (2002).

B. Mutations that Disrupt Mesenchymal/Epithelial Interactions

Mutational analysis has revealed another group of genes that influence the compartmentalization of the otocyst, but in contrast, these genes are not expressed within the otic vesicle. Instead they are expressed in the surrounding periotic mesenchyme and/or neural plate. Examples include secreted proteins such as sonic hedgehog, homeobox genes such as *Hoxa1* and *Prx1/2*, and genes involved in the retinoic acid signaling pathways.

Sonic hedgehog (*Shh*), which is secreted by the notochord and/or floor plate, is known to be important in epithelial/mesechymal interactions in

52   Quint and Steel

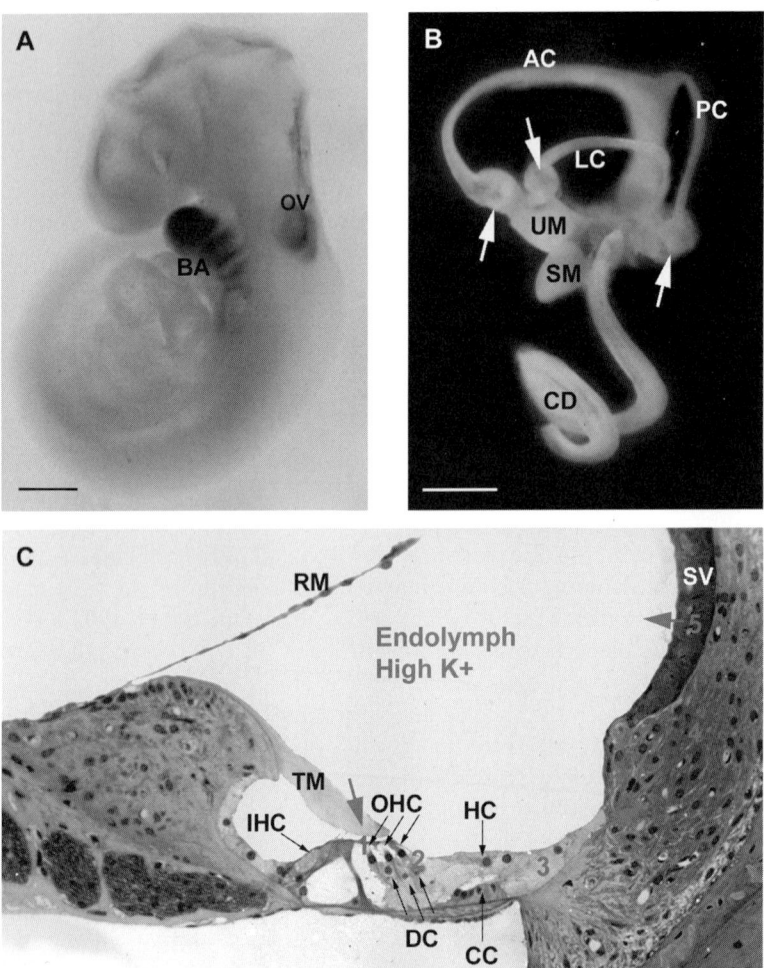

Figure 1 Components of the mouse inner ear. (A) RNA *in situ* hybridization of a mouse at E9.5 showing *Dlx5* expression in the otic vesicle (OV) and branchial arches (BA). Anterior to the left: (B) Paint-filled inner ear from a P0 mouse. The cochlear portion contains the cochlear duct (CD), whereas the vestibular portion contains the saccular macula (SM); utricular macula (UM); and lateral, posterior, and anterior semicircular canals (LC, PC, AC). Arrows point to the ampullae, containing the cristae, which are also part of the vestibular system. Anterior to the left, ventral down: (C) Semithin section of the cochlear duct from an adult mouse, stained with toluidine blue. RM, Reissner's membrane; SV, stria vascularis; TM, tectorial membrane; IHC, inner hair cell; OHC, outer hair cell; DC, Deiter's cells; HC, Hensens's cells; CC, Claudius' cells. Medial to the left. Numbering illustrates the sequence of steps in the hypothesized potassium recycling pathway, referred to in the text in detail (see Section III.D). Scale bars = 500 μm (A and B), 25 μm, (C). (See Color Insert.)

2. Mouse Genetics and Inner Ear Development

several developmental systems, such as tooth and hair follicle morphogenesis. In mice with null mutations in *Shh*, ear induction occurs normally but ventral otic derivatives such as the cochlear duct and cochleovestibular ganglion fail to develop (Liu *et al.*, 2002; Riccomagno *et al.*, 2002). This may be due to altered (reduced) *Pax2* expression (Liu *et al.*, 2002). Shh also interacts with the bone morphogenetic proteins (BMPs) during chondrogenesis (e.g. Quint *et al.*, 2002) and is expressed in regions of mesenchymal condensations around the otocyst that will form the bony capsule of the inner ear. Targeted disruption of *Shh* results in a poorly defined otic capsule and, occasionally, absent semicircular canals (Liu *et al.*, 2002).

The homeobox genes *Prx1* and *Prx2*, which are predominantly expressed in the periotic mesenchyme, are also involved in the formation of the bony labyrinth. In the *Prx2* knockout mouse there are no visible inner ear malformations, but in the *Prx1/Prx2* double knockout inner ear malformations are found that are similar to the *Prx1* single knockout but more severe (ten Berge *et al.*, 1998). In the double knockout the otic capsule is smaller than normal and the lateral semicircular canal is often absent, probably due to reduced outgrowth of the pouches that form the canals.

As a final example, the vitamin A derivative retinoic acid (RA) is known to be involved in a wide spectrum of embryonic development and is mediated through a range of nuclear receptors (RARs, RXRs), binding proteins (CRABPs), and response elements (RAREs) (see Chapter 9 by Romand). Three of the nuclear receptor (RAR) isoforms, RARα, β, and γ, are known to influence gross inner ear development. RARα is expressed in the sensory epithelium and neural components of the inner ear, whereas RARβ and RARγ are expressed in the mesenchyme-derived tissues such as the inner spiral sulcus and modiolus of the cochlea (Romand *et al.*, 1998, 2002). The RARγ and RARβ single knockout mice and the RARα and RARβ double knockout mice do not exhibit any obvious cochlear or vestibular phenotypes. However, absence of both RARα and RARγ together results in cochlear and vestibular malformations (Romand *et al.*, 2002), including a small otocyst, absent endolymphatic duct, and aborted semicircular canal formation. These malformations are similar to those found in the *Hoxa1*-deficient mouse but are more severe, suggesting that RA signaling may have several downstream targets, including the *Hoxa1* gene (Romand *et al.*, 2002).

C. Mutations that Disrupt Endolymph Homeostasis

Many of the inner ear malformations that have been described result from altered transcription factor expression (see Section II.A). However, it is also clear that disruption of the ionic composition of the endolymphatic fluid (see Fig. 1C) can have deleterious morphogenetic consequences. As an illustration, the gene pendrin (*Pds*), which underlies the human condition Pendred syndrome (Everett *et al.*, 1997), encodes an anion transporter within the inner ear, kidney, and thyroid gland. Targeted disruption of *Pds* in mice results in clear behavioral abnormalities such as circling and head bobbing, similar to other mice with known vestibular defects. Indeed, disruption of *Pds* results in the dilation of the endolymphatic duct, semicircular canals, cochlea and saccule, as well as reduced otoconia in the maculae and, later, degeneration of hair cells (visible by postnatal day 15) (Everett *et al.*, 2001). The gross malformations of the endolymphatic compartments are not visible until after E15 and the cochlear hair cells seem to develop normally prior to their degeneration. This suggests that the hair cell degeneration observed may be a secondary effect of disrupted endolymph homeostasis (see Section III.D) because hair cells require a unique ionic environment to function normally.

Two examples of proteins involved in K^+ recycling and endolymph homeostasis are the Na-K-Cl cotransporter Slc12a2, located on the basolateral side of the strial marginal cells, and the ion channel component Kcne1 (lsk), located on the apical surface of the strial marginal cells. Mutations in *Slc12a2* underlie the mutant mouse *shaker-with-syndactylism* (Dixon *et al.*, 1999), which exhibits circling behavior similar to that described in the *Pds* mutant, and mutant mice fail to produce endolymph. In mice with targeted mutations of the *Slc12a2* gene, the endolymphatic compartments collapse around the time of birth (Delpire *et al.*, 1999). The ion channel formed by Kcnq1 and Kcne1 helps to regulate the output of K^+ into the scala media, and disruption of this gene in the spontaneously occurring mutant mouse *punk rocker* and the *Kcne1* knockout mouse results in the collapse of the endolymphatic compartment (Letts *et al.*, 2000; Vetter *et al.*, 1996). In both the *shaker-with-syndactylism* (*Slc12a2*) and *punk rocker* (*Kcne1*) mouse mutants, the semicircular canals appear thinner than normal. This may result from the early collapse of the endolymphatic lumen leading to an abnormal template for bony labyrinth formation (Deol 1963; Dixon *et al.*, 1999; Letts *et al.*, 2000).

D. Mutations that Affect Sensory Patch Specification

Abnormal sensory patch specification may also lead to gross inner ear malformations. Differentiation of the sensory patches within the inner ear begins with a synchronous exit of cells from the cell cycle at around E12.5–E13.5 (Chen *et al.*, 2002; Ruben, 1967). In the cochlea, these cells form the "zone of nonproliferating cells" (ZNPC) that will form the organ of Corti (Chen et al., 2002). Cell cycle exit by the ZNPC cells appears to be promoted by the cyclin-dependent kinase inhibitor p27^{kip1} (Chen and Segil, 1999) because disruption of this gene allows cell proliferation to continue into the postnatal organ of Corti (Lowenheim et al., 1999). The postmitotic mechanism by which the sensory precursor cells of the organ of Corti are defined within the ZNPC is still not clearly understood. However, shortly after terminal mitosis, the prosensory precursor cells along the cochlear duct express Notch1 (Lanford et al., 1999) and Jag1 (Chen et al., 2002; Morrison et al., 1999). The early expression of these proteins in the presumptive organ of Corti may help to define the boundaries of the developing organ, contributing to its precise patterning (Chen et al., 2002), although later both *Notch1* and *Jag1* expression are restricted to the supporting cells. Evidence for Notch signaling in defining early boundaries of the sensory patches is supported by observations from the inner ears of the two allelic mutant mice mutated at the *Jag1* locus: headturner (Kiernan et al., 2001) and slalom (Tsai et al., 2001). In these mice some of the cristae in the vestibular system are small or absent and the patterning of the hair cells within the organ of Corti is abnormal, with fewer rows of outer hair cells and more inner hair cells. These mutants either have subsequent truncations of the posterior and anterior semicircular canals, with the truncation centered on the missing cristae and their surrounding ampullae, or they have small ampullae and small cristae. This suggests that the absence of the crista may lead to abnormal formation of the otic capsule and consequent malformation of the inner ear. It will be interesting to discover whether other genes involved in early specification of the sensory patches of the inner ear also lead to gross malformation of the labyrinth.

III. Development of the Neuroepithelium

The sensory patch primordia give rise to the organ of Corti in the cochlea and the utricle, saccule, and three cristae of the vestibular system. Within these endorgans are the sensory hair cells, so-called because of their apical projections of actin-rich stereocilia. The stereocilia are grouped into highly ordered bundles that are mechanically deflected by sound waves (in the cochlea) or body movements (in the vestibular system). Bundle deflection opens ion channels, located in the tips of the stereocilia, through which cations enter the hair cell, leading to the generation of a receptor potential. The hair cells are separated by a regular, defined arrangement of supporting cells that, at least in the organ of Corti, are highly differentiated (see Figs. 1C and 2). In fact, the organ of Corti is probably a victim of its own specialization because this is likely to be one of the factors contributing to its apparent incapacity to regenerate following trauma. Each of the endorgans

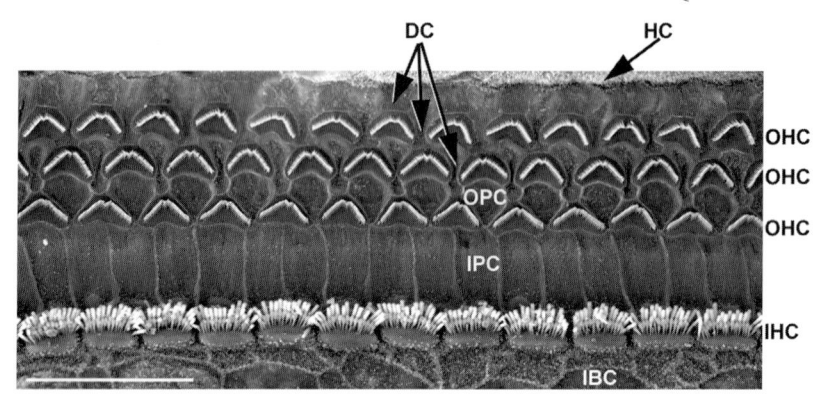

Figure 2 Scanning electron micrograph of the adult mouse organ of Corti, illustrating the three rows of outer hair cells (OHC), one row of inner hair cells (IHC) and supporting inner border cells (IBC), inner pillar cells (IPC), outer pillar cells (OPC), Deiter's cells (DC), and Hensen's cells (HC). Lateral is uppermost. Scale bar = 20 μm.

is overlain by a gelatinous extracellular matrix. In the cochlea this is the tectorial membrane, in the utricle and saccule it is the otoconial membrane, and in the cristae it is the cupula. These membranes are crucial to the sensory transduction process (see Section III.G).

A. Formation of the Sensory Patches

The formation of the intricate pattern of hair cells, support cells, and overlying membranes is encoded by many different genes, and mouse mutants are proving to be invaluable in the determination of what these genes might be. As an example, one group of genes, the *fibroblast growth factors* (*Fgfs*), which are known to be involved in many aspects of development, are also important in the formation of the sensory patches. There are at least 23 different Fgfs, all of which act through four known Fgf receptors (Fgfr1–Fgfr4) (see Pickles and Chir, 2002, and Chapter 8 by Wright and Manson). The *Fgf receptor 1* (*Fgfr1*) loss-of-function mouse exhibits a dose-dependent disruption of hair cell patterning within the organ of Corti (Pirvola *et al.*, 2002), whereas *Fgfr3* mutants do not have normal pillar cells (Colvin *et al.*, 1996). Pirvola *et al.* (2002) showed that Fgfr1 is expressed in the ventromedial wall of the otocyst at E10.5 and its expression increases with time within the organ of Corti primordium of the cochlear duct. Using an otic-specific *Fgfr1* conditional knockout they showed that outer hair cell patterning, but not differentiation, is affected by *Fgfr1* inactivation. They suggest that Fgfr1, possibly in conjunction with Fgf10, may help to define the sensory epithelial progenitor cells in the greater epithelial ridge prior to their exit from

2. Mouse Genetics and Inner Ear Development

the cell cycle. In this study the cellular patterning of the vestibular sensory patches did not seem to be affected. Disruption of Fgfs and their receptors has so far provided the best examples of genes that influence sensory patch formation; however, large-scale mutational analysis is still in its infancy and so hopefully there will be many more examples to add in the future.

B. Hair Cell Patterning in the Sensory Patches: Evidence from the Cochlea

What makes a progenitor cell turn into a hair cell? The answer to this question is one of the holy grails (there are several) in auditory research, and the quest to understand the factors that influence, control, and maintain hair cell differentiation has never been more in focus. Developmental and evolutionary biologists are interested because hair cell development in the mammal seems to parallel many of the cell fate determination mechanisms discovered in lower vertebrates such as Notch signaling in *Drosophila*. Anatomists are interested because the hair cell is such a precisely polarized cell with its own unique architecture; it is even motile. Physiologists and psychophysicists are interested because of the hair cell's amazing functional range, transducing sounds and vibrations as low in frequency as 2 Hz to well over 100 kHz in bats, and clinicians are interested because of the possibility that hair cells may be encouraged to regenerate in order to replace those lost in the diseased inner ear.

Much of what we now know about hair cell specification is reminiscent of the mechanisms of cell fate determination in lower vertebrates, in particular for example the lateral inhibition signaling pathways in *Drosophila* (reviewed in Eddison *et al.*, 2000). However, fundamental differences do seem to exist between species and so our understanding of mammalian hair cell development must ultimately be based on studies of mammalian hair cell development. Having said that, the reports during the late 1980s of the extensive capacity for hair cell regeneration in birds (Corwin and Cotanche, 1988; Ryals and Rubel, 1988) and then the regenerative limitations of mammals (Forge *et al.*, 1993; Warchol *et al.*, 1993) provoked questions about how to stimulate mammalian hair cell development. In order to induce regeneration of hair cells in the organ of Corti, we clearly needed to know more about the mechanisms that underlie hair cell specification, differentiation, and maintenance, in both the developing and mature inner ear. We now have a much clearer, although far from complete, understanding of some of these mechanisms and the molecules involved.

Based on experiments in chicks, it is now generally accepted that immature support cells and hair cells arise from the same progenitors (Fekete *et al.*, 1998). Following the specification of the prosensory region (see Section III. A), a subpopulation of cells then expresses the basic helix-loop-helix (bHLH) transcription factor Mammalian atonal homologue 1 (Math1) (see Chapter 10 by Gao). These cells are destined to become hair cells. The first evidence for the involvement of Math1 in the specification of mammalian hair cells came from a targeted deletion study where *Math1* was knocked out. *Math1* is the mouse homologue of the *Drosophila* gene *atonal* (a proneural gene) and is expressed in the hindbrain, otic vesicle (i.e., in the developing hair cells), dorsal spinal cord, and external germinal layer of the cerebellum. In *Math1* null mice ($Math1^{\beta-gal}/\beta_{-gal}$) hair cells fail to develop but supporting cells do form, and possibly in greater numbers than expected (Bermingham et al., 1999). This suggests that *Math1* might be a molecular prohair cell switch. Furthermore, overexpression of Math1 in cultures of the postnatal rat organ of Corti induces the production of extra hair cells in the adjacent epithelium (Zheng and Gao, 2000), suggesting that *Math1* is both necessary and sufficient for promoting hair cell development, at least in some cell types. Further studies using transgenic mice containing a *Math1* enhancer coupled to a reporter construct further clarified that this gene is not involved in the selection/specification of the hair cell precursors. Instead, it is only expressed in the subpopulation of postmitotic cells that are destined to become hair cells (Chen et al., 2002). Chen et al. also showed a longitudinal gradient of *Math1* expression along the developing cochlear duct similar to that reported by Bermingham et al. (1999), but in addition showed that there is a radial gradient of expression from inner to outer hair cells. This progression of expression from the base to the apex and from the inner hair cells to the outer hair cells is consistent with the normal pattern of hair cell maturation.

Specification of the hair cells by Math1 is followed by the upregulation of *Jag2* in the same cell (Lanford et al., 2000; Zine and de Ribaupierre, 2002). Because Jag2 is a ligand for the transmembrane Notch1 receptor, it binds to Notch in the adjacent (soon to be nonsensory) cell. This in turn may activate the downstream bHLH genes *Hes1* and *Hes5* (Zheng et al., 2000a; Zine and de Ribaupierre, 2002), which are homologues of the *Drosophila hairy* and *enhancer-of-split* genes and downstream targets of Notch. *Hes1* and *Hes5* may then lead to the downregulation of *Math1* and upregulation of *Notch1* in the putative nonsensory cell, thus acting as a negative reinforcement for the nonsensory fate (lateral inhibition). Evidence for this model came from the analysis of *Math1*, *Jag2*, and *Notch1* expression in *Hes1* and *Hes5* mutant mice (Zine and de Ribaupierre, 2002). The absence of *Hes1* and *Hes5* results in the overexpression of *Math1* and *Jag2*, and this fits neatly with earlier observation that mutations in the *Hes* genes result in an increase in inner (*Hes1* mutant) and outer (*Hes5* mutant) hair cells (Zine et al., 2001). In addition, Zheng et al. (2000a) showed that cotransfection of rat cochlear cultures with *Hes1* prevents the production of supernumerary hair cells mediated by ectopic *Math1* (Zheng and Gao, 2000).

It is important to remember, however, that much of the evidence for lateral inhibition pathways in the organ of Corti is circumstantial. Many of the observations described could be explained by Notch signaling having a role in determining the boundaries of the sensory patches leading to more or fewer hair cells, which could explain why often complete rows of hair cells are absent or extra rows present in the mutant mice described here.

C. Hair Cell Differentiation: Specializations of the Cytoskeleton

1. The Stereociliary Bundle

From the ENU mutagenesis screens and the spontaneously occurring mouse mutants that have been characterized, a surprising number of deaf animals exhibit primary stereociliary bundle defects. These mutants have provided key information about the factors that regulate the development of the bundle, its maintenance, and its function.

The normal development of mammalian hair cell bundle has been described elsewhere (e.g., for hamsters by Kaltenbach *et al.*, 1994; see Chapter 13 by Eatock and Hurley). The first sign of hair bundle differentiation can be detected at around E16.5 in the base of the mouse cochlea (earlier in the vestibular system). Initially the whole hair cell apex is covered with small microvilli clustered around a central kinocilium, which later migrates to the lateral side of the cell apex (Fig. 3). A subpopulation of microvilli then develop into stereocilia, with their own specialized cytoskeleton, arranged in a series of rows graded in height, increasing toward the kinocilium (see Fig. 3B). Excess microvilli are resorbed back into the hair cell and in the cochlea the kinocilium is also resorbed, leaving a defined V-or W-shaped cluster of stereocilia in the outer hair cells and a slightly crescent-shaped line of stereocilia in the inner hair cells (Fig. 4A and B). Tiny filamentous lateral linkages connect the shafts of adjacent stereocilia both within and between rows (lateral links) and connect the tips of shorter with taller stereocilia (tip links) (Fig. 4C). In the vestibular system there are also links that connect the proximal regions of the stereociliary shafts (ankle links) (reviewed in Forge and Wright, 2002). These links are probably of different composition relating to their different roles, but they seem to contribute to the rigidity of the bundle as a unit.

Although the mechanisms that regulate the growth and stepped height arrangement of the stereocilia are still not understood, several mouse mutants do seem to exhibit defects in genes probably involved in this pathway. For example, in the *Myo15* mutant, shaker2, the stereocilia, although in the correct position, fall short of their normal height (Probst *et al.*, 1998). However, abnormal amounts of actin, or cytocauds, are found in the hair cell apex (in the cuticular plate region beneath the bundle),

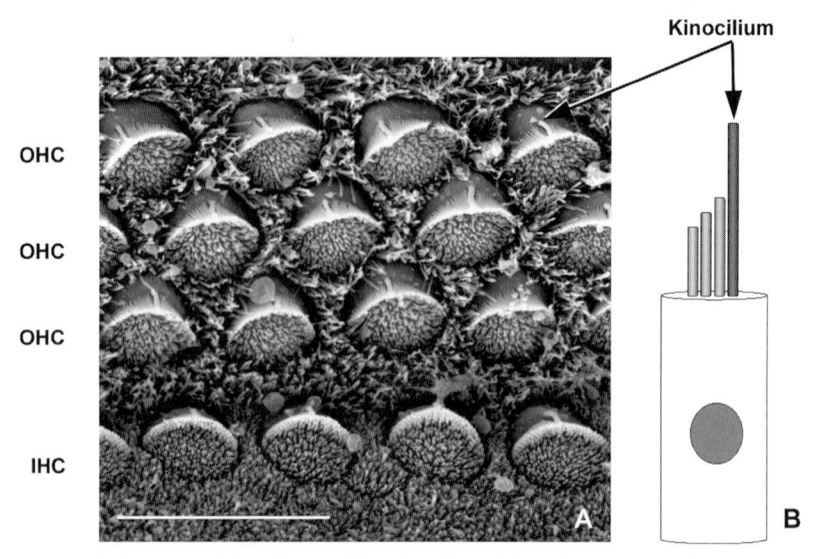

Figure 3 The immature hair cell stereociliary bundle. (A) Scanning electron micrograph of the immature mouse organ of Corti at E18.5, illustrating the three rows of outer hair cells (OHC) and one row of inner hair cells (IHC). Lateral is uppermost. Scale bar = 10 μm. (B) Schematic illustration of the immature hair cell. The hair bundle projects from the apical surface and consists of a kinocilium (arrows) and stereocilia that increase in height toward the kinocilium. Medial is to the left.

indicating that myosin XV may be important in the deployment of actin into the developing stereocilia.

a. The Establishment and Maintenance of Hair Bundle Polarity. The hair cells of the stereociliary bundle have been likened to the bristles on the thorax and wings of fruit flies and polarized cells of the mammalian respiratory tract and fallopian tubes (reviewed in Lewis and Davies, 2002). This is so because each of these cell types is an example of a polarized cell with a plane of asymmetry, or planar cell polarity (PCP), within the plane of the epithelium. This directional polarity allows, for example, mucus to be swept out of the lungs and, in the hair cells of the inner ear, movement and sound to be synchronously transduced by hair cells within the cochlear duct. PCP within the hair cell is reflected by the asymmetry of its stereocilia, which are organized in rows of graded height from one side of the cell apex (the medial side) to the other (the lateral side) (illustrated in Fig. 3B). PCP in the hair cell should not be confused with the apical–basal (bundle end–synaptic end) asymmetry of these epithelial cells.

In the organ of Corti the alignment of adjacent hair cells, all exhibiting PCP, is clearly coordinated along the length of the cochlear duct. This gives

2. Mouse Genetics and Inner Ear Development 61

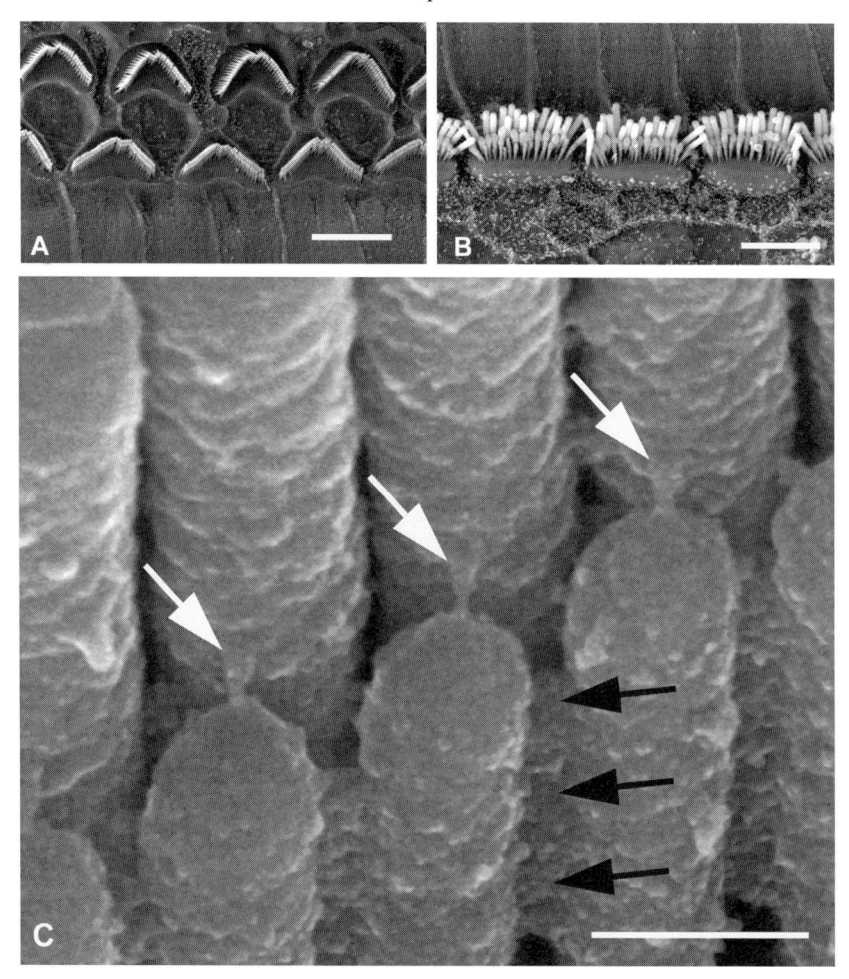

Figure 4 Scanning electron micrographs of the cochlear hair cell stereociliary bundles. (A) In the adult outer hair cell bundle, excess microvilli and the kinocilium are resorbed leaving a V- or W-shaped array of stereocilia. (B) The adult inner hair cell bundles contain a slightly crescent-shaped array of stereocilia. (C) Adjacent stereocilia are connected by lateral links (black arrows) and the tips of smaller stereocilia are connected to the shafts of taller stereocilia by tip links (white arrows). Photograph in panel C provided by D. N. Furness. Scale bars = 5 μm in A and B and 150 nm in C.

the three characteristic rows of uniformly orientated V-shaped outer hair cell bundles and one row of straight or very slightly curved inner hair cell bundles (see Fig. 4A and B). The vestibular system is also highly organized, but in a more complicated manner (see Denman-Johnson and Forge, 1999). The morphological asymmetry of the bundle allows it to have an excitatory and inhibitory direction of mechanical movement. This property is fundamental to the mechanotransduction role described later. Many of the genes known to be involved in PCP establishment in *Drosophila* (reviewed in Adler, 2002) have homologues in the mouse (reviewed in Lewis and Davies, 2002), and some are known to be expressed in the cochlea, such as *Celsr1* (one of the mouse homologues of *flamingo*) (Shima et al., 2002). These genes will therefore certainly provide good candidates for any mutant mice that might exhibit defects in PCP establishment.

The mutants currently described in the literature with stereociliary defects can tell us which genes are *not* involved in the establishment of PCP but may be involved in maintenance of the polarity. The stereociliary bundles of the myosin; cadherin; and integrin mutants, such as shaker1 (*Myo7a*; Gibson et al., 1995), Snell's waltzer (*Myo6*; Avraham et al., 1995), waltzer (*Cdh23*; Di Palma et al., 2001), Ames waltzer (*Pcdh15*; Alagramam et al., 2001), and the *Inta8* knockout (Littlewood Evans and Muller, 2000), all initially begin to differentiate normally, develop stereocilia of graded height, and establish bundles with uniform orientation. In each case, however, the bundles soon embark on a process of progressive disorganization (each mutant with its own phenotype). Progressive disorganization of the stereocilia in the *Myo6* mutant, Snell's waltzer, is associated with fusion of the stereocilia. Myosin VI is different from most other unconventional myosins described because it is a minus-end–directed protein (Wells et al., 1999). This means that it tracks along actin in the stereocilia in the direction of the pointed or minus end, which is toward the hair cell apex away from the tip of the stereocilia. It is possible therefore that this protein acts to maintain bundle organization by anchoring down the hair cell apical membrane or other stereocilia components at the level of the stereocilia rootlets (Avraham et al., 1995; Self et al., 1999).

The stereociliary disorganization that occurs in the *Myo7a* (shaker1) and *Cdh23* (waltzer) mutants (Fig. 5) may be explained by the study by Boeda et al. (2002) describing harmonin in these mutants. The harmonin gene is mutated in Usher syndrome type 1c (USH1c; Bitner-Glindzicz et al., 2000) and the harmonin protein seems to interact with myosin VIIa and cadherin 23 in the immature, developing hair bundle. Harmonin, shuttled by myosin VIIa, may bridge cadherin-23–containing lateral links to the actin core of adjacent stereocilia (Boeda et al., 2002). The absence of both harmonin and cadherin 23 in the mature stereocilia suggests that these complexes would only be present in the immature bundle, conferring early stability prior to the maturation of the stereociliary rootlet network. This may provide a mechanism by which at least the bundle integrity can be maintained following initial differentiation and explain why in the myosin VIIa and cadherin mutants the bundle becomes progressively disorganized (Alagramam et al., 2001; Di Palma et al., 2001; Kussel-Andermann et al., 2000; Self et al., 1998).

2. Mouse Genetics and Inner Ear Development 63

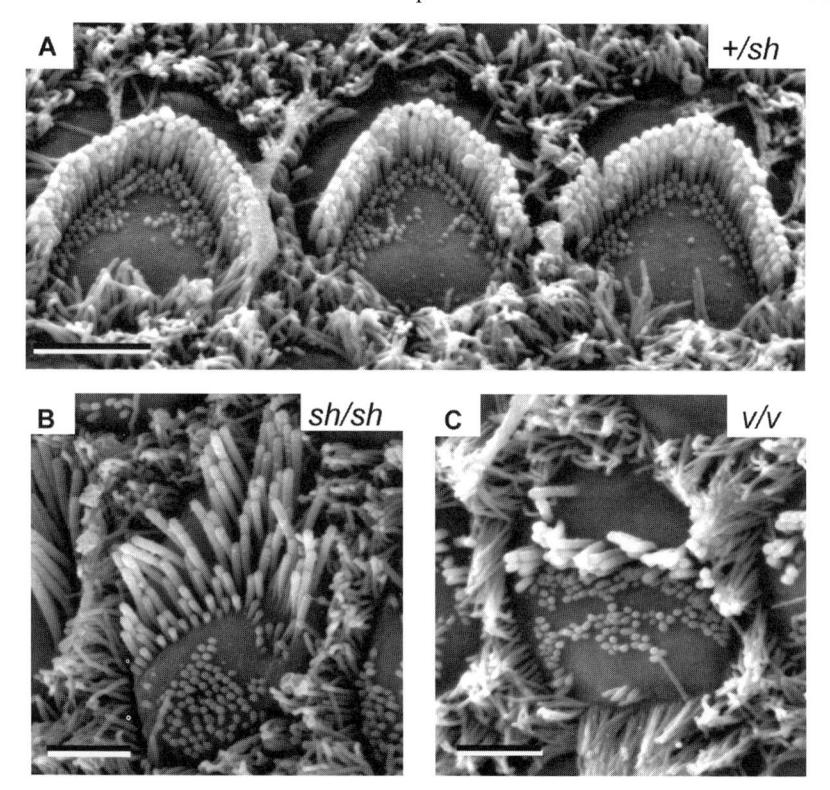

Figure 5 Scanning electron micrographs of mouse outer hair cell bundles at P4. (A) Outer hair cell bundles from a control (shaker1 heterozygote $+/Myo7a^{4626SB}$) displaying a broadly V-shaped cluster of stereocilia; note the waltzer heterozygote (not shown) is indistinguishable from the shaker1 heterozygote. (B) An outer hair cell bundle from a shaker1 homozygous mutant ($Myo7a^{4626SB}/Myo7a^{4626SB}$); although adjacent stereocilia are more separated than in the controls and have failed to form the normal V-shaped cluster, the polarity of the bundle has been established as indicated by stereocilia of increasing height toward the lateral (uppermost) edge of the bundle. (C) An outer hair cell bundle from a waltzer homozygous mutant ($Cdh23^v/Cdh23^v$). Again, polarity is established but the V-shaped arrangement of stereocilia is absent and the stereocilia are separated. Tissue samples provided by R Holme. Scale bars = 2.5 μm.

In addition to the mutants described, there are several other mouse mutants in which the affected gene has not yet been identified that also exhibit defects in the development of the stereociliary bundle. Examples published from our own laboratory (Fig. 6) include tailchaser (Kiernan *et al.*, 1999), whirler (Holme *et al.*, 2002), and tasmanian devil (Erven *et al.*, 2002). In the tailchaser mutant the stereociliary bundle of the outer hair cells fails to achieve its mature V shape. The inner hair cell bundles are also affected, albeit to a lesser extent, and hair cells from all of the inner

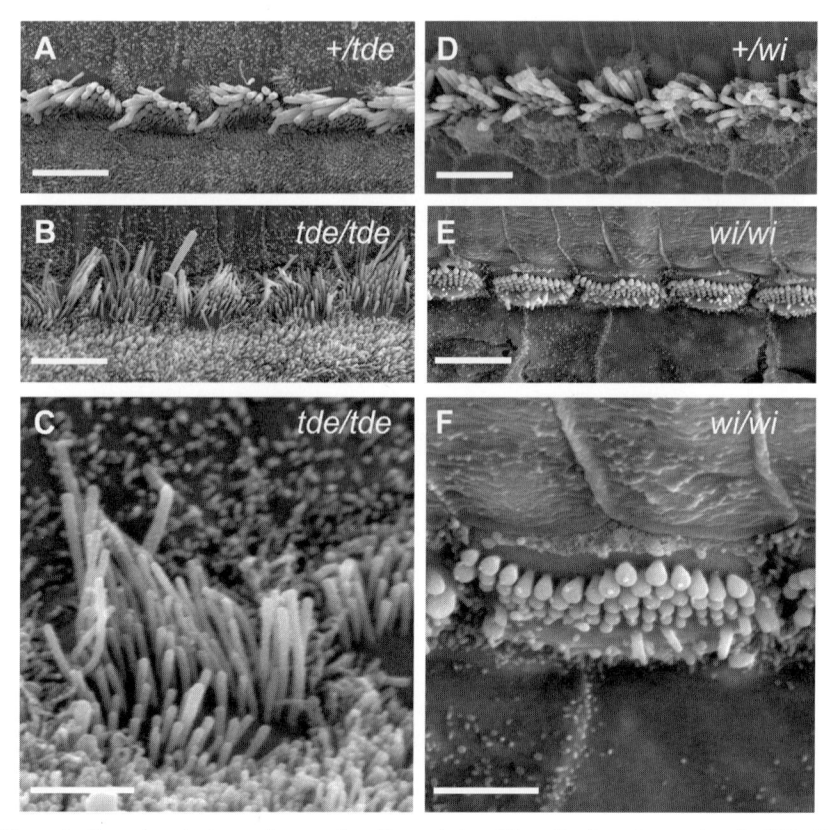

Figure 6 Scanning electron micrographs of inner hair cell bundles from tasmanian devil (A–C) aged 11 days and whirler (D–F) mouse mutants aged 30 days. The homozygote tasmanian devil stereocilia (B, C) appear thinner than the littermate (heterozygote) controls (A). In contrast, the homozygote whirler mutant stereocilia (E, F) are shorter than the littermate (heterozygote) controls (D). Tasmanian devil photographs (panels A–C) provided by A. Erven and whirler tissue samples provided by R. Holme. Scale bars = 5 µm A, B, D, and E; 2 µm, C and F.

ear sensory regions, including the vestibular system, eventually undergo progressive degeneration. It will be interesting to see if this mutation, which maps to chromosome 2 in the mouse, provides any insights into human progressive deafness once identified. The whirler mouse exhibits short stereocilia (see Fig. 6A–C), similar to the shaker2 (*Myo15*) and jerker (*Espn*) mouse mutants (Probst *et al.*, 1998; Zheng *et al.*, 2000b). The whirler mutation maps to chromosome 4. Tasmanian devil mice were generated using an insertional random mutagenesis approach. The Rous sarcoma virus (RSV) promoter driving the expression of the human placental alkaline phosphatase (RSV–PLAP) reporter gene was randomly incorporated into the genome

2. Mouse Genetics and Inner Ear Development   65

using pronuclear injection methods (Skynner et al., 1999), and in the tasmanian devil line the insert is presumed to disrupt expression of a native gene. Using fluorescence in situ hybridization (FISH) visualization of the (reporter) transgene, tasmanian devil was mapped to chromosome 5. This mutant exhibits abnormally thin stereocilia that become progressively more disorganized and floppy (see Fig. 6D–F). There is a delay in reabsorption of the redundant microvilli and later there is degeneration of the whole bundle. It has been suggested that the cross-linking of the parallel arrays of actin might be affected in this mutant (Erven et al., 2002).

In summary, all of the mouse mutants that have been described with defects in stereociliary bundle development seem to establish planar cell polarity, at least initially, and then undergo progressive disorganization of failure to complete their maturation process. The rapid disorganization of hair cell bundles in the *Myo7a* and *Cdh23* mutants suggest that these genes are important in the very early maintenance of bundle polarity.

b. The Transduction Channel. Why do the hair cell bundles have such an elaborate polarization of their stereocilia? Early physiology experiments investigating the transduction mechanism of the hair cell showed through mechanical deflection of the stereocilia that movement toward the tallest stereocilium results in cell depolarization, i.e., opening of ion channels and influx of cations, predominantly K^+. Conversely, deflection toward the shortest stereocilia results in hyperpolarization (closing of ion channels). These early studies also showed that the hair cell transduction channels actually lie somewhere within the bundle (Hudspeth and Jacobs, 1979), probably near their tips, although their location is not precisely known (reviewed in Strassmaier and Gillespie, 2002). It is likely, however, that the tip links, linking shorter with taller stereocilia (see Fig. 4C), are somehow involved in gating the channels (Assad et al., 1991; Pickles et al., 1984), and the speed at which the transduction channels seem to open supports the hypothesis that they are directly gated by mechanical elements and do not involve a second messenger system (reviewed in Strassmaier and Gillespie, 2002). Although the precise identity of the hair cell transduction channel is still unknown, several mutations in deaf mice encode ion channels located in the hair cells.

Mutations in the gene transmembrane cochlear-expressed gene 1 (*Tmc1*) underlies the deafness and beethoven mouse mutants and human nonsyndromic deafness DFNB7/B11 and DFNA36 (Kurima et al., 2002; Vreugde et al., 2002). This novel protein is predicted to have a structure similar to many known ion channels. It is selectively expressed in the hair cells, and in the deafness homozygous animals transduction is never normal (Steel and Bock, 1980). Because this protein is not related to any other known proteins, little is currently known about its role in the hair cells.

A second mouse mutant, varitint-waddler (*Va*), has a mutation in the gene encoding mucolipin 3 (*Mcoln3*; Di Palma *et al.*, 2002). This mouse is deaf and has the classic head bobbing and circling behavior described in so many other mice with vestibular defects (reviewed in Steel, 2002). *Mcoln3* shows sequence similarities with members of the transient receptor potential (TRP) superfamily, which includes the fly mechanoreceptor ion channels (reviewed in Clapham et al., 2001). The TRP ion channels have many of the physiological characteristics expected of the hair cell transduction channel, such as amiloride sensitivity and cation nonselectivity (reviewed in Strassmaier and Gillespie, 2002). Again, further analysis of this gene product is needed to determine its role in auditory transduction.

c. Recent Genetic Insights into Hair Cell Adaptation. The rigidity of the stereocilia, which is crucial for transduction, is attributed to its cytoskeleton. The stereocilia are almost entirely composed of parallel arrays of actin filaments, cross-linked by actin-binding proteins such as fimbrin (Flock *et al.*, 1982) and the recently identified espin protein, which is mutated in the spontaneous recessive mouse mutation, jerker (Zheng *et al.*, 2000b). In the jerker mouse the stereocilia begin to develop, but around the time of auditory function onset they rapidly degenerate, suggesting that espin might play a role in the maintenance of actin integrity during mechanical stimulation. Although the rigidity of the bundle alone cannot account for the sensitivity or functional dynamic range of the hair cell, another property of the bundle might be able to. That is its ability to adapt to mechanical stimulation. Although the physiological analysis of adaptation is beyond the scope of this chapter, the adaptation properties of hair cells are established during differentiation and maturation and are likely to be a property of the cell plasma membrane and cytoskeleton (Kros, *et al.*, 2002). Mouse mutants, and in particular those mice carrying mutations in the myosin gene family, have been extremely informative regarding the biological nature of the adaptation machinery. The clues seem to lie in the ultrastructural condition of mutant hair cell stereociliary bundles.

The unconventional myosin gene, *Myo7a* (described earlier), is mutated in the shaker1 mouse (Gibson *et al.*, 1995). Expression analysis of myosin Vlla in the inner ear vestibular and cochlear sensory patches shows that it is present just prior to the morphological differentiation of hair cells at around E13.5 (reviewed in Zuo, 2002) and is clearly restricted to the hair cells by E15.5. Thus myosin Vlla has proven to be a good hair cell marker (see Rivolta *et al.*, 1998, and Section III.E). Transgenic mice containing a reporter construct incorporating components of the *Myo7a* gene and promoter region have been generated (Boeda *et al.*, 2001) and may prove useful in the future for the generation of conditional transgenic mice with hair-cell–targeted mutations, or indeed for hair-cell–directed gene therapies.

Of the several mutant *Myo7a* alleles that have been described, the least severe, original shaker1 allele, results in the formation of clearly polarized hair bundles that subsequently undergo a process of stereocilia degeneration and resorption (Self *et al.*, 1998). Other alleles show more rapid bundle disorganization and some abnormal kinocilium positioning (Self *et al.*, 1998). It is known that myosin Vlla protein is found along the length of each stereocillium (Hasson *et al.*, 1997b), and Richardson *et al.* (1999) have previously shown that aminoglycoside antibiotics, which normally bind to open ion channels, do not accumulate in the stereocilia of shaker1 mutant mice (Richardson *et al.*, 1999). In addition, Kros *et al.* (2002) showed that the stereocilia of mice with *Myo7a* mutations lose their deflection sensitivity due to the loss of tension within the bundle (Kros *et al.*, 2002). Both the pharmacological and physiological data suggest that at rest and during stimulation the transducer channels do not open as normally expected. This, in conjunction with the anatomical data, suggests that properly functioning myosin Vlla is a requisite for normal transducer channel gating. This may be due to its ability to modulate the tension of the tip links (Kros *et al.*, 2002), probably indirectly by acting as an anchor between the actin core and the cell membrane along the length of the stereocilium.

Myosin Vlla may also anchor the actin cytoskeleton to the plasma membrane in regions where the lateral, interstereociliary links are found (see Fig. 4C). This may facilitate bundle integrity and maintenance of polarity. Evidence for this hypothesis comes from the identification of the transmembrane protein vezatin, which binds to the C-terminal FERM domain of myosin Vlla and also localizes to the lateral links at the base of adjacent vestibular stereocilia (ankle links). Vezatin is abnormally localized in shaker1 mutants (Kussel-Andermann *et al.*, 2000), which exhibit disorganized vestibular, as well as cochlear, hair bundles. Cadherin 23 (Di Palma *et al.*, 2001) has been proposed as a candidate for one of the lateral link proteins and members of the cadherin family, including cadherin 23 (*Cdh23*, mutated in the waltzer mouse [Di Palma *et al.*, 2001]) and protocadherin 15 (*Pcdh15*, mutated in the Ames waltzer mouse [Alagramam et al., 2001]) also result in disorganized stereocilia when disrupted. The phenotype is much like that seen in the shaker1 mice (see Fig. 5). Given that myosin Vlla and vezatin colocalize in the region of a subset of lateral links, that cadherin 23 may be a lateral link protein, and that *Cdh23* and *Pcdh15* mutants share similar disorganized stereocilia phenotypes to the *Myo7a* mutants, these groups of proteins may interact as complexes, at least in the vestibular system, to maintain bundle rigidity and polarity.

Another myosin, myosin 1c (*Myo1c*) has also been implicated more directly in the hair cell transduction adaptation process. It is present in the stereociliary hair bundle (Gillespie et al., 1993; Hasson et al., 1997a) at the upper anchor of the tip link (Garcia et al., 1998; Steyger et al., 1998) and therefore is present in the suspected region of the transducer ion channels. It is thought to restore hair bundle sensitivity during prolonged stimulation by acting as a motor that alters the tension of the gated channels, resulting in the closure of transduction channels. Using a "chemical–genetic" approach, Holt et al. (2002) showed that inhibition of myosin 1c results in the blockade of hair cell adaptation to mechanical stimulation. The key to these experiments was the generation of a mouse containing a mutated version of the *Myo1c* gene. The mutated version of the gene was highly expressed in the vestibular hair cells and was engineered to be susceptible to inhibition by N^6-modified ADP analogs. Whole cell recordings were made from vestibular hair cells and the chemical inhibitor delivered through the recording pipette and as a result adaptation was blocked. However, it is not clear from this study if myosin 1c works independently or in conjunction with myosin Vlla in this process.

2. Outer Hair Cells as Force Generators: Cells on the Move

The transduction channel adaptation described earlier renders the sensory hair cells of the inner ear exquisitely sensitive to mechanical stimuli. However, there is an ATP-independent mechanism by which stimuli are amplified, and this process is intrinsic to the outer hair cells of the organ of Corti. It is thought to increase the sensitivity of the cochlea to sound by over 100-fold and has been termed "active amplification." It has been known since the mid-1980s that outer hair cells can undergo voltage-gated conformational changes; when the outer hair cells depolarize they get shorter and when they hyperpolarize they get longer (Brownell et al., 1985; Kachar et al., 1986; reviewed in Dallos and Fakler, 2002). This knowledge led to the hypothesis that the inner hair cells perform a primary signal detection role, whereas the outer hair cells undergo rapid length changes that mechanically amplify the traveling wave of the basilar membrane in a frequency-specific (cycle-by-cycle) manner (Ashmore, 1987). Until recently, the molecular machinery that underlies outer-hair-cell–specific electromotility was not known. However, subtractive cloning between motile outer hair cells and nonmotile inner hair cells identified a cDNA unique to the outer hair cells (Zheng et al., 2000b). This protein, named prestin, is present from a few days after birth. Evidence that it was indeed the motor protein came from the analysis of mice with targeted disruption of the *prestin* gene. In the absence of prestin, the outer hair cells are nonmotile and although mechanoelectrical transduction is normal, there is a 40 dB loss of auditory sensitivity

D. Endolymph Homeostasis

Mechanoelectrical transduction by hair cells, described in the previous section, occurs by the mechanical opening of the transduction ion channels near the tips of the stereocilia and influx of (mainly) K^+ ions. In order for this to occur, there is an excess of K^+ and low levels of Na^+ in the fluid bathing the hair cell apices (i.e., the endolymph), compared with low K^+ and high Na^+ levels in intracellular fluids. This difference in ionic environments generates a potential difference across the apex of the hair cells, thus assisting the influx of cations through the transduction channels, down a chemical gradient, and resulting in hair cell depolarization. The high levels of K^+ in the endolymphatic compartment of the cochlea, which contributes to the generation of a resting potential or endocochlear potential (EP) of around 100 mV, is maintained by active pumping mechanisms. Through the generation of mice with defects in ion channels, transporters, and pumps, we are now starting to understand the ionic recycling processes that occur in the inner ear (reviewed in Steel, 1999; Steel and Kros, 2001; and Wangemann, 2002; illustrated in Fig. 1C).

The KCNQ4 ion channel is mutated in a form of dominantly inherited progressive hearing loss in humans (Kubisch et al., 1999; see Fig. 1C, Step 1). Currently there is no mouse model for this condition. Removal of K^+ from the hair cells after depolarization is thought to occur via the Kcnq4 ion channel. From here the K^+ ions are taken up by the surrounding supporting cells, possibly by the Kcc4 cotransporter encoded by the *Slc12a7* gene (Boettger et al., 2002; see Fig. 1C, Steps 2 and 3). *Slc12a7* is expressed in the supporting cells adjacent to the inner and outer hair cells of the cochlea, and mice with mutations in this gene become rapidly deaf after the onset of hearing, at around postnatal day 14 (Boettger et al., 2002). From the supporting cells, K^+ ions pass through their network of connexin-rich gap junctions into the mesenchymal fibrocytes in the spiral ligament (Kikuchi et al., 1995; see Fig. 1C, Step 4). Mutation of the transcription factor *Pou3f4*, which is expressed in the cells of the spiral ligament, results in fewer contacts between the fibrocytes and a reduced, but not absent, EP (Minowa et al., 1999).

Connexins are present in all gap junctions between the supporting cells of the inner ear. In mice, there are two independent intercellular gap junction networks: one within the neurosensory epithelium (see Fig. 1C, Steps 2 and 3) that forms around E16 and the other within the connective tissues of the fibrocytes in the spiral ligament and spiral limbus and in the basal, intermediate, and marginal cells of the stria vascularis (see Fig. 1C, Steps 4 and 5). This second network forms around birth. At least four different connexins, GJB2 (connexin 26), GJB3 (connexin 31), GJB6 (connexin 30), and GJA1 (connexin 43), have been implicated in homeostasis of the inner ear endolymph. Mutations affecting any one of these proteins in human results in hearing impairment. Mutations in *GJB2* underlie DFNB1 and are thought to account for up to 50% of all inherited cases of prelingual nonsyndromic deafness (reviewed in Petersen, 2002). Animals with a targeted mutation in *Gjb6* are viable but have no EP and their hair cells degenerate postnatally (Teubner et al., 2003). In contrast, the *Gjb2* knockout mouse is embryonic lethal and so the mechanisms that underlie this prominent form of deafness have been hard to model. However, Cohen-Salmon and colleagues have generated a mouse in which a mutation in the *Gjb2* gene has been targeted only in the epithelia of the organ of Corti (Cohen-Salmon et al., 2002). Using the *Cre-lox* system (Nagy, 2000), they have circumnavigated the problem of embryonic lethality. Briefly, two mouse transgenic lines were generated, one containing two *loxP* sites one on each side of exon II of the *Gjb2* gene and one containing the bacteriophage *P1 Cre recombinase* gene under control of the mouse *Otog* promoter. *Otog* expression is specific to epithelial cells of the otic vesicle. When these two mouse lines were mated, some of the resulting offspring contained both the *loxP*-flanked ("floxed") *Gjb2* and the otic vesicle–targeted *Cre* allele. In these double transgenic mice, the *Cre recombinase* excises the floxed exon II segment of *Gjb2* (this method is reviewed in Zuo, 2002). *Gjb2* is effectively knocked out in these animals, but only in the tissue for which the *Cre* promotor is specific, in this case the epithelial cells of the otic vesicle. In the double transgenic mice, there is an initial loss of supporting cells that are nearest to the inner hair cells. The onset of supporting cell degeneration occurs around the time of hearing onset (around P14 in mouse), when the inner hair cells first begin to transduce. Shortly after the onset of supporting cell degeneration, the inner hair cells also begin to degenerate. The inner hair cell loss may be a secondary effect due to K^+ and/or glutamate accumulation in the perilymphatic fluid bathing the hair cell bodies, which would normally be removed by the supporting cells. However, this early damage to the hair cells and supporting cells of the organ of Corti make it difficult to ask questions about the role of Gjb2 in the K^+ recycling model and so there is more to discover before the picture is complete.

From the supporting cell network and then the spiral ligament, K^+ is pumped into the marginal cells of the stria vascularis (which secretes endolymph) via the Na-K-Cl$_2$ cotransporter, encoded by *Slc12a2*, on the basolateral side of the marginal cells (described in Section II.C; see Fig. 1C, Step 5) (Delpire et al., 1999; Dixon et al., 1999). Finally, ion channels on the apical surface of the strial marginal cells, which are composed of Kcnq1 (KvLQT1) and Kcne1 (Isk), probably regulate the output of K^+ into the scala media (see Fig. 1C, Step 5). Mutation of either of these components in mice results in the disruption of endolymph homeostasis and the collapse of the endolymphatic compartment (Lee et al., 2000; Letts et al., 2000; Vetter et al., 1996; see Section II.C).

Another category of animals that exhibit defects in endolymphatic homeostasis includes mice with defects in the pigment cells of the stria vascularis and dark cells of the vestibular system. Often these mice also have anomalies associated with coat color. Mice lacking the Ephb2 receptor tyrosine kinase in the dark cells of the vestibular system exhibit the circling behaviors associated with vestibular dysfunction and have thinned semicircular canals and reduced endolymph production in the ampullae (Cowan et al., 2000). The cochlea in these animals does not seem to be affected. Other mutants in this category that also have an abnormal EP include Dominant spotting (*Kit*), Steel (*Mgf*), Microphthalmia (*Mitf*), Piebald (*Ednrb*), and Lethal spotting (*Edn3*) (reviewed in Steel et al., 2002). All four genes are known to be important for the survival of newly migrated melanocytes, derived from the neural crest (Cable et al., 1995; Steel et al., 1992). Mutations lead to a reduction or total absence of melanocytes in both the skin (causing white patches) and stria vascularis. Defects are not related to the pigment within the melanocytes because albino animals have no pigment but do have melanocytes and an EP. However, melanocytes appear to be the only cell type in the stria that expresses the *Kcnj10* gene encoding the Kir4-1 potassium channel, and it has been shown that knockout of this channel abolishes the EP (Marcus et al., 2002). This explains the requirement for melanocytes for generating an EP.

E. Survival of the Hair Cell

The hair cell is far from home after its fate has been specified. To fully mature and survive, hair cells need not only the correct environment, as illustrated by hair cell degeneration in mutant mice with homeostasis problems, but also other survival factors, some of which we are now starting to understand. One of the first examples is the transcription factor *Pou4f3* (formerly *Brn3c* or *Brn3.1*). In the developing mouse inner ear, *Pou4f3* is strongly expressed in the hair cells both of the cochlea and the vestibular system as early as E12.5 and continues into adulthood (Xiang et al., 1998). Targeted null mutation of this gene in the mouse results in an absence of hair cells in the early postnatal inner ear (Xiang et al., 1997), accompanied by vestibular dysfunction and profound deafness (Erkman et al., 1996; Xiang et al., 1997). The reason for the absence of hair cells in the *Pou4f3* null mouse is not, as was first thought, because this gene is involved in hair cell specification. *Pou4f3* is only expressed in the postmitotic nascent hair cell, beginning at the early stages of differentiation. In the null mouse hair cells begin to differentiate, as demonstrated by the presence of the early hair-cell–specific markers, myosin VI and myosin VIIa, but then degenerate via apoptosis prior to migration into the upper stratum of the epithelium and prior to the formation of a hair bundle (Xiang *et al.*, 1998). Analysis of the *Pou4f3* knockout mouse suggests that this gene is not necessary for hair cell determination, but it is necessary for some aspects of hair cell differentiation and survival, placing it downstream of *Math1* (see Section III.B).

A mutation in the equivalent gene (*POU4F3*) in humans is known to cause autosomal dominant progressive hearing loss (Vahava *et al.*, 1998). However, this mutation differs from the mouse knockout model because it is dominant and hearing loss onset is relatively late, at 18–30 years. In the human, therefore, *POU4F3* may not affect early differentiation of the hair cells (i.e., early migration and bundle formation), but it may affect their long-term survival.

More recently, in the mouse there have been several genes shown to be important for the long-term survival of hair cells, but not necessarily for their differentiation. For example, mice that have had the transcriptional repressor *Gfi1* (a homologue of the *Drosophila* gene *senseless*) knocked out exhibit apoptotic hair cells in the embryonic inner ear and a complete absence of hair cells by birth, or shortly afterward (Wallis *et al.*, 2003). In contrast to *Pou4f3* null mice, however, the hair cells in the *Gfi1* knockout do seem to be present in the upper epithelium and also seem to develop hair bundles, although no comprehensive ultrastructural analysis is yet available on these mice. This would indicate that *Gfi1* is not necessary for the gross differentiation of hair cells but may be key to their survival. However, the full role of this gene is far from clear because null mutants seem to have more fundamental hair cell organization problems, such as fewer rows of outer hair cells.

A second example is the *Barhl1* homeobox gene, expressed in vestibular and cochlear hair cells at E14.5 and continuing into postnatal stages (Li *et al.*, 2002). In the postnatal cochlea of the mutant, outer hair cells of the apical region degenerate, progressing basally, and later inner hair cells degenerate in a basal-to-apical pattern. Mice exhibit a low-frequency hearing loss at early ages. At first, the early expression of *Barhl1* was taken to suggest that it might have a role in differentiation. However, based on marker-gene analysis, light microscopy, and some ultrastructural analysis, the hair cells in null mice seem to reach a mature differentiation state prior to degenerative events.

A third example of late-onset progressive hair cell loss comes from the beethoven mouse, in which the novel gene *Tmc1* is mutated (Vreugde *et al.*, 2002; see also Section III.C.1.b). In humans, mutations in this gene underlie forms of dominant (DFNA36) and recessive (DFNB7/B11DFN) deafness (Kurima *et al.*, 2002). *Tmc1* is thought to encode a protein with several transmembrane domains, but it is not yet known what the function of the protein might be.

Understanding precisely which part of the hair cell specification, differentiation, and survival processes genes are involved in will clearly be important for the design of appropriate therapeutic approaches to deafness. One of the challenges, however, is going to be discriminating between genes that affect hair cell survival primarily, such as *Gifi1*, and genes that have a secondary effect on survival by providing the correct environment, such as *Pds* and *Gjb2*. However, the variety of patterns of hair cell loss, for whatever reason, seen in the progressive deafness mutants further illustrates the complexity and specificity of factors that may be involved in the survival of hair cells. Further analysis of these late-onset degenerative processes and their causative genes may shed more light on the factors involved in late-onset progressive deafness in humans.

F. Neurogenesis and Neural Defects

The cells forming the VIII cranial nerve that innervate the cochlear and vestibular hair cells arise from within the otic placode very early during placode invagination (reviewed in Fekete and Wu, 2002). The delamination of neuroblasts from the anteroventral otic vesicle involves two bHLH genes, *Neurogenic differentiation 3* (*Neurod3*, previously *neurogenin-1*) (Ma *et al.*, 2000) and *Neurogenic differentiation 1* (*Neurod1*, previously *NeuroD*) (Kim *et al.*, 2001; Liu *et al.*, 2000), and in the cochlea, the zinc finger gene *Gata3* (Rivolta and Holley, 1998). In mice with mutations in *Neurod3* the vestibulocochlear ganglion is absent (Ma *et al.*, 2000). There is also hair cell loss in the sensory epithelia, where the saccule seems to be most affected (Ma *et al.*, 2000).

The neurons of the vestibulocochlear ganglion also require survival factors for their full differentiation and maturation, much like the sensory hair cells. In the ganglion neurons, these factors, or neurotrophins, include nerve growth factor (Ngf), brain-derived neurotrophic factor (Bdnf), neurotrophin 5 (Ntf5), and neurotrophin 3 (Ntf3). Their receptors are neurotrophic tyrosine kinase receptor type 1 (Ntrk1, previously trkA), which is the receptor for Ngf; Ntrk2, the receptor for Bdnf and Ntf5; and Ntrk3, the receptor for Ntf3 (see Chapter 7 by Pirlova and Ylikoski). Targeted deletion of the neurotrophins and their receptors results in neural cell death; loss of both Bdnf/Ntrk2 and Ntf3/Ntrk3 results in a loss of all ganglion neurons (Ernfors *et al.*, 1995; Liebl *et al.*, 1997). For a comprehensive review of other aspects of neurogenesis, see Rubel and Fritzsch (2002) and Chapter 15 by Hashino.

G. Extracellular Membranes

A membranous extracellular matrix covers each of the sensory epithelia in the inner ear. In the cochlea this is the tectorial membrane, in the utricle and saccule this is the otoconial membrane, and in the cristae this is the cupula. Analysis of the development of these structures has been relatively neglected, but their role in sensory transduction is crucial. In each case, the tallest stereocilia of the hair cell bundles project into the overlying matrix. In the cochlea, only the outer hair cell bundles are embedded in the tectorial membrane. In each organ, the incoming mechanical stimulus (e.g., gravity or the traveling wave induced by sound) generates movement of the respective overlying membrane. This is thought to act in a shearing motion against the hair cell apices, resulting in the deflection of the rigid stereociliary bundles and activation of the transduction channels. Each membrane has its own complex architecture. The major component tends to be a collagenase-sensitive material, such as collagen II, V, or IX in the tectorial membrane, and noncollagenous glycoproteins, including α-tectorin (*Tecta*), β-tectorin (*Tectb*), and otogelin (*Otog*). Mutation of *Tecta* in the mouse results in fewer otoconia in the otoconial membranes of the maculae, and in the cochlea the tectorial membrane is detached from the organ of Corti and lacks all noncollagenous material (Legan *et al.*, 2000). These mice do not exhibit balance defects. On the other hand, mutation of *Otog*, affecting another glycoprotein, otogelin, in both a targeted disruption (Simmler *et al.*, 2000a) and a presumed allelic mutation, twister (Simmler *et al.*, 2000b), results in severe balance defects in mice. The extracellular membranes in all of the vestibular organs are detached from the underlying sensory epithelium; there are fewer otoconia in the maculae; and in the cochlea, the fibrillar network is disorganized in the tectorial membrane (Simmler *et al.*, 2000a). The tectorial membrane of mice with a targeted disruption of *Col11a2* also exhibits disorganized collagen fibrils and has a midfrequency hearing loss (McGuirt *et al.*, 1999). Mutational analysis has therefore allowed the components of these extracellular matrices to be dissected and illustrates that all of the components are necessary for the normal functioning of the membranes. Interestingly, in each case the sensory hair cells do not seem to be affected by disruption of their overlying membrane, although transduction sensitivity is reduced (Legan *et al.*, 2000).

IV. Concluding Remarks

This chapter has reviewed the advances in our understanding of inner ear development that have been most facilitated by the analysis of mutant mice. Targeted mutagenesis models (knock-ins and knockouts) are increasingly being used to study human deafness and also as tools to understand fundamental developmental processes. On the other hand, phenotype-driven mutagenesis programs have already been incredibly successful in the identification of novel genes involved in inner ear development and function. Whereas gene-driven strategies can target any part of the developmental system, many of the mutants derived from phenotype-driven screens seem to have either lateral semicircular canal truncations or neuroepithelial phenotypes. This may be a limitation of the screening process, which uses the most profound outcome of gene alteration as a measure of effect, i.e., deafness or an inability to balance correctly. It may be that, functionally speaking, an animal can appear normal following the inactivation of certain genes; it is well known that there is a large degree of functional redundancy in our genetic makeup, meaning that the function of an absent gene is often adopted by a closely related gene. Analysis of the effect of gene mutation by whichever approach is extremely exciting. Already the prestin gene has been shown to encode the protein that is the molecular motor of outer hair cell motility, and genes encoding some of the unconventional myosins such as myosin VIIa and myosin 1c are known to be crucial to the transduction process fundamental to the inner ear's sensory role. But there are still many more unknown molecules playing key roles in hearing and balance. However, the challenges of discovering these unknown molecules will be greatly facilitated by the molecular methods now in our toolbox.

References

Abdelhak, S., Kalatzis, V., Heilig, R., Compain, S., Samson, D., *et al.* (1997). A human homologue of the *Drosophila eyes absent* gene underlies branchio-oto-renal (BOR) syndrome and identifies a novel gene family. *Nat. Genet.* **15**, 157–164.

Adler, P. N. (2002). Planar signaling and morphogenesis in *Drosophila*. *Dev. Cell* **2**, 525–535.

Ahituv, N., and Avraham, K. B. (2002). Mouse models for human deafness: Current tools for new fashions. *Trends Mol. Med.* **8**, 447–451.

Alagramam, K. N., Murcia, C. L., Kwon, H. Y., Pawlowski, K. S., Wright, C. G., and Woychik, R. P. (2001). The mouse Ames waltzer hearing-loss mutant is caused by mutation of Pcdh 15, a novel protocadherin gene. *Nat. Genet.* **27**, 99–102.

Ashmore, J. F. (1987). Links Abstract A fast motile response in guinea-pig outer hair cells: The cellular basis of the cochlear amplifier. *J Physiol* **388**, 323–47.

Assad, J. A., Shepherd, G. M., and Corey, D. P. (1991). Tip-link integrity and mechanical transduction in vertebrate hair cells. *Neuron* **7**, 985–994.

Avraham, K. B., Hasson, T., Steel, K. P., Kingsley, D. M., Russell, L. B., et al. (1995). The mouse Snell's waltzer deafness gene encodes an unconventional myosin required for structural integrity of inner ear hair cells. *Nat. Genet.* **11**, 369–375.

Bermingham, N. A., Hassan, B. A., Price, S. D., Vollrath, M. A., Ben-Arie, N., et al. (1999). Math1: An essential gene for the generation of inner ear hair cells. *Science* **284**, 1837–1841.

Bitner-Glindzicz, M., Lindley, K. J., Rutland, P., Blaydon, D., Smith, V. V., et al. (2000). A recessive contiguous gene deletion causing infantile hyperinsulinism, enteropathy and deafness identifies the Usher type 1C gene. *Nat. Genet.* **26**, 56–60.

Boeda, B., El-Amraoui, A., Bahloul, A., Goodyear, R., Daviet, L., et al. (2002). Myosin VIIa, harmonin and cadherin 23, three Usher 1 gene products that cooperate to shape the sensory hair cell bundle. *EMBO J* **21**, 6689–6699.

Boeda, B., Weil, D., and Petit, C. (2001). A specific promoter of the sensory cells of the inner ear defined by transgenesis. *Hum. Mol. Genet* **10**, 1581–1589.

Boettger, T., Hubner, C. A., Maier, H., Rust, M. B., Beck, F. X., and Jentsch, T. J. (2002). Deafness and renal tubular acidosis in mice lacking the K-Cl co-transporter Kcc4. *Nature* **416**, 874–878.

Brownell, W. E., Bader, C. R., Bertrand, D., de Ribaupierre, Y. (1985). Evoked mechanical responses of isolated cochlear outer hair cells. *Science* **11**, 227 194–6.

Cable, J., Jackson, I. J., and Steel, K. P. (1995). Mutations at the W locus affect survival of neural crest-derived melanocytes in the mouse. *Mech. Dev.* **50**, 139–150.

Cantos, R., Cole, L. K., Acampora, D., Simeone, A., and Wu, D. K. (2000). Patterning of the mammalian cochlea. *Proc. Natl. Acad. Sci. USA* **97**, 11707–11713.

Carney, P. R., and Silver, J. (1983). Studies on cell migration and axon guidance in the developing distal auditory system of the mouse. *J. Comp. Neurol.* **215**, 359–369.

Chen, P., Johnson, J. E., Zoghbi, H. Y., and Segil, N. (2002). The role of Math 1 in inner ear development: Uncoupling the establishment of the sensory primordium from hair cell fate determination. *Development* **129**, 2495–2505.

Chen, P., and Segil, N. (1999). p27(Kip1) links cell proliferation to morphogenesis in the developing organ of Corti. *Development* **126**, 1581–1590.

Clapham, D. E., Runnels, L. W., and Strubing, C. (2001). The TRP ion channel family. *Nat. Rev. Neurosci.* **2**, 387–396.

Cohen-Salmon, M., Ott, T., Michel, V., Hardelin, J. P., Perfettini, I., et al. (2002). Targeted ablation of connexin26 in the inner ear epithelial gap junction network causes hearing impairment and cell death. *Curr. Biol.* **12**, 1106–1111.

Colvin, J. S., Bohne, B. A., Harding, G. W., McEwen, D. G., and Ornitz, D. M. (1996). Skeletal overgrowth and deafness in mice lacking fibroblast growth factor receptor 3. *Nat. Genet.* **12**, 390–397.

Corwin, J. T., and Cotanche, D. A. (1988). Regeneration of sensory hair cells after acoustic trauma. *Science* **240**, 1772–1774.

Cowan, C. A., Yokoyama, N., Bianchi, L. M., Henkemeyer, M., and Fritzsch, B. (2000). EphB2 guides axons at the midline and is necessary for normal vestibular function. *Neuron* **26**, 417–430.

Dallos, P., and Fakler, B. (2002). Prestin, a new type of motor protein. *Nat. Rev. Mol. Cell Biol.* **3**, 104–11.

Delpire, E., Lu, J., England, R., Dull, C., and Thorne, T. (1999). Deafness and imbalance associated with inactivation of the secretory Na-K-2Cl co-transporter. *Nat. Genet.* **22**, 192–195.

Denman-Johnson, K., and Forge, A. (1999). Establishment of hair bundle polarity and orientation in the developing vestibular system of the mouse. *J. Neurocytol.* **28**, 821–835.

Deol, M. S. (1963). The development of the inner ear in mice homozygous for shaker-with-syndactylism. *J. Embrol. Exp. Morph.* **11**, 493–512.

Di Palma, F., Belyantseva, I. A., Kim, H. J., Vogt, T. F., Kachar, B., and Noben-Trauth, K. (2002). Mutations in Mcoln3 associated with deafness and pigmentation defects in varitint-waddler (Va) mice. *Proc. Natl. Acad. Sci. USA* **99**, 14994–14999.

Di Palma, F., Holme, R. H., Bryda, E. C., Belyantseva, I. A., Pellegrino, R., et al. (2001). Mutations in Cdh23, encoding a new type of cadherin, cause stereocilia disorganization in waltzer, the mouse model for Usher syndrome type 1D. *Nat. Genet.* **27**, 103–107.

Dixon, M. J., Gazzard, J., Chaudhry, S. S., Sampson, N., Schulte, B. A., and Steel, K. P. (1999). Mutation of the Na-K-Cl co-transporter gene Slc12a2 results in deafness in mice. *Hum. Mol. Genet.* **8**, 1579–1584.

Doetschman, T., Gregg, R. G., Maeda, N., Hooper, M. L., Melton, D. W., et al. (1987). Targetted correction of a mutant HPRT gene in mouse embryonic stem cells. *Nature* **330**, 576–578.

Eddison, M., Le Roux, I., and Lewis, J. (2000). Notch signaling in the development of the inner ear: Lessons from *Drosophila*. *Proc. Natl. Acad. Sci. USA* **97**, 11692–11699.

Erkman, L., McEvilly, R. J., Luo, L., Ryan, A. K., Hooshmand, F., et al. (1996). Role of transcription factors Brn-3.1 and Brn-3.2 in auditory and visual system development. *Nature* **381**, 603–606.

Ernfors, P., Van De Water, T., Loring, J., and Jaenisch, R. (1995). Complementary roles of BDNF and NT-3 in vestibular and auditory development. *Neuron* **14**, 1153–1164.

Erven, A., Skynner, M. J., Okumura, K., Takebayashi, S., Brown, S. D., et al. (2002). A novel stereocilia defect in sensory hair cells of the deaf mouse mutant Tasmanian devil. *Eur. J. Neurosci.* **16**, 1433–1441.

Everett, L. A., Belyantseva, I. A., Noben-Trauth, K., Cantos, R., Chen, A., et al. (2001). Targeted disruption of mouse Pds provides insight about the inner-ear defects encountered in Pendred syndrome. *Hum. Mol. Genet.* **10**, 153–161.

Everett, L. A., Glaser, B., Beck, J. C., Idol, J. R., Buchs, A., et al. (1997). Pendred syndrome is caused by mutations in a putative sulphate transporter gene (PDS). *Nat. Genet.* **17**, 411–422.

Favor, J., Sandulache, R., Neuhauser-Klaus, A., Pretsch, W., Chatterjee, B., et al. (1996). The mouse Pax2(1Neu) mutation is identical to a human PAX2 mutation in a family with renal-coloboma syndrome and results in developmental defects of the brain, ear, eye, and kidney. *Proc. Natl. Acad. Sci. USA* **93**, 13870–13875.

Fekete, D. M., Muthukumar, S., and Karagogeos, D. (1998). Hair cells and supporting cells share a common progenitor in the avian inner ear. *J. Neurosci.* **18**, 7811–7821.

Fekete, D. M., and Wu, D. K. (2002). Revisiting cell fate specification in the inner ear. *Curr. Opin. Neurobiol.* **12**, 35–42.

Flock, A., Bretscher, A., and Weber, K. (1982). Immunohistochemical localization of several cytoskeletal proteins in inner ear sensory and supporting cells. *Hear. Res.* **7**, 75–89.

Forge, A., Li, L., Corwin, J. T., and Nevill, G. (1993). Ultrastructural evidence for hair cell regeneration in the mammalian inner ear. *Science* **259**, 1616–1619.

Forge, A., and Wright, T. (2002). The molecular architecture of the inner ear. *Br. Med. Bull.* **63**, 5–24.

Fortnum, H. M., Summerfield, A. Q., Marshall, D. H., Davis, A. C., and Bamford, J. M. (2001). Prevalence of permanent childhood hearing impairment in the United Kingdom and implications for universal neonatal hearing screening: Questionnaire based ascertainment study. *BMJ* **323**, 536–540.

Garcia, J. A., Yee, A. G., Gillespie, P. G., and Corey, D. P. (1998). Localization of myosin-Ibeta near both ends of tip links in frog saccular hair cells. *J. Neurosci.* **18**, 8637–8647.

Gibson, F., Walsh, J., Mburu, P., Varela, A., Brown, K. A., et al. (1995). A type VII myosin encoded by the mouse deafness gene shaker-1. *Nature* **374**, 62–64.

Gillespie, P. G., Wagner, M. C., and Hudspeth, A. J. (1993). Identification of a 120 kd hair-bundle myosin located near stereociliary tips. *Neuron* **11**, 581–594.

Gordon, J. W., Scangos, G. A., Plotkin, D. J., Barbosa, J. A., and Ruddle, F. H. (1980). Genetic transformation of mouse embryos by microinjection of purified DNA. *Proc. Natl. Acad. Sci. USA* **77**, 7380–7384.

Hadrys, T., Braun, T., Rinkwitz-Brandt, S., Arnold, H. H., and Bober, E. (1998). Nkx5-1 controls semicircular canal formation in the mouse inner ear. *Development* **125**, 33–39.

Hasson, T., Gillespie, P. G., Garcia, J. A., MacDonald, R. B., Zhao, Y., *et al.* (1997). Unconventional myosins in inner-ear sensory epithelia. *J. Cell. Biol.* **137**, 1287–1307.

Hasson, T., Walsh, J., Cable, J., Mooseker, M. S., Brown, S. D., and Steel, K. P. (1997). Effects of shaker-1 mutations on myosin-Vlla protein and mRNA expression. *Cell Motil. Cytoskeleton* **37**, 127–138.

Holme, R. H., Kiernan, B. W., Brown, S. D., and Steel, K. P. (2002). Elongation of hair cell stereocilia is defective in the mouse mutant whirler. *J. Comp. Neurol.* **450**, 94–102.

Holt, J. R., Gillespie, S. K., Provance, D. W., Shah, K., Shokat, K. M., Corey, D. P., Mercer, J. A., and Gillespie, P. G. (2002). A chemical-genetic strategy implicates myosin-1c in adaptation by hair cells. *Cell* **108**, 371–81.

Hooper, M., Hardy, K., Handyside, A., Hunter, S., and Monk, M. (1987). HPRT-deficient (Lesch-Nyhan) mouse embryos derived from germline colonization by cultured cells. *Nature* **326**, 292–295.

Hrabe de Angelis, M. H., Flaswinkel, H., Fuchs, H., Rathkolb, B., Soewarto, D., *et al.* (2000). Genome-wide, large-scale production of mutant mice by ENU mutagenesis. *Nat. Genet.* **25**, 444–447.

Hudspeth, A. J., and Jacobs, R. (1979). Stereocilia mediate transduction in vertebrate hair cells (auditory system/cilium/vestibular system). *Proc. Natl. Acad. Sci. USA* **76**, 1506–1509.

Justice, M. J., Noveroske, J. K., Weber, J. S., Zheng, B., and Bradley, A. (1999). Mouse ENU mutagenesis. *Hum. Mol. Genet.* **8**, 1955–1963.

Kachar, B., Brownell, W. E., Altschuler, R., and Fex, J. (1986). Electrokinetic shape changes of cochlear outer hair cells. *Nature* **322**, 365–8.

Kaltenbach, J. A., Falzarano, P. R., and Simpson, T. H. (1994). Postnatal development of the hamster cochlea. II. Growth and differentiation of stereocilia bundles. *J. Comp. Neurol.* **350**, 187–198.

Kaufman, M. H. (1992). "The Atlas of Mouse Development." Academic Press, San Diego, CA.

Kiernan, A. E., Ahituv, N., Fuchs, H., Balling, R., Avraham, K. B., *et al.* (2001). The Notch ligand Jagged1 is required for inner ear sensory development. *Proc. Natl. Acad. Sci. USA* **98**, 3873–3878.

Kiernan, A. E., and Steel, K. P. (2000). Mouse homologues for human deafness. *Adv. Otorhinolaryngol.* **56**, 233–243.

Kiernan, A. E., Zalzman, M., Fuchs, H., de Angelis, M. H., Balling, R., *et al.* (1999). Tailchaser (Tlc): A new mouse mutation affecting hair bundle differentiation and hair cell survival. *J. Neurocytol.* **28**, 969–985.

Kikuchi, T., Kimura, R. S., Paul, D. L., and Adams, J. C. (1995). Gap junctions in the rat cochlea: Immunohistochemical and ultrastructural analysis. *Anat. Embryol. (Berl)* **191**, 101–118.

Kim, W. Y., Fritzsch, B., Serls, A., Bakel, L. A., Huang, E. J., *et al.* (2001). NeuroD-null mice are deaf due to a severe loss of the inner ear sensory neurons during development. *Development* **128**, 417–426.

Kros, C. J., Marcotti, W., van Netten, S. M., Self, T. J., *et al.* (2002). Reduced climbing and increased slipping adaptation in cochlear hair cells of mice with Myo7a mutations. *Nat. Neurosci.* **5**, 41–47.

Kubisch, C., Schroeder, B. C., Friedrich, T., Lutjohann, B., El-Amraoui, A., *et al.* (1999). KCNQ4, a novel potassium channel expressed in sensory outer hair cells, is mutated in dominant deafness. *Cell* **96**, 437–446.

Kurima, K., Peters, L. M., Yang, Y., Riazuddin, S., Ahmed, Z. M., et al. (2002). Dominant and recessive deafness caused by mutations of a novel gene, TMC1, required for cochlear hair-cell function. *Nat. Genet.* **30**, 277–284.

Kussel-Andermann, P., El-Amraoui, A., Safieddine, S., Nouaille, S., Perfettini, I., et al. (2000). Vezatin, a novel transmembrane protein, bridges myosin VIIA to the cadherin-catenins complex. *EMBO J* **19**, 6020–6029.

Lander, E. S., Linton, L. M., Birren, B., Nusbaum, C., Zody, M. C., et al. (2001). Initial sequencing and analysis of the human genome. *Nature* **409**, 860–921.

Lanford, P. J., Lan, Y., Jiang, R., Lindsell, C., Weinmaster, G., et al. (1999). Notch signalling pathway mediates hair cell development in mammalian cochlea. *Nat. Genet.* **21**, 289–292.

Lanford, P. J., Shailam, R., Norton, C. R., Gridley, T., and Kelley, M. W. (2000). Expression of Math1 and HES5 in the cochleae of wildtype and Jag2 mutant mice. *J. Assoc. Res. Otolaryngol.* **1**, 161–171.

Lee, M. P., Ravenel, J. D., Hu, R. J., Lustig, L. R., Tomaselli, G., et al. (2000). Targeted disruption of the Kvlqt1 gene causes deafness and gastric hyperplasia in mice. *J. Clin. Invest.* **106**, 1447–1455.

Legan, P. K., Lukashkina, V. A., Goodyear, R. J., Kossi, M., Russell, I. J., and Richardson, G. P. (2000). A targeted deletion in alpha-tectorin reveals that the tectorial membrane is required for the gain and timing of cochlear feedback. *Neuron* **28**, 273–285.

Letts, V. A., Valenzuela, A., Dunbar, C., Zheng, Q. Y., Johnson, K. R., and Frankel, W. N. (2000). A new spontaneous mouse mutation in the Kcne1 gene. *Mamm. Genome* **11**, 831–835.

Lewis, J., and Davies, A. (2002). Planar cell polarity in the inner ear: How do hair cells acquire their oriented structure? *J. Neurobiol.* **53**, 190–201.

Li, C. W., Van De Water, T. R., and Ruben, R. J. (1978). The fate mapping of the eleventh and twelfth day mouse otocyst: An *in vitro* study of the sites of origin of the embryonic inner ear sensory structures. *J. Morphol.* **157**, 249–267.

Li, S., Price, S. M., Cahill, H., Ryugo, D. K., Shen, M. M., and Xiang, M. (2002). Hearing loss caused by progressive degeneration of cochlear hair cells in mice deficient for the Barhl1 homeobox gene. *Development* **129**, 3523–3532.

Lieberman, M. C., Gao, J., He, D. Z., Wu, X., Jia, S., and Zuo, J. (2002). Prestin is required for electromotility of the outer hair cell and for the cochlear amplifier. *Nature* **19**, 300–4.

Liebl, D. J., Tessarollo, L., Palko, M. E., and Parada, L. F. (1997). Absence of sensory neurons before target innervation in brain-derived neurotrophic factor-, neurotrophic 3-, and TrkC-deficient embryonic mice. *J. Neurosci.* **17**, 9113–9121.

Littlewood Evans, A., and Muller, U. (2000). Stereocilia defects in the sensory hair cells of the inner ear in mice deficient in integrin alpha8beta1. *Nat. Genet.* **24**, 424–428.

Liu, M., Pereira, F. A., Price, S. D., Chu, M. J., Shope, C., et al. (2000). Essential role of BETA2/NeuroD1 in development of the vestibular and auditory systems. *Genes Dev.* **14**, 2839–2854.

Liu, W., Li, G., Chien, J. S., Raft, S., Zhang, H., et al. (2002). Sonic hedgehog regulates otic capsule chondrogenesis and inner ear development in the mouse embryo. *Dev. Biol.* **248**, 240–250.

Liu, X. Z., Walsh, J., Mburu, P., Kendrick-Jones, J., Cope, M. J., et al. (1997a). Mutations in the myosin VIIA gene cause non-syndromic recessive deafness. *Nat. Genet.* **16**, 188–190.

Liu, X. Z., Walsh, J., Tamagawa, Y., Kitamura, K., Nishizawa, M., et al. (1997b). Autosomal dominant non-syndromic deafness caused by a mutation in the myosin VIIA gene. *Nat. Genet.* **17**, 268–269.

Lowenheim, H., Furness, D. N., Kil, J., Zinn, C., Gultig, K., et al. (1999). Gene disruption of p27(Kip1) allows cell proliferation in the postnatal and adult organ of Corti. *Proc. Natl. Acad. Sci. USA* **96**, 4084–4088.

Ma, Q., Anderson, D. J., and Fritzsch, B. (2000). Neurogenin 1 null mutant ears develop fewer, morphologically normal hair cells in smaller sensory epithelia devoid of innervation. *J. Assoc. Res. Otolaryngol.* **1**, 129–143.

Marazita, M. L., Ploughman, L. M., Rawlings, B., Remington, E., Arnos, K. S., and Nance, W. E. (1993). Genetic epidemiological studies of early-onset deafness in the U.S. school-age population. *Am. J. Med. Genet.* **46**, 486–491.

Marcus, D. C., Wu, T., Wangemann, P., and Kofuji, P. (2002). KCNJ10 (Kir4.1) potassium channel knockout abolishes endocochlear potential. *Am. J. Physiol. Cell Physiol.* **282**, C403–C407.

McGuirt, W. T., Prasad, S. D., Griffith, A. J., Kunst, H. P., Green, G. E., Melchondia, *et al.* (2001). Mutations in COL11A2 cause non-syndromic hearing loss (DFNA13). *Nat. Genet.* **23**, 413–419.

Minowa, O., Ikeda, K., Sugitani, Y., Oshima, T., Nakai, S., *et al.* (1999). Altered cochlear fibrocytes in a mouse model of DFN3 nonsyndromic deafness. *Science* **285**, 1408–1411.

Morrison, A., Hodgetts, C., Gossler, A., Hrabe de Angelis, M., and Lewis, J. (1999). Expression of Delta1 and Serrate1 (Jagged1) in the mouse inner ear. *Mech. Dev.* **84**, 169–172.

Morton, N. E. (1991). Genetic epidemiology of hearing impairment. *Ann. NY Acad. Sci.* **630**, 16–31.

Nagy, A. (2000). Cre recombinase: The universal reagent for genome tailoring. *Genesis* **26**, 99–109.

Nolan, P. M., Peters, J., Strivens, M., Rogers, D., Hagan, J., *et al.* (2000). A systematic, genome-wide, phenotype-driven mutagenesis programme for gene function studies in the mouse. *Nat. Genet.* **25**, 440–443.

Okabe, M., Ikawa, M., Kominami, K., Nakanishi, T., and Nishimune, Y. (1997). 'Green mice' as a source of ubiquitous green cells. *FEBS Lett.* **407**, 313–319.

Palmiter, R. D., Brinster, R. L., Hammer, R. E., Trumbauer, M. E., Rosenfeld, M. G., *et al.* (1982a). Dramatic growth of mice that develop from eggs microinjected with metallothionein-growth hormone fusion genes. *Nature* **300**, 611–615.

Palmiter, R. D., Chen, H. Y., and Brinster, R. L. (1982b). Differential regulation of metallothionein-thymidine kinase fusion genes in transgenic mice and their offspring. *Cell* **29**, 701–710.

Petersen, M. B. (2002). Non-syndromic autosomal-dominant deafness. *Clin. Genet.* **62**, 1–13.

Pickles, J. O., and Chir, B. (2002). Roles of fibroblast growth factors in the inner ear. *Audiol. Neurootol.* **7**, 36–39.

Pickles, J. O., Comis, S. D., and Osborne, M. P. (1984). Cross-links between stereocilia in the guinea pig organ of Corti, and their possible relation to sensory transduction. *Hear. Res.* **15**, 103–112.

Pirvola, U., Ylikoski, J., Trokovic, R., Hebert, J. M., McConnell, S. K., and Partanen, J. (2002). FGFR1 is required for the development of the auditory sensory epithelium. *Neuron* **35**, 671–680.

Pratt, T., Sharp, L., Nichols, J., Price, D. J., and Mason, J. O. (2000). Embryonic stem cells and transgenic mice ubiquitously expressing a tau-tagged green fluorescent protein. *Dev. Biol.* **228**, 19–28.

Probst, F. J., Fridell, R. A., Raphael, Y., Saunders, T. L., Wang, A., *et al.* (1998). Correction of deafness in shaker-2 mice by an unconventional myosin in a BAC transgene. *Science* **280**, 1444–1447.

Quint, E., Smith, A., Avaron, F., Laforest, L., Miles, J., *et al.* (2002). Bone patterning is altered in the regenerating zebrafish caudal fin after ectopic expression of sonic hedgehog and bmp2b or exposure to cyclopamine. *Proc. Natl. Acad. Sci. USA* **99**, 8713–8718.

Quint, E., Zerucha, T., and Ekker, M. (2000). Differential expression of orthologous Dlx genes in zebrafish and mice: Implications for the evolution of the Dlx homeobox gene family. *J. Exp. Zool.* **288**, 235–241.

Riccomagno, M. M., Martinu, L., Mulheisen, M., Wu, D. K., and Epstein, D. J. (2002). Specification of the mammalian cochlea is dependent on Sonic hedgehog. *Genes Dev.* **16**, 2365–2378.

Richardson, G. P., Forge, A., Kros, C. J., Marcotti, W., Becker, D., *et al.* (1999). A missense mutation in myosin VIIA prevents aminoglycoside accumulation in early postnatal cochlear hair cells. *Ann. NY Acad. Sci.* **884**, 110–124.

Rivolta, M. N., Grix, N., Lawlor, P., Ashmore, J. F., Jagger, D. J., and Holley, M. C. (1998). Auditory hair cell precursors immortalized from the mammalian inner ear. *Proc. R Soc. Lond. B Biol. Sci.* **265**, 1595–1603.

Rivolta, M. N., and Holley, M. C. (1998). GATA3 is downregulated during hair cell differentiation in the mouse cochlea. *J. Neurocytol.* **27**, 637–647.

Romand, R., Sapin, V., and Dolle, P. (1998). Spatial distributions of retinoic acid receptor gene trascripts in the prenatal mouse inner ear. *J. Comp. Neurol.* **393**, 298–308.

Romand, R., Hashino, E., Dolle, P., Vonesch, J. L., Chambon, P., and Ghyselinck, N. B. (2002). The retinoic acid receptors RARalpha and RARgamma are required for inner ear development. *Mech. Dev.***119**, 213–23.

Rubel, E. W., and Fritzsch, B. (2002). Auditory system development: Primary auditory neurons and their targets. *Annu. Rev. Neurosci.* **25**, 51–101.

Ruben, R. J. (1967). Development of the inner ear of the mouse: A radioautographic study of terminal mitoses. *Acta Otolaryngol. Suppl.* **220**, 1–44.

Ryals, B. M., and Rubel, E. W. (1988). Hair cell regeneration after acoustic trauma in adult Coturnix quail. *Science* **240**, 1774–1776.

Schimenti, J., and Bucan, M. (1998). Functional genomics in the mouse: Phenotype-based mutagenesis screens. *Genome Res.* **8**, 698–710.

Self, T., Mahony, M., Fleming, J., Walsh, J., Brown, S. D., and Steel, K. P. (1998). Shaker-1 mutations reveal roles for myosin VIIA in both development and function of cochlear hair cells. *Development* **125**, 557–566.

Self, T., Sobe, T., Copeland, N. G., Jenkins, N. A., Avraham, K. B., and Steel, K. P. (1999). Role of myosin VI in the differentiation of cochlear hair cells. *Dev. Biol.* **214**, 331–341.

Shima, Y., Copeland, N. G., Gilbert, D. J., Jenkins, N. A., Chisaka, O., *et al.* (2002). Differential expression of the seven-pass transmembrane cadherin genes Celsr1-3 and distribution of the Celsr2 protein during mouse development. *Dev. Dyn.* **223**, 321–332.

Simmler, M. C., Cohen-Salmon, M., El-Amraoui, A., Guillaud, L., Benichou, J. C., *et al.* (2000a). Targeted disruption of otog results in deafness and severe imbalance. *Nat. Genet.* **24**, 139–143.

Simmler, M. C., Zwaenepoel, I. I., Verpy, E., Guillaud, L., Elbaz, C., *et al.* (2000b). Twister mutant mice are defective for otogelin, a component specific to inner ear acellular membranes. *Mamm. Genome.* **11**, 961–966.

Skynner, M. J., Drage, D. J., Dean, W. L., Turner, S., Watt, D. J., and Allen, N. D. (1999). Transgenic mice ubiquitously expressing human placental alkaline phosphatase (PLAP): An additional reporter gene for use in tandem with beta-galactosidase (lacZ). *Int. J. Dev. Biol.* **43**, 85–90.

Steel, K. P. (1999). Perspectives: Biomedicine. The benefits of recycling. *Science* **285**, 1363–1364.

Steel, K. P. (2002). Varitint-waddler: A double whammy for hearing. *Proc. Natl. Acad. Sci. USA* **99**, 14613–14615.

Steel, K. P., and Bock, G. R. (1980). The nature of inherited deafness in deafness mice. *Nature* **288**, 159–161.

Steel, K. P., Davidson, D. R., and Jackson, I. J. (1992). TRP-2/DT, a new early melanoblast marker, shows that steel growth factor (c-kit ligand) is a survival factor. *Development* **115**, 1111–1119.

Steel, K. P., and Kimberling, W. (1996). Approaches to Understanding the Molecular Genetics of Hearing and Deafness. pp 10–40. Clinical Aspects of Hearing. Eds. Van De Water T. R., Popper a. N., Fay R. R. Springer Verlag New York.

Steel, K. P., Erven, A., and Kiernan, A. E. (2002). Mice as models for Human Hereditary Deafness pp 247–296, in: Genetics and Auditory Disorders. Eds. Keats B. J. B, Popper, A. N., Fay, R. R., Springer Verlag New York.

Steel, K. P., and Kros, C. J. (2001). A genetic approach to understanding auditory function. *Nat. Genet.* **27**, 143–149.

Steyger, P. S., Gillespie, P. G., and Baird, R. A. (1998). Myosin Ibeta is located at tip link anchors in vestibular hair bundles. *J. Neurosci.* **18**, 4603–4315.

Stock, D. W., Ellies, D. L., Zhao, Z., Ekker, M., Ruddle, F. H., and Weiss, K. M. (1996). The evolution of the vertebrate Dlx gene family. *Proc. Natl. Acad. Sci. USA* **93**, 10858–10863.

Strassmaier, M., and Gillespie, P. G. (2002). The hair cell's transduction channel. *Curr. Opin. Neurobiol.* **12**, 380–386.

ten Berge, D., Brouwer, A., Korving, J., Martin, J. F., and Meijlink, F. (1998). Prx1 and Prx2 in skeletogenesis: Roles in the craniofacial region, inner ear and limbs. *Development* **125**, 3831–3842.

Teubner, B., Michel, V., Pesch, J., Lautermann, J., Cohen-Salmon, M., et al. (2003). Connexin30 (Gjb6)-deficiency causes severe hearing impairment and lack of endocochlear potential. *Hum. Mol. Genet.* **12**, 13–21.

Thomas, K. R., and Capecchi, M. R. (1987). Site-directed mutagenesis by gene targeting in mouse embryo-derived stem cells. *Cell* **51**, 503–512.

Torres, M., Gomez-Pardo, E., and Gruss, P. (1996). Pax2 contributes to inner ear patterning and optic nerve trajectory. *Development* **122**, 3381–3391.

Tsai, H., Hardisty, R. E., Rhodes, C., Kiernan, A. E., Roby, P., et al. (2001). The mouse slalom mutant demonstrates a role for Jagged1 in neuroepithelial patterning in the organ of Corti. *Hum. Mol. Genet.* **10**, 507–512.

Vahava, O., Morell, R., Lynch, E. D., Weiss, S., Kagan, M. E., et al. (1998). Mutation in transcription factor POU4F3 associated with inherited progressive hearing loss in humans. *Science* **279**, 1950–1954.

Venter, J. C., Adams, M. D., Myers, E. W., Li, P. W., Mural, R. J., et al. (2001). The sequence of the human genome. *Science* **291**, 1304–1351.

Vetter, D. E., Mann, J. R., Wangemann, P., Liu, J., McLaughlin, K. J., et al. (1996). Inner ear defects induced by null mutation of the isk gene. *Neuron* **17**, 1251–1264.

Vreugde, S., Erven, A., Kros, C. J., Marcotti, W., Fuchs, H., et al. (2002). Beethoven, a mouse model for dominant, progressive hearing loss DFNA36. *Nat. Genet.* **30**, 257–258.

Wakayama, T., Perry, A. C., Zuccotti, M., Johnson, K. R., and Yanagimachi, R. (1998). Full-term development of mice from enucleated oocytes injected with cumulus cell nuclei. *Nature* **394**, 369–374.

Wallis, D., Hamblen, M., Zhou, Y., Venken, K. J., Schumacher, A., et al. (2003). The zinc finger transcription factor Gfi1, implicated in lymphomagenesis, is required for inner ear hair cell differentiation and survival. *Development* **130**, 221–232.

Wang, W., Van De Water, T., and Lufkin, T. (1998). Inner ear and maternal reproductive defects in mice lacking the Hmx3 homeobox gene. *Development* **125**, 621–634.

Wangemann, P. (2002). K+ cycling and the endocochlear potential. *Hear. Res.* **165**, 1–9.

Warchol, M. E., Lambert, P. R., Goldstein, B. J., Forge, A., and Corwin, J. T. (1993). Regenerative proliferation in inner ear sensory epithelia from adult guinea pigs and humans. *Science* **259**, 1619–1622.

Waterston, R. H., Lindblad-Toh, K., Birney, E., Rogers, J., Abril, J. F., *et al.* (2002). Initial sequencing and comparative analysis of the mouse genome. *Nature* **420**, 520–562.

Weil, D., Blanchard, S., Kaplan, J., Guilford, P., Gibson, F., *et al.* (1995). Defective myosin VIIA gene responsible for Usher syndrome type 1B. *Nature* **374**, 60–61.

Wells, A. L., Lin, A. W., Chen, L. Q., Safer, D., Cain, S. M., *et al.* (1999). Myosin VI is an actin-based motor that moves backwards. *Nature* **401**, 505–508.

Xiang, M., Gan, L., Li, D., Chen, Z. Y., Zhou, L., *et al.* (1997). Essential role of POU-domain factor Brn-3c in auditory and vestibular hair cell development. *Proc. Natl. Acad. Sci. USA* **94**, 9445–9450.

Xiang, M., Gao, W. Q., Hasson, T., and Shin, J. J. (1998). Requirement for Brn-3c in maturation and survival, but not in fate determination of inner ear hair cells. *Development* **125**, 3935–3946.

Xu, P. X., Adams, J., Peters, H., Brown, M. C., Heaney, S., and Maas, R. (1999). Eya1-deficient mice lack ears and kidneys and show abnormal apoptosis of organ primordia. *Nat. Genet.* **23**, 113–117.

Zerucha, T., Stuhmer, T., Hatch, G., Park, B. K., Long, Q., *et al.* (2000). A highly conserved enhancer in the Dlx5/Dlx6 intergenic region is the site of cross-regulatory interactions between Dlx genes in the embryonic forebrain. *J. Neurosci.* **20**, 709–721.

Zheng, J. L., and Gao, W. Q. (2000). Overexpression of Math1 induces robust production of extra hair cells in postnatal rat inner ears. *Nat. Neurosci.* **3**, 580–586.

Zheng, J. L., Shou, J., Guillemot, F., Kageyama, R., and Gao, W. Q. (2000a). Hes1 is a negative regulator of inner ear hair cell differentiation. *Development* **127**, 4551–4560.

Zheng, L., Sekerkova, G., Vranich, K., Tilney, L. G., Mugnaini, E., and Bartles, J. R. (2000b). The deaf jerker mouse has a mutation in the gene encoding the espin actin-bundling proteins of hair cell stereocilia and lacks espins. *Cell* **102**, 377–385.

Zine, A., Aubert, A., Qiu, J., Therianos, S., Guillemot, F., *et al.* (2001). Hes1 and Hes5 activities are required for the normal development of the hair cells in the mammalian inner ear. *J. Neurosci.* **21**, 4712–4720.

Zine, A., and de Ribaupierre, F. (2002). Notch/Notch ligands and Math1 expression patterns in the organ of Corti of wild-type and Hes1 and Hes5 mutant mice. *Hear. Res.* **170**, 22–31.

Zuo, J. (2002). Transgenic and gene targeting studies of hair cell function in mouse inner ear. *J. Neurobiol.* **53**, 286–305.

3

Formation of the Outer and Middle Ear, Molecular Mechanisms

Moisés Mallo
Instituto Gulbenkian de Ciência
Rua da Quinta Grande 6, 2780-156 Oeiras, Portugal

I. Introduction
II. Basic Anatomical and Embryological Overview
III. Genetic Determinants of Outer and Middle Ear Development
    A. The Outer Ear
    B. The Middle Ear
IV. Concluding Remarks
    References

The outer and middle ears collect, transmit, and amplify sound waves from the environment to stimulate sensory receptors located in the inner ear. Over the last decade enormous advances have been made in the understanding of the molecular mechanisms controlling the development of these areas. Molecular controls are set at different levels, including the production and differentiation of neural crest cells, modulation of proliferation and survival mechanisms, control of tissue interactions, and establishment of patterning processes in the branchial arches. Many of the genes involved in these processes have already been identified and their role is starting to be understood with the help of the production and analysis of relevant animal models. From these genes, a major proportion is represented by a variety of transcription factors and of signaling molecules belonging to different families and their receptors. © 2003, Elsevier Inc.

I. Introduction

The hearing apparatus of animals living in an aerial environment consists of three anatomical compartments: the outer, middle, and inner ears. The inner ear houses the sensorial receptors, the hair cells, embedded in a liquid medium known as the endolymph. The outer and middle ears are the evolutionary solution to the need for stimulating the hair cells on a liquid medium with sound waves produced in air, a medium about 1000-fold less dense than the endolymph. The functional solution consisted of recording the sound waves as vibrations on an elastic, low-mass membrane, the eardrum, or

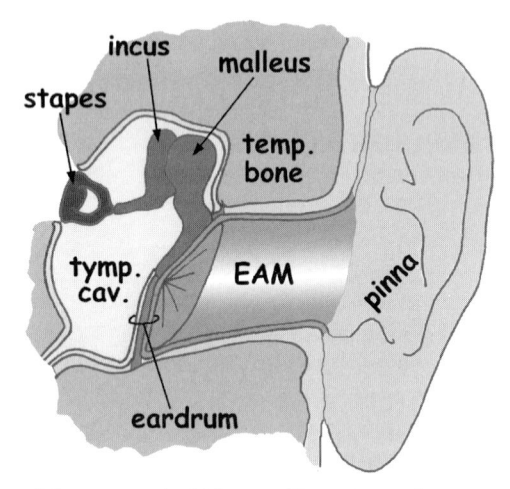

Figure 1 Anatomy of the outer and middle ears. The outer ear is composed of the pinna and the external acoustic meatus (EAM). The eardrum (or tympanic membrane) is located at the end of the EAM. It is composed of three layers provided by the epithelia of the EAM and middle ear cavity and fibrous tissue between the two epithelia. The middle ear is located on the tympanic cavity (within the temporal bone). The main components of the middle ear are the ossicles: malleus, incus, and stapes. Other components, such as muscles, nerves, and arteries, are not shown for the sake of simplicity. (See Color Insert.)

tympanic membrane, and then amplifying the sound waves for transmission to the inner ear by one or several skeletal elements that constitute the middle ear to produce excitatory waves in the endolymph. The basic functional principles are quite similar among the different phyla, but the details in the anatomy of this area can vary a lot from one animal class to another. The discussion in this chapter concentrates on the mammalian middle and outer ears because they are the best studied at the genetic and molecular levels.

II. Basic Anatomical and Embryological Overview

The basic anatomy of the outer ear is quite simple and consists of the ear pinna and the external acoustic meatus (EAM) (Fig. 1). In a schematic way, the outer ear can be seen as a funnel to bring sound waves from the environment to the eardrum, located at the end of the EAM. The eardrum can be considered as the interface between the outer and middle ears. It is a thin membrane composed of three layers: an ectodermal epithelium, provided by the EAM; a middle fibrous layer that gives elasticity to the membrane; and an inner epithelium, which is part of the lining of the middle ear cavity. The tympanic membrane is attached throughout most of its circumference to an opening in the lateral wall of the temporal bone.

3. Outer and Middle Ear Development

The mammalian middle ear (see Fig. 1) contains a chain of three ossicles located in a cavity within the temporal bone (the middle ear cavity). Two muscles (stapedial and tensor tympani) to regulate the mobility of the ossicles; the stapedial artery (in some mammals only during embryological stages); and a nerve, the corda tympani, complete the basic components of the middle ear cavity. The first ossicle, going from lateral to medial, is the malleus. It provides the physical connection to the tympanic membrane through the insertion of its manubrium between the two epithelial layers of the eardrum. In addition to the manubrium, the malleus has a head that provides the articular surface for the head of the incus; a neck that connects the head with the manubrium; and the processus brevis, a small prominence at the base of the manubrium. The incus serves as a bridge between the malleus and the third ossicle, the stapes, with which it establishes contact through its long process. The stapes is composed of two basic elements: the footplate, inserted in the oval window, and the arch, which provides the articular surface for the incudal long process.

Embryologically, all the components of the outer and middle ears develop from the first and second branchial arches (Carlson, 1999; Mallo, 1998) (Fig. 2). The ear pinna is formed around the first branchial cleft from several protuberances in the first and second arches. These protuberances, known as auricular hillocks, undergo a series of complex rearrangements to form the definitive pinna. Each of the hillocks contributes to specific areas of the pinna, those in the second arch being the ones forming most of the structure. The EAM is directly formed from the ectoderm of the cleft that separates the first and second branchial arches (first branchial cleft). The invagination of this ectoderm is controlled and coordinated by a C-shaped skeletal structure, the tympanic ring, that develops from the first arch mesenchyme (Declau *et al.*, 1989; Mallo and Gridley, 1996; Michaels and Soucek, 1989). This ring exists as an individual entity only during embryogenesis, becoming progressively integrated into the temporal bone at postnatal stages to serve as the attachment of the eardrum. The tympanic ring is formed by endomembranous ossification in a sequential fashion. An initial condensation in the proximal part of the first arch mesenchyme grows in a circumferential fashion around the first branchial cleft to invade the second arch area (Mallo and Gridley, 1996). While the ring is being formed, the tip of the cleft's ectoderm becomes associated with it all around its length and is pulled inward to form the meatal plug that will eventually become the EAM (Declau *et al.*, 1989; Mallo and Gridley, 1996; Michaels and Soucek, 1989). In addition, as the ring grows, the invaginated EAM starts to flatten down in the plane defined by the ring and becomes apposed to the endoderm of the middle ear cavity. The mesodermal tissue that remains between the two epithelia differentiates to form the fibrous layer.

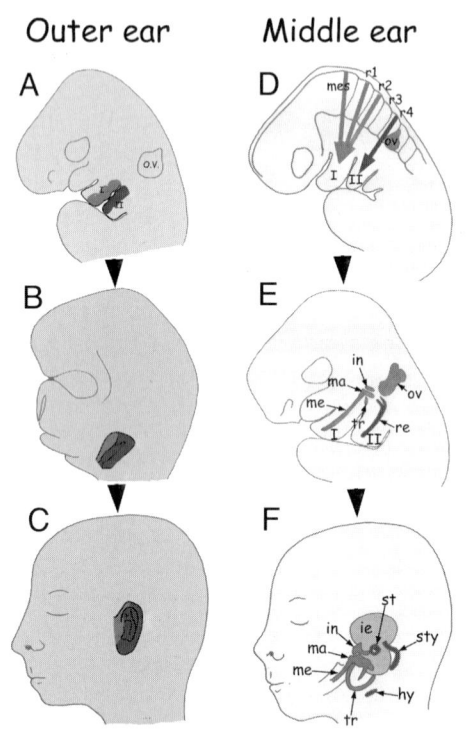

Figure 2 Embryological development of the outer and middle ears. (A) The outer ear develops from hillocks on the first (I) and second (II) branchial arches (represented in red and blue, respectively). (B, C) As development proceeds, they give rise to specific areas of the pinna. (D) The middle ear ossicles develop from the cranial neural crest. Crest cells from the caudal mesencephalon (mes) and rhombomere 1 and 2 (r1 and r2), shown in red, populate the first branchial arch (I). r4 neural crest (blue) populates the second arch (II). (E, F) From the first arch develop the Meckel's cartilage (me), malleus (ma), incus (in), and tympanic ring (tr). Skeletal development in the second arch starts with Reichert's cartilage (re), which will eventually form the stapes (st), styloid process (sty), and lesser horn of the hyoid bone (hy). The otic vesicle (ov) will give rise to the inner ear (ie). (See Color Insert.)

The middle ear ossicles are also formed from the proximal mesenchyme of the first and second arches (Carlson, 1999; Mallo, 1998). The malleus and incus develop from the first arch and the stapes from the second. The malleus and incus are formed at the proximal extremity of Meckel's cartilage (the first arch cartilage) as a condensation perpendicular to the main axis of the branchial arch (Miyake et al., 1996). This initial condensation separates into two parallel components that eventually form the two ossicles. The stapes is formed from the proximal part of the second arch primordial cartilage (known as Reichert's cartilage). The second arch cartilage also gives rise to two other skeletal elements: the styloid process and the lesser horn of the

3. Outer and Middle Ear Development 89

hyoid bone. Although not directly related to the middle ear, these structures are important for understanding some of the mutant phenotypes affecting the ear region.

Similar to the rest of the facial skeleton, the middle ear ossicles are of neural crest origin. The neural crest is a group of migratory mesenchymal cells that originate from the dorsal part of the developing neural tube, which gives rise to a wide variety of tissues and structures all over the vertebrate body (Nieto, 2001). In the cranial region (i.e., those cells arising from the most rostral areas of the neural tube), neural crest cells are also responsible for the genesis of most of the craniofacial skeleton that, unlike axial and apendicular skeleton, is not of mesodermal origin (Couly *et al.*, 1993). The neural crest cells that form the middle ear ossicles arise from specific areas of the neural tube during a precise time window. For instance, the stapes derives from crest cells migrating from the fourth rhombomere and the malleus, incus, and tympanic ring from neural crest originating in rhombomeres 1 and 2 and caudal mesencephalon (Köntges and Lumsden, 1996). In the mouse, the neural crest cells giving rise to the middle ear ossicles migrate between day 8 plus 4.5 hours and day 8 plus 7.5 hours (Mallo, 1997).

Finally, with the exception of the gonial bone, a small dermal element that will be a part of the malleus, the middle ear ossicles develop by endochondral ossification (i.e., by formation of a cartilaginous template).

III. Genetic Determinants of Outer and Middle Ear Development

A. The Outer Ear

1. The Ear Pinna

For many years malformations in the outer ear that result from genetic mutations or teratogenic processes have been described. However, insights into the molecular mechanisms of ear pinna development had to wait until the genetic bases of human syndromes affecting ear formation were identified and mutant mice for those and other genes could be generated and analyzed (Table I).

There are several human syndromes that include various degrees of alterations in the outer ears, normally associated with abnormalities in other organs. Among the most salient and best characterized of these syndromes is the branchio-oto-renal (BOR) syndrome, which combines craniofacial, ear, and kidney defects (Abdelhak *et al.*, 1997). The genetic defect in the BOR syndrome has been traced to haploinsufficiency of the *EYA1* gene (Abdelhak *et al.*, 1997). Analysis of mice homozygous for an *Eya1* null

Table I Summary of the Outer and Middle Ear Phenotypes of Various Mutants

	Outer ear	Middle ear	References
Ap2	ND[a]	Absent ossicles	Schorle et al., 1996; Zhang et al., 1996
Bmp5	Small pinna; normal EAM	Normal	Kingsley et al., 1992
Dlx1	Normal	Slightly affected stapes	Qiu et al., 1997
Dlx2	Normal	Malleus normal; incus fused to skull wall; stapes reduced	Qiu et al., 1995
Dlx5	ND	Normal malleus and incus; absent stapes in some embryos	Acampora et al., 1999; Depew et al., 1999
Dlx5/Dlx6	Absent pinna	Severely affected ossicles	Beverdam et al., 2002; Depew et al., 2002; Robledo et al., 2002
Edn1	Small pinna	All ossicles strongly affected or absent	Kurihara et al., 1994
Ednra	Small pinna	All ossicles strongly affected or absent	Clouthier et al., 1998
Ece1	Small pinna	All ossicles strongly affected or absent	Yanagisawa et al., 1998
Eya1	Small or absent pinna	Smaller and fused malleus and incus; absent stapes	Abdelhak et al., 1997; Xu et al., 1999
Fgf8[b]	ND	Vestigial malleus; absent incus; unaffected stapes	Trumpp et al., 1999
Gsc	Normal pinna; absent EAM	Malleus without manubrium	Rivera-Pérez et al., 1995, 1999; Yamada et al., 1995

Hoxa1	Normal	Slightly affected stapes[c]	Lufkin et al., 1991; Chisaka et al., 1992
Hoxa1/Hoxb1	Absent pinna; affected EAM	Hypomorphic malleus; absent stapes	Gavalas et al., 1998, 2001; Rossel and Capecchi, 1999
Hoxa2	Reduced pinna; duplicated EAM	Duplicated malleus; duplicated incus; absent stapes	Gendron-Maguire et al., 1993; Rijli et al., 1993; Mallo and Gridley, 1996
Msx1	Normal	Malleus without processus brevis	Satokata and Maas, 1994
Pbx1	Smaller pinna	Normal malleus and incus; absent stapes (transformed into other structures?)	Selleri et al., 2001
Prx1	Normal pinna; absent EAM	Malleus without manubrium; incus fused to extra cartilage; small stapes fused to extra cartilage	Martin et al., 1995; Mallo et al., 2000
Ptx1	ND	Absent gonial bone	Lanctôt et al., 1999; Szeto et al., 1999
RARs[d]	Small/absent pinna[e]	Malformed incus[f]; absent stapes[e]	Lohnes et al., 1994
Wnt1/Wnt3a	ND	Absent stapes	Ikeya et al., 1997

[a]Not determined. [b]Data on a conditional mutant in which Fgf8 was removed from the first branchial arch. [c]Two independent mutants gave slightly different middle ear phenotypes. [d]Different combinations of null mutants for *RARα*, *RARβ2*, and *RARγ*. [e]Only in *RARα/RARγ* double mutant. [f]In *RARα/RARβ2* and *RARα/RARγ* double mutants and *RARα/RARβ2/RARγ* triple mutants.

mutation showed that this gene is essential for development of different areas of the ear, including the pinna, which was malformed or absent in these mutants (Xu et al., 1999). The role of *Eya1* in the formation of this structure is still unresolved. One possible scenario is that the outer ear phenotype is a secondary consequence of the major alterations in the branchial arches also found in these mutants. Alternatively, considering that *Eya1* is expressed in cartilaginous condensations (Xu et al., 1997), the absence of pinnae could result from the inability of the ear mesenchyme to form the ear cartilage, thus hampering the growth of this structure. Indeed, the alterations observed in other skeletal structures of $Eya1^{-/-}$ embryos, including the middle ear (see later), indicate that this gene plays a role in cartilage formation. So far, it is not possible to distinguish between these possibilities because detailed analyses have been lacking.

The *short ear* (*se*) is a classic mutant strain (Lynch, 1921) that combines very small outer ears with other abnormalities in skeletal and soft tissues (Green, 1951; Green and Green, 1942). Using a positional cloning approach, the mutation was found in the *Bmp5* gene (Kingsley et al., 1992). This gene belongs to the bone morphogenetic protein (BMP) family, which has been implicated in a wide variety of developmental processes (Ducy and Karsenty, 2000; Hogan, 1996). In the case of the ear pinna, the role of *Bmp5* seems to be at relatively late stages of its development, affecting growth rather than early patterning processes. Indeed, pinna development seemingly proceeds normally until the mesenchyme starts to condense to form the ear cartilage. In addition, the already abnormal ear cartilages of *se* mice lack perichondrium (Green and Green, 1942), which very likely contributes to the retarded growth of this cartilage, resulting in the formation of very small pinnae. In keeping with this role for *Bmp5*, expression analyses indicate that it is not active during early branchial arch development in the regions that will become the pinna, but it is strongly expressed both in early skeletal condensations, including that of the ear cartilage, and in the perichondrium of differentiating cartilages (King et al., 1994).

With the advent of gene-targeting techniques, other mutations have been found to affect formation of the ear pinna. Mutations in the *Hoxa2* gene were among the first described to have a strong impact on the formation of this structure (Gendron-Maguire et al., 1993; Rijli et al., 1993). In the absence of *Hoxa2* a normal pinna fails to form and it is replaced by a protuberance with no recognizable shape. The requirement of *Hoxa2* in this process is dose dependent because intermediate phenotypes were observed in animals with reduced *Hoxa2* expression levels (Ohnemus et al., 2001). External ear defects in *Hoxa2* null mutants are detected at early developmental stages, indicating that this gene is required for the initial steps of ear pinna formation. As mentioned, the ear pinna is formed from auricular hillocks in the first and second arch. Morphological studies indicate that most of the definitive pinna derives from the second arch hillocks, whereas those in the first arch are responsible for a rather small portion of this structure (Carlson, 1999). Because *Hoxa2* is strongly expressed in the second arch (but not at all in the first) (Mallo 1997; Prince and Lumsden, 1994), it seems likely that the role of *Hoxa2* in pinna formation is related to the second-arch–derived portion. Considering that the branchial arches of *Hoxa2* null mutants are morphologically normal (Gendron-Maguire *et al.*, 1993; Rijli *et al.*, 1993), the ear pinna phenotype cannot be simply the result of the absence of second arches. *Hoxa2* has been shown to be required for proper specification of the skeleton derived from the second branchial arch, which in $Hoxa2^{-/-}$ embryos displays first-arch–like morphology (Gendron-Maguire *et al.*, 1993; Rijli *et al.*, 1993). Thus, one possibility is that *Hoxa2* plays an active role in defining the identity of the second arch hillocks. According to this hypothesis, the abnormal protuberance observed in the place of the ear pinnae of $Hoxa2^{-/-}$ embryos could represent a double set of first-arch–like hillocks. An alternative hypothesis is that the outer ear phenotype of the *Hoxa2* mutants is secondary to the major changes in skeletal patterning in the second arches of these embryos. The amount and spatial location of the second arch mesenchyme recruited to form middle ear and neck skeletal structures is very different in wildtype and mutant embryos (Kanzler *et al.*, 1998). Therefore, the absence of ear pinna in the $Hoxa2^{-/-}$ embryos could be the consequence of the abnormal recruitment of the second arch mesenchyme normally contributing to the outer ear cartilage to form middle ear skeletal structures. Consistent with the latter possibility, we have observed that transgenic embryos in which most of the second arch mesenchyme is converted into endochondral elements by ectopic *Sox9* expression (Kanzler *et al.*, 1998) have smaller ear pinnae (M. Mallo, unpublished results). To clearly differentiate between the two hypotheses, it is necessary to identify specific markers for the first and second arch hillocks.

Mutations in other *Hox* genes have also produced ear pinna phenotypes. In particular, compound $Hoxa1^{-/-}$; $Hoxb1^{-/-}$ mutants seem to lack completely this structure (Gavalas *et al.*, 1998; Rossel and Capecchi, 1999). None of these *Hox* genes is expressed in the developing branchial arches (Murphy and Hill, 1991), indicating that they have a rather indirect role in the formation of the pinna. Indeed, analysis of $Hoxa1^{-/-}$; $Hoxb1^{-/-}$ embryos showed that their second branchial arches are strongly hypomorphic as a consequence of defects in the production of migratory neural crest cells (Gavalas *et al.*, 1998; Rossel and Capecchi, 1999). It is interesting to note that no residual ear pinna was detected in these mutant embryos. This is quite surprising considering that the first branchial arch and its neural crest seem not to be affected, and it indicates that some as yet undefined interaction occurs between the first and second arches. Molecular analyses of the patterning of the ectodermal and endodermal epithelia surrounding the first and second branchial arches failed to reveal any detectable changes (Gavalas et al., 2001), indicating that epithelial alterations are not likely the mediators of these interactions.

Outer ear malformations have also been described in several other mouse mutants, including $Prx1^{-/-}$; $Prx2^{-/-}$ embryos, $Dlx5^{-/-}$; $Dlx6^{-/-}$ embryos, $Tbx1$ null mutants, and mutant embryos lacking endothelin1 signaling. In these cases, the outer ear malformation was just reported but not studied; therefore this chapter does not discuss them in more detail.

2. The External Acoustic Meatus

Although EAM and pinna development must be somehow coordinated in order to be functional, the development of these two structures seems to be regulated by independent mechanisms. This is clear from the analysis of the Gsc and the $Prx1$ mutant embryos, in which fairly normal ear pinnae coexist with absent EAMs (Mallo et al., 2000; Martin et al., 1995; Rivera-Pérez et al., 1995; Yamada et al., 1995).

As briefly described previously, development of the EAM seems to be largely dependent on the tympanic ring. Support for the functional relationship between the tympanic ring and EAM development comes essentially from mutant and teratogenic analyses showing that when the tympanic ring fails to develop (as in Gsc and $Prx1$ null mutants or in retinoic acid (RA)-treated embryos) the EAM is also missing (Mallo, 1997; Mallo et al., 2000; Martin et al., 1995; Rivera-Pérez et al., 1995; Yamada et al., 1995). Importantly, Gsc and $Prx1$ are not expressed in the epithelium of the cleft (Mallo and Gridley, 1996; Mallo et al., 2000), further highlighting the secondary nature of the absence of EAM. In addition, $Eya1$ null mutants, which have very small tympanic rings, show atresia of the EAM and malformed eardrums (Xu et al., 1999). It has also been observed in human patients that abnormal formation of the tympanic ring is clinically associated with atresia of the EAM (Lambert and Dodson, 1996). Finally, the tympanic ring duplication that occurs in $Hoxa2$ mutants results in duplicated invagination of the first branchial cleft and in a somewhat duplicated tympanic membrane (Mallo and Gridley, 1996).

Several genes that are required for the formation of the tympanic ring have already been identified. Among them the most salient are Gsc and $Prx1$ because although null mutations for these genes also produce phenotypes in other body areas, the absence of the ring occurs in the near absence of other major malformations in the ear area (Mallo et al., 2000; Martin et al., 1995; Rivera-Pérez et al., 1995; Yamada et al., 1995).

The role of Gsc in tympanic ring formation has been investigated in some detail using a chimeric analysis approach (Rivera-Pérez et al., 1999). This study revealed that Gsc may play more than one role in the formation of the tympanic ring. Although in $Gsc^{-/-}$ embryos the primordial condensation for this structure cannot be detected, Gsc-negative cells can form part of the tympanic ring condensation if Gsc-positive cells are present. In addition, at later stages the tympanic rings of these chimeras presented areas that failed to ossify, which might correspond to the ring areas populated by $Gsc^{-/-}$ cells. A possible explanation for these observations is that Gsc acts noncell autonomously to produce the ring condensation, but that it is required cell autonomously at later stages to complete the differentiation program. Whether this is indeed the case awaits identification of the molecular mediators of Gsc activity during tympanic ring formation.

A possible role for *Prx1* in the formation of the ring can be suggested on the basis of the analysis of additional malformations seen in the branchial arch skeleton of $Prx1^{-/-}$ mutants (Martin et al., 1995). In most of these malformations, an excess of endochondral ossification is evident in areas of the first and second arches. In some cases, such as for the squamous bone, endochondral structures replace skeletal elements that should develop by dermal ossification. Since these areas represent domains of high *Prx1* expression and this gene seems to be downregulated in *Sox9* (fated to cartilage)-positive areas (Mallo et al., 2000), it is possible that *Prx1* is part of a switch between dermal/endochondral fates for skeletogenic condensations. If this hypothesis is true, the absence of tympanic ring could derive from the inability of the primordial condensation to take a dermal fate, thus hindering further development of this structure. Interestingly, in $Prx1^{-/-}$ embryos, a small protuberance was described attached to the proximal part of the Meckel's cartilage, close to the area where the tympanic ring primordium should be formed (Martin et al., 1995). It can be speculated that this abnormal structure is the remnant of the primordium that underwent endochondral instead of dermal ossification and that this change in fate both blocked further growth of the structure and facilitated its fusion to a nearby cartilage (Meckel's cartilage). Although attractive, this hypothesis has not been proven and awaits experimental testing.

Despite their importance, expression of *Prx1* and *Gsc* is not sufficient for the formation of a tympanic ring in the branchial arch area since transcripts for both genes are also detected in the second branchial arches (Cserjesi et al., 1992; Gaunt et al., 1993; Kanzler et al., 1998; Kuratani et al., 1994; Mallo and Gridley, 1996). The inability of the second arch mesenchyme to differentiate into a tympanic ring is due to the expression of the *Hoxa2* gene in this area. Analysis of *Hoxa2* null mutants revealed that a tympanic ring is also formed in the second arch area (Gendron-Maguire et al., 1993; Rijli et al., 1993), indicating a dominant role of *Hoxa2* on the *Gsc* and *Prx1* activities.

Other genes and factors also required for proper tympanic ring development include *Dlx5/Dlx6* (together) (Beverdam et al., 2002; Depew et al., 2002; Robledo et al., 2002); *Ptx1* (Lanctôt et al., 1999; Szeto et al.,

1999); *Eya1* (Xu *et al.*, 1999); and several secreted factors and their receptors, including Fgf8 (Trumpp *et al.*, 1999), Endothelin 1 (*Edn1*) (Kurihara *et al.*, 1994), Endothelin receptor A (*Ednra*) (Clouthier *et al.*, 1998) and Endothelin converting enzyme 1 (*Ece1*) (Yanagisawa *et al.*, 1998). *Dlx5/Dlx6*, *Ptx1*, and *Eya1* are transcription factors whose exact role in the formation of the tympanic ring has not been investigated. In the case of the secreted molecules, a variety of data indicate that they are mediators of the epithelial–mesenchymal interactions required for patterning and morphogenetic processes in the branchial arches (Bee and Thorogood, 1980; Dunlop and Hall, 1995; Hall, 1980). Expression of both *Gsc* and *Prx1* has been shown to be dependent on these epithelial–mesenchymal interactions (Kuratani *et al.*, 1994; Tucker *et al.*, 1999), and in the specific case of *Gsc*, it has been reported that it is downregulated in the absence of Edn1 signaling (Clouthier *et al.*, 1998). Therefore, these secreted molecules are involved in modulating expression of genes known to be required for the formation of the tympanic ring. However, as is discussed in the following sections for the middle ear bones, other mechanisms, including altered proliferation or cell survival, could also be responsible for the lack of tympanic ring in these mutants.

B. The Middle Ear

1. The Ossicles

In general, the genes known to be required for ossicle development can be classified into two major groups. The first group includes those genes that control general mechanisms of endochondral ossification. Although mutations in these genes might have strong effects on middle ear development, they are not discussed in detail because of space constraints (interested readers can see Karsenty and Wagner, 2002, and Mallo, 1998). The second group includes those genes with more specific roles in the induction, patterning, and morphogenesis of the branchial arch skeleton (see Table I). Although we are referring to this second group as "quite specific," malformations in the middle ear are normally associated with defects in other branchial arch–derived structures, indicating that there may be no "ossicle-specific" mechanisms but that we should rather contemplate the formation of these structures in the wider context of the patterning of branchial arch–derived structures. In addition, middle ear phenotypes derived from specific gene mutations are often associated with malformations in other organs (and not necessarily the skeleton), indicating that some of the patterning mechanisms working in the branchial arch mesenchyme are shared by other, apparently unrelated, developmental processes.

3. Outer and Middle Ear Development

a. Factors Controlling Early Neural Crest Development. The skeletogenic mesenchyme of the branchial arches derives from the cranial neural crest (Chai *et al.*, 2000; Couly *et al.*, 1993; Jiang *et al.*, 2002; Noden, 1983). Therefore, a first group of mutations with a repercussion on middle ear skeletal development are those that interfere with neural crest development itself. Numerous studies performed mostly on chicken embryos have implicated a wide variety of genes at different stages of neural crest development (Endo *et al.*, 2002; Gammill and Bronner-Fraser, 2002; Garcia-Castro *et al.*, 2002; Liem *et al.*, 1995; Nieto, 2001). However, our knowledge about their role during formation of the middle ear ossicles is quite limited because of a dearth of genetic proof for their implication in middle ear development. *AP-2* (Schorle *et al.*, 1996; Zhang *et al.*, 1996), *Bmp2* (Kanzler *et al.*, 2000). *Hoxa1/Hoxb1* (Gavalas *et al.*, 1998; Rossel and Capecchi, 1999), and *Wnt1/Wnt3a* (Ikeya *et al.*, 1997) are among the few examples for which genetic data are already available. *AP-2* is a transcription factor expressed in premigratory and postmigratory neural crest cells (Mitchell *et al.*, 1991) that is essential for proper development of the neural crest in general. In *AP-2* null mutant embryos, neural crest cells are produced and migrate, but differentiation of postmigratory cells is compromised at some stage, resulting in strong facial malformations that include the middle ear region (Schorle *et al.*, 1996; Zhang *et al.*, 1996). The exact role of this molecule is still unresolved. *Hoxa1* and *Hoxb1* are also transcription factors implicated in the development of the second arch neural crest. Their function is required at the premigratory stage. Inactivation of *Hoxb1* by itself has no obvious negative effect on development of neural crest–derived structures (Goddard *et al.*, 1996; Studer *et al.*, 1996). In the case of *Hoxa1*, a few $Hoxa1^{-/-}$ embryos develop with minor deficiencies in the stapes (Chisaka *et al.*, 1992; Lufkin *et al.*, 1991). However, in the absence of both genes development of the middle ear structures is strongly compromised (Gavalas *et al.*, 1998; Rossel and Capecchi, 1999). Molecular analyses of such *Hoxa1/Hoxb1* double mutant embryos revealed the absence of the second arch migratory neural crest (Gavalas *et al.*, 1998, 2001; Rossel and Capecchi, 1999), in keeping with the absence of stapes in these mutants. Surprisingly, the middle ear phenotype is not restricted to the second arch but also affects first arch structures. The reason for this remains unclear, but as mentioned earlier interactions between the branchial arches have been hypothesized to explain this unexpected result (Gavalas *et al.*, 2001).

Bmp2 and *Wnt1/Wnt3a* belong to gene families that have been implicated in neural crest development in birds (García-Castro *et al.*, 2002; Liem *et al.*, 1995).

Bmp2 null mutants do not have migratory neural crest cells, and inactivation of the BMP pathway in specific areas of the neural tube resulted in a strong block in neural crest cell production from the targeted areas (Kanzler *et al.*, 2000). Phenotypic analyses further indicated that the skeletogenic neural crest population (which gives rise to the middle ear ossicles) was also affected in these embryos (Kanzler *et al.*, 2000). However, this inhibition was not restricted to this subpopulation because all of the neural crest derivatives that were tested appeared to be affected upon inhibition of BMP signaling (Kanzler *et al.*, 2000; Ohnemus *et al.*, 2002). Hence, Bmp2 may have a general role during early neural crest production, either for its induction or for its migration. Likewise, $Wnt1^{-/-}$; $Wnt3a^{-/-}$ mutant embryos showed a specific loss of the stapes without any detectable effects on other middle ear components (Ikeya *et al.*, 1997). However, the effects of this mutation are not restricted to this ossicle because other nonskeletal neural crest derivatives are also affected in these mutants. From a variety of cellular and molecular analyses it seems that the phenotype is due to proliferation defects of a specific population of cranial neural crest cells.

b. Secreted Factors and Receptors Involved in Skeletal Development in the Branchial Arches. A second group of mutations shown to have profound effects on skeletal development of the branchial arches are those in secreted molecules or their receptors acting directly on the branchial arches. In particular, genetic studies have shown that Fgf8 and Edn1 signaling are essential for development of this area. Null mutations in *Edn1*, its receptor *Ednra*, or the converting enzyme *Ece1* resulted in mice that lack most of the first and second arch derivatives, including the middle ear ossicles (Clouthier *et al.*, 1998; Kurihara *et al.*, 1994; Yanagisawa *et al.*, 1998). Also, when *Fgf8* is removed from the first arch ectoderm using a conditional mutation strategy (*Fgf8*; *Nes-cre* mice), strong hypomorphic effects were observed in first arch mesenchymal-derived structures (Trumpp *et al.*, 1999). Both *Edn1* and *Fgf8* are expressed in the branchial arch epithelia (Clouthier *et al.*, 1998; Crossley and Martin, 1995; Thomas *et al.*, 1998) (*Edn1* also has been reported to be expressed in the core of the branchial arches and in the endothelium of the blood vessels) and thus are thought to be among the most important mediators of the epithelial–mesenchymal interactions implicated in morphogenesis in the branchial arches. The roles of Fgf8 seem to be multiple. On the one hand, a strong increase in apoptosis in postmigratory neural crest cells has been reported in *Fgf8*; *Nes-cre* embryos (Trumpp *et al.*, 1999). Therefore, Fgf8 acts as a survival factor. On the other hand, Fgf8 has also been shown to be required to induce mesenchymal expression of genes that are known or suspected to be required for development of the branchial arches. A variety of *in vitro* culture studies have suggested a role for Fgf8 in the induction of genes such as *Lhx6*, *Lhx7*, *Ptx1*, *Barx1*, or *Gsc* (Bobola *et al.*, 2003; Grigoriou *et al.*, 1998; St. Amand *et al.*, 2000; Tucker *et al.*, 1998, 1999). Analysis of conditional *Fgf8* mutant embryos confirmed the absence of *Lhx6*, *Barx1*, and *Ptx1* expression in the first arch mesenchyme in the absence of Fgf8 (Bobola *et al.*, 2003; Trumpp *et al.*, 1999). This implies that Fgf8 has an impact not only on cell survival but also on the induction of other genes required for proper development of branchial arch structures. Likewise, Edn1 is involved both in proliferation/survival processes and in the control of gene expression in the branchial arch mesenchyme. Contrary to what has been described for *Fgf8* mutants, in the absence of Edn1 signaling cell death is quite mild and occurs relatively late in development (around E11.5) (Clouthier *et al.*, 2000), indicating that it may not be a major determinant of the *Edn1* phenotype. Downregulation has been reported for a variety of genes known to be involved in developmental processes in the branchial arches that are more likely mediators of the *Edn1* phenotype. For instance, transcript levels of *Gsc*, *Dlx3*, *Dlx6*, and *dHAND* are partially or completely reduced in the first and second arches of mutant embryos without Edn1 signaling (Charite *et al.*, 2001; Clouthier *et al.*, 1998, 2000; Thomas *et al.*, 1998), and *Dlx2* expression is lost from the second arches (Clouthier *et al.*, 2000). Interestingly, analysis of the *Fgf8* mutants indicated that both signaling pathways are somehow connected because epithelial expression of *Edn1* in the first arch is compromised in the absence of Fgf8 (Trumpp *et al.*, 1999).

Other signaling molecules are probably also involved in the epithelial–mesenchymal interactions required for branchial arch skeletogenic processes, most particularly members of the BMP and hedgehog families (Francis-West *et al.*, 1998). However, mutant analyses so far have not defined whether any of these molecules is important for middle ear development, particularly because of the early death of the relevant mutants. Only in the case of compound *Bmp5/Bmp7* mutants is it reasonable to expect major defects in the middle ear ossicles because of the massive apoptosis that occurs in the postmigratory neural crest of these embryos (Solloway and Robertson, 1999). The role of any of these or other signaling molecules awaits elucidation by the production and characterization of the relevant conditional mutants.

Retinoic acid (RA) signaling is also required for middle ear formation. Analysis of compound mutants for the RA receptors (*RARs*) (Lohnes *et al.*, 1994) revealed the complete absence of stapes (double mutants for *RARs* α and γ) and abnormal development of the incus ($\alpha/\beta2$ double muntant, α/γ double mutant, and $\alpha/\beta2/\gamma$ triple mutant). The latter ossicle presented an extraskeletal prominence attached to the ossicle's body, which was often fused to the alisphenoid. Quite surprisingly, however, the malleus was normal in all of the mutant combinations for *RARs*. The role of RA in the development of this area is still not clear. A wealth of experimental data, mostly resulting from exogenous RA administration during embryogenesis, indicates that high levels of this molecule can strongly affect early phases of neural crest development (Mallo, 1997; Morriss-Kay, 1993). However, it seems unlikely that the middle ear phenotypes observed in the *RAR* mutant embryos resulted from alterations in neural crest production or its early migratory properties because transcripts for these receptors were not detected in the relevant areas for neural crest production. Conversely, strong expression of different combinations of these receptors was observed in the branchial arches, thus suggesting a local RA requirement in this area (Dollé *et al.*, 1990; Ruberte *et al.*, 1990, 1991). Although the exact role of RA signaling in craniofacial and middle ear development is still unclear, it has been suggested that it might be required for proper modulation of cell death and cell fate specification processes in the postmigratory neural crest (Lohnes *et al.*, 1994).

c. Transcription Factors Involved in Branchial Arch Development. A third group of mutations with specific effects on the development of the middle ear structures includes those in a variety of transcription factors. Among them, the major subclass with proven effects on middle ear development are the homeobox-containing genes.

Hoxa2 was among the first *Hox* genes identified to be required for proper middle ear development. This gene is expressed in the second but not in the first branchial arch mesenchyme (Mallo, 1997; Prince and Lumsden, 1994) and its activity is essential for the formation of typical second arch skeletal derivatives (in the case of the middle ear ossicles, the stapes) from the second arch mesenchyme (Gendron-Maguire *et al.*, 1993; Rijli *et al.*, 1993). In *Hoxa2* null mutants these second arch derivatives are replaced by a duplicated set of first arch derivatives (malleus, incus, tympanic ring) in a mirror image disposition with respect to their first arch counterparts. This phenotype suggested some kind of "first arch ground state" in the first and second arches that is overridden by *Hoxa2* (Gendron-Maguire *et al.*, 1993; Rijli *et al.*, 1993). The role of *Hoxa2* in this process is still a matter of controversy. The classical view is *Hoxa2* acting as a selector gene. This would imply a dominant role for this gene in the patterning of branchial arch mesenchyme. In keeping with this hypothesis, widespread *Hoxa2* overexpression in the first branchial arches of chicken, *Xenopus*, and zebrafish embryos produced alterations in the first arch–derived skeleton that were scored as transformations into second arch identities (Grammatopoulos *et al.*, 2000; Hunter and Prince, 2002; Pasqualetti *et al.*, 2000). However, interpretation of these experiments is not so straightforward because equivalent experiments in the mouse or *Hoxa2* misexpression restricted to the neural crest populating the first arch resulted in the loss of the first arch skeleton instead of its transformation into second arch elements (Mallo and Brändlin, 1997; Kanzler *et al.*, 1998; Grammatopoulos *et al.*, 2000; Creuzet *et al.*, 2002). Considering that the branchial arch skeleton is neural crest derived (Chai *et al.*, 2000; Couly *et al.*, 1993; Jiang *et al.*, 2002; Noden, 1983), the latter results indicate that the information carried by the *Hoxa2*-expressing neural crest cells is not sufficient to generate a second arch pattern but requires interactions with properly expressed environmental cues. Interestingly, it has been shown that signals from the foregut endoderm play an essential role in the specification of the shape and orientation of the arch skeleton (Couly *et al.*, 2002). Therefore, it is possible that the role of *Hoxa2* is to modulate the response of the neural crest cells to signals provided by the arch environment. FGFs are likely among these signals because *Hoxa2* has been shown to physiologically block expression of *Ptx1* in the second arch by interference with FGF signaling (Bobola *et al.*, 2003).

Another homeobox-containing gene, *Msx1*, has been reported to be required for the formation of a small protuberance in the malleus, the processus brevis (Satokata and Maas, 1994). This effect is quite interesting because it is the only phenotype observed in the middle ear of these mutant embryos. The role of *Msx1* in the formation of this structure is still unknown.

Pbx1 is also required for middle ear development. The phenotype of the *Pbx1* mutants is not restricted to the branchial area and includes other body domains (Selleri *et al.*, 2001). An overview of this mutant phenotype, together with extensive biochemical studies, suggests that the *Pbx1* phenotype is somehow connected to the misfunctioning of several *Hox* genes (Mann and Chan, 1996; Selleri *et al.*, 2001). In the middle ear region the affected area is mostly the second arch. The *Pbx1* mutant phenotype in this area includes the absence of stapes and oval window and the formation of cartilage that extends from the lesser horn of the hyoid bone (also a second arch derivative) and ends close to the third second arch derivative, the styloid process, which is also somehow dismorphic. The first arch derivatives, malleus and incus, are quite normal in appearance; only a small ectopic cartilage develops close to the malleus. As discussed previously, the second arch, the main area affected in the *Ptx1* mutants, is also the domain of *Hoxa2* activity (Gendron-Maguire *et al.*, 1993; Rijli *et al.*, 1993). Some characteristics of the phenotype are similar, such as the absence of stapes and oval window, but others are clearly different. If, as the *Hox–Pbx* biochemical interactions would suggest, the *Pbx1* mutant phenotype is due to an interference with *Hoxa2* activity, this would imply that other *Pbx* or *Pbx*-like molecules must cover some of the functions of *Hoxa2*. However, it should be kept in mind that the *Hox/Pbx* interaction is just a possible explanation for the *Pbx1* mutant phenotype and it cannot be ruled out that this gene has a *Hox*-independent function in middle ear development (and that of other areas).

The *Dlx* family of genes also plays important roles in patterning processes in the craniofacial region, including those areas that produce the middle ear. *Dlx1*, *Dlx2*, *Dlx5*, and *Dlx6* have both specific and redundant roles in the development of the middle ear ossicles (Acampora *et al.*, 1999; Beverdam *et al.*, 2002; Depew *et al.*, 1999, 2002; Qiu *et al.*, 1995, 1997; Robledo *et al.*,

2002). Of all of these genes, *Dlx1* seems to have the most middle ear–restricted role, which is specific for the stapes. In $Dlx1^{-/-}$ embryos this ossicle is very dismorphic, with a complete loss of its arch (Qiu et al., 1997). The origin of this malformation has never been investigated in detail, and it is not clear whether *Dlx1* is required for skeletogenesis or if the stapedial phenotype is an indirect consequence of soft tissue malformations also observed in these mutants (Qiu et al., 1997). In particular, the absence of the stapedial artery could be instrumental in the $Dlx1^{-/-}$ phenotype. Stapes and stapedial artery have been shown to develop in a coordinated fashion (Anson and Cauldwell, 1942). Interestingly, the stapes of $Edn1^{-/-}$ embryos resembles that of $Dlx1^{-/-}$ mice (Kurihara et al., 1994), and the endothelium of the stapedial artery is a source of Edn1 (Thomas et al., 1998). Although *Dlx1* regulation by Edn1 has never been reported, other *Dlx* genes, particularly *Dlx2*, *Dlx3*, and *Dlx6*, have been shown to be targets and/or mediators of Edn1 signaling in the branchial area (Charite et al., 2001; Clouthier et al., 1998, 2000; Thomas et al., 1998), opening up the possibility that *Dlx1* function is also connected to the stapedial artery and Edn1 signaling. In support of this hypothesis, $Dlx2^{-/-}$ animals have malformations in the stapes similar to those of *Dlx1* mutants (Qiu et al., 1995, 1997), and *Dlx2* expression seems to be specifically downregulated in the second arches of *Ednra* null mutant embryos (Clouthier et al., 2000).

Unlike *Dlx1*, *Dlx2* function is not restricted to the second arch mesenchyme. Malformations in the proximal area of the first branchial arch have also been reported in $Dlx2^{-/-}$ animals (Qiu et al., 1995). In particular, the incus is attached to a cartilage in the lateral wall of the skull. In theory, considering *Dlx2* expression pattern, the activity of this gene could be required either in the neural-crest–derived mesenchyme or in the branchial arch ectoderm to modulate epithelial–mesenchymal interactions (Qiu et al., 1997; Robinson and Mahon, 1994). So far there is not enough information to discern between these possibilities, but indirect evidence suggests that *Dlx2* could play some role in restricting chondrogenic areas within the arch mesenchyme (Smith and Schneider, 1998; Thomas et al., 1997). In particular, it has been reported that in compound *Dlx1*; *Dlx2* mutants *Sox9*, an early chondrogenic marker (Bell et al., 1997; Ng et al., 1997), was upregulated in the area where molars (missing in these mutants) should develop (Thomas et al., 1997).

Recently, *Dlx5* and *Dlx6* have been shown to play major roles in patterning processes in the craniofacial area, including the ear region (Beverdam et al., 2002; Depew et al., 2002; Robledo et al., 2002). Whereas mutations in *Dlx5* had minor and variable effects on middle ear development, restricted to the absence of stapes in some exencephalic mutants (Acampora et al., 1999; Depew et al., 1999), the combined *Dlx5*/*Dlx6* mutation resulted in extensive remodeling of the facial area (Beverdam et al., 2002; Depew et al., 2002; Robledo et al., 2002). Concentrating on the middle ear, the first arch derivatives were deeply affected. The incus was still recognizable but fused to a cartilaginous structure that was interpreted as the malleus transformed into an incus-like element. The gonial bone, an endomembranous element that contributes to the mature malleus (Novacek, 1993), was also absent from these mutants. In the case of the second arch elements, the stapes could still be detected but did not have an arch. The origin of these complex malformations in the middle ear ossicles of *Dlx5/Dlx6* double mutants is not easy to explain, mainly because expression analyses indicate that these two genes are mostly expressed in the distal part of the branchial arches (Depew *et al.*, 2002) (the ossicles are proximal derivatives). Therefore, the most probable explanation is that these malformations are a secondary consequence of the extreme reorganization that occurs in the branchial-arch–derived soft and hard tissues of these mutants.

As discussed earlier, *Prx1* is involved in the development of several branchial arch skeletal derivatives, including the tympanic ring and the middle ear bones (Martin *et al.*, 1995). In the case of the latter structures, the three ossicles are affected to various extents. The only deficiency observed in the malleus of *Prx1* mutant embryos is the absence of manubrium, which, as is discussed later, is secondary to the lack of EAM (Mallo *et al.*, 2000). The incus and stapes are more strongly affected. Their morphologies are atypical, and instead of being independent elements, they are fused to abnormal cartilaginous structures. The incus is contiguous to an endochondral element that replaces the squamous bone in the wall of the skull. In addition, Reichert's cartilage seems to have developed anomalously and is still recognizable as a solid structure running from the hyoid to the incus, with which it makes contact. A remnant of the stapes can be found attached to this cartilage. During normal development, Reichert's cartilage gives rise to the styloid process, the lesser horn of the hyoid, and the stapes at three different levels of its length, the rest of it being converted into the stylohyoid ligament. As already mentioned, *Sox9* expression seems to be absent from the domains where *Prx1* expression is kept at high levels (Mallo *et al.*, 2000). It is then possible that the extra cartilages found in the *Prx1* mutants result from the inability of particular areas to downregulate *Sox9*. According to this hypothesis, *Prx1*-mediated downregulation of *Sox9*, acting together with mechanisms promoting dermal ossification, would produce the squamous bone in the area of the proximal first arch. In the second arch, in addition to the *Prx1*-dependent *Sox9* block, dermal ossification would also be prevented by *Hoxa2* activity (Kanzler *et al.*, 1998) and thus the preskeletogenic condensation would not be able to take any of the skeletogenic pathways, instead taking an alternative fate to make a ligament.

Ptx1 has also been shown to affect development of middle ear structures. In the case of the ossicle chain, this effect seems to be restricted to the only part that develops through endomembranous ossification, namely the gonial bone (Novacek, 1993), which is missing from the $Ptx1^{-/-}$ embryos (Lanctôt et al., 1999; Szeto et al., 1999). The role of *Ptx1* in the development of this or other skeletal structures is still not known. However, from the phenotypes observed in the branchial area, it seems likely that *Ptx1* plays a role in regulating some late step in the ossification processes.

Eya1 is another transcription factor essential for middle ear development. As mentioned, null mutants for this gene have phenotypes in the outer, middle, and inner ears (Xu et al., 1999). The middle ear anomalies affect the various components differently, with the constant absence of the stapes and several degrees of malformations in the malleus and incus, which are often fused. Interestingly, the stapes is already affected in the heterozygous state, failing to insert correctly into the oval window, thus generating a transmission hearing loss. It is possible that the role of *Eya1* in middle ear formation is related to some specific step of skeletogenesis because this gene is expressed in the condensing skeletogenic mesenchyme (Xu et al., 1997) and other skeletal structures are also affected in null mutant animals (Xu et al., 1999).

2. Muscles and Cavity

Not much is known about the mechanisms controlling the proper insertion of the middle ear muscles or of the making of the middle ear cavity; therefore this chapter considers them very briefly.

Experiments performed in chicken embryos (Köntges and Lumsden, 1996) indicate that the neural crest cells, although they themselves are not responsible for the formation of the facial musculature, might play an essential role in coordinating their proper insertion in the neural crest–derived skeleton. If this also holds true for mammals, it is expected that the insertion of the two muscles relevant for middle ear function might also depend on neural crest cells. Analysis of mutant embryos with middle ear malformations might help in solving this question. In addition, the mechanisms by which the neural crest cells control muscle insertion are still an unresolved issue.

The middle ear cavity derives from the first pharyngeal pouch. Available data seem to indicate that the formation of this cavity is rather passive and that when the pouch fails to form (e.g., when the first two arches are fused or when they are hypomorphic), the cavity is just not made (Mallo, 1997). As with the muscles, not much attention has been focused on this issue and therefore it is difficult to draw any conclusions. Thus, whereas in *Eya1* mutant embryos the absence of the middle ear cavity has been reported, it is unclear whether this gene plays an active role in the formation of this structure or if the phenotype is just a secondary consequence of major deformations in the branchial arch region.

3. Constructing a Functional Bridge

The main function of the middle ear is to transmit and amplify vibrations from the tympanic membrane into the inner ear. Therefore, construction of the middle ear implies not only the building of individual components but also their proper arrangement and connection with the eardrum and otic capsule.

The connection between the middle ear and the tympanic membrane is provided by the insertion of a malleal prominence known as the manubrium between the endodermal and ectodermal layers of the eardrum. A variety of genetic and *in vitro* culture experiments indicates that development of the manubrium itself depends on the EAM (Mallo *et al.*, 2000; Martin *et al.*, 1995; Rivera-Pérez *et al.*, 1995; Yamada *et al.*, 1995). In the absence of this epithelium, the manubrium is very small or absent even in otherwise fairly normal-looking ossicles. This is evident in the *Prx1* and *Gsc* mutant embryos in which, as discussed, the EAM is not formed as a consequence of the failure of the tympanic ring to develop (Mallo *et al.*, 2000; Martin *et al.*, 1995; Rivera-Pérez *et al.*, 1995; Yamada *et al.*, 1995).

The molecular mechanisms for this inducing activity are still not clear. *In vitro* recombination experiments indicate that the EAM is capable of inducing *Sox9* expression (one of the earliest markers of endochondral ossification [Bell *et al.*, 1997; Ng *et al.*, 1997]) in first branchial arch mesenchyme (Mallo *et al.*, 2000). In addition, this induction does not occur adjacent to the epithelium but at a distance, similarly to what is observed *in vivo* during manubrial induction (Mallo *et al.*, 2000). These results thus suggest that the EAM is the source of at least two kinds of signals, one able to induce chondrogenesis and another that dominantly represses the former but has a shorter range of activity. It is possible that *Prx1* plays a role in mediating the repressing activity because, both *in vivo* and in tissue recombinants, it is strongly expressed adjacent to the EAM with a pattern complementary to that of *Sox9*. Several secreted factors of the FGF and BMP families have been shown to be expressed in the EAM (Mallo *et al.*, 2000). However, their hypothetical role in the process of manubrial induction remains to be determined.

The construction of the tympanic membrane provides a paradigmatic example of morphogenesis by reciprocal tissue interactions to integrate development of different individual elements to produce a functional structure. First, the cleft ectoderm acts on the underlying mesenchyme to induce formation of the tympanic ring. This is the first step in ensuring that the EAM will form between the first and second arches, and thus within the ear pinna. The newly induced tympanic ring then grows and acts on the cleft ectoderm to produce an epithelial invagination (also known as the meatal plug), the future EAM. While invaginating, the medial surface of the cleft ectoderm secretes factors that induce chondrogenesis in the underlying mesenchyme that will eventually form the manubrium. The spatial and temporal characteristics of manubrial induction are responsible for the proper positioning of this structure, to become integrated into the mature eardrum when the endodermal and ectodermal epithelia appose each other. In summary, a cascade of properly timed epithelial–mesenchymal interactions results in both the proper location of the tympanic membrane at the end of the EAM and the attachment of the ossicles to the eardrum.

On the other side of the ossicle chain, the stapes must become inserted into the oval window. Analyses of middle ears of several mutant and teratogenically malformed embryos have started to produce a picture of the molecular determinants of the stapedial insertion into the oval window. *Hoxa2* and *Pbx1* are essential for the formation of the oval window because null mutations for these genes result in fenestraless inner ears (Gendron-Maguire *et al.*, 1993; Rijli *et al.*, 1993; Selleri *et al.*, 2001). However, it is important to point out that the absence of oval windows is not just the consequence of the failure of the stapes to develop (no stapes is formed in any of these two mutants) because inner ears with oval windows have been described in embryos that do not have the ossicle (Kanzler *et al.*, 2000; Louryan and Glineur, 1991; Mallo, 1997).

Hoxa2 is strongly expressed around the stapes in the area of oval window formation (Kanzler *et al.*, 1998). Considering that *Hox* gene expression in the neural crest is incompatible with skeletal differentiation (Couly *et al.*, 1998; Creuzet *et al.*, 2002; Grammatopoulos *et al.*, 2000), a possible role for *Hoxa2* in oval window formation is the inhibition of chondrogenesis in this area of the developing otic capsule. In the case of *Pbx1*, one possibility is that its role is connected to *Hoxa2* and that the absence of both stapes and oval window in the $Pbx1^{-/-}$ embryos results from the inability of *Hoxa2* to function properly in the absence of *Pbx1*. An alternative possibility is that the role of *Pbx1* in the formation of the oval window is independent of *Hox* gene activity. Distinguishing between these two possibilities awaits experimental evaluation.

Although the presence of a stapes is not required for the formation of an oval window, coordination between the formation of both the ossicle and the fenestra is essential for functional stapedial insertion. For instance, oval windows formed in the absence of the ossicle are smaller (Kanzler *et al.*, 2000; Louryan and Glineur, 1991; Mallo, 1997). Moreover, in some cases of partial deletion of the stapedial footplate after a retinoic acid (RA) treatment, this structure was properly located but abnormally attached to the otic cartilage (Mallo, 1997). In addition, mutations in the *Brn4* gene, which is expressed in the circumference of the oval window but not in the stapes (Phippard *et al.*, 1998), resulted in the development of stapes with an abnormal shape and with cartilaginous attachments to the window (Phippard *et al.*, 1999).

Interestingly, mutations in the *Eya1* have a strong impact on the development of the stapes and its attachment to the oval window (Xu *et al.*, 1999). As discussed, haploinsufficiency of this gene produces transmission deafness as a result of the abnormal placement of the stapes relative to the oval window. Both structures are apparently present, but an anomalous location of the VII cranial nerve seems to interfere with their proper connection. Although the primary role of *Eya1* in this process is not clear, this phenotype further highlights the importance of properly coordinating the formation of all relevant components in order to allow functional structures.

Finally, the joints between the different ossicles must also be properly established to allow their adequate movement. Little is known about how these connections are formed. Once again, *Eya1* seems to participate in this process because in $Eya1^{-/-}$ embryos fusions between the malleus and incus have been described (Xu *et al.*, 1999). It would be interesting to determine if genes that have been implicated in joint development in other body areas, such as *Wnt14* (Hartmann and Tabin, 2001), *noggin* (Brunet *et al.*, 1998), or *GDF5* (Storm and Kingsley, 1996) (or other members of these gene families), also play a role in the middle ear. Interestingly, recent data indicate that *Gdf6* is required for proper formation of the joints between the middle ear ossicles and for the attachment of the stapes in the oval window (Settle *et al.*, 2003). In particular, the articular surfaces connecting the malleus with the incus and the incus with the stapes were significantly reduced. These malformations resulted from a local reduction in cell proliferation in the developing articular surfaces of the three ossicles, which was not observed in the nonarticular areas of these skeletal elements (Settle *et al.*, 2003).

IV. Concluding Remarks

This chapter has covered in some detail the findings on the genetic determinants of outer and middle ear development. From these pages it is clear that considerable advances have been made in this field and, given the current trend, it is expected that in the following years the list of genes that play a role in the development of these areas will grow rapidly. The availability of the complete sequence of human and mouse genomes will surely help in this process because the identification of genes responsible for some human syndromes and mouse mutants affecting the outer and middle ears will be strongly facilitated. However, from the discussion presented it is also evident that little is yet known about the cellular mechanisms mediating the action of most of these genes and about their mutual interactions and their epistatic relationships. These issues, together with the understanding of how the genetic background affects the penetrance and expressivity of some of the mutant phenotypes, are among the most important challenges for the future in order to obtain a complete picture of the cellular and molecular mechanisms that govern and coordinate the construction of functional structures as complex as the mammalian outer and middle ears.

References

Abdelhak, S., Kalatzis, V., Heilig, R., Compain, S., Samson, D., et al. (1997). A human homologue of the *Drosophila eyes absent* gene underlies branchio-oto-renal (BOR) syndrome and identifies a novel gene family. *Nat. Genet.* **15**, 157–164.

Acampora, D., Merlo, G. R., Paleari, L., Zerega, B., Postiglione, M. P., et al. (1999). Craniofacial, vestibular and bone defects in mice lacking the *Distal-less*-related gene *Dlx5*. *Development* **126**, 3795–3809.

Anson, B. J., and Cauldwell, E. W. (1942). The developmental anatomy of the human stapes. *Ann. Otol. Rhinol. Laryngol.* **51**, 891–904.

Bee, J., and Thorogood, P. (1980). The role of tissue interactions in the skeletogenic differentiation of avian neural crest cells. *Dev. Biol.* **78**, 47–62.

Bell, D. M., Leung, K. K., Wheatley, S. C., Ng, L. J., Zhou, S., et al. (1997). SOX9 directly regulates the type-II collagen gene. *Nat. Genet.* **16**, 174–178.

Beverdam, A., Merlo, G. R., Paleari, L., Mantero, S., Genova, F., et al. (2002). Jaw transformation with gain of symmetry after *Dlx5/Dlx6* inactivation: Mirror of the past? *Genesis* **34**, 221–227.

Bobola, N., Carapuço, M., Ohnemus, S., Kanzler, B., Leibbrandt, A., et al. (2003). Mesenchymal patterning by *Hoxa2* requires blocking FGF-dependent activation of *Ptx1*. *Development* **130**, 3403–3414.

Brunet, L. J., McMahon, J. A., McMahon, A. P., and Harland, R. M. (1998). Noggin, cartilage morphogenesis, and joint formation in the mammalian skeleton. *Science* **280**, 1455–1457.

Carlson, B. M. (1999). "Human Embryology and Developmental Biology," 2 ed. Mosby, St. Louis.

Chai, Y., Jiang, X., Ito, Y., Bringas, P., Jr., Han, J., et al. (2000). Fate of the mammalian cranial neural crest during tooth and mandibular morphogenesis. *Development* **127**, 1671–1679.

Charité, J., McFadden, D. G., Merlo, G., Levi, G., Clouthier, D. E., et al. (2001). Role of Dlx6 in regulation of an endothelin-1-dependent, *dHAND* branchial arch enhancer. *Genes Dev.* **15**, 3039–3049.

Chisaka, O., Musci, T. S., and Capecchi, M. R. (1992). Developmental defects of the ear, cranial nerves and hindbrain resulting from targeted disruption of the mouse homeobox gene *Hox-1.6*. *Nature* **355**, 516–520.

Clouthier, D. E., Hosoda, K., Richardson, J. A., Williams, S. C., Yanagisawa, H., et al. (1998). Cranial and cardiac neural crest defects in endothelin-A receptor-deficient mice. *Development* **125**, 813–824.

Clouthier, D. E., Williams, S. C., Yanagisawa, H., Wieduwilt, M., Richardson, J. A., and Yanagisawa, M. (2000). Signaling pathways crucial for craniofacial development revealed by endothelin-A receptor-deficient mice. *Dev. Biol.* **217**, 10–24.

Couly, G. F., Coltey, P. M., and Le Douarin, N. M. (1993). The triple origin of skull in higher vertebrates: A study in quail-chick chimeras. *Development* **117**, 409–429.

Couly, G., Grapin-Botton, A., Coltey, P., Ruhin, B., and Le Douarin, N. M. (1998). Determination of the identity of the derivatives of the cephalic neural crest: Incompatibility between *Hox* gene expression and lower jaw development. *Development* **125**, 3445–3459.

3. Outer and Middle Ear Development

Couly, G., Creuzet, S., Bennaceur, S., Vincent, C., and Le Douarin, N. M. (2002). Interactions between *Hox*-negative cephalic neural crest cells and the foregut endoderm in patterning the facial skeleton in the vertebrate head. *Development* **129**, 1061–1073.

Creuzet, S., Couly, G., Vincent, C., and Le Douarin, N. M. (2002). Negative effect of *Hox* gene expression on the development of the neural crest-derived facial skeleton. *Development* **129**, 4301–4313.

Crossley, P. H., and Martin, G. R. (1995). The mouse *Fgf8* gene encodes a family of polypeptides and is expressed in regions that direct outgrowth and patterning in the developing embryo. *Development* **121**, 439–451.

Cserjesi, P., Lilly, B., Bryson, L., Wang, Y., Sassoon, D. A., and Olson, E. N. (1992). MHox: A mesodermally restricted homeodomain protein that binds an essential site in the muscle creatine kinase enhancer. *Development* **115**, 1087–1101.

Declau, F., Moeneclaey, L., and Marquet, J. (1989). Normal growth pattern of the middle ear cleft in the human fetus. *J. Laryngol. Otol.* **103**, 461–465.

Depew, M. J., Liu, J. K., Long, J. E., Presley, R., Meneses, J. J., *et al.* (1999). *Dlx5* regulates regional development of the branchial arches and sensory capsules. *Development* **126**, 3831–3846.

Depew, M. J., Lufkin, T., and Rubenstein, J. L. (2002). Specification of jaw subdivisions by *Dlx* genes. *Science* **298**, 381–385.

Dollé, P., Ruberte, E., Leroy, P., Morriss-Kay, G., and Chambon, P. (1990). Retinoic acid receptors and cellular retinoid binding proteins. I. A systematic study of their differential pattern of transcription during mouse organogenesis. *Development* **110**, 1133–1151.

Ducy, P., and Karsenty, G. (2000). The family of bone morphogenetic proteins. *Kidney Int.* **57**, 2207–2214.

Dunlop, L. L., and Hall, B. K. (1995). Relationships between cellular condensation, preosteoblast formation and epithelial-mesenchymal interactions in initiation of osteogenesis. *Int. J. Dev. Biol.* **39**, 357–371.

Endo, Y., Osumi, N., and Wakamatsu, Y. (2002). Bimodal functions of Notch-mediated signalling are involved in neural crest formation during avian ectoderm development. *Development* **129**, 863–873.

Francis-West, P., Ladher, R., Barlow, A., and Graveson, A. (1998). Signalling interactions during facial development. *Mech. Dev.* **75**, 3–28.

Gammill, L. S., and Bronner-Fraser, M. (2002). Genomic analysis of neural crest induction. *Development* **129**, 5731–5741.

García-Castro, M. I., Marcelle, C., and Bronner-Fraser, M. (2002). Ectodermal Wnt function as a neural crest inducer. *Science* **297**, 848–851.

Gaunt, S. J., Blum, M., and De Robertis, E. M. (1993). Expression of the mouse *goosecoid* gene during mid-embryogenesis may mark mesenchymal cell lineages in the developing head, limbs and body wall. *Development* **117**, 769–778.

Gavalas, A., Studer, M., Lumsden, A., Rijli, F. M., Krumlauf, R., and Chambon, P. (1998). *Hoxa1* and *Hoxb1* synergize in patterning the hindbrain, cranial nerves and second pharyngeal arch. *Development* **125**, 1123–1136.

Gavalas, A., Trainor, P., Ariza-McNaughton, L., and Krumlauf, R. (2001). Synergy between *Hoxa1* and *Hoxb1*: The relationship between arch patterning and the generation of cranial neural crest. *Development* **128**, 3017–3027.

Gendron-Maguire, M., Mallo, M., Zhang, M., and Gridley, T. (1993). *Hoxa-2* mutant mice exhibit homeotic transformation of skeletal elements derived from cranial neural crest. *Cell* **75**, 1317–1331.

Goddard, J. M., Rossel, M., Manley, N. R., and Capecchi, M. R. (1996). Mice with targeted disruption of *Hoxb-1* fail to form the motor nucleus of the VIIth nerve. *Development* **122**, 3217–3228.

Grammatopoulos, G. A., Bell, E., Toole, L., Lumsden, A., and Tucker, A. S. (2000). Homeotic transformation of branchial arch identity after *Hoxa2* overexpression. *Development* **127**, 5355–5365.

Green, E. L., and Green, M. C. (1942). The development of three manifestations of the short ear gene in the mouse. *J. Morphol.* **70**, 1–19.

Green, M. C. (1951). Further morphological effects of the short ear gene in the house mouse. *J. Morphol.* **88**, 1–22.

Grigoriou, M., Tucker, A. S., Sharpe, P. T., and Pachnis, V. (1998). Expression and regulation of *Lhx6* and *Lhx7*, a novel subfamily of LIM homeodomain encoding genes, suggests a role in mammalian head development. *Development* **125**, 2063–2074.

Hall, B. K. (1980). Tissue interactions and the initiation of osteogenesis and chondrogenesis in the neural crest-derived mandibular skeleton of the embryonic mouse as seen in isolated murine tissues and in recombinations of murine and avian tissues. *J. Embryol. Exp. Morph.* **58**, 251–264.

Hartmann, C., and Tabin, C. J. (2001). Wnt-14 plays a pivotal role in inducing synovial joint formation in the developing appendicular skeleton. *Cell* **104**, 341–351.

Hogan, B. L. (1996). Bone morphogenetic proteins in development. *Curr. Opin. Genet. Dev.* **6**, 432–438.

Hunter, M. P., and Prince, V. E. (2002). Zebrafish *hox* paralogue group 2 genes function redundantly as selector genes to pattern the second pharyngeal arch. *Dev. Biol.* **247**, 367–389.

Ikeya, M., Lee, S. M. K., Johnson, J. E., McMahon, A. P., and Takada, S. (1997). Wnt signalling required for expansion of neural crest and CNS progenitors. *Nature* **389**, 966–970.

Jiang, X., Iseki, S., Maxson, R. E., Sucov, H. M., and Morriss-Kay, G. M. (2002). Tissue origins and interactions in the mammalian skull vault. *Dev. Biol.* **241**, 106–116.

Kanzler, B., Kuschert, S. J., Liu, Y. H., and Mallo, M. (1998). *Hoxa2* restricts the chondrogenic domain and inhibits bone formation during development of the branchial area. *Development* **125**, 2587–2597.

Kanzler, B., Foreman, R. K., Labosky, P. A., and Mallo, M. (2000). BMP signaling is essential for development of skeletogenic and neurogenic cranial neural crest. *Development* **127**, 1095–1104.

Karsenty, G., and Wagner, E. F. (2002). Reaching a genetic and molecular understanding of skeletal development. *Dev. Cell* **2**, 389–406.

King, J. A., Marker, P. C., Seung, K. J., and Kingsley, D. M. (1994). *BMP5* and the molecular, skeletal, and soft-tissue alterations in *short ear* mice. *Dev. Biol.* **166**, 112–122.

Kingsley, D. M., Bland, A. E., Grubber, J. M., Marker, P. C., Russell, L. B., *et al.* (1992). The mouse *short ear* skeletal morphogenesis locus is associated with defects in a bone morphogenetic member of the TGF-β superfamily. *Cell* **71**, 399–410.

Köntges, G., and Lumsden, A. (1996). Rhomboencephalic neural crest segmentation is preserved throughout craniofacial ontogeny. *Development* **122**, 3229–3242.

Kuratani, S., Martin, J. F., Wawersik, S., Lilly, B., Eichele, G., and Olson, E. N. (1994). The expression pattern of the chick homeobox gene *gMHox* suggests a role in patterning of the limbs and face and in compartimentalization of the somites. *Dev. Biol.* **161**, 357–369.

Kurihara, Y., Kurihara, H., Suzuki, H., Kodama, T., Maemura, K., *et al.* (1994). Elevated blood pressure and craniofacial abnormalities in mice deficient in *endothelin-1*. *Nature* **368**, 703–710.

Lambert, P. R., and Dodson, E. E. (1996). Congenital malformations of the external auditory canal. *Otolaryngol. Clin. North. Am.* **29**, 741–760.

Lanctôt, C., Moreau, A., Chamberland, M., Tremblay, M. L., and Drouin, J. (1999). Hindlimb patterning and mandible development require the *Ptx1* gene. *Development* **126**, 1805–1810.

Liem, K. F., Jr., Tremml, G., Roelink, H., and Jessell, T. M. (1995). Dorsal differentiation of neural plate cells induced by BMP-mediated signals from epidermal ectoderm. *Cell* **82**, 969–979.

Lohnes, D., Mark, M., Mendelsohn, C., Dolle, P., Dierich, A., et al. (1994). Function of the retinoic acid receptors (RARs) during development (I). Craniofacial and skeletal abnormalities in RAR double mutants. *Development* **120**, 2723-2748.

Louryan, S., and Glineur, R. (1991). Mouse middle ear abnormalities following retinoic acid administration: Comparison with far/far features. *Teratology* **44**, 25A.

Lufkin, T., Dierich, A., LeMeur, M., Mark, M., and Chambon, P. (1991). Disruption of the *Hox 1.6* homeobox gene results in defects in a region corresponding to its rostral domain of expression. *Cell* **66**, 1105-1119.

Lynch, C. J. (1921). Short ears, an autosomal mutation in the house mouse. *Am. Nat.* **55**, 421-426.

Mallo, M. (1997). Retinoic acid disturbs mouse middle ear development in a stage-specific fashion. *Dev. Biol.* **184**, 175-186.

Mallo, M. (1998). Embryological and genetic aspects of middle ear development. *Int. J. Dev. Biol.* **42**, 11-22.

Mallo, M., and Brändlin, I. (1997). Segmental identity can change independently in the hindbrain and rhombencephalic neural crest. *Dev. Dyn.* **210**, 146-156.

Mallo, M., and Gridley, T. (1996). Development of the mammalian ear: Coordinate regulation of formation of the tympanic ring and the external acoustic meatus. *Development* **122**, 173-179.

Mallo, M., Schrewe, H., Martin, J. F., Olson, E. N., and Ohnemus, S. (2000). Assembling a functional tympanic membrane: Signals from the external acoustic meatus coordinate development of the malleal manubrium. *Development* **127**, 4127-4136.

Mann, R. S., and Chan, S. K. (1996). Extra specificity from extradenticle: The partnership between HOX and PBX/EXD homeodomain proteins. *Trends Genet.* **12**, 258-262.

Martin, J. F., Bradley, A., and Olson, E. N. (1995). The *paired*-like homeo box gene *Mhox* is required for early events of skeletogenesis in multiple lineages. *Genes Dev.* **9**, 1237-1249.

Michaels, L., and Soucek, S. (1989). Development of the stratified squamous epithelium of the human tympanic membrane and external canal: The origin of auditory epithelial migration. *Am. J. Anat.* **184**, 334-344.

Mitchell, P. J., Timmons, P. M., Hebert, J. M., Rigby, P. W., and Tjian, R. (1991). Transcription factor AP-2 is expressed in neural crest cell lineages during mouse embryogenesis. *Genes Dev.* **5**, 105-119.

Miyake, T., Cameron, A. M., and Hall, B. K. (1996). Stage-specific onset of condensation and matrix deposition for Meckel's and other first arch cartilages in inbreed C57BL/6 mice. *J. Craniofac. Genet. Dev. Biol.* **16**, 32-47.

Morriss-Kay, G. (1993). Retinoic acid and craniofacial development: Molecules and morphogenesis. *Bioessays* **15**, 9-15.

Murphy, P., and Hill, R. E. (1991). Expression of the mouse *labial*-like homeobox-containing genes, *Hox 2.9* and *Hox 1.6*, during segmentation of the hindbrain. *Development* **111**, 61-74.

Ng, L. J., Wheatley, S., Muscat, G. E., Conway-Campbell, J., Bowles, J., et al. (1997). SOX9 binds DNA, activates transcription, and coexpresses with type II collagen during chondrogenesis in the mouse. *Dev. Biol.* **183**, 108-121.

Nieto, M. A. (2001). The early steps of neural crest development. *Mech. Dev.* **105**, 27-35.

Noden, D. M. (1983). The role of the neural crest in patterning of avian cranial skeletal, connective and muscle tissues. *Dev. Biol.* **96**, 144-165.

Novacek, M. J. (1993). Patterns of diversity in the mammalian skull. *In* "The Skull" (J. Hanken and B. K. Hall, eds.), Vol. 2, pp. 438-545. University of Chicago Press, Chicago.

Ohnemus, S., Bobola, N., Kanzler, B., and Mallo, M. (2001). Different levels of *Hoxa2* are required for particular developmental processes. *Mech. Dev.* **108**, 135-147.

Ohnemus, S., Kanzler, B., Jerome-Majewska, L. A., Papaioannou, V. E., Boehm, T., and Mallo, M. (2002). Aortic arch and pharyngeal phenotype in the absence of BMP-dependent neural crest in the mouse. *Mech. Dev.* **119**, 127-135.

Pasqualetti, M., Ori, M., Nardi, I., and Rijli, F. M. (2000). Ectopic *Hoxa2* induction after neural crest migration results in homeosis of jaw elements in *Xenopus*. *Development* **127**, 5367–5378.

Phippard, D., Heydemann, A., Lechner, M., Lu, L., Lee, D., *et al.* (1998). Changes in the subcellular localization of the *Brn4* gene product precede mesenchymal remodeling of the otic capsule. *Hear. Res.* **120**, 77–85.

Phippard, D., Lu, L., Lee, D., Saunders, J. C., and Crenshaw, E. B., III (1999). Targeted mutagenesis of the POU-domain gene *Brn4/Pou3f4* causes developmental defects in the inner ear. *J. Neurosci.* **19**, 5980–5989.

Prince, V., and Lumsden, A. (1994). *Hoxa-2* expression in normal and transposed rhombomeres: Independent regulation in the neural tube and neural crest. *Development* **120**, 911–923.

Qiu, M., Bulfone, A., Martínez, S., Meneses, J. J., Shimamura, K., *et al.* (1995). Null mutation of *Dlx2* results in abnormal morphogenesis of proximal first and second branchial arch derivatives and abnormal differentiation in the forebrain. *Genes Dev.* **9**, 2523–2538.

Qiu, M., Bulfone, A., Ghattas, I., Meneses, J. J., Christensen, L., *et al.* (1997). Role of the *Dlx* homeobox genes in proximodistal patterning of the branchial arches: Mutations of *Dlx-1, Dlx2* and *Dlx-1* and *-2* alter morphogenesis of proximal skeletal and soft tissue structures derived from the first and second arches. *Dev. Biol.* **185**, 165–184.

Rijli, F. M., Mark, M., Lakkaraju, S., Dierich, A., Dolle, P., and Chambon, P. (1993). A homeotic transformation is generated in the rostral branchial region of the head by disruption of *Hoxa-2*, which acts as a selector gene. *Cell* **75**, 1333–1349.

Rivera-Pérez, J., Mallo, M., Gendron-Maguire, M., Gridley, T., and Behringer, R. R. (1995). *Goosecoid* is not an essential component of the mouse gastrula organizer but is required for craniofacial and rib development. *Development* **121**, 3005–3012.

Rivera-Pérez, J. A., Wakamiya, M., and Behringer, R. R. (1999). *Goosecoid* acts cell autonomously in mesenchyme-derived tissues during craniofacial development. *Development* **126**, 3811–3821.

Robinson, G. W., and Mahon, K. A. (1994). Differential and overlapping expression domains of *Dlx-2* and *Dlx-3* suggest distinct roles for *Distal-less* homeobox genes in craniofacial development. *Mech. Dev.* **48**, 199–215.

Robledo, R. F., Rajan, L., Li, X., and Lufkin, T. (2002). The *Dlx5* and *Dlx6* homeobox genes are essential for craniofacial, axial, and appendicular skeletal development. *Genes Dev.* **16**, 1089–1101.

Rossel, M., and Capecchi, M. (1999). Mice mutant for both *Hoxa1* and *Hoxb1* show extensive remodeling of the hindbrain and defects in craniofacial development. *Development* **126**, 5027–5040.

Ruberte, E., Dolle, P., Krust, A., Zelent, A., Morriss-Kay, G., and Chambon, P. (1990). Specific spatial and temporal distribution of retinoic acid receptor gamma transcripts during mouse embryogenesis. *Development* **108**, 213–222.

Ruberte, E., Dolle, P., Chambon, P., and Morriss-Kay, G. (1991). Retinoic acid receptors and cellular retinoid binding proteins. II. Their differential pattern of transcription during early morphogenesis in mouse embryos. *Development* **111**, 45–60.

St. Amand, T. R., Zhang, Y., Semina, E. V., Zhao, X., Hu, Y. P., *et al.* (2000). Antagonistic signals between BMP4 and FGF8 define the expression of *Pitx1* and *Pitx2* in mouse tooth-forming anlage. *Dev. Biol.* **217**, 323–332.

Satokata, I., and Maas, R. (1994). *Msx1* deficient mice exhibit cleft palate and abnormalities of craniofacial and tooth development. *Nat. Genet.* **6**, 348–356.

Schorle, H., Meier, P., Buchert, M., Jaenisch, R., and Mitchell, P. J. (1996). Transcription factor AP-2 essential for cranial closure and craniofacial development. *Nature* **381**, 235–238.

Selleri, L., Depew, M. J., Jacobs, Y., Chanda, S. K., Tsang, K. Y., et al. (2001). Requirement for *Pbx1* in skeletal patterning and programming chondrocyte proliferation and differentiation. *Development* **128**, 3543-3557.

Settle, S. H., Jr., Rountree, R. B., Sinha, A., Thacker, A., Higgins, K., and Kingsley, D. M. (2003). Multiple joint and skeletal patterning defects caused by single and double mutations in the mouse *Gdf6* and *Gdf5* genes. *Dev. Biol.* **254**, 116-130.

Smith, K. K., and Schneider, R. A. (1998). Have gene knockouts caused evolutionary reversals in the mammalian first arch? *BioEssays* **20**, 245-255.

Solloway, M. J., and Robertson, E. J. (1999). Early embryonic lethality in *Bmp5;Bmp7* double mutant mice suggests functional redundancy within the 60A subgroup. *Development* **126**, 1753-1768.

Storm, E. E., and Kingsley, D. M. (1996). Joint patterning defects caused by single and double mutations in members of the bone morphogenetic protein (BMP) family. *Development* **122**, 3969-3979.

Studer, M., Lumsden, A., Ariza-McNaughton, L., Bradley, A., and Krumlauf, R. (1996). Altered segmental identity and abnormal migration of motor neurons in mice lacking *Hoxb-1*. *Nature* **384**, 630-634.

Szeto, D. P., Rodriguez-Esteban, C., Ryan, A. K., O'Connell, S. M., Liu, F., et al. (1999). Role of the Bicoid-related homeodomain factor Pitx1 in specifying hindlimb morphogenesis and pituitary development. *Genes Dev.* **13**, 484-494.

Thomas, B. L., Tucker, A. S., Qui, M., Ferguson, C. A., Hardcastle, Z., et al. (1997). Role of *Dlx-1* and *Dlx-2* genes in patterning of the murine dentition. *Development* **124**, 4811-4818.

Thomas, T., Kurihara, H., Yamagishi, H., Kurifara, Y., Yazaki, Y., et al. (1998). A signaling cascade involving endothelin-1, dHAND and *Msx1* regulates development of neural-crest-derived branchial arch mesenchyme. *Development* **125**, 3005-3014.

Trumpp, A., Depew, M. J., Rubenstein, J. L. R., Bishop, J. M., and Martin, G. R. (1999). Cre-mediated gene inactivation demonstrates that FGF8 is required for cell survival and patterning of the first branchial arch. *Genes Dev.* **13**, 3136-3148.

Tucker, A. S., Matthews, K. L., and Sharpe, P. T. (1998). Transformation of tooth type induced by inhibition of BMP signaling. *Science* **282**, 1136-1138.

Tucker, A. S., Yamada, G., Grigoriou, M., Pachnis, V., and Sharpe, P. T. (1999). Fgf-8 determines rostral-caudal polarity in the first branchial arch. *Development* **126**, 51-61.

Xu, P. X., Woo, I., Her, H., Beier, D., and Maas, R. (1997). Mouse homologues of the *Drosophila eyes absent* gene require *Pax6* for expression in lens and nasal placodes. *Development* **124**, 219-231.

Xu, P. X., Adams, J., Peters, H., Brown, M. C., Heaney, S., and Maas, R. (1999). *Eya1*-deficient mice lack ears and kidneys and show abnormal apoptosis of organ primordia. *Nat. Genet.* **23**, 113-117.

Yamada, G., Mansouri, A., Torres, M., Stuart, E. T., Blum, M., et al. (1995). Targeted mutation of the murine *goosecoid* gene results in craniofacial defects and neonatal death. *Development* **121**, 2917-2922.

Yanagisawa, H., Yanagisawa, M., Kapur, R. P., Richardson, J. A., Williams, S. C., et al. (1998). Dual genetic pathways of endothelin-mediated intercellular signaling revealed by targeted disruption of endothelin converting enzyme-1 gene. *Development* **125**, 825-836.

Zhang, J., Hagopian-Donaldson, S., Serbedzija, G., Elsemore, J., Plehn-Dujowich, D., et al. (1996). Neural tube, skeletal and body wall defects in mice lacking transcription factor AP-2. *Nature* **381**, 238-241.

4

Molecular Basis of Inner Ear Induction

Stephen T. Brown, Kareen Martin, and Andrew K. Groves
Gonda Department of Cell and Molecular Biology, House Ear Institute, 2100 West Third Street, Los Angeles, California 90057

I. Introduction
II. Inner Ear Induction in Fish
   A. Molecular Inducers of the Inner Ear
   B. Genes Expressed in Presumptive Otic Tissue
III. Inner Ear Induction in Amphibians
IV. Inner Ear Induction in Birds
   A. Molecular Inducers of the Inner Ear
   B. Genes Expressed in Presumptive Otic Tissue
V. Inner Ear Induction in Mammals
   A. Molecular Inducers of the Inner Ear
   B. Genes Expressed in Presumptive Otic Tissue
VI. Perspectives on Inner Ear Induction
   A. Molecular Inducers and Tissue Sources
   B. Receptor Targets in Presumptive Otic Tissue
   C. Conservation of Genetic Networks
   D. General Placode Induction: A Two-Step Process?
VII. Conclusions
   References

    The induction of the vertebrate inner ear is a complex developmental process that has been under investigation for decades. The traditional tools of embryology used in most of these studies have provided a wealth of information on the subject; however, they are generally limited in focus to morphological features. Recent advances in molecular biology have provided the opportunity to study inner ear induction in ways not possible in previous years. The capacity for visualizing and manipulating gene expression in combination with more traditional embryological techniques has changed the focus of inner ear induction from morphology to changes in gene expression. This chapter provides a critical review of recent studies relating to the molecular induction of the inner ear. A major theme to emerge from these studies is the dependence of inner ear induction on fibroblast growth factor (FGF) signaling. The source(s) of these FGF signals is still not entirely clear, though the available data are consistent with a mesodermal and/or hindbrain origin as has been long proposed. Numerous early otic marker genes have been identified as well, and their conservation in ear induction is quite clear. Functional data regarding these genes are still largely incomplete, although a role for several of these genes in zebrafish inner ear induction has been demonstrated. As a whole, these studies have made exciting and provocative contributions to the field, thus creating a more complete and precise picture of inner ear induction. © 2003, Elsevier Inc.

I. Introduction

The vertebrate inner ear is arguably the most complex vertebrate sensory organ. In addition to the sensation of sound vibrations, the inner ear also provides the central nervous system with information regarding spatial orientation, angular momentum, and gravitational force. The formation of such an elaborate sensory structure from a nondistinct patch of ectodermal cells is a truly remarkable feat, and one that is poorly understood. Morphologically the inner ear is known to begin as an ectodermal thickening called the otic placode, which then becomes internalized to form a hollow vesicle that gradually gives rise to the cochlea, semicircular canals, and other sensory structures. The study of the molecular and genetic basis of this developmental process is in its infancy; however, much progress has been made. This chapter examines this progress with regard to the first step in inner ear development, the formation of the otic placode from undifferentiated embryonic ectoderm through the process of induction. However, before doing so it is necessary to revisit the concept of induction and how it relates to development at the molecular level.

Embryonic induction has been defined as "an interaction between an inducing and a responding tissue that alters the path of differentiation of the responding tissue" (Gurdon, 1987; Jacobson and Sater, 1988). The first step in the study of an inductive process is the identification of inducing and responding tissues. A classic and relatively straightforward example of induction is lens formation in non-neural ectoderm due to contact with underlying neuroepithelium of the optic vesicle (Chow and Lang, 2001; Grainger, 1992). By comparison, inner ear induction is less well understood since, although the responding tissue has been clearly identified as ectoderm adjacent to the developing hindbrain, the identity of ear-inducing tissues is still somewhat controversial. Nearly a century's worth of studies have generated a consensus view that both mesodermal and neural tissue nearest the presumptive ear contribute to its induction, though the relative contributions of each tissue varies depending on the experimental organism and the methodology involved. A review of these studies is not attempted here; instead interested readers are referred to excellent discussions of this topic found in Baker and Bronner-Fraser (2001) and Kiernan *et al.* (2002). A definitive demonstration of the necessity or sufficiency of any of these tissues for normal induction of the inner ear is currently lacking, and future studies involving molecular tools will help clarify the issue. For this chapter, both mesodermal and neural tissue are considered as potential sources of inducing signals as the available molecular data relating to otic placode induction are examined. This topic is addressed in Section VI in light of the molecular data for each model organism that follow in Sections II–V.

The availability of molecular tools to analyze gene expression and function has resulted in the concept of embryonic induction becoming more complicated and reductionist. One can no longer think of the appearance of a placode as a single developmental event. The entire process clearly involves the products of many genes, and multiple genetic pathways can now be viewed on a much finer scale. While morphological changes resulting from inductive processes provide the basis for studying induction, they appear well after the changes in gene expression underlying them are under way. In other words, they are indicators that molecular induction has already occurred. To understand the molecular events responsible for these changes, it is necessary to think of induction in molecular terms. From this perspective, induction is manifested by changes in gene expression caused by molecules released from inducing tissue(s), with a single change in gene expression considered the functional unit of induction. The identification of changes in the expression of individual genes and of molecules inducing these changes is the current focus of most research involving induction of the inner ear. In time these molecular events will be seen as part of larger genetic networks triggered by extrinsic molecular inducers, thereby creating a more complete picture of the molecular basis of inner ear induction.

In order to evaluate candidate inducing tissues or molecules, it is crucial to know approximately when the induction starts. Although the exact onset of otic placode induction is not known for any organism, a reasonable estimate can be generated experimentally by determining the developmental stage at which otic tissue is specified with respect to given placode characteristics. Historically, researchers had to rely on morphological landmarks such as the thickening of placodal ectoderm or its invagination to form a pit or vesicle to make such estimates. More recently, the advent of molecular markers of the otic placode has made it clear that otic placode induction begins well before it is morphologically distinct, and thus specification must be reassessed at the molecular level. In practice, specification is determined by explanting pieces of presumptive otic ectoderm of different ages and maintaining them in culture in the absence of any inducing signals such as growth factors or serum. If the explanted ectoderm has already begun to respond to inducing signals, it is possible that it will express otic markers during the culture period. Such tissue is said to be *specified*, thus providing a useful operational definition of when otic placode induction starts. A recent example of such a study has been done in the chick with regard to *Pax2*,

Sox3, BMP7, and *Notch* (Groves and Bronner-Fraser, 2000). It is important to emphasize, however, that specification can be described only with respect to the molecular markers used in the assay and that any inference of the starting point of induction must by necessity be a provisional one, subject to change by the discovery of earlier placode markers. Consequently, the onset of otic placode induction can be said to be no later than the point of specification for the earliest known ear marker.

The end of inner ear induction is a somewhat arbitrary designation, yet to limit the scope of this review we need to clarify when induction is considered more or less complete. For practical reasons we consider this to be when the otic placode has begun its transformation into the otic vesicle, whether by invagination or cavitation. In theory it is more appropriate to consider induction complete when the otic anlage is *committed* to forming a placode (or vesicle), with the point of "commitment" being determined experimentally by placing otic tissue of varying ages in an ectopic "challenging" environment and looking for placode or vesicle formation. No such experiments have been reported for fish or mammalian species; however, this has been done in chick (Groves and Bronner-Fraser, 2000) and several amphibian species (Gallagher *et al.*, 1996; Ginsburg, 1995). The results of these studies indicate that the timing of commitment to an ear fate from species to species is quite variable; therefore we have chosen to define the end of otic placode induction as the onset of the transition from placode to vesicle.

The experimental evidence described in Sections II–V reveals the genes involved in inner ear induction among different species to be highly conserved, a fact made more obvious through the simultaneous examination of ear induction in multiple species. We feel, however, that the reader would be better served by examining the available data for each model organism separately. This will provide the reader with a clearer picture of the current understanding of inner ear induction in various model organisms before addressing broader implications in Section VI. Therefore, the available molecular data regarding the inner ear induction process in fish (zebrafish and medaka), amphibians (*Xenopus laevis*), birds (chick and quail), and mammals (mouse and human) are examined in turn.

II. Inner Ear Induction in Fish

A. Molecular Inducers of the Inner Ear

1. Fgf Family Members

Many studies have implicated hindbrain tissue adjacent to presumptive otic ectoderm and subjacent mesoderm as the primary sources of ear-inducing signals. Because of their expression in the early hindbrain, the secreted

4. Molecular Basis of Inner Ear Induction   119

factors Fgf3 and Fgf8 are considered potential ear-inducing factors in fish. Both of these genes are expressed in the margin and embryonic shield at 50% epiboly with only *fgf3* maintained in cells of the prechordal hypoblast as they migrate from the margin past the otic anlagen at ~75% epiboly (Furthauer *et al.*, 1997; Phillips *et al.*, 2001; Reifers *et al.*, 1998; see Fig. 1A). At about this time expression of both *fgf3* and *fgf8* begins in the neurectoderm destined to give rise to rhombomere 4, with expression progressively increasing and becoming restricted to rhombomere 4 as the ear is induced (Furthauer *et al.*, 1997; Maroon *et al.*, 2002; Phillips *et al.*, 2001; Reifers *et al.*, 1998; see Fig. 1A).

Four recent studies in zebrafish targeted *fgf3* and/or *fgf8* using morpholino treatment or the *fgf8* mutant *acerebellar* (*ace*; Reifers *et al.*, 1998) to eliminate or reduce their activity during early ear development. Knockdown of Fgf3 alone resulted in smaller otic vesicles and decreased the expression domain of the early ear marker *pax2.1* (Leger and Brand, 2002; Maroon *et al.*, 2002; Phillips *et al.*, 2001), in addition to slightly reducing *sox9a* levels (Liu *et al.*, 2003). Expression of *pax8* was also examined with two groups reporting a decrease in the area of *pax8* expression in the ear (Leger and Brand, 2002; Phillips *et al.*, 2001) and another reporting no effect (Maroon *et al.*, 2002). Two studies (Leger and Brand, 2002; Maroon *et al.*, 2002) also demonstrated a corresponding decrease in *dlx3b* (we adopt recent changes in zebrafish gene nomenclature and refer to *dlx3* and *dlx7* as *dlx3b* and *dlx4b*, respectively), whereas another reports no change in *dlx3b* (or *dlx4b*) expression (Liu *et al.*, 2003). Analysis of *ace* mutants (Leger and Brand, 2002; Liu *et al.*, 2003; Phillips *et al.*, 2001) or embryos treated with *fgf8*-specific morpholinos (Maroon *et al.*, 2002) gave results similar to those observed in the Fgf3 knockdown studies discussed previously, with the exception of *dlx3b* expression being reduced in all cases (Leger and Brand, 2002; Liu *et al.*, 2003; Maroon *et al.*, 2002). Decreased expression domains for *eya1* and *six4.1* were seen as well (Leger and Brand, 2002).

The simultaneous targeting of *fgf3* and *fgf8* resulted in the complete absence of otic vesicle formation in a significant proportion of embryos in all four studies, whether *fgf8*-specific morpholinos or ace mutants were used (Leger and Brand, 2002; Liu *et al.*, 2003; Maroon *et al.*, 2002; Phillips *et al.*, 2001). As might be expected, the expression of numerous ear marker genes was also negatively effected. A corresponding loss of *pax2.1* expression in presumptive otic ectoderm was observed in each study that analyzed this marker (Leger and Brand, 2002; Maroon *et al.*, 2002; Phillips *et al.*, 2001). Effects on *dlx3b* expression, however, varied from study to study. In one study, expression of *dlx3b* was reported to be frequently absent in these embryos (Maroon *et al.*, 2002), while in another *dlx3b* (along with *eya1*) was reported to initiate normally with expression not maintained after placode induction (Leger and Brand, 2002). A third study (Liu *et al.*, 2003)

A   **FISH**

50% epiboly   *80% epiboly*   *18ss*

fgf3/fgf8   Furthauer et al., 1997;
fgf3   *   Leger and Brand, 2002;
fgf3/fgf8   Maroon et al., 2002;
                                                      Phillips et al., 2001;
                                                      Reifers et al., 1998

B   **CHICK**

         Stage 6   *Stage 7 (2ss)*   *Stage 9 (7ss)*

Fgf3   Mahmood et al., 1995;
                                                      A. Groves, unpublished obs.
Fgf19   ♦   Ladher et al., 2000a

Wnt8c   Ladher et al., 2000a

C   **MOUSE**

              1ss   *3ss*   *7ss*

Fgf3   Wilkinson et al., 1988
Fgf10   Wright and Mansour, 2003

　　 Shield/Germ Ring (zebrafish)
　　 Neural ectoderm
　　 Mesoderm
　 * prechordal hypoblast
　 ♦ *Fgf19* expression appears beneath presumptive otic ectoderm

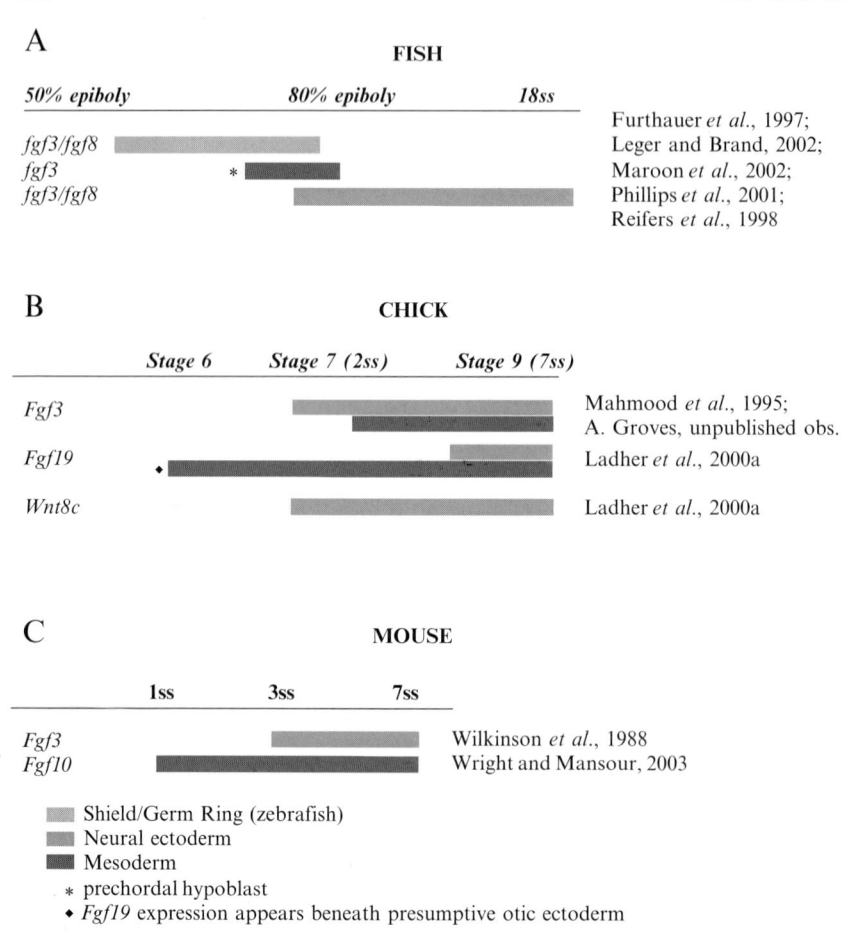

Figure 1 Summary of the temporal expression of candidate inner ear–inducing molecules in zebrafish, chick, and mouse. The approximate duration of expression for each factor in potential ear-inducing tissues is shown relative to the indicated developmental stages. Factors of neural origin are represented by blue bars, factors of mesodermal origin are represented by red bars, and factors expressed in the zebrafish shield and germ ring are represented by a green bar. The location of *Fgf19* expression in chick changes progressively from mesoderm beneath the neural plate to mesoderm beneath presumptive otic ectoderm to the ventral neural tube. The approximate time of the initial appearance of *Fgf19* expression in mesoderm beneath presumptive otic ectoderm is indicated by a diamond(♦). (See Color Insert.)

reports that the loss of Fgf3 and Fgf8 causes a slight reduction in *dlx3b* expression (in addition to a severe reduction in *sox9a* expression). It is not clear if the age of embryos stained for *dlx3b* in these studies overlap. Considering these results as a whole, it seems reasonable to speculate that the maintenance of *dlx3b* expression in the presumptive ear is Fgf dependent,

4. Molecular Basis of Inner Ear Induction

although its initiation may not be. However, further studies are needed to clarify the relationship between Fgf signaling and *dlx3b* expression in the presumptive ear of fish.

The effects on *pax8* expression in these studies also differ. Two studies report little to no *pax8* expression following *fgf3*-specific morpholino treatment in an *ace* background (Leger and Brand, 2002; Phillips *et al.*, 2001), whereas another indicates at most a slight reduction in *pax8* following treatment with morpholinos specific for both *fgf3* and *fgf8* (Maroon *et al.*, 2002). The FGF receptor inhibitor SU5402 (Mohammadi *et al.*, 1997) was also used in two of these studies to inhibit Fgf signaling as a whole during ear induction with somewhat differing results (Leger and Brand, 2002; Maroon *et al.*, 2002). Leger and Brand report a significant proportion of embryos completely lacking otic vesicles, as well as otic *pax8*, *pax2.1*, and *dlx3b* expression. In the Maroon *et al.* study, *pax2.1* expression was also inhibited, but surprisingly *pax8* was relatively unaffected. In this study there was no description of morphological effects.

The conflicting results regarding *pax8* expression following Fgf3 and Fgf8 knockdown or SU5402 treatment are quite significant. Leger and Brand conclude that *pax8* expression and induction of the ear require Fgf signaling, whereas Maroon *et al.* propose that *pax8* expression is Fgf independent. Differences in hybridization stringency could provide one possible explanation, though this was not evident with regard to other genes. It is also possible that fish treated with *fgf8*-specific morpholinos are not equivalent to *ace* mutants, which express an 83–amino acid truncated protein (Reifers *et al.*, 1998). The discrepancies seen with SU5402 treatment may be due to differences in concentration and the time of administration (60 μM SU5402 at 30% or 60% epiboly for the Maroon study vs 16 μM SU5402 at 70% epiboly for the Leger study). However, the study reporting no effect of SU5402 on *pax8* expression used a much higher concentration administered at a slightly earlier developmental stage. It is generally assumed that SU5402 specifically inhibits all four FGF receptors and no other receptor tyrosine kinases, though studies demonstrating inhibition of FGF receptors other than FGFR1 are not in the literature. In addition, it has been reported that SU5402 is a potent inhibitor of at least one non-FGF RTK, specifically the VEGF receptor FLK (Mohammadi *et al.*, 1997). Thus the specificity of SU5402 for FGF receptors is somewhat uncertain. Alternative methods that target FGF receptors specifically are necessary to resolve these seemingly contradictory results.

With regard to the studies mentioned, it is important to note that disruption of Fgf3 and Fgf8 function in zebrafish causes repatterning of the hindbrain in addition to effects on otic placode induction (Maves *et al.*, 2002; Walshe *et al.*, 2002). It is therefore not clear whether Fgf3 and Fgf8 are acting directly to induce the inner ear in zebrafish or indirectly by specifying rhombomere identities, which then provide other inducing signals. Several recent studies show the effects of repatterning of the hindbrain on induction of the ear. In one study the zebrafish *pbx2* and *pbx4* genes were disrupted, resulting in the entire hindbrain taking on the identity of rhombomere 1 and therefore lacking *fgf3* expression and *fgf8* upregulation in what would normally become rhombomere 4 (Waskiewicz *et al.*, 2002). Interestingly, the otic placode continued to be induced in approximately the normal position, suggesting that a rhombomere-specific signal may not be required from the hindbrain to induce the otic placode. In another study, retinoic acid (RA), a signaling molecule that causes the caudalization of the developing hindbrain (Maden, 2002), was used to expand the domain of *fgf3* and *fgf8* expression rostrally. This resulted in the expansion of *pax8* expression to completely encircle the anterior neural plate and the corresponding formation of ectopic otic vesicles (Phillips *et al.*, 2001). The expansion of *pax8* expression was shown to be dependent on both Fgf3 and Fgf8 since RA treatment of *ace* mutants or fish treated with *fgf3*-specific morpholinos failed to replicate these results (Phillips *et al.*, 2001). These results support the idea that Fgf3 and Fgf8 in combination are sufficient to induce ectopic *pax8* expression and otic vesicle formation from the hindbrain; however, the role of RA in this process is not clear. Ectopic expression of *fgf8* alone in the vicinity of the hindbrain has been shown to cause the induction of *pax2.1* expression; however, no ectopic vesicle formation was observed (Leger and Brand, 2002). *pax8* expression was not examined. In another study, the caudal expansion of *fgf3*, but not *fgf8*, expression in the early hindbrain was seen in *valentino* mutants homozygous for the val^{337} allele (Kwak *et al.*, 2002), a putative null allele resulting in a 130–amino acid truncated gene product (Moens *et al.*, 1998). In these embryos, patterning defects in the ear are apparent at later stages; however, the ear is induced normally and no ectopic vesicle formation is reported (Kwak *et al.*, 2002). Taken together, these studies suggest that Fgf3 and Fgf8 of hindbrain origin may be sufficient but not necessary to induce the zebrafish ear. Residual Fgf8 in the hindbrains of fish deficient for *pbx2* and *pbx4* (Waskiewicz *et al.*, 2002) may account for ear induction in those fish; however, this is not consistent with the inability of ectopic Fgf8 near the hindbrain to induce ectopic ears (Leger and Brand, 2002). The ear-inducing effects of Fgf3 and Fgf8 during normal development may well be derived from other tissue sources such as the shield or germ ring, with their expression in the hindbrain serving primarily to influence later patterning of the ear.

2. Other Potential Inducers

A possible role for mesendodermal tissue in zebrafish otic placode induction is illustrated by the loss of *one-eyed pinhead* (*oep*) gene function. The *oep* gene product is an EGF-related ligand necessary for nodal signaling and the formation of multiple tissues including prechordal mesendoderm (Gritsman *et al.*, 1999; Zhang *et al.*, 1998). Thus in embryos lacking expression of *oep* (of both maternal and zygotic origin), mesendodermal tissue normally underlying the developing ear is largely absent. One study of maternal/zygotic mutants (*mzoep*) reported otic vesicle formation that was delayed and morphologically abnormal (Leger and Brand, 2002). In these embryos *fgf3* and *fgf8* were initiated normally in the hindbrain, though only *fgf3* expression was maintained in rhombomere 4. In addition *pax8*, as well as *pax2.1* and *dlx3b*, was expressed in its normal position. These results suggest that presumptive hindbrain tissue is sufficient to induce the otic placode, though the abnormal character of its induction also suggests some contribution from mesendoderm. Also, the gross defects in these embryos as a whole make these results difficult to interpret since the hindbrain is clearly defective as well. In a similar study, zebrafish embryos treated with *oep*-specific morpholinos were used to phenocopy *mzoep* mutant embryos. The most severely affected embryos in this study lacked prechordal mesendoderm and also exhibited delayed and abnormal otic vesicle formation that was in a position posterior to its normal location (Phillips *et al.*, 2001). In contrast to the Leger and Brand study, expression of *pax8* in the otic primordia of these embryos was delayed and greatly reduced. Expression of *fgf3* and *fgf8*, although negatively affected in the shield, prechordal domain, and germring, is normal in the hindbrain of these embryos. This suggests that an as yet unknown factor of mesendodermal origin is necessary for normal *pax8* expression in the ear if in fact Fgf3 and Fgf8 contribute to ear induction via the hindbrain. Future studies involving the selective removal of prechordal mesendoderm or its putative ear-inducing factor(s) are needed since the pleiotropic nature of Oep knockdown makes interpretation difficult.

3. Summary

In summary, the studies described in Sections II.A.1 and II.A.2 support the necessity of Fgf signaling via both Fgf3 and Fgf8 for normal induction of the inner ear and several associated marker genes in fish. The sufficiency of these signals to induce the ear in a neutral environment, however, has not yet been demonstrated. It is also not clear when and where these factors perform their role(s) in ear induction. *fgf3* and *fgf8* expression in the presumptive hindbrain seems a likely source of their impact on ear induction; however, there is evidence suggesting this may not be the case. Without specifically knowing the onset of otic placode induction, it is impossible to ignore the potential for Fgf3 and Fgf8 to exert inductive effects from tissues such as the shield and germ ring during gastrulation. The experiments described do not distinguish between these sources. Methods that selectively

B. Genes Expressed in Presumptive Otic Tissue

The earliest genetic marker of the zebrafish inner ear, previously considered to be *pax8*, has been shown to be the transcription factor and forkhead gene family member *foxi1* (Solomon *et al.*, 2003). Mutation of this gene was clearly shown to be responsible for the *hearsay* mutant phenotype, a range of phenotypes with the most severely affected lacking all ear morphology. It is noteworthy that this is the first case of mutation in a single gene reported to cause the complete abrogation of ear development. Not surprisingly, several early otic markers are affected in these mutants with *pax8*, *pax2.1*, and *dlx3b* expression being absent. Effects on *dlx4b*, which is expressed in parallel with *dlx3b* in the ear (Akimenko *et al.*, 1994; Ellies *et al.*, 1997; Solomon and Fritz, 2002), were not reported. Ectopic expression of *foxi1* was also shown to induce corresponding expression of *pax8*, *dlx3b*, and *dlx4b*, though not *pax2.1*. In light of the similar ear phenotypes of *foxi1* mutant fish to those lacking *fgf3* and *fgf8*, the relationship between these genes is of considerable interest. It seems quite possible that *foxi1* expression in the presumptive ear is to some degree dependent on *fgf3* and/or *fgf8*; however, this issue was not addressed.

Four of the zebrafish *pax* genes (*pax2.1*, *pax2.2*, *pax5*, and *pax8*) are expressed in the ear at some point, with *pax8* appearing first at 80–90% epiboly, followed by *pax2.1* at the 3 somite stage (Pfeffer *et al.*, 1998). Expression of *pax2.2* and *pax5* is not seen until the 6 and 18 somite stages, respectively (Pfeffer *et al.*, 1998). There are no reported mutations or knockdown studies in zebrafish directly involving the *pax8* gene. Its ectopic expression has been accomplished via RA treatment and is dependent on the ectopic expression of *fgf3* and *fgf8* as described in Section II.A.1 (Phillips *et al.*, 2001). As described in Section II.A.2, knockdown of Oep activity can severely reduce *pax8* expression in the otic primordium despite normal expression of *fgf3* and *fgf8* in the hindbrain (Phillips *et al.*, 2001). This suggests an additional source of otic *pax8* induction of mesendodermal origin, although this can also be explained by the missing *fgf* expression in non-neural tissues (Phillips *et al.*, 2001). Foxi1 might well mediate the impact of these signals on *pax8* expression, though this issue is not yet addressed in the literature. Mutations in the *pax2.1* gene are responsible for the *no isthmus* (*noi*) phenotype in zebrafish (Brand *et al.*, 1996). Embryos homozygous for the putative null *pax2.1* allele *noi*^{tu29a} (encoding a 139-residue–truncated peptide; Lun and Brand, 1998) show no obvious defects in ear induction, though the mature ear is sometimes reduced in size (Brand *et al.*, 1996;

4. Molecular Basis of Inner Ear Induction 125

Lun and Brand, 1998; Pfeffer *et al.*, 1998). Its expression in otic ectoderm is clearly dependent on Fgf signaling and Foxi1 (see Section II.A.1 and the previous paragraph, respectively), as well as expression of *dlx3b* and *dlx4b* (Solomon and Fritz, 2002). Expression of *pax2.1* may also be influenced by Sox3 because ectopic expression of *sox3* in medaka on rare occasions (~3%) is reported to induce ectopic vesicles expressing *Pax2* in trunk ectoderm (Koster *et al.*, 2000). Loss-of-function studies involving *Pax8* alone or in combination with *noi* mutants have not been reported but should do a great deal to clarify the roles of these *pax* genes in otic placode induction in fish.

The zebrafish *dlx3b* and *dlx4b* genes are expressed in presumptive otic ectoderm after the appearance of *pax8* and just prior to *pax2.1*. The expression pattern of both genes is very similar, being apparent at ~80% epiboly in ectoderm surrounding the anterior border of the neural plate and becoming limited to the otic and olfactory placodes by the onset of somitogenesis (Akimenko *et al.*, 1994; Ellies *et al.*, 1997; Liu *et al.*, 2003; Solomon and Fritz, 2002). A chromosomal deletion in zebrafish designated *b380* has been described that removes *dlx3b*, *dlx4b*, and 13 other genes and ESTs (Fritz *et al.*, 1996; Liu *et al.*, 2003; Solomon and Fritz, 2002). When this deletion is homozygous, the otic and olfactory placodes and their derivatives fail to form entirely. Consequently, morpholinos specific for *dlx3b* and/or *dlx4b* were used to knock down the activity of these genes and assess their role in ear induction. Knockdown of Dlx4b had no detectable impact on ear formation or marker gene expression except for a slight decrease in *sox9a* and *sox9b* (Liu *et al.*, 2003; Solomon and Fritz, 2002). Knockdown of Dlx3b resulted in the formation of small otic vesicles containing one otolith with variable decreases in the expression domains of *pax2.1* (Liu *et al.*, 2003; Solomon and Fritz, 2002), *sox9a* (Liu *et al.*, 2003), and *eya1* (Solomon and Fritz, 2002). Expression of *pax8* was unchanged (Solomon and Fritz, 2002). Simultaneous knockdown of Dlx3b and Dlx4b resulted in smaller otic vesicles lacking otoliths and the absence of *pax2.1* expression. Expression of *eya1* was further decreased, though *pax8* was still unaffected (Solomon and Fritz, 2002). Thus in zebrafish *dlx3b* and *dlx4b* appear to act in a genetic pathway upstream of *pax2.1*. This pathway also appears to be dependent on Fgf signaling and *foxi1* expression as a result of their effect on *dlx3b* expression (see Section II.A.1 and this section). *pax8* may also be part of this pathway should its expression in the ear be dependent of Fgf signaling and would therefore act upstream of the *dlx* genes. These results also suggest that, although *dlx3b* and *dlx4b* play important roles in ear induction, at least one more gene contained within the *b380* deletion plays an important role in the induction of the inner ear in fish.

The *sox9a* gene is contained within the *b380* deletion and is expressed in otic ectoderm during placode induction in fish (Chiang *et al.*, 2001; Liu *et al.*, 2003). The simultaneous knockdown of Dlx3b, Dlx4b, and Sox9a by

Table I Molecular Markers of Otic Ectoderm in Approximate Order of Appearance: ZEBRAFISH

Stage		Marker	References
8.5 h	75% epiboly	*foxil*[a]	Solomon *et al.*, 2003
		pax8[a]	Pfeffer *et al.*, 1998
		sox9a[a]	Liu *et al.*, 2003
10 h	End of gastrulation	*eya1*[a]	Sahly *et al.*, 1999
		six4.1[a]	Kobayashi *et al.*, 2000
		dlx3b, dlx4b[a]	Akimenko *et al.*, 1994; Ellies *et al.*, 1997; Liu *et al.*, 2003; Solomon and Fritz, 2002
		erm[a]	Raible and Brand, 2001
		pea3[a]	Raible and Brand, 2001
		cldna, cldnb[b]	Kollmar *et al.*, 2001
11 h	3 somites	*pax2.1*[a]	Krauss *et al.*, 1991
		tbx2[a]	Ruvinsky *et al.*, 2000
12 h	6 somites	*pax2.2*[a]	Pfeffer *et al.*, 1998
		scyba[c]	Long *et al.*, 2000
13 h	8 somites	*islet1*[a]	Korzh *et al.*, 1993

[a] Nuclear factors.
[b] Tight junction components.
[c] Secreted factors.

morpholino injection caused a dramatic decrease in the size of the otic placode, and in some embryos complete loss of otic vesicle formation, thus phenocopying the *b380* deletion (Liu *et al.*, 2003). Since morpholino treatment directed at *dlx3b* and *dlx4b* was not sufficient to reproduce the *b380* phenotype (Solomon and Fritz, 2002), it would seem that *sox9a* also plays an essential role in ear induction. Injection of *dlx3b*, *dlx4b*, and *sox9a* mRNA into *b380* mutants was able to completely rescue otic vesicle formation and *pax2.1* expression in the ear. Knockdown of *sox9a* alone via morpholino injection caused a slight reduction in the domain of *dlx3b* expression and subsequent size of the otic vesicle. Expression of *sox9a* in the presumptive ear, like many other otic markers, appears to be dependent on Fgf signaling since the absence of Fgf3 and Fgf8 results in the loss of otic *sox9a* expression (Liu *et al.*, 2003).

The *eya1* gene provides another earlier ear marker in fish. Its expression during otic placode induction is similar to *dlx3b* and *dlx4b*, appearing along the anterior neural plate border, followed by expression in the otic ectoderm that continues through formation of the otic vesicle (Sahly *et al.*, 1999). The *dog-eared* mutation in zebrafish has been shown to be in the *eya1* gene and when homozygous causes vestibular defects in the mature ear. Ear induction does not appear to be affected (Whitfield et al., 2002). Initiation of eya1 expression in the ear is reported to be independent of fgf3 and fgf8 expression in zebrafish, whereas its maintenance in the ear is dependent on these factors (Leger and Brand, 2002). Its expression in the ear is reported to decrease somewhat with the knockdown of Dlx3b and Dlx4b (Solomon and Fritz, 2002) and may be upregulated by Sox3 in medaka (Koster et al., 2000). Zebrafish *six4.1* is another gene expressed in essentially the same pattern as *dlx3b*, *dlx4b*, and *eya1* with regard to ear development, being expressed in the anterior neural plate border and becoming progressively restricted to the region of the presumptive ear (Kobayashi et al., 2000). *six4.1* expression is also reported to decrease in *ace* mutants (Leger and Brand, 2002); however, no other information regarding effects on *six4.1* expression has been reported. A number of other markers are also expressed in presumptive otic ectoderm in zebrafish; however, their relation to ear development in fish is not clear. A listing of genes expressed during the early stages of ear development in zebrafish is shown in Table I.

III. Inner Ear Induction in Amphibians

Molecular studies on ear induction in amphibians are few, consisting of a pair of studies in *Xenopus laevis* involving Fgf2 and Fgf3. *Fgf2* in *Xenopus* is broadly expressed in the anterior portion of the embryo during neurula stages (Song and Slack, 1994). *Fgf3* is initially expressed in rhombomere 3 (r3) at stage 13–14, expanding to include r4 and r5 by stage 15, and becoming limited to r4 by neural tube closure (Lombardo et al., 1998). *Fgf3* is also expressed in the otic vesicle, though none was detected in presumptive otic tissue at earlier stages. In separate experiments, beads soaked in Fgf2 or Fgf3 were shown to induce ectopic vesicles when implanted in posterior ectoderm at stage 13–14 in at least 50% of embryos (Lombardo et al., 1998; Lombardo and Slack, 1998). As with other such studies, the sufficiency of these factors to induce placode characteristics is not clear because of the potential for indirect induction via tissues such as the underlying mesoderm. Studies done *in vitro* will be required to avoid this uncertainty. The necessity of these factors regarding ear induction also remains to be demonstrated.

A handful of otic marker genes have been identified in *Xenopus*. As in zebrafish and mice, *Pax8* expression is seen throughout early ear development (Heller and Brandli, 1999). Other early ear markers likely to play roles in ear induction based on their expression are *Pax2* (Heller and Brandli, 1997), *Six1* and *Six2* (Ghanbari et al., 2001; Pandur and Moody, 2000), *Eya1* (David et al., 2001), and *Sox9* (Spokony et al., 2002). For a comprehensive list of early ear markers in *Xenopus*, see Table II.

Table II Molecular Markers of Otic Ectoderm in Approximate Order of Appearance: XENOPUS

Stage		Marker	References
12	Late gastrula	*Pax8*[a]	Heller and Brandli, 1999
13		*Sox9*[a]	Spokony et al., 2002
14		*Tbx2*[a]	Takabatake et al., 2000
21	8–9 somites	*Six1, Six2*[a]	Ghanbari et al., 2001; Pandur and Moody, 2000
		Pax2[a]	Heller and Brandli, 1997
		Delta 1[b]	Schlosser and Northcutt, 2000
22	9–10 somites	*Groucho-related genes 4 and 5*[a]	Molenaar et al., 2000; Roose et al., 1998
		p21 activated kinase[c]	Islam et al., 2000

[a]Nuclear factors.
[b]Receptors.
[c]Cytoskeletal effector proteins.

IV. Inner Ear Induction in Birds

A. Molecular Inducers of the Inner Ear

Three soluble factors in the chick have been implicated in ear induction as a result of their expression in mesodermal and/or hindbrain tissues prior to the appearance of the otic placode: Fgf3, Fgf19, and Wnt8c. Several studies have been performed to address Fgf3 as a potential ear-inducing factor based on its expression in the early hindbrain. In the initial study involving Fgf3 and inner ear induction, it was demonstrated that antisense oligonucleotides specific for the first 15 nucleotides of the mouse *Fgf3* coding region prevented otic vesicle formation when introduced into explanted head tissue from stage 10 chick embryos (Represa et al., 1991). In retrospect, this study raises two concerns regarding its relevance to inner ear induction. First, it is now clear that by Hamburger Hamilton stage 10 (Hamburger and Hamilton, 1992) the otic ectoderm is already committed to an ear fate and thus induction has already occurred (Groves and Bronner-Fraser, 2000). These results are therefore more relevant to invagination of the placode than to its induction. Second, the subsequent cloning of chick *Fgf3* (Mahmood et al., 1995) revealed that the oligonucleotide used in the study contains three mismatches with regard to chick *Fgf3* and thus would not be predicted to affect chick *Fgf3* expression. A more recent study has shown that misexpression of *Fgf3* causes the formation of ectopic otic vesicles expressing the ear markers *Pax2*, *BMP7*, and *Nkx5.1*, among others (Vendrell et al., 2000). The range of ectopic vesicle formation was limited to ectoderm adjacent to the neural tube extending from just posterior to the eye to the sixth somite. Although these results suggest the sufficiency of Fgf3 to induce placodal characteristics in non-otic tissue, it is not clear if Fgf3 acts directly or indirectly. The necessity of Fgf3 for otic placode induction in the chick is also still not known. Loss-of-function experiments will be required to address this.

Fgf19 and Wnt8c have both been identified as inducers of the otic placode and a number of early otic markers in cultured explants of avian otic ectoderm (Ladher *et al.*, 2000a). Chick *Fgf19* is initially expressed in mesodermal tissue beneath the neural plate at stage 6 and expands to the mesoderm underlying the presumptive otic ectoderm. At this time *Fgf19* expression is also seen in the ventral portion of the neural tube, where it persists until stage 9, at which point all *Fgf19* expression in the vicinity of the developing ear ceases (Ladher *et al.*, 2000a; see Fig. 1B). *Wnt8c* expression with regard to induction of the ear is seen in the hindbrain from stages 7 through 9 (Ladher *et al.*, 2000a; see Fig. 1B). A series of tissue culture experiments revealed that Fgf19-soaked beads were able to induce *Wnt8c* expression along with *Pax2* and *Fgf3*, but only in tissue that contained some presumptive neural ectoderm. *Wnt8c*-expressing COS-7 cells in turn were able to induce *Fgf3* expression in stage 7 otic ectoderm. When combined, Fgf19-soaked beads and *Wnt8c*-expressing cells were able to induce placodal thickening in addition to *Fgf3*, *Nkx5.1*, *Dlx5*, and *SOHo1* expression in stage 7 otic ectoderm (Ladher *et al.*, 2000a).

The authors of this study propose a model of avian otic placode induction in which Fgf19 induces *Wnt8c* expression in the presumptive hindbrain, followed by the cooperative induction of placode formation and associated marker genes (e.g., *Fgf3*, *Pax2*, *Nkx5.1*) by both Fgf19 and Wnt8c. One difficulty with this model is the fact that expression of *Pax2* in chick otic ectoderm is seen before *Fgf19* expression spreads to the underlying mesoderm (Groves and Bronner-Fraser, 2000; Ladher *et al.*, 2000a). Although *Pax2* is the earliest known gene to mark the presumptive ear in birds, it is highly likely that earlier marker genes such as *Pax8* or *Fox* gene homologues will also be identified in the chick. An ear induction model that accounts for the appearance of *Pax2* and other putative earlier markers is therefore necessary. Recent work in our laboratory demonstrates that *Fgf3* is expressed in mesoderm underlying the presumptive placode prior to *Fgf19*, and prior to the appearance of *Pax2* in the overlying ectoderm (A. Groves, unpublished observations). It is possible that mesodermal Fgf3 (and/or some other factor) is responsible for the earliest steps in ear induction. Because of the timing of its expression, we think Fgf19 is more likely to induce later markers and/or to contribute to the maintenance of earlier ones. As described earlier in this section, Fgf3 may be sufficient to induce the same characteristics (Vendrell *et al.*, 2000), and the spatiotemporal expression of *Fgf3 in vivo* is consistent with such a role. The sufficiency of Fgf19 in combination with Wnt8c, and possibly Fgf3 by itself, to induce the otic placode in competent tissue seems clear; however, the actual role of each factor *in vivo* remains to be demonstrated.

B. Genes Expressed in Presumptive Otic Tissue

The expression of numerous avian genes within presumptive otic tissue has been demonstrated; however, no studies to date have revealed their roles in ear development. Thus for now they serve as otic markers, though the expression of such conserved genes as *Pax2, Eya1*, and *Six4* suggests roles similar to those of homologous genes in other organisms. A listing of known early otic markers in chick and/or quail is presented in Table III. The absence of avian model organisms useful for genetic studies has made it difficult to assess the roles of given genes in ear development in birds. Recent

Table III Molecular Markers of Otic Ectoderm in Approximate Order of Appearance: CHICK

Stage		Marker	References
8	4 ss	*Pax2*[a]	Groves and Bronner-Fraser, 2000
	5 ss	*Lmx1*[a]	Giraldez, 1998
	6 ss	*Sprouty2*[b]	Chambers and Mason, 2000
9	7 ss	*Gata3*[a]	Sheng and Stern, 1999
		Sox3[a]	Groves and Bronner-Fraser, 2000
		Frizzled1[c]	Stark et al., 2000
		Dlx3[a]	Brown and Groves, unpublished observations; Pera and Kessel, 1999
		Bmp7[b]	Groves and Bronner-Fraser, 2000
		Frzb1[b]	Baranski et al., 2000; Duprez et al., 1999
		SOHo1[a]	Kiernan et al., 1997
	9 ss	*Nkx5.1*[a]	Herbrand et al., 1998
10	10 ss	*Gbx2*[a]	Hidalgo-Sanchez et al., 2000
		Notch1[c]	Groves and Bronner-Fraser, 2000
		Fgf8[b]	Hidalgo-Sanchez et al., 2000
		Wnt5a[b]	Baranski et al., 2000
		Six4[a]	Esteve and Bovolenta, 1999
11	13 ss	*Frizzled2*[c]	Stark et al., 2000
		Bmp4[b]	Wu and Oh, 1996
		Delta1[c]	Wu and Oh, 1996
		Serrate1[c]	Wu and Oh, 1996
		SFrp1[b]	Esteve et al., 2000
		SFrp2[b]	Ladher et al., 2000b

[a] Nuclear factors.
[b] Secreted factors.
[c] Receptors.

advances in gene knockdown such as RNAi will hopefully prove fruitful in studies with chick embryos and allow the roles of specific genes in the development of the inner ear and other organ systems to be assessed directly.

V. Inner Ear Induction in Mammals

A. Molecular Inducers of the Inner Ear

1. FGF Family Members

The only two factors in mice thus far considered to be potential inducers of the inner ear are Fgf3 and Fgf10. Murine *Fgf3* is expressed in rhombomeres 5 and 6 of the hindbrain, as well as in the otic ectoderm at the time of placode formation (Mahmood *et al.*, 1996; McKay *et al.*, 1996; Wright and Mansour, 2003; see Fig. 1C). *Fgf3*-deficient mice have been described and show no evidence of effects on inner ear induction (Mansour *et al.*, 1993). No physical abnormalities in otic development or the adjacent hindbrain structures were observed in any embryos through E9.5, by which time the otic vesicle has formed, although abnormalities in later ear development were clearly apparent. Mice deficient in the *Kreisler* gene product, a factor necessary for proper hindbrain development (McKay *et al.*, 1994), exhibit an expanded domain of *Fgf3* expression in the hindbrain without the normal increase of expression in the region of rhombomeres 5 and 6 (McKay *et al.*, 1996). In these mice the ear is also induced normally with defects appearing later in ear development (McKay *et al.*, 1996).

Fgf10 expression has recently been demonstrated in mesoderm underlying the presumptive placode from the one- to seven- somite stages (Wright and Mansour, 2003; See Fig. 1C). This is the first report of a secreted factor in mammals being expressed in this tissue prior to the appearance of the otic placode and specific ear markers. Expression of *Fgf10* is also seen later in the otic placode, along with *Fgf3* (Pirvola *et al.*, 2000; Wright and Mansour, 2003; See Fig. 1C). Mice deficient for *Fgf10* have been generated and, unlike *Fgf3*-deficient mice, exhibit a smaller otic vesicle at E10.5 (Ohuchi *et al.*, 2000). Mice deficient for both *Fgf3* and *Fgf10* have also been generated, revealing ear phenotypes ranging from the formation of extremely small vesicles to the complete absence of otic vesicles altogether (Wright and Mansour, 2003). To test for molecular evidence of ear induction in these mice, they were examined for *Pax8*, *Pax2*, and *Gbx2* expression in otic ectoderm at the eight- somite stage. Expression of all three of these early otic marker genes was not seen in this tissue, although in many embryos their expression was seen more laterally. The hindbrains of these embryos were also examined and found to be normal in morphology and marker gene expression

(*Kreisler, HoxB1, Krox20*), suggesting that the ear phenotype is not simply an indirect consequence of hindbrain defects (Wright and Mansour, 2003). These results present the first clear demonstration that FGF signaling is necessary for ear induction in mammals and are consistent with studies involving *fgf3* and *fgf8* in zebrafish. In addition, the hindbrain and mesodermal sources of FGF3 and FGF10, respectively, provide strong evidence that inductive signals are derived from both tissues in mice.

2. FGF Receptors

FGFs exert their effects by binding to and activating subsets of the FGF receptors. Table IV lists the ability of several relevant FGFs to activate various FGF receptors. FGF3 and FGF10 have both been shown to activate FGFR1b and FGFR2b (Miralles *et al.*, 1999; Ohuchi *et al.*, 2000; Ornitz *et al.*, 1996; Pirvola *et al.*, 2000); therefore one might expect mice deficient for these receptor genes to show defects in ear induction. There are no studies providing clear evidence of a role for FGFR1 in ear induction since *FGFR1*-deficient embryos die prior to otic placode formation (Pirvola *et al.*, 2002). However, one study has reported the effects of two hypomorphic alleles of *FGFR1*, both of which result in apparently normal ear induction followed by defects in sensory cell architecture (Pirvola *et al.*, 2002). Mice completely deficient in FGFR2 (both IIIb and IIIc isoforms) die by E10.5 to E11.5, though by this stage a marked reduction in the size of the otic vesicle is apparent (Xu *et al.*, 1998). *FGFR2b*-deficient mice have been reported and also exhibited reduced otic vesicles, though less markedly so (Ohuchi *et al.*, 2000; Pirvola *et al.*, 2000; Revest *et al.*, 2001), indicating that FGFR2c may contribute to ear induction. *FGFR2c*-deficient mice have

Table IV Mammalian FGF/FGF Receptor Interactions

	FGFR1		FGFR2		FGFR3		FGFR4
Ligands	IIIb	IIIc	IIIb	IIIc	IIIb	IIIc	
FGF2	•	•		•		•	•
FGF3	•		•				
FGF8				•		•	•
FGF10	•[a]		•[a]				
FGF19							•[b]

From Ornitz *et al.*, 1996, except where noted.
[a]Miralles *et al.*, 1999.
[b]Xie *et al.*, 1999.

as yet not been reported. The milder phenotype observed in the *FGFR2* knockout mice compared with the *Fgf3/Fgf10* double knockout is likely due to the continued expression of FGFR1b, which also binds FGF3 and FGF10. Thus, contrary to the implications of results involving *FGFR1* hypomorphic alleles, FGFR1b might play a significant role in murine inner ear induction after all. *FGFR3*-deficient mice exhibited no effects on ear induction, though defects were seen at later stages in the organ of Corti (Colvin *et al.*, 1996). This is not unexpected since there is no evidence that FGF3 or FGF10 activates FGFR3.

FGFR4-deficient mice have been generated and are reported to have liver defects with no mention of defects in the ear or other organs (Yu *et al.*, 2000). This is especially relevant considering the potential role of Fgf19 in avian ear induction (Ladher *et al.*, 2000a). The *in vitro* binding characteristics of human FGF19 have been reported and reveal that FGF19 binds to FGFR4 but not to FGFR1c, FGFR2b, FGFR2c, FGFR3b, or FGFR3c (Xie *et al.*, 1999). FGFR1b was not included in this study and thus its ability to bind and be activated by FGF19 is not known. Interestingly, *Fgf15*-deficient mice (the mouse homologue of *Fgf19* has been designated *Fgf15*) do not appear to have defects in ear induction (S. Mansour, personal communication). As with many other knockout mice, functional redundancy may explain the lack of a mutant phenotype so a role for FGF15 in murine ear induction is not necessarily excluded. A more detailed examination of *FGFR4*-deficient mice with regard to ear development may prove to be informative. With regard to *Fgf* receptor gene knockouts in general, it seems clear that compound knockouts and the use of conditional alleles will be necessary to sort out the roles of different FGF receptors in murine ear induction.

B. Genes Expressed in Presumptive Otic Tissue

Two genes in the *Dlx* family, *Dlx5* and *Dlx6*, have been implicated in otic placode induction in mammals by virtue of their expression patterns. Both genes are expressed in the border region of the anterior neural plate fated to give rise to all the cranial placodes (Merlo *et al.*, 2002; Quint *et al.*, 2000; Robledo *et al.*, 2002). Unlike zebrafish *dlx3b* and *dlx4b*, which are also expressed at the neural plate border, mouse Dlx5 and Dlx6 are not upregulated in the otic or olfactory anlagen (although they are expressed in both tissues during their later development). Two studies of mice deficient for *Dlx5* show significant vestibular abnormalities; however, ear induction appears to be unaffected (Acampora *et al.*, 1999; Depew *et al.*, 1999). Mice deficient for both *Dlx5* and *Dlx6* show much more severe defects in later ear development, though the otic vesicle seems to form normally

Table V Molecular Markers of Otic Ectoderm in Approximate Order of Appearance: MOUSE

Stage		Marker	References
E8	1–7 ss	$Pax8^a$	Pfeffer et al., 1998
		$BF1^a$	Hatini et al., 1999
		$Dlx5^a$	Depew et al., 1999; Yang et al., 1998
		$Fgf3^b$	McKay et al., 1996
		$Bmp7^b$	Solloway and Robertson, 1999
E8.5	8–12 ss	$Sox2^a$	Wood and Episkopou, 1999
		$Hmx3(Nkx5.1)^a$	Rinkwitz-Brandt et al., 1995
		$c\text{-}kit^c$	Orr-Urtreger et al., 1990
		$Notch1^c$	Reaume et al., 1992
E9	13–20 ss	$Ngn1^a$	Ma et al., 1998
		$NeuroD^a$	Ma et al., 1998
		$Activin\ receptor\ 1^c$	Yoshikawa et al., 2000
		$Bmp4^b$	Morsli et al., 1998
		$Delta1^c$	Ma et al., 1998; Morrison et al., 1999

[a] Nuclear factors.
[b] Secreted factors.
[c] Receptors.

(Merlo et al., 2002; Robledo et al., 2002). Thus DLX5 and DLX6 do not appear to be necessary for induction of the mouse inner ear.

Knockout studies in mice have been reported for a number of other relevant transcription factors. *Foxi1* null mice have no detectable defects in development of the inner ear until E13.5, after which gross defects in various inner ear structures become apparent (Hulander et al., 1998, 2003). This is in contrast to zebrafish *hearsay* mutants, which are also deficient in Foxi1 and fail to initiate ear development altogether in a significant proportion of embryos (Solomon et al., 2003). This difference may be due to the later onset of *Foxi1* expression during murine ear development, which appears to be after the otic placode has formed (Hulander et al., 1998). *Pax8*-deficient mice show no defects in ear development, possibly due to compensation by PAX2 (Mansouri et al., 1998). *Pax2*-deficient mice fail to form the cochlea and spiral ganglion; however, induction of the ear seems to be unaffected (Torres et al., 1996). Interestingly, these mice show no changes in the expression of *Eya1* (Heanue et al., 2002), a gene commonly regulated by *Pax* gene products in other developmental systems. Compensation by PAX8, however, is a possibility. Mice deficient for both *Pax2* and *Pax8* should be more informative because potential compensation between these genes will be avoided.

Inner ear development in *Eya1*-deficient mice appears to initiate normally but is arrested at the otic vesicle stage (Xu et al., 1999). Not surprisingly, expression of both *Pax2* and *Pax8* in these mice is unaffected (Xu *et al.*, 1999). Mice deficient for *Six4* exhibit no defects in inner ear development (Ozaki *et al.*, 2001). This may be due to compensation by *Six1*, which has a similar expression pattern (Ohto *et al.*, 1998; Oliver *et al.*, 1995). *Six1*-deficient mice have been described; however, no defects in development of the inner ear have been reported (Laclef *et al.*, 2003). Finally, *Gbx2*-deficient mice undergo normal ear induction with vestibular defects appearing later in development (Wassarman *et al.*, 1997). Other ear markers likely to play roles in ear induction based on their expression patterns have not yet been knocked out. For a detailed listing of genes expressed during early ear development in mice, see Table V.

VI. Perspectives on Inner Ear Induction

A. Molecular Inducers and Tissue Sources

The experimental evidence described in Sections II–V implicates several secreted factors as inducers of the otic placode. With one exception, all of these factors are members of the FGF family of growth factors. In fish these are Fgf3 and Fgf8; in amphibians Fgf2 and Fgf3; in birds Fgf3, Fgf19, and Wnt8c; and in mammals FGF3 and FGF10. For a visual summary of the temporal expression of these genes, see Fig. 1. In assessing the contributions of each factor to the inductive process it helps to think in terms of necessity and sufficiency. A gene, or more specifically the product it encodes, is considered *necessary* if the phenotype under consideration (e.g., marker gene expression, morphological character) changes as a result of its absence. In zebrafish and mice, gene knockout and knockdown strategies show that respective pairs of FGFs, though partially redundant, are necessary for otic placode induction in these organisms. In amphibians and birds, experiments establishing the necessity of specific factors have not been reported. A gene product is considered *sufficient* if it directly causes a phenotypic change in tissue that would otherwise remain unchanged, an assessment best made with isolated tissue explants. The sufficiency to directly induce otic placode formation or marker gene expression has been demonstrated only in chick with regard to Fgf19 and Wnt8c. Clearly, further experimentation establishing the necessity and sufficiency of given factors is required since the necessity *and* sufficiency of any given factor has not been established. Assessing the necessity and sufficiency of the factors mentioned will not exclude the existence of other as yet undiscovered factors that participate in the induction process; however, it would provide a clear method of assessing the contribution of each factor. With a more complete picture of otic placode induction in these terms we will move closer to a molecular model of this process with regard to extrinsic factors and the tissues from which they originate.

In light of recent molecular data, the uncertain issue of the contributions of mesodermal and neural tissue to otic placode induction can be looked at from a new perspective. The necessity of *Fgf3* and *Fgf10* together for otic placode induction in mice clearly suggests contributions from both tissues since during the induction process they are expressed in presumptive hindbrain and mesodermal tissue, respectively. The same might be said with regard to zebrafish studies involving *fgf3* and *fgf8*, which are both expressed in the presumptive hindbrain and mesendodermal tissues. However, the tissue source(s) of their inducing activity is still not clear. Although the hindbrain is strongly implicated, *fgf3* and *fgf8* may also influence ear induction from the shield or germring. The effects of *oep* perturbation on otic placode induction also imply an ear-inducing signal(s) of mesendodermal origin in fish. Pleiotropic effects of this mutation involving the hindbrain and other tissue, however, create uncertainty. In chick, expression patterns of *Fgf3*, *Fgf19*, and *Wnt8c*, combined with *in vitro* studies, clearly suggest that both mesodermal and hindbrain tissue contribute to induction of the otic placode; however, the potential role of mesodermal factors expressed prior to *Fgf19* (such as *Fgf3*) is not clear. It remains possible that the hindbrain may not be necessary for otic placode induction in birds. Overall, these molecular data support the notion that both mesodermal and neural tissues contribute to otic placode induction, though the matter is still somewhat uncertain and may differ among various classes of organisms.

B. Receptor Targets in Presumptive Otic Tissue

There are four different FGF receptor genes, three of which produce alternatively spliced variants containing one of two distinct FGF-binding domains (IgIIIb or IgIIIc), with only one known gene product derived from the *FGFR4* gene (Powers *et al.*, 2000). *FGFR1* and *FGFR2* also express additional splice variants that are not addressed here. Given the apparent role of FGFs in otic placode induction, it would be expected that their target receptors be expressed in presumptive otic tissue. However, this has not been confirmed since FGF receptor expression data in various model organisms are not complete and typically do not distinguish between different receptor subtypes. With this in mind, the available expression data suggest that FGF receptor targets for the FGFs identified in Table 4 are for the most part expressed in presumptive otic tissue (Sleptsova-Friedrich *et al.*, 2001; Thisse *et al.*, 1995; Walshe and Mason, 2000; Wright and Mansour, 2003). One possible exception is the lack of FGFR4 (the only known Fgf19 target) expression in chick otic ectoderm (Marcelle *et al.*, 1994). However, because of the ability

4. Molecular Basis of Inner Ear Induction

of Fgf19 to induce otic markers in culture (Ladher et al., 2000a), it seems likely that another FGF receptor mediates this effect. One such candidate is FGFR1b since its ability to be activated by FGF19 is not known.

There is a tendency to view FGF receptors as "On/Off" switches with differing affinities for the various FGFs; however, in reality FGF signaling appears to be much more complicated. The activation of FGF receptor tyrosine kinase activity requires not only FGF binding but also heparan sulfate binding and dimerization with another FGF receptor (Powers et al., 2000). It is not known if the signal transmitted by an activated FGF receptor differs depending on its dimerization partner or the FGF molecule bound; therefore the level of signaling complexity is potentially quite large. One way to possibly simplify this issue is to address the functional equivalence of FGFs having the same receptor specificity. This could be accomplished by replacing one FGF with another sharing the same receptor targets. For example, in mice it seems clear that FGF3 and FGF10 together are necessary for normal otic placode induction (Wright and Mansour, 2003), and both appear to have the same receptor targets, FGFR1b and FGFR2b (see Table 4). Their functional equivalence could be tested by the creation of knock-in mice with the coding exons of *Fgf10* replaced by the coding region for *Fgf3* (or vice versa). In such a mouse *Fgf3* would theoretically be expressed as would *Fgf10* (provided there are no intronic regulatory sequences) and would serve to stimulate FGFR1b and FGFR2b in place of FGF10 throughout development, in addition to its endogenous function. If these mice were to develop in a completely normal fashion, this would imply the functional equivalence of FGF3 and FGF10 and suggest that the crucial aspect of FGF signaling is the receptor being activated. Defective development in any context would suggest that, despite having the same receptor targets, FGF3 and FGF10 are functionally distinct.

Another issue worth noting with regard to FGF signaling is the general assumption that FGFs and their receptors are essentially interchangeable from species to species. Although this appears to be true for the most part, there is at least one exception, that being the different binding characteristics of Fgf3 from *Xenopus* compared with Fgf3 from zebrafish or mice (Kiefer et al., 1996), and there are likely more. Most FGF receptor/ligand studies involve genes of mammalian origin and may not necessarily apply to species in other classes. In addition, it remains possible that these *in vitro* studies may not completely correlate with FGF/FGFR function *in vivo*. Nevertheless, similar studies with FGF/FGFR homologues in chick, zebrafish, and *Xenopus* would help clarify matters.

The only non-FGF factor as yet shown to have a role in otic placode induction is Wnt8c in chick. The targets of Wnt factors are receptors in the *Frizzled* family. The only reported expression of *Frizzled* genes in presumptive otic tissue indicate that in chick *Fz-1* is expressed beginning at the seven-somite stage (stage 9), just as *Wnt8c* expression in the hindbrain is shutting down (Stark *et al.*, 2000). Expression of *Fz-2* and *Fz-7* appear after otic placode induction is complete (Stark *et al.*, 2000). Since *Fz-1* and *Wnt8c* expression appear to overlap very briefly and relatively late in the induction process, it would seem likely that another receptor is responsible for its proposed role in placode induction. The effects of Wnt8c on stage 7 otic ectoderm *in vitro* (Ladher *et al.*, 2000a) certainly suggest as much.

C. Conservation of Genetic Networks

The identification of genes associated with inner ear induction in numerous organisms has made it clear that gene homologues appear to function in similar contexts in each organism. We have already seen this with regard to FGF signaling and we also see numerous examples with regard to the expression of genes that mark the presumptive ear. In each organism examined thus far we see very early otic expression of *Pax8* with the exception of the chick. In this case, it is almost certain that avian *Pax8* will soon be isolated and seen to play a similar role in ear development since antibodies specific for mouse PAX8 are seen to stain presumptive otic tissue in birds (K. Martin and A. Groves, unpublished observations). Another *Pax* gene, *Pax2*, is expressed in the ear in all organisms studied. Other genes consistently seen to mark early ear development in a variety of organisms include members of the *Eya, Six, Gbx*, and *Sox* gene families. Other genes have thus far been examined in only one species so their functional conservation, though likely, remains to be seen.

The expression of members of the *Pax, Six*, and *Eya* gene families during early ear development suggests the conservation of a genetic network involving these genes, a network that is seen in numerous other developmental systems. This network was originally described with regard to *Drosophila* eye development (for review see Wawersik and Maas, 2000) and is seen in vertebrate eye, kidney, and muscle development as well (Heanue *et al.*, 1999; Vainio and Lin, 2002; Wawersik and Maas, 2000). We have seen that members of the *Pax, Six*, and *Eya* gene families are expressed during the early development of the inner ear in all organisms examined. Genes in the *Dach* family, which are downstream members of this genetic network in other developmental systems, provide further evidence of conservation of this genetic network since they are also expressed during later inner ear development (Davis *et al.*, 1999, 2001a,b; Heanue *et al.*, 1999, 2002; Vainio and Lin, 2002; Wawersik and Maas, 2000). The genetic relationships among these genes, all of which are transcriptional regulators, have been made clear regarding *Drosophila* eye development. The *Pax* gene homologues *eyeless* (*ey*) and *twin of eyeless* (*toy*) are necessary for the subsequent

4. Molecular Basis of Inner Ear Induction   139

expression of the *Eya* and *Six* gene homologues *eyes absent* (*eya*) and *sine oculis* (*so*), respectively (Wawersik and Maas, 2000). *eya* and *so* in turn regulate the expression of *dachshund* (*dac*), a homologue of vertebrate *Dach* genes. The expression patterns of vertebrate *Pax, Six, Eya,* and *Dach* genes in ear, kidney, and muscle development are completely consistent with this genetic hierarchy, and their genetic relationships appear to be conserved. Thus, it seems clear that this genetic network has been retained for use in ear development. Other genes expressed in the ear are likely to be part of conserved genetic networks; however, they have yet to be identified as such.

One apparent exception to the functional conservation of genes involved in otic placode induction is the role of *Dlx* genes. The *Dlx* gene family consists of six family members (*Dlx1–Dlx6*) in most organisms, with zebrafish having two additional members (Panganiban and Rubenstein, 2002). Most *Dlx* genes have been cloned and examined in multiple developmental contexts, and *Dlx* genes are clearly involved in ear development in every organism examined (Baker and Bronner-Fraser, 2001). The only organism for which *Dlx* genes have a clear role in otic placode induction, however, is zebrafish. Expression of the zebrafish *dlx3b* and *dlx4b* genes is seen throughout the border region of the anterior neural plate, the region destined to give rise to all of the cranial placodes. This expression pattern is also conserved in birds, amphibians, and mammals, with *Dlx5* and *Dlx6* fulfilling this role (Luo *et al.*, 2001; Pera *et al.*, 1999; Quint *et al.*, 2000; S. Brown and A. Groves, unpublished observations). During subsequent development, expression of *Dlx* genes at the neural plate border gradually recedes to the most anterior regions, with upregulation in presumptive olfactory and otic ectoderm occurring only in zebrafish. This may explain the severe effects on otic placode induction in zebrafish embryos treated with *dlx3b/dlx4b*-specific morpholinos that are not apparent in *Dlx5/Dlx6* double knockout mice. Thus it appears that, unlike in fish, in mammals *Dlx* genes are not essential for normal otic placode induction. Although the *Dlx* genes have all been cloned in mice and zebrafish, this is not the case with *Xenopus* and chick. Cloning of the remaining genes in these organisms, followed by a complete analysis of *Dlx* gene expression, will be necessary to determine if *Dlx* genes have a possible role in placode induction in amphibians and birds.

D. General Placode Induction: A Two-Step Process?

The nature of placode induction in general is a topic of ongoing debate. The point of contention is the existence, and hence requirement, of a preliminary developmental step conferring general placode competence on tissue that will subsequently give rise to specific sensory and neurogenic placodes. Specifically, this tissue is the ectoderm bordering the anterior neural plate during late gastrulation ("preplacodal" ectoderm), which will give rise to the olfactory, lens, trigeminal, otic, and epibranchial placodes. Several observations suggest that preplacodal ectoderm may acquire a general and perhaps necessary placode competence prior to the unique formation and development of each placode. One such observation is seen in mice, humans, and several amphibian and fish species as the appearance of a uniformly thickened preplacodal ectoderm that selectively recedes around each presumptive placode as it becomes visible (Baker and Bronner-Fraser, 2001). This morphological observation certainly reflects an underlying molecular process occurring throughout preplacodal ectoderm; however, it is clearly not essential for placode formation in all species since in organisms such as chick and zebrafish, no such thickening occurs (Baker and Bronner-Fraser, 2001).

Another observation suggestive of a state of general placode competence is the common expression pattern of multiple genes specifically marking the preplacodal ectoderm in all vertebrate organisms studied. Genes expressed in this manner include members of the *Dlx*, *Msx*, *Six*, *Eya*, and *BMP* gene families (Baker and Bronner-Fraser, 2001; Streit, 2002). One might predict that the removal of products from a given gene family would impact the formation of all placodes if these gene products contribute to the formation and/or maintenance of a general state of placode competence. To see the effects of this we can simply revisit the experiments in zebrafish and mice regarding *Dlx* genes. Recall that the *b380* deletion in zebrafish removes *dlx3b*, *dlx4b*, and a number of other genes, resulting in the complete absence of the olfactory and otic placodes and their derivatives. Selective knockdown of Dlx3b and Dlx4b also had a significant effect on placode formation and development (although it was not sufficient to completely reproduce the b380 phenotype). The impact of the loss of *dlx3b* and *dlx4b* expression or the *b380* deletion on the induction of other cranial placodes was not addressed, thus making it unclear if placode induction was affected as a whole. It could be argued that the impact of Dlx3b and Dlx4b deficiency on olfactory and otic placode induction was due to the loss of these gene products at the time of their upregulation in the presumptive olfactory and otic regions rather than their absence from the preplacodal domain. The effects of *Dlx5* and *Dlx6* inactivation in mice support this argument. The *Dlx* genes expressed in preplacodal ectoderm in mice (and also birds and amphibians) are *Dlx5* and *Dlx6*, and though mice deficient for both of these genes have numerous skeletal defects, no evidence has been reported relating to abnormal placode induction, otic or otherwise. Overall, the results of these two studies are not conclusive since *Dlx* genes may still contribute to the putative generation of general placode competence without being necessary for it to occur. Several other genes with preplacodal expression patterns such as *Six4* and *Eya1* have been knocked out in mice; however, the capacity for compensation by genes from the same family make it difficult to speculate on their roles in general placode induction.

A third observation suggesting a state of general placode competence is of an experimental nature. Unspecified epiblast tissue from stage 4 quail embryos is competent to form an otic placode and express associated marker genes when transplanted in place of host tissue normally giving rise to these structures (Groves and Bronner-Fraser, 2000). However, the same tissue is not competent to express otic marker genes such as *Pax2* and *Pax8* when exposed to Fgf2 *in vitro*. In contrast, presumptive trigeminal ectoderm (contained within preplacodal ectoderm) is competent in such experiments (K. Martin and A. Groves, unpublished observations). This difference might well be due to the expression of genes characteristic of the preplacodal domain in trigeminal ectoderm that are lacking in stage 4 epiblast. If this is true, then the same result would be expected if any preplacodal ectoderm were to be substituted for trigeminal ectoderm. These experiments have not yet been done but will no doubt be of great interest. If all ectoderm of preplacodal origin is competent to express characteristic placode marker genes when exposed to various inducing factors, it may be possible to reproduce this state by overexpressing preplacodal marker genes in tissue such as stage 4 epiblast.

In summary, there are several lines of evidence suggesting an initial step of general placode competence preceding specific placode induction, though they are merely suggestive and not sufficient to prove that such a state exists. Further experimentation addressing the issue directly is clearly required. For additional information on this topic the reader is referred to excellent reviews by Baker and Bronner-Fraser (2001), Begbie and Graham (2001), and Graham and Begbie (2000).

VII. Conclusions

The studies described in this chapter reveal that the process of inner ear induction shares significant similarities among different species. FGF signaling appears to be crucial for normal ear induction in all species examined, although the source of its action is still not clear. In mouse and chick these signals are likely to originate from cranial paraxial mesoderm, as well as from the hindbrain. In zebrafish, FGF signaling may potentially influence ear induction from the shield, germ ring, and/or the hindbrain. Although FGF signals from the hindbrain in each of these organisms are clearly important for early ear development, it is still not certain that these signals contribute to the induction of the inner ear. The conservation of genes expressed in presumptive otic tissue, many of which are also dependent on FGF signals, is readily apparent. The roles of most otic marker genes during ear induction are still unclear; however, much ground has been gained in this area. The increasing use of comparative studies in different species is continuing to provide new and important contributions to the understanding of inner ear induction. We predict that the coming years will provide answers to the remaining questions surrounding inner ear induction through a combination of classical embryological techniques, molecular biology, and genetics.

Acknowledgments

The authors would like to thank Andres Collazo and Olivier Bricaud for their fruitful discussions and valuable advice during the course of preparing this chapter. We also thank the National Institutes of Health (NIDCD Grant 1 R03 DC05566-1) for financial support during the writing of this chapter and our continuing work on avian inner ear induction.

References

Acampora, D., Merlo, G. R., Paleari, L., Zerega, B., Postiglione, M. P., *et al.* (1999). Craniofacial, vestibular and bone defects in mice lacking the Distal-less-related gene Dlx5. *Development* **126,** 3795–3809.

Akimenko, M. A., Ekker, M., Wegner, J., Lin, W., and Westerfield, M. (1994). Combinatorial expression of three zebrafish genes related to distal-less: Part of a homeobox gene code for the head. *J. Neurosci.* **14,** 3475–3486.

Baker, C. V., and Bronner-Fraser, M. (2001). Vertebrate cranial placodes I. Embryonic induction. *Dev. Biol.* **232,** 1–61.

Baranski, M., Berdougo, E., Sandler, J. S., Darnell, D. K., and Burrus, L. W. (2000). The dynamic expression pattern of frzb-1 suggests multiple roles in chick development. *Dev. Biol.* **217,** 25–41.

Begbie, J., and Graham, A. (2001). The ectodermal placodes: A dysfunctional family. *Philos. Trans. R. Soc. Lond. B Biol. Sci.* **356,** 1655–1660.

Brand, M., Heisenberg, C. P., Jiang, Y. J., Beuchle, D., Lun, K., *et al.* (1996). Mutations in zebrafish genes affecting the formation of the boundary between midbrain and hindbrain. *Development* **123,** 179–190.

Chambers, D., and Mason, I. (2000). Expression of sprouty2 during early development of the chick embryo is coincident with known sites of FGF signalling. *Mech. Dev.* **91,** 361–364.

Chiang, E. F., Pai, C. I., Wyatt, M., Yan, Y. L., Postlethwait, J., and Chung, B. (2001). Two sox9 genes on duplicated zebrafish chromosomes: Expression of similar transcription activators in distinct sites. *Dev. Biol.* **231,** 149–163.

Chow, R. L., and Lang, R. A. (2001). Early eye development in vertebrates. *Annu. Rev. Cell Dev. Biol.* **17,** 255–296.

Colvin, J. S., Bohne, B. A., Harding, G. W., McEwen, D. G., and Ornitz, D. M. (1996). Skeletal overgrowth and deafness in mice lacking fibroblast growth factor receptor 3. *Nat. Genet.* **12,** 390–397.

4. Molecular Basis of Inner Ear Induction

David, R., Ahrens, K., Wedlich, D., and Schlosser, G. (2001). *Xenopus Eya1* demarcates all neurogenic placodes as well as migrating hypaxial muscle precursors. *Mech. Dev.* **103**, 189–192.

David, R. J., Shen, W., Heanue, T. A., and Mardon, G. (1999). Mouse Dach, a homologue of *Drosophila dachshund*, is expressed in the developing retina, brain and limbs. *Dev. Genes Evol.* **209**, 526–536.

Davis, R. J., Shen, W., Sandler, Y. I., Amoui, M., Purcell, P., *et al.* (2001a). Dach1 mutant mice bear no gross abnormalities in eye, limb, and brain development and exhibit postnatal lethality. *Mol. Cell Biol.* **21**, 1484–1490.

Davis, R. J., Shen, W., Sandler, Y. I., Heanue, T. A., and Mardon, G. (2001b). Characterization of mouse Dach2, a homologue of *Drosophila dachshund*. *Mech. Dev.* **102**, 169–179.

Depew, M. J., Liu, J. K., Long, J. E., Presley, R., Meneses, J. J., *et al.* (1999). Dlx5 regulates regional development of the branchial arches and sensory capsules. *Development* **126**, 3831–3846.

Duprez, D., Leyns, L., Bonnin, M. A., Lapointe, F., Etchevers, H., *et al.* (1999). Expression of Frzb-1 during chick development. *Mech. Dev.* **89**, 179–183.

Ellies, D. L., Stock, D. W., Hatch, G., Giroux, G., Weiss, K. M., and Ekker, M. (1997). Relationship between the genomic organization and the overlapping embryonic expression patterns of the zebrafish dlx genes. *Genomics* **45**, 580–590.

Esteve, P., and Bovolenta, P. (1999). cSix4, a member of the six gene family of transcription factors, is expressed during placode and somite development. *Mech. Dev.* **85**, 161–165.

Esteve, P., Morcillo, J., and Bovolenta, P. (2000). Early and dynamic expression of cSfrp1 during chick embryo development. *Mech. Dev.* **97**, 217–221.

Fritz, A., Rozowski, M., Walker, C., and Westerfield, M. (1996). Identification of selected gamma-ray induced deficiencies in zebrafish using multiplex polymerase chain reaction. *Genetics* **144**, 1735–1745.

Furthauer, M., Thisse, C., and Thisse, B. (1997). A role for FGF-8 in the dorsoventral patterning of the zebrafish gastrula. *Development* **124**, 4253–4264.

Gallagher, B. C., Henry, J. J., and Grainger, R. M. (1996). Inductive processes leading to inner ear formation during *Xenopus* development. *Dev. Biol.* **175**, 95–107.

Ghanbari, H., Seo, H. C., Fjose, A., and Brandli, A. W. (2001). Molecular cloning and embryonic expression of *Xenopus* Six homeobox genes. *Mech. Dev.* **101**, 271–277.

Ginsburg, A. S. (1995). Determination of the labyrinth in different amphibian species and its correlation with determination of the other ectoderm derivatives. *Roux's Arch. Dev. Biol.* **204**, 351–358.

Giraldez, F. (1998). Regionalized organizing activity of the neural tube revealed by the regulation of lmx1 in the otic vesicle. *Dev. Biol.* **203**, 189–200.

Graham, A., and Begbie, J. (2000). Neurogenic placodes: A common front. *Trends Neurosci.* **23**, 313–316.

Grainger, R. M. (1992). Embryonic lens induction: Shedding light on vertebrate tissue determination. *Trends Genet.* **8**, 349–355.

Gritsman, K., Zhang, J., Cheng, S., Heckscher, E., Talbot, W. S., and Schier, A. F. (1999). The EGF-CFC protein one-eyed pinhead is essential for nodal signaling. *Cell* **97**, 121–132.

Groves, A. K., and Bronner-Fraser, M. (2000). Competence, specification and commitment in otic placode induction. *Development* **127**, 3489–3499.

Gurdon, J. B. (1987). Embryonic induction—molecular prospects. *Development* **99**, 285–306.

Hamburger, V., and Hamilton, H. L. (1992). A series of normal stages in the development of the chick embryo. 1951. *Dev. Dyn.* **195**, 231–272.

Hatini, V., Ye, X., Balas, G., and Lai, E. (1999). Dynamics of placodal lineage development revealed by targeted transgene expression. *Dev. Dyn.* **215**, 332–343.

Heanue, T. A., Davis, R. J., Rowitch, D. H., Kispert, A., McMahon, A. P., et al. (2002). Dach1, a vertebrate homologue of *Drosophila dachshund*, is expressed in the developing eye and ear of both chick and mouse and is regulated independently of Pax and Eya genes. *Mech. Dev.* **111,** 75–87.

Heanue, T. A., Reshef, R., Davis, R. J., Mardon, G., Oliver, G., et al. (1999). Synergistic regulation of vertebrate muscle development by Dach2, Eya2, and Six1, homologs of genes required for *Drosophila eye* formation. *Genes Dev.* **13,** 3231–3243.

Heller, N., and Brandli, A. W. (1997). *Xenopus Pax-2* displays multiple splice forms during embryogenesis and pronephric kidney development. *Mech. Dev.* **69,** 83–104.

Heller, N., and Brandli, A. W. (1999). *Xenopus Pax-2/5/8* orthologues: Novel insights into Pax gene evolution and identification of Pax-8 as the earliest marker for otic and pronephric cell lineages. *Dev. Genet.* **24,** 208–219.

Herbrand, H., Guthrie, S., Hadrys, T., Hoffmann, S., Arnold, H. H., et al. (1998). Two regulatory genes, cNkx5-1 and cPax2, show different responses to local signals during otic placode and vesicle formation in the chick embryo. *Development* **125,** 645–654.

Hidalgo-Sanchez, M., Alvarado-Mallart, R., and Alvarez, I. S. (2000). Pax2, Otx2, Gbx2 and Fgf8 expression in early otic vesicle development. *Mech. Dev.* **95,** 225–229.

Hulander, M., Kiernan, A. E., Blomqvist, S. R., Carlsson, P., Samuelsson, E. J., et al. (2003). Lack of pendrin expression leads to deafness and expansion of the endolymphatic compartment in inner ears of Foxi1 null mutant mice. *Development* **130,** 2013–2025.

Hulander, M., Wurst, W., Carlsson, P., and Enerback, S. (1998). The winged helix transcription factor Fkh10 is required for normal development of the inner ear. *Nat. Genet.* **20,** 374–376.

Islam, N., Poitras, L., and Moss, T. (2000). The cytoskeletal effector xPAK1 is expressed during both ear and lateral line development in *Xenopus*. *Int. J. Dev. Biol.* **44,** 245–248.

Jacobson, A. G., and Sater, A. K. (1988). Features of embryonic induction. *Development* **104,** 341–359.

Kiefer, P., Mathieu, M., Mason, I., and Dickson, C. (1996). Secretion and mitogenic activity of zebrafish FGF3 reveal intermediate properties relative to mouse and *Xenopus* homologues. *Oncogene* **12,** 1503–1511.

Kiernan, A. E., Nunes, F., Wu, D. K., and Fekete, D. M. (1997). The expression domain of two related homeobox genes defines a compartment in the chicken inner ear that may be involved in semicircular canal formation. *Dev. Biol.* **191,** 215–229.

Kiernan, A. E., Steel, K. P., Fekete, D. M. (2002) Development of the Mouse Inner Ear. *In* "Mouse Development: Patterning, Morphogenesis and Organogenesis," J. Rossant and P. P. L. Tam, eds. pp. 539–566. Academic Press, San Diego.

Kobayashi, M., Osanai, H., Kawakami, K., and Yamamoto, M. (2000). Expression of three zebrafish Six4 genes in the cranial sensory placodes and the developing somites. *Mech. Dev.* **98,** 151–155.

Kollmar, R., Nakamura, S. K., Kappler, J. A., and Hudspeth, A. J. (2001). Expression and phylogeny of claudins in vertebrate primordia. *Proc. Natl. Acad. Sci. USA* **98,** 10196–10201.

Korzh, V., Edlund, T., and Thor, S. (1993). Zebrafish primary neurons initiate expression of the LIM homeodomain protein Isl-1 at the end of gastrulation. *Development* **118,** 417–425.

Koster, R. W., Kuhnlein, R. P., and Wittbrodt, J. (2000). Ectopic Sox3 activity elicits sensory placode formation. *Mech. Dev.* **95,** 175–187.

Krauss, S., Johansen, T., Korzh, V., and Fjose, A. (1991). Expression of the zebrafish paired box gene pax[zf-b] during early neurogenesis. *Development* **113,** 1193–1206.

Kwak, S. J., Phillips, B. T., Heck, R., and Riley, B. B. (2002). An expanded domain of fgf3 expression in the hindbrain of zebrafish valentino mutants results in mis-patterning of the otic vesicle. *Development* **129,** 5279–5287.

4. Molecular Basis of Inner Ear Induction

Laclef, C., Hamard, G., Demignon, J., Souil, E., Houbron, C., and Maire, P. (2003). Altered myogenesis in Six1-deficient mice. *Development* **130**, 2239–2252.

Ladher, R. K., Anakwe, K. U., Gurney, A. L., Schoenwolf, G. C., and Francis-West, P. H. (2000a). Identification of synergistic signals initiating inner ear development. *Science* **290**, 1965–1967.

Ladher, R. K., Church, V. L., Allen, S., Robson, L., Abdelfattah, A., *et al.* (2000b). Cloning and expression of the Wnt antagonists Sfrp-2 and Frzb during chick development. *Dev. Biol.* **218**, 183–198.

Leger, S., and Brand, M. (2002). Fgf8 and Fgf3 are required for zebrafish ear placode induction, maintenance and inner ear patterning. *Mech. Dev.* **119**, 91–108.

Liu, D., Chu, H., Maves, L., Yan, Y. L., Morcos, P. A., *et al.* (2003). Fgf3 and Fgf8 dependent and independent transcription factors are required for otic placode specification. *Development* **130**, 2213–2224.

Lombardo, A., Isaacs, H. V., and slack, J. M. (1998). Expression and functions of FGF-3 in *Xenopus* development. *Int. J. Dev. Biol.* **42**, 1101–1107.

Lombardo, A., and Slack, J. M. (1998). Postgastrulation effects of fibroblast growth factor on *Xenopus* development. *Dev. Dyn.* **212**, 75–85.

Long, Q., Quint, E., Lin, S., and Ekker, M. (2000). The zebrafish scyba gene encodes a novel CXC-type chemokine with distinctive expression patterns in the vestibulo-acoustic system during embryogenesis. *Mech. Dev.* **97**, 183–186.

Lun, K., and Brand, M. (1998). A series of no isthmus (noi) alleles of the zebrafish pax2.1 gene reveals multiple signaling events in development of the midbrain-hindbrain boundary. *Development* **125**, 3049–3062.

Luo, T., Matsuo-Takasaki, M., Lim, J. H., and Sargent, T. D. (2001). Differential regulation of Dlx gene expression by a BMP morphogenetic gradient. *Int. J. Dev. Biol.* **45**, 681–684.

Ma, Q., Chen, Z., del Barco Barrantes, I., de la Pompa, J. L., and Anderson, D. J. (1998). Neurogenin1 is essential for the determination of neuronal precursors for proximal cranial sensory ganglia. *Neuron* **20**, 469–482.

Maden, M. (2002). Retinoid signalling in the development of the central nervous system. *Natl. Rev. Neurosci.* **3**, 843–853.

Mahmood, R., Kiefer, P., Guthrie, S., Dickson, C., and Mason, I. (1995). Multiple roles for FGF-3 during cranial neural development in the chicken. *Development* **121**, 1399–1410.

Mahmood, R., Mason, I. J., and Morriss-Kay, G. M. (1996). Expression of Fgf-3 in relation to hindbrain segmentation, otic pit position and pharyngeal arch morphology in normal and retinoic acid-exposed mouse embryos. *Anat. Embryol. (Berl)* **194**, 13–22.

Mansour, S. L., Goddard, J. M., and Capecchi, M. R. (1993). Mice homozygous for a targeted disruption of the proto-oncogene int-2 have developmental defects in the tail and inner ear. *Development* **117**, 13–28.

Mansouri, A., Chowdhury, K., and Gruss, P. (1998). Follicular cells of the thyroid gland require Pax8 gene function. *Nat. Genet.* **19**, 87–90.

Marcelle, C., Eichmann, A., Halevy, O., Breant, C., and Le Douarin, N. M. (1994). Distinct developmental expression of a new avian fibroblast growth factor receptor. *Development* **120**, 683–694.

Maroon, H., Walshe, J., Mahmood, R., Kiefer, P., Dickson, C., and Mason, I. (2002). Fgf3 and Fgf8 are required together for formation of the otic placode and vesicle. *Development* **129**, 2099–2108.

Maves, L., Jackman, W., and Kimmel, C. B. (2002). FGF3 and FGF8 mediate a rhombomere 4 signaling activity in the zebrafish hindbrain. *Development* **129**, 3825–3837.

McKay, I. J., Lewis, J., and Lumsden, A. (1996). The role of FGF-3 in early inner ear development: An analysis in normal and kreisler mutant mice. *Dev. Biol.* **174**, 370–378.

McKay, I. J., Muchamore, I., Krumlauf, R., Maden, M., Lumsden, A., and Lewis, J. (1994). The kreisler mouse: A hindbrain segmentation mutant that lacks two rhombomeres. *Development* **120,** 2199–2211.

Merlo, G. R., Paleari, L., Mantero, S., Zerega, B., Adamska, M., *et al.* (2002). The Dlx5 homeobox gene is essential for vestibular morphogenesis in the mouse embryo through a BMP4-mediated pathway. *Dev. Biol.* **248,** 157–169.

Miralles, F., Czernichow, P., Ozaki, K., Itoh, N., and Scharfmann, R. (1999). Signaling through fibroblast growth factor receptor 2b plays a key role in the development of the exocrine pancreas. *Proc. Natl. Acad. Sci. USA* **96,** 6267–6272.

Moens, C. B., Cordes, S. P., Giorgianni, M. W., Barsh, G. S., and Kimmel, C. B. (1998). Equivalence in the genetic control of hindbrain segmentation in fish and mouse. *Development* **125,** 381–391.

Mohammadi, M., McMahon, G., Sun, L., Tang, C., Hirth, P., *et al.* (1997). Structures of the tyrosine kinase domain of fibroblast growth factor receptor in complex with inhibitors. *Science* **276,** 955–960.

Molenaar, M., Brian, E., Roose, J., Clevers, H., and Destree, O. (2000). Differential expression of the Groucho-related genes 4 and 5 during early development of *Xenopus laevis*. *Mech. Dev.* **91,** 311–315.

Morrison, A., Hodgetts, C., Gossler, A., Hrabe de Angelis, M., and Lewis, J. (1999). Expression of Delta1 and Serrate1 (Jagged1) in the mouse inner ear. *Mech. Dev.* **84,** 169–172.

Morsli, H., Choo, D., Ryan, A., Johnson, R., and Wu, D. K. (1998). Development of the mouse inner ear and origin of its sensory organs. *J. Neurosci.* **18,** 3327–3335.

Ohto, H., Takizawa, T., Saito, T., Kobayashi, M., Ikeda, K., and Kawakami, K. (1998). Tissue and developmental distribution of Six family gene products. *Int. J. Dev. Biol.* **42,** 141–148.

Ohuchi, H., Hori, Y., Yamasaki, M., Harada, H., Sekine, K., *et al.* (2000). FGF10 acts as a major ligand for FGF receptor 2 IIIb in mouse multi-organ development. *Biochem. Biophys. Res. Commun.* **277,** 643–649.

Oliver, G., Wehr, R., Jenkins, N. A., Copeland, N. G., Cheyette, B. N., *et al.* (1995). Homeobox genes and connective tissue patterning. *Development* **121,** 693–705.

Ornitz, D. M., Xu, J., Colvin, J. S., McEwen, D. G., MacArthur, C. A., *et al.* (1996). Receptor specificity of the fibroblast growth factor family. *J. Biol. Chem.* **271,** 15292–15297.

Orr-Urtreger, A., Avivi, A., Zimmer, Y., Givol, D., Yarden, Y., and Loani, P. (1990). Developmental expression of c-kit, a proto-oncogene encoded by the W locus. *Development* **109,** 911–923.

Ozaki, H., Watanabe, Y., Takahashi, K., Kitamura, K., Tanaka, A., *et al.* (2001). Six4, a putative myogenin gene regulator, is not essential for mouse embryonal development. *Mol. Cell. Biol.* **21,** 3343–3350.

Pandur, P. D., and Moody, S. A. (2000). *Xenopus* Six1 gene is expressed in neurogenic cranial placodes and maintained in the differentiating lateral lines. *Mech. Dev.* **96,** 253–257.

Panganiban, G., and Rubenstein, J. L. (2002). Developmental functions of the Distal-less/Dlx homeobox genes. *Development* **129,** 4371–4386.

Pera, E., and Kessel, M. (1999). Expression of DLX3 in chick embryos. *Mech. Dev.* **89,** 189–193.

Pera, E., Stein, S., and Kessel, M. (1999). Ectodermal patterning in the avian embryo: Epidermis versus neural plate. *Development* **126,** 63–73.

Pfeffer, P. L., Gerster, T., Lun, K., Brand, M., and Busslinger, M. (1998). Characterization of three novel members of the zebrafish Pax2/5/8 family: Dependency of Pax5 and Pax8 expression on the Pax2.1 (noi) function. *Development* **125,** 3063–3074.

Phillips, B. T., Bolding, K., and Riley, B. B. (2001). Zebrafish fgf3 and fgf8 encode redundant functions required for otic placode induction. *Dev. Biol.* **235,** 351–365.

4. Molecular Basis of Inner Ear Induction

Pirvola, U., Spencer-Dene, B., Xing-Qun, L., Kettunen, P., Thesleff, I., et al. (2000). FGF/FGFR-2(IIIb) signaling is essential for inner ear morphogenesis. *J. Neurosci.* **20,** 6125–6134.

Pirvola, U., Ylikoski, J., Trokovic, R., Hebert, J. M., McConnell, S. K., and Partanen, J. (2002). FGFR1 is required for the development of the auditory sensory epithelium. *Neuron* **35,** 671–680.

Powers, C. J., McLeskey, S. W., and Wellstein, A. (2000). Fibroblast growth factors, their receptors and signaling. *Endocr. Relat. Cancer* **7,** 165–197.

Quint, E., Zerucha, T., and Ekker, M. (2000). Differential expression of orthologous Dlx genes in zebrafish and mice: Implications for the evolution of the Dlx homeobox gene family. *J. Exp. Zool.* **288,** 235–241.

Raible, F., and Brand, M. (2001). Tight transcriptional control of the ETS domain factors Erm and Pea3 by Fgf signaling during early zebrafish development. *Mech. Dev.* **107,** 105–117.

Reaume, A. G., Conlon, R. A., Zirngibl, R., Yamaguchi, T. P., and Rossant, J. (1992). Expression analysis of a Notch homologue in the mouse embryo. *Dev. Biol.* **154,** 377–387.

Reifers, F., Bohli, H., Walsh, E. C., Crossley, P. H., Stainier, D. Y., and Brand, M. (1998). Fgf8 is mutated in zebrafish acerebellar (ace) mutants and is required for maintenance of midbrain-hindbrain boundary development and somitogenesis. *Development* **125,** 2381–2395.

Represa, J., Leon, Y., Miner, C., and Giraldez, F. (1991). The int-2 proto-oncogene is responsible for induction of the inner ear. *Nature* **353,** 561–563.

Revest, J. M., Spencer-Dene, B., Kerr, K., De Moerlooze, L., Rosewell, I., and Dickson, C. (2001). Fibroblast growth factor receptor 2-IIIb acts upstream of Shh and Fgf4 and is required for limb bud maintenance but not for the induction of Fgf8, Fgf10, Msx1, or Bmp4. *Dev. Biol.* **231,** 47–62.

Rinkwitz-Brandt, S., Justus, M., Oldenettel, I., Arnold, H. H., and Bober, E. (1995). Distinct temporal expression of mouse Nkx-5.1 and Nkx-5.2 homeobox genes during brain and ear development. *Mech. Dev.* **52,** 371–381.

Robledo, R. F., Rajan, L., Li, X., and Lufkin, T. (2002). The Dlx5 and Dlx6 homeobox genes are essential for craniofacial, axial, and appendicular skeletal development. *Genes Dev.* **16,** 1089–1101.

Roose, J., Molenaar, M., Peterson, J., Hurenkamp, J., Brantjes, H., et al. (1998). The *Xenopus* Wnt effector XTcf-3 interacts with Groucho-related transcriptional repressors. *Nature* **395,** 608–612.

Ruvinsky, I., Oates, A. C., Silver, L. M., and Ho, R. K. (2000). The evolution of paired appendages in vertebrates: T-box genes in the zebrafish. *Dev. Genes Evol.* **210,** 82–91.

Sahly, I., Andermann, P., and Petit, C. (1999). The zebrafish eya1 gene and its expression pattern during embryogenesis. *Dev. Genes. Evol.* **209,** 399–410.

Schlosser, G., and Northcutt, R. G. (2000). Development of neurogenic placodes in *Xenopus laevis*. *J. Comp. Neurol.* **418,** 121–146.

Sheng, G., and Stern, C. D. (1999). Gata2 and Gata3: Novel markers for early embryonic polarity and for non-neural ectoderm in the chick embryo. *Mech. Dev.* **87,** 213–216.

Sleptsova-Friedrich, I., Li, Y., Emelyanov, A., Ekker, M., Korzh, V., and Ge, R. (2001). fgfr3 and regionalization of anterior neural tube in zebrafish. *Mech. Dev.* **102,** 213–217.

Solloway, M. J., and Robertson, E. J. (1999). Early embryonic lethality in Bmp5;Bmp7 double mutant mice suggests functional redundancy within the 60A subgroup. *Development* **126,** 1753–1768.

Solomon, K. S., and Fritz, A. (2002). Concerted action of two dlx paralogs in sensory placode formation. *Development* **129,** 3127–3136.

Solomon, K. S., Kudoh, T., Dawid, I. B., and Fritz, A. (2003). Zebrafish foxi1 mediates otic placode formation and jaw development. *Development* **130,** 929–940.

Song, J., and Slack, J. M. (1994). Spatial and temporal expression of basic fibroblast growth factor (FGF-2) mRNA and protein in early *Xenopus* development. *Mech. Dev.* **48**, 141–151.

Spokony, R. F., Aoki, Y., Saint-Germain, N., Magner-Fink, E., and Saint-Jeannet, J. P. (2002). The transcription factor Sox9 is required for cranial neural crest development in *Xenopus*. *Development* **129**, 421–432.

Stark, M. R., Biggs, J. J., Schoenwolf, G. C., and Rao, M. S. (2000). Characterization of avian frizzled genes in cranial placode development. *Mech. Dev.* **93**, 195–200.

Streit, A. (2002). Extensive cell movements accompany formation of the otic placode. *Dev. Biol.* **249**, 237–254.

Takabatake, Y., Takabatake, T., and Takeshima, K. (2000). Conserved and divergent expression of T-box genes Tbx2-Tbx5 in *Xenopus*. *Mech. Dev.* **91**, 433–437.

Thisse, B., Thisse, C., and Weston, J. A. (1995). Novel FGF receptor (Z-FGFR4) is dynamically expressed in mesoderm and neurectoderm during early zebrafish embryogenesis. *Dev. Dyn.* **203**, 377–391.

Torres, M., Gomez-Pardo, E., and Gruss, P. (1996). Pax2 contributes to inner ear patterning and optic nerve trajectory. *Development* **122**, 3381–3391.

Vainio, S., and Lin, Y. (2002). Coordinating early kidney development: Lessons from gene targeting. *Natl. Rev. Genet.* **3**, 533–543.

Vendrell, V., Carnicero, E., Giraldez, F., Alonso, M. T., and Schimmang, T. (2000). Induction of inner ear fate by FGF3. *Development* **127**, 2011–2019.

Walshe, J., Maroon, H., McGonnell, I. M., Dickson, C., and Mason, I. (2002). Establishment of hindbrain segmental identity requires signaling by FGF3 and FGF8. *Curr. Biol.* **12**, 1117–1123.

Walshe, J., and Mason, I. (2000). Expression of FGFR1, FGFR2 and FGFR3 during early neural development in the chick embryo. *Mech. Dev.* **90**, 103–110.

Waskiewicz, A. J., Rikhof, H. A., and Moens, C. B. (2002). Eliminating zebrafish pbx proteins reveals a hindbrain ground state. *Dev. Cell* **3**, 723–733.

Wassarman, K. M., Lewandoski, M., Campbell, K., Joyner, A. L., Rubenstein, J. L., *et al.* (1997). Specification of the anterior hindbrain and establishment of a normal mid/hindbrain organizer is dependent on Gbx2 gene function. *Development* **124**, 2923–2934.

Wawersik, S., and Maas, R. L. (2000). Vertebrate eye development as modeled in *Drosophila*. *Hum. Mol. Genet.* **9**, 917–925.

Whitfield, T. T., Riley, B. B., Chiang, M. Y., and Phillips, B. (2002). Development of the zebrafish inner ear. *Dev. Dyn.* **223**, 427–458.

Wilkinson, D. G., Peters, G., Dickson, C., and McMahon, A. P. (1988). Expression of the FGF-related proto-oncogene int-2 during gastrulation and neurulation in the mouse. *EMBO J.* **7**, 691–695.

Wood, H. B., and Episkopou, V. (1999). Comparative expression of the mouse Sox1, Sox2 and Sox3 genes from pre-gastrulation to early somite stages. *Mech. Dev.* **86**, 197–201.

Wright, T. J., and Mansour, S. L. (2003). Fgf3 and Fgf10 are required for mouse otic placode induction. *Development* **130**, 3379–3390.

Wu, D. K., and Oh, S. H. (1996). Sensory organ generation in the chick inner ear. *J. Neurosci.* **16**, 6454–6462.

Xie, M. H., Holcomb, I., Deuel, B., Dowd, P., Huang, A., *et al.* (1999). FGF-19, a novel fibroblast growth factor with unique specificity for FGFR4. *Cytokine* **11**, 729–735.

Xu, P. X., Adams, J., Peters, H., Brown, M. C., Heaney, S., and Maas, R. (1999). Eya1-deficient mice lack ears and kidneys and show abnormal apoptosis of organ primordia. *Nat. Genet.* **23**, 113–117.

Xu, X., Weinstein, M., Li, C., Naski, M., Cohen, R. I., *et al.* (1998). Fibroblast growth factor receptor 2 (FGFR2)-mediated reciprocal regulation loop between FGF8 and FGF10 is essential for limb induction. *Development* **125**, 753–765.

4. Molecular Basis of Inner Ear Induction

Yang, L., Zhang, H., Hu, G., Wang, H., Abate-Shen, C., and Shen, M. M. (1998). An early phase of embryonic Dlx5 expression defines the rostral boundary of the neural plate. *J. Neurosci.* **18,** 8322–8330.

Yoshikawa, S. I., Aota, S., Shirayoshi, Y., and Okazaki, K. (2000). The ActR-I activin receptor protein is expressed in notochord, lens placode and pituitary primordium cells in the mouse embryo. *Mech. Dev.* **91,** 439–444.

Yu, C., Wang, F., Kan, M., Jin, C., Jones, R. B., *et al.* (2000). Elevated cholesterol metabolism and bile acid synthesis in mice lacking membrane tyrosine kinase receptor FGFR4. *J. Biol. Chem.* **275,** 15482–15489.

Zhang, J., Talbot, W. S., and Schier, A. F. (1998). Positional cloning identifies zebrafish one-eyed pinhead as a permissive EGF-related ligand required during gastrulation. *Cell* **92,** 241–251.

5

Molecular Basis of Otic Commitment and Morphogenesis: A Role for Homeodomain-Containing Transcription Factors and Signaling Molecules

Eva Bober,[1] Silke Rinkwitz,[2] and Heike Herbrand[3]
[1]Institute of Physiological Chemistry
Martin-Luther University Halle-Wittenberg
Holly Strasse 1, D-06097
Halle, Germany
[2]Institute of Biology and Environmental Sciences, Neurogenetics
Carl von Ossietzky University
Carl von Ossietzky Strasse 9-11, D-26129
Oldenburg, Germany
[3]Institute of Immunology
Medical School Hannover
Feodor-Lynen Strasse 21, D-30625
Hannover, Germany

I. Introduction
II. Patterning of the Inner Ear Anlagen
   A. Different States of Commitment
   B. Homeobox-Containing Genes and Pattern Formation
   C. Domains and Borders: Correlation to Gene Expression Pattern
III. Transcription Factors and Diffusible Signals: A Complex Network
   A. Fibroblast Growth Factors Control Induction and Pattern Formation
   B. The Role of Bone Morphogenetic Protein Signals in Semicircular Canal Formation and Establishing Gene Expression Patterns
IV. Patterning and Morphogenesis
   A. Homeobox Genes Patterning the Hindbrain Are Essential for Inner Ear Development
   B. *Pax2*, *Nkx5-1*, and *Dlx5*: How Patterns Relate to Morphogenesis
   C. Formation of Semicircular Canals Depends on Intimate Control of Proliferation and Apoptosis
V. Conclusions
   References

This chapter reviews the patterning processes in the developing inner ear. Pattern formation commences in the forming otic placode and thus takes place in parallel with and is influenced by the ear induction. The commitment steps that were inferred from classical transplantation and explantation experiments are now assumed to be directly related to pattern formation and are addressed first. The next part concentrates on transcription factors, which are regionally expressed at the early stages of otic development (placode, pit, vesicle) and divides these morphologically homogeneous structures into fields or compartments committed to form individual parts of the future inner ear. Furthermore, the role of regulatory loops between signaling molecules and transcription factors during pattern formation are discussed. The last part discusses the importance of patterning genes for the morphogenetic processes that shape the inner ear. Here, data on inactivation of some of the patterning genes (*Nkx5-1*, *Dlx5*, *Pax2*, and others) in the mouse are reviewed. © 2003, Elsevier Inc.

I. Introduction

Early stages of inner ear development include the induction of competent ectoderm to form the otic placode, immediately followed by patterning processes that divide the invaginating placode and later the otic pit and vesicle into several territories with different fates. These territories are marked by expression domains of genes specifying future inner ear structures. A prominent role is played here by transcription factors encoding homeodomain-containing proteins. Numerous investigations on otic development have led to the identification of several genes expressed during early developmental stages. The function of some of these genes has been studied by gain-and loss-of-function approaches using different animal models and has revealed aspects of molecular mechanisms underlying the individual steps of otic commitment and morphogenesis.

In this chapter we summarize current data on the expression and function of genes patterning the otocyst. We review the recent experiments that address the possible regulatory loops between transcription factors and diffusible signaling molecules, which are essential for establishing gene expression patterns within the developing otocyst. We discuss the implications of these findings for the previously proposed compartment and boundary models. As with other developing organs or body parts, patterning of the inner ear anlagen is a prerequisite for its future morphogenesis. Therefore, in the last part some examples are presented on how patterning genes can influence the formation of individual parts or structures of the future inner ear.

II. Patterning of the Inner Ear Anlagen

A. Different States of Commitment

The original concept that a given organ develops in successive steps as induction, pattern formation, differentiation, and morphogenesis occurring in a timely ordered sequence has been significantly modified. Experiments

5. Homeobox Genes and Inner Ear Development

analyzing the early steps of inner ear development in different vertebrate species have made clear that inner ear induction is not a single all-or-none decision but consists of several events that may proceed in parallel and sometimes independently of each other (Groves and Bronner-Fraser, 2000; for a review see Noramly and Grainger, 2002, and Chapter 4 by Brown *et al.*). Similarly, no clear-cut border can be drawn between induction and patterning. In fact, patterning commences well before ear induction is completed, and some ear-inducing signals might be engaged in patterning as well. The term *patterning* describes division of the organ anlage in discrete territories that acquire different states of commitment to future fates. According to the definition proposed by Slack (1991), three phases of commitment can be recognized. The first, known as specification, can be functionally defined as an ability of a specified region to develop outside of the embryo in a neutral environment. For example, ear ectoderm can be specified to form an otocyst. The next step of commitment is known as determination. In functional terms, determination describes the ability to form a given structure in ectopic positions within the embryo. The last commitment step is the potential to form a full range of derivatives in an appropriate environment. Individual commitment steps have been described for *Xenopus*, chick, and mouse otocysts by cell lineage tracing and/or generation of fate maps (Brigande *et al.*, 2000a; Kil and Collazo, 2001, 2002; Li *et al.*, 1978; Streit, 2002).

What are the early specification/determination events in the developing ear placode? One of the critical determination steps is the ability of the placode to invaginate and form the otic pit, which eventually closes into a vesicular structure (otocyst). Although many genes are known to be expressed during these early stages, it is still not clear what molecular mechanisms are responsible for the formation of vesicle. In order to define the required environmental and temporal factors more precisely, we performed a series of transplantations of chick otic placodes of different ages (HH stage 8–18) into various locations of the embryo (Herbrand *et al.*, 1998; Herbrand *et al.*, manuscript in prep.). It appears that placodes become fully determined to form vesicles relatively late, around HH stage 13, when transplants give rise to vesicles in all tested locations within the embryo (Table I). The results suggest that determination takes place in a gradual fashion because the ability to form vesicles is strongly dependent on the local environment. Placodes as young as HH stage 8 (4–6 somites) form vesicles when transplanted along the neural tube rostral to their normal position at the level of the first two rhombomeres or caudal midbrain. At stage 10 (10–11 somites) the majority of transplanted placodes is able to form vesicles when transferred along the neural tube either rostral or further caudal to their endogenous position, thus defining a more advanced determination step. In a putatively less permissive or even hostile environment such as the wing buds, the first formation (i.e., determination) of vesicles has been seen in placodes (or rather otic pits) of HH stage 13 (18–19 somites).

Table I Commitment Steps of Chick Otic Placodes

Age of placodes HH stage	Specification/determination features	References
8	Gene expression: *Pax2, Sox3, BMP7, Notch*, vesicle formation occasionally?	Groves and Bronner-Fraser, 2000
8–9+	Vesicle formation rostral, regionalized *Nkx5*-expression	Bober and Herbrand, upubl.; Herbrand et al., 1998
10–12	Vesicle formation rostral and caudal; gene expression partially preserved	
13–14	Vesicle formation in all locations (rostral, caudal, trunk, wing buds); gene expression; ED formation in rostral positions	Herbrand et al., 1998; Swanson et al., 1990; Waddington, 1937

What are the molecular mechanisms responsible for these consecutive determination steps? From a variety of previous and more recent experiments it is clear that signals arising from an underlying mesoderm and adjacent hindbrain are necessary for placode formation (see reviews by Fritzsch et al., 1998, and Torres and Giraldez, 1998, for references). It is also obvious that many of these signals are not only important to induce the otic placode but also for placode commitment to proceed. Such signals have been addressed extensively in several reviews (Baker and Bronner-Fraser, 2001; Noramly and Grainger, 2002; Rinkwitz et al., 2001). We therefore concentrate on signals that play a role in placode commitment steps in addition to their proposed role in the ear induction.

In addition to signals from outside the otic vesicle, we were interested in identification of genes whose expression coincides with vesicle specification. One candidate gene, *Nkx5-1*, is expressed exclusively in the ear anlagen: the otic placode, pit, and vesicle in all vertebrate species we have tested thus far (Fig. 1A–C). The only additional expression domain of *Nkx5-1* is the lateral line in fish and presumably aquatic amphibia (Adamska et al., 2001b). In transplantation experiments we followed *Nkx5-1* expression under the assumption that its function is intrinsically connected to vesicle formation. However, contrary to our expectation, *Nkx5-1* expression only roughly correlates with the onset of placode specification to form vesicle and does not seem to be necessary for vesicle formation (see Table I and Fig. 1D–L). The same finding seems to be true for other genes such as *Pax2, Dlx5*, and *SOHo* expressed in the placode (Herbrand et al., 1998; Herbrand et al., manuscript in preparation). All of these genes appear to be involved in the subdivision of the otocyst into discrete territories. Since the majority of such patterning genes encode homeodomain-containing proteins, a short summary of the main characteristics of the homeobox gene family follows.

5. Homeobox Genes and Inner Ear Development

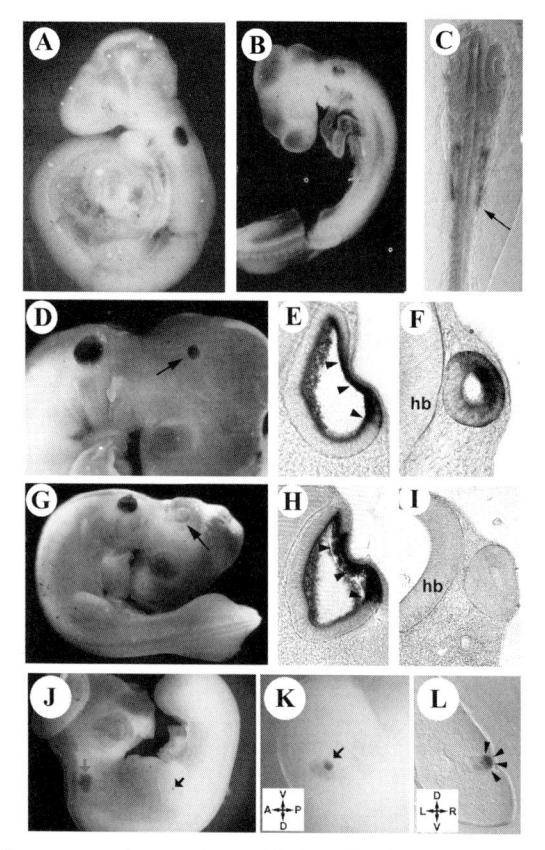

Figure 1 *Nkx5-1* gene expression correlates with determination to form otic vesicle. Vertebrate embryos were hybridized as whole mounts with an antisense RNA probe complementary to the *Nkx5-1* mRNA sequence. The blue staining marks *Nkx5-1* expression in the otocysts; the only additional hybridizing structures are the anlagen of the lateral line marked in C by an arrow. A shows a 10.5 days 'p. c. mouse embryo, B a 3-day-old chick embryo, and C a medaka fish embryo. D–L show *Nkx5-1 in situ* hybridizations on chick embryos with additional otic placodes transplanted to ectopic locations. D demonstrates ectopic vesicle (arrow) positive for *Nkx5-1*, which developed after a placode from a 12-somite–stage donor embryo was transplanted at the level of the midbrain. E and F show transversal sections through the control: endogenous vesicle (E) and the ectopic one (F). Note that *Nkx5-1* expression is restricted to the lateral wall as depicted by arrowheads in E. G shows an outcome of a similar experiment as presented in D–F but with younger placodes transplanted at the 8-somite–stage. Note that the ectopic vesicle has formed but it does not express *Nkx5-1* (arrow in G); transversal sections through the endogenous and ectopic vesicles of the embryo presented in G are shown in H and I, respectively. J shows an embryo with a placode of 14-somite–stage transplanted into the wing bud (black arrow); a red arrow points at the endogenous vesicle. K and L show higher magnifications of the ectopic vesicle in the wing bud; notice a regionally restricted *Nkx5-1* expression (arrowheads in L). hb, hindbrain; A, anterior; P, posterior; V, ventral; D, dorsal.

A

```
                    Drosophila ┌── Antp-C ──┐┌── BX-C ──┐
                               lab pb    Dfd Scr Antp  Ubx abd-A Abd-B
                    HOM-C   ■ ■    ■ ■ ■   ■ ■ ■
```

Hox-A chr 7	■ ■ ■ ■ ■	■ ■ ■ ■
Hox-B chr 17	■ ■ ■ ■ ■ ■	■ ■
Hox-C chr 12	■ ■	■ ■ ■ ■ ■
Hox-D chr 2	■ ■ ■	■ ■ ■ ■ ■ ■
Paralogs	1 2 3 4 5 6	7 8 9 10 11 12 13

hindbrain ← | → trunk
3' ←――――――――――――――――――― 5'

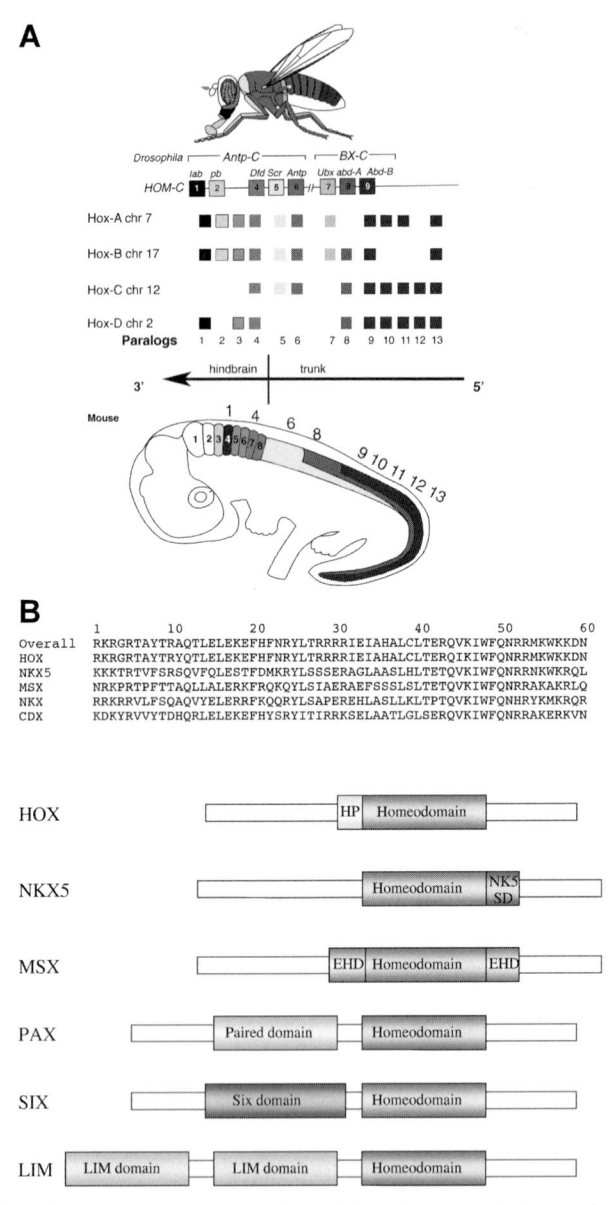

Mouse

B

```
              1         10        20        30        40        50        60
Overall   RKRGRTAYTRAQTLELEKEFHFNRYLTRRRRIEIAHALCLTERQVKIWFQNRRMKWKKDN
HOX       RKRGRTAYTRYQTLELEKEFHFNRYLTRRRRIEIAHALCLTERQIKIWFQNRRMKWKKDN
NKX5      KKKTRTVFSRSQVFQLESTFDMKRYLSSSERAGLAASLHLTETQVKIWFQNRRNKWKRQL
MSX       NRKPRTPFTTAQLLALERKFRQKQYLSIAERAEFSSSLSLTETQVKIWFQNRRAKAKRLQ
NKX       RRKRRVLFSQAQVYELERRFKQQRYLSAPEREHLASLLKLTPTQVKIWFQNHRYKMKRQR
CDX       KDKYRVVYTDHQRLELEKEFHYSRYITIRRKSELAATLGLSERQVKIWFQNRRAKERKVN
```

HOX — HP | Homeodomain

NKX5 — Homeodomain | NK5 SD

MSX — EHD | Homeodomain | EHD

PAX — Paired domain | Homeodomain

SIX — Six domain | Homeodomain

LIM — LIM domain | LIM domain | Homeodomain

Figure 2 Homeobox-containing genes in *Drosophila* and mouse. A shows schematically the organization of the *HOM-C* homeobox gene cluster in *Drosophila* and the four paralogous groups (A, B, C, D) of the corresponding *Hox* genes in the mouse. An arrow indicates the relative position within the genomic cluster with genes expressed caudally placed at the 5'-end and the more rostrally expressed genes at the 3'-end. In the scheme of the mouse embryo the segmentation of the hindbrain into eight rhombomeres is depicted. The colors of the individual

B. Homeobox-Containing Genes and Pattern Formation

Pattern formation was recognized as an orderly appearance of specific structures along the embryonic axes long before the underlying molecular mechanisms were identified. As has been gradually revealed over time, two groups of molecules, transcription factors and signaling molecules, control pattern formation and the subsequent morphogenesis. Homeobox-containing genes were among the first identified transcriptional regulators that play crucial roles during embryonic development. They were discovered in homeotic mutants of the fruit fly *Drosophila melanogaster*. Homeotic mutations lead to the transformation of some body parts into other similar but different parts. The most impressive examples are the four-winged fly caused by a deletion of the ultrabithorax gene and the antennapedia mutation, where the ectopic expression of the *Antennapedia* gene results in a partial transformation of the head segment into the thoracic segment with legs instead of antennae growing out of the head socket (Lewis, 1978). Both genes are located in a genomic cluster of homeobox genes called *HOM-C* or *Hox*-gene cluster (see later). All seven members of the *HOM-C* cluster are expressed in the developing embryo according to their position within the cluster; genes placed at the 3′-region are expressed earlier and more rostrally than genes located farther 5′ (Fig. 2A). The pivotal role of *HOM-C* genes in the establishment of segmental identities along the anteroposterior axis of the embryo is attributed to their ability to bind specific DNA-regulatory sequences and regulate transcription of target genes. All genes contain a highly conserved 180 bp homeobox sequence motive encoding a common protein domain, the homeodomain (Fig. 2B). This domain consists of 60 amino acids and forms a conserved helix-turn-helix structure, which binds the DNA-consensus sequence of target genes. According to this structural feature, the homeodomain-containing proteins are also called homeoproteins.

However, mutations in homeobox genes do not always lead to a homeotic type of phenotype (see later).

The discovery that homeobox genes are extremely well conserved is a true milestone in the molecular research on development. Homeobox genes have been found in all eukaryotic organisms studied so far. Any multicellular organism, from worm to man, appears not only to possess homologs of the *HOM-C* family members but also to preserve their genomic cluster organization and at least some of their functional features and regulatory

---

genes are used to mark the rostral limits of their expression in the embryo. In B, the consensus protein sequence of the homeodomain (overall) is shown in a comparison to the Hox and other non-Hox homeodomains (Nkx5, Msx, Nkx, and Cdx); below the sequence the structural features of the Hox and non-Hox homeoproteins are shown schematically.

mechanisms. One of the best studied homeobox-gene complexes in vertebrates is the Hox genes that pattern the embryo along the anteroposterior axis caudal to the midbrain (Glover, 2001; Kammermeier and Reichert, 2001; Lumsden and Krumlauf, 1996; Maconochie et al., 1996; see also Fig. 2A). *Hox* genes represent homologues of the individual genes of the *HOM-C* complexes in *Drosophila* and are numbered according to their position within the cluster, starting with *Hox1* at the 3′ end, which marks the most rostral *Hox* gene expression and is followed by consecutive numbers of *Hox* genes placed farther 5′. In higher vertebrates the *Hox* gene cluster is present four times in a haploid genome, probably due to gene duplication events during evolution. Thus, each vertebrate homologue of a particular *Drosophila* gene is present as a group of a maximum of four so-called paralogues. Paralogous genes are marked by letters a–d (see Fig. 2A). Several inactivating mutations of different *Hox* genes have been generated in the mouse. Inactivation of genes with the more rostral position in the *Hox* cluster leads to changes of identity of the segmental hindbrain structures, the rhombomeres, resembling the transformations observed in *Drosophila* homeotic mutations. However, the high sequence conservation among the paralogous genes allows functional compensation within the group. Therefore, more significant phenotypic defects are observed when two or more paralogous genes are inactivated simultaneously. Genes of the first paralogous group (*Hoxa1* and *Hoxb1*) seem to be responsible for specifying rhombomeres 4 and 5, which in turn are important for inner ear formation (see Section IV.A). Mice that are deficient for each of these genes show malformations of inner ear development, albeit with a variable expressivity and an incomplete penetrance (Chisaka et al., 1992; Lufkin et al., 1991). In double mutant mice deficient for both *Hoxa1* and *Hoxb1* genes, inner ear phenotype is completely penetrant (Gavalas et al., 1998). Despite intensive research, the molecular basis of *Hox* gene functions remains elusive. It was assumed that *Hox* genes are responsible for transcriptional activation of a whole set of so-called realizators. However, only a few genes have been identified as direct targets of Hox transcription factors. It has been proposed that a specific developmental program may be exerted by activation of single-target genes rather than a whole battery of regulators. The complex response to the action of a single gene that leads to formation of complete parts of the body might be determined by the molecular memory and the "tissue context" of the responding cell (Lohmann and McGinnis, 2002). Apparently, intensive research has to be performed to fully evaluate the functional properties of the *Hox*-gene family.

Besides the *Hox* homeobox-containing genes, a high number of non-*Hox* homeobox genes has been identified (see Fig. 2B). The non-*Hox* genes are involved in the regulation of a variety of other developmental processes; many

5. Homeobox Genes and Inner Ear Development

of them play important roles in essential cellular processes, such as proliferation, differentiation, cell communication, stress response, and apoptosis, and are often involved in cancerogenesis (Abate-Shen, 2002; Cillo *et al.*, 2001; Owens and Hawley, 2002). This chapter discusses the role of members of three different homeobox gene families, the *Pax*-, *NK*-, and *Dlx*-gene families, during the inner ear patterning and morphogenesis.

Pax genes were isolated on the basis of sequence homology to *Drosophila*-paired box–containing homeobox genes (see Fig. 2C). The nine *Pax* genes in vertebrates (*Pax1–Pax9*) can be subdivided into six classes according to their paired box sequence similarity, the presence or absence of a partial or complete paired-class homeobox, and a conserved octapeptide (Gruss and Walther, 1992; Walther *et al.*, 1991). All *Pax* genes play important regulatory roles in many aspects of organogenesis, with *Pax6* being the best studied example. *Pax6* exerts a highly conserved function as one of the key regulators of eye development (Czerny *et al.*, 1999; Simpson and Price, 2002). Inactivating mutations of *Pax* genes in the mouse usually lead to strong dominant or semidominant phenotypes. At least two *Pax* gene mutations, *Pax3* and *Pax2*, are known to affect inner ear development (Epstein *et al.*, 1991; Torres *et al.*, 1996).

The *NK* or *Nkx* gene family represents a heterogeneous group of genes containing a homeobox similar to the homeoboxes of *Drosophila NK* genes (Harvey, 1996). Several of the *NK* genes are essential for development of different organs, as has been demonstrated by gene inactivation experiments. The most interesting phenotypes in mutants deficient in individual *NK/Nkx* genes include *NK2* (*tinman*) in *Drosophila* and *Hox11*, *Nkx2–5*, and *Nkx5/Hmx* genes in the mouse. Deletion of each gene results either in a complete lack of development of a particular organ or in its severe morphological malformation (Azpiazu and Frasch, 1993; Hadrys *et al.*, 1998; Prall *et al.*, 2002; Roberts *et al.*, 1994; Wang *et al.*, 1998).

Similar to *Pax* and *NK* genes, the *Dlx*-gene family comprises genes originally isolated in *Drosophila* and, as do *Pax* and *NK* genes, *Dlx* genes play an essential role in organogenesis. The founder of the gene family, the *Drosophila distal-less* (*Dll*) gene is required for the formation of the distal portion of the legs, antennae, and mouthparts (Cohen *et al.*, 1989). Multiple homologues have been isolated in other species. In vertebrates *Dlx* genes are expressed in various locations and appear to play different roles during embryonic development (Sumiyama *et al.*, 2002). Later we describe a role of *Dlx5* in the formation of vestibular structures. As is the case for the *hox* genes, targets of other homeodomain transcription factors are largely unknown. Interestingly, interactions with members of the BMP family were demonstrated for two different subgroups of the *NK* family, *Nkx2* and *Nkx5*, and also for *Dlx* genes (Monzen *et al.*, 2002; see also Section II.C).

C. Domains and Borders: Correlation to Gene Expression Pattern

During commitment of placodal cells to inner ear fates, different parts of the developmental program are activated in subsets of cells. In analogy to other morphogenetic processes (e.g., the imaginal discs of appendages in *Drosophila*), it was proposed that the inner ear anlage is divided into distinct territories that are destined to form specific parts of the future inner ear (Fekete, 1996). According to this hypothesis, forming territories are marked by specific expression of regulatory genes, typically transcription factors. The borders of such compartments are predicted to represent cell lineage restriction territories and prevent mixing of cells between compartments. They might also constitute a source of new signals acting as morphogens that induce new structures or subterritories. Consistent with this theoretical expectation many asymmetrically expressed genes have been identified in early otic structures (for reviews see Noramly and Grainger, 2002, and Rinkwitz *et al.*, 2001). In addition, fate maps and lineage tracing have been performed to correlate gene expression patterns with fate maps and cell lineage data (Brigande *et al.*, 2000a; Kil and Collazo, 2002).

Nkx5-1 and *SOHO* genes encode homeodomain-containing transcription factors that show regionally restricted expression patterns in the otocyst and represent examples of compartment-restricted genes. Both genes are activated within the developing otic placode and display highly dynamic expression patterns during placode and otic pit stages. Figure 3A–C exemplifies expression of the *SOHO* gene in placodes and two different otic pit stages. In the placode *SOHO* is activated in the posterior part, directly adjacent to the neural tube (see Fig. 3A). During the invagination process *SOHO* transcripts gradually relocate to a more rostral and lateral position, farther away from the neural tube (Fig. 3B and C). *Nkx5-1* also shows a regionally restricted expression pattern that changes dynamically during placode invagination (Herbrand *et al.*, 1998). Therefore, it is difficult to assign transcriptional activities of each gene to specific compartments during placode or otic pit stages. With the closure of the otic vesicle both genes become confined to the dorsolateral wall of the otocyst, where a smaller *Nkx5-1* domain seems to be completely included within the broader *SOHO* expression region (Fig. 3D and E). At this stage several other genes show complementary or partially overlapping expression patterns, as demonstrated for *Pax2* in Fig. 2F.

Two independently established fate maps revealed a great extent of cell mixing during placode and otic pit stages, which corroborates the dynamic gene expression patterns observed during this time period and argues against the existence of static compartments during these stages of development (Kil and Collazo, 2001; Streit, 2002). Another fate map analysis performed at the

5. Homeobox Genes and Inner Ear Development 161

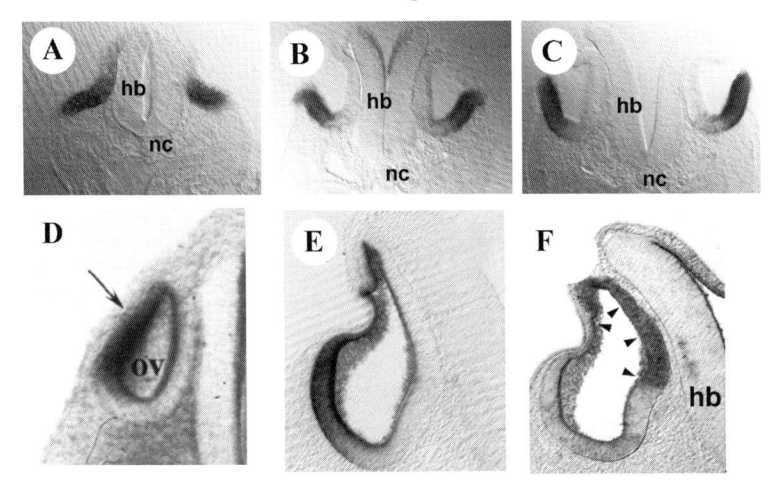

Figure 3 Examples of regionally restricted gene expression domains in the developing otocyst. A–C show transversal sections through chick embryos of HH stages 10, 13, and 14 at the level of the otic placode (A), the otic pit (B), and the closed otic vesicle (C) hybridized with *SOHo*. In D–F cross-sections through otic vesicles at HH stage 17 *in situ* hybridized with *Nkx5-1* (D), *SOHo* (E), and *Pax2* (arrows in F) are shown. hb, hindbrain; nc, notochord; OV, otic vesicle.

time of the vesicle closure revealed restrictions of cell intermingling along two borders: anteroposterior (A/P) and mediolateral (M/L) (Brigande *et al.*, 2000a,b). The first border was proposed to divide the vesicle mediolaterally at the basis of the forming endolymphatic duct. *In situ* hybridization presented by Brigande *et al.* (2000a) correlates this border with sharp *SOHO* and *Pax2* domains. Our data suggest some overlap of both gene expression domains in the closed vesicle and suggest placing the border between *Pax2* and *Nkx5-1* expression domains (see Fig. 3E and F). Therefore, the Nkx5-1 domain could in fact represent the lateral compartment as inferred from the fate mapping studies, whereas *Pax2* would mark the medioventral part of the vesicle. We show in Section IV.A that the function of these two transcription factors correlates with the predicted fate of otocyst regions marked by their expression.

In summary, the comparative analyses of fate maps and gene expression data suggest that compartments are separated by dynamic borders that change their position during development until they reach the transitory fixation, probably after closure of the otic vesicle. It is evident that more functional investigations are needed to understand the molecular circuits laying down the blueprint for inner ear morphogenesis.

III. Transcription Factors and Diffusible Signals: A Complex Network

A. Fibroblast Growth Factors Control Induction and Pattern Formation

Members of the fibroblast growth factor (FGF) family have been postulated to be essential for inner ear induction and its correct morphogenesis (Fritzsch *et al.*, 1998; Mahmood *et al.*, 1996; Mansour *et al.*, 1993; McKay *et al.*, 1996; Torres and Giraldez, 1998; see Chapter 8 by Wright and Mansour). More recently, different FGFs were demonstrated to induce ectopic vesicles in *Xenopus* and chick embryos (Adamska *et al.*, 2001a; Lombardo and Slack, 1998; Vendrell *et al.*, 2000). Three different FGFs were used in these studies, and the inductory or morphogenetic potential of individual FGFs appeared different in each investigation. In *Xenopus*, beads soaked with FGF2 and FGF3 were implanted posteriorly in neural plate embryos. Both factors induced ectopic vesicles with comparable efficiency (Lombardo and Slack, 1998). In a study by Vendrell *et al.* (2000), retrovirus-mediated overexpression of FGF2 and FGF3 in HH stage 8–10 chick embryos was investigated, but only FGF3 was able to induce vesicles, whereas FGF2 showed no effects. In both studies ectopic vesicles revealed signs of initial pattern formation, as demonstrated by dorsal *Wnt3a* expression in *Xenopus* and regionally restricted expression of *Pax2*, *cek8/EphA4*, *BMP7*, and *Nkx5-1* in chick. The third study analyzed the effects of FGF2- and FGF8-soaked beads implanted at HH stage 8–10 close to the endogenous otic vesicles in chick (Adamska *et al.*, 2001a). In contrast to the Vendrell paper, FGF2 could efficiently induce ectopic otic structures, as was the case in the *Xenopus* system. However, no ear induction was seen with FGF8-soaked beads implanted near the endogenous otocysts. Interestingly, in this study expression of *Nkx5-1*, *Pax2*, and *SOHO* was detected in only a small percentage of ectopic otocysts, whereas the *Dlx5* expression was absent in ectopic ears, indicating that only parts of the patterning program were activated. The variability of results and the differences between species and experimental designs did not allow assignment of a specific function to particular FGFs. Since FGF receptors are able to bind different FGFs, the growth factors used to induce otic vesicles in the experiments described previously might mimic the activity of still another member of the FGF family, which acts under physiological conditions *in vivo*. In the ear anlagen and in the adjacent hindbrain several FGFs and their receptors are expressed; therefore, it appears likely that not a single FGF but a combination of them synergistically induce otic fates *in vivo* (Karabagli *et al.*, 2002; Pickles and Chir, 2002; Pirvola *et al.*, 2000). Interestingly, experiments in fish using different FGF mutants or morpholino antisense oligonucleotides inactivating

5. Homeobox Genes and Inner Ear Development

FGF3 and FGF8 alone or in combination suggest that these two factors act in concert to exert full function (Maroon *et al.*, 2002; Phillips *et al.*, 2001; see Chapter 12 by Riley). In addition, FGF19 expression has been demonstrated to coincide with the previously predicted inducing territory (Ladher *et al.*, 2000). FGF19 appears to be present at the correct time point in the precise mesodermal region that was proposed more than 50 years ago to send the initial ear-inducing signal. Furthermore, FGF19 seems to induce the expression of and act synergistically with wnt-8c, which is expressed in the neural ectoderm. Both factors likely collaborate to induce ear placodes and expression of a relatively broad range of molecules (i.e., *Pax2*, *Nkx5-1*, FGF3, *Dlx5*, *SOHO*) in the preneural stage ectoderm (Ladher *et al.*, 2000).

As mentioned, FGFs are believed to influence not only ear induction but also further morphogenesis of the ear vesicle, but little information is available about the exact role of FGFs in these processes. Despite this shortcoming, two main morphogenetic events are known to be under FGF control: the delamination of neuroblasts with the subsequent formation of the otic ganglion and the establishment of specific expression patterns within the otocyst. Yet, the knowledge of which aspect of the patterning program is under FGF control remains elusive. Contradictory results were reported on the effects of FGF2 on gene expression patterns within the endogenous otic vesicles. Adamska *et al.* (2001a) demonstrated activation of *Nkx5-1*, *SOHO*, and *Pax2* after implantation of FGF2-soaked beads close to the otocyst. *Dlx5* and *BMP4* were unaffected by ectopic FGF signals. In contrast, viral overexpression of FGF2 did not influence transcription of *Pax2* or *lmx-1*, whereas upregulation of those genes was apparent when an FGF3-expressing virus was used (Vendrell *et al.*, 2000). Different transcriptional responses in these two investigations might be due to differences in expression levels and location of the ectopic signal. Such an explanation is supported by data indicating the dependence of transcriptional response on the location of the ectopic FGF source (Adamska *et al.*, 2001a). To further complicate the situation, the expression domains of these two FGFs did not correlate with a specific patterning function. In contrast, a tightly restricted location of FGF8 expression was found in the chick otic placode and later within the otocyst. FGF8 is activated for a short time of 4–8 hours at the anterior aspect of the otic placode. Since FGF8 appearance in this region immediately precedes activation of the *Nkx5-1* gene, it has been proposed that this particular member of the FGF family might represent the primary signal responsible for activation of a gradient-like distribution of *Nkx5-1* in the otic placode (Adamska *et al.*, 2001a). Another pulse of FGF8 expression was observed in a small ventromedial region of the otocyst where neuroblasts delaminate to form the otic ganglion (Adamska *et al.*, 2001a; Hidalgo-Sanchez, 2000). Data from the FGF8 zebrafish mutant, ace, support a role for FGF8 in otic ganglion formation (Adamska *et al.*, 2000). Although delamination of cells can still be morphologically discerned in ace mutants, these cells fail to activate *Nkx5-1* transcription, which is normally present in ganglion-forming cells at this stage of development. Thus, FGF8 may regulate *Nkx5-1* transcription in the placode and in delaminating ganglion cells in two different vertebrate species, the chick and the zebrafish.

B. The Role of Bone Morphogenetic Protein Signals in Semicircular Canal Formation and Establishing Gene Expression Patterns

Bone morphogenetic proteins (BMPs) constitute a subgroup of the TGFβ superfamily and exert various heterogeneous effects during development (for reviews see Balemans and Hul, 2002, and Hogan, 1996). Experiments testing the function of BMPs in chick embryos have revealed that BMPs play an essential role in the formation of semicircular canals (Chang *et al.*, 1999; Gerlach *et al.*, 2000). In both studies BMPs or their antagonist (noggin) were placed in the vicinity of the inner ear by different methods (retrovirus, soaked beads, or cell pellets). Noggin strongly inhibited the formation of semicircular canals. It was even possible to block outpouching of individual canals dependent on the location of noggin sources (Gerlach *et al.*, 2000). On the other hand, BMPs alone were not able to induce ectopic canals. Since noggin is able to bind several members of the BMP family and inhibits their action, it is not clear which BMP is critical for canal development *in vivo*. Expression studies revealed that BMP4, 5, and 7 are expressed in different patterns in the placode and the otocyst, whereas BMP2 is activated at later stages, after the otocyst has undergone further morphogenesis (Chang *et al.*, 2002; Oh *et al.*, 1996). It seems likely that the three BMPs expressed at the early stages are involved in patterning processes leading to morphogenesis of the semicircular canals (see later). BMP2 was proposed to be responsible for the continual outgrowth of the semicircular canals after their initial formation (Chang *et al.*, 1999).

Despite the growing experimental evidence of a critical role for BMP signaling in various aspects of inner ear development, only a few downstream targets have been identified in the inner ear. Chang *et al.* (1999) documented that BMPs are important to maintain *Msx1* and *p75NGFR* expression in some sensory organs (cristae) because expression of these genes decreased after exposure to an ectopic noggin source. We have tested expression of several ear patterning genes after implantation of cell aggregates producing either BMP2 or noggin. BMPs seem to activate *Nkx5-1*, *Dlx5*, and *SOHO* genes, which are all involved in canal formation. Interestingly, *Pax2*, a gene essential for cochlear morphogenesis but dispensable for semicircular canal formation (Torres *et al.*, 1996; see also later), was not activated within the ear, although *Pax2* expression was increased in other tissues in response to

5. Homeobox Genes and Inner Ear Development

```
         ┌──→ Nkx5-1 ←──┐
         │              │
         │  ──→ Dlx5 ┤  │
  BMP2 ──┤              ├── FGF2 / 8
         │    Pax2 ←──  │
         │              │
         └──→ SOHo ←────┘
```

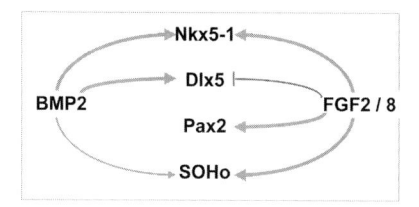

Figure 4 FGFs and BMPs control expression of inner ear patterning genes. Schematic illustration of positive and negative influences of BMP and FGF signals on expression of homeobox genes *Nkx5-1*, *Dlx5*, *SOHo*, and *Pax2* in the developing otocyst.

ectopic BMP signals. These data suggest that BMP signaling controls activities of homeobox genes involved in the formation of semicircular canals, thus supporting the experimental evidence in chick for the essential role of BMPs in canal morphogenesis presented earlier. In fact, in contrast to previous experiments (Gerlach et al., 2000) we were able to induce supernumerary canal-like structures through ectopic BMP sources (Herbrand et al., manuscript in prep.).

In contrast to the consistent activation of vestibular genes by BMPs, FGFs do not seem to affect any specific group of patterning genes. Figure 4 schematically depicts the action of these two gene families on otocyst patterning. Since some of classical FGF knockouts are early lethal or without phenotype, the generation of conditional and combinatorial knockouts will likely be needed in order to inactivate more than one gene at a chosen time point of development.

IV. Patterning and Morphogenesis

A. Homeobox Genes Patterning the Hindbrain Are Essential for Inner Ear Development

A critical dependence of inner ear development on signals from neighboring tissues was most extensively studied in the case of hindbrain influences. Phenotypes of several classical inner ear mutations such as fidget or kreisler were ascribed to the primary developmental defects within the adjacent hindbrain many years ago (Rinkwitz et al., 2001, and references therein). In the early 1990s the importance of the hindbrain for inner ear morphogenesis was documented on the molecular level. The first steps of hindbrain development include formation of segmental structures, so-called rhombomeres, which acquire specific identities and strict borders, preventing cell mixing between adjacent rhombomeres. Interestingly, these borders correlate with anterior borders of expression of *Hox* genes from the first paralogous group of the *Hox*-gene complex (see Fig. 2A). In analogy to *Drosophila* genes of the *HOM-C* complex, it was originally postulated that this specific *Hox* expression pattern indicates its function in establishing and maintaining the rhombomeres, as is the case for the homologous homeotic selector genes specifying individual segments in *Drosophila* embryo. In fact, inactivation of the *Hoxa-1* gene by homologous recombination in the mouse confirmed this assumption (Chisaka *et al.*, 1992; Lufkin *et al.*, 1991). *Hoxa-1* is expressed in the neural tube before the onset of segmentation with an anterior limit of expression coinciding with a presumptive rhombomere 3/4 border. The rhombomeric organization in the *Hoxa-1* null mice is disrupted, and as demonstrated by histological changes and changes in the distribution of the rhombomere-specific marker genes, rhombomere 4 is markedly reduced, whereas rhombomere 5 is almost completely absent (Mark *et al.*, 1993). Although *Hoxa-1* is never expressed in the cells forming the inner ear, inner ear development is severely disrupted in *Hoxa-1* null mice. As already mentioned, gene duplications in higher vertebrates led to formation of paralogous genes exerting overlapping functions.

Such functional redundancy is also observed for two genes of the first paralogous group, *hoxa-1* and *hoxb-1*, so that double null mice mutants appear to be more affected than singular mutations. The inner ears of *Hoxa-1/Hoxb-1* null mutants appear as cystic structures without any recognizable vestibular or cochlear parts (Gavalas *et al.*, 1998). In spite of earlier postulations that intact hindbrain is crucial for the induction of the otic placode, this event seems to take place normally in rhombomere 4/5–deficient mutants. The otic vesicle is also formed but is displaced further away from the hindbrain as compared to the wild type. In the meantime more gene mutations originally affecting hindbrain segmentation (e.g., the *kreisler* gene) have been shown to be involved in the inner ear morphogenesis, confirming that an intact identity of rhombomeres 4 and 5 is essential for ear development (Cordes and Barsh, 1994; Frohman *et al.*, 1993; McKay *et al.*, 1994).

B. *Pax2, Nkx5-1,* and *Dlx5:* How Patterns Relate to Morphogenesis

The previously proposed role of compartments and boundaries formation within the otocyst to lay down a "blueprint" for morphogenesis of the inner ear was tested *in vivo* in mouse knockout models (Anagnostopoulos, 2002; Brigande *et al.*, 2000b; Fekete, 1999). It appears that in some cases inactivation of genes (such as *FGF3* or *Eya-1*) leads to broad changes of inner ear morphology. These genes are expressed at very early stages before subdivision of the placode and thus are unlikely to take part in compartmentalization of the otocyst. On the other hand, at least three examples have been described where the affected structures of the inner ear corresponded to a

5. Homeobox Genes and Inner Ear Development 167

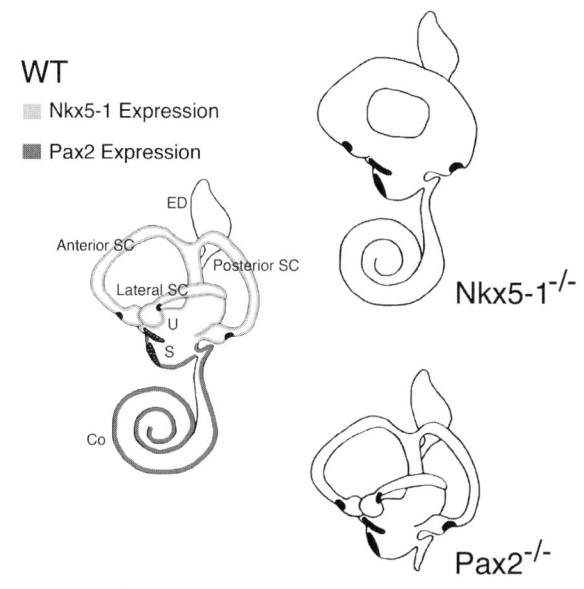

Figure 5 Knockout ears after inactivation of otocyst patterning genes *Nkx5-1* and *Pax2*. On the left a schematic drawing of an adult inner ear shows in two different colors the structures developing from different otocyst parts that originally expressed either *Nkx5-1* (dorsolateral otocyst, blue) or *Pax2* (ventromedial otocyst, red). Two schemes on the right exemplify the affected structures of the inner ears of *Nkx5-1* and *Pax2* deficient mice (see text for further explanations).

specific gene expression pattern in the otocyst and respective fate maps. Figure 5 describes in a simplified fashion such roles for *Pax2*, *Nkx5-1*, and *Dlx5*. *Nkx5-1* and *Dlx5* were both assumed to mark the dorsolateral compartment of the otocyst based on their expression patterns. Mice deficient for either of these genes show very similar malformation of the semicircular canals. In the most severely affected individuals canal formation is almost completely blocked. Despite similarities in expression pattern and phenotypic changes in knockouts, each gene appears to act in an independent pathway since *Nkx5-1* expression is unaffected in $Dlx5^{-/-}$ otocysts and vice versa (Acampora et al., 1999; Merlo et al., 2002). Also, *Nkx5-1* activity appears to be regulated by FGF signals (particularly FGF8 as highlighted earlier), whereas *Dlx5* does not respond to FGF signaling (Adamska et al., 2001a). Interestingly, both genes seem to be included into the regulatory loop existing between BMP signals and homeobox genes (see Section IV.C for further discussion). Another transcription factor, *Pax2*, is expressed in the medial wall of the otocyst in a fashion complementary to that of *Nkx5-1* (Herbrand et al., 1998; Rinkwitz-Brandt et al., 1996). Fate mapping of different parts of the explanted mouse otocysts revealed different developmental potentials of these two otocyst regions (Li *et al.*, 1978). The dorsolateral otocyst (expressing *Nkx5-1*) gave rise predominantly to semicircular canals and other vestibular structures, whereas the medioventral (*Pax2*-expressing domain) gave rise to predominantly auditory structures. In agreement with expression pattern and fate mapping data, *Pax2* was demonstrated to be essential for the morphogenesis of the cochlea. Inactivation of *Pax2* resulted in agenesis of the entire auditory apparatus, whereas the vestibular part developed normally (Torres *et al.*, 1996). An almost complementary phenotype appeared after inactivation of two dorsomedially expressed genes, *Nkx5-1* and *Dlx5*. In both mutants vestibular structures were affected but cochlear development was normal (Acampora *et al.*, 1999; Hadrys *et al.*, 1998).

C. Formation of Semicircular Canals Depends on Intimate Control of Proliferation and Apoptosis

Morphogenesis of semicircular canals begins with the formation of outpouches of the dorsolateral wall of the otocyst. By mechanisms that are not completely understood the opposite walls within the outpouching region are brought together and set up the so-called canal plate as shown schematically in Fig. 6A. The canals are then formed by resorption of the central regions of the canal plates. Apoptosis of canal plate cells in the chick or recruitment of cells back into the otic epithelium in mouse have been proposed as underlying mechanisms, although it remains open whether both processes play a role in canal plate removal in each species (Fekete *et al.*, 1997; Martin and Swanson, 1993). The later steps involve a continual outgrowth and further shaping of the canals. Knockouts of five different genes, *Nkx5-1*, *netrin*, *Otx-1*, *Dlx5*, and *Nor-1*, as well as many other mutants, have demonstrated the different potential roles for these genes during early semicircular canal morphogenesis in mice (Acampora *et al.*, 1999; Hadrys *et al.*, 1998; Merlo *et al.*, 2002; Morsli *et al.*, 1999; Ponnio *et al.*, 2002; Salminen *et al.*, 2000). In some of these knockouts additional data highlighting possible underlying mechanisms were generated.

Inactivation of netrin-1, a molecule otherwise known from axonal guidance regulation (Forcet *et al.*, 2002), leads to malformations of semicircular canals strongly resembling the phenotypic changes of *Nkx5-1* and *Dlx5* mutants. However, netrin-1 is not required for *Nkx5-1* expression, as indicated by its unchanged expression in netrin-1 deficient mice (Salminen *et al.*, 2000). In contrast, the size of the *netrin-1* expression domain (which is normally confined to the part of epithelium forming the canal plates) is significantly reduced in *Nkx5-1* knockout animals (Fig. 6B and C). It has been proposed that netrin-1 induces proliferation of mesenchymal cells adjacent

5. Homeobox Genes and Inner Ear Development   169

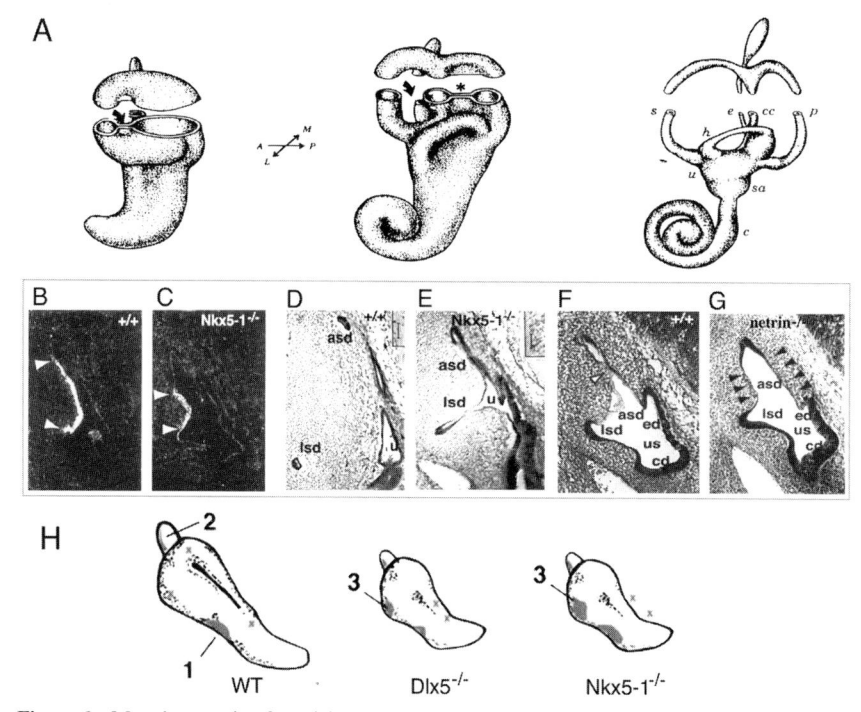

Figure 6 Morphogenesis of semicircular canals. The role of Nkx5-1, Dlx5, and netrin. In A three consecutive stages during morphogenesis of semicircular canals are schematically drawn. Below, sections of inner ears of wild type (B) and Nkx5-1 deficient (C) mice hybridized *in situ* with a *netrin* probe are shown; the size of the *netrin* expression domain is indicated by arrowheads. D and E show sections of wild type (D) and Nkx5-1 deficient (E) mice illustrating that the resorbtion of the canal plates does not take place in the Nkx5-1 mutant. F and G show similar sections from the wild type and netrin mutant mice with the region of the canal plate forming normally in the wild type (a white arrowhead in F) while it fails to form in the netrin mutant (arrowheads in G). In H advanced stages of otic vesicles from wild type (left, WT), Dlx5, and Nkx5-1 deficient mice are schematically presented. Apoptotic regions are indicated in green. The two hot spots of apoptosis (1 and 2) in the wild type are diminished in the Dlx5 mutant and increased in the Nkx5-1 mutant. An additional apoptotic region arises in both mutants (3). asd, anterior semicircular duct; lsd, lateral semicircular duct; u, ventricle; ed, endolymphatic duct; cd, cochlear duct; us, utriculo-saccular space.

to the canal plate epithelium and the proliferating mesenchyme "pushes" the otocyst walls together to form the canal plate. In *netrin-1$^{-/-}$* mice this proliferation is markedly reduced so that the formation of a proper canal plate does not occur (Fig. 6F and G; Salminen *et al.*, 2000). Interestingly, a survival role of netrin-1 was proposed in the developing nervous system. Netrin-1 increases survival of neurons by binding to their receptors and preventing receptor-induced apoptosis (Llambi *et al.*, 2001). Netrin-1 may also induce removal of canal plate cells by intercalation through influencing cell movements and rearrangement of extracellular matrix. The exact netrin-1 function in the process of canal formation has yet to be evaluated.

Nkx5-1 function seems to be confined to a slightly later developmental stage of canal formation than proposed for netrin-1. In *Nkx5-1* mutants, and similarly in *Dlx5* mutants, canal plates were formed but their resorption was either incomplete or delayed (Fig. 6D and E; Merlo et al., 2002; Salminen et al., 2000). This malresorption could be due to changes in *BMP4* expression observed in both mutants. In *Nkx5-1* mutant mice, both *BMP4* expression domains in the otocyst are weaker and broader with no sharp boundaries as compared to the wildtype. In *Dlx5* mutants, *BMP4* expression is almost completely abolished. In both mutants, patterns of proliferating and apoptotic areas are also significantly changed (Fig. 6H). BMPs are involved in regulated cell death processes in many organ systems in the developing embryo as has been demonstrated in the case of eye development (Trousse et al., 2001). Distinct regions of proliferation and cell death have also been demonstrated in the developing chick inner ear, including a "hotspot" at the developing canal plate region (Lang et al., 2000). Therefore it is plausible that BMP4 regulates canal plate resorption by inducing cell death in this region. Significantly, similar interactions between BMP signaling and the action of another member of the *Nkx* gene family were described during heart and gut development (Jamali et al., 2001; Smith et al., 2000). In summary, we propose the following sequence of events during canal morphogenesis: (1) Locally proliferating mesenchyme pushes the opposing vesicle walls together. Netrin-1 seems to be a critical factor controlling this step, whereas Nkx5-1 might regulate the size of the netrin-1 expression domain. (2) Resorption of canal plates occurs. This step might be regulated by homeobox genes such as *Nkx5-1* and *Dlx5*. Their interactions with BMP signaling possibly define areas of cell death. In addition, recruitment of canal plate cells into the epithelium may also play a role in canal morphogenesis. (3) A continual growth of formed semicircular canals is regulated by orphan nuclear receptor Nor-1 (Ponnio et al., 2002).

V. Conclusions

The following axioms might be formulated. Pattern formation in the developing otocyst is established by a concert action of signaling molecules and transcription factors. It leads to division of the otic vesicle into compartments determined to form different structures of the inner ear. Compartments are marked by expression of specific sets of transcription factors, many of them encoded by homeobox genes. Signals that control pattern formation are either derived from neighboring tissues or are generated autonomously by the otocyst. Two hindbrain segments, rhombomeres 4 and 5, are essential for inner ear morphogenesis. Therefore, disruption of *hox* genes responsible for rhombomere formation or other critical rhombomere-specific genes will critically affect ear development.

The importance of otocyst-patterning genes such as *Pax2*, *Nkx5-1*, and *Dlx5* for the morphogenesis of individual structures of the inner ear such as the cochlea and semicircular canals has been demonstrated *in vivo* by inactivation of these genes in the mouse. Although the exact molecular role has still to be elucidated, existence of regulatory loops between the homeodomain–containing transcription factors Dlx5 and Nkx5-1 and members of the BMP family of signaling molecules was documented to be essential for canal morphogenesis. However, the further delineation of the gene function requires identification of their target genes and a functional assessment of their regulatory roles. It can be expected that the extensive knowledge gained by genomic and proteomic approaches will help to identify sets of regulators and realizators involved in pattern formation and morphogenesis of inner ear structures.

Acknowledgments

We would like to thank Thorsten Hadrys for help with Fig. 2 and Thomas Braun for critically reading the chapter and sparking interesting discussions. The research of E. B. was supported by Deutsche Forschungsgemeinschaft and the Boehringer Ingelheim Fonds.

References

Abate-Shen, C. (2002). Deregulated homeobox gene expression in cancer: Cause or consequence? *Nature Rev* **2**, 777–785.

Acampora, D., Merlo, G. R., Paleari, L., Zerega, B., Postiglione, M. P., *et al.* (1999). Craniofacial, vestibular and bone defects in mice lacking the Distal-less-related gene Dlx5. *Development* **126**, 3795–3809.

Adamska, M., Herbrand, H., Adamski, M., Krüger, M., Braun, T., and Bober, E. (2001a). FGFs control patterning of the inner ear but are not able to induce the full ear program. *Mech. Dev.* **109**, 303–313.

Adamska, M., Leger, S., Brand, M., Hadrys, T., Braun, T., and Bober, E. (2000). Inner ear and lateral expression of a zebrafish Nkx5-1 gene and its downregulation in the ears of FGF8 mutant, ace. *Mech. Dev.* **97**, 161–165.

Adamska, M., Wolff, A., Kreusler, M., Wittbrodt, J., Braun, T., and Bober, E. (2001b). Five Nkx5 genes show differential expression patterns in anlagen of sensory organs in medaka. Insight into the evolution of the gene family. *Dev. Genes Evol.* **211**, 338–349.

Anagnostopoulos, A. V. (2002). A compedium of mouse knockouts with inner ear defects. *Trends Genet* **18**, S21–S38.

Azpiazu, N., and Frasch, M. (1993). Tinman and bagpipe: Two homeobox genes that determine cell fates in the dorsal mesoderm of *Drosophila*. *Genes Dev.* **7**, 1325–1340.

Baker, C. V., and Bronner-Fraser, M. (2001). Vertebrate cranial placodes I. Embryonic induction. *Dev. Biol.* **232**, 1–61.

Balemans, W., and Hul, W. V. (2002). Extracellular regulation of BMP signaling in vertebrates: A cocktail of modulators. *Dev. Biol.* **250**, 231–250.

Brigande, J. V., Iten, L. E., and Fekete, D. (2000a). A fate map of chick otic cup closure reveals lineage boundaries in the dorsal otocyst. *Dev. Biol.* **227**, 256–270.

Brigande, J. V., Kiernan, A., Gao, X., Iten, L. E., and Fekete, D. M. (2000b). Molecular genetics of pattern formation in the inner ear: Do compartments boundaries play a role? *Proc. Natl. Acad. Sci. USA* **97**, 11700–11706.

Chang, W., Nunes, F. D., De Jesus-Escobar, J. M., Harland, R., and Wu, D. (1999). Ectopic noggin blocks sensory and nonsensory organ morphogenesis in the chicken inner ear. *Dev. Biol.* **216**, 369–381.

Chang, W., ten Dijke, P., and Wu, D. (2002). BMP pathways are involved in otic capsule formation and epithelial-mesenchymal signaling in the developing chicken inner ear. *Dev. Biol.* **251**, 380–394.

Chisaka, O., Musci, T. S., and Capecchi, M. R. (1992). Developmental defects of the ear, cranial nerves and hindbrain resulting from targeted disruption of the mouse homeobox gene Hox-1.6. *Nature* **355**, 516–520.

Cillo, C., Cantile, M., Faiella, A., and Boncinelli, E. (2001). Homeobox genes in normal and malignant cells. *J. Cell. Physiol.* **188**, 161–169.

Cohen, S. M., Brönner, G., Küttner, F., Jürgens, G., and Jäckle, H. (1989). Distal-less encodes a homeodomain protein required for limb development in *Drosophila*. *Nature* **338**, 432–434.

Cordes, S. P., and Barsh, G. S. (1994). The mouse segmentation gene kr encodes a novel basic domain-leucin zipper transcription factor. *Cell* **79**, 1025–1034.

Czerny, T., Halder, G., Kloter, U., Souabni, A., Gehring, W. J., and Busslinger, M. (1999). Twin of eyeless, a second Pax-6 gene of *Drosophila*, acts upstream of eyeless in the control of eye development. *Mol. Cell* **3**, 297–307.

Epstein, D. J., Vekemans, M., and Gros, P. (1991). Splotch (Sp2H), a mutation affecting development of the mouse neural tube, shows a deletion within the paired homeodomain of Pax-3. *Cell* **67**, 767–774.

Fekete, D. M. (1996). Cell fate specification in the inner ear. *Curr. Opin. NeuroBiol.* **12**, 35–42.

Fekete, D. M. (1999). Development of the vertebrate ear: Insights from knockouts and mutants. *Trends Neurosci.* **22**, 263–269.

Fekete, D. M., Homburger, S. A., Waring, M. T., Riedl, A. E., and Garcia, L. F. (1997). Involvement of programmed cell death in morphogenesis of the vertebrate inner ear. *Development* **124**, 2451–2461.

Forcet, C., Stein, E., Pays, L., Corset, V., Llambi, F., et al. (2002). Netrin-1-mediated axon outgrowth requires deleted in colorectal cancer-dependent MAPK activation. *Nature* **417**, 443–447.

Fritzsch, B., Barald, K. F., and Lomax, M. I. (1998). Early embryology of the vertebrate inner ear. "Development of the Auditory Systems" (E. W. Rubel, A. N. Popper and R. R. Fay Eds.), 80–145. New York: Springer Verlag.

Frohman, M. A., Martin, G. R., Cordes, S. P., Halamek, L. P., and Barsh, G. S. (1993). Altered rhombomere-specific gene expression and hyoid bone differentiation in the mouse segmentation mutant, kreisler (kr). *Development* **117**, 925–936.

Gavalas, A., Studer, M., Lumsden, A., Rilji, F. M., Krumlauf, R., and Chambon, P. (1998). Hoxa1 and Hoxb1 synergize in patterning the hindbrain, cranial nerves and second pharyngeal arch. *Development* **125**, 1123–1136.

Gerlach, L. M., Hutson, M. R., Germiller, J. A., Ngyuen-Luu, D., Victor, J. C., and Barald, K. (2000). Addition of the BMP antagonist, noggin, disrupts avian inner ear development. *Development* **127**, 45–54.

Glover, J. C. (2001). Correlated patterns of neuron differentiation and Hox gene expression in the hindbrain: A comparative analysis. *Brain Res. Bull.* **55**, 683–693.

Groves, A. K., and Bronner-Fraser, M. (2000). Competence, specification and commitment in otic placode induction. *Development* **127**, 3489–3499.

Gruss, P., and Walther, C. (1992). Pax in development. *Cell* **69**, 719–722.

Hadrys, T., Braun, T., Rinkwitz-Brandt, S., Arnold, H. H., and Bober, E. (1998). Nkx5-1 controls semicircular canal formation in the mouse inner ear. *Development* **125**, 33–39.

Harvey, R. P. (1996). Nk-2 homeobox genes and heart development. *Dev. Biol.* **178**, 203–216.

Herbrand, H., Guthrie, S., Hadrys, T., Hoffmann, S., Arnold, H. H., *et al.* (1998). Two regulatory genes, cNkx5-1 and cPax2, show different responses to local signals during otic placode and vesicle formation in the chick embryo. *Development* **125**, 645–654.

Hidalgo-Sanchez, M., Alvarado-Mallart, R., and Alvarez, I. S. (2000). Pax2, otx2, gbx2 and fgf8 expression in early otic vesicle development. *Mech. Dev.* **95**, 225–229.

Hogan, B. L. M. (1996). Bone morphogenetic proteins in development. *Curr. Opin. Genet. Dev.* **6**, 432–438.

Jamali, M., Karamboulas, C., Rogerson, P. J., and Skerjanc, I. S. (2001). BMP signaling regulates Nkx5-2 activity during cardiomyogenesis. *FEBS Lett.* **509**, 126–130.

Kammermeier, L., and Reichert, H. (2001). Common developmental genetic mechanisms for patterning invertebrate and vertebrate brains. *Brain. Res. Bull.* **55**, 675–682.

Karabagli, H., Karabagli, P., Ladher, R. K., and Schoenwolf, G. C. (2002). Comparison of the expression patterns of several fibroblast growth factors during chick gastrulation and neurulation. *Anat. Embryol. (Berl).* **205**, 365–370.

Kil, S. H., and Collazo, A. (2001). Origins of inner ear sensory organs revealed by fate map and time-lapse analyses. *Dev. Biol.* **233**, 365–379.

Kil, S. H., and Collazo, A. (2002). A review of inner ear fate maps and cell lineage studies. *J. NeuroBiol.* **53**, 129–142.

Ladher, R. K., Anakwe, K. U., Gurney, A. L., Schoenwolf, G. C., and Francis-West, P. H. (2000). Identification of synergistic signals initiating inner ear development. *Science* **290**, 1965–1967.

Lang, H., Bever, M. M., and Fekete, D. M. (2000). Cell proliferation and cell death in the developing chick inner ear: Spatial and temporal patterns. *J. Comp. Neurol.* **417**, 205–220.

Lewis, E. B. (1978). A gene complex controlling segmentation in *Drosophila*. *Nature* **276**, 565–570.

Li, C. W., Van de Water, T. R., and Ruben, R. J. (1978). The fate mapping of the eleventh and twelfth day mouse otocyst: An in vitro study of the sites of origin of the embryonic inner ear sensory structures. *J. Morphol.* **157**, 249–268.

Llambi, F., Causeret, F., Bloch-Gallego, E., and Mehlen, P. (2001). Netrin-1 acts as a survival factor via its receptors UNC5H and DCC. *EMBO J.* **20**, 2715–2722.

Lombardo, A., and Slack, J. M. (1998). Postgastrulation effects of fibroblast growth factor on *Xenopus* development. *Dev. Dyn.* **212**, 75–85.

Lohmann, I., and McGinnis, W. (2002). Hox genes: It's all a matter of context. *Curr. Biol.* **12**, R514–516.

Lufkin, T., Dierich, A., LeMeur, M., Mark, M., and Chambon, P. (1991). Disruption of the Hox-1.6 homeobox gene results in defects in a region corresponding to its rostral domain of expression. *Cell* **66**, 1105–1119.

Lumsden, A., and Krumlauf, R. (1996). Patterning the vertebrate neuraxis. *Science* **274**, 1109–1113.

Maconochie, M., Nonchev, S., Morrison, A., and Krumlauf, R. (1996). Paralogous Hox genes: Function and regulation. *Annu. Rev. Genet.* **30**, 529–556.

Mahmood, R., Mason, I., and Morris-Kay, G. M. (1996). Expression of FGF3 in relation to hindbrain segmentation, otic pit position and pharyngeal arch morphology in normal and retinoic acid-exposed mouse embryos. *Anat. Embryol.* **194**, 13–22.

Mansour, S. L., Goddard, J. M., and Capecchi, M. R. (1993). Mice homozygous for a targeted disruption of the proto-oncogene int-2 have developmental defects in the tail and inner ear. *Development* **117**, 13–28.

Mark, M., Lufkin, T., Vonesch, J. L., Ruberte, E., Olivo, J. C., *et al.* (1993). Two rhombomeres are altered in Hoxa-1 mutant mice. *Development* **119**, 319–338.

Maroon, H., Waalshe, J., Mahmood, R., Kiefer, P., Dickson, C., and Mason, I. (2002). Fgf3 and fgf8 are required for formation of the otic placode and vesicle. *Development* **129**, 2099–2108.

Martin, P., and Swanson, G. J. (1993). Descriptive and experimental analysis of the epithelial remodelings that control semicircular canal formation in the developing mouse inner ear. *Dev. Biol.* **159**, 549–558.

McKay, I. J., Lewis, J., and Lumsden, A. (1996). The role of FGF3 in early inner ear development: An analysis in normal and kreisler mutant mice. *Dev. Biol.* **174**, 370–378.

McKay, I. J., Muchamore, L., Krumlauf, R., Maden, M., Lumsden, A., and Lewis, J. (1994). The kreisler mouse: A hindbrain segmentation mutant that lacks two rhombomeres. *Development* **120**, 2199–2211.

Merlo, G. R., Paleari, L., Mantero, S., Zerega, B., Adamska, M., *et al.* (2002). The Dlx5 homeobox gene is essential for vestibular morphogenesis in the mouse embryo through a BMP4-mediated pathway. *Dev. Biol.* **248**, 157–169.

Monzen, K., Nagai, R., and Komuro, I. (2002). A role for bone morphogenetic protein signaling in cardiomyocyte differentiation. *Trends Cardiovasc. Med.* **12**, 263–269.

Morsli, H., Tuorto, F., Choo, D., Postiglione, M. P., Simeone, A., and Wu, D. K. (1999). Otx1 and Otx2 activities are required for the normal development of the mouse inner ear. *Development* **126**, 2335–2343.

Noramly, S., and Grainger, R. M. (2002). Determination of the embryonic inner ear. *J. NeuroBiol.* **53**, 100–128.

Oh, S. H., Johnson, R., and Wu, D. K. (1996). Differential expression of bone morphogenetic proteins in the developing vestibular and auditory sensory organs. *J. Neurosci.* **16**, 6463–6475.

Owens, B. M., and Hawley, R. G. (2002). Hox and non-Hox homeobox genes in leukemic hematopoiesis. *Stem Cells* **20**, 364–379.

Phillips, B. T., Bolding, K., and Riley, B. B. (2001). Zebrafish fgf3 and fgf8 encode redundant functions required for otic placode induction. *Dev. Biol.* **235**, 351–365.

Pickles, J. O., and Chir, B. (2002). Roles of fibroblast growth factors in the inner ear. *Audiol. Neurotol.* **7**, 36–39.

Pirvola, U., Spencer-Dene, B., Xing-Qun, L., Kettunen, P., Thesleff, I., *et al.* (2000). FGF/FGFR-2(IIIb) signaling is essential for inner ear morphogenesis. *J. Neurosci.* **20**, 6125–6134.

Ponnio, T., Burton, Q., Pereira, F. A., Wu, D. K., and Coneely, O. M. (2002). The nuclear receptor Nor-1 is essential for proliferation of the semicircular canals of the mouse inner ear. *Mol. Cell. Biol.* **22**, 935–945.

Prall, O. W., Elliott, D. A., and Harvey, R. P. (2002). Developmental paradigms in heart disease: Insights from tinamn. *Ann. Med.* **34**, 148–156.

Rinkwitz, S., Bober, E., and Baker, R. (2001). Development of the vertebrate inner ear. *Ann. NY Acad. Sci.* **942**, 1–14.

Rinkwitz-Brandt, S., Arnold, H. H., and Bober, E. (1996). Regional expression of Nkx5–1, Nkx5–2, Pax2 and sek genes during inner ear development. *Hear. Res.* **99**, 129–138.

Roberts, C. W., Shutter, J. R., and Korsmayer, S. J. (1994). Hox11 controls the genesis of the spleen. *Nature* **368**, 747–749.

Salminen, M., Meyer, B. I., Bober, E., and Gruss, P. (2000). Netrin 1 is required for semicircular canal formation in the mouse inner ear. *Development* **127**, 13–22.

Simpson and Price (2002). Pax6: A pleiotrophic player in development. *Bioessays* **24**, 1041–1051.

Slack, J. M. W. (1991). "From Eggs to Embryo." Cambridge University Press, Cambridge.

Smith, D. M., Nielsen, C., Tabin, C. J., and Roberts, D. J. (2000). Roles of BMP signaling and Nkx2.5 in patterning at the chick midgut-foregut boundary. *Development* **127**, 3671–3681.

Streit, A. (2002). Extensive cell movements accompany formation of the otic placode. *Dev. Biol.* **249**, 237–254.

Sumiyama, K., Irvine, S. Q., Stock, D. W., Weiss, K. M., Kawasaki, K., *et al.* (2002). Genomic structure and functional control of the Dlx3–7 bigene cluster. *Proc. Natl. Acad. Sci. USA* **99**, 780–785.

Swanson, G. J., Howard, M., and Lewis, J. (1990). Epithelial autonomy in the development of the inner ear of a bird embryo. *Dev. Biol.* **137**, 243–257.

Torres, M., and Giraldez, F. (1998). The development of the vertebrate inner ear. *Mech. Dev.* **71**, 5–21.

Torres, M., Gomez-Pardo, E., and Gruss, P. (1996). Pax2 contributes to inner ear patterning and optic nerve trajectory. *Development* **121**, 4057–4065.

Trousse, F., Esteve, P., and Bovolenta, P. (2001). BMP4 mediates apoptotic cell death in the developing chick eye. *J. Neurosci.* **21**, 1292–1301.

Vendrell, V., Carnicero, E., Giraldez, F., Alonso, M. T., and Schimmang, T. (2000). Induction of inner ear fate by FGF3. *Development* **127**, 2011–2019.

Waddington, C. H. (1937). The determination of the auditory placode in the chick. *J. Exp. Biol.* **14**, 1435–1449.

Walther, C., Guenet, J. L., Simon, D., Deutch, V., Jostes, B., Goulding, M. D., Plachov, D., Balling, R., and Gruss, P. (1991). Pax: A murine multigene family of paired-box containing genes. *Genomics* **11**, 423–434.

Wang, W., Van De Water, T., and Lufkin, T. (1998). Inner ear and maternal defects in mice lacking the Hmx3 homeobox gene. *Development* **125**, 621–634.

6

Growth Factors and Early Development of Otic Neurons: Interactions between Intrinsic and Extrinsic Signals

Berta Alsina,[1] Fernando Giraldez,[1] and Isabel Varela-Nieto[2]
[1]DCEXS-Universitat Pompeu Fabra, Dr Aiguader 80, 08003 Barcelona, Spain
[2]Instituto de Investigaciones Biomédicas "Alberto Sols", CSIC-UAM
Arturo Duperier 4, Madrid 28029, Spain

I. Introduction
II. Early Development of Cochlear (Auditory) and Vestibular Neurons
   A. A Common Scheme for Vertebrate Neurogenesis
   B. Inner Ear Neurogenesis
III. Extrinsic Factors in Inner Ear Neurogenesis: Fibroblast Growth Factor, Nerve Growth Factor, and Insulin-Like Growth Factor-1 Families of Growth Factors
   A. The Family of Fibroblast Growth Factors
   B. The Family of Nerve Growth Factors
   C. The Insulin-Like Growth Factors System
   D. Other Diffusible Factors
IV. Conclusions
   References

The interplay between intrinsic and extrinsic factors is essential for the transit into different cell states during development. Secreted factors mediate crucial steps in development such as cell growth and survival, bias between self-renewal and differentiation, or choices between different cell fates. Three families of growth factors emerge as critical for early steps in otic neurogenesis: fibroblast growth factors (FGFs); nerve growth factor (NGF)-related neurotrophins (NTs); and insulin-related factors, particularly insulin-like growth factor-1 (IGF-1).

Otic neurons connect the sensory mechanoreceptors of the ear, the hair cells, with their targets in the central nervous system (CNS). The generation of otic neurons is a sequential process that includes first the specification of otic precursors in the otic epithelium, second the delamination of epithelial neuroblasts to form the cochleovestibular ganglion (CVG), third the proliferative expansion of ganglionar neuroblasts, and finally the differentiation of neurons that innervate back the vestibular and cochlear (auditory) sensory organs. Each step is characterized by the expression of a particular set of transcription factors and differentially regulated by growth factor signaling, based on which a simple model for early otic neurogenesis is proposed. © 2003, Elsevier Inc.

I. Introduction

Otic neurons innervate hair cells located in the auditory and vestibular sensory organs of the inner ear. Local mechanical perturbations are transduced by hair cells into synaptic potentials, which elicit the activation of auditory (cochlear) and vestibular neurons that project toward the homonymous central nuclei, the first input station of the auditory pathway. Auditory neurons inform the brain of the intensity and spectral properties of sound, and vestibular neurons carry information about position, velocity, and acceleration. The result of nervous processing is our perception of sound and equilibrium, as well as many functionally relevant reflexes of enormous complexity. The inner ear derives from the otic placode, an ectodermal thickening that develops adjacent to the hindbrain (Torres and Giraldez, 1998; see Chapter 4 by Groves *et al.*). The otic placode invaginates and pinches off the ectoderm to form the otic vesicle, which is a transient structure that undergoes multiple developmental changes associated with cell proliferation, differentiation, and death. This results in the ear labyrinth: the cochlea, the utricle and saccule, and the semicircular canals, each containing their corresponding sensory organs (Cantos *et al.*, 2000; Fekete and Wu, 2002). The otic vesicle shows developmental autonomy, meaning that it can be explanted from the embryo and it exhibits patterning, morphogenesis, and cell diversification, including the genesis of mechanotransducing hair cells and neurons (Swanson *et al.*, 1990). The whole system is both intricate and simple, i.e., it contains morphological complexity but based on a limited number of cell types, which are originated from the otic vesicle. The generation of otic neurons is a sequential process. First, otic neurons are specified in the otic epithelium, then neuronal precursors delaminate to form the cochleovestibular ganglion, the CVG, where they proliferate, and then differentiate and innervate back the vestibular and cochlear (auditory) sensory organs. The CVG is a transient condensation of cells that house the nascent delaminating neuroblasts. Further in development, it generates separated cochlear and vestibular ganglions. The term cochlea strictly applies to the coiled auditory apparatus, which is characteristic of some but not all vertebrates. However, it is used extensively as synonymous for auditory. We shall keep the name CVG, *sensu lato*, referring to the auditory and vestibular ganglion of different species.

All of these features make the ear a very attractive model system for studying basic problems of development, including how neurons are generated in specific regional patterns, how signaling factors control early neurogenesis, and how immature neuroblasts transit into fully functional neurons. The possibility of applying genetic and molecular approaches to the study of this model has given a new impulse to its understanding.

This chapter concentrates on the initial steps of development of otic neurons, from the specification of neural fate in the otic epithelium cells to neuronal differentiation within the CVG. Other relevant aspects of the development of auditory and vestibular neurons—target-dependent survival, innervation, and differentiation—are addressed in other chapters of this book and in reviews (Fritzsch and Beisel, 2001; Rubel and Fritzsch, 2002; see also Adam *et al.*, 1998, and Eddison *et al.*, 2000, for a general discussion of hair cell specification). Our first task is to review the cellular and molecular information about neural determination and differentiation in the ear and to define cell states and the transition steps from precursors to neurons (Fig. 1). In parallel, we review the information available on proneural and neurogenic genes in the ear (Table I). Our second task is to address the role of diffusible factors in the control of ear neurogenesis by reviewing the effects of FGFs, the NGF family of neurotrophins, and insulin-like growth factor-1 (IGF-1) (Fig. 2 and Table II). There is growing evidence that these families of intercellular signaling molecules participate in the regulation of early stages of neural development by mediating cell-to-cell interactions. Diffusible factors are locally synthesized during development and elicit a network of interconnected signaling pathways leading to specific gene transcription. Transcription factors in turn modulate the expression of factors and key signaling proteins in a cycle that finally instructs the cell to proliferate, die, or differentiate.

II. Early Development of Cochlear (Auditory) and Vestibular Neurons

Neurons that populate the CVG derive from the otic placode (D'Amico-Martel and Noden, 1983). Neurogenic placodes are local epithelial condensations that have evolved with vertebrates and account for the concentration of specialized sense organs in the head (for a review, see Baker and Bronner-Fraser, 2001). Classical studies established that, unlike the dual origin—placodal and neural crest—of most cranial ganglia, CVG sensory neurons have an exclusive placodal origin (D'Amico-Martel and Noden, 1983). Neuronal fate is specified in the otic placode as discussed later. Neuroblasts then delaminate from the otic vesicle to form the CVG, where they differentiate into mature neurons. At some stage along this process, the neural-committed

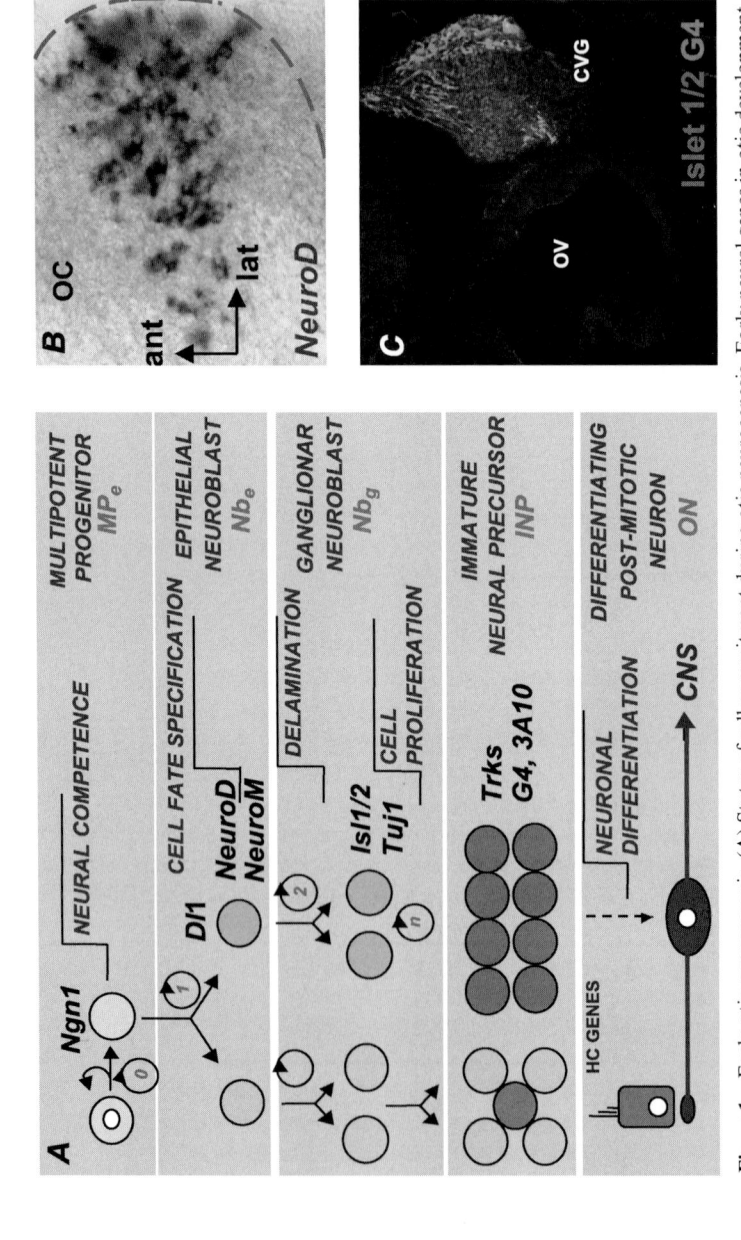

Figure 1 Early otic neurogenesis. (A) States of cell commitment during otic neurogenesis. Early neural genes in otic development. The neural lineage is labeled in different tones of blue and the sensory lineage in different tones of yellow. (B) Epithelial neuroblasts singled out by NeuroD expression. *In situ* hybridization with a NeuroD probe in HH stage 14 chick otic cup. (C) Islet-1 and G4 expression in ganglionar neuroblasts. Double immunocytological recognition with an anti-Islet-1/2 antibody and an anti-G4 antibody. (See Color Insert).

Table I Cell States and Molecular Markers during Early Neurogenesis

	Epithelial multipotent progenitor cell MP$_e$	Epithelial neuroblast Nb$_e$	Ganglionar neuroblast Nb$_g$	Immature neural precursor INP
Ngn 1	nd	nd	−	−
NeuroD/M	−	+	+	−
Delta1	nd	+	−	−
Notch[a]	nd	−	−	nd
Islet-1/2	−	−	+	−
Tuj1	−	−	+	+/−
G4	−	−	−	+
3A10	−	−	−	+
TrkB/TrkC	−	−	+	+
PCNA/BrdU	+	−	+	−

Data compiled from expression studies in different animal species (see text for references). The list does not pretend to be comprehensive. [a]Notch stands for *activated* Notch; positive expression (+), lack of expression (−), nd, Not determined; −, lack of expression; ±, positive expression.

[a]*Notch* stands for activated Notch.

population expands, and further, it diversifies into two basic cell types, I and II, and two different identities, vestibular and auditory (cochlear).

A. A Common Scheme for Vertebrate Neurogenesis

General principles for the understanding of molecular mechanisms underlying sensory organ development start to be elucidated after insight is gained from fundamental studies in *Drosophila* (Campuzano and Modolell, 1992). Proneural and neurogenic genes are central elements in the process of determination and selection of neural precursors. Two main classes of proneural genes in *Drosophila* are the *achaete-scute* complex (*asc*), mainly associated with determination of external sense organs and central nervous system (CNS) progenitors, and the *atonal* proneural gene (*ato*), related to chordotonal, proprioceptive sense organs and photoreceptors. Proneural genes are tissue-specific basic helix-loop-helix proteins (bHLH proteins) that bind to a common DNA sequence called the E-box sequence, forming heterodimers with E-proteins (Bertrand *et al.*, 2002, Lee, 1997). In *Drosophila*, proneural genes are expressed in clusters of cells from which individual neuronal precursor cells are singled out. Genetic studies have demonstrated that they are necessary and sufficient for the specification and determination of neural precursors from the ectoderm (Jimenez and Modolell, 1993). Once proneural genes establish a neuronal competent region, neuronal precursors emerge by a process of lateral inhibition mediated by the Delta–Notch transmembrane

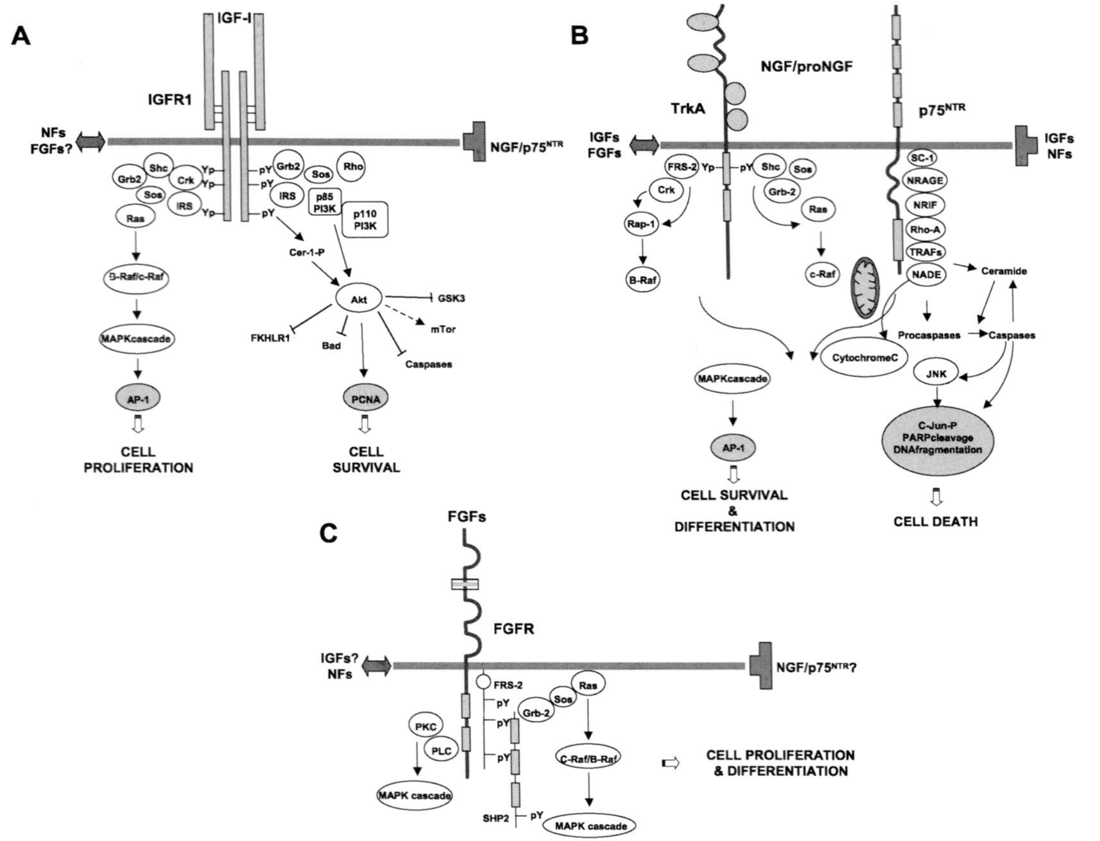
A
IGF-I
IGFR1
NFs
FGFs?
NGF/p75NTR
Grb2
Shc
Crk
Sos
IRS
Ras
Yp
pY
Grb2
Sos
Rho
IRS
p85
PI3K
p110
PI3K
Cer-1-P
B-Raf/c-Raf
MAPKcascade
AP-1
Akt
GSK3
mTor
FKHLR1
Bad
Caspases
PCNA
CELL
PROLIFERATION
CELL
SURVIVAL
B
NGF/proNGF
TrkA
p75NTR
IGFs
FGFs
IGFs
NFs
FRS-2
Crk
Rap-1
B-Raf
Shc
Sos
Grb-2
Ras
c-Raf
SC-1
NRAGE
NRIF
Rho-A
TRAFs
NADE
Ceramide
Procaspases
Caspases
CytochromeC
JNK
MAPKcascade
AP-1
C-Jun-P
PARPcleavage
DNAfragmentation
CELL SURVIVAL
&
DIFFERENTIATION
CELL DEATH
C
FGFs
FGFR
IGFs?
NFs
NGF/p75NTR?
FRS-2
pY
Grb-2
Sos
Ras
PKC
PLC
MAPK cascade
C-Raf/B-Raf
SHP2
MAPK cascade
CELL PROLIFERATION
& DIFFERENTIATION

6. Growth Factors and Early Development of Otic Neurons

proteins. Notch–ligands, Dl, are expressed in those cells that are committed to becoming neurons and inhibiting neuronal fate in neighboring cells. This is carried out by activation of Notch receptors and Notch effectors such as Hairy (Hry) or Enhancer of Split (E[spl]), which in turn repress neuronal fate. Therefore another crucial element in the operation of proneural genes is the Delta–Notch pathway that contributes to establishing cell asymmetry and diversification within an equally competent domain.

The vertebrate homologues of *Drosophila* proneural genes fall in two main groups according to their structure: the *ASC*-related genes (*Mash, Cash, Zash,* and *Xash*) and the *ATONAL*-related genes, which in turn are grouped into subfamilies on the basis of their amino acid sequence similarities— *MATH, Neurogenin* (*Ngn*), *NeuroD,* and *Olig* (see Bertrand *et al.*, 2002, for a review). Also, the Delta–Notch pathway has several vertebrate homologues for the ligands Dl, Serrate (Ser), and Jagged; the Notch receptors (N); and the Notch targets Hes (Hairy and E[spl] homologue) and Id (Justice and Jan, 2002; Lee, 1997). The Notch-signaling pathway is modulated at different levels and displays cross talk with other signaling pathways (Panin and Irvine, 1998).

Figure 2 Intracellular signaling pathways activated by NGF, IGF-1, and FGFs. Diagrams illustrate the signaling pathways activated by extrinsic factors to balance biological processes underlying inner ear neurogenesis. Double-headed arrows indicate positive or equivalent interactions among factors acting on inner ear development, whereas blocked arrows indicate negative interactions. (A) Cell signaling by IGF-1. The (IGF1Rs) are $\alpha_2\beta_2$ heterotetramers. The extracellular glycosylated subunits bind the IGFs. The β subunit of this receptor is a transmembrane polypeptide with a highly conserved tyrosine kinase catalytic domain. Binding of receptor results in its autophosphorylation on tyrosine residues and subsequent phosphorylation of intracellular proteins. Among these are the insulin receptor substrates (IRS-1/3) that phosphorylate the regulatory subunit of phosphoinositide 3-kinase (PI3-K). This in turn activates Akt, a serine/threonine protein kinase that activates FKHR-L1 (forkhead transcription factor), inactivates Bad by phosphorylation, and blocks procaspase 9, leading to the inactivation of apoptosis pathways and, consequently, cell survival. A parallel and connected signal transduction pathway activated by IGF-1 binding is the cascade including Ras and Raf family members that leads to activation of the mitogen-activated protein kinases cascade and cellular proliferation. Arrows denote facilitating action, whereas crossbars indicate inhibitory influence. (B) Cell signaling by NGF. NGF can exert different effects on target inner ear neurons depending on the receptor to which it binds. NGF and its related NTs promote cell proliferation, survival, and differentiation of neurons expressing high-affinity tyrosine kinase Trk receptors. In the absence of Trk receptors, binding to the low-affinity p75^{NTR} receptors promotes cell death. Signaling through p75^{NTR} requires the binding of cytoplasmic proteins, which is followed by an increase in ceramide levels and the activation of downstream targets such as JNK and caspases, among others; JNK; and c-Jun-N-terminal kinase. (C) Cell signaling by FGFs. General or specific inner ear information on the specific signaling pathways activated by each of the FGFs is not available. There is a general agreement on what FGF receptors 1–3 dimerize following activation by FGFs, which is followed by tyrosine phosphorylation of FRS2 and the activation of the Ras, Raf, MAPK cascade, phospholipase C (PLC), and protein kinase C (PKC), leading to the control of cell proliferation and differentiation.

Table II Actions of Diffusible Factors during Inner Ear Neurogenesis

	Proliferation	Survival	Apoptosis	Migration	Early differentiation
IGF-1	+	+	−	−	+
Insulin	+	nd	nd	nd	−
IGF2	+	nd	nd	nd	nd
NT[a]	+	+	+	nd	+
FGF2	+	nd	nd	+	+
FGF8	−	nd	nd	nd	nd
FGF10	−	−	−	−	+

Data compiled from transgenic mice and from *in vitro* experiments performed either in otic explants or in cultures of dissociated cells of rodent or chicken inner ear (see text for references). The list does not pretend to be comprehensive. +, Positive modulation; −, negative modulation; nd, not determined.

[a]Neurotrophin (NT) actions on apoptosis are mediated by p75^{NTR} low-affinity receptor (see text).

In vertebrates, proneural genes act on neural fate determination and neural differentiation, but also they are involved in neuronal identity (Brunet and Ghysen, 1999). The Delta–Notch pathway is also required for vertebrate neurogenesis and refines the proneural potential by establishing (asymmetrical) lineages from common progenitors (Justice and Jan, 2002; Myat et al., 1996; Landford et al., 1999). However, in vertebrates the scenario is more complex than that described in *Drosophila*. First, given members of the family have multiple roles, such as cell fate determination, cell differentiation, or neuronal identity (Brunet and Ghysen, 1999). Second, there are cooperative or synergistic functions among different genes (Bertrand et al., 2002). Third, family members have redundant functions (Bertrand et al., 2002). Finally, the expression of these genes is also regulated by extracellular signals and diffusible factors (Raballo et al., 2000; Vaccarino et al., 1999). Not a minor added complexity is the fact that even in related species there are unexpected differences in the expression and/or role of particular genes or gene subfamilies on the background of a conserved mechanism (Manzanares et al., 2001).

B. Inner Ear Neurogenesis

What is known about the molecular basis of ear neurogenesis? Unfortunately, the understanding of the molecular basis of ear neurogenesis is in a very primitive state, and even the basic cellular processes are poorly understood. Most information comes from chick and mouse embryos, to which we refer unless otherwise stated. The first visible output of otic neurogenesis

6. Growth Factors and Early Development of Otic Neurons

is the delamination of neural cells (otic neuroblasts, see later) from the otic vesicle to populate the CVG. However, most likely cell diversification starts very early in otic development, at the otic placode stage. This is indicated by the expression in the otic placode of both *Ngn1* (but not *Ngn2* or the Aschlike *Mash* and *Cash* genes) and *Dll* (Abu-Elmagd et al., 2001; Adam et al., 1998; Riley et al., 1999; Lewis et al., 1998; Haddon et al., 1998). The expression of these genes persists throughout the otic cup and the otic vesicle, but none appears to be expressed in delaminating neuroblasts or the CVG. Therefore, it can be assumed that the early common expression domain of *Ngn1* and *Dll* corresponds to the prospective neural sensory domain, which is the region of the otic vesicle epithelium committed to generate otic neurons and sensory organs. This early expression of neural genes is anticipated by the expression of FGF10 (Alsina and Giraldez, unpublished results) and by the expression of the cell adhesion molecule BEN (Goodyear et al., 2001).

It is believed that both neurons and sensory cells derive from a common progenitor (an epithelial multipotent progenitor cell, MP_e; see Fig. 1A), but direct evidence is not available (see Lang and Fekete, 2001). It is not known whether the neural sensory domain is compartmentally restricted (Brigande et al., 2000), nor if there is a correspondence between the site of generation of otic neurons and the location of their epithelial targets (Lang and Fekete, 2001).

Evidence for an early role of the proneural gene *Ngn1* in neuronal determination comes from the analysis of the mouse null mutant of *Ngn1*, which exhibits a loss of ganglion neurons (Ma et al., 1999, 2000). *Ngn1* deletion also causes a reduction in the number of hair cells, which could be due to the existence of a common step in gene activation for the determination of both neuronal and receptor cell lineages.

This suggests that the newly generated neuroblasts (Nb_es) remain in cell division arrest before delamination, in spite of the proliferative capacity that Nb_es display once in the CVG (see later).

Delamination of otic neuronal precursors starts at the otic cup stage and ends by the late otic vesicle. The time-course of delamination was extensively documented in the chick embryo by D'Amico-Martel and Noden (1983) and Hemond and Morest (1991). Neuroblasts start to emerge at stage 12–13 (Hamburger and Hamilton, 1992) and the peak of migration occurs at the closure of the otic cup into the early otic vesicle, at stages 16–17, which correspond to E9.5 and E10.5–12.5 in the mouse (Liu et al., 2000). The LIM homeodomain transcription factor Islet-1/2 is expressed in few cells within the epithelium, but it is intensely expressed in cells located at the sites of delamination and in those that have delaminated (see Fig. 1C; Adam et al., 1998; Camarero et al., 2003). The Lim homeodomain gene family of transcription factors is known to be crucial for specification of neuronal identity in neural tube–derived neurons, but there are no functional studies on the role of Islet-1/2 in the inner ear (Ericson et al., 1992; Hobert and Westphal, 2000). Early ganglion cells also express other markers such as the neuron-specific βIII-tubulin Tuj1 (Camarero et al., 2003; Memberg and Hall, 1995), FGF10 in the mouse but not in the chick (Pirvola et al., 2000), N-CAM (Richardson et al., 1987), or Lunatic fringe (Cole et al., 2000).

As mentioned previously, neuronal precursors in the CVG develop into neurons that innervate back the sensory patches developed in parallel in the otic vesicle. D'Amico-Martel (1982) carried out a very detailed analysis of the temporal and regional patterns of cell division in the chick CVG and mapped the sequence of generation of postmitotic neurons. Final mitoses start during delamination and proceed until the vestibular and auditory ganglions are well developed, well after delamination has ceased. Both migration and terminal mitosis occur earlier in the vestibular than in the cochlear–auditory ganglion (D'Amico-Martel, 1982; Hemond and Morest, 1991; Ruben, 1967). This pattern parallels the rostrocaudal (rostral being distal to the otic vesicle) sequence of maturation of CVG neurons and the earlier differentiation of vestibular neurons (Hemond and Morest, 1991; Whitehead and Morest, 1985). Hemond and Morest (1991) noted that most migrating cells are still intermitotic and, therefore, CVG neurons differentiate only after migration and proliferation of precursors. The occurrence of cell division in cells that are committed to the neural fate has been demonstrated directly (Adam et al., 1998; Begbie et al., 2001, 2002), as has been their dependence on IGF-1 activity (Camarero et al., 2003). The implication of these studies is that delaminating cells constitute a cell population that is committed to neural fate but is still proliferative and that it is not until a yet unknown number of cell divisions have occurred that they are allowed to differentiate, i.e., it constitutes a transit-amplifying cell population. We can identify this third cellular state as a ganglionar neuroblast (Nb_g) characterized by the expression of *NeuroD*, Islet-1/2, Tuj1, and proliferation markers such as the proliferating cell nuclear antigen (PCNA) and phosphohistone 3 (Camarero et al., 2003). As discussed later, ganglionar neuroblasts form a transit-amplifying cell population that depends on IGF1 for survival, expansion, and differentiation.

After cell division has reached an end, neuroblasts become postmitotic and transit into immature neurons, which then differentiate and extend projections toward both their peripheral and central targets (Whitehead and Morest, 1985). These cells are immature neuronal precursors (INPs), a fourth cellular state characterized as being postmitotic but not yet differentiated. These precursors downregulate the expression of most of the early neural genes and start to express another set of markers, which are related with neurite extension and survival. Among the latter are the fasciculin G4 (see Fig. 1C; Camarero et al., 2003), the neurofilament-associated antigen 3A10 (Adam et al., 1998), and TrkB/TrkC neurotrophin receptors (Brumwell et al., 2000; Kim et al., 2001; Alsina, Varela-Nieto, and Giraldez, unpublished observations).

6. Growth Factors and Early Development of Otic Neurons

This cellular state extends until the final maturation that generates otic neurons (ONs), which express the whole panoply of ionic currents, synaptic receptors, and neurotransmitters. This final step of maturation occurs later during development. For example, in the chick first terminal mitoses occur by day 3, whereas the first indication of expression of specific ionic currents and synaptic contacts take place by day 7 (Valverde *et al.*, 1992; Whitehead and Morest, 1985). IGF-1 and FGF2 may be critical for the transition between the proliferative state and the initiation of neuronal differentiation and neurite outgrowth (Brumwell *et al.*, 2000; Camarero *et al.*, 2003).

A view on the process of otic neurogenesis is summarized in Fig. 1A. The different cellular states leading to neuron generation exhibit a distinct combination of molecular markers (see Table I) and a specific dependence on growth factors for survival, proliferation, and differentiation. To date there has not been a molecular label for the multipotent precursor (MP$_e$) and, as mentioned previously, the possibility that neural and sensory cell types derive from a common precursor has not been demonstrated directly. The epithelial neuroblast (Nb$_e$) is characterized by the expression of *NeuroD* and *NeuroM*, and its differentiation is probably dependent on the local expression of FGF10 (Alsina and Giraldez, unpublished results). Upon delamination, the Nb$_g$ expresses Islet-1/2 and Tuj1 and is dependent on IGF1 for survival and proliferation. The expansion of Nb$_g$ leads to a population of immature neural precursors, which are dependent on FGFs and neurotrophins for survival and differentiation (see later). The scheme of Fig. 1 can be used as a working framework to understand both cell-autonomous processes and the action signals in the generation of otic neurons to establish the cellular states and the coordinated participation of transcription and diffusible factors in the generation of otic neurons. It could be expected that diffusible factors produced locally during development modulate neurogenesis at different stages: first, at the stage of patterning of the neural sensory domain; second, during lineage decision steps by shifting gene expression toward different cellular fates; third, at the proliferative expansion required for given cell lineages, by controlling cell division, survival, or both; and finally, at the point of cell differentiation, either in an instructive or a permissive manner.

III. Extrinsic Factors in Inner Ear Neurogenesis: Fibroblast Growth Factor, Nerve Growth Factor, and Insulin-Like Growth Factor-1 Families of Growth Factors

The specification of cell fate is the result of the interaction between two complementary sets of determinative factors, those secreted (extrinsic) signals that are present in the environment of the cell and those that operate in a cell-autonomous manner (intrinsic) (Edlund and Jessell, 1999). The final number of neurons depends on a dynamic balance between self-renewal, differentiation, and death of the precursor population. This balance is modulated by factors that provide positive or inhibitory signals to meet the demands of tissue development. Exit from the cell cycle needs to be finely coordinated with the process of differentiation because lack of coordination may lead to apoptosis (Sommer and Rao, 2002). The neuronal precursors generated in the inner ear successively transit into different cell states, which are determined by the expression of different proneural genes and by their responsiveness to various extrinsic signals. We now summarize the available data on the effects of three families of diffusible factors that have been reported to regulate cell state transitions during inner ear neurogenesis. Further information on the functions of FGFs and neurotrophins during inner ear development can be found in other chapters of this book.

A. The Family of Fibroblast Growth Factors

Fibroblast growth factors (FGFs) and receptors (FGFRs) play multiple roles in cell communication during embryonic development (see also Chapter 8 by Wright and Mansour for a general view of FGFs in otic development). To date, 23 different FGFs have been discovered, and they have been numbered according to their order of identification (Ornitz and Itoh, 2001). The main structural feature of FGFs is a central core constituted by a highly homologous sequence of 140 amino acids. Another common feature is their strong affinity for heparin that facilitates the interaction with heparin-like glycosaminoglycans in the extracellular matrix (Burke et al., 1998). FGFs bind to four classes of tyrosine kinase receptors, FGFR1–4, that mediate their effects. FGFRs contain an extracellular domain with three Ig loops, a transmembrane domain, and an intercellular kinase domain. Various isoforms of FGFR2 (FGFR2IIIb or FGFR2IIIc) are generated by alternative splicing with different Ig domains and specificity for particular FGFs. Binding of FGFs to membrane receptors elicits a number of intracellular signaling pathways that are depicted in Fig. 2 (Powers et al., 2000; Ridyard and Robbins, 2003). Comparison of expression patterns of the different FGFRs shows that FGFR1 expression is highly conserved among different vertebrates, FGFR2 has many features that are conserved and some that are divergent, and FGFR4 is highly divergent (Golub et al., 2000). Typically, FGFs have mitogenic and chemotactic activities on mesodermal and ectodermal cells, but their effects are tremendously diverse during development and in adult life. It has been reported that FGFs have critical roles in neural induction (reviewed by Wilson and Edlund, 2001), patterning (Cole et al., 2000), cell migration, proliferation, and survival (Raballo et al., 2000; Vaccarino et al., 1999).

The role of FGFs in inner ear induction has been studied extensively in different animal species (see Chapter 4 by Groves *et al.*), but more recently attention has been focused on the role of FGFs in other processes of inner ear development, including otic neurogenesis (see also Chapter 8 by Wright and Mansour). Several FGFs are expressed in the otic primordium at the time of neurogenesis, but species differences are striking (Karabagli *et al.*, 2002).

FGF2 is expressed in the ectoderm, including the otic placode, and throughout the otic vesicle stage (Vendrell *et al.*, 2000). Ectopic application of FGF2 and FGF8 by means of microbeads enhances transcription of several patterning genes in the chick (Adamska *et al.*, 2001). This study showed that FGF2 and FGF8 cause an enlargement of the CVG, which is in agreement with previous results reported by Hossain *et al.*, (1996) showing that FGF2 increases migration and differentiation of CVG neurons (see also Adamska *et al.*, 2001; Zheng *et al.*, 1997; Zhou *et al.*, 1996). Interestingly, FGF2 upregulates *TrkB*, promotes brain-derived neurotrophic factor (BDNF) action in auditory neurons (Brumwell *et al.*, 2000), and shows a sequential interaction between FGF2, BDNF, NT-3, and their receptors that define so-called critical periods of development of ganglion neurons (Hossain *et al.*, 2002).

FGF3 was first localized in the neural sensory epithelium of the otic vesicle and in the CVG in the mouse (Wilkinson *et al.*, 1989) and *Xenopus* (Tannahill *et al.*, 1992), but not in the chick embryo (Mahmood *et al.*, 1995). The null mutation of *FGF3* in mouse causes reduction of the CVG ganglion on the background of morphogenetic defects (Mansour *et al.*, 1993).

Chick *FGF8* is expressed in a medial anterior–posterior band in the otic vesicle and is complementary to *Otx2* laterally and *Gbx2* medially. Delaminating neuroblasts are observed at the boundary between *FGF8* and *Otx2* domains (Hidalgo-Sanchez *et al.*, 2000). *FGF8* was detected in the otic vesicle and later in the CVG and nonsensory epithelium (Colvin *et al.*, 1999). Loss of function of *FGF8* in zebrafish (the *ace*, acerebellar mutant) and the knockdown of *FGF3* with antisense morpholinos cause alterations in the induction and formation of the otic vesicle, with small otic ganglia and defects in sensory organs (Kwak *et al.*, 2002; Leger and Brand, 2002; Phillips *et al.*, 2001).

FGF10 is a 208–amino acid glycoprotein that possesses a signal sequence with potential for cell export and intercellular signaling (Yamasaki *et al.*, 1996). The study of *FGF10* expression in the mouse showed mRNA expression in the otic placode and otic vesicle within the proneural–sensory domain epithelium and also in CVG neurons (Pirvola *et al.*, 2000). The mouse *knockout* of *FGF10* shows a reduction in the number of otic neurons and defects in the sensory epithelium, along with vestibular defects (Pauley *et al.*, 2003). Based on binding studies, FGFR2(IIIb) is thought to be the potential target of FGF10, although FGFR2 is also activated by FGF1, 3, and 7 and FGF10 *also* binds to FGFR1 (Powers *et al.*, 2000). *FGF10* and *FGFR2(IIIb)* expression domains in the early otic vesicle are complementary in the mouse and *FGFR2* is expressed dorsally in the prospective nonsensory epithelium, whereas *FGF10* and *FGF3* are coexpressed in the ventral domain. This discovery led to the suggestion that FGFs and FGFRs may act in a paracrine manner to provide an explanation to the alterations caused by the deletion of *FRFR2(IIIb)* in the sensory epithelium and the CVG. *FGFR2(IIIb)* null mutants fail to form the CVG, but the mechanism is unknown. It does not seem to be initial patterning because main axes and broad domains are still present in the null animals. Disruption of *FGFR1* signaling in the otic epithelium causes defects in the organ of Cortic with no reported effect on the CVG. These defects have been explained by a reduced proliferation of the precursor pool that gives rise to the auditory sensory epithelium (Pirvola *et al.*, 2002).

In the chick *FGF10*, but not *FGF3*, is also expressed in the proneural sensory domain of the otic placode. This occurs well before typical patterning genes such as *Pax-2* or *Nkx5.1* are regionally restricted, indicating that the neural sensory domain is specified very early, probably before further regional specification. Recent experiments have shown that *FGF10* overexpression causes the induction of *NeuroD*, but not *Delta1* or *Hes5*, and that FGF receptor inhibition suppresses *NeuroD*, indicating that FGF signaling may be required for commitment of otic epithelial cells to neural fate (results unpublished).

In summary, FGF2, 3, 8, and 10 may have different but complementary roles in the inner ear, acting at different steps during neurogenesis (see Table II). FGF3 and 10 seem to be more related to specification and early differentiation of neurons, whereas FGF2 and 8 appear to be related to migration and late differentiation, although different FGFs mimicking each others' functions are sometimes difficult to exclude. It will be interesting to explore the existence of FGF cooperation or looping like that in other model systems (Kettunen *et al.*, 2000; Xu *et al.*, 1998).

Interactions of FGFs with other families of factors in developing otic neurons have not been explored. Drago *et al.* (1991) showed that FGF2 stimulates proliferation and differentiation of neuroepithelial precursors of cerebral cortex and that this effect requires endogenous IGF1. On the other hand, FGFs are able to induce *Trk*-receptor expression in the ear, suggesting that FGFs may facilitate the neuronal response to neurotrophins (Brumwell *et al.*, 2000; see Fig. 2).

B. The Family of Nerve Growth Factors

NGF and the related neurotrophins, brain-derived neurotrophic factor (BDNF) and neurotrophin-3 (NT-3), are a family of secreted proteins structurally related that are essential for the development and functioning of the nervous system (Bibel and Barde, 2000; Huang and Reichardt, 2001). The role of neurotrophins in neuronal survival is mainly mediated by the activation of high-affinity tyrosine kinase Trk receptors (Lee *et al.*, 2001; Lee and Kim, 2001; see Fig. 2 for signaling pathways). Neurotrophins also bind the low-affinity p75 neurotrophin receptor (p75^{NTR}) that is a modulator of survival and death decisions (Barrett, 2000; Casaccia-Bonnefil *et al.*, 1999; Chao and Bothwell, 2002; Lee *et al.*, 2001; (see Fig. 2). The p75^{NTR} receptor belongs to the superfamily of death receptors. It possesses a death domain that upon binding to ligand recruits a specific subset of docking proteins that elicit the intracellular responses leading to cell death (Hempstead, 2002; Liepinsh *et al.*, 1997). *p75^{NTR} knockout* mice present severe nervous system defects, but the impact of p75^{NTR} deficiency on the inner ear has not been evaluated (Lee *et al.*, 2001; von Schack *et al.*, 2001). On the contrary, there is compiling evidence showing that *knockout* mice for BDNF, NT-3, or their high-affinity TrkB and TrkC receptors present important defects in the inner ear as discussed in detail in other chapters and reviewed by Fritzsch *et al.* (1997). In the developing inner ear of mice and chicken there is an early expression of the receptors for neurotrophins (Bernd and Li, 1999; Bernd and Represa, 1989; Represa *et al.*, 1991; Fritzsch *et al.*, 1997; Pirvola *et al.*, 1997). Indeed, neurotrophin dependence of cochlear and vestibular neurons is preceded by the expression of specific receptors, which is one of the earliest signs of cochlear and vestibular identity. Earlier work *in vitro* studied the role of neurotrophins in primary cultures of neurons or in organotypic cultures of explanted CVG, developing cochlear ganglion (CG), and vestibular ganglion (VG) (see Table II, Represa *et al.*, 1991a and 1993). These studies pointed out that neurotrophins promote either cell proliferation of neuronal precursors or neuronal differentiation, depending on the stage of development of the chicken otocysts. By following this experimental approach it was shown that CG depends on NT-3, whereas VG depends on BDNF for neuronal survival and differentiation in chicken, mouse, and humans (Adamson *et al.*, 2002; Avila *et al.*, 1993; Pirvola *et al.*, 1994; Sokolowski, 1997; Vega *et al.*, 1999). These data were confirmed later when the phenotype of single and double mutant mice for each of the factors or receptors was reported (reviewed by Fritzsch *et al.*, 1997). The role for NGF as an inductor of apoptosis during nervous system development and maturation has been evidenced (Frade and Barde, 1998; Majdan *et al.*, 1997). More recently it has been reported that apoptosis is induced by the unprocessed factor, pro-NGF (Hempstead, 2002; Lee *et al.*, 2001; Lee and Kermani, 2001). NGF induces apoptotic cell death in organotypic cultures of otic vesicles and in the CVG through binding to p75^{NTR} (Frago *et al.*, 1998; Sanz *et al.*, 1999b) and, interestingly, NGF-induced cell death occurs in specific areas of the otic vesicle and CVG. The presence of p75^{NTR} has been extensively reported at different stages of inner ear development in several animal species. *p75^{NTR}* expression has been associated also with an early specification of vestibular sensory patches in the chicken embryo (Schecterson and Bothwell, 1994; von Bartheld et al., 1991; Wu and Oh, 1996). The precise signaling pathway(s) used by p75^{NTR} to activate cell death remain unclear. They may involve the generation of ceramide (Dobrowsky et al., 1994; Frago et al., 1998; Hirata et al., 2001), and the activation of Jun-N-terminal kinase (Harrington et al., 2002; Sanz et al., 1999a), caspase cascade (Frago et al., 2003; Gu et al., 1999; Troy et al., 2002), and cyclin-dependent kinases (Frade, 2000). The proapoptotic actions of NGF/p75^{NTR} are strictly controlled during development by survival factors, as shown in Fig. 2 and discussed in detail later.

C. The Insulin-Like Growth Factors System

IGF-1 and 2, insulin, their cellular receptors, and the specific IGF binding proteins (IGFBP1–6) form the insulin-related growth factor system. Related factors and receptors have been described in invertebrates (reviewed in Varela-Nieto et al., 2003). IGFs are produced locally during development by many tissues and have autocrine and paracrine functions. Local expression is maintained postnatally in certain tissues, such as the brain, but the endocrine contribution of the liver increases. The cellular actions of these factors are mediated by binding to membrane-bound tyrosine kinase receptors. The classic receptors include the insulin receptor and the IGF type 1 receptor (IGF1R). Both receptors are heterotetramers conformed by two extracellular α subunits (binding activity) and two transmembrane β subunits (tyrosine kinase and signaling activities) (see Fig. 2). A second class of IGF receptor, the cation-independent, mannose-6-phosphate monomeric IGF type 2 receptor (IGF2R), has factor clearance functions and to date has no known signaling activity. Insight into the specific role of each receptor during development has been provided by the analysis of mutant mice (Efstratiadis, 1998; Nakae et al., 2001). Finally, binding of IGFs to their specific receptors is modulated by the IGF binding proteins (IGFBP).

The elements of the IGF system are expressed in most areas of the nervous system of birds and mammals during development. As development proceeds to adulthood, higher levels of expression are associated with areas that maintain plasticity. In the early developing chicken inner ear there is expression of IGF-1, insulin, insulin receptor, and IGFR1 in the otic epithelium and in the CVG (Leon et al., 1999; Varela-Nieto et al., 2003). IGF1 is also expressed during maturation of the rodent auditory system and in adult hair cells (Lee and Cotanche, 1996; Saffer et al., 1996). During the early postnatal period of mouse inner ear development, from postnatal day P5 to P20, IGF1 is expressed in the organ of Corti and cochlear ganglia (Camarero et al., 2001). The cochlear ganglia also express IGF2 and IGFR1 (Camarero and

Varela-Nieto, unpublished observations). There is no information available on the local expression pattern of other elements of the IGF system in the developing inner ear. The analysis of a human fetal cochlear cDNA library has indicated the presence of IGF-1 and IGFBP1, 3, and 5. Analysis of gene expression profiles of the rat cochlea has evidenced the presence of IGF2 and IGFBP2 and 6 (Cho et al., 2002; Resendes et al., 2002).

Studies on primary cell cultures and genetically modified animal models have shown that IGF-1 is essential for the normal development and function of the vertebrate nervous system. Furthermore, IGF1 is currently accepted as a neurotrophic factor, and it is being considered for treatment of neurodegenerative diseases (Dore et al., 1997; Savage et al., 2001). The actions of IGF-1 in the nervous system include control of cell size, stimulation of cell proliferation and survival, increase of neuronal differentiation, axonogenesis, myelination, and finally modulation of synaptogenesis and neurotransmission (reviewed in Varela-Nieto et al., 2003). The importance of IGF-1 in ear development is stressed by the fact that a deficiency in IGF-1 results in sensorineural deafness in humans (Woods et al., 1996, 1997). Besides, the lack of IGF-1 in mice severely affects postnatal survival, differentiation, and maturation of the cochlear ganglion cells and causes abnormal innervation of the sensory cells in the organ of Corti (Camarero et al., 2001, 2002).

1. Actions of IGF-1 in Cell Growth, Proliferation, and Survival

In cultured chicken otic vesicle and CVG, IGF-1 is a survival and growth factor capable of blocking apoptosis (see Table II) (Frago et al., 1998, 2003; Leon et al., 1995, 1998; Sokolowski, 1997). IGF-1 binding to its high-affinity tyrosine kinase receptor increases the levels of the lipid mediator inositol phosphoglycan; activates the Raf/MAPK cascade; and induces the expression of Fos, Jun, and the proliferating cell nuclear antigen (PCNA), leading to cell growth (Frago et al., 2000; Leon et al., 1995, 1998; Sanz et al., 1999a,b). On the other hand, IGF-1 promotes cell survival during development and protects otic explants from cell death caused by factor deprivation or in response to NGF/p75^{NTR}. IGF-1 exerts its antiapoptotic effect in the chicken otic vesicle by the activation of the Akt/protein kinase B pathway and the modulation of the levels of the proapoptotic lipid mediator ceramide (Frago et al., 2003). Akt downstream signaling diverges depending on the cellular context and apoptotic stimuli. IGF-1 prevents caspase-3 activation, Jun N-terminal kinase activation, and p53 transcriptional activity; induces dephosphorylation of Bad; and modulates Bax and Bcl2 levels (see Fig. 2) (reviewed in Bevan, 2002). In the developing inner ear, cell death contributes to morphogenesis and to cell number control in the otic epithelium and within the CVG (Alvarez, 1989; Alvarez and Navascues, 1990; Fekete et al., 1997; Sanz et al., 1999a); therefore, the balance between proliferation and programmed cell death is carefully regulated. The signaling pathways regulating neuronal death *in vivo* are not fully understood and, in this context, the developing inner ear offers a physiologically relevant model for the study of signaling networks during development. Interestingly, IGF-1 and NGF/p75^{NTR} signaling pathways interact at different levels, being the apoptosis induced by NGF, balanced by survival signals, and activated by IGF-1, among other factors (see Fig. 2).

Endogenous IGF-1 is essential for the generation of the CVG in chicken embryos. By interfering with the actions of the endogenous factor in otic vesicle explants, we have recently shown that in the absence of IGF-1, the CVG size is reduced or the CVG is not even formed. Associated with this phenotype are an increase in cell death, a reduction in cell proliferation, and a reduction in the levels of expression of neuroblasts and neuronal markers (Islet-1/2 antibody, Tuj1, and G4). In contrast, NeuroD expression is not altered, suggesting that IGF-1 is not acting in the specification of neuroblast precursors within the epithelium. NeuroD null mice present a severe loss of CVG neurons during development, and it is worth noting that insulin is one of the transcriptional targets of NeuroD (Alvarez and Navascues, 1990; Kaneto *et al.*, 2002; Kim *et al.*, 2001). CVG neurogenesis and early differentiation are recovered by treatment with the exogenous factor, which is also able to induce proliferation and differentiation in cultured CVG explants (Camarero *et al.*, 2003). Interestingly, the early actions of IGF-1 on cell proliferation are mimicked by other factors of the family, insulin and IGF-2, but later actions on cell differentiation are specific for IGF-1 (Camarero *et al.*, 2003; Leon *et al.*, 1998).

The analysis of the *Igf1* null mice cochlea confirmed the importance of IGF1 for the correct maturation of the inner ear. P20 *Igf1* null mice show reduced size of the cochlea, cochlear ganglia, and mean size of cochlear ganglia sensory neurons. *Igf1* null mice present a general, although organ-specific, growth retardation (Cheng and Feldman, 1998; Liu *et al.*, 1993, 1998; Powell-Braxton *et al.*, 1993). On the other hand, transgenic mice overexpressing IGF1 present increased cell size and decreased apoptosis, among other alterations in the CNS (Behringer *et al.*, 1990; Carson *et al.*, 1993; Ye *et al.*, 1995). However, the ear phenotype of these mutants has not yet been studied. In invertebrates, the IGF system also participates in the control of cell size (Garofalo, 2002; Oldham and Hafen, 2003).

In addition to cell size alterations, *Igf1* null and transgenic mice have important changes in brain size that are due not only to alterations in neuron growth and size but also to changes in the proliferation rate of neuronal precursors. In the mouse cochlea, however, IGF-1 does not have an impact on cell number at P5, suggesting that other members of the family, possibly IGF-2, compensate IGF-1 actions during early development (Burns and Hassan, 2001; Efstratiadis, 1998; Camarero, Vigil and Varela-Nieto, unpublished results). In contrast, during the mouse inner ear postnatal period of maturation, from birth to P20, the neurons of the cochlear ganglia become strictly dependent on IGF-1 and its deficit causes retarded maturation and decreased cell survival (Camarero *et al.*, 2001, 2002). About 3500 cochlear neurons are lost during this period in the cochlear ganglia of null mice compared with the wild-type control. Cell loss occurs by apoptosis, which is associated with an increase in the activated form of caspase-3. It decreases from the base to the apex of the cochlea from P5 to P20, which suggests that the more mature neurons are more affected by IGF-1 deficit (Camarero *et al.*, 2001, 2002).

2. Actions of IGF-1 on Cell Differentiation and Myelination

IGFs enhance the differentiation of specific sets of neural populations (reviewed in Varela-Nieto *et al.*, 2003). IGF-1 deficiency causes decreased neuronal differentiation in the mouse cochlear ganglion and an aberrant synaptogenesis at the organ of Corti, suggesting that IGF-1 is required to reach full function. IGF-1 also plays a significant role in myelination during nervous system development. Total myelin content is increased in the brain of IGF-1 transgenic mice, and it is significantly reduced in all brain regions of developing IGF-1 null mice (Beck *et al.*, 1995; Carson *et al.*, 1993; Cheng and Feldman, 1998; Gao *et al.*, 1999; Ye *et al.*, 1995). In the inner ear, IGF-1 deficiency is associated with a sustained deficit in the cochlear nerve and ganglia myelination (Camarero *et al.*, 2001, 2002). Damage of cochlear ganglion neurons may be deleterious to hair cells and vice versa (Duan *et al.*, 2000). Therefore, the alterations observed in cochlear neurons of *Igf1* null mice could compromise midterm survival of hair cells of the organ of Corti.

3. Actions of IGF-1 on Neuroprotection and Regeneration

Hearing and balance impairment caused by hair cell loss or dysfunction is a high-prevalence multifactorial disease that currently has no restorative treatment available. IGF-1 and insulin alone or in combination with other growth and neurotrophic factors protects otic cells from ototoxic damage and promotes hair cell regeneration (see Chapter 14 by Ryan). When considering insulin actions, it should be noted that insulin, a typical component of cell culture media, effectively binds the high affinity receptor for IGF-1; therefore, many of the reported actions of insulin could in fact be due to the activation of IGFR1. In the chick and rodent inner ear, IGFs increase vestibular hair cell proliferation and differentiation and protect cochlear hair cells from ototoxic damage (Duan *et al.*, 2000; Kanzaki *et al.*, 2002; Kopke *et al.*, 2001; Kuntz and Oesterle, 1998; Malgrange *et al.*, 2002; Oesterle *et al.*,

1997; Romand and Chardin, 1999; Stacey and McLean, 2000; Staecker and Van De Water, 1998). The potential of IGF-1 for the treatment of neurodegenerative diseases together with the reported actions of IGF-1 in inner ear development and function support the hypothesis that this factor is a good candidate for inner ear regeneration therapy.

D. Other Diffusible Factors

Other families of factors have a potential importance in early development of neural and sensory structures of the ear. Information on these factors is still very limited and this chapter briefly enumerates some of them.

Sonic hedgehog (Shh) is secreted by the notocord and/or the floor plate; has a transmembrane receptor, Patched; and has been reported to regulate auditory cell fate in the mouse inner ear. Mice mutant for *Shh* present an altered morphogenesis of the cochlear duct and the CVG is not formed. In addition, Hedgehog signaling is required for correct anterior–posterior patterning in the zebra fish otic vesicle (Hammond *et al.*, 2003; Riccomagno *et al.*, 2002; Stone *et al.*, 1996).

Glial cell line–derived neurotrophic factor (GDNF) promotes survival of specific subsets of neurons, including inner ear sensory neurons (see also Chapter 7 by Pirvola and Ylikoski). GDNF actions are mediated by the Ret receptor tyrosine kinase that associates to the coreceptor GFRα1 (Trupp *et al.*, 1995). *Ret* expression is modulated during chick development (Hashino *et al.*, 1999b), and developing vestibular ganglion neurons become sensitive to GDNF after target innervation (Hashino *et al.*, 1999a). Actions of GDNF seem to be very important after target innervation (Hashino *et al.*, 1999a).

Bone morphogenetic protein-4 (Bmp4) is expressed during multiple stages of development in the chick inner ear in association with presumptive sensory organ and semicircular canal formation (Cole *et al.*, 2000; Mowbray *et al.*, 2001; Wu and Oh, 1996). Nogging blockade of the BMP signal produces anomalies in semicircular canal formation and structural patterning of the cristae (Chang *et al.*, 1999). Bmp4 expression is tightly regulated by the homeobox gene *Dlx5*; absence of this pathway is associated in the mouse embryo with abnormal vestibular morphogenesis (Chang *et al.*, 1999; Merlo *et al.*, 2002). Bmp4 and 7 are expressed early in the otic placode and otic vesicle (Oh *et al.*, 1996; Wu and Oh, 1996), but no information is yet available on the effects of Bmp on otic neurons.

Ciliary-derived neurotrophic factor (CNTF) and other yet undetermined factors may be important for early differentiation and survival as suggested by experiments with otic-conditioned media (Bianchi and Cohan, 1993; Bianchi *et al.*, 1998).

IV. Conclusions

Ear neurons are specified early in otic development at the otic placode stage. Neurons are originated from proliferation of a multipotent precursor. Neuronal precursors are determined by the expression of *Ngn1* and singled out by the subsequent operation of the Delta–Notch signaling pathway. The proliferative and undifferentiated state of the progenitor is maintained by unknown signals. Nascent neuroblasts in the epithelium express the proneural genes NeuroD and NeuroM; epithelial neuroblasts delaminate generating ganglionar neuroblasts that populate the cochleovestibular ganglion. This is a transit-amplifying cell population that typically expresses Islet-1/2 and Tuj1 and shows a strict dependence on IGF-1. Ganglionar neuroblasts develop into immature neural precursors that express TrkB and TrkC neurotrophin receptors and depend on neurotrophins for final differentiation and survival. FGFs, IGF-1, and neurotrophins interact to ensure the full differentiation program of ear neurons.

Acknowledgments

This work was supported in part by grants PM99-0111, BMC2001-2132-C02-02, BMC2002-00355, and G03 from the Dirección General de Investigación, Ciencia y Tecnología and Instituto de Salud Carlos III. Islet-1/2 monoclonal antibody was obtained from the Developmental Studies Hybridoma Bank under the auspices of the National Institute of Child Health and Human Development and maintained by the University of Iowa, Department of Biological Sciences, Iowa City, Iowa. We thank Dr. Guadalupe Camarero for sharing unpublished data and Drs. Thomas Schimmang and Yolanda Leon for critical comments on the manuscript.

References

Abu-Elmagd, M., Ishii, Y., Cheung, M., Rex, M., Le Rouedec, D., and Scotting, P. J. (2001). cSox3 expression and neurogenesis in the epibranchial placodes. *Dev. Biol.* **237**, 258–269.

Adam, J., Myat, A., Le Roux, I., Eddison, M., Henrique, D., *et al.* (1998). Cell fate choices and the expression of Notch, Delta and Serrate homologues in the chick inner ear: Parallels with *Drosophila* sense-organ development. *Development* **125**, 4645–4654.

Adamska, M., Herbrand, H., Adamski, M., Kruger, M., Braun, T., and Bober, E. (2001). FGFs control the patterning of the inner ear but are not able to induce the full ear program. *Mech. Dev.* **109**, 303–313.

Adamson, C. L., Reid, M. A., and Davis, R. L. (2002). Opposite actions of brain-derived neurotrophic factor and neurotrophin-3 on firing features and ion channel composition of murine spiral ganglion neurons. *J. Neurosci.* **22**, 1385–1396.

Alvarez, I. S., Martin-Partido, G., Rodriguez-Gallardo, L., Gonzalez-Ramos, C., and Navascues, J. (1989). Cell proliferation during early development of the chick embryo otic

anlage: Quantitative comparison of migratory and nonmigratory regions of the otic epithelium. *J. Comp. Neurol.* **290**, 278–288.
Alvarez, I. S., and Navascues, J. (1990). Shaping, invagination, and closure of the chick embryo otic vesicle: Scanning electron microscopic and quantitative study. *Anat. Rec.* **228**, 315–326.
Avila, M. A., Varela-Nieto, I., Romero, G., Mato, J. M., Giraldez, F., et al. (1993). Brain-derived neurotrophic factor and neurotrophin-3 support the survival and neuritogenesis response of developing cochleovestibular ganglion neurons. *Dev. Biol.* **159**, 266–275.
Baker, C. V., and Bronner-Fraser, M. (2001). Vertebrate cranial placodes I. Embryonic induction. *Dev. Biol.* **232**, 1–61.
Barrett, G. L. (2000). The p75 neurotrophin receptor and neural apoptosis. *Prog. Neurobiol.* **61**, 205–229.
Beck, K. D., Powell-Braxton, L., Widmer, H. R., Valverde, J., and Hefti, F. (1995). Igf1 gene disruption results in reduced brain size, CNS hypomyelination, and loss of hippocampal granule and striatal parvalbumin-containing neurons. *Neuron* **14**, 717–730.
Begbie, J., Ballivef, M., and Graham, A. (2002). Early steps in the production of sensory neurons by the neurogenic placodes. *Mol. Cell Neurosci.* **21**, 502–511.
Begbie, J., and Graham, A. (2001). The ectodermal placodes: A dysfunctional family. *Philos. Trans. R. Soc. Lond. B Biol. Sci.* **356**, 1655–1660.
Behringer, R. R., Lewin, T. M., Quaife, C. J., Palmiter, R. D., Brinster, R. L., and D'Ercole, A. J. (1990). Expression of insulin-like growth factor I stimulates normal somatic growth in growth hormone-deficient transgenic mice. *Endocrinology* **127**, 1033–1040.
Bernd, P., and Li, R. (1999). Differential expression of trkC mRNA in the chicken embryo from gastrulation to development of secondary brain vesicles. *Brain Res. Dev. Brain Res.* **116**, 205–209.
Bernd, P., and Represa, J. (1989). Characterization and localization of nerve growth factor receptors in the embryonic otic vesicle and cochleovestibular ganglion. *Dev. Biol.* **134**, 11–20.
Bertrand, N., Castro, D. S., and Guillemot, F. (2002). Proneural genes and the specification of neural cell types. *Nat. Rev. Neurosci.* **3**, 517–530.
Bevan, P. (2002). Insulin signalling. *J. Cell Sci.* **114**, 1429–1430. (See poster on cell science at a glance, http://www.biologists.com/jcs.)
Bianchi, L. M., and Cohan, C. S. (1993). Effects of the neurotrophins and CNTF on developing statoacoustic neurons: Comparison with an otocyst-derived factor. *Dev. Biol.* **159**, 353–365.
Bianchi, L. M., Dolnick, R., Medd, A., and Cohan, C. S. (1998). Developmental changes in growth factors released by the embryonic inner ear. *Exp. Neurol.* **150**, 98–106.
Bibel, M., and Barde, Y. A. (2000). Neurotrophins: Key regulators of cell fate and cell shape in the vertebrate nervous system. *Genes Dev.* **14**, 2919–2937.
Brigande, J. V., Kiernan, A. E., Gao, X., Iten, L. E., and Fekete, D. M. (2000). Molecular genetics of pattern formation in the inner ear: Do compartment boundaries play a role? *Proc. Natl. Acad. Sci. USA* **97**, 11700–11706.
Brumwell, C. L., Hossain, W. A., Morest, D. K., and Bernd, P. (2000). Role for basic fibroblast growth factor (FGF-2) in tyrosine kinase (TrkB) expression in the early development and innervation of the auditory receptor: In vitro and in situ studies. *Exp. Neurol.* **162**, 121–145.
Brunet, J. F., and Ghysen, A. (1999). Deconstructing cell determination: Proneural genes and neuronal identity. *Bioessays* **21**, 313–318.
Burke, D., Wilkes, D., Blundell, T. L., and Malcolm, S. (1998). Fibroblast growth factor receptors: Lessons from the genes. *Trends Biochem. Sci.* **23**, 59–62.
Burns, J. L., and Hassan, A. B. (2001). Cell survival and proliferation are modified by insulin-like growth factor 2 between days 9 and 10 of mouse gestation. *Development* **128**, 3819–3830.

Camarero, G., Avendano, C., Fernandez-Moreno, C., Villar, A., and Contreras, J. (2001). Delayed inner ear maturation and neuronal loss in postnatal Igf-1-deficient mice. *J. Neurosci.* **21**, 7630–7641.

Camarero, G., Leon, Y., Villar, A., Gorospe, I., De Pablo, F., *et al.* (2003). Insulin-like growth factor 1 is required for survival of transit-amplifying neuroblasts and differentiation of otic neurons. *Dev. Biol.* (In press.)

Camarero, G., Villar, M. A., Contreras, J., Fernandez-Moreno, C., Pichel, J. G., *et al.* (2002). Cochlear abnormalities in insulin-like growth factor-1 mouse mutants. *Hear. Res.* **170**, 2–11.

Campuzano, S., and Modolell, J. (1992). Patterning of the *Drosophila* nervous system: The achaete-scute gene complex. *Trends Genet.* **8**, 202–208.

Cantos, R., Cole, L. K., Acampora, D., Simeone, A., and Wu, D. K. (2000). Patterning of the mammalian cochlea. *Proc. Natl. Acad. Sci. USA* **97**, 11707–11713.

Carson, M. J., Behringer, R. R., Brinster, R. L., and McMorris, F. A. (1993). Insulin-like growth factor I increases brain growth and central nervous system myelination in transgenic mice. *Neuron* **10**, 729–740.

Casaccia-Bonnefil, P., Gu, C., Khursigara, G., and Chao, M. V. (1999). p75 neurotrophin receptor as a modulator of survival and death decisions. *Microsc. Res. Tech.* **45**, 217–224.

Chang, W., Nunes, F. D., Jesus-Escobar, J. M., Harland, R., and Wu, D. K. (1999). Ectopic noggin blocks sensory and nonsensory organ morphogenesis in the chicken inner ear. *Dev. Biol.* **216**, 369–381.

Chao, M. V., and Bothwell, M. (2002). Neurotrophins: To cleave or not to cleave. *Neuron* **33**, 9–12.

Cheng, H. L., and Feldman, E. L. (1998). Bidirectional regulation of p38 kinase and c-Jun N-terminal protein kinase by insulin-like growth factor-I. *J. Biol. Chem.* **273**, 14560–14565.

Cho, Y., Gong, T. W., Stover, T., Lomax, M. I., and Altschuler, R. A. (2002). Gene expression profiles of the rat cochlea, cochlear nucleus, and inferior colliculus. *J. Assoc. Res. Otolaryngol.* **3**, 54–67.

Cole, L. K., Le Roux, I., Nunes, F., Laufer, E., Lewis, J., and Wu, D. K. (2000). Sensory organ generation in the chicken inner ear: Contributions of bone morphogenetic protein 4, serrate 1, and lunatic fringe. *J. Comp. Neurol.* **424**, 509–520.

Colvin, J. S., Feldman, B., Nadeau, J. H., Goldfarb, M., and Ornitz, D. M. (1999). Genomic organization and embryonic expression of the mouse fibroblast growth factor 9 gene. *Dev. Dyn.* **216**, 72–88.

D'Amico-Martel, A. (1982). Temporal patterns of neurogenesis in avian cranial sensory and autonomic ganglia. *Am. J. Anat.* **163**, 351–372.

D'Amico-Martel, A., and Noden, D. M. (1983). Contributions of placodal and neural crest cells to avian cranial peripheral ganglia. *Am. J. Anat.* **166**, 445–468.

Dobrowsky, R. T., Werner, M. H., Castellino, A. M., Chao, M. V., and Hannun, Y. A. (1994). Activation of the sphingomyelin cycle through the low-affinity neurotrophin receptor. *Science* **265**, 1596–1599.

Dore, S., Kar, S., and Quirion, R. (1997). Rediscovering an old friend, IGF-I: Potential use in the treatment of neurodegenerative diseases. *Trends Neurosci.* **20**, 326–331.

Drago, J., Murphy, M., Carroll, S. M., Harvey, R. P., and Bartlett, P. F. (1991). Fibroblast growth factor-mediated proliferation of central nervous system precursors depends on endogenous production of insulin-like growth factor I. *Proc. Natl. Acad. Sci. USA* **88**, 2199–2203.

Duan, M., Agerman, K., Ernfors, P., and Canlon, B. (2000). Complementary roles of neurotrophin 3 and a N-methyl-D-aspartate antagonist in the protection of noise and aminoglycoside-induced ototoxicity. *Proc. Natl. Acad. Sci. USA* **97**, 7597–7602.

Eddison, M., Le Roux, I., and Lewis, J. (2000). Notch signaling in the development of the inner ear: Lessons from *Drosophila*. *Proc. Natl. Acad. Sci. USA* **97**, 11692–11699.

Edlund, T., and Jessell, T. M. (1999). Progression from extrinsic to intrinsic signaling in cell fate specification: A view from the nervous system. *Cell* **96**, 211–224.

Efstratiadis, A. (1998). Genetics of mouse growth. *Int. J. Dev. Biol.* **42**, 955–976.

Ericson, J., Thor, S., Edlund, T., Jessell, T. M., and Yamada, T. (1992). Early stages of motor neuron differentiation revealed by expression of homeobox gene Islet-1. *Science* **256**, 1555–1560.

Fekete, D. M., Homburger, S. A., Waring, M. T., Riedl, A. E., and Garcia, L. F. (1997). Involvement of programmed cell death in morphogenesis of the vertebrate inner ear. *Development* **124**, 2451–2461.

Fekete, D. M., and Wu, D. K. (2002). Revisiting cell fate specification in the inner ear. *Curr. Opin. NeuroBiol.* **12**, 35–42.

Frade, J. M. (2000). Unscheduled re-entry into the cell cycle induced by NGF precedes cell death in nascent retinal neurones. *J. Cell. Sci.* **113**, 1139–1148.

Frade, J. M., and Barde, Y. A. (1998). Nerve growth factor: Two receptors, multiple functions. *Bioessays* **20**, 137–145.

Frago, L. M., Camerero, G., Canon, S., Paneda, C., Sanz, C., *et al.* (2000). Role of diffusible and transcription factors in inner ear development: Implications in regeneration. *Histol. Histopathol.* **15**, 657–666.

Frago, L. M., Canon, S., De La Rosa, E. J., Leon, Y., and Varela-Nieto, I. (2003). Programmed cell death in the developing inner ear is balanced by nerve growth factor and insulin-like growth factor I. *J. Cell Sci.* **116**, 475–486.

Frago, L. M., Leon, Y., De La Rosa, E. J., Gomez-Munoz, A., and Varela-Nieto, I. (1998). Nerve growth factor and ceramides modulate cell death in the early developing inner ear. *J. Cell Sci.* **111**, 549–556.

Fritzsch, B., and Beisel, K. W. (2001). Evolution and development of the vertebrate ear. *Brain Res. Bull.* **55**, 711–721.

Fritzsch, B., Silos-Santiago, I., Bianchi, L. M., and Farinas, I. (1997). The role of neurotrophic factors in regulating the development of inner ear innervation. *Trends Neurosci.* **20**, 159–164.

Gao, W. Q., Shinsky, N., Ingle, G., Beck, K., Elias, K. A., and Powell-Braxton, L. (1999). IGF-I deficient mice show reduced peripheral nerve conduction velocities and decreased axonal diameters and respond to exogenous IGF-I treatment. *J. Neurobiol.* **39**, 142–152.

Garofalo, R. S. (2002). Genetic analysis of insulin signaling in *Drosophila*. *Trends Endocrinol. Metab.* **13**, 156–162.

Golub, R., Adelman, Z., Clementi, J., Weiss, R., Bonasera, J., and Servetnick, M. (2000). Evolutionarily conserved and divergent expression of members of the FGF receptor family among vertebrate embryos, as revealed by FGFR expression patterns in *Xenopus*. *Dev. Genes Evol.* **210**, 345–357.

Goodyear, R. J., Kwan, T., Oh, S. H., Raphael, Y., and Richardson, G. P. (2001). The cell adhesion molecule BEN defines a prosensory patch in the developing avian otocyst. *J. Comp. Neurol.* **434**, 275–288.

Gu, C., Casaccia-Bonnefil, P., Srinivasan, A., and Chao, M. V. (1999). Oligodendrocyte apoptosis mediated by caspase activation. *J. Neurosci.* **19**, 3043–3049.

Haddon, C., Jiang, Y. J., Smithers, L., and Lewis, J. (1998). Delta-Notch signalling and the patterning of sensory cell differentiation in the zebrafish ear: Evidence from the mind bomb mutant. *Development* **125**, 4637–4644.

Hamburger, V., and Hamilton, H. L. (1992). A series of normal stages in the development of the chick embryo. 1951. *Dev. Dyn.* **195**, 231–272.

Hammond, K. L., Loynes, H. E., Folarin, A. A., Smith, J., and Whitfield, T. T. (2003). Hedgehog signalling is required for correct anteroposterior patterning of the zebrafish otic vesicle. *Development* **130**, 1403–1417.

Harrington, A. W., Kim, J. Y., and Yoon, S. O. (2002). Activation of Rac GTPase by p75 is necessary for c-jun N-terminal kinase-mediated apoptosis. *J. Neurosci.* **22**, 156–166.

Hashino, E., Dolnick, R. Y., and Cohan, C. S. (1999a). Developing vestibular ganglion neurons switch trophic sensitivity from BDNF to GDNF after target innervation. *J. Neurobiol.* **38**, 414–427.

Hashino, E., Johnson, E. M., Jr., Milbrandt, J., Shero, M., Salvi, R. J., and Cohan, C. S. (1999b). Multiple actions of neurturin correlate with spatiotemporal patterns of Ret expression in developing chick cranial ganglion neurons. *J. Neurosci.* **19**, 8476–8486.

Hemond, S. G., and Morest, D. K. (1991). Ganglion formation from the otic placode and the otic crest in the chick embryo: Mitosis, migration, and the basal lamina. *Anat. Embryol. (Berl)* **184**, 1–13.

Hempstead, B. L. (2002). The many faces of p75NTR. *Curr. Opin. Neurobiol.* **12**, 260–267.

Hidalgo-Sanchez, M., Alvarado-Mallart, R., and Alvarez, I. S. (2000). Pax2, Otx2, Gbx2 and Fgf8 expression in early otic vesicle development. *Mech. Dev.* **95**, 225–229.

Hirata, H., Hibasami, H., Yoshida, T., Ogawa, M., Matsumoto, M., *et al.* (2001). Nerve growth factor signaling of p75 induces differentiation and ceramide-mediated apoptosis in Schwann cells cultured from degenerating nerves. *Glia* **36**, 245–258.

Hobert, O., and Westphal, H. (2000). Functions of LIM-homeobox genes. *Trends Genet.* **16**, 75–83.

Hossain, W. A., Brumwell, C. L., and Morest, D. K. (2002). Sequential interactions of fibroblast growth factor-2, brain-derived neurotrophic factor, neurotrophin-3, and their receptors define critical periods in the development of cochlear ganglion cells. *Exp. Neurol.* **175**, 138–151.

Hossain, W. A., Zhou, X., Rutledge, A., Baier, C., and Morest, D. K. (1996). Basic fibroblast growth factor affects neuronal migration and differentiation in normotypic cell cultures from the cochleovestibular ganglion of the chick embryo. *Exp. Neurol.* **138**, 121–143.

Huang, E. J., and Reichardt, L. F. (2001). Neurotrophins: Roles in neuronal development and function. *Annu. Rev. Neurosci.* **24**, 677–736.

Itoh, M., Kim, C. H., Palardy, G., Oda, T., Jiang, Y. J., *et al.* (2003). Mind bomb is a ubiquitin ligase that is essential for efficient activation of notch signaling by delta. *Dev. Cell* **4**, 67–82.

Jimenez, F., and Modolell, J. (1993). Neural fate specification in *Drosophila*. *Curr. Opin. Genet. Dev.* **3**, 626–632.

Justice, N. J., and Jan, Y. N. (2002). Variations on the Notch pathway in neural development. *Curr. Opin. NeuroBiol.* **12**, 64–70.

Kaneto, H., Sharma, A., Suzuma, K., Laybutt, D. R., Xu, G., *et al.* (2002). Induction of c-Myc expression suppresses insulin gene transcription by inhibiting NeuroD/BETA2-mediated transcriptional activation. *J. Biol. Chem.* **277**, 12998–13006.

Kanzaki, S., Beyer, L. A., Canlon, B., Meixner, W. M., and Raphael, Y. (2002). The cytocaud: A hair cell pathology in the waltzing Guinea pig. *Audiol. Neurootol.* **7**, 289–297.

Karabagli, H., Karabagli, P., Ladher, R. K., and Schoenwolf, G. C. (2002). Comparison of the expression patterns of several fibroblast growth factors during chick gastrulation and neurulation. *Mech. Dev.* **78**, 159–163.

Kettunen, P., Laurikkala, J., Itaranta, P., Vainio, S., Itoh, N., and Thesleff, I. (2000). Associations of FGF-3 and FGF-10 with signaling networks regulating tooth morphogenesis. *Dev. Dyn.* **219**, 322–332.

Kim, W. Y., Fritzsch, B., Serls, A., Bakel, L. A., Huang, E. J., *et al.* (2001). NeuroD-null mice are deaf due to a severe loss of the inner ear sensory neurons during development. *Development* **128**, 417–426.

Kopke, R. D., Jackson, R. L., Li, G., Rasmussen, M. D., Hoffer, M. E., et al. (2001). Growth factor treatment enhances vestibular hair cell renewal and results in improved vestibular function. Proc. Natl. Acad. Sci. USA **98**, 5886–5891.

Kuntz, A. L., and Oesterle, E. C. (1998). Transforming growth factor alpha with insulin stimulates cell proliferation in vivo in adult rat vestibular sensory epithelium. J. Comp. Neurol. **399**, 413–423.

Kwak, S. J., Phillips, B. T., Heck, R., and Riley, B. B. (2002). An expanded domain of fgf3 expression in the hindbrain of zebrafish valentino mutants results in mis-patterning of the otic vesicle. Development **129**, 5279–5287.

Lanford, P. J., Lan, Y., Jiang, R., Lindsell, C., Weinmaster, G., et al. (1999). Notch signalling pathway mediates hair cell development in mammalian cochlea. Nat. Genet. **21**, 289–292.

Lang, H., and Fekete, D. M. (2001). Lineage analysis in the chicken inner ear shows differences in clonal dispersion for epithelial, neuronal, and mesenchymal cells. Dev. Biol. **234**, 120–137.

Lee, F. S., Kim, A. H., Khursigara, G., and Chao, M. V. (2001). The uniqueness of being a neurotrophin receptor. Curr. Opin. Neurobiol. **11**, 281–286.

Lee, J. E. (1997). Basic helix-loop-helix genes in neural development. Curr. Opin. Neurobiol. **7**, 13–20.

Lee, K. H., and Cotanche, D. A. (1996). Potential role of bFGF and retinoic acid in the regeneration of chicken cochlear hair cells. Hear. Res. **94**, 1–13.

Lee, R., Kermani, P., Teng, K. K., and Hempstead, B. L. (2001). Regulation of cell survival by secreted proneurotrophins. Science **294**, 1945–1948.

Leger, S., and Brand, M. (2002). Fgf8 and Fgf3 are required for zebrafish ear placode induction, maintenance and inner ear patterning. Mech. Dev. **119**, 91–108.

Leon, Y., Sanz, C., Frago, L. M., Camarero, G., Canon, S., et al. (1999). Involvement of insulin-like growth factor-I in inner ear organogenesis and regeneration. Horm. Metab. Res. **31**, 126–132.

Leon, Y., Sanz, C., Giraldez, F., and Varela-Nieto, I. (1998). Induction of cell growth by insulin and insulin-like growth factor-I is associated with Jun expression in the otic vesicle. J. Comp. Neurol. **398**, 323–332.

Leon, Y., Vazquez, E., Sanz, C., Vega, J. A., Mato, J. M., et al. (1995). Insulin-like growth factor-I regulates cell proliferation in the developing inner ear, activating glycosyl-phosphatidylinositol hydrolysis and Fos expression. Endocrinology **136**, 3494–3503.

Lewis, A. K., Frantz, G. D., Carpenter, D. A., de Sauvage, F. J., and Gao, W. Q. (1998). Distinct expression patterns of notch family receptors and ligands during development of the mammalian inner ear. Mech. Dev. **78**, 159–163.

Liepinsh, E., Ilag, L. L., Otting, G., and Ibanez, C. F. (1997). NMR structure of the death domain of the p75 neurotrophin receptor. EMBO J. **16**, 4999–5005.

Liu, J. L., Grinberg, A., Westphal, H., Sauer, B., Accili, D., et al. (1998). Insulin-like growth factor-I affects perinatal lethality and postnatal development in a gene dosage-dependent manner: Manipulation using the Cre/loxP system in transgenic mice. Mol. Endocrinol. **12**, 1452–1462.

Liu, J. P., Baker, J., Perkins, A. S., Robertson, E. J., and Efstratiadis, A. (1993). Mice carrying null mutations of the genes encoding insulin-like growth factor I (Igf-1) and type 1 IGF receptor (Igf1r). Cell **75**, 59–72.

Liu, M., Pleasure, S. J., Collins, A. E., Noebels, J. L., Naya, F. J., et al. (2000). Loss of BETA2/NeuroD leads to malformation of the dentate gyrus and epilepsy. Proc. Natl. Acad. Sci. USA **97**, 865–870.

Ma, Q., Anderson, D. J., and Fritzsch, B. (2000). Neurogenin 1 null mutant ears develop fewer, morphologically normal hair cells in smaller sensory epithelia devoid of innervation. *J. Assoc. Res. Otolaryngol.* **1**, 129–143.

Ma, Q., Fode, C., Guillemot, F., and Anderson, D. J. (1999). Neurogenin 1 and neurogenin 2 control two distinct waves of neurogenesis in developing dorsal root ganglia. *Genes Dev.* **13**, 1717–1728.

Mahmood, R., Kiefer, P., Guthrie, S., Dickson, C., and Mason, I. (1995). Multiple roles for FGF-3 during cranial neural development in the chicken. *Development* **121**, 1399–1410.

Majdan, M., Lachance, C., Gloster, A., Aloyz, R., Zeindler, C., et al. (1997). Transgenic mice expressing the intracellular domain of the p75 neurotrophin receptor undergo neuronal apoptosis. *J. Neurosci.* **17**, 6988–6998.

Malgrange, B., Rigo, J. M., Coucke, P., Thiry, M., Hans, G., et al. (2002). Identification of factors that maintain mammalian outer hair cells in adult organ of Corti explants. *Hear. Res.* **170**, 48–58.

Mansour, S. L., Goddard, J. M., and Capecchi, M. R. (1993). Mice homozygous for a targeted disruption of the proto-oncogene int-2 have developmental defects in the tail and inner ear. *Development* **117**, 13–28.

Manzanares, M., Locascio, A., and Nieto, M. A. (2001). The increasing complexity of the Snail gene superfamily in metazoan evolution. *Trends Genet.* **17**, 178–181.

Memberg, S. P., and Hall, A. K. (1995). Dividing neuron precursors express neuron-specific tubulin. *J. Neurobiol.* **27**, 26–43.

Merlo, G. R., Paleari, L., Mantero, S., Zerega, B., Adamska, M., et al. (2002). The Dlx5 homeobox gene is essential for vestibular morphogenesis in the mouse embryo through a BMP4-mediated pathway. *Dev. Biol.* **248**, 157–169.

Mowbray, C., Hammerschmidt, M., and Whitfield, T. T. (2001). Expression of BMP signalling pathway members in the developing zebrafish inner ear and lateral line. *Mech. Dev.* **108**, 179–184.

Myat, A., Henrique, D., Ish-Horowicz, D., and Lewis, J. (1996). A chick homologue of Serrate and its relationship with Notch and Delta homologues during central neurogenesis. *Dev. Biol.* **174**, 233–247.

Nakae, J., Kido, Y., and Accili, D. (2001). Distinct and overlapping functions of insulin and IGF-I receptors. *Endocr. Rev.* **22**, 818–835.

Oesterle, E. C., Tsue, T. T., and Rubel, E. W. (1997). Induction of cell proliferation in avian inner ear sensory epithelia by insulin-like growth factor-I and insulin. *J. Comp. Neurol.* **380**, 262–274.

Oh, S. H., Johnson, R., and Wu, D. K. (1996). Differential expression of bone morphogenetic proteins in the developing vestibular and auditory sensory organs. *J. Neurosci.* **16**, 6463–6475.

Oldham, S., and Hafen, E. (2003). Insulin/IGF and target of rapamycin signaling: A TOR de force in growth control. *Trends Cell Biol.* **13**, 79–85.

Ornitz, D. M., and Itoh, N. (2001). Fibroblast growth factors. *Genome Biol.* **2**, 3005.

Panin, V. M., and Irvine, K. D. (1998). Modulators of Notch signaling. *Semin. Cell Dev. Biol.* **9**, 609–617.

Pauley, S., Wright, T. J., Pirvola, U., Ornitz, D. M., Beisel, K., and Fritzsch, B. (2003). Expression and function of FGF10 in mamalian inner er development. *Dev. Dyn.* **227**, 203–215.

Phillips, B. T., Bolding, K., and Riley, B. B. (2001). Zebrafish fgf3 and fgf8 encode redundant functions required for otic placode induction. *Dev. Biol.* **235**, 351–365.

Pirvola, U., Arumae, U., Moshnyakov, M., Palgi, J., Saarma, M., and Ylikoski, J. (1994). Coordinated expression and function of neurotrophins and their receptors in the rat inner ear during target innervation. *Hear. Res.* **75**, 131–144.

Pirvola, U., Hallbook, F., Xing-Qun, L., Virkkala, J., Saarma, M., and Ylikoski, J. (1997). Expression of neurotrophins and Trk receptors in the developing, adult, and regenerating avian cochlea. *J. Neurobiol.* **33**, 1019–1033.

Pirvola, U., Spencer-Dene, B., Xing-Qun, L., Kettunen, P., Thesleff, I., *et al.* (2000). FGF/FGFR-2(IIIb) signaling is essential for inner ear morphogenesis. *J. Neurosci.* **20**, 6125–6134.

Pirvola, U., Ylikoski, J., Trokovic, R., Hebert, J. M., McConnell, S. K., and Partanen, J. (2002). FGFR1 is required for the development of the auditory sensory epithelium. *Neuron* **35**, 671–680.

Powell-Braxton, L., Hollingshead, P., Warburton, C., Dowd, M., Pitts-Meek, S., *et al.* (1993). IGF-I is required for normal embryonic growth in mice. *Genes Dev.* **7**, 2609–2617.

Powers, C. J., McLeskey, S. W., and Wellstein, A. (2000). Fibroblast growth factors, their receptors and signaling. *Endocr. Relat. Cancer* **7**, 165–197.

Raballo, R., Rhee, J., Lyn-Cook, R., Leckman, J. F., Schwartz, M. L., and Vaccarino, F. M. (2000). Basic fibroblast growth factor (Fgf2) is necessary for cell proliferation and neurogenesis in the developing cerebral cortex. *J. Neurosci.* **20**, 5012–5023.

Represa, J., Avila, M. A., Miner, C., Giraldez, F., Romero, G., *et al.* (1991a). Glycosylphosphatidylinositol/inositol phosphoglycan: A signaling system for the low-affinity nerve growth factor receptor. *Proc. Natl. Acad. Sci. USA* **88**, 8016–8019.

Represa, J., Avila, M. A., Romero, G., Mato, J. M., Giraldez, F., and Varela-Nieto, I. (1993). Brain-derived neurotrophic factor and neurotrophin-3 induce cell proliferation in the cochleovestibular ganglion through a glycosyl-phosphatidylinositol signaling system. *Dev. Biol.* **159**, 257–265.

Represa, J., Van De Water, T. R., and Bernd, P. (1991b). Temporal pattern of nerve growth factor receptor expression in developing cochlear and vestibular ganglia in quail and mouse. *Anat. Embryol. (Berl)* **184**, 421–432.

Resendes, B. L., Robertson, N. G., Szustakowski, J. D., Resendes, R. J., Weng, Z., and Morton, C. C. (2002). Gene discovery in the auditory system: Characterization of additional cochlear-expressed sequences. *J. Assoc. Res. Otolaryngol.* **3**, 45–53.

Riccomagno, M. M., Martinu, L., Mulheisen, M., Wu, D. K., and Epstein, D. J. (2002). Specification of the mammalian cochlea is dependent on Sonic hedgehog. *Genes Dev.* **16**, 2365–2378.

Richardson, G. P., Crossin, K. L., Chuong, C. M., and Edelman, G. M. (1987). Expression of cell adhesion molecules during embryonic induction. III. Development of the otic placode. *Dev. Biol.* **119**, 217–230.

Ridyard, M. S., and Robbins, S. M. (2003). Fibroblasts growth factor-2-induced signaling through lipid raft-associated fibroblast growth factor receptor substrate 2 (FRS2). *J. Biol. Chem.* **278**, 13803–13809.

Riley, B. B., Chiang, M., Farmer, L., and Heck, R. (1999). The deltaA gene of zebrafish mediates lateral inhibition of hair cells in the inner ear and is regulated by pax2.1. *Development* **126**, 5669–5678.

Romand, R., and Chardin, S. (1999). Effects of growth factors on the hair cells after ototoxic treatment of the neonatal mammalian cochlea in vitro. *Brain Res.* **825**, 46–58.

Rubel, E. W., and Fritzsch, B. (2002). Auditory system development: Primary auditory neurons and their targets. *Annu. Rev. Neurosci.* **25**, 51–101.

Ruben, R. J. (1967). Development of the inner ear of the mouse: A radioautographic study of terminal mitoses. *Acta Otolaryngol.* Suppl. **220**, 1–44.

Saffer, L. D., Gu, R., and Corwin, J. T. (1996). An RT-PCR analysis of mRNA for growth factor receptors in damaged and control sensory epithelia of rat utricles. *Hear. Res.* **94**, 14–23.

6. Growth Factors and Early Development of Otic Neurons

Sanz, C., Leon, Y., Cañon, S., Alvarez, L., Giraldez, F., and Varela-Nieto, I. (1999a). Pattern of expression of the jun family of transcription factors during the early development of the inner ear: Implications in apoptosis. *J. Cell Sci.* **112**, 3967–3974.

Sanz, C., Leon, Y., Troppmair, J., Rapp, U. R., and Varela-Nieto, I. (1999b). Strict regulation of c-Raf kinase levels is required for early organogenesis of the vertebrate inner ear. *Oncogene* **18**, 429–437.

Savage, M. O., Camacho-Hubner, C., Dunger, D. B., Ranke, M. B., Ross, R. J., and Rosenfeld, R. G. (2001). Is there a medical need to explore the clinical use of insulin-like growth factor I? *Growth Horm. IGF Res.* **11 Suppl A**, S65–S69.

Schecterson, L. C., and Bothwell, M. (1994). Neurotrophin and neurotrophin receptor mRNA expression in developing inner ear. *Hear. Res.* **73**, 92–100.

Sokolowski, B. H. (1997). Quantitative analysis of long-term survival and neuritogenesis in vitro: Cochleovestibular ganglion of the chick embryo in BDNF, NT-3, NT-4/5, and insulin. *Exp. Neurol.* **145**, 1–15.

Sommer, L., and Rao, M. (2002). Neural stem cells and regulation of cell number. *Prog. NeuroBiol.* **66**, 1–18.

Stacey, D. J., and McLean, W. G. (2000). Cytoskeletal protein mRNA expression in the chick utricle after treatment in vitro with aminoglycoside antibiotics: Effects of insulin, iron chelators and cyclic nucleotides. *Brain Res.* **871**, 319–332.

Staecker, H., and Van De Water, T. R. (1998). Factors controlling hair-cell regeneration/repair in the inner ear. *Curr. Opin. Neurobiol.* **8**, 480–487.

Stone, D. M., Hynes, M., Armanini, M., Swanson, T. A., *et al.* (1996). The tumour-suppressor gene patched encodes a candidate receptor for Sonic hedgehog. *Nature* **384**, 129–134.

Swanson, G. J., Howard, M., and Lewis, J. (1990). Epithelial autonomy in the development of the inner ear of a bird embryo. *Dev. Biol.* **137**, 243–257.

Tannahill, D., Isaacs, H. V., Close, M. J., Peters, G., and Slack, J. M. (1992). Developmental expression of the *Xenopus* int-2 (FGF-3) gene: Activation by mesodermal and neural induction. *Development* **115**, 695–702.

Torres, M., and Giraldez, F. (1998). The development of the vertebrate inner ear. *Mech. Dev.* **71**, 5–21.

Troy, C. M., Friedman, J. E., and Friedman, W. J. (2002). Mechanisms of p75-mediated death of hippocampal neurons. Role of caspases. *J. Biol. Chem.* **277**, 34295–34302.

Trupp, M., Ryden, M., Jornvall, H., Funakoshi, H., Timmusk, T., *et al.* (1995). Peripheral expression and biological activities of GDNF, a new neurotrophic factor for avian and mammalian peripheral neurons. *J. Cell Biol.* **130**, 137–148.

Vaccarino, F. M., Schwartz, M. L., Raballo, R., Rhee, J., and Lyn-Cook, R. (1999). Fibroblast growth factor signaling regulates growth and morphogenesis at multiple steps during brain development. *Curr. Top. Dev. Biol.* **46**, 179–200.

Valverde, M. A., Sheppard, D. N., Represa, J., and Giraldez, F. (1992). Development of Na(+)- and K(+)-currents in the cochlear ganglion of the chick embryo. *Neuroscience* **51**, 621–630.

Varela-Nieto, I., de la Rosa, E. J., Valenciano, A. I., and Leon, Y. (2003). Cell death in the nervous system: Lessons from insulin and insulin-like growth factors. *Mol. Neurobiol.* **28**, 1–27.

Vega, J. A., San José, I., Cabo, R., Rodriguez, S., and Represa, J. (1999). Trks and p75 genes are differentially expressed in the inner ear of human embryos. What may Trks and p75 null mutant mice suggest on human development? *Neurosci. Lett.* **272**, 103–106.

Vendrell, V., Carnicero, E., Giraldez, F., Alonso, M. T., and Schimmang, T. (2000). Induction of inner ear fate by FGF3. *Development* **127**, 2011–2019.

von Bartheld, C. S., Patterson, S. L., Heuer, J. G., Wheeler, E. F., Bothwell, M., and Rubel, E. W. (1991). Expression of nerve growth factor (NGF) receptors in the developing inner ear of chick and rat. *Development* **113**, 455–470.

von Schack, D., Casademunt, E., Schweigreiter, R., Meyer, M., Bibel, M., and Dechant, G. (2001). Complete ablation of the neurotrophin receptor p75NTR causes defects both in the nervous and the vascular system. *Nat. Neurosci.* **4**, 977–978.

Whitehead, M. C., and Morest, D. K. (1985). The development of innervation patterns in the avian cochlea. *Neuroscience* **14**, 255–276.

Wilkinson, D. G., Bhatt, S., and McMahon, A. P. (1989). Expression pattern of the FGF-related proto-oncogene int-2 suggests multiple roles in fetal development. *Development* **105**, 131–136.

Wilson, S. I., and Edlund, T. (2001). Neural induction: Toward a unifying mechanism. *Nat. Neurosci.* **4 Suppl**, 1161–1168.

Woods, K. A., Camacho-Hubner, C., Barter, D., Clark, A. J., and Savage, M. O. (1997). Insulin-like growth factor I gene deletion causing intrauterine growth retardation and severe short stature. *Acta Paediatr. Suppl.* **423**, 39–45.

Woods, K. A., Camacho-Hubner, C., Savage, M. O., and Clark, A. J. (1996). Intrauterine growth retardation and postnatal growth failure associated with deletion of the insulin-like growth factor I gene. *N. Engl. J. Med.* **335**, 1363–1367.

Wu, D. K., and Oh, S. H. (1996). Sensory organ generation in the chick inner ear. *J. Neurosci.* **16**, 6454–6462.

Xu, X., Weinstein, M., Li, C., Naski, M., Cohen, R. I., *et al.* (1998). Fibroblast growth factor receptor 2 (FGFR2)-mediated reciprocal regulation loop between FGF8 and FGF10 is essential for limb induction. *Development* **125**, 753–765.

Yamasaki, M., Miyake, A., Tagashira, S., and Itoh, N. (1996). Structure and expression of the rat mRNA encoding a novel member of the fibroblast growth factor family. *J. Biol. Chem.* **271**, 15918–15921.

Ye, P., Carson, J., and D'Ercole, A. J. (1995). In vivo actions of insulin-like growth factor-I (IGF-I) on brain myelination: Studies of IGF-I and IGF binding protein-1 (IGFBP-I) transgenic mice. *J. Neurosci.* **15**, 7344–7356.

Zheng, J. L., Helbig, C., and Gao, W. Q. (1997). Induction of cell proliferation by fibroblast and insulin-like growth factors in pure rat inner ear epithelial cell cultures. *J. Neurosci.* **17**, 216–226.

Zhou, X., Hossain, W. A., Rutledge, A., Baier, C., and Morest, D. K. (1996). Basic fibroblast growth factor (FGF-2) affects development of acoustico-vestibular neurons in the chick embryo brain in vitro. *Hear. Res.* **101**, 187–207.

7

Neurotrophic Factors during Inner Ear Development

Ulla Pirvola and Jukka Ylikoski
Institute of Biotechnology and Department of Otolaryngology
University of Helsinki, 00014 Helsinki, Finland

I. Neurotrophic Factors
II. The Neurotrophin System
   A. Neurotrophic Hypothesis
   B. Expression of Neurotrophins and Trk Receptors in the Embryonic Inner Ear
   C. Inner Ear Phenotype of Neurotrophin and Trk Null Mutant Mice
   D. Neurotrophin and Trk Induction in the Early-Developing Inner Ear
   E. Neurotrophins during Postnatal Inner Ear Development
   F. Truncated Trk Receptors during Inner Ear Development
   G. The p75 Receptor
III. The Glial Cell Line–Derived Neurotrophic Factor Family
   A. Glial Cell Line–Derived Neurotrophic Factor and Its Receptors in the Developing Inner Ear
IV. Conclusions
   References

    Neurotrophins are crucial for the development and maintenance of distinct sets of neurons of the central and peripheral nervous systems. Two members of this family, neurotrophin-3 (NT-3) and brain-derived neurotrophic factor (BDNF), are expressed in the developing cochlear and vestibular sensory epithelia. Their cognate tyrosine kinase receptors belonging to the Trk family are expressed in the sensory neurons of the inner ear ganglia. Knockout studies have shown the key role of the neurotrophin system in promoting survival and proper innervation of the embryonic inner ear neurons of the mouse. Recent evidence indicates that although both neurotrophins act as survival factors *in vivo*, BDNF is especially important in the regulation of innervation. Glial cell line–derived neurotrophic factor (GDNF) is a member of another family of neurotrophic factors. *Gdnf*, *Nt-3*, and *Bdnf* expressions show dynamic changes in the early postnatal cochlea of rodents concomitant with the rearrangement of the innervation pattern to the hair cells, suggesting a role for these factors in the remodeling process. The distinct expression of the p75 neurotrophin receptor in specific nonneuronal regions of the rodent inner ear suggests novel functions for this receptor, independent of the Trk receptors, perhaps in stimulating cellular apoptosis as recent data on other model systems suggest. © 2003, Elsevier Inc.

I. Neurotrophic Factors

Neurotrophic factors are defined as endogenous soluble proteins regulating survival, growth, morphological plasticity and synthesis, or synthesis of proteins for differentiated functions of neurons (Hefti et al., 1993). They include the neurotrophins (NTs), ciliary neurotrophic factor, epidermal growth factor, fibroblast growth factors (FGFs), insulin-like growth factor (IGF), and members of the glial cell line–derived neurotrophic factor (GDNF) family. In the developing inner ear, by far the best characterized of these factors are the NTs. Their role in the embryonic inner ear of the mouse has been conclusively shown by gene targeting technology. This chapter concentrates on the role of NTs in the rodent inner ear. In this chapter data on GDNF are also discussed, although much less is known about its role and the role of its receptors in the inner ear. IGF and FGFs and their function in the developing inner ear are discussed in other chapters.

II. The Neurotrophin System

A. Neurotrophic Hypothesis

Classic studies by Hamburger (1934) and colleagues have shown that target cells of innervating neurons produce trophic signals, which promote and maintain neuronal survival and regulate successful innervation. These pioneering studies have demonstrated that neurons of the peripheral nervous system (PNS) are initially produced in vast excess. A proportion of sensory, sympathetic, and motor neurons die during normal development. This naturally occurring death has been termed programmed cell death (PCD) (reviewed by Oppenheim, 1991). Today it is known that PCD proceeds through apoptosis—a gene-regulated mode of cellular death with distinct morphological manifestations, such as chromatin condensation and DNA fragmentation. The early studies laid the backround for the now well-established neurotrophic theory stating that target-derived trophic molecules control the size of the neuronal population, which innervates the target field (reviewed by Korsching, 1993). Analyses of transgenic mice have provided strong evidence for this hypothesis (reviewed by Huang and Reichardt, 2001).

The prototypical NT, nerve growth factor (NGF), was discovered in the mid-twentieth century (reviewed by Levi-Montalcini, 1967). Structurally and functionally highly homologous molecules—BDNF, NT-3, and neurotrophin-4/5 (NT-4/5) (vertebrates)—were characterized much later (reviewed by Bibel and Barde, 2000; Huang and Reichardt, 2001). All NTs share similar biochemical properties. They are small, basic, and secreted proteins. They exist as homodimers and are synthesized as larger proforms, which are proteolytically cleaved to mature forms. Two types of classic *in vitro* assays, the neuronal survival assay using dissociated sensory neurons (Davies and Lindsay, 1985) and the neurite outgrowth assay using ganglion explants cultured in a three-dimensional collagen matrix (Ebendal, 1989), demonstrated first the neuronal survival- and differentiation-promoting activities of NTs. These assays have shown that NTs act on different but partially overlapping neuronal populations. NTs mediate their effects through two types of receptors, the Trk tyrosine kinase receptors (TrkA, TrkB, TrkC) and the neurotrophin receptor p75. The latter belongs to the superfamily of tumour necrosis factor receptors.

B. Expression of Neurotrophins and Trk Receptors in the Embryonic Inner Ear

An early study showed NGF-like immunoreactivity in the rat cochlear hair cells during the first postnatal week (Després *et al.*, 1988). Since later *in situ* hybridization studies have not detected *Ngf* mRNA in the inner ear sensory epithelia, it is possible that this NGF antibody partially cross-reacts with the later characterized NT-3 and/or BDNF. On the other hand, NGF-like immunostaining was not found in the cochlea after the first postnatal week, although *Nt-3* transcripts mRNAs have been detected in the auditory hair cells (HCs) at later stages (Ylikoski *et al.*, 1993). *In situ* hybridizations have shown the expression of *Nt-3* (Figs. 1, 2A and B) and *Bdnf* but not *Ngf* or *Nt-4/5* in the inner ear sensory epithelia, the onset of these expressions starting already in the sensory patches of the midgestational otic vesicle (embryonic day 10 in the mouse). During late embryogenesis and the neonatal period, *Nt-3* and *Bdnf* expressions show distinct spatial segregation, *Nt-3* being expressed in the organ of Corti of the cochlear duct and in the utricular and saccular maculae (see Figs. 1 and 2B), whereas *Bdnf* is found in all sensory epithelia of the developing ear, including the ampullary cristae. NT-3 is the primary NT in the developing and adult cochlea and BDNF is the primary NT in the vestibular organs. In the extending cochlear duct, at stages before the onset of morphological differentiation, *Nt-3* is strongly expressed in the greater epithelial ridge, which comprises the precursor cells of the organ of Corti (from embryonic day 13 onward in the mouse) (see Fig. 2A). In the cochlea at late embryogenesis and neonatal stages, *Nt-3* is expressed in the in the HCs as well as in the adjacent supporting cells. Levels of *Nt-3* expression are more prominent in the differentiating supporting cells than in the differentiating HCs. *Bdnf* expression is confined to HCs in both the developing cochlea and the vestibular organs (Ernfors *et al.*, 1992; Pirvola *et al.*, 1992, 1994; Schecterson and Bothwell, 1994; Wheeler *et al.*, 1994; Ylikoski *et al.*, 1993). *Nt-3* and *Bdnf* expression patterns have been

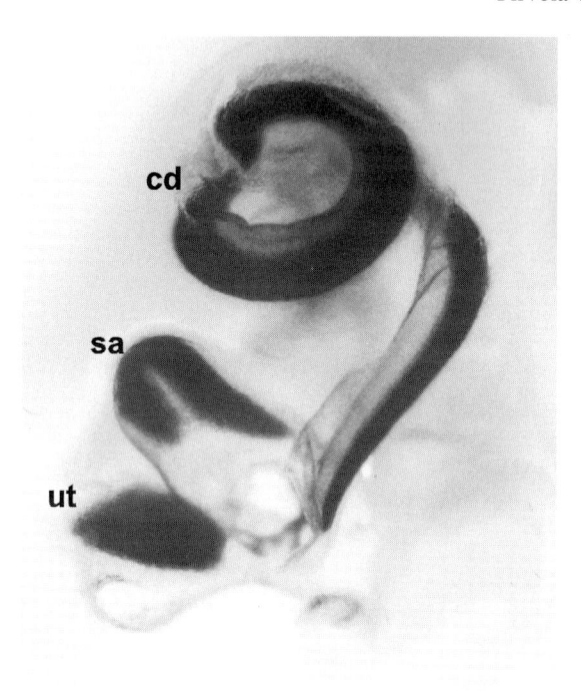

Figure 1 *Nt-3* expression in the inner ear sensory epithelia of an *Nt-3* knockout mouse at birth as revealed by *lacZ* staining. We are grateful to Dr. L. F. Reichardt for donating the *Nt-3* mutant mouse line to us. Cd, cochlear duct; sa, saccular macula; ut, utricular macula. (See Color Insert.)

recently verified by histochemistry using mice in which the *lacZ* reporter gene is inserted into the *Bdnf* and *Nt-3* loci (Farinas et al., 2001).

Consistent with the NT expression patterns, mRNAs coding for the cognate receptors for NT-3 (TrkC) (Fig. 2C) and BDNF (TrkB) have been detected in the sensory neurons of the inner ear ganglia throughout development and during adult life (Ernfors et al., 1992; Pirvola et al., 1994; Schecterson and Bothwell, 1994; Ylikoski et al., 1993). *TrkB* and *TrkC* mRNAs are coexpressed in the cochlear and vestibular neurons (Pirvola et al., 1994; Ylikoski et al., 1993), an observation that was later verified by immunohistochemistry (Farinas et al., 2001). As would be expected from this coexpression, a synergistic (overlapping) rather than additive effect on neuronal survival has been observed when both NTs are added to the culture medium (Mou et al., 1997; Pirvola et al., 1994). The fact that NGF does not have a significant effect on inner ear neurons was actually an expected result since expression of its high-affinity receptor *TrkA* was not found in these neurons, with the exception of a weak and transient expression in the

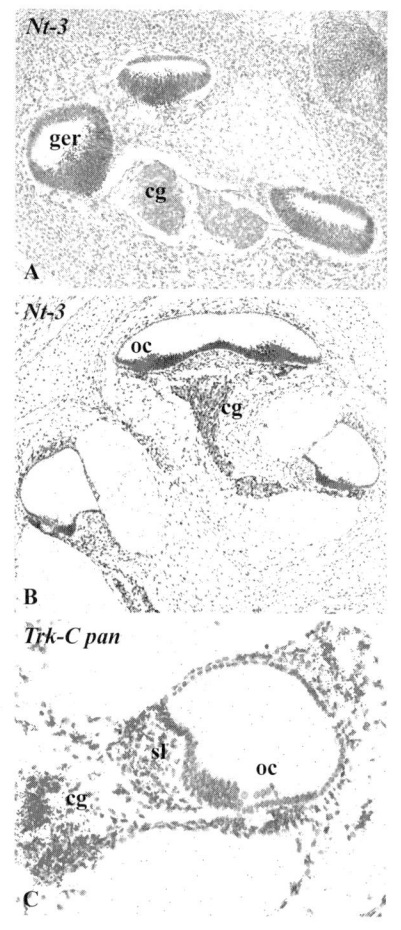

7. Neutrophic Factors during Inner Ear Development 211

Figure 2 *Nt-3* and *TrkC* expressions in the embryonic cochlea of the mouse as revealed by *in situ* hybridization. (A) At embryonic day 13.5, the greater epithelial ridge (ger) of the nascent cochlear duct shows *Nt-3* expression. (B) At birth, *Nt-3* is expressed in the organ of Corti (oc) of the cochlear duct. (C) The *pan-TrkC* probe detects both the catalytic and truncated forms of *TrkC*. Based on the use of isoform-specific probes, the full-length (catalytic) isoform is expressed exclusively in the neurons of the cochlear ganglion (cg). Truncated *TrkC* is expressed, in addition to the ganglion, in mesenchymal regions such as the spiral limbus (sl) and the domain next to the lateral epithelial wall. (See Color Insert.)

early-developing cochleovestibular ganglion (embryonic day 11 in the mouse) (Farinas *et al.*, 2001; Pirvola *et al.*, 1994; Schecterson and Bothwell, 1994). Furthermore, NGF had been earlier shown to act on neural crest–derived, but not on placodal-derived neurons (Davies and Lindsay, 1985). Taken together, these initial studies showed that NT-3 and BDNF and their TrkC and TrkB tyrosine kinase receptors promote survival of embryonic inner ear neurons and elicit neurite outgrowth from the inner ear ganglia *in vitro*.

It is interesting that BDNF is the primary NT in the inner ears of birds and reptiles, whereas NT-3 is nearly absent from their ears, during both development and adulthood (Don *et al.*, 1997; Hallböök *et al.*, 1993; Hallböök and Fritzsch, 1997; Pirvola *et al.*, 1997). As it does in mammals, HC-derived BDNF acts on the innervating sensory neurons of these species. Thus, it appears that the evolution of the coiling hearing organ is associated with the introduction of a novel NT, NT-3.

C. Inner Ear Phenotype of Neurotrophin and Trk Null Mutant Mice

The development of gene-targeting technology provided a crucial tool to test the initial suggestions that NT-3, BDNF, and their Trk receptors are required for the establishment of the afferent innervation of the inner ear. Table I shows the percentages of lost inner ear neurons reported by different research groups in *Nt-3*, *Bdnf*, *TrkB*, and *TrkC* knockout mice, as well as in *Nt-3/Bdnf* and *TrkB/TrkC* double mutants. The analyses have been performed during neonatal life. The inner ear has been found to be one of the most severely affected organs in these mutants. Most dramatically, in *Nt-3/Bdnf* and *TrkB/TrkC* double null mice, all inner ear neurons are lost (Ernfors *et al.*, 1995; Silos-Santiago *et al.*, 1997). Results from the analysis of mutant phenotypes are in line with the expression data, showing that the survival of

Table I Inner Ear Neuronal Losses in *Neurotrophin* and *Trk*-Inactivated Mice as Analyzed at Neonatal Stages

Ganglion	Nt-3−/−	Bdnf−/−	Nt-3−/− Bdnf−/−	Ngf−/−	Nt-4/5−/−	TrkC−/−	TrkB−/−	TrkC−/− TrkB−/−	TrkA−/−
Cochlear	85[a,b]	7[b]	100[b]	ND	NO[c,d]	50–70[e,f]	15[e,f]	100[g]	NO[h]
Vestibular	20–35[a,b]	80[c,i,j]	100[b]	ND	NO[c,d]	15[e,f,k]	56[e,f]	100[g]	NO[h]

[a]Farinas *et al.*, 1994.
[b]Ernfors *et al.*, 1995.
[c]Conover *et al.*, 1995.
[d]Liu *et al.*, 1995.
[e]Schimmang *et al.*, 1995.
[f]Minichiello *et al.*, 1995.
[g]Silos-Santiago *et al.*, 1997.
[h]Huang and Reichardt, 2001.
[i]Ernfors *et al.*, 1994.
[j]Bianchi *et al.*, 1996.
[k]Tessarollo *et al.*, 1997.
ND, Not done; *NO*, not observed.

cochlear neurons depends primarily on NT-3, whereas vestibular neurons depend especially on BDNF. Regarding innervation to the vestibular organs, as would be expected from the expression patterns, *Bdnf* and *TrkB* inactivations lead to complete loss of afferent innervation to the cristae and a reduced and aberrant innervation to the maculae (Bianchi *et al.*, 1996; Fritzsch *et al.*, 1995; Schimmang *et al.*, 1995).

Although different research groups are uniform regarding the extent of neuronal loss in the cochlear (about 85% in *Nt-3* null mutants and 25% in *Bdnf* null mutants) and vestibular ganglia (about 5% in *Nt-3* mutants and 80% in *Bdnf* mutants), there are different interpretations of the effects of the gene inactivations on the innervation patterns to the cochlear HCs and on the survival of the two types of cochlear neuron subclasses. Type I neurons comprise 90–95% of cochlear ganglion neurons and they innervate the inner hair cells (IHCs). The remaining population consists of small-diameter type II neurons, which innervate the outer hair cells (OHCs). Ernfors and collegues have suggested that innervation of type I neurons to IHCs and the survival of these neurons depend on NT-3/TrkC interaction, whereas innervation of type II neurons to OHCs and their survival is selectively dependent on BDNF/TrkB signaling (Agerman *et al.*, 2003; Ernfors *et al.*, 1995; Minichiello *et al.*, 1995; Schimmang *et al.*, 1995). In contrast, Fritzsch and collegues have not found differential dependence of the two cochlear neuron populations on different NTs (Bianchi *et al.*, 1996; Coppola *et al.*, 2001; Farinas *et al.*, 2001; Fritzsch *et al.*, 1995, 1997, 1998). They have found that the mutations cause neuronal loss in longitudinal gradients along the cochlear duct, disruption of NT-3/TrkC signaling having the strongest effect in the basal coil and *Bdnf* and *TrkB* inactivations in the apex. Fritzsch and collegues suggest that these defects are caused by spatiotemporal gradients in *Nt-3* and *Bdnf* expression levels, *Nt-3* being strongly expressed in the basal turn of the developing cochlea and *Bdnf* in the apex (Bianchi *et al.*, 1996; Farinas *et al.*, 2001; Fritzsch *et al.*, 1997). Studies using *lacZ* knock-in mice in the *Nt-3* and *Bdnf* loci support this suggestion of apical–basal differences in *Nt* expressions (Farinas *et al.*, 2001). Despite the differences in the interpretation of the innervation defects in the mutants, both groups agree that since TrkB and TrkC receptors are coexpressed in the cochlear neurons, the specificity is achieved by spatiotemporal differences in ligand expression levels.

Interesting approaches have been used to further dissect NT functions *in vivo*. The *Nt-3* gene has been replaced by *Bdnf* (Coppola *et al.*, 2001; Farinas *et al.*, 2001) and, vice versa, *Bdnf* by *Nt-3* (Agerman *et al.*, 2003). Thus, in these strategies signaling by one Trk receptor is eliminated, whereas signaling by another Trk is preserved. However, it should be noted that although NT-3 preferentially signals through TrkC, it can also signal to some extent through TrkA and TrkB, as shown directly in other peripheral ganglia (Farinas et al., 1998; Huang et al., 1999; Tessarollo et al., 1997). Both gene replacement strategies result in almost complete restoration of cochlear neuron numbers and target innervation. Also in the vestibular system, *Bdnf* replacement by *Nt-3* rescues a large number of vestibular neurons from death (Agerman et al., 2003). Taken together, these data suggest that both BDNF and NT-3 can promote survival of inner ear neurons *in vivo*, these results being in line with the data showing that TrkB and TrkC are coexpressed in the inner ear neurons. Interestingly, in contrast to neuronal death, *Bdnf* replacement by *Nt-3* results in complete lack of innervation to the ampullary cristae and only a partial rescue of innervation to the utricular and saccular maculae. Thus, BDNF/TrkB interaction appears to be essential for terminal innervation and synaptogenesis in the vestibular system (Agerman et al., 2003). These results are likely to be due to the fact that TrkB and TrkC activate partially different intracellular signaling events (Postigo et al., 2002). Also, the analyses of cochleas of mutants in which *Nt-3* is replaced by *Bdnf* point to the role of BDNF in the growth of nerve terminals into contact with HCs (Coppola et al., 2001). These data are consistent with the fact that *Bdnf* expression is confined to HCs, whereas *Nt-3* is also expressed in the noninnervated supporting cells of the cochlea. Although it was first suggested that *TrkC* null mutation and *TrkB/TrkC* double null mutation impair auditory HC development (Minichiello et al., 1995; Schimmang et al., 1995), later analyses indicate that anterograde signaling from the cochlear sensory neurons does not have a major impact on HC development (Farinas et al., 2001; Fritzsch et al., 1997).

D. Neurotrophin and Trk Induction in the Early-Developing Inner Ear

NT/Trk interactions seem not to act on the precursors of inner ear neurons. These precursors delaminate from the otocyst epithelium and migrate to form the cochleovestibular ganglion. They do not express *TrkB* or *TrkC* mRNAs (Farinas et al., 2001; Pirvola et al., 1994; Schecterson and Buthwell, 1994), and no direct defects have been documented at the level of these precursors in *Nt* or *Trk* mutant mice. Initial expression of the full-length (catalytic) isoforms of *TrkB* and *TrkC* receptors is detected at the initial stages of cochleovestibular ganglion compaction, when neuronal differentiation starts (Pirvola et al., 1994). In agreement, analyses of the inner ears of *Nt-3*, *Bdnf*, *TrkB*, and *TrkC*$^{-/-}$ mice show that cochleovestibular neurons are initially formed, but they die apoptotically between E13.5 and E15.5 (Bianchi et al., 1996; Farinas et al., 2001; Schimmang et al., 1995).

Basic helix-loop-helix (bHLH) transcription factors are crucial for early neurogenesis in the PNS (Ma et al., 1998). Of the bHLH factors, *neurogenin1* (*Ngn1*) inactivation blocks the delamination of neuronal precursors from the neurogenic region of the otic vesicle, resulting in a complete failure of inner ear neurons to develop (Ma *et al.*, 1998, 2000). Neurogenins are upstream activators of another bHLH gene, *NeuroD*, which promotes the withdrawal of precursors from the cell cycle and their differentiation into neurons (Lee *et al.*, 1995). *NeuroD* null mice lack most inner ear neurons. This defect results from impaired delamination of neuronal precursors from the otic epithelium and, since a small neuronal population is initially formed, also from the failure of the early neurons to survive. Interestingly, *TrkB* and *TrkC* expressions are downregulated in *NeuroD* mutants (Kim *et al.*, 2001; Liu *et al.*, 2000), suggesting that these genes belong to the same genetic pathway.

First signs of *Nt-3* and *Bdnf* expressions can be seen in the otic vesicle before neurites have penetrated into the epithelium. Since also *Trk* receptors start to be expressed in the neurons of the cochleovestibular ganglion before target innervation, the upregulation of these expressions seems to be independent of each other. Very little is known about the mechanisms regulating induction of NTs in the embryonic ear. In other systems, WNT factors have been shown to induce *Nt-3* expression (Krylova *et al.*, 2002; Patapoutian *et al.*, 1999). A challenge for inner ear researchers is to reveal the genetic pathways composed of intrinsic (transcription factors) and extrinsic (diffusible molecules) factors, which regulate specification toward the neuronal lineage, neuroblast differentiation, and neuronal survival and differentiation.

E. Neurotrophins during Postnatal Inner Ear Development

Nts and *Trk* receptors continue to be expressed in the postnatal inner ear (Wheeler *et al.*, 1994; Ylikoski *et al.*, 1993). In the postnatal vestibular organs, these genes show similar expression patterns to those shown during embryogenesis. In the cochlea, *Nt-3* and *Bdnf* expressions show distinct rearrangements during the first and second postnatal week. By the end of the second postnatal week, *Nt-3* expression, which is found in the IHCs, OHCs, and supporting cells of the neonatal cochlea, is shifted to the IHCs only. This expression persists during adulthood (Ylikoski *et al.*, 1993). During the second postnatal week, *Bdnf* expression is turned off in the cochlear HCs. This time period when qualitative and quantitative changes occur in *Nt* expression coincides with the remarkable remodeling of the cochlear innervation in which the afferent innervation is shifted mainly to the IHCs and in which the majority of efferent fibers originating from the brainstem terminate on the OHCs (Lenoir *et al.*, 1980). Importantly, all these changes shortly precede the onset of hearing function. Thus, the expression data suggest that BDNF and NT-3 are involved in the remodeling of cochlear innervation and synaptogenesis (Wiechers *et al.*, 1999; Ylikoski *et al.*, 1993), perhaps together with GDNF (Ylikoski *et al.*, 1998). In this function, BDNF, which is expressed exclusively in HCs, might be particularly important, taking into account the knowledge that it influences activity-dependent synaptic rearrangements elsewhere in the developing nervous system (Cabelli *et al.*, 1995; Cohen-Cory and Fraser, 1995). Because inactivation of the genes of the NT/Trk system has a strong effect on the embryonic inner ear and since most transgenic mouse lines of this signaling system die around birth, novel methods are required to elucidate function of NTs in the postnatal inner ear. One future method is the conditional spatiotemporally controlled mutagenesis using *Cre-loxP* technology and a tamoxifen-inducible Cre-ER^T, a fusion between Cre recombinase and a mutated hormone-binding domain of the estrogen receptor (Metzger and Chambon, 2001).

F. Truncated Trk Receptors during Inner Ear Development

All *Trk* genes are alternatively spliced, leading to a variety of functionally different receptor proteins. Truncated *TrkB* and *TrkC* isoforms contain identical extracellular and transmembrane domains but different intracellular domains as compared with their full-length counterparts. In truncated *Trk* isoforms, the intracellular tyrosine kinase domain is replaced by a short sequence having no homology to any known protein motifs. This does not affect the ability of truncated Trks to bind NTs. Full-length and truncated *Trk* isoforms are coexpressed in many neurons. Truncated *Trks* are prominently expressed in many nonneuronal cell types where full-length variants are not usually expressed (Barbacid, 1994).

We have investigated the expression of truncated *TrkB* and *TrkC* receptor variants in the inner ear. During embryogenesis and early postnatal life, truncated *TrkB* and *TrkC* are expressed in the cochleovestibular ganglion, in distinct mesenchymal areas surrounding the otic epithelium, and in restricted regions of the otic epithelium. Notably, we have not found expression of truncated (or full-length) *Trk* receptors in the cochlear HCs at any stage (Fig. 2C) (Pirvola *et al.*, 1994; Xing-Qun *et al.*, 1999; Ylikoski *et al.*, 1993). The functional significance of truncated Trk isoforms in the inner ear is unknown. Also, in other tissues, despite the many suggested functions, direct evidence of their roles *in vivo* remains to be shown. Suggested functions include a dominant negative inhibition when truncated and full-length isoforms are coexpressed (neurons), regulation of the availability of NTs to responsive cells, and modulation of neuritic growth (reviewed by Bibel and Barde, 2000).

G. The p75 Receptor

p75 is a pan-NT receptor. It has been suggested to increase the affinity and specificity of NT/Trk interactions, to stimulate axonal growth, and to promote retrograde transport of NTs. Intriguingly, in addition to mediating cell survival when coexpressed with Trk receptors, p75 activation by NTs can induce apoptotic death of neuronal and nonneuronal cells. NTs induce apoptosis through p75 in cells not expressing Trk receptors. Most intriguingly, recent evidence shows that the historical designation of the p75 receptor as a "low-affinity" NT receptor is not entirely appropriate. It has been discovered that the precursor forms of NGF and BDNF bind to p75 with much higher affinity than the mature NTs and that the pro-NTs stimulate p75-mediated apoptosis more efficiently than do their mature forms (Lee *et al.*, 2001).

During otocyst stages, p75 is expressed in distinct domains of the otic epithelium, in the surrounding mesenchyme, and in the cochleovestibular ganglion. During late embryogenesis, it is found in the neurons of the cochlear and vestibular ganglia. Interestingly, p75 shows transient expression in the early-differentiating pillar cells of the organ of Corti during a few days before birth. During postnatal life, inner ear neurons continue to express this receptor (Després *et al.*, 1991; Gestwa *et al.*, 1999; Pirvola *et al.*, 1994, 2002; Schecterson and Bothwell, 1994; von Bartheld *et al.*, 1991; Ylikoski *et al.*, 1993). No reports exist on the inner ear phenotype of p75 null mutant mice. Studies using organotypic cultures of early-developing chick inner ears have shown that exogenous NGF stimulates p75-mediated apoptosis in the otic epithelium and cochleovestibular ganglion (Frago *et al.*, 1998, 2002). Further studies are needed to find out the *in vivo* relevance of this observation. The transient expression of p75 in late-embryonic pillar cells is interesting. Since these supporting cells have not been shown to undergo apoptosis during the period of differentiation, the functional significance of this expression remains obscure.

III. The Glial Cell Line–Derived Neurotrophic Factor Family

GDNF was originally found to be a trophic factor for midbrain dopaminergic neurons (Lin *et al.*, 1993). Later GDNF was shown to have distinct effects on other neuronal populations, during both embryogenesis and postnatal life. GDNF also has important roles outside the nervous system, especially in the developing kidney. Three related members have been discovered: neurturin, artemin, and persephin (reviewed by Airakasinen and Saarma, 2002). All of these factors are distant members of the TGFβ superfamily. GDNF family members promote survival of specific neuronal populations in both the central nervous system (CNS) and the PNS, except persephin, which has not been shown to act on PNS neurons (Milbrandt et al., 1998). GDNF family ligands bind to a receptor complex consisting of a signaling component and a ligand-binding component. The former molecule is the c-RET proto-oncogene, a receptor tyrosine kinase, and the latter is a member of the family of the glycosyl phosphatidylinositol-anchored GDNF receptor α (GFRα1–3) (Airaksinen and Saarma, 2002). GDNF signals preferentially through GFRα1/c-RET.

A. Glial Cell Line–Derived Neurotrophic Factor and Its Receptors in the Developing Inner Ear

We have found strong expression of *c-Ret* in the early-developing rodent inner ear (our unpublished observations). *c-Ret* mRNA is expressed in distinct domains of the otocyst epithelium and in the cochleovestibular ganglion. However, the functional significance of this expression remains unresolved since no morphological changes can be seen in the inner ears of *c-Ret* null mutant mice, as analyzed at birth, at the stage when these mutants die (our unpublished observations). *c-Ret* expression is downregulated in the late-embryonic inner ear. At this stage and during postnatal life, we have not detected *c-Ret* mRNA in the inner ear by *in situ* hybridization (in the same sections as, e.g., the neighboring geniculate ganglion serves as a positive control). In contrast to *c-Ret*, *Gdnf* starts to be expressed in the cochlear HCs at the end of the first postnatal week. This upregulation shows a distinct basal-to-apical coil progression, corresponding to the spatiotemporal gradient of maturation of the organ of Corti. During the second postnatal week, *Gdnf* is expressed in the IHCs and OHCs. Thereafter, the expression is shifted to the IHCs only, where it persists throughout adulthood (Ylikoski et al., 1998). We have found that GDNF efficiently promotes survival of dissociated cochlear neurons of 1-week-old rats. In addition, local delivery of GDNF into the traumatized cochlea of adult guinea pigs prevents neuronal loss (Ylikoski et al., 1998). Based on the expression pattern, in addition to promoting survival of postnatal inner ear neurons, GDNF, together with NTs, might be involved in the remodeling of cochlear innervation and in synaptogenesis. This may be analogous to GDNF's role in the early postnatal neuromuscular junction, where it has been shown to stimulate axonal branching and synaptic remodeling (Keller-Peck et al., 2001). Since we have not detected *c-Ret* mRNA in the inner ear neurons (except at the earliest developmental stages), the question arises as to what is the signaling receptor for GDNF. Cochlear neurons express *GFRα1* (Ylikoski et al., 1998). In neuronal cell lines and in cultured *c-Ret*$^{-/-}$ primary neurons, GFRα1 stimulation by GDNF activates several intracellular signaling events (Poteryaev et al., 1999; Trupp et al., 1997). However, because there is no direct evidence for GDNF signaling in $Ret^{-/-}$ mice, the physiological relevance of c-RET–independent signaling *in vivo* remains to be shown. Another possibility is the existence of a new, still uncharacterized transmembrane receptor for GDNF–GFRα1 (reviewed by Airaksinen and Saarma, 2002). Thus, how GDNF acts on the inner ear neurons is currently unclear.

IV. Conclusions

Neurotrophic factors produced by the inner ear sensory epithelia act against the developmentally regulated apoptotic death of cochlear and vestibular neurons as shown in transgenic mouse models and *in vitro* assays. Neurotrophic factors can also promote survival of adult inner ear neurons, as demonstrated by the delivery of trophic factors into traumatized ears. Although the role of NTs in the innervation process has been established, their specific role in synaptogenesis and their possible role as chemoattractants guiding the primary neurites toward the target field need more clarification. Taking into account the recent progress in the identification of genes that are essential for HC formation during embryogenesis, high hopes exist that this knowledge—perhaps in conjunction with stem cell approaches—could be used to generate new HCs in the adult mammalian inner ear. To be functional, new HCs must establish proper neuronal connections and they might also be needed to prevent neuronal degeneration following traumas. Knowledge of the molecular regulation of NT gene expression in the embryonic precursor cells and of the innervation process during development are likely to be important in understanding how the neuritic regrowth into contact with regenerated HCs is achieved.

Acknowledgment

Our work is supported by the Sigrid Jusélius Foundation.

References

Agerman, K., Hjerling-Leffler, J., Blanchard, M. P., Scarfone, E., Canlon, B., *et al.* (2003). BDNF gene replacement reveals multiple mechanisms for establishing neurotrophin specificity during sensory nervous system development. *Development* **130**, 1479–1491.
Airaksinen, M. S., and Saarma, M. (2002). The GDNF family: Signalling, biological functions and therapeutic value. *Nat. Rev. Neurosci.* **3**, 383–394.
Barbacid, M. (1994). The Trk family of neurotrophin receptors. *J. Neurobiol.* **25**, 1386–1403.
Bianchi, L. M., Conover, J. C., Fritzsch, B., DeChiara, T., Lindsay, R. M., and Yancopoulos, G. D. (1996). Degeneration of vestibular neurons in late embryogenesis of both heterozygous and homozygous BDNF null mutant mice. *Development* **122**, 1965–1973.

Bibel, M., and Barde, Y. A. (2000). Neurotrophins: Key regulators of cell fate and cell shape in the vertebrate nervous system. *Genes Dev.* **14**, 2919–2937.

Cabelli, R. J., Hohn, A., and Shatz, C. J. (1995). Inhibition of ocular dominance column formation by infusion of NT-4/5 or BDNF. *Science* **267**, 1662–1666.

Cohen-Cory, S., and Fraser, S. E. (1995). Effects of brain-derived neurotrophic factor on optic axon branching and remodelling in vivo. *Nature* **378**, 192–196.

Conover, J. C., Erickson, J. T., Katz, D. M., Bianchi, L. M., Poueymirou, W. T., *et al.* (1995). Neuronal deficits, not involving motor neurons, in mice lacking BDNF and/or NT4. *Nature* **375**, 235–238.

Coppola, V., Kucera, J., Palko, M. E., Martinez-De Velasco, J., Lyons, W. E., *et al.* (2001). Dissection of NT3 functions in vivo by gene replacement strategy. *Development* **128**, 4315–4327.

Davies, A. M., and Lindsay, R. M. (1985). The cranial sensory ganglia in culture: Differences in the response of placode-derived and neural crest-derived neurons to nerve growth factor. *Dev. Biol.* **111**, 62–72.

Després, G., Giry, N., and Romand, R. (1988). Immunohistochemical localization of nerve growth factor-like protein in the organ of Corti of the developing rat. *Neurosci. Lett.* **85**, 5–8.

Després, G., Hafidi, A., and Romand, R. (1991). Immunohistochemical localization of nerve growth factor receptor in the cochlea and in the brainstem of the perinatal rat. *Hear. Res.* **52**, 157–166.

Don, D. M., Newman, A. N., Micevych, P. E., and Popper, P. (1997). Expression of brain-derived neurotrophic factor and its receptor mRNA in the vestibuloauditory system of the bullfrog. *Hear. Res.* **114**, 10–20.

Ebendal, T. (1989). Use of collagen gels to bioassay nerve growth factor activity. *In* "Nerve Growth Factors" (R. A. Rush, ed.), pp. 81–93. Wiley & Sons Ltd, New York.

Ernfors, P., Lee, K. F., and Jaenisch, R. (1994). Mice lacking brain-derived neurotrophic factor develop with sensory deficits. *Nature* **368**, 147–150.

Ernfors, P., Merlio, J. P., and Persson, H. (1992). Cells expressing mRNA for neurotrophins and their receptors during embryonic rat development. *Eur. J. Neurosci.* **4**, 1140–1158.

Ernfors, P., Van De Water, T., Loring, J., and Jaenisch, R. (1995). Complementary roles of BDNF and NT-3 in vestibular and auditory development. *Neuron* **14**, 1153–1164.

Farinas, I., Jones, K. R., Backus, C., Wang, X. Y., and Reichardt, L. F. (1994). Severe sensory and sympathetic deficits in mice lacking neurotrophin-3. *Nature* **369**, 658–661.

Farinas, I., Jones, K. R., Tessarollo, L., Vigers, A. J., Huang, E., *et al.* (2001). Spatial shaping of cochlear innervation by temporally regulated neurotrophin expression. *J. Neurosci.* **21**, 6170–6180.

Farinas, I., Wilkinson, G. A., Backus, C., Reichardt, L. F., and Patapoutian, A. (1998). Characterization of neurotrophin and Trk receptor functions in developing sensory ganglia: Direct NT-3 activation of TrkB neurons in vivo. *Neuron* **21**, 325–334.

Frago, L. M., Canon, S., de la Rosa, E., León, Y., and Varela-Nieto, I. (2002). Programmed cell death in the developing inner ear is balanced by nerve growth factor and insulin-like growth factor I. *J. Cell Sci.* **116**, 475–486.

Frago, L. M., León, Y., de la Rosa, E., Gómez-Munoz, A., and Varela-Nieto, I. (1998). Nerve growth factor and ceramides modulate cell death in the early developing inner ear. *J. Cell Sci.* **111**, 549–556.

Fritzsch, B., Barbacid, M., and Silos-Santiago, I. (1998). The combined effects of trkB and trkC mutations on the innervation of the inner ear. *Int. J. Dev. Neurosci.* **16**, 493–505.

Fritzsch, B., Farinas, I., and Reichardt, L. F. (1997). Lack of neurotrophin 3 causes losses of both classes of spiral ganglion neurons in the cochlea in a region-specific fashion. *J. Neurosci.* **17**, 6213–6225.

Fritzsch, B., Silos-Santiago, I., Smeyne, R., Fagan, A. M., and Barbacid, M. (1995). Reduction and loss of inner ear innervation in trkB and trkC receptor knockout mice: A whole mount DiI and scanning electron microscopic analysis. *Aud. Neurosci.* **1**, 401–417.

Gestwa, G., Wiechers, B., Zimmermann, U., Praetorius, M., Rohboch, K., et al. (1999). Differential expression of trkB.T1 and trkB.T2, truncated trkC, and p75^{NGFR} in the cochlea prior to hearing function. *J. Comp. Neurol.* **414**, 33–49.

Hallböök, F., and Fritzsch, B. (1997). Distribution of BDNF and trkB mRNA in the otic region of 3.5 and 4.5 day chick embryos as revealed with a combination of in situ hybridization and tract tracing. *Int. J. Dev. Biol.* **41**, 725–732.

Hallböök, F., Ibanez, C. F., Ebendal, T., and Persson, H. (1993). Cellular localization of brain-derived neurotrophic factor and neurotrophin-3 mRNA expression in the early chicken embryo. *Eur. J. Neurosci.* **5**, 1–14.

Hamburger, V. (1934). The effects of wing bud extirpation on the development of the central nervous system in chick embryos. *J. Exp. Zool.* **68**, 449–494.

Hefti, F., Denton, T. L., Knusel, B., and Lapchak, P. A. (1993). Neurotrophic factors: What are they and what are they doing? *In* "Neurotrophic Factors" (S. E. Loughlin and J. H. Fallon, eds.), pp. 25–49. Academic Press, San Diego.

Huang, E. J., and Reichardt, L. F. (2001). Neurotrophins: Roles in neuronal development and function. *Annu. Rev. Neurosci.* **24**, 677–736.

Huang, E. J., Wilkinson, G. A., Farinas, I., Backus, C., Zang, K., et al. (1999). Expression of Trk receptors in the developing mouse trigeminal ganglion: In vivo evidence from NT-3 activation of TrkA and TrkB in addition to TrkC. *Development* **126**, 2191–2203.

Keller-Peck, C. R., Feng, G., Sanes, J. R., Yan, Q., Lichtman, J. W., and Snider, W. D. (2001). Glial cell line-derived neurotrophic factor administration in postnatal life results in motor unit enlargement and continuous synaptic remodeling at the neuromuscular junction. *J. Neurosci.* **21**, 6136–6146.

Kim, W. Y., Fritzsch, B., Serls, A., Bakel, L. A., Huang, E. J., et al. (2001). NeuroD-null mice are deaf due to a severe loss of the inner ear sensory neurons during development. *Development* **128**, 417–426.

Korsching, S. (1993). The neurotrophic factor concept: A reexamination. *J. Neurosci.* **13**, 2739–2748.

Krylova, O., Herreros, J., Cleverley, K. E., Ehler, E., Henriquez, J. P., et al. (2002). WNT-3, expressed by motoneurons, regulates terminal arborization of neurotrophin-3-responsive spinal sensory neurons. *Neuron* **35**, 1043–1056.

Lee, J. E., Hollenberg, S. M., Snider, L., Turner, D. L., Lipnick, N., and Weintraub, H. (1995). Conversion of *Xenopus* ectoderm into neurons by NeuroD, a basic helix-loop-helix protein. *Science* **268**, 836–844.

Lee, R., Kermani, P., Teng, K. K., and Hempstead, B. L. (2001). Regulation of cell survival by secreted proneurotrophins. *Science* **294**, 1945–1948.

Lenoir, M., Shnerson, A., and Pujol, R. (1980). Cochlear receptor development in the rat with emphasis on synaptogenesis. *Anat. Embryol.* **160**, 253–262.

Levi-Montalcini, R. (1967). The nerve growth factor 35 years later. *Science* **237**, 1154–1162.

Lin, L. F., Doherty, D. H., Lile, J. D., Bektesh, S., and Collins, F. (1993). GDNF: A glial cell line-derived neurotrophic factor for midbrain dopaminergic neurons. *Science* **260**, 1130–1132.

Liu, M., Pereira, F. A., Price, S. D., Chu, M. J., Shope, C., et al. (2000). Essential role of BETA2/NeuroD1 in development of the vestibular and auditory systems. *Genes Dev.* **14**, 2839–2854.

Liu, X., Ernfors, P., Wu, H., and Jaenisch, R. (1995). Sensory but not motor neuron deficits in mice lacking NT4 and BDNF. *Nature* **375**, 238–241.

Ma, Q., Anderson, D. J., and Fritzsch, B. (2000). Neurogenin1 null mutant ears develop fewer, morphologically normal hair cells in smaller sensory epithelia devoid of innervation. *J. Assoc. Res. Otolaryngol.* **1**, 129–143.

Ma, Q., Chen, Z., del Barco Barrantes, I., de la Pompa, J. L., and Anderson, D. J. (1998). Neurogenin1 is essential for the determination of neuronal precursors for proximal cranial sensory ganglia. *Neuron* **20**, 469–482.

Metzger, D., and Chambon, P. (2001). Site- and time-specific gene targeting in the mouse. *Methods* **24**, 71–80.

Milbrandt, J., de Sauvage, F. J., Fahrner, T. J., Baloh, R. H., Leitner, M. L., *et al.* (1998). Persephin, a novel neurotrophic factor related to GDNF and neurturin. *Neuron* **20**, 245–253.

Minichiello, L., Piehl, F., Vazquez, E., Schimmang, T., Hokfelt, T., *et al.* (1995). Differential effects of combined trk receptor mutations on dorsal root ganglion and inner ear sensory neurons. *Development* **121**, 4067–4075.

Mou, K., Hunsberger, C. L., Cleary, J. M., and Davis, R. L. (1997). Synergistic effects of BDNF and NT-3 on postnatal spiral ganglion neurons. *J. Comp. Neurol.* **386**, 529–539.

Oppenheim, R. W. (1991). Cell death during development of the nervous system. *Annu. Rev. Neurosci.* **14**, 453–501.

Patapoutian, A., Backus, C., Kispert, A., and Reichardt, L. F. (1999). Regulation of neurotrophin-3 expression by epithelial-mesenchymal interactions: The role of Wnt factors. *Science* **283**, 1180–1183.

Pirvola, U., Arumae, U., Moshnyakov, M., Palgi, J., Saarma, M., and Ylikoski, J. (1994). Coordinated expression and function of neurotrophins and their receptors in the rat inner ear during target innervation. *Hear. Res.* **75**, 131–144.

Pirvola, U., Hallbook, F., Xing-Qun, L., Virkkala, J., Saarma, M., and Ylikoski, J. (1997). Expression of neurotrophins and Trk receptors in the developing, adult, and regenerating avian cochlea. *J. Neurobiol.* **33**, 1019–1033.

Pirvola, U., Ylikoski, J., Palgi, J., Lehtonen, E., Arumae, U., and Saarma, M. (1992). Brain-derived neurotrophic factor and neurotrophin 3 mRNAs in the peripheral target fields of developing inner ear ganglia. *Proc. Natl. Acad. Sci. USA* **89**, 9915–9919.

Pirvola, U., Ylikoski, J., Trokovic, R., Hebert, J. M., McConnell, S. K., and Partanen, J. (2002). FGFR1 is required for the development of the auditory sensory epithelium. *Neuron* **35**, 671–680.

Postigo, A., Calella, A. M., Fritzsch, B., Knipper, M., Katz, D., *et al.* (2002). Distinct requirements for TrkB and TrkC signaling in target innervation by sensory neurons. *Genes Dev.* **16**, 633–645.

Poteryaev, D., Titievsky, A., Sun, Y. F., Thomas-Crusells, J., Lindahl, M., *et al.* (1999). GDNF triggers a novel ret-independent Src kinase family-coupled signaling via a GPI-linked GDNF receptor alpha1. *FEBS Lett.* **463**, 63–66.

Schimmang, T., Minichiello, L., Vazquez, E., San Jose, I., Giraldez, F., *et al.* (1995). Developing inner ear sensory neurons require TrkB and TrkC receptors for innervation of their peripheral targets. *Development* **121**, 3381–3391.

Schecterson, L. C., and Bothwell, M. (1994). Neurotrophin and neurotrophin receptor mRNA expression in the developing inner ear. *Hear. Res.* **73**, 92–100.

Silos-Santiago, I., Fagan, A. M., Garber, M., Fritzsch, B., and Barbacid, M. (1997). Severe sensory deficits but normal CNS development in newborn mice lacking TrkB and TrkC tyrosine protein kinase receptors. *Eur. J. Neurosci.* **9**, 2045–2056.

Tessarollo, L., Tsoulfas, P., Donovan, M. J., Palko, M. E., Blair-Flynn, J., *et al.* (1997). Targeted deletion of all isoforms of the trkC gene suggests the use of alternate receptors by its ligand neurotrophin-3 in neuronal development and implicates trkC in normal cardiogenesis. *Proc. Natl. Acad. Sci. USA* **94**, 14776–14781.

7. Neutrophic Factors during Inner Ear Development

Trupp, M., Belluardo, N., Funakoshi, H., and Ibánez, C. F. (1997). Complementary and overlapping expression of glial cell line-derived neurotrophic factor (GDNF), *c*-ret protooncogene, and GDNF receptor-α indicates multiple mechanisms of trophic actions in the adult rat CNS. *J. Neurosci.* **17**, 3554–3567.

von Bartheld, C. S., Patterson, S. L., Heuer, J. G., Wheeler, E. F., Bothwell, M., and Rubel, E. W. (1991). Expression of nerve growth factor (NGF) receptors in the developing inner ear of chick and rat. *Development* **113**, 455–470.

Wheeler, E. F., Bothwell, M., Schecterson, L. C., and von Bartheld, C. S. (1994). Expression of BDNF and NT-3 mRNA in hair cells of the organ of Corti: Quantitative analysis in developing rats. *Hear. Res.* **73**, 46–56.

Wiechers, B., Gestwa, G., Mack, A., Carroll, P., Zenner, H. P., and Knipper, M. (1999). A changing pattern of brain-derived neurotrophic factor expression correlates with the rearrangement of fibers during cochlear development of rats and mice. *J. Neurosci.* **19**, 3033–3042.

Xing-Qun, L., Pirvola, U., Saarma, M., and Ylikoski, J. (1999). Neurotrophic factors in the auditory periphery. *Ann. N. Y. Acad. Sci.* **884**, 292–304.

Ylikoski, J., Pirvola, U., Moshnyakov, M., Palgi, J., Arumae, U., and Saarma, M. (1993). Expression patterns of neurotrophin and their receptor mRNAs in the rat inner ear. *Hear. Res.* **65**, 69–78.

Ylikoski, J., Pirvola, U., Virkkala, J., Suvanto, P., Xing-Qun, L., *et al.* (1998). Guinea pig auditory neurons are protected by glial cell line-derived growth factor from degeneration after noise trauma. *Hear. Res.* **124**, 17–26.

8

FGF Signaling in Ear Development and Innervation

Tracy J. Wright and Suzanne L. Mansour
Department of Human Genetics, University of Utah, Salt Lake City, Utah 84112

I. Introduction
  A. Fibroblast Growth Factor Signaling
II. Fibroblast Growth Factor Signaling in Ear Development
  A. Fibroblast Growth Factor Signaling in Outer Ear Development
  B. Fibroblast Growth Factor Signaling in Middle Ear Development
  C. Fibroblast Growth Factor Signaling in Inner Ear Development
III. Conclusions and Future Directions
  References

The 22 mammalian fibroblast growth factors (FGFs) acting through seven main receptor tyrosine kinase isoforms play roles in controlling the growth, differentiation, motility, and survival of cells during development. A variety of experimental approaches have begun to reveal the ways in which FGF signaling is used to control the development of the outer, middle, and inner ears. Structurally related groups of FGFs have similar receptor binding specificity *in vitro* and in cultured cells. For some aspects of ear development, both the relevant FGF ligand and receptor is known. In other cases, only one member of the pair has been determined, but plausible candidates can be proposed based on expression data. Despite the large number of distinct roles for FGF signaling that have been uncovered in the developing auditory system to date, it is clear that we have only begun to scratch the surface of this topic. Future studies aimed at uncovering redundant functions of FGFs and unraveling the later functions of FGFs that are required for early stages of development will help in defining the full complement of roles for FGF signaling in otic development. © 2003, Elsevier Inc.

I. Introduction

The mammalian peripheral auditory system has three components: the outer (external), middle, and inner ears. The outer ear collects sound, funneling it to the tympanic membrane. Vibrations of the tympanic membrane are transmitted through the bones of the middle ear and delivered via the oval window of the otic capsule to the sensory epithelium of the inner ear. The sensory hair cells convert the vibrations into electrical impulses that are transmitted to the central nervous system via the bipolar sensory neurons of the cochlear division of the eighth cranial ganglion. The three components of the peripheral auditory system have separate, but closely apposed, embryological origins (Kaufman and Bard, 1999; Sadler, 1995; Sulik and Cotanche, 1995). Thus, it is likely that there is tight developmental coordination between the three parts to ensure efficient functioning of the mature auditory system. Indeed, there is evidence that proper formation of the external auditory meatus, a component of the external ear, depends on normal development of the tympanic ring, a middle ear component (Mallo and Gridley, 1996), and that signaling from the external auditory meatus is required for normal development of the manubrium of the malleus, a middle ear ossicle (Mallo *et al.*, 2000). Definitive evidence of signaling between the developing middle and inner ears is lacking, but normal morphogenesis of the inner ear does depend on signals from the mesenchymal tissues within which the middle ear bones are developing (Phippard *et al.*, 1999; ten Berge *et al.*, 1998). However, this chapter discusses the development of each ear component separately, and with reference only to potential roles for the fibroblast growth factor (FGF) signaling system in controlling each component's development. Studies in the mouse are emphasized because this is the only mammalian system in which it is possible to systematically assess gene function by targeted mutagenesis. Where relevant, studies utilizing other important model systems, primarily the chick and zebrafish, are highlighted. FGF signaling is clearly involved in the development of each of the three parts of the peripheral auditory system. Relatively few reports of roles for FGF signaling in the development of the external and middle ears have appeared. By contrast, studies of FGF signaling during inner ear development are numerous. Despite the large number of distinct roles for FGF signaling that have been uncovered to date, it is clear just from a cursory examination of the expression patterns of *Fgf* and *Fgf* receptor genes in the developing otic region that this topic will continue to add to our understanding of auditory development for many years to come.

A. Fibroblast Growth Factor Signaling

1. Fibroblast Growth Factors

Fibroblast growth factors were originally identified and isolated based on their mitogenic effects on cells in culture (Gospodarowicz, 1974). Many studies have revealed additional cellular functions for FGFs, including the promotion of cellular differentiation, motility, and survival. The FGFs comprise a family of small (17–35 kD), typically secreted, highly basic proteins that have strong affinity for the heparan sulfate moiety present on proteoglycans resident in the extracellular matrix. FGFs share a conserved core of approximately 120 amino acids that assumes a three-dimensional structure similar to that of the cytokine interleukin-1β (Fig. 1A, for recent reviews see Dono, 2003; Ford-Perriss *et al.*, 2001; Ornitz, 2000). Phenotypic analysis of mice bearing targeted mutations in 15 different *Fgf* genes has begun to reveal the ways in which the cellular functions of FGFs are utilized during the normal development of a wide variety of tissues and organs, including the three parts of the ear (Ornitz and Itoh, 2001; Pickles and Chir, 2002; Powers *et al.*, 2000). In addition, studies of humans and mice bearing dominant gain-of-function mutations in FGF receptor (*Fgfr*) genes have shown that constitutive or ectopic activation of FGF signaling has especially severe consequences for skeletal development (McIntosh *et al.*, 2000; Ornitz and Marie, 2002). Some of these *Fgfr* mutations are also associated with anomalous development and/or function of the various components of the ear (Orvidas *et al.*, 1999; Vallino-Napoli, 1996).

Gene discovery efforts and the near completion of the human and mouse genomes have revealed that FGFs are encoded by 22 distinct genes. These can be visualized in their chromosomal context by using several dierent browsers (e.g., *http://www.ensembl.org/*, *http://genome.ucsc.edu/*, *http://www.ncbi.nlm.nih.gov/mapview/*). The 10 *Fgf* genes described so far in chick and the 4 *Fgf* genes found in zebrafish are all clearly orthologous to *Fgf* genes in mammals, suggesting that there may be no more new FGFs to be described in the organisms typically used in studies of the development and innervation of ear. *Fgf* genes typically have a three-exon structure, with some showing evidence of alternative splicing that does not affect the coding sequences (Brookes *et al.*, 1989; Mansour and Martin, 1988). One exception is the *Fgf8* gene, which is alternatively spliced within the coding region to generate 7 distinct mRNAs in mice and 4 in humans. These encode FGF8 isoforms that vary within the amino terminal region and which may mediate distinct biological responses, although little is known about their relative expression profiles *in vivo* (Blunt *et al.*, 1997; MacArthur *et al.*, 1995; A. Joyner, personal communication).

2. Fibroblast Growth Factor Receptors

FGFs signal by binding to and activating FGF receptors, which are single-pass transmembrane proteins with three immunoglobulin-like (Ig) domains for ligand binding and an intracellular tyrosine kinase domain for signaling (Fig. 1B and C). Mammals have four *Fgfr* genes, which generate many different protein isoforms through the use of alternative splicing. Most attention has been directed to the alternative splicing events that generate the "b" and "c" isoforms of the third Ig domain of FGF receptors 1, 2, and 3. The *Fgfr4* gene does not possess the "b" exon and thus produces only the

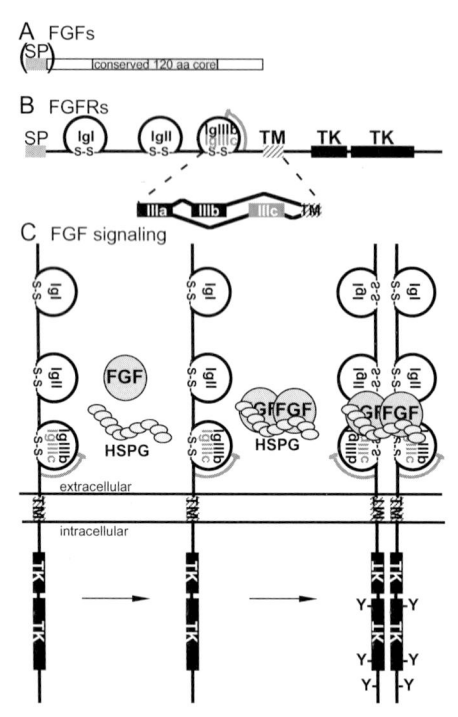

Figure 1 FGFs, FGF receptors, and FGF signaling. (A) Schematic diagram of FGFs. Secreted FGFs have a signal peptide (SP). FGFs share a core homology region of approximately 120 amino acids. The white boxes indicate amino- and carboxy-terminal regions that are unique to each FGF. (B) Schematic diagram of FGF receptors (FGFRs) and the alternative splicing event that leads to production of "b" and "c" isoforms of FGFR1, FGFR2, and FGFR3. The FGFRs have a signal peptide (SP), three immunoglobulin (Ig) domains, a transmembrane domain (TM, hatched box) and a split tyrosine kinase domain (TK, filled boxes). The IgIII domain is encoded by three exons, IIIa (black box), IIIb (black box), and IIIc (gray box). Splicing of exon IIIa to exon IIIb yields mRNAs encoding the "b" type isoforms and splicing of exon IIIa to exon IIIc yields mRNAs encoding the "c" type isoforms. (C) FGF signaling. A generic FGFR receptor is depicted embedded within the cell membrane, with a nearby FGF (gray-filled circle) and heparan sulfate proteoglycan (HSPG, connected ovals). Following binding of the FGFs to HSPG, the dimerized ligand complex binds to and dimerizes the receptor. This leads to activation of receptor tyrosine autophosphorylation (depicted using Y), as well as phosphorylation of signaling intermediates (not shown).

equivalent of a "c" isoform. The "b" and "c" isoforms of FGF receptors are significant because they provide much of the FGF ligand binding specificity. Thus, there are 7 major receptor isoforms to potentially interact with 22 different ligands. As it turns out, most FGFs bind to and activate more than one but not all FGF receptors. Indeed, a phylogenetic tree of the 22 FGFs shows that highly related ligands have similar receptor binding specificity

(Fig. 2). These ligand–receptor interactions can have different affinities, leading ultimately to different signal strengths (Beer et al., 2000; Konishi et al., 2000; Ornitz et al., 1996; Xu et al., 2000; Yamashita et al., 2002). Thus, it appears that FGF signaling can potentially be regulated qualitatively by the local availability of appropriate ligand–receptor combinations and quantitatively by the affinity of particular ligand–receptor combinations. Although these ligand–receptor binding and activity studies are typically carried out *in vitro* or by using artificial cell culture systems, their relevance to FGF signaling *in vivo* generally (Colvin et al., 1996; Deng et al., 1996; Liu et al., 2002; Ohbayashi et al., 2002), and in the ear in particular, is increasingly apparent (Mansour et al., 1993; Pauley et al., 2003; Pirvola et al., 2000; Wright and Mansour, 2003).

3. Fibroblast Growth Factor Signal Transduction

FGFs signal from one cell to another by binding first to heparan sulfate proteoglycans in the extracellular matrix. This interaction promotes binding of the FGF ligand to the extracellular immunoglobulin-like domains of its cell surface receptor and promotes dimerization of the ligand–receptor complexes. Once dimerized, the intracellular tyrosine kinase domain of the receptor is autophosphorylated and also phosphorylates several different signal adaptors at the top of a variety of intracellular signaling pathways (Fig. 1C; Klint and Claesson-Welsh, 1999; Ornitz and Marie, 2002; Powers et al., 2000). One of the best understood of these is the RAS–MAPK pathway. In this case, phosphorylation of the FGF receptor initiates a cytoplasmic cascade of sequential protein kinase phosphorylations that culminates in the activation of the extracellular regulated kinase (ERK) mitogen activated protein (MAP) kinase, which in turn phosphorylates transcription factors that change patterns of gene expression in the responding cell (Chang and Karin, 2001; Kolch, 2000; Pearson et al., 2001). Among the transcriptional targets of FGF signaling through this pathway are at least three distinct classes of feedback inhibitors: Sprouty (Furthauer et al., 2001; Hacohen et al., 1998; Minowada et al., 1999), Sef (Furthauer et al., 2002; Kovalenko et al., 2003; Tsang et al., 2002), and dual-specificity protein phosphatases (Camps et al., 2000; Keyse, 2000) that act at different levels of the pathway. The FGFs that are not secreted (FGFs11–14) cannot bind to FGF receptors and they signal by binding to tissue-specific intracellular scaffold proteins, such as IB2, that recruit specific MAPK signaling modules (Schoorlemmer and Goldfarb, 2001).

FGF signaling is generally used to communicate between epithelial and mesenchymal tissues (paracrine signaling) (Hogan, 1999; Martin, 1998; Thesleff and Mikkola, 2002) but in some cases is thought to operate via an autocrine mechanism (Fedorov et al., 2001; Sakaue et al., 2002; Ying et al.,

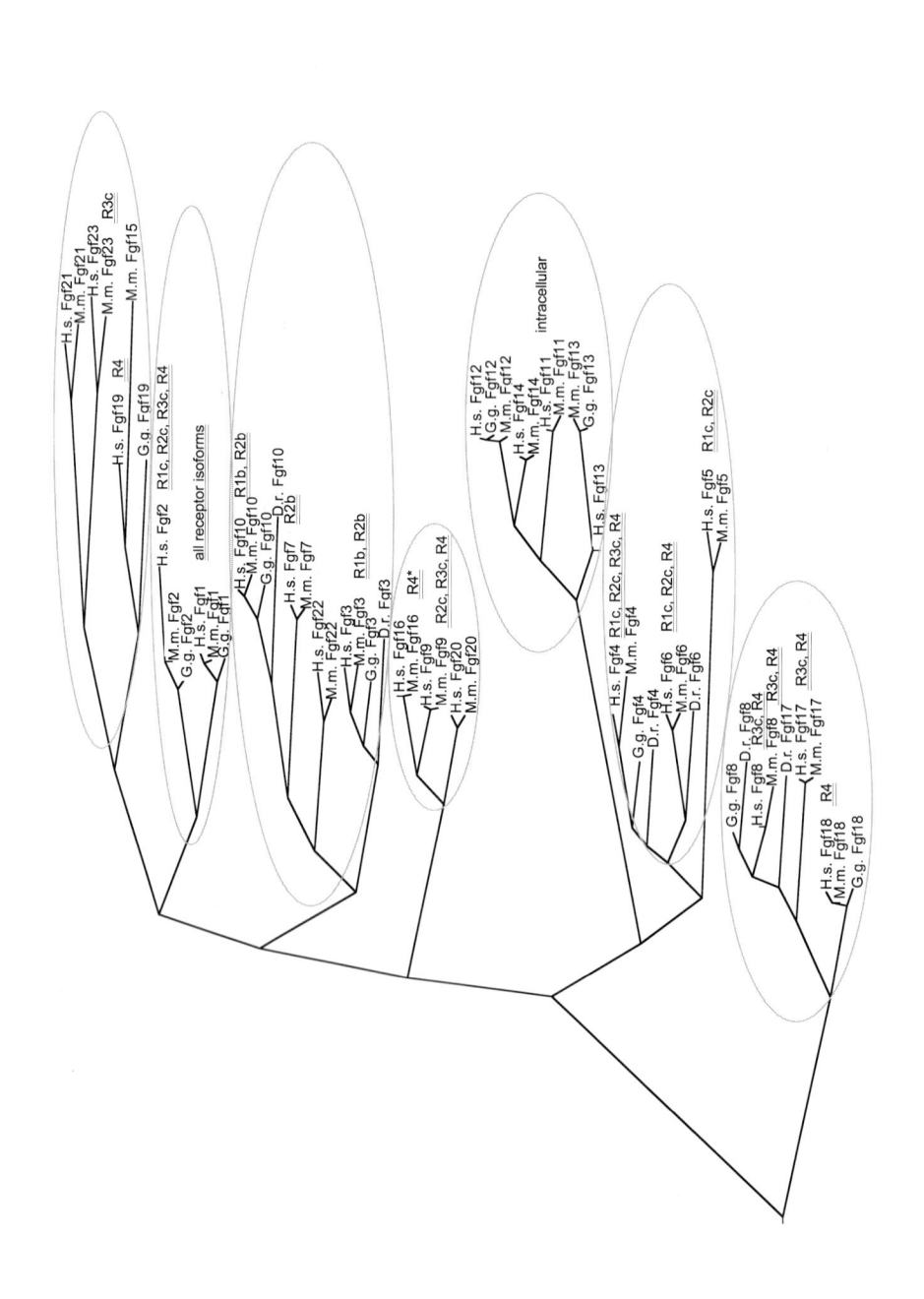

2003). Regardless of the mechanism, it is important to appreciate that FGFs do not act on cells that are located very far from their sites of synthesis. Roehl and Nusslein-Volhard (2001) showed that ectopic induction of zebrafish *Fgf8* from a point source was able to activate one target gene, *Pea3*, within 3–4 cell diameters and another target gene, *Erm*, within 7–8 cell diameters from the source. Therefore, in order to make predictions about possible functions of particular FGFs based on expression data or to interpret the results of inactivating a particular ligand or receptor, it is necessary also to understand the relative location of the receptors that interact with the ligand in question. Most of the available expression data on FGFs and their receptors are highly dispersed. Several surveys have been reported (Bachler and Neubuser, 2001; Karabagli *et al.*, 2002a,b; Kettunen *et al.*, 1998; Ozawa *et al.*, 1996; Walshe and Mason, 2000; Wilke *et al.*, 1997), though none focus particularly on the ear. In addition to localizing active ligand–receptor pairs, a complete understanding of the roles of FGF signaling in any developmental process requires understanding the regulatory inputs (both positive and negative) into the pathway, as well as the ultimate targets of signaling.

In the following sections we focus first on genetic studies implicating FGF8 and FGFR1 in both outer and middle ear development. Next, we consider four FGFs—FGF3, 8, 10, and 19—that individually and in combination have been implicated by genetic ablation, overexpression, and collagen gel culture studies in the development of the inner ear and its innervation. In the case of FGF3 and FGF10, it is clear that both molecules signal through FGFR2b and possibly also through FGFR1b. FGF8 may signal through an undetermined receptor in one context and through FGFR3 at another. In the case of FGF19, the receptor used *in vivo* has not been determined, but *in vitro* studies point to FGFR4. A mechanism for the negative regulation of FGF signaling by Sprouty4 during zebrafish inner ear development has

---

Figure 2 Structurally related FGFs have similar receptor binding properties. Phylogenetic tree showing relationships of human (H.s.), mouse (M.m.), chick (G.g.), and zebrafish (D.r.) FGFs. Groups of similar FGFs are enclosed within gray ovals. The highest-affinity receptor(s) determined for individual FGFs is/are listed to the right of the FGFs. When more than one receptor isoform is indicated, it is likely that there is no significant difference in the affinity of the ligand for the different receptors. Receptor specificity determined for a ligand from one species is likely to apply to the orthologous ligands of other species. Receptor binding specificity has not been determined for FGF15, FGF20, FGF21, FGF22, and FGF23. The asterisk beside the FGF16 result indicates that the ligand was prepared using a rat rather than a mouse cDNA. It should be noted that except for FGF1–9, which were tested at the same time on the same panel of receptor-expressing cell lines (Ornitz *et al.*, 1996), the relative activities of the different FGFs are largely unknown. Protein sequences were retrieved from GenBank and aligned using MegAlign (DNASTAR) software. Receptor binding specificities were compiled from the literature (Beer *et al.*, 2000; Konishi *et al.*, 2000; Mathieu *et al.*, 1995; Ornitz *et al.*, 1996; Xu *et al.*, 2000; Yamashita *et al.*, 2002).

been demonstrated, and expression analysis of a number of additional negative regulators strongly suggests candidates that may be acting in otic development in other species.

II. Fibroblast Growth Factor Signaling in Ear Development

A. Fibroblast Growth Factor Signaling in Outer Ear Development

The outer or external ear consists of the auricle (pinna), which is unique to mammals; the external acoustic meatus (auditory canal); and the ectodermal layer of the trilaminar tympanic membrane (eardrum). The pinna derives from ectoderm and underlying mesenchyme located on both the anterior (branchial arch 1, mandibular arch) and posterior (branchial arch 2, hyoid arch) sides of the first branchial cleft, whereas the auditory canal and ectodermal layer of the tympanic membrane derive from the ectoderm of the cleft itself (Kaufman and Bard, 1999; Sadler, 1995; Wright, 1997; see Chapter 3 by Mallo). Of these three parts, there is published genetic evidence from the mouse of a primary role for FGF signaling only for the development of the pinna. It should be noted, however, that pinna abnormalities are more readily scored in mouse models than are auditory canal or tympanic membrane abnormalities. In addition, patients with Crouzon syndrome, which is caused by gain-of-function mutations in *Fgfr2*, may present with microtia and atresia of the auditory canal (Lambert and Dodson, 1996; OMIM #123500, *http://www.ncbi.nlm.nih.gov:80/entrez/dispomim.cgi?id=123500*). Thus, it would not be surprising if additional existing mouse *Fgf* and *Fgf* receptor mutants were eventually found to have abnormalities involving the auditory canal and/or tympanic membrane.

In the mouse, the complete absence of *Fgfr1* is lethal at gastrulation, before the earliest stages of development of any parts of the ear are initiated (Deng *et al.*, 1994; Yamaguchi *et al.*, 1994). Mice that are homozygous for a hypomorphic *Fgfr1* allele (*Fgfr1^{n7}*) that produces 20% of the wildtype level of *Fgfr1* mRNA, however, have a greatly reduced pinna. These defects are not seen in mice homozygous for an *Fgfr1b* isoform-specific deletion (Partanen *et al.*, 1998), suggesting that the "c" isoform is required for pinna development. *Fgfr1* transcripts (isoforms not determined) are expressed quite widely in both the branchial arch ectoderm and mesenchyme that will give rise to the pinna (Trokovic *et al.*, 2003). Based on genetic studies, one candidate for an FGFR1 ligand involved in pinna formation is FGF8, the gene for which is expressed in branchial arch ectoderm and endoderm (Crossley and Martin, 1995; Lin *et al.*, 2002).

Support for this hypothesis comes from analysis of mice that carry a hypomorphic allele of *Fgf8* (*Fgf8^{neo}*) in combination with a null allele of

Fgf8. This genotype results in a reduction in the pinna similar to that seen in homozygous *Fgfr1^{n7}* mice (Abu-Issa et al., 2002). Because elimination of *Fgf8* specifically from branchial arch ectoderm does not appear to cause pinna defects (Trumpp et al., 1999), this ligand may signal to a mesenchymal receptor from pharyngeal pouch endoderm rather than from the arch ectoderm. One caveat to this proposal of FGF8/FGFR1 signaling, however, is that none of the FGF8 isoforms are thought to activate either of the FGFR1 isoforms (Blunt et al., 1997; MacArthur et al., 1995; Ornitz et al., 1996). Thus, the FGF8 signal could actually be mediated by a different receptor, such as FGFR2, which is also expressed in the arch mesenchyme (Trokovic et al., 2003). Perhaps a better candidate ligand for FGFR1 in pinna formation is FGF4, which does activate FGFR1c (Ornitz et al., 1996) and the gene for which is expressed in pharyngeal pouch endoderm just prior to the initiation of pinna development (Niswander and Martin, 1992). Hypomorphic alleles of *Fgf4* also cause reductions in the pinna (A. Moon, personal communication). Unfortunately, neither the developmental origins nor the tissues critical for either of these potential ligand/receptor pairs have been established. It is attractive to speculate, however, that signaling from FGF ligands expressed by the pharyngeal pouch endoderm to mesenchymal FGF receptors initiates the mesenchymal swellings that presage pinna development. See Fig. 3A for a summary diagram illustrating the *Fgfs*, *Fgfrs*, and tissues that play roles in mouse outer ear development.

The avian external ear does not include a pinna, but roles for FGF signaling in the development of the external auditory canal and tympanic membrane can be suggested based on the expression patterns of *Fgfs* and their receptors in the vicinity of the developing external ear in chick embryos. In many instances these are similar to the expression patterns of the mouse homologues, providing additional clues to their potential involvement in mammalian external ear development. *Fgf3, 4, 8, 12, 14,* and *18* can be detected in the chick pharyngeal endodermal pouches. *Fgfs3, 8,* and *18* can also be detected in the branchial arch ectoderm (Karabagli et al., 2002a,b; Mahmood et al., 1995; Stolte et al., 2002). Prior to and during external ear development *Fgfr1, 2,* and *3* are all expressed in the branchial arches, although the specific isoforms have not been distinguished so it is not yet possible to use this information to propose specific ligand–receptor interactions that might be involved in outer ear development. *Fgfr1* is expressed more strongly in the arch ectoderm and endoderm of the pharyngeal pouches than in the arch mesenchyme. *Fgfr2* is expressed in the branchial arch ectoderm and *Fgfr3* is expressed relatively early in the pharyngeal endoderm. *Fgfr3* is not, however, detected in the branchial arch mesenchyme until slightly later in development (Walshe and Mason, 2000; Wilke et al., 1997).

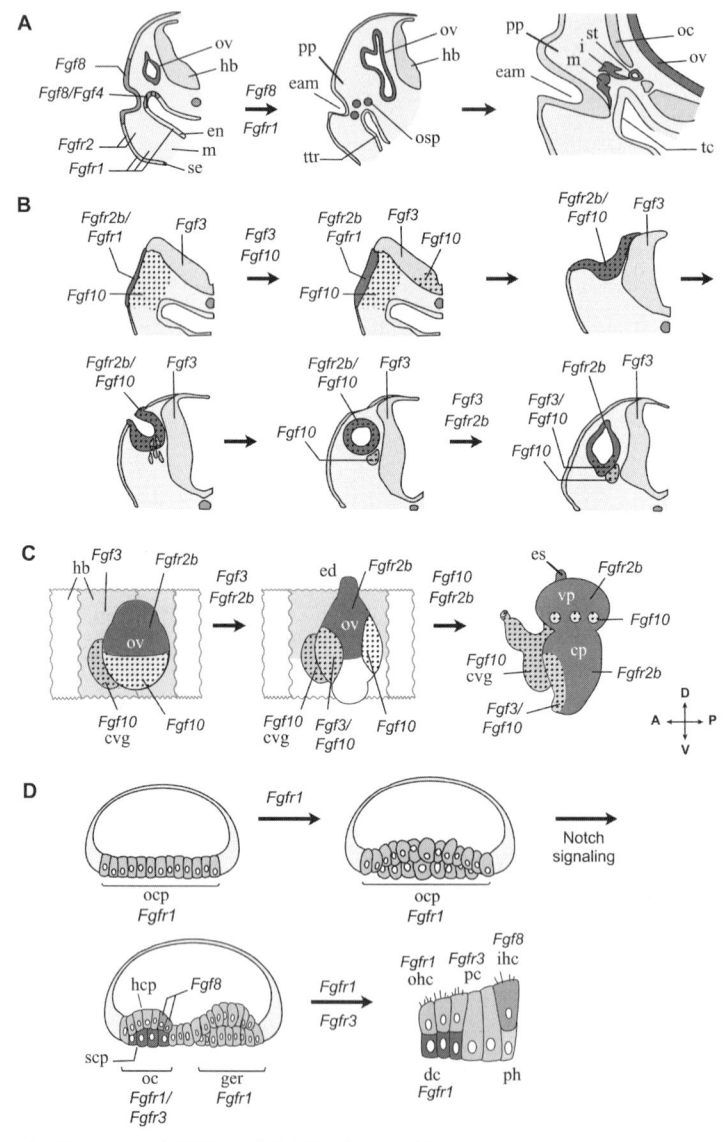

Figure 3 Summary of FGFs and FGFRs involved in mouse ear development. (A) Outer and middle ear development. The left diagram depicts the patterns of gene expression in the otic region immediately prior to the initiation of outer and middle ear development. *Fgfr1*, expressed throughout the surface ectoderm (se), mesenchyme (m), and endoderm (en), is required both for pinna and ossicle development. *Fgf8* (green highlight), expressed in the first branchial cleft of the se and in the pharyngeal pouch en, is also required for both pinna and ossicle development. *Fgf4*, expressed in the pharyngeal pouch en, is required for pinna development, but potential

B. Fibroblast Growth Factor Signaling in Middle Ear Development

The middle ear is composed of a cavity formed from the first pharyngeal pouch endoderm, which also gives rise to the endodermal layer of the tympanic membrane. In mammals, the middle ear cavity contains three ossicles, the malleus, incus, and stapes, which are formed from neural crest–derived mesenchyme of the first (malleus and incus) and second (stapes) branchial

---

roles in middle ear development have not been examined. A role for *Fgfr2* in se and m has not been established but is suggested because FGF8 is not a good ligand for FGFR1 isoforms but does activate FGFR2c. The middle and right diagrams show the progressive FGF signal-dependent development of the outer and middle ears. Additional abbreviations: eam, external auditory meatus; hb, hindbrain; i, incus; m, malleus; oc, otic capsule; osp, ossicle precursors; ov, otic vesicle; pp, pinna primordium; st, stapes; tc, tympanic cavity; ttr, tubotympanic recess. (B) The earliest stages of inner ear development showing the expression patterns of *Fgf* and *Fgfr* genes required for otic vesicle formation. The otic placode, which thickens, invaginates and forms a closed vesicle, is shown in red. The neurons of the eighth ganglion, which derive from the otic epithelium, are shown in pink. The hindbrain and non-otic ectoderm are blue; the endoderm is yellow; and the mesenchyme is not colored, except in the region of *Fgf10* expression (dotted). Other regions of *Fgf10* expression are also dotted and retain their assigned colors. *Fgf3* and *Fgfr2b* are generally expressed throughout the tissues indicated, except at the last stage depicted, in which *Fgf3* is confined with *Fgf10* to the most ventral region of the vesicle, the rest of which expresses *Fgfr2b*. *Fgf3* expressed in the hindbrain and *Fgf10* expressed in the mesenchyme, acting through *Fgfr2b* (and potentially *Fgfr1*) expressed in the otic placode, are required redundantly for induction of the placode. All components of this signaling system continue to be expressed in changing patterns after the placode is induced, but specific roles for FGF signaling in invagination and vesicle formation have not been established. See text for a description of the roles of FGF signaling in the early stages of zebrafish development. (C) FGF signaling in the early morphogenesis of the otic vesicle. Lateral views of the otic vesicle (ov) and cochleovestibular ganglion (cvg, pink) from shortly after vesicle closure through endolymphatic duct (ed) induction and the earliest stages of semicircular canal formation. *Fgf3* (blue), expressed in the hindbrain (hb), acting through *Fgfr2b* (red) expressed in the dorsal half of the ov is required for ed induction. Later functions for *Fgf3* expressed in the anterior ov are possible but not yet explored. *Fgf10* (dotted) is expressed throughout this period in the cvg but does not play a unique role in its formation. *Fgf10*, expressed initially in the ventral ov, then in both the anterior and posterior ov, and eventually in the developing sensory patches of the cochlear primordium (cp) and vestibular primordia (vp) acts through *Fgfr2b*, expressed in nonsensory tissue, and may be required for the initial (dorsal) outpouching of the semicircular canal plates but is not required for (ventral) cochlear development. (D) FGF signaling in the proliferation and differentiation of the organ of Corti (oc). Sequential diagrams depict sections taken through the developing cochlear duct in which signaling through FGFR1, activated by an unknown ligand, stimulates proliferation of undifferentiated organ of Corti precursors (ocp). Notch signaling is then required for the cell fate decision separating hair cell precursors (hcp) from supporting cell precursors (scp). Cells in both the primitive oc and greater epithelial ridge (ger) express *Fgfr1*, whereas only cells in the primitive oc coexpress *Fgfr3*. Signaling through FGFR1, stimulated by an unknown ligand, is required for differentiation of the hcps to outer hair cells (ohc). Signaling through FGFR3, possibly stimulated by FGF8, is required for the differentiation of scps to pillar cells (pc). Other abbreviations: dc, Deiter's cell; ph; phalangeal cell. (See Color Insert.)

arches (Kaufman and Bard, 1999; Mallo, 2001, Sadler, 1995; see Chapter 3 by Mallo). Chicks, like most other nonmammalian species, have a single ossicle (the columella), which is derived from second arch neural crest (Van De Water *et al.*, 1980). The other skeletal element associated with the middle ear is the tympanic ring, which supports the tympanic membrane and is derived from first arch mesenchyme. This bone is eventually incorporated into the temporal bone of the skull. The muscles associated with the malleus (tensor tympani) and stapes (stapedius) are derived from first and second branchial arch mesoderm and are innervated by the trigeminal and facial nerves, respectively (Kaufman and Bard, 1999; Mallo, 2001; Sadler, 1995). An excellent review of the details of middle ear development is provided in Chapter 3 by Mallo.

There is clear genetic evidence of a role for FGF signaling in the development of the mouse middle ear bones. As mentioned earlier, *Fgf8* is expressed in a dynamic pattern in the ectoderm and endoderm of branchial arches 1 and 2 (Crossley and Martin, 1995; Lin *et al.*, 2002). Like *Fgfr1* null embryos, *Fgf8* null embryos do not survive past gastrulation (Sun *et al.*, 1999). Mice that carry one hypomorphic (reduced function) and one null allele of *Fgf8* show variable reductions or loss of the malleus, incus, and typmpanic ring, all of which derive from neural crest mesenchyme in branchial arch 1 (Abu-Issa *et al.*, 2002; Frank *et al.*, 2002). Furthermore, analysis of mouse embryos in which *Fgf8* expression is normal during gastrulation but is removed specifically from first arch ectoderm revealed loss of the tympanic ring and incus and reduction of the malleus. This phenotype is caused by excessive cell death in the mutant mesenchyme, indicating a paracrine role for ectodermal FGF8 signals in mesenchymal cell survival (Trumpp *et al.*, 1999). These authors also showed that expression of several branchial arch 1 mesenchyme transcription factor genes, including *Barx1*, *Pax9*, *Lhx6*, and *Gsc*, are affected by the loss of ectodermal *Fgf8*. Whether any of these genes are primary targets of FGF8 signals remains to be determined, but it is interesting to note that *Gsc* and *Pax9* null mutants also have middle ear ossicle and typmpanic ring defects (Peters *et al.*, 1998; Rivera-Perez *et al.*, 1995; Yamada *et al.*, 1995).

How is this signal mediated? As indicated earlier, FGF8 isoforms activate the "c" isoforms of FGF receptors 2 and 3 and also activate FGFR4 (Blunt *et al.*, 1997; MacArthur *et al.*, 1995; Ornitz *et al.*, 1996). There is very little published information on isoform-specific expression of the *Fgf* receptor genes in the branchial arches, but *Fgfr2* (isoforms not distinguished) is clearly present in branchial arch mesenchyme, whereas *Fgfr3* and *Fgfr4* are not present in this tissue (Trokovic *et al.*, 2003; Wright, Hatch, and Mansour, unpublished data). An *Fgfr2c*-specific loss of function mutant has been generated. These mice have defects of bone formation, but unfortunately there has been no description yet of their middle ears (Eswarakumar *et al.*, 2002).

Several lines of evidence point to an important role for *Fgfr1c* in middle ear development. *Fgfr1* is expressed both in branchial arch ectoderm and in the neural crest–derived mesenchyme that migrates into the branchial arches to form the middle ear bones. Reduction in the level of *Fgfr1* expression, through the use of hypomorphic alleles, causes reduction or elimination of all three middle ear bones (Trokovic *et al.*, 2003). As noted for the external ear defects, no such effect is seen with an *Fgfr1b* isoform-specific deletion allele (Partanen *et al.*, 1998), suggesting that the "c" isoform is required for ossicle development. The stapes phenotype seen in animals homozygous for the *Fgfr1^{n7}* hypomorphic allele correlates with a reduction in the amount of neural crest cells that migrate into the second arch. Surprisingly, elimination of *Fgfr1* specifically in the neural crest has no effect on the development of any of the ossicles (Trokovic *et al.*, 2003), suggesting that *Fgfr1* (presumably the "c" isoform) expression is required in some tissue other than the neural crest, perhaps in one of the branchial arch epithelia. Thus, it will be of interest to eliminate *Fgfr1* specifically in the branchial arch ectoderm and pharyngeal pouch endoderm to determine which tissue is responsible for the middle ear defects observed by Trokovic and colleagues. The only clues as to an FGF that could be signaling to FGFR1c expressed in branchial arch epithelia come from a survey of chick *Fgfs* that found *Fgf2* to be localized to branchial arch mesenchyme (Karabagli *et al.*, 2002b). Whether this is also the case for mouse is not known. See Fig. 3A for a summary diagram illustrating the relationships between *Fgf* and *Fgfr* gene expression and the tissues involved in middle ear formation.

C. Fibroblast Growth Factor Signaling in Inner Ear Development

1. Induction and Formation of the Otic Placode

The inner ear, including sensory, nonsensory, and neuronal cells of the eighth (cochleovestibular) ganglion, derives from a patch of ectodermal tissue, termed the otic placode, which is located adjacent to the developing hindbrain. Cells that are competent to form the otic placode are initially present in a large swath of surface ectoderm extending anteriorly to posteriorly from the forebrain to the region where the first somite will form. As development proceeds the preplacodal tissue is progressively restricted to the region lateral to hindbrain rhombomeres 5 and 6. The placode itself is first evident morphologically as a region of thickened ectoderm. This occurs at slightly different somite stages in the different experimental species that are used in studies of ear development. This topic has been reviewed (Baker and Bronner-Fraser, 2001; Groves, 2003; Kiernan *et al.*, 2002; Noramly and Grainger, 2002; Whitfield *et al.*, 2002) and Chapter 4 by Groves *et al.* contains the most recent description of otic placode development.

a. Effects of Ectopic FGF Expression on Preplacodal Ectoderm. A variety of experimental approaches have led to the conclusion that FGF signaling plays a critical role in the induction of the otic placode. Tissue ablation and transplantation experiments have shown that the hindbrain adjacent to the placodal region and the mesoderm/mesenchyme underlying the prospective placodal tissue are required for placodal induction (reviewed in Baker and Bronner-Fraser, 2001; Groves, 2003; Noramly and Grainger, 2002). FGF signals are sufficient to mimic many aspects of this inductive event. For example, application of heparin beads soaked with FGF2 to neural plate stage *Xenopus* embryos induces ectopic vesicles of variable size that express the otic marker gene *XWnt3a*. In some cases the expression pattern of this gene is dorsally restricted as it is in the normal otic vesicle (Lombardo and Slack, 1998). It is not clear whether FGF2 could be an endogenous inducer of the placode because expression of its gene in both inducing tissues is very weak in chick (Karabagli *et al.*, 2002a) and undetectable in mouse (Wright and Mansour, unpublished data). *Fgf3*, however, which is expressed in the hindbrain of all species analyzed, is a better candidate for an endogenous otic inducer (Mahmood *et al.*, 1995, 1996; Maroon *et al.*, 2002; McKay *et al.*, 1996; Phillips *et al.*, 2001; Tannahill *et al.*, 1992; Wilkinson *et al.*, 1988). Infection of chick embryonic hindbrain and ectoderm at stages 8–10 with an *Fgf3*-expressing herpes simplex virus vector induces ectopic otic-like vesicles that express a variety of marker genes, including *Pax2*, with appropriate regional restrictions (Vendrell *et al.*, 2000). *Fgf8* is coexpressed with *Fgf3* in the zebrafish hindbrain (Maves *et al.*, 2002; Walshe *et al.*, 2002). Although ectopic overexpression of *Fgf8* in zebrafish embryos expands the otic placode, as marked by *Pax2.1* expression, it is unable to induce ectopic vesicles (Leger and Brand, 2002).

In addition to hindbrain-expressed *Fgf* genes, there are also mesodermal/mesenchymal *Fgfs*, such as *Fgf3* (Karabagli *et al.*, 2002a) and *Fgf19* (Ladher *et al.*, 2000) in the chick and *Fgf10* in the mouse (Wright and Mansour, 2003). Application of FGF19-soaked beads in combination with WNT8c-expressing cells to chick stage 5 (preplacodal) anterior cephalic ectoderm in collagen gel culture has been shown to induce expression of a panoply of otic marker genes, suggesting that FGF19 could participate in otic induction (Ladher *et al.*, 2000).

Fgf8 has also been detected in the chick otic placode itself (Hidalgo-Sanchez *et al.*, 2000). Implantation of FGF8- (or FGF2)-coated beads into the vicinity of the developing chick otic placode induces otic marker genes, including *Pax2*. Some ectopic vesicles are also observed, but these do not express all otic markers and do not develop normally (Adamska *et al.*, 2001). Although all of the mentioned overexpression studies suggest that placodal tissue has the capacity to respond to FGF signals by initiating otic development, they beg the question of whether any of the applied FGFs are endogenous otic inducers.

b. Effects of Blocking FGF Signaling on Otic Placode Development. A specific small molecule inhibitor of the tyrosine kinase activity of all FGF receptors, SU5402 (Mohammadi *et al.*, 1997), has been used to demonstrate that otic placode induction does in fact require FGF signaling. Zebrafish embryos soaked in SU5402 during preplacodal stages fail to form an otic vesicle and do not express the otic placode marker genes, *Pax2.1*, *Pax8*, and *Dlx3* (Leger and Brand, 2002; Maroon *et al.*, 2002).

Several different groups have provided genetic evidence that *Fgf3* and *Fgf8* are the specific *Fgf* genes that are required for otic placode induction in zebrafish embryos. *Fgf8* (*ace*) mutant embryos treated with *Fgf3* morpholinos or wildtype embryos treated with both *Fgf3* and *Fgf8* morpholinos fail to form otic vesicles (Leger and Brand, 2002; Maroon *et al.*, 2002; Phillips *et al.*, 2001). Both of these genes are expressed in hindbrain rhombomere 4, and their absence also causes major abnormalities of hindbrain patterning, making it uncertain whether one or both of these FGFs act directly or indirectly on the prospective otic placode (Maves *et al.*, 2002; Walshe *et al.*, 2002). Furthermore, the identities of FGF receptor isoforms expressed by the zebrafish hindbrain and prospective placode have not been determined, although it seems that *Fgfr2* (isoform[s] not determined) is expressed in several hindbrain rhombomeres (Tonou-Fujimori *et al.*, 2002). Thus, it is not yet clear whether either FGF3 or FGF8 can signal directly to the preplacodal ectoderm. Regardless of whether the mechanism is direct or indirect, it appears that FGF signals are required only for appropriate induction of gene expression in the ectoderm, not for proliferation or survival of the ectodermal cells (Leger and Brand, 2002; Maroon *et al.*, 2002).

Genetic evidence of a requirement for FGF signaling in normal otic induction in the mouse is now available. Similarly to zebrafish *Fgf8*/*Fgf3* mutant/morphant zebrafish, mice that are homozygous for null alleles of both *Fgf3* and *Fgf10* fail to form otic vesicles and have aberrant patterns of otic placode marker gene expression, but ectodermal cell proliferation and survival are not affected. Genes encoding both of the receptors that are activated by FGF3 and FGF10 (*Fgfr2b* and *Fgfr1b*) are expressed in the prospective placodal tissue just as it begins to thicken and, in contrast to the zebrafish *Fgf3*/*Fgf8* mutants, hindbrain patterning in the mouse *Fgf3*/*Fgf10* double mutants is normal, suggesting that these two FGFs act directly on the otic placode. Because *Fgf3* is expressed in the hindbrain and *Fgf10* is expressed in the mesenchyme underlying the prospective placode in mouse embryos, it is possible, though not assured, that these two genes may provide the ultimate signal necessary for normal otic placode formation (Wright and Mansour, 2003). It is notable that neither an *Fgfr2b* isoform-specific mutation nor an *Fgfr2* dominant-negative transgene causes otic phenotypes that are as severe as the *Fgf3/Fgf10* double mutant phenotype (Celli *et al.*, 1998; Pirvola *et al.*, 2000). As discussed in more detail later, the *Fgfr2b* mutant mice form small otic vesicles that go on to develop abnormally. This result suggests the possibility that *Fgfr1b* may provide some redundancy for *Fgfr2b* during placode induction. Because an *Fgfr1b* isoform-specific mutation has been generated (Partanen *et al.*, 1998), it will be possible to directly address this possibility. See Fig. 3B for a diagram summarizing the roles of *Fgf* and *Fgfr* genes in the early development of the mouse inner ear.

The question of whether *Fgf8* might also play a role in mouse otic placode induction has yet to be addressed. Mouse *Fgf8* is not expressed in the hindbrain but is instead found briefly in the preplacodal ectoderm and mesenchyme and is expressed in the pharyngeal endoderm over a longer period of time (Crossley and Martin, 1995), as is also the case in chick embryos (Hidalgo-Sanchez *et al.*, 2000; Karabagli *et al.*, 2002b; Stolte *et al.*, 2002). Thus, it is possible that *Fgf8* could signal directly to the placode in either an autocrine or paracrine mode from the placode itself or from the mesenchyme, respectively. Alternatively, it might have an indirect effect on otic development by signaling from either the ectoderm or endoderm to activate expression of mesenchymal inducers, such as FGF10. Because *Fgf8* null embryos die of gastrulation defects prior to otic placode induction (Sun *et al.*, 1999), determining the potential role of *Fgf8* in otic induction will have to await analysis of *Fgf8* conditional mutants (Meyers *et al.*, 1998; Moon and Capecchi, 2000). Localization of candidate receptors (FGFR2c, FGFR3c, FGFR4) for an FGF8 signal at placodal stages could also help in formulating an appropriate model for FGF8 function in placodal development.

c. Downstream Consequences of FGF Signaling. Although it is clearly established that FGF signaling is required for otic induction in fish (see Chapter 12 by Kiley) and mice, and the same is likely to be true for other species, the molecular consequences of FGF signaling to the preplacodal ectoderm are only beginning to be investigated. As indicated earlier, loss of FGF3 and FGF8 function in zebrafish or loss of *Fgf3* and *Fgf10* in mice results in the loss of ectodermal expression of a variety of marker genes encoding transcription factors, including *Pax2*, *Pax8*, and *Dlx3/Dlx5* (zebrafish/mouse). Some of these genes could be direct targets of FGF signaling, but others might be indirect targets that are activated by the primary transcriptional targets of FGF signaling. A study in zebrafish has begun to divide some of the placodal transcription factors required for ear development into FGF-dependent and FGF-independent classes. Using both germline mutations and morpholinos to block gene expression, as well as injections of mRNAs to rescue particular genes, Liu *et al.* (2003) showed that in addition

8. FGF Signaling in Ear Development and Innervation

to some of the previously determined genes, placodal *Sox9a* expression is also reduced when *Fgf3* and *Fgf8* expression is blocked. *Sox9a*, in turn, is required for expression of *Dlx3b* (*Dlx3* in older literature) and *Dlx4b* (*Dlx7* in older literature), suggesting that *Sox9a* may be a direct target of FGF signaling but that neither of the *Dlx* genes is a primary target of FGF signals. The intracellular signaling cascade responsible for linking the FGF signals to *Sox9a* was not determined. It is possible, however, that the RAS/MAPK pathway controls at least part of the response of ectodermal cells to FGF signaling because transcripts for the RAS–MAPK phosphorylation targets, ERM and PEA3, are absent from the otic region of SU5402-treated zebrafish embryos (Maroon *et al.*, 2002; Raible and Brand, 2001; Roehl and Nusslein-Volhard, 2001) and from *Fgf8* (*ace*) mutants (Raible and Brand, 2001; Roehl and Nusslein-Volhard, 2001). Furthermore, it is known that mouse *Sox9* induction by FGF signaling in another context is mediated by the RAS–MAPK cascade (Murakami *et al.*, 2000).

When *Fgf3* and *Fgf10* are both inactivated in mice, *Pax2* expression is absent from the entire ectoderm and three other marker genes, *Dlx5*, *Pax8*, and *Gbx2*, are excluded from the dorsal-most ectoderm (Wright and Mansour, 2003). One possible interpretation of this result is that expression of all four genes in the dorsal ectoderm is dependent on FGF signals and that there are additional (unknown) signals that are sufficient for induction of *Dlx5*, *Pax8*, and *Gbx2* expression in the more ventral regions of the ectoderm. Additional support for transcriptional induction of *Pax2* and *Dlx5* by FGFs comes from the previously mentioned studies in which FGF beads or FGF viruses were applied to chick ectodermal tissues and induction of *Pax2* and *Dlx5* expression was observed (Adamska *et al.*, 2001; Ladher *et al.*, 2000; Vendrell *et al.*, 2000). Alternatively, it could be that FGF3 and FGF10 signals are required for *Pax2* induction and that *Pax2*, in turn, is required in the dorsal but not in the ventral region to activate the other three genes. Additional studies will clearly be required to sort out the direct vs indirect targets of FGF signals to the placode.

Sprouty (*Spry*) genes are also transcriptional targets of FGF signaling through the RAS/MAPK pathway and encode inhibitors of FGF signaling (Gross *et al.*, 2001; Hacohen *et al.*, 1998; Hanafusa *et al.*, 2002). Thus, ectopic overexpression of *Spry* genes inhibits FGF-dependent processes and mimics FGF loss-of-function phenotypes (Minowada *et al.*, 1999). Indeed, when *Spry4*, which is inducible by both *Fgf3* and *Fgf8*, is ectopically expressed in zebrafish embryos, the otic vesicle is reduced in size, although not eliminated (Furthauer *et al.*, 2001). Mice have similar *Spry* genes, and expression of *Spry1*, *2*, and *4* has been reported during several phases of otic development (de Maximy *et al.*, 1999; Zhang *et al.*, 2001), suggesting that these genes may also play roles in limiting FGF signaling during mouse otic development.

In addition, the connection between FGF signals, changes in gene expression, and the morphological changes underlying the thickening and invagination of placodal tissue have not been tackled. One might expect that these morphological changes could involve alterations in cell–cell adhesion. In this respect it is interesting to note that one of the targets of FGFR1 signaling during gastrulation is the cell adhesion molecule, E-cadherin (Ciruna and Rossant, 2001), which is expressed throughout the forming placode (Richardson et al., 1987). Presumably it will be possible to address systematically the downstream targets of FGF signaling during placodal development by comparing gene expression profiles in wildtype placodal tissue with those of ectodermal tissue in which FGF signaling has been blocked, either chemically or genetically.

2. Otic Vesicle Formation

Shortly after the otic placode forms in frogs, chicks, and mice, it begins to invaginate, forming the otic cup. This is followed by closure of the cup to form a roughly spherical vesicle known as the otocyst, or otic vesicle, a process that may have parallels to primary neurulation (Mansour and Schoenwolf, 2003). Zebrafish have a different mode of vesicle formation. In this species, placodal cells form an aggregate that undergoes cavitation to form a vesicle, much as occurs during neurulation in zebrafish and in secondary neurulation in higher vertebrates (Haddon and Lewis, 1996; Whitfield et al., 2002; see Chapter 12 by Riley). At least two *Fgf* genes, *Fgf10* (Karabagli et al., 2002a; Pirvola et al., 2000; Wright and Mansour, 2003) and *Fgf16* (Hatch, Wright, and Mansour, unpublished data), as well as *Fgf* receptor genes, including *Fgfr2b* (Pirvola et al., 2000; Wright and Mansour, 2003) and *Fgfr3c* (Hatch, Wright, and Mansour, unpublished data), continue to be expressed in the mouse otic cup prior to vesicle formation, but it is not known whether FGF signaling plays a role in the cup-to-vesicle transition.

3. Otic Ganglion Formation and Otic Vesicle Morphogenesis

During the period of otic cup/vesicle formation, the otic epithelium delaminates neuroblasts from the anteroventral region (Altman and Bayer, 1982; Carney and Silver, 1983). These aggregate to form the eighth cranial or otic ganglion, which eventually splits into a cochlear and a vestibular component. Once the otic vesicle is formed, it initiates both cellular differentiation and morphogenesis. The first step of otic morphogenesis in frogs, chicks, and mice is to pinch off the dorsomedial region of the otic vesicle to form the endolymphatic duct/sac anlage. Zebrafish do not form an endolymphatic duct until after the rest of otic morphogenesis is complete (Bever and Fekete, 2002). In mice and chicks, the ventral portion of the otic epithelium begins to elongate, forming the cochlear (auditory) duct. Fish and amphibia do not form a cochlea but instead house their auditory apparatus in the saccule (Bever and Fekete, 2002; Paterson, 1948). The dorsal portion of the mouse and chick otic vesicle forms the vestibular system, composed of three mutually perpendicular semicircular ducts, the utricle, and the saccule. The first step in forming the semicircular ducts is a localized evagination of the otic epithelium. The first outpouching occurs dorsolaterally to form the vertical canal plate, which gives rise sequentially to the superior (anterior) and posterior semicircular ducts. The second outpouching occurs laterally to form the lateral (horizontal) canal plate, which gives rise to the lateral semicircular duct. After the evagination of the canal pouches, the two epithelial sheets detach from the surrounding mesenchyme and approach one another prior to fusion in the central region of the outpouching (Kiernan *et al.*, 2002; Morsli *et al.*, 1998). Zebrafish semicircular canal formation differs slightly in that there are no outpouchings of the epithelium from the otic vesicle. Instead, the epithelium protrudes inward at four sites into the vesicle lumen. Opposing protrusions eventually form the equivalent of a fusion plate (Whitfield *et al.*, 2002). In mice and zebrafish, the central cells are removed, perhaps by moving back into the duct epithelium (Martin and Swanson, 1993; Waterman and Bell, 1984). In chicks and frogs, the fused area undergoes apoptosis (Fekete *et al.*, 1997; Haddon and Lewis, 1996).

4. Fibroblast Growth Factor Signaling in Otic Ganglion Formation and Inner Ear Morphogenesis

a. *Fgf3*. Mice that are homozygous for a null allele of *Fgf3* have inner ear morphogenesis and otic ganglion formation defects that are highly variable (Mansour *et al.*, 1993). In addition, after many generations of backcrossing, they also exhibit variably small otic vesicles (Wright and Mansour, unpublished). The variability reflects both incomplete penetrance, i.e., not every homozygous mutant animal is affected, and variable expressivity, i.e., the two ears in a single animal can be affected to vastly different extents. Indeed, it is possible to identify individuals with one ear that is entirely normal and another ear that is completely abnormal as assessed by morphological and functional criteria. These findings suggest the possibility that *Fgf3* might be functionally redundant with another *Fgf* gene, and that explanation has been strengthened by the analysis of the *Fgf3/Fgf10* double mutants described previously.

The morphogenetic defects appear to be initiated by failure of endolymphatic duct formation. This leads in some cases to cystic development of the otic epithelium, with no clear evidence for formation of any recognizable cochlear or vestibular compartments or any sensory cell development, similar to the ears described for the *kreisler* (*MafB*) and *HoxA1* mouse mutants (Chisaka *et al.*, 1992; Deol, 1964). The otic ganglion in *Fgf3* mutants is also abnormally small, but a detailed description of its development has not been reported. Both of these features of the *Fgf3* null phenotype could be correlated with the expression pattern of *Fgf3* expression in the period immediately prior to the initial phenotypic effects (Mansour *et al.*, 1993). The failure of the endolymphatic duct to form correlates with the loss of *Fgf3* expression in hindbrain rhombomeres 5 and 6, which are located immediately adjacent to the dorsomedial site on the otic vesicle from which the endolymphatic duct emerges (Mahmood *et al.*, 1996; Mansour *et al.*, 1993; McKay *et al.*, 1996; Wilkinson *et al.*, 1988). This region of the otic vesicle expresses *Fgfr2b*, which encodes an FGF3 receptor (Pirvola *et al.*, 2000; Wright, Hatch, and Mansour, unpublished). The otic ganglion defects correlate with the loss of *Fgf3* expression from the neurogenic (anteroventral) region of the otic vesicle (McKay *et al.*, 1996; Pirvola *et al.*, 2000; Wilkinson *et al.*, 1989) and suggest a role for FGF3 in promoting the proliferation or migration of neuronal precursors. Because *Fgfr2b* transcripts appear to be excluded from this region (Pirvola *et al.*, 2000), it is not clear how the FGF3 signal is received. One possibility is that *Fgfr1b* might be expressed in the neurogenic region of the vesicle. Alternatively, it is possible that there is a reciprocal signaling loop involving receptors in the adjacent mesenchyme that turn on signaling molecules that communicate back to the neurogenic region. Low-level expression of *Fgfr2b* in the mesenchyme is consistent with this idea, but it is not known what signal it might induce to communicate to the developing neuroblasts (Pirvola *et al.*, 2000). The small vesicle size could actually be a consequence of an earlier role in placode development. It would be interesting to generate a conditional *Fgf3* allele in order to determine whether the different sites of *Fgf3* expression are actually responsible for the distinct mutant phenotypes or whether all of the phenotypes have a single origin. Furthermore, *Fgf3* is also expressed at later stages in the developing sensory epithelium and a conditional mutant would also permit assessment of its potential role in sensory development, independent of its morphogenetic role in endolymphatic duct formation.

Zebrafish embryos with reduced levels of FGF3 have inner ear phenotypes that are similar in some respects to mice that lack *Fgf3*. There are no *Fgf3* germline mutations available in zebrafish, but FGF3 function has been reduced by injecting morpholinos into embryos at the 1–2 cell stage. The treatment leads to the development of variably small otic vesicles (Leger and Brand, 2002; Phillips *et al.*, 2001). Otic vesicle size is reduced by approximately 30–35% (Maroon *et al.*, 2002) and molecular analysis detects a reduction, but not an absence, of expression of the otic placode marker genes *Pax8*, *Pax2.1*, *Dlx3*, *Pax5*, *MshD*, *MsxC*, *Otx1*, *Gsc*, and *Neurogenin 1* (Leger and Brand, 2002; Maroon *et al.*, 2002; Phillips *et al.*, 2001), arguing that

Fgf3 on its own plays a role in determining the size of the placodal region in zebrafish. The variability of this phenotype is probably due to the redundancy with *Fgf8* that was described earlier. The otic phenotype of zebrafish *Fgf3* morphants appears to initiate earlier than that of most mouse embryos missing *Fgf3* only. Removal of just one copy of *Fgf10* from mouse *Fgf3* mutant homozygotes, however, causes a reduction in the size of the otic vesicle (Wright and Mansour, 2003), similar to that seen in zebrafish *Fgf3* morphants. This suggests that although the basic requirement for FGF signaling in otic placode induction is the same for both species, there may be (relatively small) species-to-species variation in the spatial, temporal, and quantitative regulation of *Fgf* expression, all of which contribute to the phenotypic differences observed when only a single *Fgf* is depleted.

Although it is difficult to analyze late otic phenotypes in morpholino-treated embryos due to depletion of the morpholinos, Leger and Brand (2002) noted an absence of all epithelial protrusions of the semicircular canals in some FGF3-depleted embryos. This morphogenetic phenotype is not likely to have anything to do with endolymphatic duct formation, as proposed for the mouse, because the zebrafish endolymphatic duct develops after the semicircular canals (Bever and Fekete, 2002). Another similarity between the mouse *Fgf3* mutants and the zebrafish *Fgf3* morphants is the failure of or reduction in sensory hair cell production (Kwak *et al.*, 2002). It is not clear, however, whether these phenotypes reveal specific roles for *Fgf3* in epithelial budding to form the semicircular canals and in hair cell specification or whether they are secondary consequences of the initial small otic vesicle phenotype.

b. *Fgf10*. Subsequent to the preplacodal mesenchymal expression described previously, *Fgf10* is expressed throughout the mouse and chick otic cup (Karabagli *et al.*, 2002a; Pirvola *et al.*, 2000; Wright and Mansour, 2003). In the mouse, it persists in the otic vesicle, but its expression is progressively excluded from the nonsensory regions, eventually being concentrated in all of the prospective sensory patches, in the otic epithelium, and in the immature and mature neurons of the otic ganglion (Pauley *et al.*, 2003; Pirvola *et al.*, 2000).

Fgf10 null homozygotes have normal otic vesicle and endolymphatic duct formation, but the observation that they have small otic capsules suggested that they might also have inner ear defects (Ohuchi *et al.*, 2000). Indeed, an analysis of inner ear morphogenesis in *Fgf10* mutants confirms this point (Pauley *et al.*, 2003). The major defect in *Fgf10* mutant ears is a failure to form the posterior semicircular canal and its crista. There are also major dysmorphologies of the anterior and horizontal canals, but their crista can be recognized even though they are smaller than normal. In addition, the saccule and utricle are slightly reduced in size. The cochlea is remarkably normal. One possible explanation of the mutant phenotype is that FGF signaling, primarily initiated by FGF10, is required for the evagination of the canal pouches. The variability in the formation of rudimentary pouches, leading to aberrant formation of the anterior and lateral canals, might be due to redundancy provided by another FGF, such as FGF3, which is also expressed in the prospective sensory patches (Pirvola *et al.*, 2000; Wilkinson *et al.*, 1989) and which interacts with the same receptor isoforms as FGF10 (Mathieu *et al.*, 1995; Ornitz *et al.*, 1996). Because otic vesicle development is entirely blocked in standard *Fgf3/Fgf10* double mutant embryos (Wright and Mansour, 2003), conditional alleles will be needed to address this possibility.

Innervation of *Fgf10* mutant ears is also abnormal. There are no nerve fibers in the area where the missing posterior canal would have been located, and the supply of fibers to the anterior and horizontal cristae is reduced. There is, however, no defect in the initial formation of the sensory neurons, suggesting that the abnormalities of innervation are secondary to the loss of target sensory cells (Pauley *et al.*, 2003). Thus, the *Fgf10* single mutant phenotype does not reveal a role for FGF10 in the development of the otic ganglion, but as we see later, the phenotype of mice that lack the receptor for FGF10 strongly implicates this ligand in otic neurogenesis.

It is interesting to note that the primary defects in both *Fgf3* and *Fgf10* mutants appear to involve the budding of the otic epithelium to form the endolymphatic duct and canal pouches, respectively. *Fgf10* is also required for the budding morphogenesis of the limb and lung (Min *et al.*, 1998; Sekine *et al.*, 1999), and this aspect of FGF signaling in vertebrates is similar to the role of FGF (branchless) signaling in *Drosophila* tracheal development (Sutherland *et al.*, 1996).

c. ***Fgf8.*** The redundant role that *Fgf8* and *Fgf3* play in zebrafish otic placode induction has been discussed previously. However, analysis of otic development in *Fgf8* (*ace*) mutants has shown that this gene may also play a role in later otic development in this species (Leger and Brand, 2002). Like the zebrafish *Fgf3* morphants, *ace* mutants at 20–24 h of development have small otic vesicles and the number of cells expressing molecular markers for the anteromedial, dorsal, and posterior domains is reduced (Kwak *et al.*, 2002; Leger and Brand, 2002; Phillips *et al.*, 2001). Additional later aspects of this phenotype include a small otic ganglion; mislocalization of the sensory patches, especially the three cristae; and a substantial (>50%) reduction in the number of hair cells (Kwak *et al.*, 2002; Leger and Brand, 2002). It is possible that these defects occur secondarily to the reduced otic vesicle size. However, *Fgf8* is expressed in an anterior patch of the vesicle from which the anterior macula will develop, and more weakly and transiently at the posteromedial pole of the vesicle. From 48 h, *Fgf8* transcripts can be detected in the lumenal cell layer of the anterior macula containing the hair cells, in the cristae, and in the epithelial protrusions of the semicircular canal system. Therefore, it is possible that the later otic defects seen in *ace* mutants are due to a combination of direct and indirect effects of the lack of *Fgf8* on otic vesicle development and sensory hair cell differentiation. These data from the zebrafish suggest strongly that analysis of FGF8 function in conditional mouse mutants (Meyers *et al.*, 1998; Moon and Capecchi, 2000) will also reveal roles for this ligand in mouse otic morphogenesis.

d. *Fgfr2b*. Since both FGF3 and FGF10 signal most strongly through FGFR2b, it is not surprising that the otic defects seen in *Fgfr2b* mutants are more severe and penetrant than those of either the *Fgf3* or *Fgf10* mutants (Pirvola *et al.*, 2000; Revest *et al.*, 2001). In most cases, the membranous labyrinth of ears analyzed just prior to birth has the strong *Fgf3* phenotype, consisting of cystic development with no sensory organ development (Pirvola *et al.*, 2000). At early stages of development, the otic vesicles are small and lack an endolymphatic duct. This phenotype might be explained by the loss of FGFR2b from the dorsomedial region of the otic vesicle, where it could receive FGF3 signals from the adjacent hindbrain rhombomeres 5 and 6 and FGF10 signals from the medial region of the vesicle to induce endolymphatic duct morphogenesis. In addition, formation of the otic ganglion is rudimentary, though not entirely blocked in *Fgfr2b* mutants. In this case, it appears that there is some initial delamination and migration of neuronal precursors but that FGF signaling through FGFR2b is necessary for survival of the neurons of the ganglion. This phenotype is somewhat more difficult to understand. The *Fgf3* and *Fgf10* genes are both expressed in the neurogenic region of the vesicle, but *Fgfr2b* expression is excluded from that region. As suggested in the discussion of the *Fgf3* mutant phenotype, FGF3 and FGF10 expressed in the neurogenic region may signal to FGFR2b in the adjacent regions of the vesicle, and this may in turn induce another signal that impinges upon the delaminating neuroblasts to promote their survival.

Additional support for the role of *Fgfr2b* in otic morphogenesis comes from studies of transgenic mice expressing a soluble dominant negative form of FGFR2b. Although not examined in detail, the transgenic mice appear to have inner ears that are very similar to those of the *Fgfr2b* mutants (Celli *et al.*, 1998). See Fig. 3C for a diagram summarizing the roles of *Fgf* and *Fgfr* genes in the early morphogenesis of the mouse inner ear.

5. Development of the Sensory Epithelium

The otic vesicle epithelium gives rise to all of the sensory and nonsensory cells of the mature inner ear. Hair cells and supporting cells within the sensory epithelia of the avian inner ear share a common precursor (Fekete et al., 1998; Stone and Rubel, 2000). As discussed in more detail later, proliferation of these precursor cells appears to be mediated in part by FGF signals. The bipotential precursor cells undergo their final mitoses during the same period when the epithelium is undergoing its complex morphogenesis (Ruben, 1967). The initial decision of precursor cells to adopt a sensory vs a nonsensory fate is controlled by the Notch signaling system (Bryant et al., 2002; Eddison et al., 2000; Fekete and Wu, 2002). The genetic cascades involved in subsequent stages of the differentiation of hair cells are beginning to be elucidated, but to date there is no experimental evidence for specific roles for FGF signaling in hair cell fate specification per se. There may, however, be a role for FGF signaling to promote outer hair cell differentiation. In addition to hair cells, there are many types of supporting cells in the mature sensory epithelium, but relative to hair cells, few details of the genetic programs contributing to their differentiation are known. In general, however, it appears that normal differentiation of supporting cells depends on normal differentiation of hair cells because mutations in hair cell–specific genes, such as *Pou4f3*, that disrupt hair cell differentiation also affect supporting cell differentiation (Erkman et al., 1996; Xiang et al., 1997). Expression of several FGFs has been reported in developing hair cells, suggesting the possibility that FGF signaling could provide signals for the differentiation of supporting cells. Indeed, as discussed in the next section, it appears that differentiation of at least one supporting cell type, the pillar cell, depends on FGF signaling.

a. Role of FGF Signaling in the Proliferation of Hair Cell Precursors and the Differentiation of Outer Hair Cells. *Fgfr1* is expressed throughout inner ear development. It can be detected in early somite stage mouse embryos in the prospective placode (Wright and Mansour, 2003). At the early otocyst stage, *Fgfr1* transcripts are found in the periotic mesenchyme, where they persist for the entire course of otic development. As the cochlea elongates ventrally and differentiates, *Fgfr1* transcripts are seen first in the ventral wall and then slightly later in the thickened greater epithelial ridge, as well as in the developing organ of Corti, eventually becoming concentrated in outer hair cells and Deiters' (supporting) cells (Pirvola et al., 2002).

Complete elimination of *Fgfr1* function is lethal during the early postimplantation to gastrulation period of mouse development (Deng et al., 1994; Yamaguchi et al., 1994); therefore otic-specific functions of *Fgfr1* were evaluated using hypomorphic (Partanen et al., 1998) and conditional mutants (Trokovic et al., 2003). Pirvola et al. (2002) showed that the gross morphogenesis of the inner ears of newborn mice with either 10% or 20% of normal *Fgfr1* levels is normal; however, cellular patterning in the organ of Corti, but not in the vestibular sensory organs, is disrupted. In both hypomorphic mutants, the third row of outer hair cells is missing. In the animals with only 10% of normal *Fgfr1* levels, the basal half of the organ of Corti is even more severely affected, with patches rather than rows of sensory cells and supporting cells. In newborn animals devoid of otic epithelial *Fgfr1*, the cochlear phenotype is similar to but even more severe than that seen in the ears with 10% of normal *Fgfr1*. There are very few outer hair cells, with most being located in the apical half of the cochlea, and the patches of sensory tissue that do form are composed mainly of patches of inner hair cells and supporting cells.

Analysis of the development of the conditional *Fgfr1* null ears and the expression of molecular markers does not suggest that this phenotype is related to a failure to specify hair cell fate; rather, the defect may be explained by a specific reduction in the proliferation of precursor cells that give rise to both the sensory and supporting cells of the organ of Corti (Pirvola *et al.*, 2002). Because none of the phenotypes described are evident in *Fgfr1b* mutants, it is likely that the defects are due to loss of FGFR1c. Taken together, it appears that FGF signaling through FGFR1c plays a dose-dependent role in regulating the number of cells that give rise to the sensory epithelium. The fact that some hair cells still form in the absence of *Fgfr1* suggests the possibility that either there are FGF-independent mechanisms of generating an initial pool of hair cell precursors or that there is some redundancy provided by a different FGF receptor. The milder phenotype seen in the hypomorphic mutants suggests that FGFR1c is also needed in the organ of Corti, after precursor cells have proliferated and made the supporting cell vs hair cell fate choice, to stimulate differentiation of outer hair cells. The FGF ligands involved in these two functions of FGFR1 are not known. None of the *Fgfs* that are known to be expressed in the vicinity of the *Fgfr1*-expressing otic epithelial cells (*Fgf3*, *Fgf8*, and *Fgf10*) are thought to have significant affinity for FGFR1c. Additional studies of the expression and function of FGFs that do activate FGFR1c will be informative.

b. FGF Signals in the Differentiation of Pillar Cells. Two lines of evidence support a specific role for FGF signaling in the differentiation of pillar cells, one of the supporting cell types present in the organ of Corti. Mice that are homozygous for a null allele of *Fgfr3* have no auditory brainstem response and normal middle ears, suggesting profound sensorineural hearing loss (Colvin *et al.*, 1996). The only morphological abnormality apparent in the *Fgfr3* null mutant cochleae is a complete absence of pillar cells and the tunnel of Corti, a space that the pillar cells ordinarily surround and that separates the inner hair cells from the outer hair cells. Instead, there are two extra rows of "Deiters'-like" cells underneath the hair cells, suggesting the possibility that these unusual cells may share a common origin with Deiters' cells and that they failed to receive an FGF signal necessary for appropriate differentiation as pillar cells. Normal expression of FGFR3, initially in the band of cells that will differentiate as pillar cells, Deiters' cells, or outer hair cells, and subsequently in the pillar cells themselves, is consistent with this interpretation (Mueller et al., 2002; Pirvola et al., 1995). These studies did not determine which isoforms of FGFR3 are expressed by the developing sensory epithelium. They do, however, raise the question of why differentiation of the Deiters' cells and outer hair cells is not affected by the *Fgfr3* mutation. One possibility is that the unaffected cell types may also express another FGF receptor that can be activated by the same ligand that activates FGFR3 in pillar cells (receptor redundancy). Consistent with this possibility, *Fgfr1* (isoform not determined) is detectable in developing Deiters' and outer hair cells, but not in pillar cells (Pirvola et al., 2002). Another possible interpretation is that FGFR3 signaling is not required for differentiation of Deiters' and outer hair cells.

Studies of mouse embryonic cochleae cultured *in vitro* in the presence of the FGF receptor inhibitor, SU5402, have added additional insight into the role of FGF signaling in pillar cell development (Mueller et al., 2002). Cochleae were explanted at embryonic day 13 or 14, cultured for 1 day, and then treated with the inhibitor. This treatment resulted in a remarkably accurate phenocopy of the *Fgfr3* null mutant, reinforcing the conclusion that FGF signaling is required for pillar cell differentiation. In addition, the results of varying timing of initiating inhibitor treatment and the quantity of inhibitor applied showed that FGF signaling is required continuously in a dose-dependent fashion throughout the period of pillar cell development. Pillar cells can even differentiate after transient inhibition of FGF signaling, and submaximal concentrations of the inhibitor only partially inhibit differentiation of pillar cells. Because *Fgfr1*, which is expressed by developing Deiters' cells and outer hair cells, was presumably also inhibited by this treatment, these results, taken together with the *Fgfr1* mutant studies described previously (Pirvola et al., 2002), suggest that Deiters' cell differentiation does not depend on FGF signaling and that the FGFR1-dependent signal for differentiation of outer hair cells may occur prior to the stage at which the cultured cochleae were treated with the inhibitor. In addition, Mueller et al. (2002) showed that FGF signaling, induced using FGF2 protein, is sufficient to promote production of supernumerary pillar cells. It should be noted that FGF2, which is not expressed in the developing cochlea (Luo et al., 1993), is not likely to provide the endogenous differentiation signal. Instead, FGF8, which is expressed by the developing inner hair cells (Pirvola et al., 2002) and which efficiently activates FGFR3c (Ornitz et al., 1996) and/or a related FGF (e.g., FGF17 or FGF18), is the likely endogenous ligand responsible for promoting pillar cell differentiation. Studies of an *Fgf8* conditional allele, inactivated using an otic epithelial CRE-driver such as $Foxg1(BF-1)^{Cre}$ (Hebert and McConnell, 2000), could clarify this issue, unless *Fgf8* also plays an earlier role in otic epithelial development. Additional *Fgfs* that could potentially play roles in the development and differentiation of supporting cells include *Fgf3* and *Fgf10*, which have both been detected in developing hair cells (Pirvola et al., 2000, 2002; Wilkinson et al., 1989). To date, their specific roles in the development of the sensory epithelium have not been investigated. See Fig. 3D for a diagram summarizing the roles of *Fgf* and *Fgfr* genes in the proliferation and differentiation of the organ of Corti.

III. Conclusions and Future Directions

As we have seen, FGF signaling is already known to play important roles in virtually every aspect of the development and innervation of the ear (summarized in Fig. 3). *Fgf8* and *Fgfr1*, which are not likely to comprise a ligand/receptor pair, have roles in both outer and middle ear development in the mouse. Hindbrain-expressed *Fgf3* and mesenchyme-expressed *Fgf10* acting through placodal *Fgfr2b* and possibly *Fgfr1b* are required for induction of the otic placode in mouse. FGF signaling is also required for placode induction in fish, but in this case the Fgfs are both expressed in hindbrain and the receptors have not been established. *Fgf3* and *Fgf10*, probably acting through *Fgfr2b*, also play later roles in mouse endolymphatic duct formation and canal plate outpouching, respectively. Finally, FGF signals are required for the initial proliferation of organ of Corti precursors, as well as for outer hair cell differentiation. Finally, an FGF8/FGFR3 interaction is apparently required for pillar cell differentiation.

That said, our knowledge of the true scope of FGF signaling in these processes is rudimentary. For obvious reasons, the most detailed studies have involved the *Fgfs* and *Fgfrs* that were among the first to be discovered. As we gain a better understanding of the expression patterns of all of these genes during the development of the various experimental organisms, we will be in a better position to carry out investigations of their functions in the ear. We have begun to appreciate the issues of both ligand and receptor redundancy that can initially mask important aspects of gene function in any developmental processes. The ability to create double mutants and to apply multiple FGFs to tissues in culture will continue to help in unraveling the complexities of redundancy. Conversely, early defects or even lethality can confound our genetic understanding of late functions of FGFs and their receptors. Two areas of research will contribute to overcoming these difficulties. The emerging technology for conditional gene inactivation is beginning to allow us to control the timing and specific location of gene disruption. In addition, *in vitro* culture systems suitable for the various stages of ear development are improving steadily. Together, these technologies promise to allow a finer dissection of the roles of FGF signaling in all stages of ear development. Finally, our knowledge of the downstream genetic pathways controlled by FGF signaling and our understanding of the ways in which FGF signaling is integrated with other signaling pathways to control ear development and function remain to be elucidated.

Acknowledgments

Research in the authors' laboratory is supported by the NIH/NIDCD and the Deafness Research Foundation. We thank Diana Lim for her tireless efforts to produce Figure 3.

References

Abu-Issa, R., Smyth, G., Smoak, I., Yamamura, K., and Meyers, E. N. (2002). Fgf8 is required for pharyngeal arch and cardiovascular development in the mouse. *Development* **129**, 4613–4625.

Adamska, M., Herbrand, H., Adamski, M., Kruger, M., Braun, T., and Bober, E. (2001). FGFs control the patterning of the inner ear but are not able to induce the full ear program. *Mech. Dev.* **109**, 303–313.

Altman, J., and Bayer, S. (1982). Development of the cranial nerve ganglia and related nuclei in the rat. *Adv. Anat. Embryol. Cell Biol.* **74**, 1–90.

Bachler, M., and Neubuser, A. (2001). Expression of members of the Fgf family and their receptors during midfacial development. *Mech. Dev.* **100**, 313–316.

Baker, C. V., and Bronner-Fraser, M. (2001). Vertebrate cranial placodes I. Embryonic induction. *Dev. Biol.* **232**, 1–61.

Beer, H. D., Vindevoghel, L., Gait, M. J., Revest, J. M., Duan, D. R., *et al.* (2000). Fibroblast growth factor (FGF) receptor 1-IIIb is a naturally occurring functional receptor for FGFs that is preferentially expressed in the skin and the brain. *J. Biol. Chem.* **275**, 16091–16097.

Bever, M. M., and Fekete, D. M. (2002). Atlas of the developing inner ear in zebrafish. *Dev. Dyn.* **223**, 536–543.

Blunt, A. G., Lawshe, A., Cunningham, M. L., Seto, M. L., Ornitz, D. M., and MacArthur, C. A. (1997). Overlapping expression and redundant activation of mesenchymal fibroblast growth factor (FGF) receptors by alternatively spliced FGF-8 ligands. *J. Biol. Chem.* **272**, 3733–3738.

Brookes, S., Smith, R., Casey, G., Dickson, C., and Peters, G. (1989). Sequence organization of the human *int*-2 gene and its expression in teratocarcinoma cells. *Oncogene* **4**, 429–436.

Bryant, J., Goodyear, R. J., and Richardson, G. P. (2002). Sensory organ development in the inner ear: Molecular and cellular mechanisms. *Br. Med. Bull.* **63**, 39–57.

Camps, M., Nichols, A., and Arkinstall, S. (2000). Dual specificity phosphatases: A gene family for control of MAP kinase function. *FASEB J.* **14**, 6–16.

Carney, P. R., and Silver, J. (1983). Studies on cell migration and axon guidance in the developing distal auditory system of the mouse. *J. Comp. Neurol.* **215**, 359–369.

Celli, G., LaRochelle, W. J., Mackem, S., Sharp, R., and Merlino, G. (1998). Soluble dominant-negative receptor uncovers essential roles for fibroblast growth factors in multi-organ induction and patterning. *EMBO J.* **17**, 1642–1655.

Chang, L., and Karin, M. (2001). Mammalian MAP kinase signalling cascades. *Nature* **410**, 37–40.

8. FGF Signaling in Ear Development and Innervation

Chisaka, O., Musci, T. S., and Capecchi, M. R. (1992). Developmental defects of the ear, cranial nerves and hindbrain resulting from targeted disruption of the mouse homeobox gene *Hox-1.6. Nature* **355**, 516–520.

Ciruna, B., and Rossant, J. (2001). FGF signaling regulates mesoderm cell fate specification and morphogenetic movement at the primitive streak. *Dev. Cell* **1**, 37–49.

Colvin, J. S., Bohne, B. A., Harding, G. W., McEwen, D. G., and Ornitz, D. M. (1996). Skeletal overgrowth and deafness in mice lacking fibroblast growth factor receptor 3. *Nat. Genet.* **12**, 390–397.

Crossley, P. H., and Martin, G. R. (1995). The mouse *Fgf8* gene encodes a family of polypeptides and is expressed in regions that direct outgrowth and patterning in the developing embryo. *Development* **121**, 439–451.

de Maximy, A. A., Nakatake, Y., Moncada, S., Itoh, N., Thiery, J. P., and Bellusci, S. (1999). Cloning and expression pattern of a mouse homologue of *Drosophila* sprouty in the mouse embryo. *Mech. Dev.* **81**, 213–216.

Deng, C., Wynshaw-Boris, A., Zhou, F., Kuo, A., and Leder, P. (1996). Fibroblast growth factor receptor 3 is a negative regulator of bone growth. *Cell* **84**, 911–921.

Deng, C. X., Wynshaw-Boris, A., Shen, M. M., Daugherty, C., Ornitz, D. M., and Leder, P. (1994). Murine FGFR-1 is required for early postimplantation growth and axial organization. *Genes Dev.* **8**, 3045–3057.

Deol, M. S. (1964). The abnormalities of the inner ear in *kreisler* mice. *J. Embryol. Exp. Morph.* **12**, 475–490.

Dono, R. (2003). Fibroblast growth factors as regulators of central nervous system development and function. *Am. J. Physiol. Regul. Integr. Comp. Physiol.* **284**, R867–881.

Eddison, M., Le Roux, I., and Lewis, J. (2000). Notch signaling in the development of the inner ear: Lessons from *Drosophila. Proc. Natl. Acad. Sci. USA* **97**, 11692–11699.

Erkman, L., McEvilly, R. J., Luo, L., Ryan, A. K., Hooshmand, F., et al. (1996). Role of transcription factors Brn-3.1 and Brn-3.2 in auditory and visual system development. *Nature* **381**, 603–606.

Eswarakumar, V. P., Monsonego-Ornan, E., Pines, M., Antonopoulou, I., Morriss-Kay, G. M., and Lonai, P. (2002). The IIIc alternative of Fgfr2 is a positive regulator of bone formation. *Development* **129**, 3783–3793.

Fedorov, Y. V., Rosenthal, R. S., and Olwin, B. B. (2001). Oncogenic Ras-induced proliferation requires autocrine fibroblast growth factor 2 signaling in skeletal muscle cells. *J. Cell. Biol.* **152**, 1301–1305.

Fekete, D. M., Homburger, S. A., Waring, M. T., Riedl, A. E., and Garcia, L. F. (1997). Involvement of programmed cell death in morphogenesis of the vertebrate inner ear. *Development* **124**, 2451–2461.

Fekete, D. M., Muthukumar, S., and Karagogeos, D. (1998). Hair cells and supporting cells share a common progenitor in the avian inner Ear. *J. Neurosci.* **18**, 7811–7821.

Fekete, D. M., and Wu, D. K. (2002). Revisiting cell fate specification in the inner ear. *Curr. Opin. Neurobiol.* **12**, 35–42.

Ford-Perriss, M., Abud, H., and Murphy, M. (2001). Fibroblast growth factors in the developing central nervous system. *Clin. Exp. Pharmacol. Physiol.* **28**, 493–503.

Frank, D. U., Fotheringham, L. K., Brewer, J. A., Muglia, L. J., Tristani-Firouzi, M., et al. (2002). An Fgf8 mouse mutant phenocopies human 22q11 deletion syndrome. *Development* **129**, 4591–4603.

Furthauer, M., Lin, W., Ang, S. L., Thisse, B., and Thisse, C. (2002). Sef is a feedback-induced antagonist of Ras/MAPK-mediated FGF signalling. *Nat. Cell Biol.* **4**, 170–174.

Furthauer, M., Reifers, F., Brand, M., Thisse, B., and Thisse, C. (2001). sprouty4 acts in vivo as a feedback-induced antagonist of FGF signaling in zebrafish. *Development* **128**, 2175–2186.

Gospodarowicz, D. (1974). Localisation of a fibroblast growth factor and its effect alone and with hydrocortisone on 3T3 cell growth. *Nature* **249**, 123–127.

Gross, I., Bassit, B., Benezra, M., and Licht, J. D. (2001). Mammalian sprouty proteins inhibit cell growth and differentiation by preventing ras activation. *J. Biol. Chem.* **276**, 46460–46468.

Groves, A. K. (2003). Induction of the otic placode. *In* "Springer Handbook of Auditory Research" (D. K. Wu and M. W. Kelley, Eds.). Springer-Verlag (in press).

Hacohen, N., Kramer, S., Sutherland, D., Hiromi, Y., and Krasnow, M. A. (1998). *sprouty* encodes a novel antagonist of FGF signaling that patterns apical branching of the *Drosophila* airways. *Cell* **92**, 253–263.

Haddon, C., and Lewis, J. (1996). Early ear development in the embryo of the zebrafish. *Danio rerio. J. Comp. Neurol.* **365**, 113–128.

Hanafusa, H., Torii, S., Yasunaga, T., and Nishida, E. (2002). Sprouty1 and Sprouty2 provide a control mechanism for the Ras/MAPK signalling pathway. *Nat. Cell Biol.* **4**, 850–858.

Hebert, J. M., and McConnell, S. K. (2000). Targeting of cre to the Foxg1 (BF-1) locus mediates loxP recombination in the telencephalon and other developing head structures. *Dev. Biol.* **222**, 296–306.

Hidalgo-Sanchez, M., Alvarado-Mallart, R., and Alvarez, I. S. (2000). Pax2, Otx2, Gbx2 and Fgf8 expression in early otic vesicle development. *Mech. Dev.* **95**, 225–229.

Hogan, B. L. (1999). Morphogenesis. *Cell* **96**, 225–233.

Karabagli, H., Karabagli, P., Ladher, R. K., and Schoenwolf, G. C. (2002a). Comparison of the expression patterns of several fibroblast growth factors during chick gastrulation and neurulation. *Anat. Embryol. (Berl)* **205**, 365–370.

Karabagli, H., Karabagli, P., Ladher, R. K., and Schoenwolf, G. C. (2002b). Survey of fibroblast growth factor expression during chick organogenesis. *Anat. Rec.* **268**, 1–6.

Kaufman, M. H., and Bard, J. B. L. (1999). The eye and the ear. *In* "The Anatomical Basis of Mouse Development," pp. 194-208. Academic Press, San Diego.

Kettunen, P., Karavanova, I., and Thesleff, I. (1998). Responsiveness of developing dental tissues to fibroblast growth factors: Expression of splicing alternatives of FGFR1, -2, -3, and of FGFR4; and stimulation of cell proliferation by FGF-2, -4, and -9. *Dev. Genet.* **22**, 374–385.

Keyse, S. M. (2000). Protein phosphatases and the regulation of mitogen-activated protein kinase signalling. *Curr. Opin. Cell Biol.* **12**, 186–192.

Kiernan, A. E., Steel, K. P., and Fekete, D. M. (2002). Development of the mouse inner ear. *In* "Mouse Development: Patterning Morphogenesis and Organogenesis" (J. Rossant and P. Tam, Eds.), pp. 539–566. Academic Press, San Diego.

Klint, P., and Claesson-Welsh, L. (1999). Signal transduction by fibroblast growth factor receptors. *Front Biosci.* **4**, D165–177.

Kolch, W. (2000). Meaningful relationships: The regulation of the Ras/Raf/MEK/ERK pathway by protein interactions. *Biochem. J.* **351 Pt 2**, 289–305.

Konishi, M., Mikami, T., Yamasaki, M., Miyake, A., and Itoh, N. (2000). Fibroblast growth factor-16 is a growth factor for embryonic brown adipocytes. *J. Biol. Chem.* **275**, 12119–12122.

Kovalenko, D., Yang, X., Nadeau, R. J., Harkins, L. K., and Friesel, R. (2003). Sef inhibits fibroblast growth factor signaling by inhibiting FGFR1 tyrosine phosphorylation and subsequent ERK activation. *J. Biol. Chem.* **24**, 24.

Kwak, S. J., Phillips, B. T., Heck, R., and Riley, B. B. (2002). An expanded domain of fgf3 expression in the hindbrain of zebrafish valentino mutants results in mis-patterning of the otic vesicle. *Development* **129**, 5279–5287.

Ladher, R. K., Anakwe, K. U., Gurney, A. L., Schoenwolf, G. C., and Francis-West, P. H. (2000). Identification of synergistic signals initiating inner ear development. *Science* **290**, 1965–1967.

Lambert, P. R., and Dodson, E. E. (1996). Congenital malformations of the external auditory canal. *Otolaryngol. Clin. North Am.* **29**, 741–760.

Leger, S., and Brand, M. (2002). Fgf8 and Fgf3 are required for zebrafish ear placode induction, maintenance and inner ear patterning. *Mech. Dev.* **119**, 91–108.

Lin, W., Furthauer, M., Thisse, B., Thisse, C., Jing, N., and Ang, S. L. (2002). Cloning of the mouse Sef gene and comparative analysis of its expression with Fgf8 and Spry2 during embryogenesis. *Mech. Dev.* **113**, 163–168.

Liu, D., Chu, H., Maves, L., Yan, Y. L., Morcos, P. A., *et al.* (2003). Fgf3 and Fgf8 dependent and independent transcription factors are required for otic placode specification. *Development* **130**, 2213–2224.

Liu, Z., Xu, J., Colvin, J. S., and Ornitz, D. M. (2002). Coordination of chondrogenesis and osteogenesis by fibroblast growth factor 18. *Genes Dev.* **16**, 859–869.

Lombardo, A., and Slack, J. M. W. (1998). Postgastrulation effects of fibroblast growth factor on *Xenopus* development. *Dev. Dyn.* **212**, 75–85.

Luo, L., Koutnouyan, H., Baird, A., and Ryan, A. F. (1993). Acidic and basic FGF mRNA expression in the adult and developing rat cochlea. *Hear. Res.* **69**, 182–193.

MacArthur, C. A., Lawshe, A., Xu, J., Santos-Ocampo, S., Heikinheimo, M., *et al.* (1995). FGF-8 isoforms activate receptor splice forms that are expressed in mesenchymal regions of mouse development. *Development* **121**, 3603–3613.

Mahmood, R., Kiefer, P., Guthrie, S., Dickson, C., and Mason, I. (1995). Multiple roles for FGF-3 during cranial neural development in the chicken. *Development* **121**, 1399–1410.

Mahmood, R., Mason, I. J., and Morriss-Kay, G. M. (1996). Expression of *Fgf-3* in relation to hindbrain segmentation, otic pit position and pharyngeal arch morphology in normal and retinoic acid-exposed mouse embryos. *Anat. Embryol.* **194**, 13–22.

Mallo, M. (2001). Formation of the middle ear: Recent progress on the developmental and molecular mechanisms. *Dev. Biol.* **231**, 410–419.

Mallo, M., and Gridley, T. (1996). Development of the mammalian ear: Coordinate regulation of formation of the tympanic ring and the external acoustic meatus. *Development* **122**, 173–179.

Mallo, M., Schrewe, H., Martin, J. F., Olson, E. N., and Ohnemus, S. (2000). Assembling a functional tympanic membrane: Signals from the external acoustic meatus coordinate development of the malleal manubrium. *Development* **127**, 4127–4136.

Mansour, S. L., Goddard, J. M., and Capecchi, M. R. (1993). Mice homozygous for a targeted disruption of the proto-oncogene *int-2* have developmental defects in the tail and inner ear. *Development* **117**, 13–28.

Mansour, S. L., and Martin, G. R. (1988). Four classes of mRNA are expressed from the mouse *int-2* gene, a member of the FGF gene family. *EMBO J.* **7**, 2035–2041.

Mansour, S. L., and Schoenwolf, G. C. (2003). Morphogenesis of the inner ear. *In* "Springer Handbook of Auditory Research" (D. K. Wu and M. W. Kelley, Eds.). Springer-Verlag (in press).

Maroon, H., Walshe, J., Mahmood, R., Kiefer, P., Dickson, C., and Mason, I. (2002). Fgf3 and Fgf8 are required together for formation of the otic placode and vesicle. *Development* **129**, 2099–2108.

Martin, G. R. (1998). The roles of FGFs in the early development of vertebrate limbs. *Genes Dev.* **12**, 1571–1586.

Martin, P., and Swanson, G. J. (1993). Descriptive and experimental analysis of the epithelial remodellings that control semicircular canal formation in the developing mouse inner ear. *Dev. Biol.* **159**, 549–558.

Mathieu, M., Chatelain, E., Ornitz, D., Bresnick, J., Mason, I., *et al.* (1995). Receptor binding and mitgenic properties of mouse fibroblast growth factor 3. *J. Biol. Chem.* **270**, 24197–24230.

Maves, L., Jackman, W., and Kimmel, C. B. (2002). FGF3 and FGF8 mediate a rhombomere 4 signaling activity in the zebrafish hindbrain. *Development* **129**, 3825–3837.

McIntosh, I., Bellus, G. A., and Jab, E. W. (2000). The pleiotropic effects of fibroblast growth factor receptors in mammalian development. *Cell Struct. Funct.* **25**, 85–96.

McKay, I. J., Lewis, J., and Lumsden, A. (1996). The role of FGF-3 in early inner ear development: An analysis in normal and *kreisler* mutant mice. *Dev. Biol.* **174**, 370–378.

Meyers, E. N., Lewandoski, M., and Martin, G. R. (1998). An *Fgf8* mutant allelic series generated by Cre- and Flp-mediated recombination. *Nat. Genet.* **18**, 136–141.

Min, H., Danilenko, D. M., Scully, S. A., Bolon, B., Ring, B. D., et al. (1998). Fgf-10 is required for both limb and lung development and exhibits striking functional similarity to *Drosophila* branchless. *Genes Dev.* **12**, 3156–3161.

Minowada, G., Jarvis, L. A., Chi, C. L., Neubuser, A., Sun, X., et al. (1999). Vertebrate Sprouty genes are induced by FGF signaling and can cause chondrodysplasia when overexpressed. *Development* **126**, 4465–4475.

Mohammadi, M., McMahon, G., Sun, L., Tang, C., Hirth, P., et al. (1997). Structures of the tyrosine kinase domain of fibroblast growth factor receptor in complex with inhibitors. *Science* **276**, 955–960.

Moon, A. M., and Capecchi, M. R. (2000). *Fgf8* is required for outgrowth and patterning of the limbs. *Nat. Genet.* **26**, 455–459.

Morsli, H., Choo, D., Ryan, A., Johnson, R., and Wu, D. K. (1998). Development of the mouse inner ear and origin of its sensory organs. *J. Neurosci.* **18**, 3327–3335.

Mueller, K. L., Jacques, B. E., and Kelley, M. W. (2002). Fibroblast growth factor signaling regulates pillar cell development in the organ of corti. *J. Neurosci.* **22**, 9368–9377.

Murakami, S., Kan, M., McKeehan, W. L., and de Crombrugghe, B. (2000). Up-regulation of the chondrogenic Sox9 gene by fibroblast growth factors is mediated by the mitogen-activated protein kinase pathway. *Proc. Natl. Acad. Sci. USA* **97**, 1113–1118.

Niswander, L., and Martin, G. R. (1992). Fgf-4 expression during gastrulation, myogenesis, limb and tooth development in the mouse. *Development* **114**, 755–768.

Noramly, S., and Grainger, R. M. (2002). Determination of the embryonic inner ear. *J. Neurobiol.* **53**, 100–128.

Ohbayashi, N., Shibayama, M., Kurotaki, Y., Imanishi, M., Fujimori, T., et al. (2002). FGF18 is required for normal cell proliferation and differentiation during osteogenesis and chondrogenesis. *Genes Dev.* **16**, 870–879.

Ohuchi, H., Hori, Y., Yamasaki, M., Harada, H., Sekine, K., et al. (2000). FGF10 acts as a major ligand for FGF receptor 2 IIIb in mouse multi-organ development. *Biochem. Biophys. Res. Commun.* **277**, 643–649.

Ornitz, D. M. (2000). FGFs, heparan sulfate and FGFRs: Complex interactions essential for development. *Bioessays* **22**, 108–112.

Ornitz, D. M., and Itoh, N. (2001). Fibroblast growth factors. *Genome Biol.* **2**.

Ornitz, D. M., and Marie, P. J. (2002). FGF signaling pathways in endochondral and intramembranous bone development and human genetic disease. *Genes Dev.* **16**, 1446–1465.

Ornitz, D. M., Xu, J., Colvin, J. S., McEwen, D. G., MacArthur, C. A., et al. (1996). Receptor specificity of the fibroblast growth factor family. *J. Biol. Chem.* **271**, 15292–15297.

Orvidas, L. J., Fabry, L. B., Diacova, S., and McDonald, T. J. (1999). Hearing and otopathology in Crouzon syndrome. *Laryngoscope* **109**, 1372–1375.

Ozawa, K., Uruno, T., Miyakawa, K., Seo, M., and Imamura, T. (1996). Expression of the fibroblast growth factor family and their receptor family genes during mouse brain development. *Brain Res. Mol. Brain Res.* **41**, 279–288.

Partanen, J., Schwartz, L., and Rossant, J. (1998). Opposite phenotypes of hypomorphic and Y766 phosphorylation site mutations reveal a function for fgfr1 in anteroposterior patterning of mouse embryos. *Genes Dev.* **12**, 2332–2344.

Paterson, N. F. (1948). The development of the inner ear of *Xenopus laevis*. *Proc. Zool. Soc. Lond.* **119**, 269–291.

Pauley, S., Wright, T. J., Pirvola, U., Ornitz, D. M., Beisel, K., and Fritzsch, B. (2003). Expression and function of FGF10 in mammalian inner ear development. *Dev. Dyn.* **227**, 203–215.

Pearson, G., Robinson, F., Beers Gibson, T., Xu, B. E., Karandikar, M., *et al.* (2001). Mitogen-activated protein (MAP) kinase pathways: Regulation and physiological functions. *Endocr. Rev.* **22**, 153–183.

Peters, H., Neubuser, A., Kratochwil, K., and Balling, R. (1998). Pax9-deficient mice lack pharyngeal pouch derivatives and teeth and exhibit craniofacial and limb abnormalities. *Genes Dev.* **12**, 2735–2747.

Phillips, B. T., Bolding, K., and Riley, B. B. (2001). Zebrafish fgf3 and fgf8 encode redundant functions required for otic placode induction. *Dev. Biol.* **235**, 351–365.

Phippard, D., Lu, L., Lee, D., Saunders, J. C., and Crenshaw, E. B., III. (1999). Targeted mutagenesis of the POU-domain gene *Brn4/Pou3f4* causes developmental defects in the inner ear. *J. Neurosci.* **19**, 5980–5989.

Pickles, J. O., and Chir, B..(2002). Roles of fibroblast growth factors in the inner ear. *Audiol. Neurootol.* **7**, 36–39.

Pirvola, U., Cao, Y., Oellig, C., Suoqiang, Z., Pettersson, R. F., and Ylikoski, J. (1995). The site of action of neuronal acidic fibroblast growth factor is the organ of Corti of the rat cochlea. *Proc. Natl. Acad. Sci. USA* **92**, 9269–9273.

Pirvola, U., Spencer-Dene, B., Xing-Qun, L., Kettunen, P., Thesleff, I., *et al.* (2000). FGF/FGFR-2(IIIb) signaling is essential for inner ear morphogenesis. *J. Neurosci.* **20**, 6125–6134.

Pirvola, U., Ylikoski, J., Trokovic, R., Hebert, J. M., McConnell, S. K., and Partanen, J. (2002). FGFR1 is required for the development of the auditory sensory epithelium. *Neuron* **35**, 671–680.

Powers, C. J., McLeskey, S. W., and Wellstein, A. (2000). Fibroblast growth factors, their receptors and signaling. *Endocr. Relat. Cancer* **7**, 165–197.

Raible, F., and Brand, M. (2001). Tight transcriptional control of the ETS domain factors Erm and Pea3 by Fgf signaling during early zebrafish development. *Mech. Dev.* **107**, 105–117.

Revest, J. M., Spencer-Dene, B., Kerr, K., De Moerlooze, L., Rosewell, I., and Dickson, C. (2001). Fibroblast growth factor receptor 2-IIIb acts upstream of Shh and Fgf4 and is required for limb bud maintenance but not for the induction of Fgf8, Fgf10, Msx1, or Bmp4. *Dev. Biol.* **231**, 47–62.

Richardson, G. P., Crossin, K. L., Chuong, C. M., and Edelman, G. M. (1987). Expression of cell adhesion molecules during embryonic induction. III. Development of the otic placode. *Dev. Biol.* **119**, 217–230.

Rivera-Perez, J. A., Mallo, M., Gendron-Maguire, M., Gridley, T., and Behringer, R. R. (1995). Goosecoid is not an essential component of the mouse gastrula organizer but is required for craniofacial and rib development. *Development* **121**, 3005–3012.

Roehl, H., and Nusslein-Volhard, C. (2001). Zebrafish pea3 and erm are general targets of FGF8 signaling. *Curr. Biol.* **11**, 503–507.

Ruben, R. J. (1967). Development of the inner ear of the mouse: A radioautographic study of terminal mitoses. *Acta Otolaryngol.* **220**, 4–44.

Sadler T. W. (1995) Ear. *In* "Langman's Medical Embryology," pp. 347–357. Williams & Wilkins, Baltimore.

Sakaue, H., Konishi, M., Ogawa, W., Asaki, T., Mori, T., *et al.* (2002). Requirement of fibroblast growth factor 10 in development of white adipose tissue. *Genes Dev.* **16**, 908–912.

Schoorlemmer, J., and Goldfarb, M. (2001). Fibroblast growth factor homologous factors are intracellular signaling proteins. *Curr. Biol.* **11**, 793–797.

Sekine, K., Ohuchi, H., Fujiwara, M., Yamasaki, M., Yoshizawa, T., et al. (1999). Fgf10 is essential for limb and lung formation. *Nat. Genet.* **21**, 138–141.

Stolte, D., Huang, R., and Christ, B. (2002). Spatial and temporal pattern of Fgf-8 expression during chicken development. *Anat. Embryol. (Berl)* **205**, 1–6.

Stone, J. S., and Rubel, E. W. (2000). Temporal, spatial, and morphologic features of hair cell regeneration in the avian basilar papilla. *J. Comp. Neurol.* **417**, 1–16.

Sulik, K. K., and Cotanche, D. A. (1995). Embryology of the ear. *In* "Hereditary Hearing Loss and Its Syndromes" (R. J. Gorlin, H. V. Toriello, and M. M. Cohen, Eds.), pp. 22–42. Oxford University Press, New York.

Sun, X., Meyers, E. N., Lewandoski, M., and Martin, G. R. (1999). Targeted disruption of Fgf8 causes failure of cell migration in the gastrulating mouse embryo. *Genes Dev.* **13**, 1834–1846.

Sutherland, D., Samakovlis, C., and Krasnow, M. A. (1996). Branchless encodes a *Drosophila* FGF homolog that controls tracheal cell migration and the pattern of branching. *Cell* **87**, 1091–1101.

Tannahill, D., Isaacs, H. V., Close, M. J., Peters, G., and Slack, J. M. W. (1992). Developmental expression of the *Xenopus int-2* (FGF-3) gene: Activation by mesodermal and neural induction. *Development* **115**, 695–702.

ten Berge, D., Brouwer, A., Korving, J., Martin, J. F., and Meijlink, F. (1998). *Prx1* and *Prx2* in skeletogenesis: Roles in the craniofacial region, inner ear and limbs. *Development* **125**, 3831–3842.

Thesleff, I., and Mikkola, M. (2002). The role of growth factors in tooth development. *Int. Rev. Cytol.* **217**, 93–135.

Tonou-Fujimori, N., Takahashi, M., Onodera, H., Kikuta, H., Koshida, S., et al. (2002). Expression of the FGF receptor 2 gene (fgfr2) during embryogenesis in the zebrafish *Danio rerio*. *Gene Expr. Patterns* **2**, 183–188.

Trokovic, N., Trokovic, R., Mai, P., and Partanen, J. (2003). Fgfr1 regulates patterning of the pharyngeal region. *Genes Dev.* **17**, 141–153.

Trumpp, A., Depew, M. J., Rubenstein, J. L., Bishop, J. M., and Martin, G. R. (1999). Cremediated gene inactivation demonstrates that FGF8 is required for cell survival and patterning of the first branchial arch. *Genes Dev.* **13**, 3136–3148.

Tsang, M., Friesel, R., Kudoh, T., and Dawid, I. B. (2002). Identification of Sef, a novel modulator of FGF signalling. *Nat. Cell Biol.* **4**, 165–169.

Vallino-Napoli, L. D. (1996). Audiologic and otologic characteristics of Pfeiffer syndrome. *Cleft Palate Craniofac. J.* **33**, 524–529.

Van De Water, T. R., Maderson, P. F. A., and Jaskoll, T. F. (1980). The morphogenesis of the middle and external ear. *Birth Defects Orig. Artic. Ser.* **16**, 147–180.

Vendrell, V., Carnicero, E., Giraldez, F., Alonso, M. T., and Schimmang, T. (2000). Induction of inner ear fate by FGF3. *Development* **127**, 2011–2019.

Walshe, J., Maroon, H., McGonnell, I. M., Dickson, C., and Mason, I. (2002). Establishment of hindbrain segmental identity requires signaling by FGF3 and FGF8. *Curr. Biol.* **12**, 1117–1123.

Walshe, J., and Mason, I. (2000). Expression of FGFR1, FGFR2 and FGFR3 during early neural development in the chick embryo. *Mech. Dev.* **90**, 103–110.

Waterman, R. E., and Bell, D. H. (1984). Epithelial fusion during early semicircular canal formation in the embryonic zebrafish, Brachydanio rerio. *Anat. Rec.* **210**, 101–114.

Whitfield, T. T., Riley, B. B., Chiang, M. Y., and Phillips, B. (2002). Development of the zebrafish inner ear. *Dev. Dyn.* **223**, 427–458.

Wilke, T. A., Gubbels, S., Schwartz, J., and Richman, J. M. (1997). Expression of fibroblast growth factor receptors (FGFR1, FGFR2, FGFR3) in the developing head and face. *Dev. Dyn.* **210**, 41–52.

8. FGF Signaling in Ear Development and Innervation

Wilkinson, D. G., Bhatt, S., and McMahon, A. P. (1989). Expression pattern of the FGF-related proto-oncogene *int-2* suggests multiple roles in fetal development. *Development* **105**, 131–136.

Wilkinson, D. G., Peters, G., Dickson, C., and McMahon, A. P. (1988). Expression of the FGF-related proto-oncogene *int-2* during gastrulation and neurulation in the mouse. *EMBO J.* **7**, 691–695.

Wright, C. G. (1997). Development of the human external ear. *J. Am. Acad. Audiol.* **8**, 379–382.

Wright, T. J., and Mansour, S. L. (2003). *Fgf3* and *Fgf10* are required for mouse otic placode induction. *Development* **130**, 3379–3390.

Xiang, M., Gan, L., Chen, Z. Y., Zhou, L., O'Malley, M. W., *et al.* (1997). Essential role of POU-domain factor Brn-3c in auditory and vestibular hair cell development. *Proc. Natl. Acad. Sci. USA* **94**, 9445–9450.

Xu, J., Liu, Z., and Ornitz, D. M. (2000). Temporal and spatial gradients of Fgf8 and Fgf17 regulate proliferation and differentiation of midline cerebellar structures. *Development* **127**, 1833–1843.

Yamada, G., Mansouri, A., Torres, M., Stuart, E. T., Blum, M., *et al.* (1995). Targeted mutation of the murine goosecoid gene results in craniofacial defects and neonatal death. *Development* **121**, 2917–2922.

Yamaguchi, T. P., Harpal, K., Henkemeyer, M., and Rossant, J. (1994). fgfr-1 is required for embryonic growth and mesodermal patterning during mouse gastrulation. *Genes Dev.* **8**, 3032–3044.

Yamashita, T., Konishi, M., Miyake, A., Inui, K., and Itoh, N. (2002). Fibroblast growth factor (FGF)-23 inhibits renal phosphate reabsorption by activation of the mitogen-activated protein kinase pathway. *J. Biol. Chem.* **277**, 28265–28270.

Ying, Q. L., Stavridis, M., Griffiths, D., Li, M., and Smith, A. (2003). Conversion of embryonic stem cells into neuroectodermal precursors in adherent monoculture. *Nat. Biotechnol.* **21**, 183–186.

Zhang, S., Lin, Y., Itaranta, P., Yagi, A., and Vainio, S. (2001). Expression of Sprouty genes 1, 2 and 4 during mouse organogenesis. *Mech. Dev.* **109**, 367–370.

9

The Roles of Retinoic Acid during Inner Ear Development

Raymond Romand
Institut Clinique de la Souris and Institut de Génétique et de Biologie
Moléculaire et Cellulaire, 67404 Illkirch Cedex, France

I. Introduction
II. Metabolism of Retinoids and Their Receptors
III. Expression of Retinoic Acid Metabolic Enzymes and Receptors in the Developing Inner Ear
IV. Is Retinoic Acid Involved in Otic Placode Induction?
V. Retinoids as Morphogens during Early Embryogenesis
   A. The Otic Capsule
   B. Retinoic Acid: The Neuroectoderm and the Otocyst
VI. Retinoic Acid and Patterning Genes during Inner Ear Development
VII. Retinoic Acid and Hair Cell Differentiation
VIII. Concluding Remarks
   References

    The pleiotropic functions of retinoic acid (RA), a diffusible signaling molecule that acts at various stages during vertebrate embryogenesis and morphogenesis, have been studied extensively. The embryopathy generated by RA excess or deficiency affects several organs, including the inner ear. During the first stages of inner ear embryogenesis, RA may be indirectly involved in the induction of the otic placode, a process that appears to be due to complex mechanisms. RA is also implicated in several processes during inner ear ontogenesis, such as the mesenchyme–otic epithelium interactions that lead to otic capsule formation. Since RA synthesizing and metabolizing enzymes, along with RA receptors, are differentially expressed during early inner ear morphogenesis, a direct involvement of RA (especially in the morphogenesis of the labyrinthic epithelia) appears likely. However, current models also favor some indirect actions due to downstream gene regulation within the hindbrain. Later, RA is likely to intervene in the control of cell proliferation and differentiation, thus regulating hair cell differentiation from progenitor cells in the cochlea. The presence of retinoid receptors, along with RA metabolic enzymes, in the inner ear at the time of birth and during postnatal life suggests that the functional involvement of this signaling molecule goes beyond fetal development. © 2003, Elsevier Inc.

I. Introduction

The first morphological sign of inner ear development is an ectodermal thickening close to each side of the rhombencephalon that yields the so-called otic placode, from which all components of the inner ear will later derive. Several molecular markers have been found to be specifically expressed in the chick before the cranial ectoderm becomes specified to form the otic placode (Groves and Bronner-Fraser, 2000), and signals from different tissues contribute to the induction of the placode (Baker and Bronner-Fraser, 2001). The placodal stage is followed by the invagination of the otic pit, which eventually forms a closed vesicle or otocyst in which cell fates are more or less determined (Fritzsch et al., 1998; Torres and Giraldez, 1998), leading to the hypothesis of compartment boundaries of cell fate specification within the otocyst (Fekete and Wu, 2002; Kil and Collazo, 2002). From this stage, the beginning of inner ear organ diversification occurs, followed by morphogenesis and cell differentiation leading to two sensory systems, the vestibular and the cochlear organs. A growing number of gene products, such as transcription factors, hormones, adhesion molecules, and growth factors, which may act at various steps of inner ear embryogenesis and patterning, are being characterized (Rivolta, 1997; Torres and Giraldez, 1998). One of these factors is retinoic acid (RA), an active derivative of vitamin A (retinol), which regulates gene expression via the activation of specific receptors that belong to the nuclear receptor superfamily (Chambon, 1996). The expression of various proteins can be triggered or altered in the context of a differentiation program initiated by RA, which can also promote growth factor dependence (reviewed by McCaffery and Dräger, 2000). RA is indispensable for the normal development of several organs (Mark et al., 1998; Creech-Kraft and Willhite, 1996), including the ear as revealed by its teratogenic effects when administered during particular times of development (Frenz et al., 1996). A number of experimental results have demonstrated the involvement of RA on otic capsule chondrogenesis (Frenz and Liu, 1998), as well as cell differentiation (Kelley et al., 1993; Represa et al., 1990) and morphogenesis (Choo et al., 1998; Romand et al., 2002) of the inner ear. Recent reports of differential gene expression patterns of RA receptors (Raz and Kelley, 1999; Romand et al., 1998), cellular retinoid binding proteins (Romand et al., 2000), and synthetic or catabolic enzymes (Romand et al., 2003) may therefore be reexamined considering the possible functions of this molecule during inner ear ontogenesis. This chapter considers the roles of RA during three stages of inner ear development: (1) during the otocyst stage, when complex interactions between the periotic mesenchyme and the otic epithelium in relation to other molecules or tissues (such as the hindbrain) occur; (2) after the otocyst stage, when RA may be involved in proper patterning of inner ear structures;

9. Retinoic Acid and the Inner Ear 263

(3) when RA could induce hair cell differentiation, perhaps followed by (an) as yet uncharacterized role(s) in postnatal inner ear function.

II. Metabolism of Retinoids and Their Receptors

The metabolism of vitamin A, or retinol, leads to a number of compounds called retinoids, of which RA is the most biologically active. Retinoids are mainly obtained from two sources: retinol from animal meat and β-carotene from plants (Thurnham and Northrop-Clewes, 1999). Retinol can be considered a hormone because it is stored as retinyl esters in the liver and released in the blood, where it is bound to a specific carrier protein, the retinol binding protein, which may interact with a cell surface receptor to facilitate the transfer of retinol into target cells (Fig. 1; Gottesman *et al.*, 2001, and references therein).

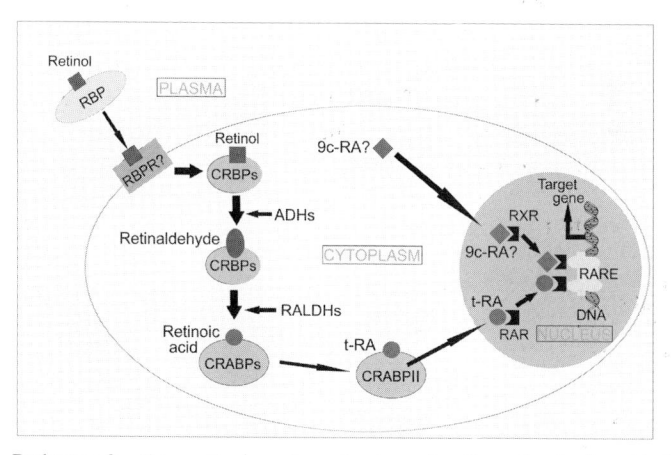

Figure 1 Pathways for the synthesis and mechanism of action of retinoic acid. Retinol in plasma is bound to a retinol binding protein (RBP) and transported toward target cells. This complex might interact with a retinol binding protein receptor (RBPR?), facilitating the transfer of retinol into the cell. In the cytoplasm, retinol is bound to two molecules known as cellular retinol binding proteins (CRBPs). Retinol is converted into retinoic acid (RA) in two successive steps by two different sets of enzymes: aldehyde dehydrogenases (ADHs) and retinaldehyde dehydrogenases (RALDHs). Then, RA interacts with two cellular RA binding proteins (CRABPs) where the complex CRABP2-RA could be internalized into the nucleus. In the nucleus, RA effects are mediated by specific receptors that in turn modulate gene expression. Two families of nuclear receptors are involved: the retinoid acid receptors (RARs), activated both by *all-trans* RA (t-RA) and *9-cis* RA (9c-RA), and the retinoid X receptors (RXRs), activated exclusively by 9c-RA. RARs and RXRs act on RA response genes by heterodimerization in binding to specific sequences known as RA response elements (RAREs). (See Color Insert.)

After being taken into the cell, retinol is bound with high affinity by cellular retinol binding proteins (CRBP1 and 2), (see Fig. 1). Proteins that bind RA with high affinity (cellular RA binding proteins: CRABP1 and 2) have also been characterized. In addition to a general role in solubilizing and stabilizing their hydrophobic and labile ligands in aqueous solution spaces, these retinoid binding proteins may have distinct functions in the regulation of synthesis, storage, transport, metabolism, and action on particular retinoids with which they are associated (Ghyselinck et al., 1999; Gottesman et al., 2001; Noy, 2000).

With the possible involvement of CRBPs, retinol is converted intracellularly into RA by two oxidative reactions (see Fig. 1). The oxidation of retinol into retinaldehyde is catalyzed by members of the alcohol dehydrogenase (ADH) family, whereas the conversion of retinaldehyde into RA is principally achieved by three enzymes usually known as retinaldehyde dehydrogenases (RALDH) 1, 2, and 3 (Duester, 2000; Edenberg, 2000; see Fig. 1), although a recent nomenclature change has designated them as ALDH1A1, A2, and A3 (Duester, 2000). A fourth retinaldehyde dehydrogenase, which may preferentially generate *9-cis* RA (see later), has also been characterized (Lin et al., 2003). Three members of the cytochrome P450 superfamily, CYP26A1, B1, and C1, specifically metabolize RA into more polar (4-hydroxy and 4-oxo) derivatives. The fact that targeted disruption of *Cyp26A1* in the mouse leads to a lethal malformative phenotype that mimics several of the known teratogenic effects of RA is consistent with the idea that these enzymes are required for tissue-specific catabolism of endogenous RA (Abu-Abed et al., 2001; Sakai et al., 2001; Stoilov, 2001). Interestingly, the *Cyp26A1*$^{-/-}$ phenotype is partly rescued by heterozygous disruption of the *Raldh2* gene, which decreases the amounts of RA synthesized within the embryo (Niederreither et al., 2002). These genetic approaches show that both the synthesis and metabolism of RA need to be tightly regulated to allow normal development.

Retinoids exert most of their effects by binding to specific receptors that modulate target gene expression. These receptors are members of the "steroid/thyroid hormone" superfamily of nuclear receptors (Evans, 1988; Mangelsdorf et al., 1995; Robinson-Rechavi et al., 2003). The diversity of RA-induced signaling can be explained by the existence of several receptors, which fall into two families: the retinoic acid receptors (RARs), which are activated both by *all-trans* and *9-cis* RA, and the "retinoid X" receptors (RXRs), which are activated exclusively by *9-cis* RA, but could merely function *in vivo* as "silent" (unliganded) dimerization partners (for review, see Chambon, 1996). There are three types of RARs (RARα, β, and γ) and three types of RXRs (RXRα, β, and γ), each existing as at least two major N-terminal isoform variants due to alternative splicing and/or promoter usage of the corresponding genes. The complexity of retinoid signaling arises from the fact that each of the RARs can heterodimerize with any of the RXRs, *in vitro* and *in vivo* (Kastner *et al.*, 1995, 1997). RXRs are also heterodimerization partners for several other nuclear receptors, including thyroid hormone (TR), vitamin D (VDR) of peroxisome proliferator-activated (PPAR) receptors, and orphan receptors (Robinson-Rechavi *et al.*, 2003). RAR/RXR heterodimers therefore act on a wide range of target genes by binding to regulatory sequences called RA response elements (RAREs) (see Fig. 1). The pleiotropic effects of retinoids are further exemplified by the findings that, according to their liganded or unliganded state, RARs and RXRs interact, respectively, with various "coactivator" and "corepressor" proteins (Chambon, 1996).

It has long been established that vitamin A intake is important for reproduction; embryonic and fetal development; growth; vision; cardiac system, nervous system, and other tissue maintenance throughout life (for review, see Wolf, 1984). Normal embryonic development is dependent on an adequate supply of vitamin A (Wilson *et al.*, 1953) such that both excess and deficiency of retinoids elicit abnormal development (Maden, 1994). The concept of vitamin A as a differentiating agent started from the early studies of vitamin A–deficient rats (Wolbach, 1954). A number of review articles discuss the current knowledge on the contribution of retinoids to early embryogenesis (Morriss-Kay and Ward, 1999; Ross *et al.*, 2000), cardiogenesis (Smith and Dickman, 1997), *Hox* gene regulation (Marshall *et al.*, 1996), nervous system development (Maden *et al.*, 1998; Malik *et al.*, 2000), retinal development (Dräger and McCaffery, 1997), and craniofacial development (Brickell and Thorogood, 1997).

Various animal models of the RA embryopathy have been described (Jarvis *et al.*, 1990; Morriss, 1972; Willhite *et al.*, 1986; Yasuda *et al.*, 1986). Both clinical and experimental observations have shown that an excess of retinoids produces inner ear malformations (Frenz *et al.*, 1996; Jarvis *et al.*, 1990; Willhite *et al.*, 1986). An excess of retinoids can induce embryopathy in the three compartments of the ear: the external ear (Wei *et al.*, 1999), the middle ear (Mallo, 1997), and the inner ear (Burk and Willhite, 1992; Frenz *et al.*, 1996). This chapter focuses on the inner ear and its surrounding otic capsule. Developmental aspects of the outer and middle ear in relationship with retinoids are discussed in Chapter 3 by Mallo.

III. Expression of Retinoic Acid Metabolic Enzymes and Receptors in the Developing Inner Ear

RA signaling needs to be tightly controlled during normal development through stage- and tissue-specific expression of its synthesizing and metabolizing enzymes and its nuclear receptors. The teratogenic effects of excess RA

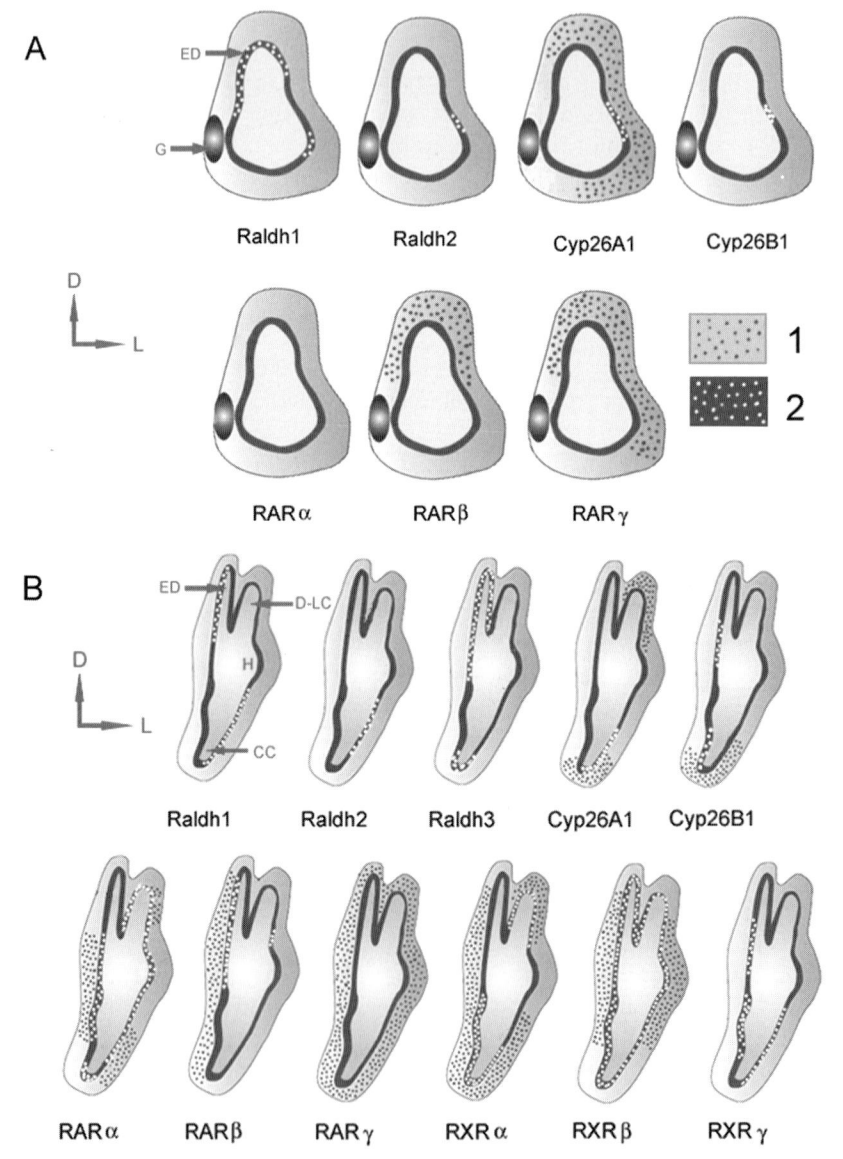

Figure 2 Distribution of synthetic and catabolic enzymes and retinoid receptor transcripts in the embryonic ear. (A) Transcript distribution of retinaldehyde dehydrogenase enzymes (*Raldh1*, *Raldh2*), catabolic enzymes (*Cyp26A1*, *Cyp26B1*), and RA receptors (RARα, RARβ, and RARγ) in E10.5 mouse. Synthetic enzyme transcripts (*Raldh1* and *2*) are mainly restricted to some regions of the otic epithelium, whereas the catabolic enzyme (*Cyp26A1*) is largely present in the mesenchyme. Only two RA receptors (RARβ and RARγ) are detected at this stage of development and restricted to mesenchyme tissues mainly around the presumptive endocochlear duct (ED). VII–VIII ganglion complex (G). (B) E12.5 stage. *Raldh1* and *2*

9. Retinoic Acid and the Inner Ear

are likely to result from disturbances in the regulated distribution of embryonic RA. In this context, detailed analyses of the transcript distributions of the various *Raldh, Cyp26*, and RAR/RXR genes may give insight into the normal role(s) of RA during morphogenesis and cell differentiation in the inner ear (Romand *et al.*, 2002, 2003).

Identification of enzymes with physiological roles in RA synthesis is important for understanding how regulated ligand synthesis contributes to retinoid signaling. For example, findings have indicated that RALDH2 does play a critical role during early embryonic development (Niederreither *et al.*, 1999). A good quantitative correlation between biochemically determined levels of RALDH activity and RA levels in corresponding tissue samples has been proved to exist (McCaffery and Dräger, 1994). Other synthetic enzymes may be involved by producing RA locally in several sensory regions, including the otic vesicle (Mic *et al.*, 2000). None of the three murine *Raldh* genes investigated (*Raldh1, 2*, and *3*) were found to be expressed in the periotic mesenchyme at embryonic day 10.5 (E10.5). However, *Raldh1* transcripts are observed in the presumptive endocochlear duct and in a restricted region of the ventrolateral part of the otocyst (Fig. 2A). The *Raldh2* gene is first expressed in a mediodorsal domain of the otocyst at E10.5 (Niederreither *et al.*, 1997; Romand *et al.*, 2001), whereas *Raldh3* is already present in the otic vesicle at E9.5 and E10.5 (Grün *et al.*, 2000; Mic *et al.*, 2000). At E12.5, no *Raldh* gene expression is seen in the periotic mesenchyme. *Raldh1* and *Raldh3* are expressed in the endolymphatic duct and the ventral region of the otocyst, whereas *Raldh2* is restricted in the ventrolateral region of the otic epithelium (Fig. 2B).

In contrast, the catabolic enzyme *Cyp26A1* is abundantly expressed in the mesenchyme surrounding the otocyst at E10.5 and is restricted to a lateral domain of the otic epithelium, whereas *Cyp26B1* is only present in a very localized region of the lateral otic epithelium (see Fig. 2A). Two days later, *Cyp26A1* gene expression is observed in mesenchymal tissues surrounding the developing anterior and posterior semicircular canals (SCCs) and in the ventrolateral region of the otic epithelium (see Fig. 2B). *Cyp26B1* presents a similar transcript distribution in the ventral part of the otic mesenchyme

---

transcripts are present in the epithelium of the endolymphatic duct (ED) and the presumptive cochlear canal (CC). *Cyp26* gene expression is mainly observed in restricted regions of the ventral otic epithelium and mesenchyme tissues around the presumptive cochlear canal and restricted in the dorsoanterior semicircular canals for Cyp26A1. All retinoid receptors are abundant in the mesenchyme except RXRγ, which is restricted to the otic epithelium. RARs and RXRs can be observed in some regions of the otic epithelium, except RARγ, which is restricted to the surrounding mesenchyme. 1, Distribution of transcripts in mesenchyme and surrounding tissues of the otocyst; 2, distribution of transcripts in the otic epithelium. (See Color Insert.)

and in a specific domain of expression localized in the medial region of the otic epithelium (see Fig. 2B).

Thus, gene expression studies have revealed mostly exclusive—and relatively complementary—inner ear distributions of *Raldh*-expressing cells, which are restricted to specific regions of the otic epithelium, and *Cyp26*-expressing cells, mainly found in periotic mesenchyme at E10.5 and in the ventral region of the otic epithelium and periotic mesenchyme at E12.5 (see Fig. 2). These patterns are clearly consistent with the idea that synthetic enzymes could act to generate local sources of RA, whereas CYP26-expressing cells may act as a dispersed sink, as discussed by Stoilov (2001). Complementary distributions between putative RA sources and sinks have also been suggested to operate during early chick embryogenesis (Swindell *et al.*, 1999) along the anteroposterior axis of the mouse embryo (Sakai *et al.*, 2001) and during hindbrain (Abu-Abed *et al.*, 2001; Maden, 2002) and retinal patterning (McCaffery *et al.*, 1999; Mey *et al.*, 2001). The complementary patterns of *Raldh* and *Cyp26* gene expression at the level of the inner ear—assuming that mRNA levels are translated into comparable levels of functional enzymes—suggest that finely tuned levels of RA are necessary to achieve its correct development. This is also consistent with the fact that RA excess generates inner ear embryopathies.

If RALDH and CYP26 enzymes play key roles for RA homeostasis (e.g., McCaffery and Dräger, 2000), retinoid receptors are essential for transducing RA signals during development (e.g., Kastner *et al.*, 1997). At E10.5 in the mouse, none of the RAR gene transcripts are present in the otic epithelium, whereas RARβ and RARγ are mainly observed in mesenchymal cells around the presumptive endolymphatic duct (see Fig. 2A). Two days later, all three RAR gene transcripts are expressed according to different spatial distributions (Fig. 2B). RARα is observed in most of the otic epithelium and specific mesenchymal regions, whereas RARβ is mainly expressed in the medial part of the otocyst, both in the epithelium and nearby mesenchyme, and RARγ gene expression is restricted to the periotic mesenchyme. RXRα transcripts exhibit a widespread distribution in mesenchymal tissues and otic epithelium, except in the presumptive endocochlear duct (see Fig. 2B). RXRβ is largely expressed throughout the otic epithelium, as well as in the lateral and medial, but not the ventral, mesenchyme. RXRγ transcripts are restricted to some areas of the otic epithelium (see Fig. 2B). Taken together, it appears that the retinoid receptors are expressed in distinct but largely overlapping regions of the E12.5 periotic mesenchyme, with the exception of RXRγ whose expression is epithelial specific. Among these receptors, only RARγ is expressed throughout the entire periotic mesenchyme (see Fig. 2B). From these distributions, we could suggest a major involvement of RARα and the three RXRs in transducing the RA signal in the sensory epithelium of the ventral part of the otocyst that corresponds to the presumptive cochlea and saccule, whereas RARγ and RXRα may be mainly involved in RA signaling during otic capsule formation.

Mapping of the expression patterns of genes encoding the retinoid receptors and the enzymes related to RA homeostasis may give some clues about RA function in specific organs during embryogenesis. However, one should be cautious about drawing conclusions based on RNA (or protein) distributions since, for example, the murine *Crabp* and *Crbp* genes presented interesting, differential transcript distributions throughout development of the inner ear, but the targeted disruptions of *Crbp1* and combined disruption of the two *Crabps* did not result in any detectable consequence on inner ear morphogenesis or adult cochlear function (Romand *et al.*, 2000). Moreover, despite extensive expression of RXR genes in inner ear tissues (Dollé *et al.*, 1994; Raz and Kelley, 1999; Romand *et al.*, 1998), no specific detectable morphological abnormality was observed in the corresponding compound mutant mice (our unpublished observations), and no functional impairment was detected from auditory brainstem-evoked potential recordings (Barros *et al.*, 1998).

IV. Is Retinoic Acid Involved in Otic Placode Induction?

The otic placode is believed to be induced by initial signals from mesoderm and nearby neural tissues (Baker and Bronner-Fraser, 2001; Noramly and Grainger, 2002; Streit, 2002). Early in development, the hindbrain is divided into a series of compartments called rhombomeres, which play pivotal roles in the differentiation and segregation of neurons along the anteroposterior axis (Lumsden and Krumlauf, 1996). Secreted factors may originate from the hindbrain at the level of the presumptive rhombomeres 5 and 6 (r5–6), which are adjacent to the prospective otic placode ectoderm (Van de Water *et al.*, 1992). Wilkinson *et al.* (1989) suggested that fibroblast growth factor 3 (FGF3) might be one such signal. It was indeed demonstrated that otocyst formation in chick embryos can be blocked *in vitro* using FGF3-blocking antibodies or antisense oligonucleotides (Represa *et al.*, 1991). Targeted disruption of the *Fgf3* gene in the mouse did not confirm an inductive role of FGF3 in otocyst formation. Rather, its function appears to be critical for later steps of morphogenesis and differentiation (Mahmood *et al.*, 1995; Mansour *et al.*, 1993; McKay *et al.*, 1996; Pirvola *et al.*, 2000). FGF3 might regulate otocyst morphogenesis through FGFR2 since this receptor is observed in the dorsal part of the otic vesicle as early as E8 in mouse embryos (Pirvola *et al.*, 2000) and its expression persists during subsequent development (Pickles, 2001). Interestingly, as pointed out by Fritzsch *et al.* (1998), there is a striking correlation between the timing of

Fgf3 expression in r5–6 and the early development of the inner ear (Mahmood *et al.*, 1996).

Recent data from the analysis of retinoid receptor mutant mice suggest that RA signaling is required for the specification of the otogenic field (Dupé *et al.*, 1999; Wendling *et al.*, 1999, 2001). Dupé *et al.* (1999) described otocyst abnormalities in RARα/RARβ double mutant embryos. In these mutants, small supernumerary otocyst-like structures (sometimes fused to the "main" otocyst, which was somewhat shifted rostrally) were formed along the caudal hindbrain neuroepithelium. This phenotype is reminiscent of that described in vitamin A–deficient (VAD) rat embryos (White *et al.*, 1998). Other retinoid receptor mutants displayed the same type of otic abnormalities, such as RXRα/RXRβ (Wendling *et al.*, 1999) and RARα/RARγ double mutants (Wendling *et al.*, 2001). It is interesting to note that all of these mouse mutants present alterations of rhombomere organization, similar to VAD rat embryos (White *et al.*, 1998). Because *Fgf3* expression has not been analyzed in the RAR/RXR compound mutants, it is unclear if a possible deregulation of its expression may be involved in the induction of the ectopic otocyst rudiments. Such a possibility appears likely, at least in the case of the RARα/RARβ compound mutants, which, according to other molecular markers, exhibit a markedly enlarged r5–6 territory (Dupé *et al.*, 1999).

In the chick it is known that ectopically expressing *Fgf3* in competent ectoderm *via* a viral vector induces otic placodes and activates the intrinsic gene expression program required for the correct development of the inner ear (Vendrell *et al.*, 2000). In zebrafish, RA treatment induces expanded domains of *fgf3* and *fgf8* expression, resulting in an ectopic expression of *pax8* and the formation of supernumerary and/or ectopic otic vesicles (Phillips *et al.*, 2001). In the same study, it is also shown that antisense morpholino oligomers directed against *fgf3* can block the effects of exogenous RA, thereby preventing the formation of the ectopic or supernumerary otic vesicles. Detailed discussions on the effects of Fgfs in zebrafish otic placode induction can be found in review articles (Noramly and Grainger *et al.*, 2002; Whitfield *et al.*, 2002; see Chapter 4 by Brown *et al.*, and Chapter 12 by Riley). According to Whitfield *et al.* (2002), RA would be required in zebrafish for normal Fgf3 expression, which in turn mediates induction of the placode.

It is interesting to point out that an excess of RA generates surpernumerary or ectopic otic vesicles in zebrafish, whereas in mice it is the opposite, corresponding to an RA deficiency. Currently it is difficult to explain this difference, which could be related to time and duration of treatment, as well as dosage, although an indirect action of RA on downstream genes may generate a different response on the ectoderm. RA deficiency made by RARα/β and RXRα/β double mutant mice enlarges the posterior expression of

Kreisler after r4 (Dupé *et al.*, 1999) as for the *Fgf3* gene expression observed after the targeted deletion of *Raldh2* in mouse (Niederreither *et al.*, 2000). These data correspond well to the observation that KREISLER is required for the upregulation of FGF3 in r5–6 (McKay *et al.*, 1996). This posterior expression of FGF3 might induce transiently supernumerary otic vesicles in competent ectoderm as observed in VAD rat and retinoid receptor mutant mice (Dupé *et al.*, 1999; Wendling *et al.*, 1999, 2001; White *et al.*, 1998), suggesting that RA might have an influence on surface ectoderm competence to form sensory placodes, although several other factors can be involved in this inductive function (Baker and Bronner-Frazer, 2001).

V. Retinoids as Morphogens during Early Embryogenesis

A. The Otic Capsule

The inner ear appears as an ectodermal thickening (otic placode) located lateral to prospective r5 and r6. Later the otic placode invaginates to form the otic pit, which subsequently closes off from the surface ectoderm to form the otic vesicle (otocyst) (Anniko, 1983). The otic vesicle is surrounded by periotic mesenchyme tissues from which a cartilaginous capsule derives, later forming the so-called otic capsule. This otic capsule forms a protective shell around the inner ear. Numerous clinical observations in humans, as well as experimental findings, have revealed the effects of RA on the otic capsule (Frenz *et al.*, 1996; Lim, 1977), which can be related to a more general action of this hormone on craniofacial development (Morriss, 1972; Sulik, 1986). The otic capsule is of neural crest origin (Couly *et al.*, 1993) and derives from the second arch (Mallo, 1997). Because it is generally accepted for the development of the skeletal elements from the neural crest cells, epithelial–mesenchymal interactions are of crucial importance (Le Douarin, 1982). It is suggested that reciprocal interactions between otic epithelium and the surrounding mesenchyme are also essential for the morphogenesis of the otic capsule and the membranous labyrinth (Frenz and Van de Water, 1991; Noden and Van de Water, 1992; Van de Water, 1983). These interactions are essential for aggregation and differentiation of periotic cartilage precursors, which form the otic capsule around the inner ear (McPhee and Van de Water, 1986; Van de Water, 1983). *In vivo* and *in vitro* experiments show that exposure to an excess of *all-trans* RA during a specific time window is teratogenic for the otic capsule (Frenz *et al.*, 1996) and inhibitory for chondrogenesis (Frenz and Liu, 1997).

Experimental evidence from RA teratogenesis suggests that the presence of ligand-activated RAR and/or the inappropriate gene expression of RARs inhibits chondrogenesis (Underhill and Weston, 1998). RARα and RARγ

null mutants display some of the defects resulting from VAD, whereas double mutants display numerous congenital defects including those of the fetal VAD syndrome (Cash et al., 1997; Lohnes et al., 1994; Mendelsohn et al., 1994a). These findings demonstrate first that RA is the active vitamin A metabolite during development and second that this hormone is involved in the organogenesis of craniofacial structures including the inner ear and the otic capsule.

Although the distribution study of the different genes related to RA metabolism was made after the critical period of the otic capsule induction, it is noteworthy to mention that several of the RA-related genes are already present at the otocyst stage. At E10.5 *Crabps* and *Crbps* transcripts are not seen in periotic mesenchyme except for *Crabp2* present in the basal region of the otic epithelium and in the apical domain, including the endolymphatic duct for *Crbp1* (Romand et al., 2000). However, single- and double-directed mutations of *Crabps* and *Crbp1* seem to be dispensable for the otic capsule morphogenesis; therefore these proteins do not appear to be essential for the RA transduction pathways in the inner ear (Romand et al., 2000). No synthetic enzyme transcript (*Radh1*, *2*, and *3*) is observed in the periotic mesenchyme between E10.5 and E12.5, although two of these enzymes are present in some regions of the otic epithelium. Alternatively, the catabolic enzyme *Cyp26A1* is largely represented in periotic mesenchyme, whose distribution is more or less complementary to *Raldhs* (see Fig. 2). Therefore, the presence of both types of enzymes may be required for a fine-tuning of RA concentration during the otic capsule morphogenesis. This could explain why experimental modification of this balance led to atresia of the capsule (Frenz and Liu, 1997; Frenz et al., 1996).

Another set of genes is important for mediating the RA effects (i.e., the retinoid receptors) that is present at E12.5, where RARα is seen in the otic epithelium and RARβ and RARγ in the periotic mesenchyme (Dollé et al., 1990; Romand et al., 2002; see Fig. 2), and later at E12.5 RARs and RXRs except RXRγ are expressed in the mesenchyme (Ruberte et al., 1990). Deletion of one of these genes does not affect the otic capsule pattern, whereas double-directed mutation of RARα/RARγ greatly affects the pattern as well as the membranous labyrinth development (Romand et al., 2002). These observations suggest that each individual RAR may not be required for the otic capsule formation because there is sufficient functional redundancy between RARs and their isoforms as revealed by knockout studies (Lohnes et al., 1993; Mendelsohn et al., 1994b); however, these genes may not appear redundant in normal development (Ross et al., 2000). The hypomorphic otic capsule observed in the RARαRARγ phenotype, which corresponds to RA deficiency or VAD, fits well with retinoic acid–induced embryopathy reported in the inner ear where the cochlear region is missing or poorly developed and the otic capsule is hypomorphic (Burk and Willhite, 1992; Frenz et al., 1996).

During the process of early otic development, which involves mesenchymal/epithelial interactions, several types of signals have been implicated in this process as equally suggested for other organs (Johnson and Tabin, 1997). Induction between mesenchyme and adjacent epithelia underlies local axis formation through differentiation and morphogenesis (Creazzo et al., 1998; Johnson and Tabin, 1997; La Mantia et al., 2000). There are several interacting signals present in the epithelia or mesenchyme of the otocyst, including RA as described earlier. There are equally several morphogenetic proteins (BMPs), and FGFs, and other factors such as transforming growth factor β1 (TGFβ1), and sonic hedgehog (SHH) (Frenz et al., 1994; Liu et al., 2002; Morsli et al., 1998; Pirvola et al., 2000).

The RAR/RXR heterodimer functions by definition as a transcription factor to regulate gene expression (Kastner et al., 1997). In the embryo several transcription factors are under the control of RA, such as *Hox* genes (Rijli and Chambon, 1997), signaling molecules, and enzymes (reviewed by McCaffery and Dräger, 2000). For instance, *Bmp4* is one of the earliest genes to be expressed in the otic epithelium of the chick at the time of the pit closure (Wu and Oh, 1996). *Bmp4*, like other *Bmps*, are found in the inner ear at different stages of development and in several other locations; therefore different functions for these diffusible molecules might be foreseen (Morsli et al., 1998; Gerlach et al., 2000; Oh et al., 1996; Takemura et al., 1996; von Bubnoff and Cho, 2001). BMP pathways have an important function in early chondrogenesis at the epithelial–mesenchyme interaction levels, where BMP signaling is involved in otic capsule chondrogenesis (Chang et al., 2002; Liu et al., 2003). It has been demonstrated that RA downregulates *Bmp4* transcription in otocyst cells, a direct effect mediated by RARs on a novel promoter within the second intron of the *Bmp4* gene (Thompson et al., 2003). RA may also act on other downstream genes such as *Shh* encoding for a secreted protein that might be involved in several morphogenic functions. It has been suggested that SHH is involved in the epithelial–mesenchymal interactions governing organ development (Sasagawa et al., 2002), particularly for correct anteroposterior patterning of the zebrafish otic vesicle (Hammond et al., 2003). This gene regulates otic capsule chondrogenesis and inner ear development in the mouse embryo (Liu et al., 2002), where SHH signaling has a direct impact on periotic mesenchyme via two potential targets, *Tbx1* and *Brn4* (Riccomagno et al., 2002). Targeted mutation of *Shh* produces otic capsule hypoplasia reminiscent of RARα/RARγ double mutation where abnormal chondrogenesis is also observed (Romand et al., 2002). Interestingly enough, this gene presents to its promoter a RARE, suggesting that it could be a downstream effector of RARs (Chang et al., 1997). RA may also interfere indirectly with other molecules such as growth factors involved in the otic epithelium–periotic mesenchyme interactions such as FGFs secreted by the hindbrain (Frenz and Liu, 1998; Maden et al., 1996), involved in different stages of inner ear development (Mansour et al., 1993; Pirvola et al., 2000).

Early during the otocyst development, most genes involved in the RA metabolism are present along with retinoid receptors (see Fig. 2). Embryopathy on the inner ear after retinoid treatment or receptor deletion as outlined earlier suggests an important role for these molecules in the development of the otic capsule. These effects are restricted in time on the otic epithelium–periotic mesenchyme interactions. However, the consequences are not restricted to the otic capsule since VAD and RAR mutants display important abnormalities during the inner ear morphogenesis due to two kinds of signals: one from the interactions between otic epithelium and its surrounding periotic mesenchyme and a second between neural tissue and the otic ectoderm (Dupé et al., 1999; Mark et al., 1993).

The embryonic inner ear sensory epithelium surrounded by the presumptive otic capsule comprises two sensory components: the dorsal region housing the presumptive vestibular sensory system and the ventral region with the auditory receptors. During early development two groups of signals shape the inner ear, one coming from the complex interaction between otic epithelium and its surrounding periotic mesenchyme as described earlier and the other from the hindbrain, although other signals may originate from other sources (Riccomagno et al., 2002). In reality the picture is certainly more complex, and these signals may influence each other in an unknown fashion. We should not forget neural crest–derived cells. Although few in number, they are crucial for the statoacoustic function, particularly the cochlea (Price and Fisher, 2001), though they do not seem crucial for inner ear patterning and cell differentiation. An important set of genes and growth factors are involved in the early inner ear morphogenesis (reviewed by Fritzsch et al., 1998; Noramly and Grainger, 2002; Rivolta, 1997; Torres and Giraldez, 1998). Some of them could be involved later in several roles during sensory organ ontogenesis, such as cell proliferation, and then in cell differentiation. RA is one of these factors, playing several functions through the complex combinatorial actions of its receptors and downstream control of other genes (McCaffery and Dräger, 2000).

B. Retinoic Acid: The Neuroectoderm and the Otocyst

From several observations (reviewed by Baker and Bronner-Fraser, 2001; Fritzsch et al., 1998), and more specifically from mutant data, it is known that inherited neural tube defects are associated with ear abnormalities (Deol, 1964, 1983; Frohman, 1993). For instance, in *Kreisler* mutant mouse, the *Fgf3* gene expression is very low (Frohman et al., 1993; McKay et al., 1994), whereas the otocyst develops but fails to undergo normal development

9. Retinoic Acid and the Inner Ear

and differentiation (Deol, 1964). Analysis of posterior brain development (hindbrain) in the *Kreisler* mutant shows that the sequential rhombomere organization is disrupted and r5 and r6 are lost (Manzanares *et al.*, 1999; McKay *et al.*, 1994). A set of genes called *Hox* or *Homeobox* genes are required for the specification and/or maintenance of specific rhombomeres and therefore the correct patterning of the hindbrain (Cordes, 2001; Deschamps *et al.*, 1999; Lumsden and Krumlauf, 1996). Deletion of some *Hox* genes and particularly *Hoxa1* induces a marked hypoplasia of the membranous labyrinth and the otic capsule (Chisaka *et al.*, 1992; Lufkin *et al.*, 1991; Mark *et al.*, 1993). This inner ear hypoplasia is related in part to alterations of r4–6, as seen from the modifications of *Hox* transcript expression (Dollé *et al.*, 1993) and hindbrain dismorphogenesis (Barrow *et al.*, 2000; Mark *et al.*, 1993). Moreover, in *Hoxa1* and *Hoxb1* mutant mice in which r5 is absent, the vestibular and cochlear regions are absent or underdeveloped (Gavalas *et al.*, 1998).

Studies of VAD animal and vitamin A excess experiments revealed altered temporal and spatial expression of RA regulated genes, notably the *Hox* genes (Gavalas, 2002; Kessel and Gruss, 1991; Maden, 2002; Wendling *et al.*, 2001). The caudal part of the hindbrain patterning, it is revealed, involves graded responses to RA signaling (Begemann and Meyer, 2001; Dupé and Lumsden, 2001). The *Hoxa1* mutation effects observed in inner ear morphogenesis (Chisaka *et al.*, 1992; Mark *et al.*, 1993) can be mimicked in part by RA excess (Frenz *et al.*, 1996) or RA deficiency as produced by RARα/RARγ double mutant (Romand *et al.*, 2002) where this mutation will affect mainly r5 and r6 (Wendling *et al.*, 2001). The inner ear hypoplasia is mediated by the RA-dependant activation where *Hoxa1* possesses a RARE on its promoter (Dupé *et al.*, 1997; Langston and Gudas, 1992), and therefore *Hoxa1* RNA is induced by RA (LaRosa and Gudas, 1988). Then in part the inner ear dismorphogenesis observed after RA excess or deprivation can be related to abnormal rhombomere segmentation that impairs the inner ear induction and/or development. The role of *Hoxa1* was demonstrated with a subteratogenic dose of RA that overrides the loss of function of *Hoxa1* mutant, which rescued some of the patterning defects in the hindbrain and their consequences in inner ear development (Pasqualetti *et al.*, 2001; Fig. 3). This experiment is important because it demonstrates several aspects in relation to RA function during hindbrain early development and inner ear induction and patterning, which can be summarized as follows:

- *Hoxa1* is transiently expressed in the hindbrain and not in the otocyst.
- The targeted inactivation of *Hoxa1* in mice results in severe abnormalities of the otic capsule and the inner ear.

- A single maternal administration of a low dose of RA is sufficient to compensate the requirement for *Hoxa1* function at two levels: (1) to rescue the signaling mechanisms required for otocyst specification and (2) to induce long-term inner ear morphological changes.
- The rescue is time dependent and specific to both vestibular and cochlear defects in mutant fetuses without affecting the development of the wildtypes. This rescue does not concern the hindbrain derivatives such as the cranial nerve nuclei.
- *Fgf3* might be the direct target of RA action in *Hoxa1* mutant embryos and may mediate inner ear rescue since it is known that *Fgf3* promoter is responsive to RA (Murakami *et al.*, 1999).

This rescue effect may be mediated by retinoid receptors, which are present in the presumptive hindbrain ectoderm and adjacent mesenchyme and are

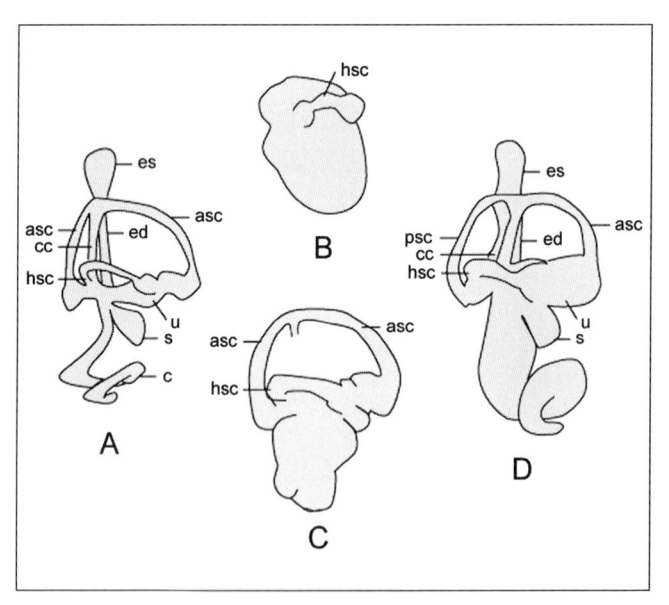

Figure 3 Morphogenesis of the inner ear in *Hoxa1* mouse mutant at E16.0 and its rescue by RA. (A) Schematic representation control of the mouse inner ear. (B, C) The most common phenotypes observed in the *Hoxa1* mutant with only the horizontal canal visible as an identifiable structure. Some mutants displayed a less severe dysmorphic phenotype as shown in C with the semicircular canals present. (D) Rescue of the inner ear from a treated mutant at E8.75 with *all-trans* RA. The two main components of the inner ear are present, as well as their spatial organization. asc, Anterior semicircular canal; c, cochlear canal; cc, common crus; ed, endolymphatic duct; es, endolymphatic sac; hsc, horizontal semicircular canal; psc, posterior semicircular canal; s, saccule; u, utricule. Modified from Pasqualetti, M., Neun, R., Davenne, M., and Rijli, F. M. (2001). Retinoic acid rescues inner ear defects in Hoxa1 deficient mice. *Nat. Genet.* **29**, 34–39.

9. Retinoic Acid and the Inner Ear

observed *Raldh1* in primitive streak mesoderm during early embryogenesis at E7.5 (Ang and Duester, 1997).

Several other observations related to synthetic and metabolic RA enzymes strengthen the function of this hormone on the hindbrain and the inner ear. An excess or a depletion of RA has dramatic effects on embryonic development; therefore, it is expected that a precise control of the RA distribution and concentration is necessary for normal embryonic development. This control can be undertaken by a balance of enzymes mediating RA synthesis and catabolism (McCaffery and Dräger, 2000; Swindell *et al.*, 1999). When comparing the spatial distribution of *Raldh2* transcripts in the posterior somitic mesoderm (Niederreither *et al.*, 1997) with those of *Cyp26A1*, the catabolic enzyme transcripts that are restricted to r2 at E8, this differential spatial distribution illustrates the complementary distribution of these enzymes on the hindbrain (Abu-Abed *et al.*, 2001, Fig. 4).

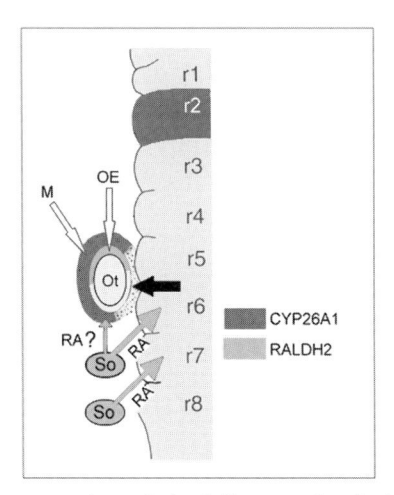

Figure 4 Schematic representation of the influence of retinoid pathways on otocyst development. Two complementary pathways may be related to inner ear ontogenesis. One is from outside the otocyst with retinoic acid (RA) synthesized by the retinaldehyde dehydrogenase 2 (RALDH2) from the somite mesoderme (So), which influences the posterior region of the hindbrain such as rhombomeres 5 and 6 (r5–r6) and also may be directly the otocyst (RA?). The catabolic enzyme CYP26A1 presents a complementary distribution in r2 that may contribute to the generation of a precise balance of RA into the posterior hindbrain. RA is known to be crucial for r5 and r6 specification through *hoxa1*, which in turn seems to be required indirectly for the normal development of the otocyst (black arrow) through a cascade of factors in which FGF3 might be involved. The second pathway is inside the presumptive inner ear where the synthetic and catabolic enzymes are present in a complementary fashion, respectively, in the otic epithelium (OE) and the mesenchyme (M). It is possible that the retinoid pathway in the prospective inner ear is independent from outside otocyst RA and functions without the influence of neighboring regions. (See Color Insert.)

The complementary spatial distribution for the synthetic and catabolic enzyme genes is also observed in the otocyst a few days later (see Figs. 2 and 4), suggesting that a control mechanism exists for modulating RA signaling during inner ear development. Complementary expression patterns are not restricted to the inner ear. The same observations were observed in the developing limb (Fujii *et al.*, 1997) and in the developing retina (McCaffery *et al.*, 1999). Disruption of RA synthesis by targeted mutation of *Raldh2* (Niederreither *et al.*, 2000), corresponding to an RA deprivation, modifies the region-specific gene expression patterns of *Pax2* and *Nkx5.1* (*Hmx3*), two genes involved in inner ear early morphogenesis (Merlo *et al.*, 2002; Rinkwitz-Brandt *et al.*, 1995; Torres *et al.*, 1996).

RA may then intervene during the early inner ear ontogenesis either by its disposal regulated by synthetic and catabolic enzymes or by the presence of its receptors that can mediate its effects on the periotic mesenchyme and the otic epithelium. Whatever the exact relations between these two structures, it seems that RA acts on both components with the possibility of complex interactions, along with other factors, leading to cell proliferation, differentiation, and morphogenesis of the labyrinthic epithelia.

VI. Retinoic Acid and Patterning Genes during Inner Ear Development

After the otic vesicle stage, the presumptive labyrinthic epithelia undergo an important proliferative stage before differentiation. Cell proliferation can be modulated *in vitro* by growth factors (Miner *et al.*, 1988; Represa *et al.*, 1988). In such assays, RA induces a precocious *in vitro* differentiation of several chick inner ear components, including early sensory and supporting epithelia (Léon *et al.*, 1995; Represa *et al.*, 1990). This effect is mediated by inhibiting the mitogen-induced *c-fos* expression that is transient and stage dependent during the otic vesicle development, suggesting that *c-fos* is a target for RA. This observation has been the basis of subsequent experiments investigating the possible involvement of RA in auditory sensory epithelium differentiation (Kelley *et al.*, 1993). This aspect is discussed later.

As in many other systems, the effects of RA on the developing inner ear are both time and dosage dependent (Choo *et al.*, 1998; Frenz *et al.*, 1996; White *et al.*, 1998). In the hindbrain as well, the two types of abnormal rhombomeric pattern seen, respectively, in RARα/RARβ and RARα/RARγ double null mutant mice reflect two distinct windows of RA action ($\sim$E7.5 and E8.5, respectively) because these phenotypes can be produced in cultured wildtype mouse embryos using a synthetic RAR antagonist at the corresponding stages (Wendling *et al.*, 2001). Furthermore, RA rescue of the cochlea defect in *Hoxa-1*$^{-/-}$ mutant mice is precisely time dependent

(Pasqualetti et al., 2001). This time window of RA effects on inner ear morphogenesis is also observed with chick embryos in vivo (Choo et al., 1998).

This temporal aspect has been investigated in detail in the chicken embryo at E2.5 by implanting a resin exchange bead saturated with RA directly into the otocyst (Choo et al., 1998). The outcome of this experiment revealed that the vestibular structures are more susceptible to RA than cochlear structures and that these effects are dose dependent. The SCCs are particularly perturbed, especially the posterior canal, and this is due to the initial outgrowth of the canal plate that can be controlled by BMPs, and more specifically by BMP2 (Chang et al., 2002). However, *Bmp4* transcripts are also present in some putative SCCs such as the superior canal in the same species (Cole et al., 2000; Gerlach et al., 2000) at the time of RA treatment, which could correspond particularly to defects observed on this SCC as a result of activation of *Bmp4* promoter by RA (Thompson et al., 2003). However, this explanation does not hold for sensory epithelia where *Bmp4* is more specifically expressed in both chicken and mouse (Cole et al., 2000; Morsli et al., 1998) because RA seems to have little effect on such sensory structures during chick inner ear morphogenesis (Choo et al., 1998). Another gene, *Msx-1*, which is a sensory epithelium marker in the chick inner ear and is expressed early in development (Wu and Oh, 1996), possesses a RARE in its promoter, suggesting that it could also be a mediator of RA effects.

The differential sensitivity reported between vestibular and cochlear structures in the case of RA excess (Choo et al., 1998) was not seen in mice deficient in RA signaling through RARα/RARγ compound deletion (Romand et al., 2002). In this case, at E10.5 the endocochlear canal is absent and the otocyst is hypomorphic, although a few days later some SCCs are present. Such mutants were studied at E18.5; despite phenotypic inner ear variations between animals the cochlear part is always the most affected of both inner ear structures as observed in experiments with an excess or a depletion of RA (Burk and Willhite, 1992; Frenz et al., 1996; Padmanabhan et al., 1990; Romand et al., 2002). Several different explanations could be given for the discrepancies between these observations. Sensitivity may be different among components of the inner ear between the two species based on subtle variations in molecular mechanisms. For example, although the formation of SCCs in the mouse and chicken is essentially similar (Fekete et al., 1997; Martin and Swanson, 1993), the mechanism of epithelium resorption seems to be different. Another explanation is also possible, that the temporal window of RA exposition in the chick is limited compared with targeted mutations. In the mouse it is known that RA deficiency resulting from targeted disruption of *Raldh2* affects early gene expression in the otic placode (Niederreither et al., 2000). Moreover, treatment with pan–RAR antagonists at E7.0 induces a loss of r5–6 identities, suggesting an early action of RA on hindbrain patterning (Wendling *et al.*, 2001). Finally, the experiment in chicken (Choo *et al.*, 1998) corresponds to an RA excess on the prospective inner ear, whereas mutant mice (Romand *et al.*, 2002) mimic a state of RA deficiency. These sets of experimental conditions may well explain the difference in inner ear component susceptibility where RA could act through different molecular pathways involving mesenchyme–epithelium and hindbrain–otocyst interactions or both at different time windows.

Some other effects may be foreseen regarding the role of RA on inner ear morphogenesis through downstream action on several genes that possess a regulatory element in their promoters such as *Bmp4*, as discussed earlier. One such gene could be *Shh*, whose targeted mutation produces an important agenesis of the otic capsule and the inner ear, especially at the cochlear level (Liu *et al.*, 2002; Riccomagno *et al.*, 2002), reminiscent of RARα/RARγ mutation (Romand *et al.*, 2002) except for the endocochlear duct, which is present at E12.5 in the *Shh* mutant (Liu *et al.*, 2002). Moreover, several genes involved in cochlear cell fates are targets of SHH signaling (Riccomagno *et al.*, 2002), where *Shh* may be required to maintain the *Pax2* expression for the cochlear duct. From the marked decrease in PAX2 protein expression in *Shh* mutants, it was suggested that *Pax2* could be a downstream target of *Shh*. It is interesting to point out that in *Raldh2* depleted embryos, *Pax2* is not expressed (Niederreither *et al.*, 2000) and is downregulated by teratogenic doses of RA during craniofacial morphogenesis (Helms *et al.*, 1997). From these observations, it is tempting to suggest a cascade of signals where dosage-dependent RA could switch on or switch off directly or indirectly *Shh*, which in turn would modify downstream target genes such as *Pax2*, a regulatory gene for inner ear early patterning.

Other genes have been suggested as possible downstream effectors of RA, such as *Otx2*, a bicoid-like transcription factor, expressed in the ventral region of the otocyst and important in the patterning of the cochlea (Cantos *et al.*, 2000). RA administration to embryonic explants or RA-deficient embryos by *Raldh2* deletion modify *Otx2* gene expression (Ang *et al.*, 1994; Niederrheither *et al.*, 1999).

VII. Retinoic Acid and Hair Cell Differentiation

Hair cell differentiation is controlled by a sequence of complex molecular and cellular signals (reviewed by Bryant *et al.*, 2002). The observation that exogenous RA added to chick otocyst *in vitro* induces differentiation of hair cells (Represa *et al.*, 1990) was the starting point suggesting that this hormone could be involved in hair cell differentiation in the inner ear. This observation was supported at that time by the discovery of RAR and RXR transcripts in the inner ear (Dollé *et al.*, 1990, 1994; Ruberte *et al.*, 1990), as

9. Retinoic Acid and the Inner Ear

well as *Crabps* (Dollé *et al.*, 1990). Equally it was demonstrated that CRBP1 and CRABP1 proteins and RA were present in the fetal cochlea and regulated during development until birth (Kelley *et al.*, 1993, Ylikoski *et al.*, 1994). Also it is worthwhile to mention that RA synthetic enzymes are observed in the cochlea after birth (Wagner *et al.*, 2002). Exogenous RA in embryonic organ of Corti explants has been shown to induce hair cell overproduction (Kelley *et al.*, 1993). From these observations, it has been suggested that RA signaling might play a role in the determination of the size of the population of prosensory cells becoming hair cells in the organ of Corti. Despite this very interesting hypothesis, current understanding is that cellular retinoid binding proteins are dispensable for RA function under normal feeding conditions (Ghyselinck *et al.*, 1999; Gorry *et al.*, 1994; Lampron *et al.*, 1995). For example, despite extensive distributions of *Crabp 1* and *2 Crbp 1* gene expressions in the otocyst at E10.5 and in the cochlea before birth, deletion of these genes did not induce impairment of the auditory function in adult mice (Romand *et al.*, 2000). Moreover, *in vitro* E16 rat cochleas in culture media without RA and growth factors produced a large number of supernumerary outer hair cells (Abdouh *et al.*, 1994). RARα was reported to be expressed in the cochlear epithelia in mouse fetuses (Dollé *et al.*, 1990, 1994; Ruberte *et al.*, 1990), and its expression was subsequently found to be present in hair cells (Raz and Kelley, 1999; Romand *et al.*, 1998). It is worthwhile to mention that knockout of any of the RARs, as well as the compound mutation of RARα/RARβ and RARβ/RARγ, did not impair the development of the organ of Corti, whereas the inner ear hypoplasia observed in RARα/RARγ mutant did not prevent the differentiation of few vestibular hair cells. From this last double mutation it can be suggested that the absence of auditory hair cells would be more related to abnormal periotic mesenchyme condensation in the cochlea, thereby preventing the normal development of the sensory epithelium (Romand *et al.*, 2002).

An *in vitro* experiment using an inhibitor of RA synthesis or an RARα-specific antagonist induced a significant decrease in the number of hair cells within the developing cochlea, suggesting that RA is necessary for the normal development of the organ of Corti (Raz and Kelley, 1999). Although RA seems to be necessary for the cochlear sensory epithelia during development, the exact mechanism of RA function may not be as simple as first thought. The RA pathways in the cochlea are surprisingly complex (Romand *et al.*, 2003) considering that this hormone is synthesized in several regions and early during inner ear embryogenesis (see Fig. 2) and retinoid receptors are present in several tissues of the vestibular and cochlear components of the inner ear. At the time of differentiation of cell types in the cochlear epithelium, from E14 onward retinoid receptors gene expressions are seen in most cell types, except RARβ and RXRγ, where RARα as well as RXRβ are strongly expressed in the organ of Corti. Genes of synthetic enzymes such as *Raldh1* are observed in two regions close to outer hair cells. *Raldh2* is restricted to the stria vascularis and Reissner's membrane, whereas *Raldh3* is only present in the spiral ganglion (Romand et al., 2001, 2003; Wagner et al., 2002). *Cyp26B1* expression at E18.5 is observed in two regions: Kölliker's organ and cells close to outer hair cells.

It has been suggested that RALDH1 and RALDH3 are involved in RA production in the developing cochlea since the expression domain of the RA-responsive reporter transgene is not altered in *Raldh2* mutant embryos at E14.5 (Niederreiter et al., 2002). Alternatively, CYP26s are known to be involved in the inactivation of RA (Abu-Abed et al., 2001; Duester, 2000; Fujii et al., 1997). Therefore, the complementary distribution of synthetic and catabolic enzymes restricted to specific cell populations could be responsible for the required precise modulation of RA necessary for cochlear development and function. Such complementary domains of distribution of these two types of enzymes are also found in the developing retina (Mey et al., 2001), as well as in several other organs (Abu-Abed et al., 2001; Sakai et al., 2001). It should be noted that *Raldh1* and *Cyp26B1*, whose gene expressions are very close to hair cells at E18.5, could suggest an important role for this hormone at this late stage of development, which is no longer related to the specific differentiation of prosensory cells into hair cells. RA produced by *Raldh2* in the stria vascularis and in Reissner's membrane may act as a diffusible factor, which in turn through its receptors might regulate downstream genes in several cell types, leading to unknown RA-regulated functions in the cochlea after birth. Two main targets of RA could be suggested: the stria vascularis, whose function is essential for normal sensory transduction (Price and Fisher, 2001), and the organ of Corti relative to its innervation. However, the exact function of RA on these cochlear cell populations remains to be characterized and further analyses are required to probe the role of this hormone in the inner ear.

VIII. Concluding Remarks

Conclusions derived from present data in trying to dissect the RA pathways in the inner ear arise from three sets of observations: (1) gene expression patterns of enzymes involved in RA metabolism and retinoid receptors, (2) *in vitro* or *in vivo* experiments creating excess or depletion of RA, and (3) targeted deletion of specific genes that very often mimic either an excess or a deficiency of RA. Such manipulations can induce specific morphological phenotypes that are not always easy to explain from a functional point of view because of the RA pleiotropic action on numerous targets. Moreover, many developmental studies were stopped at birth, rendering it impossible to observe a full phenotype. However, sometimes with adult mutants a specific phenotype is difficult to observe, for example the single retinoid receptor deletions due to a well-known redundancy between metabolic enzymes and the retinoid receptors, as well as cellular retinoid binding proteins. In this last example, despite a large distribution of these cellular retinoid binding protein genes during development in the inner ear and subsequent formulated hypotheses, it is interesting to underline that auditory thresholds in adult mice were not modified in these compound mutants. This emphasizes the necessity to go beyond morphological examinations in order to decipher the real functional role of a specific gene in the statoacoustical system, especially for those involved in cell differentiation. From present observations it could be possible to foresee a functional role for RA during the postnatal stage in mammals that could include the adult function of the vestibular and cochlear systems, along with perhaps new roles for RA beyond those related to development.

Acknowledgment

I thank Pascal Dollé for critical reading of the manuscript and Norbert Ghyselinck for fruitful comments.

References

Abdouh, A., Després, G., and Romand, R. (1994). Histochemical and scanning electron microscopic studies of supernumerary hair cells in embryonic rat cochlea *in vitro*. *Brain Res.* **660,** 181–191.

Abu-Abed, S., Dollé, P., Metzger, D., Beckett, B., Chambon, P., and Petkovich, M. (2001). The retinoic acid-metabolizing enzyme, CYP26A1, is essential for normal hindbrain patterning, vertebral identity, and development of posterior structures. *Genes Dev.* **15,** 226–240.

Ang, H. E., and Duester, G. (1997). Initiation of retinoid signaling in primitive streak mouse embryos: Spatiotemporal expression patterns of receptors and metabolic enzymes for ligand synthesis. *Dev. Dyn.* **208,** 536–543.

Ang, S. L., Conlon, R. A., Jin, O., and Rossant, J. (1994). Positive and negative signals from mesoderm regulate the expression of Otx2 in ectoderm explants. *Development* **120,** 2979–2989.

Anniko, M. (1983). Embryonic development of vestibular sense organs and their innervation. In "Development of Auditory and Vestibular Systems" (R. Romand, Ed.), pp. 375–423. Academic Press, New York.

Baker, C. V., and Bronner-Fraser, M. (2001). Vertebrate cranial placodes 1. Embryonic induction. *Dev. Biol.* **232,** 1–61.

Barros, A. C., Erway, L. C., Krezel, W., Curran, T., Kastner, P., *et al.* (1998). Absence of thyroid hormone receptors β-retinoid X receptor interactions in auditory function and in the pituitary-thyroid axis. *Neuroreport* **9,** 2933–2937.

Barrow, J. R., Stadler, H. S., and Capecchi, M. R. (2000). Roles of Hoxa1 and Hoxa2 in patterning the early hindbrain of the mouse. *Development* **127,** 933–944.

Begemann, G., and Meyer, A. (2001). Hindbrain patterning revisited: Timing and effects of retinoic acid signalling. *Bioessays* **23,** 981–986.

Brickell, P., and Thorogood, P. (1997). Retinoic acid and retinoic acid receptors in craniofacial development. *Cell Dev. Biol.* **8,** 437–443.

Bryant, J., Goodyear, R. J., and Richardson, G. P. (2002). Sensory organ development in the inner ear: Molecular and cellular mechanisms. *Br. Med. Bull.* **63,** 39–57.

Burk, D. T., and Willhite, C. C. (1992). Inner ear malformations induced by isotretinoin in hamster fetuses. *Teratology* **46,** 147–157.

Cantos, R., Cole, L. K., Acampora, D., Simeone, A., and Wu, D. K. (2000). Patterning of the mammalian cochlea. *Proc. Natl. Acad. Sci. USA* **97,** 11707–11713.

Cash, D. E., Bock, C., Schughart, K., Linney, E., and Underhill, T. M. (1997). Retinoic acid receptor a function in vertebrate limb skeletogenesis: A modulator of chondrogenesis. *J. Cell Biol.* **136,** 445–457.

Chambon, P. (1996). A decade of molecular biology of retinoid acid receptors. *FASEB J.* **10,** 940–954.

Chang, B. E., Blader, P., Fischer, N., Ingham, P. W., and Strahle, U. (1997). Axial HNF3beta and retinoic acid receptors are regulators of the zebrafish sonic hedgehog promoter. *EMBO J.* **16,** 3955–3964.

Chang, W., ten Dijke, P., and Wu, D. K. (2002). BMP pathways are involved in otic capsule formation and epithelial mesenchymal signaling in the developing chicken inner ear. *Dev. Biol.* **251,** 380–394.

Chisaka, O., Musci, T. S., and Capecchi, M. R. (1992). Developmental defects of the ear, cranial nerves and hindbrain resulting from targeted disruption of the mouse homeobox *Hox-1.6*. *Nature* **355,** 516–520.

Choo, D., Sanne, J. L., and Wu, D. K. (1998). The differential sensitivities of inner ear structures to retinoic acid during development. *Dev. Biol.* **204,** 136–150.

Cole, L. K., Le Roux, I., Nunes, F., Laufer, E., Lewis, J., and Wu, D. K. (2000). Sensory organ generation in the chicken ear: Contribution of *Bone morphogenetic protein 4, Serrate1* and *Lunatic Fringe*. *J. Comp. Neurol.* **424,** 509–520.

Cordes, S. P. (2001). Molecular genetics of cranial nerve development in mouse. *Nature Rev. Neurosci.* **2,** 611–623.

Couly, G. F., Coltey, P. M., and Le Douarin, N. M. (1993). The triple origin of skull in higher vertebrates: A study in quail-chick chimeras. *Development* **117,** 409–429.

Creazzo, T. L., Godt, R. E., Leatherbury, L., Conway, S. J., and Kirby, M. L. (1998). Role of cardiac neural crest cells in cardiovascular development. *Annu. Rev. Physiol.* **60,** 267–286.

Creech-Kraft, J., and Willhite, C. C. (1996). Retinoids in abnormal and normal embryonic development. *In* "Environmental Toxicology and Human Development" (S. Kacew and S. Lambert, Eds.), pp. 15–49. Taylor and Francis, Washington.

Deol, M. S. (1964). The abnormalities of the inner ear in kreisler mice. *J. Embryol. Exp. Morph.* **12,** 475–490.

Deol, M. S. (1983). Development of auditory and vestibular systems in mutant mice. *In* "Development of Auditory and Vestibular Systems" (R. Romand, Ed.), pp. 309–333. Academic Press, New York.

Deschamps, J., Van den Akker, E., Forlani, S., De Graaff, W., Oosterveen, T., *et al.* (1999). Initiation, establishment and maintenance of *Hox* gene expression patterns in the mouse. *Int. J. Dev. Biol.* **43,** 635–650.

Dollé, P., Fraulob, V., Kastner, Ph., and Chambon, P. (1994). Developmental expression of murine retinoid X receptor (RXR) genes. *Mech. Dev.* **45,** 91–104.

Dollé, P., Lufkin, T., Krumlauf, R., Mark, M., Duboule, D., and Chambon, P. (1993). Local alterations of *Krox-20* and *Hox* gene expression in the hindbrain suggest lack of rhombomeres 4 and 5 in homozygote null *Hoxa-1* (*Hox-1.6*) mutant embryos. *Proc. Natl. Acad. Sci. USA* **90,** 7666–7670.

Dollé, P., Ruberte, E., Leroy, P., Morriss-Kay, G., and Chambon, P. (1990). Retinoic acid receptors and cellular retinoid binding proteins. I. A systematic study of their differential pattern of transcription during mouse organogenesis. *Development* **110,** 1133–1151.

Dräger, U. C., and McCaffery, P. (1997). Retinoic acid and development of the retina. *Prog. Retin. Eye Res.* **16,** 323–346.

Duester, G. (2000). Families of retinoid dehydrogenases regulating vitamin A function. Production of visual pigment and retinoic acid. *Eur. J. Biochem.* **267,** 4315–4324.

Dupé, V., Davenne, M., Brocard, J., Dollé, P., and Mark, M. (1997). In vivo functional analysis of the *Hoxa-1* 3′ retinoic acid response element (3′RARE). *Development* **124,** 339–410.

Dupé, V., Ghyselinck, N. B., Wendling, O., Chambon, P., and Mark, M. (1999). Key roles of retinoic acid receptors alpha and beta in the patterning of the caudal hindbrain, pharyngeal arches and otocyst in the mouse. *Development* **126,** 5051–5059.

Dupé, V., and Lumsden, A. (2001). Hindbrain patterning involves graded responses to retinoic acid signalling. *Development* **128,** 2199–2208.

Edenberg H. J. (2000) Regulation of the mammalian alcohol dehydrogenase genes. *In* "Progress in Nucleic Acid Research and Molecular Biology," Vol. 64, pp. 295–341. Academic Press, San Diego.

Evans, R. M. (1988). The steroid and thyroid hormone receptors superfamily. *Science* **240,** 889–895.

Fekete, D. M., Homburger, S. A., Waring, M. T., Riedl, A. E., and Garcia, L. E. (1997). Involvement of programmed cell death in morphogenesis of the vertebrate inner ear. *Development* **124,** 2451–2461.

Fekete, D. M., and Wu, K. W. (2002). Revisiting cell fate specification in the inner ear. *Curr. Opin. Neurobiol.* **12,** 35–42.

Frenz, D. A., Liu, W., Galinovic-Schwartz, V., and Van de Water, T. R. (1996). Retinoic acid-induced embryopathy of the mouse inner ear. *Teratology* **53,** 292–303.

Frenz, D. A., and Liu, W. (1997). Effect of retinoic acid on otic capsule chondrogenesis in high-density culture suggests disruption of epithelial-mesenchyme interactions. *Teratology* **56,** 233–240.

Frenz, D. A., and Liu, W. (1998). Role of FGF3 in otic capsule chondrogenesis in vitro: An antisense oligonucleotide approach. *Growth Factors* **15,** 173–182.

Frenz, D. A., Liu, W., William, J. D., Hatcher, V., Galinovic-Schwartz, V., *et al.* (1994). Induction of chondrogenesis: Requirement for synergistic interaction of basic fibroblast growth factor and transforming growth factor-beta. *Development* **120,** 415–424.

Frenz, D. A., and Van de Water, T. R. (1991). Epithelial control of periotic mesenchyme chondrogenesis. *Dev. Biol.* **144,** 38–46.

Fritzsch, B., Barald, K. F., and Lomax, M. I. (1998). Early embryology of the vertebrate ear. *In* "Development of the Auditory System" (E. D. Rubel, A. R. Popper and R. R. Fay, Eds.), pp. 80–145. Springer-Verlag, New York.

Frohman, M. A., Martin, G. R., Cordes, S. P., Halamek, L. P., and Barsh, G. S. (1993). Altered rhombomere-specific gene expression and hyoid bone differentiation in the mouse segmentation mutant, kreisler (kr). *Development* **117,** 925–936.

Fujii, H., Sato, T., Kaneko, S., Gotoh, O., Fijii-Kuriyama, Y., *et al.* (1997). Metabolic inactivation of retinoic acid by a novel P450 differentially expressed in developing mouse embryos. *EMBO J.* **16,** 4163–4173.

Gavalas, A. (2002). ArRAnging the hindbrain. *Trends Neurosci.* **25,** 61–64.

Gavalas, A., Studer, M., Lumsden, A., Rijli, F. M., Krumlauf, R., and Chambon, P. (1998). Hoxal and Hoxbl synergize in patterning the hindbrain, cranial nerves and second pharyngeal arch. *Development* **125,** 1123–1136.

Gerlach, L. M., Hutson, M. R., Germiller, J. A., Nguyen-Luu, D., Victor, J. C., and Barald, K. F. (2000). Addition of the BMP4 antagonist, noggin, disrupts avian inner ear development. *Development* **127**, 45–54.

Ghyselinck, N. B., Bavik, C., Sapin, V., Mark, M., Bonnier, D., *et al.* (1999). Cellular retinol-binding protein I is essential for vitamin A homeostasis. *EMBO J.* **18**, 4903–4914.

Gorry, P., Lufkin, T., Dierich, A., Rochette-Egly, C., Décimo, D., *et al.* (1994). The cellular retinoic acid binding protein I is dispensable. *Proc. Natl. Acad. Sci. USA* **91**, 9032–9036.

Gottesman, M. E., Quadro, L., and Blaner, W. S. (2001). Studies of vitamin A metabolism in mouse model system. *Bio essays* **23**, 409–419.

Groves, A. K., and Bronner-Fraser, M. (2000). Competence, specification and commitment in otic placode induction. *Development* **127**, 3489–3499.

Grün, F., Hirose, Y., Kawauchi, S., Ogura, T., and Umesono, K. (2000). Aldehyde dehydrogenase 6, a cytolosic retinaldehyde dehydrogenase prominently expressed in sensory neuroepithelia during development. *J. Biol. Chem.* **275**, 41210–41218.

Hammond, K. L., Loynes, H. E., Folarin, A. A., Smith, J., and Whitfield, T. T. (2003). Hedgehog signalling is required for correct anteroposterior patterning of the zebrafish otic vesicle. *Development* **130**, 1403–1417.

Helms, J. A., Kim, C. H., Hu, D., Minkoff, R., Thaller, C., and Eichele, G. (1997). *Sonic hedgehog* participates in craniofacial morphogenesis and is down-regulated by teratogenic doses of retinoic acid. *Dev. Biol.* **187**, 25–35.

Jarvis, B. L., Johnston, M. C., and Sulik, K. K. (1990). Congenital malformations of the external, middle, and inner ear produced by isotretinoin exposure in mouse embryos. *Otolaryngol. Head Neck Surg.* **102**, 391–401.

Johnson, R. L., and Tabin, C. J. (1997). Molecular models for vertebrate limb development. *Cell* **90**, 979–990.

Kastner, P., Mark, M., and Chambon, P. (1995). Nonsteroid nuclear receptors: What are genetics studies telling us about their role in real life. *Cell* **83**, 859–869.

Kastner, P., Mark, M., Ghyselinck, N., Krezel, W., Dupe, V., *et al.* (1997). Genetic evidence that the retinoid signal is transduced by heterodimeric RXR/RAR functional units during mouse development. *Development* **124**, 313–326.

Kelley, M. W., Xiao-Mei, X., Wagner, M. A., Warchol, M. E., and Corwin, J. T. (1993). The developing organ of Corti contains retinoic acid and forms supernumerary hair cells in response to exogenous retinoid acid in culture. *Development* **119**, 1041–1053.

Kessel, M., and Gruss, P. (1991). Homeotic transformations of murine vertebrae and concomitant alteration of Hox codes induced by retinoic acid. *Cell* **67**, 89–104.

Kil, S. H., and Collazo, A. (2002). A review of inner ear fate maps and cell lineage studies. *J. Neurobiol.* **53**, 129–142.

La Mantia, A. S., Bhasin, N., Rhodes, K., and Heemskerk, J. (2000). Mesenchymal/epithelial induction mediates olfactory pathway formation. *Neuron* **28**, 411–425.

Lampron, C., Rochette-Egly, C., Gorry, P., Dollé, P., Mark, M., *et al.* (1995). Mice deficient in cellular retinoic acid binding protein II (CRABPII) or in both CRABPI and CRABPII are essentially normal. *Development* **121**, 539–548.

Langston, A. W., and Gudas, L. J. (1992). Identification of a retinoic acid responsive enhancer 3′ of the murine homeobox gene Hox-1.6. *Mech. Dev.* **38**, 217–228.

La Rosa, G. J., and Gudas, L. J. (1988). An early effect of retinoic acid cloning of an mRNA (ERA-1) exhibiting rapid and protein synthesis-independent induction during teratocarcinoma stem cell differentiation. *Proc. Natl. Acad. Sci. USA* **85**, 329–333.

Le Douarin, N. M. (1982). "The Neural Crest." Cambridge University Press, Cambridge.

Léon, Y., Sanchez, J. A., Miner, C., Ariza-McNaughton, L., Represa, J. J., and Giraldez, F. (1995). Developmental regulation of Fos-protein during proliferative growth of the otic vesicle and its relation to differentiation induced by retinoic acid. *Dev. Biol.* **167**, 75–86.

Lim, D. J. (1977). Histology of the developing inner ear. Normal anatomy and developmental anomalies. *In* "Hearing Loss in Children" (B. F. Jaffee, Ed.), pp. 27–50. University Park Press, Baltimore.

Lin, M., Zhang, M., Abraham, M., Smith, S. M., and Napoli, J. L. (2003). Mouse RALDH4: molecular cloning, cellular expression, and activity in 9-cis-retinoic acid biosynthesis in intact cells. *J. Biol. Chem.* **278**, 9856–9861.

Liu, W., Li, G., Chien, J. S., Raft, S., Zhang, H., *et al.* (2002). Sonic hedgehog regulates otic capsule chondrogenesis and inner ear development in the mouse embryo. *Dev. Biol.* **248**, 240–250.

Liu, W., Oh, S. H., Kang, Y. K., Li, G., Doan, T. M., *et al.* (2003). Bone morphogenetic protein 4 (BMP4): A regulator of capsule chongrogenesis in the developing mouse inner ear. *Dev. Dyn.* **226**, 427–438.

Lohnes, D., Kastner, P., Dierich, A., Mark, M., LeMeur, M., and Chambon, P. (1993). Function of retinoic acid receptor γ (RARγ) in the mouse. *Cell* **73**, 643–658.

Lohnes, D., Mark, M., Mendelsohn, C., Dollé, P., Dierich, A., *et al.* (1994). Function of the retinoic acid receptors (RARs) during development. (I) Craniofacial and skeletal abnormalities in RAR double mutants. *Development* **120**, 2723–2748.

Lufkin, T., Dierich, A., LeMeur, M., Mark, M., and Chambon, P. (1991). Disruption of the *Hox-1.6* homeobox gene results in defect in a region corresponding to its rostral domain of expression. *Cell* **66**, 1105–1119.

Lumsden, A., and Krumlauf, R. (1996). Patterning the vertebrate neuraxis. *Science* **274**, 1109–1115.

Maden, M. (1994). Role of retinoids in embryonic development. *In* "Vitamin A in Health and Disease" (R. Blomhoff, Ed.), pp. 289–322. Dekker, New York.

Maden, M. (2002). Retinoid signalling in the development of the central nervous system. *Nat. Rev.* **3**, 843–853.

Maden, M., Gale, E., Kosteskii, I., and Zile, M. (1996). Vitamin A-deficient quail embryos have half a hindbrain and other neural defects. *Curr. Biol.* **6**, 417–426.

Maden, M., Gale, E., and Zile, M. (1998). The role of vitamin A in the development of the central nervous system. *J. Nutr.* **128**, 471S–475S.

Mahmood, R., Kiefer, P., Guthrie, S., Dickson, C., and Mason, I. (1995). Multiple roles for FGF-3 during cranial neural development in the chicken. *Development* **121**, 1399–1410.

Mahmood, R., Mason, I. J., and Morriss-Kay, G. M. (1996). Expression of *Fgf-3* in relation to hindbrain segmentation, otic pit position and pharyngeal arch morphology in normal and retinoic acid-exposed mouse embryos. *Anat. Embryol.* **194**, 13–22.

Malik, M. A., Blusztajn, J. K., and Greenwood, C. E. (2000). Nutrients as trophic factors in neurons and the central nervous system: Role of retinoic acid. *J. Nutr. Biochem.* **11**, 2–13.

Mallo, M. (1997). Retinoic acid disturbs mouse middle ear development in a stage-dependent fashion. *Dev. Biol.* **184**, 175–186.

Mangelsdorf, D. J., Thummel, C., Beato, M., Herrlich, P., Schutz, G., *et al.* (1995). The nuclear receptor superfamily: The second decade. *Cell* **83**, 835–839.

Mansour, S. L., Goddard, J. M., and Capecchi, M. R. (1993). Mice homozygotous for a targeted disruption of the proto-oncogene int-2 have development defects in the tail and inner ear. *Development* **117**, 13–28.

Manzanares, M., Trainor, P. A., Nonchev, S., Ariza-McNaughton, L., Brodie, J., *et al.* (1999). The role of kreisler in segmentation during hindbrain development. *Dev. Biol.* **211**, 220–237.

Mark, M., Ghyselinck, N. B., Kastner, P., Dupé, V., Wendling, O., *et al.* (1998). Mesectoderm is a major target of retinoic acid action. *Eur. J. Oral. Sci.* **106**, 24–31.

Mark, M., Lufkin, T., Vonesch, J. L., Ruberte, E., Olivo, J. C., *et al.* (1993). Two rhombomeres are altered in *Hoxa-1* mutant mice. *Development* **119**, 319–338.

Marshall, H., Morrison, A., Studer, M., Popperl, H., and Krumlauf, R. (1996). Retinoids and Hox genes. *FASEB J.* **10**, 969–978.

Martin, P., and Swanson, G. I. (1993). Descriptive and experimental analysis of the epithelial remodelling that control semicircular canal formation in the developing mouse inner ear. *Dev. Biol.* **159,** 549–558.

McKay, I. J., Lewis, J., and Lumsden, A. (1996). The role of FGF-3 in early inner ear development. An analysis in normal and *kreisler* mutant mice. *Development* **174,** 370–378.

McKay, I. J., Muchamore, I., Krumlauf, R., Maden, M., Lumsden, A., and Lewis, J. (1994). The *kreisler* mouse: A hindbrain segmentation mutant that lacks two rhombomeres. *Development* **120,** 2199–2211.

McCaffery, P., and Dräger, U. C. (1994). Hot spots of retinoic acid synthesis in the developing spinal cord. *Proc. Natl. Acad. Sci. USA* **91,** 7194–7197.

McCaffery, P., and Dräger, U. C. (2000). Regulation of retinoic acid signaling in the embryonic nervous system: A master differentiation factor. *Cytokine Growth Factor Rev.* **11,** 233–249.

McCaffery, P., Wagner, E., O'Neil, J., Petkovich, M., and Dräger, U. C. (1999). Dorsal and ventral retinoic territories defined by retinoic acid synthesis break-down and nuclear receptor expression. *Mech. Dev.* **85,** 203–214.

McPhee, J. R., and Van de Water, T. R. (1986). Epithelial-mesenchymal tissue interactions guiding otic capsule formation: The role of the otocyst. *J. Embryol. Exp. Morphol.* **97,** 1–24.

Mendelsohn, C., Lohnes, D., Décimo, D., Lufkin, T., LeMeur, M., *et al.* (1994a). Function of the retinoic acid receptors (RARs) during development. (II) Multiple abnormalities at various stages of organogenesis in RAR double mutants. *Development* **120,** 2749–2771.

Mendelsohn, C., Mark, M., Dollé, P., Dierich, A., Gaub, M. P., *et al.* (1994b). Retinoic acid receptor β2 (RARβ2) null mutant mice appear normal. *Dev. Biol.* **166,** 246–258.

Merlo, G. R., Paleari, L., Mantero, S., Zerega, B., Adamska, M., *et al.* (2002). The Dlx5 homeobox gene is essential for vestibular morphogenesis in the mouse embryo through a BMP4-mediated pathway. *Dev. Biol.* **248,** 157–169.

Mey, J., McCaffery, P., and Klemeit, M. (2001). Sources and sink of retinoic acid in the embryonic chick retina: Distribution of aldehyde dehydrogenase activities, CRABP-I, and sites of retinoic acid inactivation. *Dev. Brain Res.* **127,** 135–148.

Mic, F. A., Molotkov, A., Fan, X., Cuenca, A. E., and Duester, G. (2000). RALDH3, a retinaldehyde dehydrogenase that generates retinoic acid, is expressed in the ventral retina, otic vesicle and olfactory pit during mouse development. *Mech. Dev.* **97,** 227–230.

Miner, C., Represa, J. J., Barbosa, E., and Giraldez, F. (1988). Protein-kinase C is required during the early development of the inner ear in culture. *Roux Arch. Dev. Biol.* **197,** 294–297.

Morriss, G. M. (1972). Morphogenesis of the malformations induced in rat embryos by maternal hypervitaminosis. *Am. J. Anat.* **113,** 241–250.

Morriss-Kay, G. M., and Ward, S. J. (1999). Retinoids and mammalian development. *Int. Rev. Cytol.* **188,** 73–131.

Morsli, H., Choo, D., Ryan, A., Johnson, R., and Wu, D. K. (1998). Development of the mouse inner ear and origin of its sensory organs. *J. Neurosci.* **18,** 3327–3335.

Murakami, A., Thurlow, J., and Dickson, C. (1999). Retinoic acid-regulated expression of fibroblast growth factor 3 requires the interaction between a novel transcription factor and GATA-4. *J. Biol. Chem.* **274,** 17242–17248.

Niederreither, K., Abu-Abed, S., Schuhbaur, B., Petkovich, M., Chambon, P., and Dollé, P. (2002). Genetic evidence that oxidative derivatives of retinoic acid are not involved in retinoid signaling during mouse development. *Nat. Genet.* **31,** 84–88.

Niederreither, K., McCaffery, P., Dräger, U. C., Chambon, P., and Dollé, P. (1997). Restricted expression and retinoic acid-induced downregulation of the retinaldyde dehydrogenase type 2 (RALDH2) gene during mouse development. *Mech. Dev.* **62,** 67–78.

Niederreither, K., Subbarayan, V., Dollé, P., and Chambon, P. (1999). Embryonic retinoic acid synthesis is essential for early mouse post-implantation development. *Nat. Gen.* **21,** 444–448.

Niederreither, K., Vermot, J., Schuhbaur, B., Chambon, P., and Dollé, P. (2000). Retinoic acid synthesis and hindbrain patterning in the mouse embryo. *Development* **127**, 75–85.
Noden, D. M., and Van de Water, T. R. (1992). Genetic analysis of mammalian ear development. *Trends Neurosci.* **15**, 235–237.
Noramly, S., and Grainger, R. M. (2002). Determination of the embryonic inner ear. *J. Neurobiol.* **53**, 100–128.
Noy, N. (2000). Retinoid-binding proteins: Mediators of retinoid action. *Biochem. J.* **348**, 481–495.
Oh, S. H., Johnson, R., and Wu, D. K. (1996). Differential expression of bone morphogenetic proteins in the developing vestibular and auditory sensory organs. *J. Neurosci.* **15**, 6463–6475.
Padmanabhan, R., Vaidya, E. H., and Abu-Alattan, A. A. F. (1990). Malformation of the ear induced by maternal exposure to retinoic acid in the mouse fetuses. *Teratology* **42**, 25A.
Pasqualetti, M., Neun, R., Davenne, M., and Rijli, F. M. (2001). Retinoic acid rescues inner ear defects in Hoxa1 deficient mice. *Nat. Genet.* **29**, 34–39.
Phillips, B. T., Bolding, K., and Riley, B. B. (2001). Zebrafish *fgf3* and *fgf8* encode redundant functions required for otic placode induction. *Dev. Biol.* **235**, 351–365.
Pickles, J. O. (2001). The expression of fibroblast growth factors and their receptors in the embryonic and neonatal mouse inner ear. *Hear. Res.* **155**, 54–62.
Pirvola, U., Spencer-Dene, B., Xing-Qun, L., Kettunen, P., Thesleff, I., *et al.* (2000). FGF/FGFR-2(IIIb) signaling is essential for inner ear morphogenesis. *J. Neurosci.* **20**, 6125–6134.
Price, E. R., and Fisher, D. E. (2001). Sensorineural deafness and pigmentation genes: Melanocytes and the *Mitf* transcriptional network. *Neuron* **30**, 15–18.
Raz, Y., and Kelley, M. W. (1999). Retinoic acid signaling is necessary for the development of the organ of Corti. *Dev. Biol.* **213**, 180–193.
Represa, J., León, Y., Miner, C., and Giraldez, F. (1991). The int-2 proto-oncogene is responsible for induction of the inner ear. *Nature* **353**, 561–563.
Represa, J., Miner, C., Barbosa, E., and Giraldez, F. (1988). Bombesin and other growth factors activate cell proliferation in chick embryo otic vesicles in culture. *Development* **102**, 87–96.
Represa, J., Sanchez, A., Miner, C., Lewis, J., and Giraldez, F. (1990). Retinoic acid modulation of the early development of the inner ear is associated with the control of *c-fos* expression. *Development* **110**, 1081–1090.
Riccomagno, M. M., Martinu, L., Mulheisen, M., Wu, D. K., and Epstein, D. J. (2002). Specification of the mammalian cochlea is dependent of Sonic hedgehog. *Genes Dev.* **26**, 2365–2378.
Rijli, F. M., and Chambon, P. (1997). Genetic interactions of hox genes in limb development: Learning from compound mutants. *Curr. Opin. Genet. Dev.* **7**, 481–487.
Rinkwitz-Brandt, S., Justus, M., Oldenettel, I., Arnold, H. H., and Bober, E. (1995). Distinct temporal expression of mouse *Nkx5-1* and *Nkx-2* homeobox genes during brain and ear development. *Mech. Dev.* **52**, 371–381.
Rivolta, M. (1997). Transcription factors in the ear: Molecular switches for development and differentiation. *Audiol. Neurootol.* **2**, 36–49.
Robinson-Rechavi, M., Garcia, H. E., and Laudet, V. (2003). The nuclear receptor superfamily. *J. Cell Sci.* **116**, 585–586.
Romand, R., Albuisson, E., Niederreither, K., Fraulob, V., Chambon, P., and Dollé, P. (2001). Specific expression of the retinoic acid-synthesizing enzyme RALDH2 during mouse inner ear development. *Mech. Dev.* **106**, 185–189.
Romand, R., Hashino, E., Dollé, P., Vonesh, J. L., Chambon, P., and Ghyselinck, N. B. (2002). The retinoid acid receptors RARα and RARγ are required for inner ear development. *Mech. Dev.* **119**, 213–223.

Romand, R., Niederreither, K., Abu-Abed, S., Petkovich, M., Fraulob, V., et al. (2003). Complementary expression patterns of retinois acid-synthesizing and-metabolizing enzymes in prenatal mouse inner ear structures. *Gene Exp. Patt.* (imprint)

Romand, R., Sapin, V., and Dollé, P. (1998). Spatial distributions of retinoic acid receptor gene transcripts in the prenatal mouse inner ear. *J. Comp. Neurol.* **393,** 298–308.

Romand, R., Sapin, V., Ghyselinck, N., Avan, P., Le Calvez, S., et al. (2000). Spatio-temporal distribution of cellular retinoid binding protein gene transcripts in the developing and the adult cochlea. Morphological and functional consequences in CRABP- and CRBP-null mutant mice. *Eur. J. Neurosci.* **12,** 2793–2804.

Ross, S. A., McCaffery, P. J., Drager, U. C., and DeLuca, L. M. (2000). Retinoids in embryonal development. *Physiol. Rev.* **80,** 1021–1054.

Ruberte, E. P., Dollé, P., Krust, A., Zelent, A., Morriss-Kay, G., and Chambon, P. (1990). Specific spatial and temporal distribution of retinoid acid receptors gamma transcripts during mouse embryogenesis. *Development* **108,** 213–222.

Sakai, Y., Meno, C., Fujii, H., Nishino, J., Shiratori, H., et al. (2001). The retinoic-inactivating enzyme CYP26 is essential for establishing an uneven distribution of retinoic acid along the antero-posterior axis within the mouse embryo. *Genes Dev.* **15,** 213–225.

Sasagawa, S., Takabatake, T., Takabatake, Y., Muramatsu, T., and Takeshima, K. (2002). Axes establishment during eye morphogenesis in *Xenopus* by coordinate and antagonistic actions of BMP4, Shh, and RA. *Genesis* **33,** 86–96.

Smith, S. M., and Dickman, E. D. (1997). New insights into retinoid signalling in cardiac development. *Trends Cardiovasc. Med.* **7,** 53–58.

Stoilov, I. (2001). Cytochrome P450s: Coupling development and environment. *Trends Genet.* **17,** 629–632.

Streit, A. (2002). Extensive cell movements accompany formation of the otic placode. *Dev. Biol.* **249,** 237–254.

Sulik, K. K. (1986). Isoretinoin embryopathy and the cranial neural crest: An in vivo and in vitro study. *J. Craniofac. Genet. Dev. Biol.* **6,** 211–222.

Swindell, E. C., Thaller, C., Sockanathan, S., Petkovich, M., Jessel, T. M., and Eichele, G. (1999). Complementary domains of retinoic acid production and degradation in the early chick embryos. *Dev. Biol.* **216,** 282–296.

Takemura, T., Sakafumi, S., Takebayashi, K., Umemoto, M., Nakase, T., et al. (1996). Localization of bone morphogenetic protein-4 messenger RNA in developing mouse cochlea. *Hear. Res.* **95,** 26–32.

Thompson, D. L., Gerlach-Bank, L. M., Barald, K. F., and Koening, R. J. (2003). Retinoic acid repression of bone morphogenetic protein 4 in inner ear development. *Mol. Cell Biol.* **23,** 2277–2286.

Thurnham, D. I., and Northrop-Clewes, C. A. (1999). Optimal nutrition: Vitamin A and the carotenoids. *Proc. Nutr. Soc.* **58,** 449–457.

Torres, M., and Giraldez, F. (1998). The development of the vertebrate inner ear. *Mech. Dev.* **71,** 5–21.

Torres, M., Gômez-Pardo, E., and Gruss, P. (1996). *Pax2* contribute to inner ear patterning and optic nerve trajectories. *Development* **121,** 4057–4065.

Underhill, T. M., and Weston, A. D. (1998). Retinoids and their receptors in skeletal development. *Microsc. Res. Tech.* **43,** 137–155.

Van de Water, T. R. (1983). Embryogenesis of the inner ear: "In vitro studies" In "Development of Auditory and Vestibular Systems" (R. Romand, Ed.), pp. 337–374. Academic Press, New York.

Van de Water, T. R., Frenz, D. A., Giraldez, F., Represa, J., Lefebvre, P. P., et al. (1992). Growth factors and development of the stato-acoustic system. In "Development of Auditory and Vestibular Systems 2" (R. Romand, Ed.), pp. 1–32. Elsevier, Amsterdam.

Vendrell, V., Carnicero, E., Giraldez, F., Alonso, M. T., and Shimmang, T. (2000). Induction of inner ear fate by FGF3. *Development* **127,** 2011–2019.

von Bubnoff, A., and Cho, K. W. Y. (2001). Intracellular BMP signalling regulation in vertebrates: Pathway or network? *Dev. Biol.* **239,** 1–14.

Wagner, E., Luo, T., and Dräger, U. C. (2002). Retinoic acid synthesis in the postnatal mouse brain marks distinct developmental stages and functional systems. *Cerebral Cortex* **12,** 1244–1253.

Wei, X., Makori, N., Peterson, P. E., Hummler, H., and Hendrickx, A. G. (1999). Pathogenesis of retinoic acid-induced ear malformations in a primate model. *Teratology* **60,** 83–92.

Wendling, O., Chambon, P., and Mark, M. (1999). Retinoid X receptors are essential for early mouse development and placentogenesis. *Proc. Natl. Acad. Sci. USA* **96,** 547–551.

Wendling, O., Ghyselink, N. B., Chambon, P., and Mark, M. (2001). Roles of retinoic acid receptors in early embryonic morphogenesis and hindbrain patterning. *Development* **128,** 2031–2038.

White, J. C., Shankar, V. N., Highland, M., Epstein, M. L., DeLuca, H. F., and Clagett-Dame, M. (1998). Defects in embryonic hindbrain development and fetal resorption resulting from vitamin A deficiency in the rat are prevented by feeding pharmacological levels of all-trans-retinoic acid. *Proc. Natl. Acad. Sci. USA* **95,** 13459–13464.

Whitfield, T. T., Riley, B. B., Chiang, M. Y., and Philipps, B. (2002). Development of the zebrafish inner ear. *Dev. Dyn.* **223,** 427–458.

Wilkinson, D. G., Bhatt, S., and McMahon, A. P. (1989). Expression pattern of the FGF-related proto-oncogene int-2 suggests multiple roles in fetal development. *Development* **105,** 131–136.

Willhite, C. C., Hill, R. M., and Irving, D. W. (1986). Isotretinoid-induced craniofacial malformations in human and hamsters. *J. Craniofac. Gen. Dev. Biol. Suppl.* **2,** 193–209.

Wilson, J. G., Roth, C. B., and Warkany, J. (1953). An analysis of the syndrome of malformations induced by maternal vitamin A deficiency. Effects of restoration of vitamin A at various times during gestation. *Am. J. Anat.* **92,** 89–217.

Wolbach, S. B. (1954). Effects of vitamin A deficiency and hypervitaminosis A in animals. *In* "The Vitamins" (W. H. Sebrell and R. S. Harris, Eds.), Vol. 1 pp. 106–136. Academic Press, New York.

Wolf, G. (1984). Multiple functions of vitamin A. *Physiol. Rev.* **64,** 873–938.

Wu, D. K., and Oh, S. H. (1996). Sensory organ generation in the chick inner ear. *J. Neurosci.* **15,** 6456–6462.

Yasuda, Y., Okamoto, M., Konishi, H., Matsuo, T., Kihara, T., and Tanimura, T. (1986). Developmental anomalies induced by all-*trans* retinoic acid in fetal mice. I. Macroscopic findings. *Teratology* **34,** 37–49.

Ylikoski, J., Pirvola, U., and Eriksson, U. (1994). Cellular retinol binding protein type I is prominently and differentially expressed in the sensory epithelium of the rat cochlea and vestibular organs. *J. Comp. Neurol.* **349,** 596–602.

10
Hair Cell Development in Higher Vertebrates

Wei-Qiang Gao
Department of Molecular Oncology Genentech, Inc., South San Francisco, California 94080

I. Morphogenesis of the Mammalian Inner Ear
II. Control of Hair Cell Differentiation by Specific Genes
   A. Notch Signaling Is Important for Hair Cell Fate Determination
   B. Specific Basic Helix-Loop-Helix Transcription Factors Are Essential for Regulation of Initial Hair Cell Differentiation
   C. Specific Growth Factors and Cell Cycle Regulators that Influence Proliferation of Hair Cell Progenitors Also Affect Hair Cell Differentiation
   D. Further Maturation and Maintenance of Hair Cells Depend on Expression of Other Transcription Factors
III. Production and Regeneration of New Hair Cells in Mature Inner Ears
   A. Production of New Hair Cells in Mature Inner Ears
   B. Regeneration of New Hair Cells in Gentamicin-Damaged Mature Inner Ears
IV. Conclusion
   References

Mammalian inner ear morphogenesis and production of hair cells are influenced by local cellular interactions and regulated by specific genes. In particular, Notch–Notch ligand interaction–mediated lateral inhibition plays an important role in the determination of hair cell fate. Inactivation or disruption of Notch signaling can lead to the production of supernumerary hair cells. Specific basic helix-loop-helix (bHLH) transcript factors have been shown to act downstream of Notch signaling and are essential for initial hair cell differentiation. Whereas *Math1*, a mouse homologue of the *Drosophila* gene *atonal*, is a positive regulator of hair cell differentiation, *Hes1* and *Hes5*, mammalian hairy and enhancer of split homologues, act as negative regulators. These bHLH transcription factors are expressed in the inner ear epithelial region at the time when hair cell differentiation occurs. Whereas targeted deletion of *Math1* gene leads to failure of hair cell differentiation, misexpression of *Math1* induces production of extra hair cells, and cotransfection of *Hes1* and *Math1* in postnatal rat cochlear explant cultures results in an inhibition on hair cell differentiation induced by *Math1*. In addition, specific growth factors and cell cycle regulators that influence proliferation of sensory epithelial progenitors may also affect hair cell differentiation. Furthermore, other transcription factors, including Brn3c, Barhl1, and Gfi1, are required for further maturation and maintenance of hair cells. Understanding of hair cell differentiation mechanisms may provide useful hints for stimulating hair cell regeneration in mature inner ears, which could eventually be helpful for both recovery of hearing and recovery from balance impairment induced by hair cell loss. © 2003, Elsevier Inc.

I. Morphogenesis of the Mammalian Inner Ear

Hair cells are mechanosensory cells located in mammalian inner ears that convert sound or body motion signals into electrochemical energy. The bony labyrinth of the mammalian inner ear develops from the otic placode (Van de Water, 1983) and consists of the cochlea and vestibular end organs, including the utricle, saccule, and three semicircular canals. Each of these structures contains a sensory epithelium in which hair cells and supporting cells are located. The complex morphogenesis of the inner ear is likely directed by a series of cell fate specifications, controlled by specific genes, and influenced by local cell–cell interactions (Fekete, 1996). Classic tritiated thymidine incorporation studies have shown that hair cells become postmitotic between E11.5 and E17.5, with a peak at E13.5 in rodents (Ruben, 1967; Sans and Chat, 1982). Hair cells in the mammalian vestibular end organs appear to derive from the progenitor cells or supporting cells located within the sensory epithelium (Forge *et al.*, 1993; Kuntz and Oesterle, 1998; Li and Forge, 1997; Warchol *et al.*, 1993; Zheng and Gao, 1997) in a similar manner as in birds and lower vertebrates (Corwin and Cotanche, 1988; Fekete *et al.*, 1998; Ryals and Rubel, 1988). However, the exact origin of mammalian cochlear hair cells is still unclear because no cell lineage studies using lineage tracers have been performed. Although in avian and lower vertebrate systems it is believed that hair cells and supporting cells derive from a common progenitor cell (Fekete *et al.*, 1998), whether this is true in mammals remains to be determined. Previous histological studies suggest that during embryogenesis the inner and outer hair cells in the mammalian cochlea probably derive from the greater epithelial ridge (GER) and the lesser epithelial ridge (LER) cells, respectively (see Lim and Rueda, 1992). Recent work on expression patterns and misexpression of genes controlling hair cell differentiation has provided direct evidence for this model (Gao, 2001, 2002; Shou *et al.*, 2003; Zheng and Gao, 2000; Zheng *et al.*, 2000; Zine *et al.*, 2001).

II. Control of Hair Cell Differentiation by Specific Genes

A number of genes have been implicated in ear morphogenesis and hair cell differentiation (Fekete, 1996, 1999). These include various transcription factors, secreted factors, receptor tyrosine kinases, cyclin-dependent kinase inhibitors (Chen and Segil, 1999; Lowenheim et al., 1999), and membrane-bound signaling proteins such as Notch and Notch ligands (Eddison et al., 2000; Kiernan et al., 2001; Lanford et al., 1999; Lewis, 1998; Lewis et al., 1998; Stone and Rubel, 1999; Zine et al., 2000). Whereas many of these genes work at an earlier stage of ear morphogenesis, Notch signaling molecules and specific bHLH transcription factors are essential for initial hair cell differentiation. This chapter mainly focuses on the genes involved in the process of hair cell differentiation. The roles of Notch signaling, several growth factors, one specific cell cycle regulator, and three other important transcription factors are briefly summarized. However, experiments that demonstrate how specific bHLH transcription factors regulate initial hair cell differentiation are discussed in detail. These bHLH transcription factors include *Math1* (Akazawa et al., 1995), a homologue of the *Drosophila* gene *atonal*, and *Hes1* and *Hes5*, *mammalian hairy and enhancer of split* homologues.

A. Notch Signaling Is Important for Hair Cell Fate Determination

Among many genes mentioned in the previous section, the Notch family of receptors and ligands are likely candidates for genes that regulate hair cell fate determination because they have been shown to be expressed in numerous types of epithelial cells, where they mediate a variety of cell interactions specifying cell fate during embryogenesis (Artavanis-Tsakonas et al., 1999). Originally identified as a *Dorsophila* mutation displaying abnormal differentiation of neural vs. epidermal cell lineages (Welshons, 1965), the Notch receptors are generally associated with nonterminally differentiated, often proliferating cell populations. Activation of Notch signaling influences the ability of these cells to acquire or maintain a particular cell fate (Artavanis-Tsakonas et al., 1999). So far, four vertebrate Notch genes have been identified: Notch1/TAN1, Notch2, Notch3, and Notch4/int3. Each of these Notch genes is widely expressed throughout development and has both overlapping and unique expression patterns relative to the others (Lindsell et al., 1996; Williams and Lardelli, 1995). These genes appear to bind nonpreferentially to each of the identified Notch ligands, the Delta/Serrate/Lag2 (DSL) family of ligands. Similar to the Notch receptors, DSL ligands are transmembrane proteins originally identified in *Drosophila* as Delta and Serrate. Six vertebrate DSL family genes have been identified, including the Serrate-like homologues, Jagged1 and Jagged2 (Lindsell et al., 1995; Valsecchi et al., 1997), and the Delta-like homologues, Delta, Delta2, Delta3, and Dll4 (Gray et al., 1999; Joutel and Tournier-Lasserve, 1998; Shutter et al., 2000).

Studies using *in situ* hybridization, immunohistochemistry, and transgenic approaches have shown that Notch and Notch ligands are expressed in the developing inner ear sensory epithelium (Lanford et al., 1999; Lewis et al.,

1998; Morrison et al., 1999). Disruption or inactivation of Notch signaling can lead to abnormal production of hair cells. For example, when Jagged2 gene, encoding one of the Notch ligands, is deleted, supernumerary inner and outer hair cells are produced in the cochlea (Lanford et al., 1999). Whereas Jagged2 is expressed in developing hair cells, the Lunatic fringe gene that interferes in Notch signaling is expressed in cochlear nonsensory supporting cells. Lunatic fringe has been shown to suppress the production of extra inner hair cells in Jagged2 mutant mice (Zhang et al., 2000). Formation of extra hair cells is also observed when antisense oligonucleotides against Notch1 or Jagged1, which interferes normal Notch signaling, are added to the cultures of developing cochlear explants (Zine et al., 2000). Targeted deletion of Jagged1 gene results in defects in inner ear development and abnormal hair cell numbers (Kiernan et al., 2001). In addition, involvement of Notch signaling in hair cell fate determination has been reported in zebrafish and avian ears. For instance, in the mind bomb zebrafish mutant in which there is a defect in normal Notch signaling, hair cell number increases dramatically in the sensory patches, whereas supporting cells are completely missing (Haddon et al., 1998). In chicken ears, where hair cell regeneration occurs spontaneously following injury, Delta1 mRNA levels are elevated in progenitor cells during DNA synthesis. Such upregulation persists in newly generated hair cells, but Delta1 expression becomes downregulated in cells that do not adapt the hair cell fate (Stone and Rubel, 1999). Consistent with these findings, Delta1 is also observed to be expressed transiently in the nascent hair cells in developing chicken ears but to disappear as they mature (Adam et al., 1998). These studies together confirm that appropriate Notch signaling is important for production of the proper number of hair cells during inner ear development.

B. Specific Basic Helix-Loop-Helix Transcription Factors Are Essential for Regulation of Initial Hair Cell Differentiation

Specific transcription factors are shown to be expressed in the inner ear epithelium at the time when hair cell differentiation occurs (Bermingham et al., 1999; Chen et al., 2002; Zheng et al., 2000; Zine et al., 2001). They are demonstrated to be either antagonizers or effectors of Notch signaling pathway and to affect cell fate determination of specific cells (Ohtsuka et al., 1999; Zheng et al., 2000). For example, *Math1* is a positive regulator for differentiation of cerebellar granule neurons (Ben-Arie et al., 1997), dorsal commissural interneurons (Helms and Johnson, 1998), and intestinal secretory epithelial cells (Yang et al., 2001). *Hes1*, on the other hand, is a negative regulator of the differentiation of developing neurons (Ishibashi et al., 1995; Nakamura et al., 2000) and intestinal epithelial cells (Jensen et al.,

2000). *Hes1* and *Math1* can antagonize each other both biochemically (Akazawa *et al.*, 1995) and functionally (Zheng *et al.*, 2000).

1. Math1 Is a Positive Regulator of Hair Cell Differentiation

a. Targeted Gene Deletion of *Math1* Leads to a Failure of Hair Cell Differentiation. When the *Math1* gene is deleted, no recognizable hair cells are observed in the inner ear, including the cochlea and vestibular endorgans of the mutant mice (Bermingham *et al.*, 1999). The phenotypes of *Math1* null mutants are different from those of the Brn3c knockout mice, in which hair cells are absent in mature mice (Erkman *et al.*, 1996; Xiang *et al.*, 1997b) but are formed initially and degenerate gradually afterward (Xiang *et al.*, 1998). The lack of hair cells in the *Math1* knockout mice is confirmed by immunocytochemical hair cell markers and electron microscopy (Bermingham *et al.*, 1999). These loss-of-function experiments indicate that *Math1* is required for hair cell differentiation in mammalian inner ears.

b. Misexpression of Math1 Induces Production of Extra Cochlear Hair Cells. When the cochlear explants prepared from neonatal rats are transfected with a Math1-expressing vector, pRK5–Math1–EGFP, a large number of extra hair cells are produced in the cultures (Fig. 1A–C; see Zheng and Gao, 2000), as evidenced by double labeling with anti-myosin VIIA antibody, a hair cell–specific marker (Hasson *et al.*, 1995). These supernumerary hair cells are found ectopically in the GER, which is medial to the organ of Corti (OC) where four rows of normal hair cells are located (see Fig. 1A–C). In sharp contrast, none of the EGFP-positive cells become hair cells (Fig. 1D–F) in control cultures transfected with the plasmid-expressing EGFP (Murone *et al.*, 1999). To confirm whether these ectopically induced hair cells (EHC) grow stereociliary bundles, a unique feature of inner ear hair cells, the cultures are triple labeled with a biotinylated lectin, peanut agglutinin, a selective stereociliary bundle marker in the inner ear (Zheng and Gao, 1997a). This lectin highlights not only the stereociliary bundles of normal hair cells in the OC but also those of EHC in the GER (arrowheads in Fig. 1G–I). In contrast, in the cultures transfected with the control plasmid, pRK5–EGFP, peanut agglutinin labeling is restricted to the OC hair cells and is not seen in the GER (Zheng and Gao, 2000). The production of the EHC is robust (see GER area in Fig. 1A) and there are as many as 502 EHCs generated in one of the cultures.

c. Ectopic Hair Cells Are Derived from GER Cells. The identity of the GER cells is further verified in cross sections of the cultures fixed at 1–2 days after transfection. The transfected cells (EGFP-positive) are medial to the inner hair cells in OC with an approximately four-cell distance (Fig. 2A

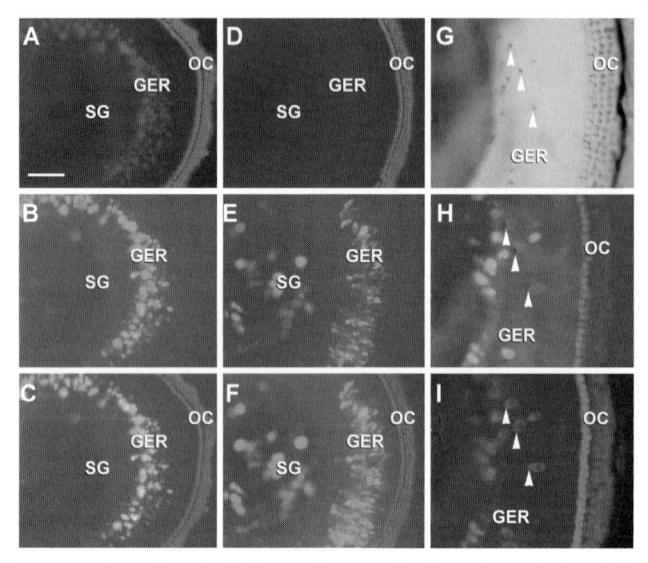

Figure 1 Overexpression of *Math1* leads to robust production of extra hair cells in the GER of P0 rat cochlear explant cultures. (A–C) Myosin VIIa (red, A) and EGFP (green, B) double immunocytochemistry (double exposure, C) of a culture transfected with the pRK5–Math1–EGFP plasmid. (D–F) Myosin VIIa (red, D) and EGFP (green, E) double immunocytochemistry (double exposure, F) of a control culture transfected with the pRK5–EGFP plasmid. (G–I) Triple labeling (lectin labeling in G, myosin and EGFP double labeling in H, and myosin VIIa labeling in I) of a culture transfected with the pRK5–Math1–EGFP plasmid shows the presence of stereociliary bundles (arrowheads) in the hair cells produced in the GER region. GER, Greater epithelial ridge; OC, organ of Corti; SG, spiral ganglion. Bar: 60 μm for A–F; 30 μm for G–I. Modified with permission from *Nature* from Zheng, J., and Gao, W. Q. (2000). Overexpression of Math1 induces robust production of extra hair cells in postnatal rat inner ears. *Nat. Neurosci.* **3,** 580–586. (See Color Insert.)

and B; see Zheng and Gao, 2000). These GER cells show a small cell body with processes extending vertically from the basement membrane to the epithelial lumen (Lim and Rueda, 1992; Sobkowicz *et al.*, 1975). Their radial pattern can be observed in the whole mount cultures because of the tilted angle (see Fig. 1E). Examination of cross sections of the cultures fixed at 6 days following transfection with the pRK5–Math1–EGFP plasmid shows that the transfected GER cells have acquired hair cell phenotypes, including expression of myosin VIIa (Fig. 2C–D) and a pear-shaped morphology with a big nucleus in the bottom of the cells but no long processes (see arrows in Fig. 2E). In contrast, those GER cells transfected with the pRK5–EGFP plasmid retain the radial morphology with a small nucleus located in the middle of an elongated, slim cell body (see arrows in Fig. 2F). Confocal microscopic analysis of the cultures reveals that many of the EHC have detached their cell bodies from the basement membrane and translocated to

10. Mammalian Hair Cell Development

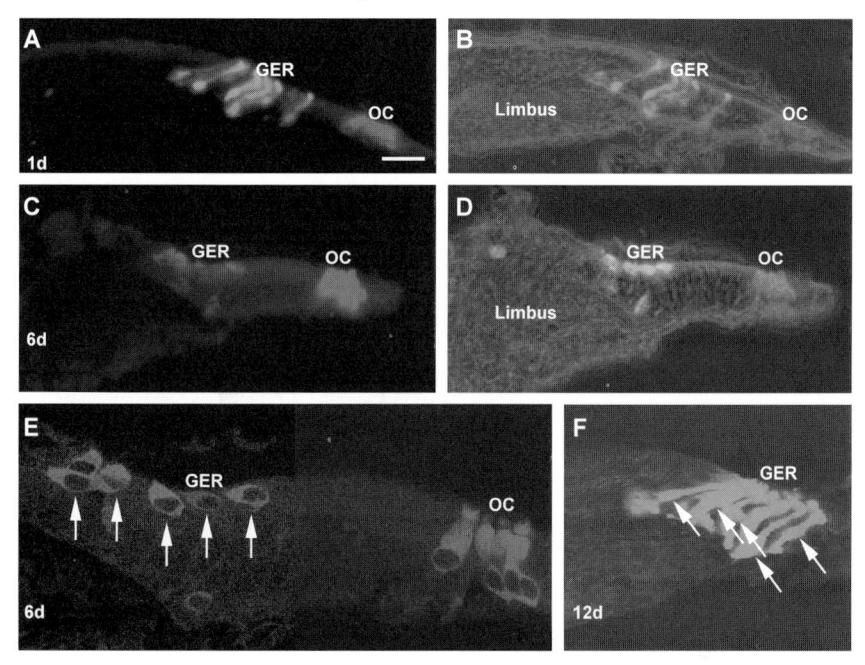

Figure 2 Morphological and immunocytochemical conversion of the GER cells into hair cells. Cross sections of cultures at 1 day (A, B) and 6 days (C, D) after transfection with the pRK5–Math1–EGFP plasmid, respectively. (A, B) Myosin VIIa (red) and EGFP (green) double labeling (A) and superimposed image of A and phase-contrast image from the same area, respectively, verify that the transfected cells are GER epithelial cells that display a radial pattern from the base to the roof and show that these cells are myosin VIIa negative at 1 day following transfection. The transfected cells are not located in the limbus. (C, D) Myosin VIIa labeling (C) and myosin VIIa, EGFP, and phase-contrast superimposed image (D) show that the transfected GER cells have converted into hair cells at 6 days after transfection. (E) High magnification confocal image of anti-myosin VIIa antibody labeling shows the EHC in the GER and normal hair cells in the OC region in C. The arrows in E indicate the large nuclei in the bottom of the EHC. (F) High-magnification confocal image of the EGFP-positive cells in the GER region of a culture transfected with the pRK5–EGFP plasmid at 12 days following transfection. The radial morphology of the transfected GER cells is retained throughout the 12-day culture period. The arrows in F indicate the small nuclei located in the middle of the radial GER cells. Bar: 40 μm for A–D, 20 μm for E–F. Modified with permission from *Nature* from Zheng, J., and Gao, W. Q. (2000). Overexpression of Math1 induces robust production of extra hair cells in postnatal rat inner ears. *Nat. Neurosci.* **3**, 580–586. (See Color Insert.)

the surface of GER even though some of them are not yet as well organized as those normal OC hair cells (see Fig. 2E).

d. Ectopically Induced Hair Cells Display Ultrastructural Features of Hair Cells. Ultrastructural examination of the transfected cultures confirms that the EHC have acquired hair cell features (Zheng and Gao, 2000). Like hair cells in the OC, EHC show light cytoplasm and large cell bodies (Fig. 3). They look healthy with a large nucleus in the bottom of the cell and have many mitochondria in the cytoplasm (see Fig. 3). Whereas normal GER cells have their nuclei in the deep layers (or near the basement membrane) and extend elongated processes with dark cytoplasm to the luminal surface where microvilli can be seen, the EHC are observed in the upper layer of the medial side of the GER (see Fig. 3). Consistent with our confocal microscopy (see Fig. 2E), some of the EHC have moved their cell bodies to the surface and display clearly cuticular plates and short stereociliary bundles (see arrows in Fig. 3). All of these characteristics confirm their hair cell identity.

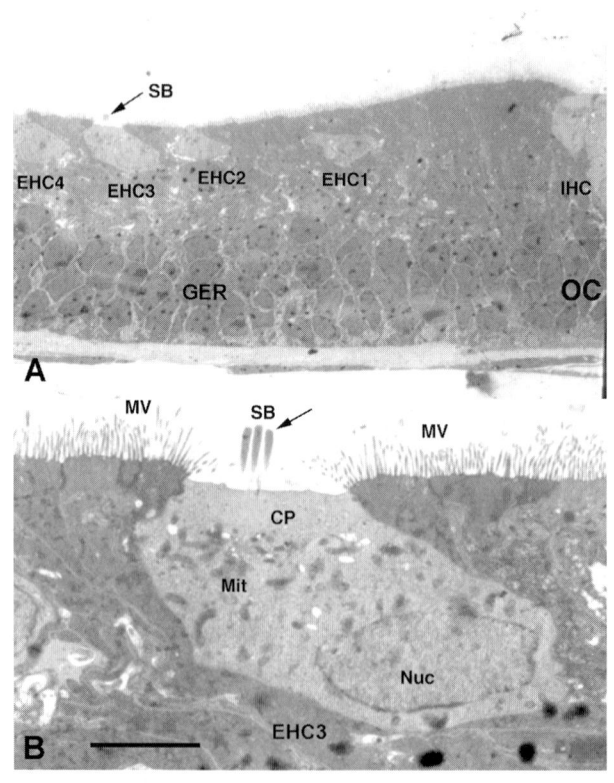

Figure 3 Ultrastructural analysis of the ectopically induced hair cells. (A) Low-magnification image of the GER region of the culture 12 days after transfection with the pRK5–Math1–EGFP plasmid shows the presence of four EHC in the medial area of the GER. (B) High-magnification image of the EHC3 in A shows the presence of cuticular plate and stereociliary bundles (arrows). Many black spots in the figure are artifact due to tissue processing and lead citrate staining. EHC, Ectopically induced hair cells; IHC, inner hair cells; Mit, mitochondria; MV, microvilli; Nuc, nucleus: SB, stereociliary bundles. Bar: 5 μm for A; 1 μm for B. Modified with permission from *Nature* from Zheng, J., and Gao, W. Q. (2000). Overexpression of Math1 induces robust production of extra hair cells in postnatal rat inner ears. *Nat. Neurosci.* **3**, 580–586.

10. Mammalian Hair Cell Development

e. Induction of Hair Cell Differentiation by *Math1* Is specific. The conversion of the GER cells into hair cells in the presence of *Math1* is specific because all of the transfected nonneuronal cells in the SG and the connective tissue fail to differentiate into hair cells. In addition, when neural cells in cerebellar slices prepared from neonatal rats are transfected with the same pRK5–Math1–EGFP plasmid under the same experimental conditions, they remain myosin VIIa negative and do not assume hair cell morphology (Zheng and Gao, 2000). Therefore, Math1 expression is sufficient to induce transdifferentiation in certain cell contexts (GER cells) but not in others (spiral ganglion, connective tissue, or nonear tissue). Moreover, overexpression of another transcription factor gene, Brn3c, which has been previously implicated in hair cell development (Erkman *et al.*, 1996; Xiang *et al.*, 1997a; 1998), in GER cells fails to induce hair cell differentiation. These results are consistent with the finding that Brn3c affects later differentiation and maturation of hair cells (Xiang *et al.*, 1998)

f. LER Cells Can also Be Induced to Become Hair Cells. We have recently generated an adenovirus-expressing human homologue of *Math1*— *Hath1*. To test whether *Hath1* has the same ability as *Math1* (Zheng and Gao, 2000) to induce hair cell differentiation, we have examined neonatal rat cochlear explant cultures infected with the adenovirus coexpressing *Hath1* and *EGFP* (Shou *et al.*, 2003). Consistent with previous electroporation experiments (Zheng and Gao, 2000), many of the infected GER cells are double labeled by anti-myosin VIIa antibody (Gao, 2001) when the cultures are fixed at 3–6 days following viral infection. Interestingly, we have found that in addition to GER cells, infected cells in the LER region that is lateral to the organ of Corti (OC) and spans about 10–14 cell widths in the superficial layer are also able to convert into hair cells (Gao, 2001, 2002; Shou *et al.*, 2003). These postnatal LER cells would normally give rise to Hensen's and Claudius' cells (Lim and Rueda, 1992) as development proceeds. None of the GER or LER cells in any of the 12 control cultures infected with the ad-EGFP virus are stained with anti-myosin VIIa antibody, indicating failure to become hair cells. These recent experiments indicate that the human homologue of *Math1* can also induce differentiation of hair cells and that LER cells have the same potential as GER cells to transdifferentiate into hair cells (Shou *et al.*, 2003). These findings provide direct evidence that GER and LER in the cochlea can act as hair cell progenitors.

g. *Math1* Facilitates Hair Cell Differentiation in Utricles. When transfection with pRK5–EGFP plasmid is performed in partially dissociated epithelial sheet cultures prepared from P3 rat utricle cells (Corwin *et al.*, 1996; Zheng *et al.*, 1997), which contain essentially supporting cells and hair cells, very few hair cells are transfected by lipofection (Zheng and Gao, 2000), presumably because of their special architecture with supporting cells. However, transfection occurs in some of the supporting cells. Whereas only 3 out of 1726 EGFP-positive cells (0.17% ± 0.09) differentiate into hair cells in 18 cultures transfected with the pRK5–EGFP plasmid, 187 out of 2607 EGFP-positive cells (7.17% ± 1.00) become hair cells in 30 cultures transfected with the pRK5–Math1–EGFP plasmid, as revealed by anti-myosin VIIa antibody double labeling (Zheng and Gao, 2000) or another vestibular hair cell marker, anti-calertinin antibody (Zheng and Gao, 1997a). These results suggest that *Math1* expression in the utricular macula facilitates the conversion of supporting cells into hair cells (Zheng and Gao, 2000).

h. *Math1* Uncouples the Establishment of the Sensory Primordium from Hair Cell Fate Determination. It has been demonstrated that *Math1* is expressed after the establishment of a zone of nonproliferating cells, which defines where the cochlear sensory epithelium is during embryogenesis (Chen *et al.*, 2002). Therefore, *Math1* is not required for the establishment of the postmitotic sensory patch but rather is essential for the differentiation of sensory hair cells within the established postmitotic sensory patch. This model is further supported by the finding that this postmitotic sensory patch is formed normally in the *Math1* null mutants (Chen *et al.*, 2002).

2. *Hes1* Is a Negative Regulator of Hair Cell Differentiation

a. Targeted Disruption of the *Hes1* Gene Leads to Formation of Extra Inner Hair Cells in the Cochlea. To determine whether *Hes1*, a know negative regulator for neuronal differentiation (Ishibashi *et al.*, 1995; Nakamura *et al.*, 2000), is important for hair cell differentiation, we have examined the phenotype of *Hes1* null mutant inner ears (Zheng *et al.*, 2000). As shown in Fig. 4A, examination of cochlear surface preparations obtained from E17.5 $Hes1^{-/-}$ mice reveals the presence of supernumerary inner hair cells (arrows in Fig. 4C). There is also a mild increase in the number of inner hair cells in the $Hes1^{+/-}$ mice, suggesting a gene–dosage effect (arrows in Fig. 4B). In contrast, there are four rows of hair cells, including one row of inner hair cells and three rows of outer hair cells, in the cochlear surface preparations dissected from E17.5 wildtype mice (see Fig. 4A). Cell counts performed from cochlear surface preparations of different genotypes show a significant increase in the number of doublet inner hair cells along 1 mm length starting from the basal end of the cochlea in the $Hes1^{-/-}$ mice (17.92 ± 3.69 s.e.m., n = 13) as compared to the $Hes1^{+/+}$ (0.20 + 0.16, n = 8), or $Hes1^{+/-}$ litter mates (5.00 ± 0.55, n = 15, p < 0.01 between $Hes1^{-/-}$ and $Hes1^{+/+}$ mice). There is also a statistical difference between $Hes1^{+/-}$ and $Hes1^{+/+}$ mice (p < 0.01).

10. Mammalian Hair Cell Development 303

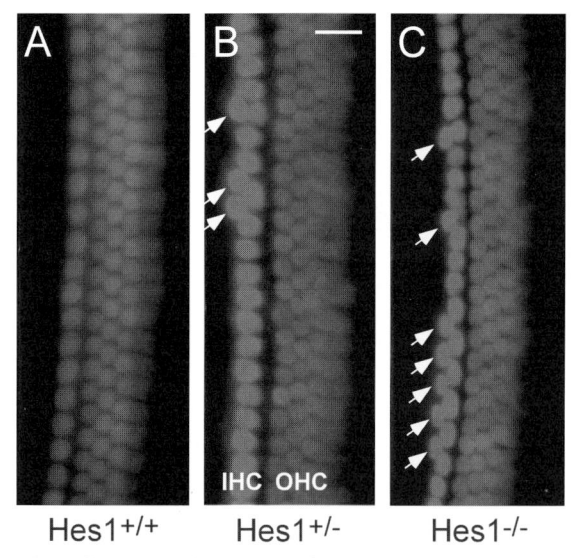

Figure 4 Formation of extra cochlear inner hair cells in *Hes1*-deficient mice. Myosin VIIa immunostaining of the cochlear surface preparations obtained from E17.5 $Hes1^{+/+}$ (A), $Hes1^{+/-}$ (B), and $Hes1^{-/-}$ mice (C). Arrows indicate the presence of extra inner hair cells. Bar: 12 μm. Modified with permission from Zheng, J., Shou, J., Guillemot, F., Kageyama, R., and Gao, W. Q. (2000). Hes1 is a negative regulator of inner ear hair cell differentiation. *Development* **127**, 4551–4560.

In contrast, no statistical difference is found in outer hair cell numbers among the three types of animals ($Hes1^{-/-}$, 63.69 ± 1.82, n = 13; $Hes1^{+/-}$, 58.87 ± 2.18, n = 15; $Hes1^{+/+}$, 58.25 ± 2.18, n = 8, p = 0.058 between $Hes1^{-/-}$ and $Hes1^{+/+}$ mice).

Examination of paraffin sections of the cochlear tissue prepared from $Hes1^{-/-}$ and $Hes1^{+/+}$ mice shows that although there are three outer hair cell rows and only one inner hair cell row in $Hes1^{+/+}$ preparations (Fig. 5A), doublet inner hair cells are seen in $Hes1^{-/-}$ preparations (arrows in Fig. 5B). Similarly, scanning electron microscopy reveals that instead of the regular four rows of hair cells observed in $Hes1^{+/+}$ mice (Fig. 5C), there are extra inner hair cells produced in the tissue prepared from $Hes1^{-/-}$ mice (arrows in Fig. 5D). These results therefore confirm our observations with anti-myosin VII antibody labeling (see Fig. 4) in cochlear surface preparations.

b. *Hes1* Also Influences Hair Cell Production in the Utricle. To determine whether hair cell production in the vestibular organs is also regulated by *Hes1*, we have performed serial cryostat sections of the utricles prepared from the $Hes1^{-/-}$, $Hes1^{+/+}$, and $Hes1^{+/-}$ mice at E17.5 as previously described (Zheng and Gao, 1997). We then immunostained these sections with

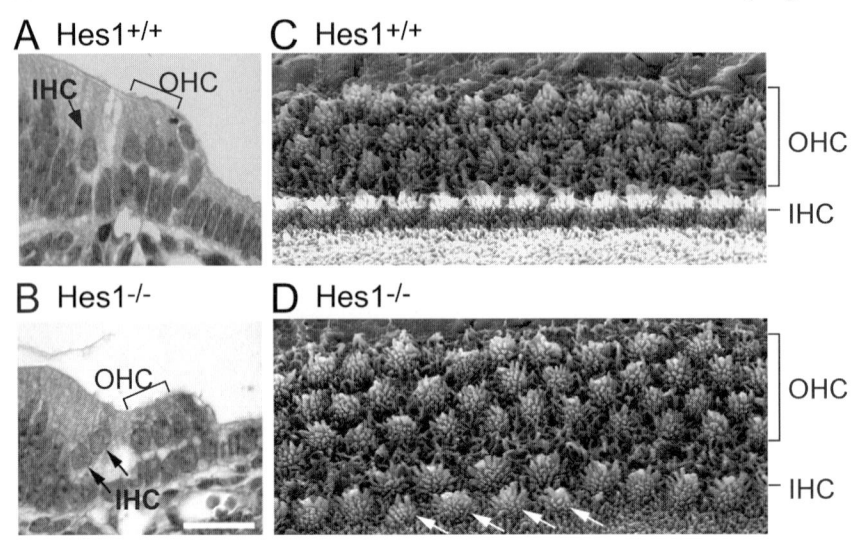

Figure 5 Confirmation of the presence of extra cochlear inner hair cells with paraffin sectioning and scanning electron microscopy of $Hes1^{-/-}$ mice. (A, B) Paraffin sections of $Hes1^{+/+}$ and $Hes1^{-/-}$ cochleae, respectively. Arrows in B show the presence of doublet inner hair cells. (C, D). Scanning electron microscopic micrographs of $Hes1^{+/+}$ and $Hes1^{-/-}$ cochlear surface preparations, respectively. Because of the initial fixation in 10% formalin, the ultrastructural preservation was not optimal. Arrows in D show the presence of several extra inner hair cells. IHC is inner hair cells; OHC is outer hair cells. Bar in B: 20 μm for A, B; 15 μm for C, D. Modified with permission from Zheng, J., Shou, J., Guillemot, F., Kageyama, R., and Gao, W. Q. (2000). Hes1 is a negative regulator of inner ear hair cell differentiation. *Development* **127**, 4551–4560.

anti-myosin VIIa antibody and performed total cell counts. We have found a significant increase in the total number of hair cells in the utricles of $Hes1^{-/-}$ mice (3193.50 ± 143.94, s.e.m., n = 4, p < 0.05) as compared to those in the $Hes1^{+/+}$ mice (2347.17 ± 80.99, n = 6) (Zheng *et al.*, 2000). The number of hair cells in $Hes1^{+/-}$ is in-between (2594.5 ± 163.19, n = 4), but the difference was not statistically significant relative to that of the $Hes1^{+/+}$ mice (p = 0.17, see Zheng *et al.*, 2000).

c. *Hes1* and *Hes5* Are Differentially Expressed in the Inner Ear. To find out the cellular expression patterns of the *Hes1* gene in the inner ear, we have performed nonradioactive RNA *in situ* hybridization with sections prepared from E17.5 rat inner ear tissue. As shown in Fig. 6A1 and 6A2, specific labeling is seen in the supporting cell layer but not in the hair cell layer of the vestibular sensory epithelium (Zheng *et al.*, 2000). Double labeling the sections with anti-myosin VIIa antibody, a hair cell–specific marker (Hasson *et al.*, 1995; Xiang *et al.*, 1997a), confirms that hair cells are devoid of labeling (see Fig. 6A2). Nonsensory epithelial cells in the transitional zone and in

10. Mammalian Hair Cell Development 305

the roof show no signals (see Fig. 6A1). In the cochlea, *Hes1* signal is seen in the GER and LER areas that are adjacent to inner and outer hair cells, respectively (Fig. 6A3 and 6A4). The sensory epithelium in which hair cells and supporting cells such as Deiters' cells and pillar cells are located show minimal labeling (see Fig. 6A4). Double labeling the sections with antimyosin VIIa antibody confirms that hair cells are essentially devoid of *Hes1* signal (see Fig. 6A4).

Given that *Hes1* is expressed in both LER and GER but extra inner hair cells are only formed in the GER, we wondered whether expression of other genes such as *Hes5* may also play a role in the control of hair cell differentiation. *Hes5* has been shown to be another negative regulator of neurogenesis (Ohtsuka *et al.*, 1999). We carried out *Hes5 in situ* hybridization on E17.5 rat inner ear tissue and have found an overlapping but distinct expression pattern as compared to that of *Hes1* (Zheng *et al.*, 2000). In the cochlea, *Hes5* signal is observed in the LER and supporting cells, including Deiters' cells and pillar cells in the sensory epithelium, but not in the GER (Fig. 6B3, 6B4). In the utricle, strong *Hes5* signal is seen in the supporting cells (Fig. 6B1). However, unlike *Hes1* that is expressed in supporting cells throughout the sensory epithelium, *Hes5* is expressed at high levels in the supporting cells in the striola region. Minimal signals or much lower levels are seen in the nonstriola region (Fig. 6B1, 6B2).

Our *Hes1/Hes5 in situ* hybridization results suggest that *Hes1* and *Hes5* may act together as negative regulators of hair cell differentiation. Whereas *Hes1* clearly contributes to a mechanism that fine-tunes the number of inner hair cells as it is expressed in the GER and LER cells, its role in regulating the number of outer hair cells is less clear and may be masked by the cooperative influence of *Hes5* as *Hes5* is expressed in the cochlear sensory epithelium and the LER (Zheng *et al.*, 2000). This model has been confirmed by Zine *et al.* (2001). In addition, Zine *et al.* (2001) show that there is a significant increase in the number of vestibular hair cells and cochlear outer hair cells in the *Hes5* null mutant mice. The formation of supernumerary hair cells was more profound in the Hes1/Hes5 double knockout mice than in *Hes1* single knockout mutants (Zine *et al.*, 2001). Collectively, these results suggest that *Hes1* and *Hes5* participate together for the control of inner ear hair cell production (Zheng *et al.*, 2000; Zine *et al.*, 2001), similar to their role during neuronal differentiation (Ohtsuka *et al.*, 1999).

d. *Hes1* Prevents Hair Cell Differentiation Induced by *Math1*. In an attempt to understand the mechanisms by which *Hes1* influences hair cell differentiation and whether there is any functional interaction between *Hes1* and *Math1*, we have cotransfected postnatal rat cochlear explant cultures with an equal amount of *Hes1*-expressing (pSV2CMV–Hes1) and *Math1*-expressing plasmids (pRK5–Math1–EGFP) and compared them to cultures

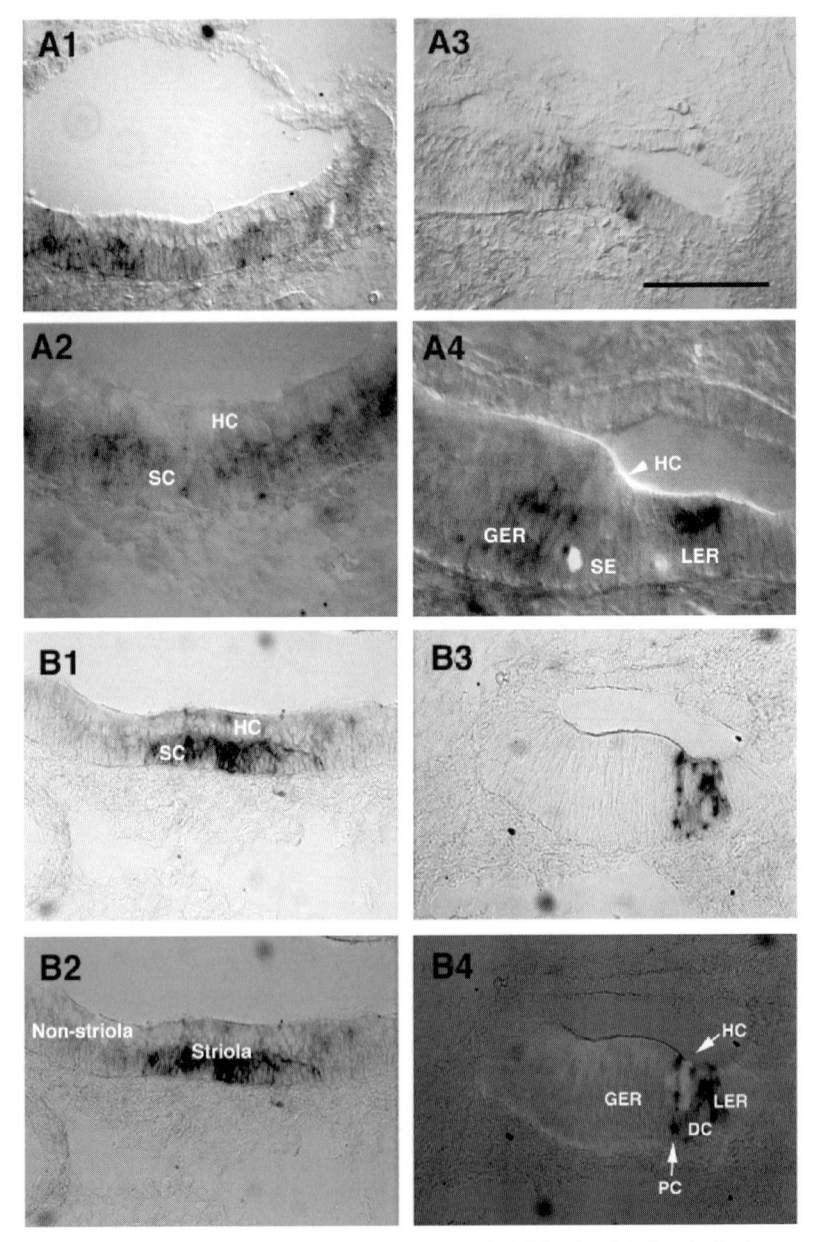

Figure 6 Nonradioactive *Hes1* and *Hes5* RNA *in situ* hybridization labeling in the inner ear. (A) Low- and high-magnification images of *Hes1* expression in E17.5 rat utricular (A1, A2) and cochlear (A3, A4) sections, respectively. (A2, A4) Myosin VIIa immunocytochemical labeling (mediated by Texas-red conjugated secondary antibody) of the sections shown in A1, A3, respectively. Note that specific *Hes1* signals are seen in supporting cell (SC) layer but not hair

transfected with pRK5–Math1–EGFP plasmid only or cotransfected with a mixture of pSV2CMV–Hes1 and pRK5–EGFP plasmids (Zheng et al., 2000). Our experiments indicate that *Math1* overexpression leads to robust production of ectopic hair cells in the GER region (Fig. 7A–C) as shown previously (Zheng and Gao, 2000). However, the induction of hair cell differentiation by *Math1* expression can be dramatically prevented by coexpression of *Hes1* (Fig. 7D–F). Overexpression of *Hes1* alone, on the other hand, does not show any apparent effects and none of the *Hes1* transfected cells becomes myosin VIIa positive (Fig. 7G–I). Previously we showed that there is a morphological change of the GER cells in the presence of *Math1* conversion from an elongated, process-bearing morphology to a pear-shaped morphology (see Fig. 7C) (Zheng and Gao, 2000). Consistently, the majority of the *Hes1/Math1* cotransfected GER cells show a process-bearing morphology (in Fig. 7D–F) similar to those of normal GER cells (see Fig. 7G–I), suggesting that the presence of *Hes1* also prevents the GER cells from undergoing a morphological change. Quantitative analysis of the cultures transfected with various vectors reveals that whereas 98.6% ± 0.6 of the *Math1* transfected GER cells become hair cells, only about 8.9% ± 1.1 of the *Hes1/Math1* cotransfected GER cells can convert into hair cells (Zheng et al., 2000). All of the *Hes1* transfected GER cells remain myosin VIIa negative and do not differentiate into hair cells. The antagonizing function of *Hes1* and *Math1* might be due to direct interactions between them because a previous biochemical study suggests so (Akazawa et al., 1995). Quantitative real-time PCR analysis with RNA extracted from the entire inner ear labyrinth tissue prepared from E13.5, E15.5, E17.5, P0, P5, P15, and adult mice indicates that *Hes1* is expressed as early as E13.5 and its expression becomes elevated around birth and is maintained in the adult. In contrast, *Math1* is expressed mainly at embryonic and early postnatal stages and its expression is greatly downregulated and becomes minimal in the adult (Zheng et al., 2000).

---

cell (HC) layer of the utricular sensory epithelium. In the cochlea, *Hes1* signal is seen in the GER and LER cells but is minimal in the sensory epithelium (SE). Hair cells (red labeling in A2, A4) are devoid of *Hes1* signal. Hybridizing the sections with sense control probes under the same experimental conditions does not show any staining (data not shown). (B) *Hes5* expression in E17.5 rat utricular (B1, B2) and cochlear (B3, B4) sections, respectively. (B2, B4) Myosin VIIa immunocytochemical labeling (mediated by Texas-red secondary antibody) of the sections shown in B1 and B3, respectively. Note that specific *Hes5* signals are seen in Deiters' cells (DC), pillar cells (PC), and the LER cells in the cochlea (B4) and in striola supporting cells (SC) in the utricle (B2). Hair cells (red labeling in B2, B4), the connective tissue, and nonsensory epithelial cells (not shown) are devoid of *Hes5* signal. Bar: 50 μm for A1, A3, B1–B4; 25 μm for A2, A4. Modified with permission from Zheng, J., Shou, J., Guillemot, F., Kageyama, R., and Gao, W. Q. (2000). Hes1 is a negative regulator of inner ear hair cell differentiation. *Development* **127**, 4551–4560. (See Color Insert.)

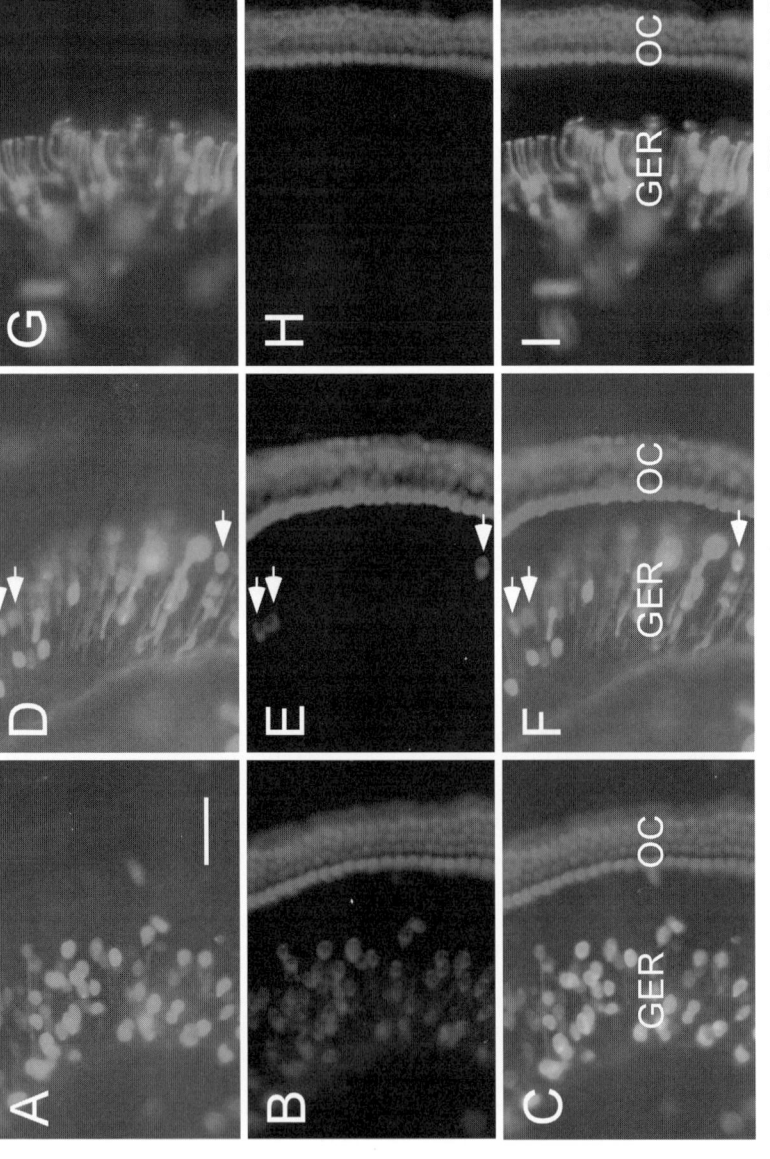

Figure 7 *Hes1* blocks hair cell differentiation induced by *Math1*. EGFP (green, A, D, G) and myosin VIIa (red, B, E, H) double immunocytochemistry (double exposure, C, F, I) of the cultures transfected with the pRK5–Math1–EGFP

C. Specific Growth Factors and Cell Cycle Regulators that Influence Proliferation of Hair Cell Progenitors Also Affect Hair Cell Differentiation

In addition to the bHLH transcription factors, specific growth factors such as heregulin, bFGF, and BMP4 may also affect production of hair cells. RT–PCR and immunostaining show that heregulin and its binding receptors are expressed in the inner ear sensory epithelium (Zhang et al., 2002; Zheng et al., 1999). Experiments with embryonic rat inner ear explant cultures demonstrate that heregulin affects hair cell differentiation by enhancing proliferation of hair cell progenitors (Gao et al., 1999). This model is supported by the observation that there is precocious differentiation of hair cells in inner ear explants prepared from the mutant mouse embryos in which *Her3* gene encoding the binding receptor for heregulin is deleted (Zheng et al., unpublished observations). Similarly, FGF receptors are expressed in the inner ear epithelium (Pirvola et al., 1995, 2002; Zheng et al., 1997). FGF signaling is required for the development of the auditory sensory epithelium (Pirvola et al., 2002; see Chapter 8 by Wright and Mansour). Full inactivation of Fgfr1 in the inner ear epithelium by Foxg1-Cre–mediated deletion results in a dramatic reduction in the number of auditory hair cells, which can be attributed to the reduced progenitor cell proliferation in the early cochlear duct (Pirvola et al., 1995). FGF signaling is also shown to regulate pillar cell development in the organ of Corti (Mueller et al., 2002). Differentiation of pillar cells is dependent on continuous activation of FGFR3. Treatment of the cochlear explant cultures with FGF2 results in a significant increase in the number of pillar cells and a small increase in the number of inner hair cells. In addition, BMP4 is expressed in the sensory epithelial patches of semicircular canals and in the supporting cells in the cochlea (Oh et al., 1996). Treatment with a BMP4 antagonist, Noggin, blocks sensory and nonsensory organ morphogenesis and disrupts morphological development of the semicircular canals (Chang et al., 1999; Gerlach et al., 2000).

Cell cycle regulator p27(Kip1), a cyclin-dependent kinase inhibitor that functions as an inhibitor of cell cycle progression, is found to be selectively expressed in the supporting-cell population of the mature OC (Chen and

---

plasmid (A–C), a mixture of equal amounts of pSVCMV–Hes1 and pRK5–Math1–EGFP plasmid (D–F) and a mixture of pSVCMV–Hes1 and pRK5–EGFP plasmid at a ratio of 5:1 (G–I), respectively. Note that whereas virtually all *Math1* transfected GER cells become hair cells (A–C), only a very small number of GER cells in the cultures cotransfected with *Hes1* and *Math1* (arrows in D–F) are able to differentiate into hair cells. All *Hes1* transfected GER cells remain myosin VIIa negative (G–I). Bar: 25 μm. Modified with permission from Zheng, J., Shou, J., Guillemot, F., Kageyama, R., and Gao, W. Q. (2000). Hes1 is a negative regulator of inner ear hair cell differentiation. *Development* **127,** 4551–4560. (See Color Insert.)

Segil, 1999; Lowenheim *et al.*, 1999). Targeted deletion of the p27(Kip1) gene leads to a prolonged proliferation of the sensory cell progenitors and the production of supernumerary hair cells and supporting cells in the cochlea (Chen and Segil, 1999; Lowenheim *et al.*, 1999). p27(Kip1) null mutant mice are severely hearing impaired (Chen and Segil, 1999; Lowenheim *et al.*, 1999). Therefore, cell cycle regulators such as p27(Kip1) can sufficiently affect proliferation of hair cell progenitors and generation of hair cells during development.

D. Further Maturation and Maintenance of Hair Cells Depend on Expression of Other Transcription Factors

Following hair cell fate determination or the initial hair cell differentiation described previously, a few other transcription factors are reported to be required for the further maturation and survival of hair cells. These include POU domain transcription factor Brn3c (Xiang *et al.*, 1998), homeobox gene Barhl1 (Li *et al.*, 2002), and zinc finger transcription factor Gfi1 (Wallis *et al.*, 2003). When any one of the three genes is knocked out, although hair cells are formed initially in the inner ear, the newly formed hair cells soon undergo a gradual degeneration and lead to hearing loss (Li *et al.*, 2002; Xiang *et al.*, 1998; Wallis *et al.*, 2003). These three genes are probably downstream of *Math1*. Based on immunohostochemical staining it has been demonstrated in the *Math1* null mutant that expression of Gfi1 protein is greatly reduced. Therefore, following the initial differentiation step, these other transcription factors are likely necessary for the later stages of hair cell differentiation, maturation, and maintenance.

III. Production and Regeneration of New Hair Cells in Mature Inner Ears

It is important to note that hair cell loss due to noise and ototoxic damage is one of the major causes of hearing and balance impairments. Although new hair cells can be produced spontaneously in mature bird and lower vertebrate ears (Corwin and Cotanche, 1988; Jones and Corwin, 1996; Ryals and Rubel, 1988), there has so far been no report showing an effective way to stimulate hair cell regeneration in mature mammalian cochleae (see Chapter 14 by Ryan). Understanding the mechanisms of hair cell differentiation would be helpful for us to stimulate hair cell regeneration following injury and could eventually lead to a therapeutic treatment of hearing and balance impairments.

A. Production of New Hair Cells in Mature Inner Ears

Although we have shown that overexpression of *Math1* can induce production of extra hair cells in immature cochleae, it is unclear whether overexpression of *Math1* in the mature mammalian inner ear can induce production of new hair cells. In rodents the inner ear tissue undergoes dramatic structural, immunocytochemical, and physiological changes within the first 3 weeks after birth (Lim and Rueda, 1992; Walsh and Romand, 1992). These changes include apoptosis (Zheng and Gao, 1997), maturation of stereociliary bundles (Lim and Rueda, 1992), functional synaptogenesis (Walsh and Romand, 1992), and acquisition of specific ionic channels (Rusch *et al.*, 1998) in hair cells. The cells in the sensory epithelia of mature mammalian inner ears become highly differentiated and show reduced response to growth factors (Gu *et al.*, 1997). Consequently, the determination of whether cells in the mature ear have the same potential as the immature ear to differentiate into hair cells is eagerly avaited (Cho, 2000; Fekete, 2000; Larkin, 2000; Seppa, 2000).

We have recently performed experiments to determine whether overexpression of *Hath1*, a human *atonal* homologue, would induce production of new hair cells in mature inner ears (Gao, 2002). Because viral gene transfer is more practical than electroporation (Zheng and Gao, 2000) for potential therapeutic applications, we have designed an adenoviral vector, ad-Hath1–EGFP, to misexpress *Hath1* (Gao, 2001, 2002).

When the adult rat utricular whole mount cultures are infected with the ad-Hath1–EGFP virus, many cells fluoresce green when live cultures are observed under UV light 1 day following viral infection. When the infected cultures were fixed and double labeled with anti-myosin VIIa antibody at 3 days following viral infection, examination of the whole mount preparations under a fluorescence microscope by focusing up and down revealed that no infected cells in the cultures were myosin VIIa positive, even though a large number of cells were infected (Gao, 2002). However, when the cultures were examined at 7–18 days following infection with the ad-Hath1–EGFP virus, the majority of infected cells in the sensory epithelium had become myosin VIIa positive (Gao, 2002; Shou *et al.*, 2003). On average, 76 of 107 (approximately 71%) infected supporting cells per utricle become hair cells, based on cell counts from 16 of 52 randomly selected cultures fixed at 7–18 days after infection. In sharp contrast, all of the infected cells in each of the 16 control cultures infected with the ad-EGFP virus remained negative to myosin VIIa antibody and failed to convert into hair cells at 12 days after infection (Gao, 2002; Shou *et al.*, 2003). The conversion of supporting cells into hair cells in the mature utricular macula was further confirmed with immunohistochemistry of cross sections of the utricular whole mount cultures with other hair cell markers. The newly generated hair cells grow stereociliary bundles (Gao,

2002; Shou et al., 2003), in addition to conversion into a well-defined paired or columnar hair cell morphology. In contrast, those cells infected with the ad-EGFP maintain the large size of their cell body and show irregular shape with numerous processes.

An important extension of the work described above is a very late study accomplished by Raphael and his co-workers (Kawamoto et al., 2003). They show very lately that that adenoviral gene transfer of *Math1* can result in appearance of immature hair cells in the organ of Corti in mature guinea pigs *in vivo*. The putative new hair cells are found adjacent to the organ of Corti in the areas which are derived from GER and LER (Zheng and Gao, 2000; Shou et al., 2003). Furthermore, Afferent auditory nerve fibers were seen to reach the vicinity of the putative new hair area. Thus, the mature mammalian cochlea retains the competence to generate new hair cells *in vivo*.

B. Regeneration of New Hair Cells in Gentamicin-Damaged Mature Inner Ears

We have also determined whether the mature inner ear epithelium has the capacity to generate new hair cells following injury. We treated the adult rat utricular macula with gentamicin, which is an ototoxin that kills the majority of hair cells in the mature ear (Warchol et al., 1993), and subsequently infected the tissue using the ad-Hath1–EGFP virus. Consistent with experiments with normal utricular macula tissue, we found that many of the infected cells (approximately 70% of the total 205 infected cells) in the sensory epithelium are myosin VIIa positive 12 days after infection (Gao, 2002; Shou et al., 2003). In contrast, no ad-EGFP–infected cells reacted with the anti-myosin VIIa antibody in the gentamicin-treated control cultures (Gao, 2002; Shou et al., 2003).

IV. Conclusion

Studies over the past several years have helped us make significant progress in our understanding of the development of inner ear hair cells in higher vertebrates. Local cell–cell interactions and specific genes play important roles in the control of the proliferation of progenitor cells in the sensory epithelium and differentiation of hair cells. A model of cochlear hair cell development is shown in Fig. 8. Although specific growth factors such as heregulin, bFGF, and BMP4 may influence proliferation of progenitor cells in the sensory epithelium around a stage of E11.5, cell cycle regulator *p27(Kip1)* regulates the terminal mitosis process between E11.5 and E14.5. Selection of hair cell fate or initial differentiation occurs from E14.5 to E15.5, which

10. Mammalian Hair Cell Development

```
Proliferation of  ← Heregulin signaling
progenitor cells  ← FGF signaling
E11.5             ← BMP signaling

Terminal mitosis  ← p27(Kip1)

E14.5

Selection of hair cell fate or  ├── Hes1/Hes5 ← Notch signaling
initial differentiation              ⊥
(expression of myosin VIIa)       Math1

E15.5

Maturation ← Brn3c, Barhl1, Gfi1
Growth of stereociliary bundles,
expression of several calcium binding proteins,
production of specific ionic channels,
and establishment of synaptic connections

P21
Functional Hair Cells
```

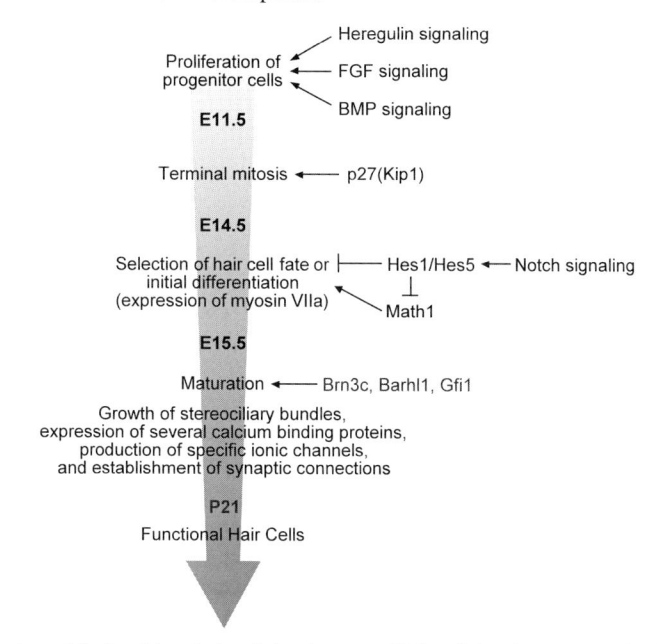

Figure 8 A model of cochlear hair cell development. Hair cell development can be divided into several stages: proliferation of progenitor cells (before E11.5), terminal mitosis (from E11.5 to E14.5), selection of hair cell fate or initial differentiation (from E14.5 to E15.5), and maturation or later differentiation and maintenance (from E15.5 to P21 and adult). Whereas specific growth factors such as heregulin, bFGF, and BMP4 may influence proliferation of progenitors, *p27(Kip1)* regulates the terminal mitosis process. Hair cell fate determination or initial differentiation marked by expression of myosin VIIa is generally regulated by Notch signaling in which *Hes1* and *Hes5* are effector genes. During this differentiation process, *Math1* is shown to be sufficient to induce hair cell differentiation and *Hes1* and *Hes5* act as negative regulators of hair cell differentiation by antagonizing *Math1*. Maturation or later differentiation of hair cells by growth of stereociliary bundles, expression of several calcium binding proteins, production of specific ionic channels, and establishment of synaptic connections occurs from E15.5 to P21. Brn3C, Barhl1, and Gfi1 transcription factors (at least) have been shown to play important roles during this later differentiation/maturation/maintenance process.

is regulated by Notch signaling in which *Hes1/Hes5* are effector genes. It has been clearly demonstrated that whereas *Math1* is a positive regulator, *Hes1/Hes5*, downstream target genes of Notch signaling pathway, act as negative regulators of hair cell differentiation by antagonizing *Math1*. There is a continuing maturation of hair cells by growth of stereociliary bundles (Lim and Rueda, 1992; Zheng and Gao, 1997b), expression of several calcium binding proteins (Zheng and Gao, 1997), production of specific ionic channels (Rusch *et al.*, 1998), and establishment of synaptic connections (Walsh and Romand, 1992). Three other transcription factors, Brn3C, Barhl1, and Gfi1, appear to be required for this late differentiation/maturation process

(Li et al., 2002; Wallis et al., 2003; Xiang et al., 1998). By P21 hair cells are mature and become functional (Walsh and Romand, 1992). Continuing maintenance of functional hair cells is critical for our normal hearing.

By taking advantage of our understanding of the hair cell differentiation process, efforts have been made to stimulate regeneration of new hair cells in mature inner ears. Earlier work in several laboratories had focused on mechanisms controlling cell proliferation (Chen and Segil, 1999; Lowenheim et al., 1999; Navaratnam et al., 1996) and had identified several mitogenic growth factors that can stimulate proliferation of supporting cells (Corwin et al., 1996; Gu et al., 1996; Kuntz and Oesterle, 1998; Lambert, 1994; Montcouquiol and Corwin, 2001; Oesterle, 1998; Oesterle et al., 1997; Yamashita and Oesterle, 1995; Zheng et al., 1997, 1999) in the inner ear sensory epithelium. Lately, more attention has been paid to genes that control hair cell differentiation. Based on our findings that *Math1* is sufficient to induce hair cell differentiation (Zheng and Gao, 2000), gene transfer experiments have been done in the mature inner ear tissue. The mature vestibular endorgans, at least, are shown to be capable of producing a large number of new hair cells in both intact and ototoxin-damaged tissue via expression of the human homologue of *Math1* (Gao, 2002; Shou et al., 2003). The finding that both GER and LER cells can be converted into hair cells in the immature cochlea *Math1* (Gao, 2002; Shou et al., 2003) provides an important insight for potential hair cell regeneration in the mature cochlea. Although some of the GER cells are believed to undergo apoptosis during maturation of the cochlear tissue (Lim and Rueda, 1992), the rest becomes inner sulcus cells, interdental cells, and other epithelial cells. The majority of the postnatal LER cells become Hensen's and Claudius' cells that remain in the mature cochlea. These data suggest that inner sulcus, interdental, Hensen's and Claudius' cells might be good target cells that can be induced to transdifferentiate into new hair cells in the mature cochlea. Indeed, a recent study in mature guinea pigs *in vivo* (Kawamoto et al., 2003) provides evidence for this notion. On the other hand, it needs to be pointed out that the studies described have so far focused on production and regeneration of hair cells using morphological and immunohistochemical approaches. In the near future it will be necessary to determine whether the ectopic hair cells induced by *Math1* expression are actually functional and to answer questions including whether they are innervated by spiral ganglion neurons, whether they express appropriate ionic channels and other required gene products, and whether they are competent for mechanotransduction. Nevertheless, these recent results are very encouraging. A combination of mitogenic supporting cell growth factors and upregulation and downregulation of specific genes involved in the control of hair cell differentiation could be a more effective way to stimulate hair cell regeneration in the inner ear, which could eventually be helpful for the treatment of hearing and balance disorders caused by hair cell loss.

Acknowledgments

I thank J. Lisa Zheng and Jianyong Shou for their contributions to this work, Ching Ching Leow for critical reading of the manuscript, and Allison Bruce for preparation of the figures.

References

Adam, J., Myat, A., Le Roux, I., Eddison, M., Henrique, D., et al. (1998). Cell fate choices and the expression of Notch, Delta and Serrate homologues in the chick inner ear: Parallels with *Drosophila* sense-organ development. *Development* **125**, 4645–4654.

Akazawa, C., Ishibashi, M., Shimizu, C., Naknish, S., and Kageyama, R. (1995). A mammalian helix-loop-helix factor structurally related to the product of *Drosophila* proneural gene atonal is a positive transcriptional regulator expressed in the developing nervous system. *J. Biol. Chem.* **270**, 8730–8738.

Artavanis-Tsakonas, S., Rand, M. D., and Lake, R. J. (1999). Notch signaling: Cell fate control and signal integration in development. *Science* **284**, 770–776.

Ben-Arie, N., Bellen, H., Armstrong, D., McCall, A., Gordadze, P., et al. (1997). Math1 is essential for genesis of cerebellar granule neurons. *Nature* **390**, 169–172.

Bermingham, N. A., Hassan, B. A., Price, S. D., Vollrath, M. A., Ben-Arie, N., et al. (1999). Math1: An essential gene for the generation of inner ear hair cells. *Science* **284**, 1837–1841.

Chang, W., Nunes, F., De Jesus-Escobar, J., Harland, R., and Wu, D. (1999). Ectopic noggin blocks sensory and nonsensory organ morphogenesis in the chicken inner ear. *Dev. Biol.* **216**, 369–381.

Chen, P., Johnson, J., Zoghbi, H., and Segil, N. (2002). The role of Math1 in inner ear development: Uncoupling the establishment of the sensory primordium from hair cell fate determination. *Development* **129**, 2495–2505.

Chen, P., and Segil, N. (1999). p27(Kip1) links cell proliferation to morphogenesis in the developing organ of Corti. *Development* **126**, 1581–1590.

Cho, A. (2000). Gene therapy could aid hearing. *ScienceNOW* **518**, 1.

Corwin, J., and Cotanche, D. (1988). Regeneration of sensory hair cells after acoustic trauma. *Science* **240**, 1772–1774.

Corwin, J. T., Warchol, M. E., Saffer, L. D., Finley, J. E., Gu, R., and Lamber, P. R. (1996). Growth factors as potential drugs for the sensory epithelia of the ear. *Ciba Foundation Symposium* **196**, 167–182, discussion 182–187.

Eddison, M., Le Roux, I., and Lewis, J. (2000). Notch signaling in the development of the inner ear: Lessons from *Drosophila*. *Proc. Natl. Acad. Sci. USA* **97**, 11692–11699.

Erkman, L., McEvilly, R. J., Luo, L., Ryan, A. K., Hooshmand, F., et al. (1996). Role of transcription factors Brn-3.1 and Brn-3.2 in auditory and visual system development. *Nature* **381**, 603–606.

Fekete, D. (2000). Making sense of making hair cells. *Trends Neurosci.* **23**, 386.

Fekete, D. M. (1996). Cell fate specification in the inner ear. *Curr. Opin. NeuroBiol.* **6**, 533–541.

Fekete, D. M. (1999). Development of the vertebrate ear: Insights from knockouts and mutants. *TINS* **22**, 263–269.

Fekete, D. M., Muthukumar, S., and Karagogeos, D. (1998). Hair cells and supporting cells share a common progenitor in the avian inner ear. *J. Neurosci.* **18**, 7811–7821.

Forge, A., Li, L., Corwin, J. T., and Nevill, G. (1993). Ultrastructural evidence for hair cell regeneration in the mammalian inner ear. *Science* **259**, 1616–1619.

Gao, W. Q. (2001) Math1, hair cell fate determination, and potential hair cell regeneration. Molecular Biology of Hearing and Deafness Conference, p. 89. Bethesda, Md (abstract).

Gao, W. Q. (2002). Robust hair cell regeneration in the mature mammalian inner ear by adenoviral expression of Hath1. The 36th Karolinska Institutet Nobel Conference "To Restore Hearing," p. 17. June 9–13, Krusenberg, Sweden (abstract).

Gao, W. Q., Zheng, J., Lewis, A., and Frantz, G. (1999) Heregulin promotes regenerative proliferation in rat utricular sensory epithelium following ototoxic damage and influences hair cell differentiation during development. Association for Research in Otolaryngology Meeting **22,** 224 (abstract).

Gerlach, L., Hutson, M., Germiller, J., Nguyen-Luu, D., Victor, J., and Barald, K. (2000). Addition of the BMP4 antagonist, noggin, disrupts avian inner ear development. *Development* **127,** 45–54.

Gray, G., Mann, R., Mitsiadis, E., Henrique, D., Carcangiu, M., *et al.* (1999). Human ligands of the Notch receptor. *Am. J. Pathol.* **154,** 785–794.

Gu, R., Marchonni, M., and Corwin, J. T. (1996) Glial growth factor enhances supporting cell proliferation in rodent vestibular epithelia cultured in isolation. Society for Neuroscience Annual Meeting **22,** 520 (abstract).

Gu, R., Marchionni, M., and Corwin, J. (1997) Age-related decreases in proliferation within isolated mammalian vestibular epithelia cultured in control and glial growth factor2 media. Association for Research in Otolaryngology Meeting **20,** 98 (abstract).

Haddon, C., Jiang, Y. J., Smithers, L., and Lewis, J. (1998). Delta-Notch signalling and the patterning of sensory cell differentiation in the zebrafish ear: Evidence from the mind bomb mutant. *Development* **125,** 4637–4644.

Hasson, T., Heintzelman, M. B., Santos-Sacchi, J., Corey, D. P., and Mooseker, M. S. (1995). Expression in cochlea and retina of myosin VIIa, the gene product defective in Usher syndrome type 1B. *Proc. Nat. Acad. Sci. USA* **92,** 9815–9819.

Helms, A., and Johnson, J. (1998). Progenitors of dorsal commissural interneurons are defined by MATH1 expression. *Development* **125,** 919–928.

Ishibashi, M., Ang, S., Shiota, K., Nakanishi, S., Kageyama, R., and Guillemot, F. (1995). Targeted disruption of mammalian hairy and enhancer of split homolog-1 (HES-1) leads to up-regulation of neural helix-loop-helix factors, premature neurogenesis, and severe neural tube defects. *Genes Dev.* **15,** 3136–3148.

Jensen, J., Pedersen, E., Galante, P., Hald, J., Heller, R., *et al.* (2000). Control of endodermal endocrine development by Hes-1. *Nat. Genet.* **24,** 36–44.

Jones, J. E., and Corwin, J. T. (1996). Regeneration of sensory cells after laser ablation in the lateral line system: Hair cell lineage and macrophage behavior revealed by time-lapse video microscopy. *J. Neurosci.* **16,** 649–662.

Joutel, A., and Tournier-Lasserve, E. (1998). Notch signalling pathway and human diseases. *Semin. Cell Dev. Biol.* **9,** 619–625.

Kawamoto, K., Ishimoto, S., Minoda, R., Brough, D. E., and Raphael, Y. (2003). Math1 gene transfer generates new cochlear hair cells in mature guinea pigs *in vivo*. *J. Neurosci.* **23,** 4395–4400.

Kiernan, A., Ahituv, N., Fuchs, H., Balling, R., Avraham, K., *et al.* (2001). The Notch ligand Jagged1 is required for inner ear sensory development. *Proc. Natl. Acad. Sci. USA* **98,** 3873–3878.

Kuntz, A. L., and Oesterle, E. C. (1998). Transforming growth factor alpha with insulin stimulates cell proliferation in vivo in adult rat vestibular sensory epithelium. *J. Comp. Neurol.* **399,** 413–423.

Lambert, P. (1994). Inner ear hair cell regeneration in a mammal: Identification of a triggering factor. *Laryngoscope* **104,** 701–717.

Lanford, P. J., Lan, Y., Jiang, R., Lindsell, C., Weinmaster, G., *et al.* (1999). Notch signalling pathway mediates hair cell development in mammalian cochlea. *Nat. Genet.* **21,** 289–292.

Larkin, M. (2000). Can lost hearing be restored? *Lancet* **356,** 744.

Lewis, A. K., Frantz, G. D., Carpenter, D. A., de Sauvage, F. J., and Gao, W. Q. (1998). Distinct expression patterns of notch family receptors and ligands during development of the mammalian inner ear. *Mech. Dev.* **78,** 159–163.

Lewis, J. (1998). Notch signalling and the control of cell fate choices in vertebrates. *Semin. Cell Dev. Biol.* **9,** 583–589.

Li, L., and Forge, A. (1997). Morphological evidence for supporting cell to hair cell conversion in the mammalian utricular macula. *Int. J. Dev. Neurosci.* **15,** 433–446.

Li, S., Price, S. M., Cahill, H., Ryugo, D. K., Shen, M. M., and Xiang, M. (2002). Hearing loss caused by progressive degeneration of cochlear hair cells in mice deficient for the Barhl1 homeobox gene. *Development* **129,** 3523–3532.

Lim, D., Rueda, J. (1992) Structural development of the cochlea. "Development of auditory and vestibular system 2" (R. Romand, ed.), pp. 33–58. Elsevier, New York.

Lindsell, C. E., Boulter, J., DiSibio, G., Gossler, A., and Weinmaster, G. (1996). Expression patterns of jagged, Delta1, Notch1, Notch2, and Notch3 genes identify ligand-receptor pairs that may function in neural development. *Mol. Cell Neurosci.* **8,** 14–27.

Lindsell, C. E., Shawber, C. J., Boulter, J., and Weinmaster, G. (1995). Jagged: A mammalian ligand that activates Notch1. *Cell* **80,** 909–917.

Lowenheim, H., Furness, D. N., Kil, J., Zinn, C., Gultig, K., *et al.* (1999). Gene disruption of p27(Kip1) allows cell proliferation in the postnatal and adult organ of Corti. *Proc. Nat. Acad. Sci. USA* **96,** 4084–4088.

Montcouquiol, M., and Corwin, J. (2001). Brief treatments with forskolin enhance s-phase entry in balance epithelia from the ears of rats. *J. Neurosci.* **21,** 947–982.

Morrison, A., Hodgetts, C., Gossler, A., Hrabe de Angelis, M., and Lewis, J. (1999). Expression of Delta1 and Serrate1 (Jagged1) in the mouse inner ear. *Mech. Dev.* **84,** 169–172.

Mueller, K., Jacques, B., and Kelley, M. (2002). Fibroblast growth factor signaling regulates pillar cell development in the organ of Corti. *J. Neurosci.* **22,** 9368–9377.

Murone, M., Rosenthal, A., and de Sauvage, F. J. (1999). Sonic hedgehog signaling by the patched-smoothened receptor complex. *Curr. Biol.* **28,** 76–84.

Nakamura, Y., Sakakibara, S. I., Miyata, T., Ogawa, M., Shimazaki, T., *et al.* (2000). The bHLH gene Hes1 as a repressor of the neuronal commitment of CNS stem cells. *J. Neurosci.* **20,** 283–293.

Navaratnam, D. S., Su, H. S., Scott, S. P., and Oberholtzer, J. C. (1996). Proliferation in the auditory receptor epithelium mediated by a cyclic AMP-dependent signaling pathway. *Nat. Med.* **2,** 1136–1139.

Oesterle, E. C., Tsue, T. T., and Rubel, E. W. (1997). Induction of cell proliferation in avian inner ear sensory epithelia by insulin-like growth factor-I and insulin. *J. Comp. Neurol.* **380,** 262–274.

Oh, S. H., Johnson, R., and Wu, D. K. (1996). Differential expression of bone morphogenetic proteins in the developing vestibular and auditory sensory organs. *J. Neurosci.* **16,** 6463–6475.

Ohtsuka, T., Ishibashi, M., Gradwohl, G., Nakanishi, S., Guillemot, F., and Kageyama, R. (1999). Hes1 and Hes5 as Notch effectors in mammalian neuronal differentiation. *EMBO J.* **18,** 2196–2207.

Pirvola, U., Gao, Y., Oellig, C., Suoqiang, Z., Pettersson, R., and Ylikoski, J. (1995). The site of action of neuronal acidic fibroblast growth factor is the organ of Corti of rat cochlea. *Proc. Natl. Acad. Sci. USA* **92,** 9269–9273.

Pirvola, U., Ylikoski, J., Trokovic, R., Hebert, J., McConnell, S., and Partanen, J. (2002). FGFR1 is required for the development of the auditory sensory epithelium. *Neuron* **35,** 671–680.

Ruben, R. J. (1967). Development of the inner ear of the mouse: A radioautographic study of terminal mitosis. *Acta. Otolaryngol. Suppl.* **220,** 1–44.

Rusch, A., Lysakowski, A., and Eatock, R. (1998). Postnatal development of type I and type II hair cells in the mouse utricle: Acquisition of voltage-gated conductances and differentiated morphology. *J. Neurosci.* **18,** 7487–7501.

Ryals, B. M., and Rubel, E. W. (1988). Hair cell regeneration after acoustic trauma in adult Coturnix quail. *Science* **240,** 1774–1776.

Sans, A., and Chat, M. (1982). Analysis of temporal and spatial patterns of rat vestibular hair cell differentiation by tritiated thymidine radioautography. *J. Comp. Neurol.* **6,** 1–8.

Seppa, N. (2000). New inner ear hair cells grow in rat tissue. *Science News* **157,** 342.

Shou, J., Zheng, J., and Gao, W. Q. (2003). Robust generation of new hair cells in the mature mammalian inner ear by adenoviral expression of Hath1. *Mol. Cell. Neurosci.* **23,** 169–179.

Shutter, J., Scully, S., Fan, W., Richards, W., Kitajewski, J., *et al.* (2000). Dl14, a novel Notch ligand expressed in arterial endothelium. *Genes Dev.* **14,** 1313–1318.

Sobkowicz, H. M., Bereman, B., and Rose, J. E. (1975). Organotypic development of the organ of Corti in culture. *J. Neurocytol.* **4,** 543–572.

Stone, J. S., and Rubel, E. W. (1999). Delta1 expression during avian hair cell regeneration. *Development* **126,** 961–973.

Valsecchi, V., Ghezzi, C., Ballabio, A., and Rugarli, E. I. (1997). JAGGED2: A putative Notch ligand expressed in the apical ectodermal ridge and in sites of epithelial-mesenchymal interactions. *Mech. Dev.* **69,** 203–207.

Van de Water, T. R. (1983). "Embryogenesis of the Inner Ear: In Vitro Studies," pp. 337–374. Academic Press, New York.

Wallis, D., Hamblen, M., Zhou, Y., Venken, K. J., Schumacher, A., *et al.* (2003). The zinc finger transcription factor Gfi1, implicated in lymphomagenesis, is required for inner ear hair cell differentiation and survival. *Development* **130,** 221–232.

Walsh, E., Romand, R. (1992) Functional development of the cochlea and the cochlear nerve. *In* "Development of Auditory and Vestibular System 2" (R. Romand, ed.). Elsevier, New York.

Warchol, M. E., Lambert, P. R., Goldstein, B. J., Forge, A., and Corwin, J. T. (1993). Regenerative proliferation in inner ear sensory epithelia from adult guinea pigs and humans. *Science* **259,** 1619–1622.

Welshons, W. J. (1965). Analysis of a gene in *Drosophila*. *Science* **150,** 1122–1129.

Williams, R. U. L., and Lardelli, M. (1995). Complementary and combinatorial patterns of Notch gene family expression during early mouse development. *Mech. Dev.* **53,** 357–368.

Xiang, M., Gan, L., Li, D., Chen, Z. Y., Zhou, L., *et al.* (1997a). Essential role of POU-domain factor Brn-3c in auditory and vestibular hair cell development. *Proc. Nat. Acad. Sci. USA* **94,** 9445–9450.

Xiang, M., Gao, W. Q., Hasson, T., and Shin, J. J. (1998). Requirement for Brn-3c in maturation and survival, but not in fate determination of inner ear hair cells. *Development* **125,** 3935–3946.

Xiang, M. Q., Gan, L., Li, D. Q., Chen, Z. Y., Zhou, L. J., *et al.* (1997b). Essential role of POU-domain factor Brn-3c in auditory and vestibular hair cell development. *Proc. Natl. Acad. Sci. USA* **94,** 9445–9450.

Yamashita, H., and Oesterle, E. C. (1995). Induction of cell proliferation in mammalian inner-ear sensory epithelia by transforming growth factor alpha and epidermal growth factor. *Proc. Nat. Acad. Sci. USA* **92,** 3152–3155.

Yang, Q., Bermingham, N., Finegold, M., and Zoghbi, H. (2001). Requirement of Math1 for secretory cell lineage commitment in the mouse intestine. *Science* **294,** 2155–2158.

Zhang, M., Ding, D., and Salvi, R. (2002). Expression of heregulin and ErbB/Her receptors in adult chinchilla cochlear and vestibular sensory epithelium. *Hear. Res.* **169,** 56–68.

10. Mammalian Hair Cell Development

Zhang, N., Martin, G., Kelley, M., and Gridley, T. (2000). A mutation in the Lunatic fringe gene suppresses the effects of a Jagged2 mutation on inner hair cell development in the cochlea. *Curr. Biol.* **10,** 659–662.

Zheng, J., and Gao, W. Q. (2000). Overexpression of Math1 induces robust production of extra hair cells in postnatal rat inner ears. *Nat. Neurosci.* **3,** 580–586.

Zheng, J., Shou, J., Guillemot, F., Kageyama, R., and Gao, W. Q. (2000). Hes1 is a negative regulator of inner ear hair cell differentiation. *Development* **127,** 4551–4560.

Zheng, J. L., Frantz, G., Lewis, A. K., Sliwkowski, M., and Gao, W. Q. (1999). Heregulin enhances regenerative proliferation in postnatal rat utricular sensory epithelium after ototoxic damage. *J. Neurocytol.* **28,** 901–912.

Zheng, J. L., and Gao, W. Q. (1997). Analysis of rat vestibular hair cell development and regeneration using calretinin as an early marker. *J. Neurosci.* **17,** 8270–8282.

Zheng, J. L., Helbig, C., and Gao, W. Q. (1997). Induction of cell proliferation by fibroblast and insulin-like growth factors in pure rat inner ear epithelial cell cultures. *J. Neurosci.* **17,** 216–226.

Zine, A., Aubert, A., Qiu, J., Therianos, S., Guillemot, F., *et al.* (2001). Hes1 and Hes5 activities are required for the normal development of the hair cells in the mammalian inner ear. *J. Neurosci.* **21,** 4712–4720.

Zine, A., Van De Water, T., and de Ribaupierre, F. (2000). Notch signaling regulates the pattern of auditory hair cell differentiation in mammals. *Development* **127,** 3373–3383.

11

Cell Adhesion Molecules during Inner Ear and Hair Cell Development, Including Notch and Its Ligands

Matthew W. Kelley
Section on Developmental Neuroscience
National Institute on Deafness and Other Communication Disorders
National Institutes of Health
Rockville, Maryland, 20850

I. Introduction
II. Adhesion Molecules
   A. Cell Adhesion Molecules
   B. Cadherins
   C. Integrins
   D. Other Adhesion Molecules
III. Adhesion Molecules and Development of the Inner Ear
   A. Expression of Adhesion Molecules in the Otocyst
   B. Adhesion Molecules and Differentiation of Sensory Epithelia
   C. Adhesion and Neurite Growth in the Inner Ear
   D. Adhesion Molecules and Development of Stereocilia Bundles
IV. Summary
   References

The vertebrate inner ear develops from a population of placodally derived epithelial cells that initially invaginate to form the otocyst. As development continues, subsets of cells within this population become specified to develop as specific structures within the ear. Subsequently, individual cells undergo morphogenetic and differentiative changes related to the development of specific regions of the ear. The factors that regulate the different developmental events that are required for formation of a complete inner ear are still not completely understood. However, there is growing evidence to suggest that cell–cell adhesion and cell adhesion molecules may play key roles in several aspects of inner ear development.

This chapter summarizes the molecular attributes of the known families of cell adhesion molecules and reviews the existing knowledge about the role of cell adhesion in inner ear development. The roles of cell adhesion in regionalization of the otocyst, cellular differentiation, and neurite extension are reviewed. In addition, the recently demonstrated role of a subset of adhesion molecules, although not necessarily cell–cell adhesion itself, in the development and orientation of hair cell stereociliary bundles is discussed. The role of adhesion molecules in stereociliary bundle development is even more intriguing in light of the fact that many of these molecules were identified as genetic mutations that lead to non-syndromic forms of deafness. These results highlight the importance of adhesion molecules in inner ear development and suggest that the developmental challenges that exist within the inner ear may have resulted in the development of novel functions for adhesion molecules. © 2003, Elsevier Inc.

I. Introduction

The vertebrate inner ear is composed of a diverse group of complex structures that includes the semicircular canals, endolymphatic duct, vestibular and auditory sensory epithelia, nonsensory epithelial regions, and vestibular and auditory ganglia. Virtually all of the cells that give rise to these structures derive from the population of epithelial cells that initially comprise the otocyst. During the course of embryogenesis, cells located within the otocyst will be influenced by both intrinsic and extrinsic factors that will regulate multiple aspects of the development of these cells, including cellular proliferation, progressive restriction and ultimate determination of cell fate and differentiation, complex morphogenetic changes, and apoptotic cell death. Despite the recent increase in the number of publications examining different aspects of the development of the inner ear, our understanding of the factors that regulate and coordinate the generation of this elaborate structure is still in its infancy.

A key factor in the formation of any epithelial structure is the ability of individual cells to form adhesive contacts with other cells and with the underlying extracellular matrix. Such adhesive contacts obviously play key roles in the development of specific three-dimensional structures, but more recent results have suggested that cell–cell and cell–basement membrane interactions can also signal within specific cells through the activation of intracellular cascades to initiate cellular responses, including proliferation, determination of cell fate, differentiation, morphogenesis, and both intra- and inter-cellular fasciculation. Classically, cell–cell adhesive interactions were thought to be mediated by either calcium-dependent cadherins or calcium-independent cell adhesion molecules (CAMs), whereas cell–basement membrane interactions were mediated by integrins. However, more recent studies have expanded the spectrum of molecules that can regulate cell adhesion to include claudins, desmosomal cadherins, and protocadherins. Not surprisingly, cell–cell adhesion and adhesion molecules have been shown to play a role in the development of a functional auditory system. But our understanding of the comprehensive role of adhesion in the different steps that must occur between otocyst and mature inner ear is still extremely limited. In this chapter, I will briefly summarize the different families of molecules that have been shown to play a role in cell adhesion and then review the existing data on the role of adhesion in the development of different aspects of the inner ear.

II. Adhesion Molecules

A. Cell Adhesion Molecules

The first family of molecules shown to play a role in cell adhesion was the immunoglobulin-related CAMs (Brackenbury *et al.*, 1977; Cunningham *et al.*, 1987). Since that time, more than 1000 papers, including many excellent reviews, have been published examining different aspects of cellular adhesion and CAM expression and function. Therefore, only a brief overview of different CAMs will be presented here. CAMs are transmembrane proteins that are characterized by the presence of structural motifs that were first identified in immunoglobulin (Ig) proteins (Edelman and Crossin, 1991; Williams and Barclay, 1988; Crossin and Krushel, 2000). These Ig motifs are located in the extracellular domains of all CAMs and play a key role in mediating homophilic and heterophilic binding between CAM molecules on adjacent cells (Cole and Burg, 1989; Cole *et al.*, 1986; Friedlander *et al.*, 1994) (Fig. 1). The number of Ig motifs is variable between different CAMs and can vary from as few as one to as many as six (Crossin and Krushel, 2000). In addition to the Ig motif, many CAMs also contain multiple fibronectin (fn) type III–like repeats that are also located in the extracellular region of the cell. A specific role for the fn repeat region has not been determined but it has been suggested that this region may play a role in clustering of CAM molecules within a single cell (Peck and Walsh, 1993; Holm *et al.*, 1995; Silletti *et al.*, 2000). Most CAMs contain a single transmembrane region followed by a cytoplasmic tail that may be involved in cellular signaling. The cytoplasmic tails of most CAMs lack obvious catalytic or enzymatic activity; however, as will be discussed later, the biological effects of these molecules appear to extend beyond simple cellular adhesion. These results suggest that second messengers may play a role in the mediation of intracellular signaling by CAMs.

Neural cell adhesion molecule (NCAM) was the first CAM to be fully characterized in terms of protein and crystal structure (D'Eustachio *et al.*, 1985; Barthels *et al.*, 1987; Kasper *et al.*, 2000). Subsequent studies have identified a large number of molecules that can be classified as CAMs (as many as 100 in mammals) (Williams and Barclay, 1988), and these have been placed into multiple subfamilies based on a similar overall structure but

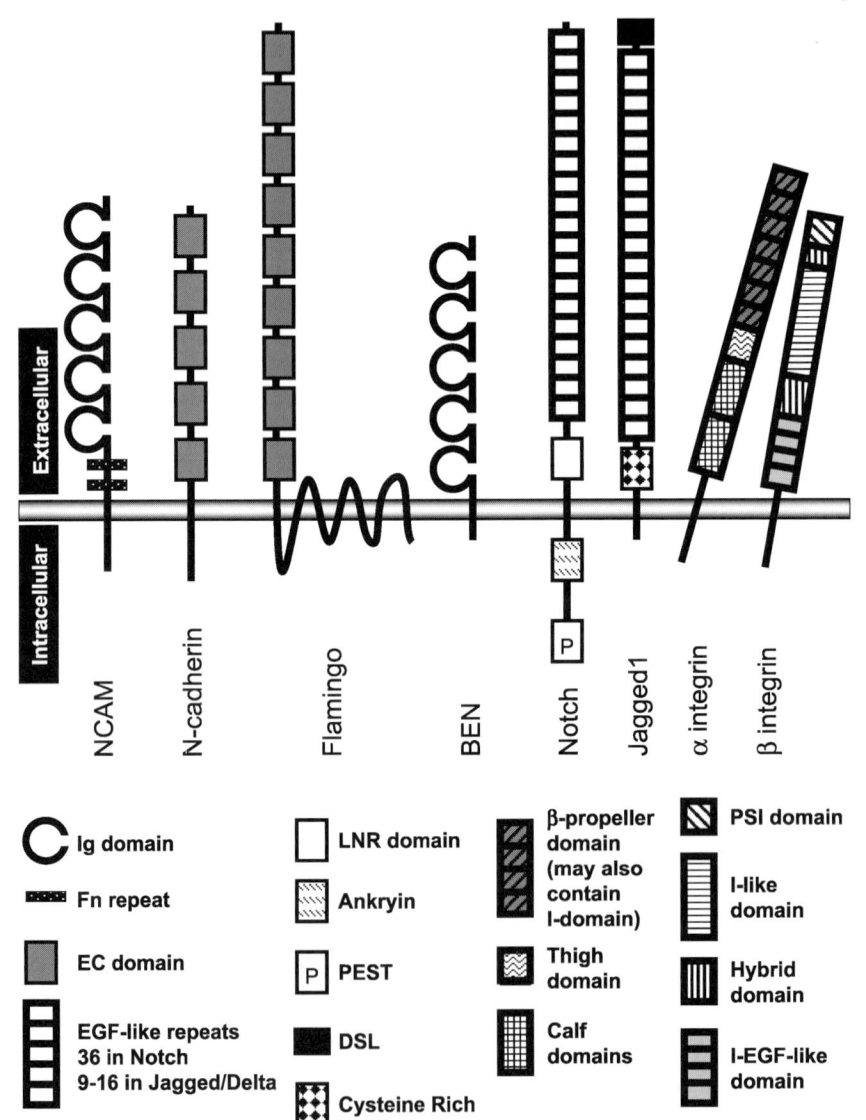

Figure 1 Domain structure of adhesion molecules. Structures for representative adhesion molecules are illustrated. Specific domains are noted in the legend. Heterophilic binding of integrins occurs between the β-propeller domain of α integrin and the I-like domain of β integrins. Some integrins may contain an I (inserted)-domain within the β-propeller. If the I-domain is present, then this is the primary site of binding to the I-like domain in the β integrin. See text and reviews cited for further details. LNR: conserved Lin12/Notch/Glp-1 region; DSL: conserved Delta/Serrate/LAG-2 domain; PSI: conserved plexin/semaphorin/integrin domain; I-EGF–like: EGF-like repeats that are unique to integrins.

variability in the number of Ig and fn domains. NCAM contains 5 Ig repeats and 2 fn repeats and so is a member of the 5/2 subfamily. Examples of other CAM subfamilies are the 5/6 subfamily (including L1), the 6/4 subfamily (including Contactin), and the 4/6 subfamily (including DCC [deleted in colorectal cancer]). The TAG1 subfamily is unique in that it lacks the transmembrane domain and is either anchored to the membrane through a glycosylphosphatidylinositol (GPI) moiety or exists in a soluble form (Furley *et al.*, 1990; Zuellig *et al.*, 1992). An intriguing aspect of the NCAM molecule is the presence of the 10 amino acid long variable alternatively spliced exon (VASE) that changes the fourth Ig domain to a variable Ig domain (Small and Akeson, 1990; Doherty *et al.*, 1992; Walsh *et al.*, 1992). In neurons, the ratio of NCAM molecules that contain VASE motif increases with developmental time (Reyes *et al.*, 1993; Linnemann *et al.*, 1993), and it has been suggested that inclusion of VASE may lead to a decrease in cellular responsiveness to NCAM (Walsh *et al.*, 1992; Liu *et al.*, 1993; Saffell *et al.*, 1994). VASE exons have not been described for any other CAMs, suggesting that this domain may be unique to NCAM. Finally, the activity of NCAM has also been shown to be regulated by the presence or absence of variable length chains of sialic acid (PSA) (Hoffman and Edelman, 1983; Sadoul *et al.*, 1983). Loss of PSA increases the adhesiveness of NCAM molecules (Walsh and Doherty, 1997); however, this effect appears to be related to a loss of plasticity, at least in the nervous system (Muller *et al.*, 1994). Similar to VASE, the presence of PSA also appears to be a unique aspect of the NCAM molecule.

NCAM was initially identified based on its ability to induce homophilic binding between cells (Hoffman and Edelman, 1983). Subsequent studies have demonstrated that this basic interaction plays a role in a number of different developmental events, including neuronal migration, and axon outgrowth, and fasciculation (Rutishuaser *et al.*, 1978; Thanos *et al.*, 1984; Fischer *et al.*, 1986; Doherty *et al.*, 1990; Drazba and Lemmon, 1990; Tomasiewicz *et al.*, 1993; Rabinowitz *et al.*, 1996). In addition, CAMs have been implicated in synaptic plasticity, axon guidance, cellular morphogenesis, and long-term potentiation (Harrelson and Goodman, 1988; Muller *et al.*, 1994; Tang *et al.*, 1994; Cremer *et al.*, 1998; Haag *et al.*, 1999; Lustig *et al.*, 1999). The observation that CAMs play a role in such a broad spectrum of developmental events, along with in vitro results demonstrating that soluble forms of some CAMs can mimic the effects of membrane-bound versions (Williams *et al.*, 1994; Doherty *et al.*, 1995), has led to the suggestion that the effects of CAMs are not limited to adhesion. In fact, binding of NCAM or L1 has been shown to lead to increases in intracellular calcium (Doherty *et al.*, 1991; Assou, 1992; Appel *et al.*, 1995), and CAM–dependent neurite outgrowth is inhibited by antagonism of nonreceptor tyrosine kinases such as fyn (Fgr/Yes related novel protein) (Beggs *et al.*, 1994;

1997). These results demonstrate diverse roles of CAMs in multiple aspects of development, as well as indicating that these effects are mediated through both adhesive and nonadhesive mechanisms.

B. Cadherins

The cadherins are an ancient family of adhesion molecules that are characterized by a conserved structure that includes the presence of a unique cadherin motif, also called an extracellular cadherin (EC) motif domain, in the extracellular region of the protein (Fig. 1) (Vleminckx and Kemler, 1999; Yagi and Takeichi, 2000). The number of EC domains varies among different members of the family from as few as two to as many as 27. Each EC domain contains specific, negatively charged sequence motifs that play a role both in homophilic binding between cadherin molecules, as well as in binding of Ca^{2+}. As their name implies, binding of Ca^{2+} plays an important role in cadherin function, and the results of a number of studies have demonstrated that homophilic cadherin binding is dependent on the presence of Ca^{2+} (first described by Takeichi et al., 1981).

Cadherins have been found throughout the metazoans with an estimated 80 different cadherins in most mammalian species (Yagi and Takeichi, 2000). Based on their structure, these have been separated into various subfamilies, including but not limited to classic, fat-like (characterized by as many as 34 EC domains), desmosomal, and protocadherins. The classic cadherins were the first family to be identified even though they represent a relatively small percentage of the total number of cadherins. Classic cadherins are characterized by a cytoplasmic domain containing p120 and α-catenin/plakoglobin binding sites, a single transmembrane domain, and five EC domains (Steinberg and McNutt, 1999). Fat-like cadherins, such as Fat and Dachsous, contain an overall structure that is similar to classic cadherins but with a greater number of EC domains (Bryant et al., 1988; Mahoney et al., 1991; Clark et al., 1995). Desmosomal cadherins, desmoglein and desmocollin, have a structure that is similar to the classic cadherins, but these molecules act as the adhesion molecules for desmosomes (reviewed in Garrod et al., 2002). Unlike classic cadherins, desmosomal cadherin binding is heterophilic rather than homophilic.

Protocadherins are a recently identified subfamily that can contain as many as 60 members in a single species (Wu and Maniatis, 1999; Wu et al., 2001). They are similar to classic cadherins in that they usually contain between five and seven EC domains and a single transmembrane domain, but their cytoplasmic domains are divergent from the classic cadherins. The specific functions of these divergent cytoplasmic domains are still somewhat unclear, but specific domains have been shown to interact with fyn tyrosine kinase (a member of the src family of cytoplasmic tyrosine kinases) (Resh, 1998), protein phosphatase 1, the Disabled 1 adaptor protein and the histone-binding protein TAF1 (TATA box-binding protein associated factor I) (Kohmura et al., 1998; Yoshida et al., 1999; Homayouni et al., 2001; Heggem and Bradley, 2003). The name protocadherin implies that this subfamily may represent the most ancestral branch of the cadherin family, and the demonstration that protocadherins are expressed in *Caenorhabditis elegans* and *Drosophila melanogaster* supports that suggestion (Sano et al., 1993). However, there appear to be a limited number of proteins that are orthologous across species, suggesting that gene duplication may have played a key role in the expansion of the family (Frank and Kemler, 2002).

Finally, a relatively small family of seven-pass transmembrane cadherins has also been identified (Hadjantonakis et al., 1997, 1998; Usui et al., 1999). At present this family is composed of four members, *Drosophila* flamingo and its three vertebrate homologs, Celsr1-3 (cadherin EGF LAG seven-pass receptor) (Formstone and Little, 2001). However, there is considerable interest in these molecules because the seven-pass transmembrane portion of these molecules has some similarity to the serpentine G-protein receptors, suggesting that this molecule could signal through a G-protein–coupled pathway (Hadjantonakis et al., 1998). In addition, flamingo and Celsr1 have been shown to act as planar polarity genes (Usui et al., 1999; Curtin et al., 2003), suggesting that this class of cadherins may be required for the generation of polarized epithelia, although the specific role of these molecules has not been determined yet.

Cadherins were initially identified based on their ability to mediate cell–cell adhesion and to regulate adhesion-related events such as cell sorting and neurite outgrowth (Steinberg and McNutt, 1999). Subsequent studies demonstrated that cadherins can also influence other aspects of cell biology, largely through their intracellular domains (Braga, 2002). The intracellular domains of classic cadherins and desmosomal cadherins contain α-catenin/plakoglobin binding domains that can act to link cadherins to the adherens junction and through that junction to the cytoskeleton. This interaction can obviously have profound effects on different aspects of cellular morphology including apical–basal polarization. Moreover, because β-catenin, an important member of the adherens junction, can also act as a transcription factor under some circumstances (Gottardi and Gombiner, 2001), the gain or loss of adherens junctions can have more far reaching effects, such as epithelial-to-mesenchymal transitions, changes in the proliferative state of cells, or effects on the determination of cell fate (Marrs and Nelson, 1996; Knudsen et al., 1998).

C. Integrins

The final major group of cell adhesion molecules are the integrins. Unlike CAMs and cadherins, integrins mediate adhesion between cells and the extracellular matrix (ECM) (Damsky and Ilic, 2002; Milner and Campbell, 2002). Integrins also differ from the other major adhesion molecules in that they form heterodimers that are comprised of a single α subunit and a single β subunit (Fig. 1). Both subunits are type I transmembrane glycoproteins with large extracellular domains, a single transmembrane region, and, a relatively short intracellular domain. The extracellular domains of both subunits form the integrin head piece that comprises the ligand binding domain (Takagi and Springer, 2002). At present, 16 different α subunits and 8 different β subunits have been identified in mammals. These subunits are used to form at least 24 different heterodimers. Each heterodimer recognizes specific ligands, with some dimers recognizing a broad range of ECM molecules, such as fibronectin, laminin, or collagen, while other dimers only bind to specific ECM matrices or to the immunoglobulin-related intercellular adhesion molecule-1 such as ICAM-1, VCAM, or MadCAM (van de Stolpe and van der Saag, 1996; Bokel and Brown, 2002).

The effects of integrins are mediated through a number of different mechanisms. Since most cells are polarized such that only a limited percentage of the cell comes in contact with the basement membrane, integrins become differentially localized to the region of the cell that is in contact with the ECM (Kraynov et al., 2000; Etienne-Manneville and Hall, 2001; Del Pozo et al., 2002). Integrins can interact with the cytoskeleton through the cytoplasmic domain of the β-subunits, which bind through talin to the actin cytoskeleton (reviewed in Liu et al., 2000). In addition, activation of integrins has been shown to be required for maintained activation of different types of growth factors and may play a role in the differential localization of growth factors within regions of the cell that are located adjacent to the ECM. Integrins can also signal through other pathways, such as focal adhesion kinase or small GTPases, largely through the localization of these factors to specific regions of a cell (Damsky and Ilic, 2002; Braga, 2002).

D. Other Adhesion Molecules

In addition to the major families of adhesion molecules reviewed earlier, there are several other molecules that have been shown to play a role in adhesion, but do not belong specifically to one of the major families. I will discuss two in particular because of their potential role in development of the ear.

11. Cell Adhesion and Development of the Ear

1. BEN/DM–GRASP/SC-1

BEN (bursal epithelium and neurons) is a glycoprotein that was originally identified in chick as a result of an antibody screen (Pourquie et al., 1990). Around the same time, the molecule was also identified by several other laboratories and given the names DM-GRASP (dorsal funiculus and ventral midline-immunoglobulin–like restricted axonal surface protein) and SC-1 (secreted protein acidic and rich in cysteine [SPARC]–related protein) (Johnston et al., 1990; Burns et al., 1991; Tanaka et al., 1991); however, for the purposes of this review it will be referred to as BEN. BEN is an approximately 100 kD protein that is a member of the immunoglobulin superfamily. It is characterized by an extracellular domain that contains V-type and C2-type Ig repeats followed by a single transmembrane domain and a short cytoplasmic domain (Fig. 1). BEN has been shown to bind both homophilically (DeBernardo and Chang, 1995; Corbel et al., 1996) and heterophilically with Ng-CAM, the chick homolog for L1 (Bowen et al., 1995; De Bernardo and Chang, 1996). Homologs for BEN have been identified in fish and mammals, including mice, rats, and humans (Laessing et al., 1994; Campbell et al., 1994). In the developing nervous system, BEN is expressed on subsets of growing neurons, including most peripheral neurons, motor neurons in the spinal cord, and sensory neurons in the dorsal root ganglion (Pourquie et al., 1992; Pollerberg and Mack, 1994). Expression is observed both in the cell body and along developing axons, leading to the suggestion that BEN may play a role in axon outgrowth, guidance, or fasciculation. BEN is also expressed in non-neuronal cells, including myeloid and erythroid cells in the hematopoietic system and immature thymocytes in the immune system (Pourquie et al., 1992).

As discussed, the binding characteristics of BEN suggest a role in cell–cell adhesion. In addition, *in vitro* studies have demonstrated that disruption of homophilic binding of BEN using specific monoclonal antibodies resulted in inhibition of retinal axon outgrowth on existing axons but not on ECM (Pollerberg and Mack, 1994). These results further support a role in cell–cell adhesion and fasciculation. Finally, inhibition of BEN signaling in developing pancreatic cells results in disruptions in the ras and JNK pathways (Stephan et al., 1999), suggesting a possible role in intracellular signaling, although this has not been demonstrated conclusively.

2. Notch

The Notch molecule was originally identified in *Drosophila* as a result of mutations that led to noticeable wing notches in hypomorphic animals (Gaiano and Fishell, 2002). Subsequently, Notch was determined to be a transmembrane protein that is expressed at the cell surface. Notch molecules are characterized by an extracellular domain that contains between 29 and

36 epidermal growth factor (EGF)-like repeats and three Notch/LIN-12 (lineage defective) repeats, a single transmembrane domain, and an intracellular domain that contains ankyrin repeats and a PEST (proline [P], glutamic acid [E], serine [S], and threonine [T]) sequence (Fig. 1). Notch is an ancient molecule that is expressed in all metazoan animals. Most vertebrates appear to contain multiple *notch* genes; for example, there are four *notch* genes in mammals. The Notch molecule is a receptor that is dependent on ligand binding for activation. The ligands for Notch can be divided into two categories: those that are similar in structure to Delta, and those that are similar in structure to Serrate/Jagged (Fleming, 1998). However, in either case, the overall molecular structure is similar, containing an extracellular domain with multiple EGF repeats, a transmembrane domain, and a short intracellular domain (Fig. 1).

The effects of Notch in both embryos and adults have been well documented. Activation of Notch has been shown to play a key role in the determination of cell fate and cellular patterning in many systems, including the ear (Gaiano and Fishell, 2002; Lanford *et al.*, 1999; Haddon *et al.*, 1998, 1999; Riley *et al.*, 1999; Zine *et al.*, 2000; Kiernan *et al.*, 2001; Tsai *et al.*, 2001). However, for the purposes of this review I will only consider Notch/Notch-ligand interactions in terms of their potential role in cellular adhesion. In fact, the interactions between Notch and Notch-ligands were initially determined based on their ability to mediate clustering of two populations of cells, one transfected with Notch and the other transfected with Delta (Fehon *et al.*, 1990). Subsequent studies have demonstrated that these interactions occur through specific EGF repeats on both Notch and Notch-ligands (Rebay *et al.*, 1991). More recently, *in vitro* studies have demonstrated specific modulation of adhesive behaviors in both keratinocytes and endothelial cells expressing specific Notch or Notch-ligand contructs (Lowell and Watt, 2001; Lindner *et al.*, 2001; Leong *et al.*, 2002). In addition, Notch signaling plays a key role in boundary formation during somitogenesis (Pourquie, 2000). Although this effect could be regulated through changes in cell fate, it is also possible that the role of Notch may include regulation of cell sorting through relative changes in adhesion (Henry *et al.*, 2001).

III. Adhesion Molecules and Development of the Inner Ear

A. Expression of Adhesion Molecules in the Otocyst

The inner ear develops from the otic placode, a thickened region of cranial ectoderm, that invaginates to form an otic cup and ultimately pinches off from the surface ectoderm to form an otic vesicle (Rinkwitz *et al.*, 2001; Bryant *et al.*, 2002). Once the vesicle is formed, an elaborate series of morphogenetic changes lead to the development of all the different epithelial and afferent neuronal structures within the ear. It has been suggested that the positions of specific sensory patches or structures may be regulated through the formation of molecular boundaries (Brigande et al., 2000; Fekete and Wu, 2002). At a minimum, the otocyst epithelial cells must become subdivided into neuronal progenitors, prosensory epithelial cells that will give rise to sensory patches, and nonsensory epithelial cells that will give rise to the epithelial cells that line most of the membraneous labyrinth. Following the formation of these subdivisions, the different regions of the labyrinth must undergo complex morphogenetic changes related to the formation of the different aspects of the membraneous labyrinth. These changes are mediated through multiple cellular events, including proliferation, migration, morphogenesis, differentiation, neurite outgrowth, and apoptosis. As discussed, cellular adhesion has been shown to play a role in mediating a number of these cellular events in other systems, suggesting that cell adhesion molecules may also be important for the development of the inner ear.

At present, the number of studies examining expression of adhesion molecules in the ear is small, especially in comparison with the growing number of adhesion molecules. Moreover, most of the studies that have been conducted have focused on later development of the ear, rather than on the earliest stages of otocyst formation. The majority of studies on early development of the ear have been conducted in the chick, and therefore, the description that follows will focus on that system, with comments regarding the expression in other species included where appropriate. The otic placode first begins to develop around Hamburger-Hamilton (HH) stage 9 as a thickening of the surface ectoderm. Just prior to invagination, uniform expression of *notch1* is initiated throughout the placode (Groves and Bronner-Fraser, 2000). In addition, NCAM, E-cadherin, N-cadherin, and B-cadherin are also broadly expressed in the placode at this stage; however, in contrast with *notch1*, the level of expression for these molecules is indistinguishable from the expression in the surrounding ectoderm (Richardson et al., 1987; Raphael et al., 1988; Murphy-Erdosh et al., 1994; Leon et al., 1997; Hrynkow et al., 1998b). Similar patterns of expression have been observed for *notch1*, E-cadherin, N-cadherin and P-cadherin (which may be homologous to chick B-cadherin) in the invaginating mouse otocyst (Nose and Takeichi, 1986; Weinmaster et al., 1991; Kimura et al., 1995).

The otocyst continues to invaginate to form the otic cup by HH stage 14 and the fully enclosed otocyst by HH stage 15. During this time, all cells continue to express *notch1*, NCAM, E- cadherin, N-cadherin, and B-cadherin. In addition, beginning near the end of stage 11, subsets of cells within the otocyst begin to express a ligand for Notch, either *delta1* or *serrate1* (Adam et al., 1998). *Delta1* expression is restricted to a small number of cells located in the anterior region of the otic cup. In contrast, *serrate1* is expressed more broadly in cells located at the medial and posterior rims of the invaginating otic cup. Around the same time, a patch of cells that express BEN is observed in the ventromedial aspect of the otocyst along the length of the anteroposterior axis (Adam et al., 1998; Goodyear et al., 2001). There are conflicting reports regarding the timing of onset of BEN relative to *delta1*. Adam et al. (1998) reported that the onset of BEN expression trailed *delta1* by 6 to 7 hours; however, the pattern of expression reported by Goodyear et al. (2001) suggests that the onset of BEN could coincide with *delta1*. It is also not clear whether the region of BEN expression overlaps with *serrate1*, but their patterns of expression at later developmental time points would certainly suggest an overlap at this point.

By HH stage 15, expression of BEN and NCAM has become restricted to a large patch of cells in the ventromedial aspect of the otocyst while E-cadherin and N-cadherin continue to be uniformly expressed throughout the otocyst (Richardson et al., 1987; Raphael et al., 1988; Goodyear et al., 2001). The BEN/NCAM-positive ventromedial region of the otocyst correlates with the site of delamination of developing neuroblasts and the region of expression of *delta1*. Delaminated neuroblasts are also positive for BEN, NCAM, and N-cadherin but do not express *delta1* (Richardson et al., 1987; Raphael et al., 1988; Adam et al., 1998; Goodyear et al., 2001). This observation led Adam et al. (1998) to suggest that *delta1* might be expressed in uncommitted progenitor cells but downregulated in cells that had become committed to develop as neuroblasts. By HH stage 19, NCAM and BEN are restricted to a ventromedial patch located directly adjacent to the developing ganglion and to a dorsal region that corresponds to the developing endolymphatic duct. E-cadherin and N-cadherin continue to be uniformly expressed throughout the otocyst. Cells within the ganglion continue to express BEN, NCAM, and N-cadherin. In addition, by HH stage 24 these cells begin to express L1 (Richardson et al., 1997; Goodyear et al., 2001). Further development of the ganglion will be discussed in a subsequent section.

As development proceeds, the otocyst becomes regionalized into sensory and nonsensory regions. In general, expression of *serrate1*, BEN, NCAM, and N-cadherin is maintained in regions of the epithelium that will develop as sensory patches (Richardson et al., 1987; Raphael et al., 1988; Adam et al., 1998; Goodyear et al., 2001). In contrast, E-cadherin and B-cadherin expression is downregulated in developing sensory patches but is maintained in regions that will develop as nonsensory regions (Raphael et al., 1988; Murphy-Erdosh et al., 1994). The patterns of expression for *jagged1* (the mammalian homolog of *serrate1*), NCAM, N-cadherin, and E-cadherin appear similar in the mouse otocyst, although existing data for early time points are very limited (Nose and Takeichi, 1986; Terkelson et al., 1989; Kimura et al., 1995; Morrison et al., 1999). In addition, *cadherin-11* has been shown to be expressed in the lateral aspect of the otocyst of the mouse at E9 (Kimura *et al.*, 1995).

Recently, in-depth studies of the expression of *serrate1* and BEN have been completed. The results of these studies provide greater details regarding the expression of these genes and their relationship with the formation of sensory versus nonsensory patches. By HH stage 13, soon after otic placode invagination, *serrate1* is present in anterior and posterior foci with weaker expression in a group of cells connecting the two (Cole *et al.*, 2000). As development continues, weak expression of *serrate1* is maintained throughout the ventral otocyst, while the foci of *serrate1* split into patches of cells that correlate with the developing sensory patches. As a result, by HH stage 21 foci of expression of *serrate1* are present in the developing anterior and posterior cristae as well as in the developing basilar papilla. In the saccule, there is expression of *serrate1* in the ventral portion but not in the dorsal region. By HH stage 26 expression of *serrate1* is restricted to the developing sensory patches. Throughout this developmental period, *notch1* continues to be expressed in all epithelial cells within the otocyst and developing ear, although there is a reduction in expression in the developing sensory epithelia.

The pattern of expression for BEN appears to be similar to *serrate1* although they have not been directly compared beyond HH stage 21 (Adam *et al.*, 1998). In addition, BEN is expressed in the presumptive endolymphatic duct, whereas expression of *serrate1* is never observed in this region of the otocyst (Goodyear *et al.*, 2001). By HH stage 22, BEN is expressed in two distinct vertromedial regions. The first is an anterior region that extends from the developing anterior cristae to the developing basilar papilla and the second is a smaller region that appears to include only the presumptive posterior cristae. BEN expression is maintained in regions that will develop as sensory patches but is downregulated in regions of the epithelium that will develop as nonsensory cells.

The expression of NCAM also appears to mirror *serrate1*; however, data to confirm this pattern of expression are not as complete as for BEN. NCAM is expressed in the region of the otocyst that gives rise to developing neuroblasts, even after delamination is complete, and is maintained in regions that will develop as sensory patches (Richardson *et al.*, 1987). As discussed, NCAM is also expressed in the developing endolymphatic duct at HH stage 21, but this expression appears to be downregulated as development proceeds. Expression of N-cadherin also appears to be similar to *serrate1*, NCAM, and BEN; however, it is not clear whether N-cadherin is expressed in the developing endolymphatic duct (Raphael *et al.*, 1988).

Information on the expression of integrins within the developing otocyst is considerably more limited. Davies and Holley (2002) examined the expression of integrin α3 and α6 and found both to be expressed throughout the mouse otocyst at E10.5. However, an increased region of expression of α6 was noted in the neuroblastic anteriolateral region. Double-labeling with the neuroblast marker NeuroD demonstrated that integrin α6 and NeuroD are co-expressed in cells located within the otocyst but neuroblast cells that had already delaminated were positive for either integrin α6 or NeuroD but not both. Since NeuroD expression persists in developing ganglion cells, the authors suggest that delaminating neuroblasts may downregulate expression of integrin α6. They further suggest that the integrin α6-positive/NeuroD-negative cells located in the ganglion may not be derived from the otocyst; however, further work is clearly required to determine the origin of these cells. Integrin β4 was weakly detected at the same stage in the medial aspect of the otocyst (Davies and Holley, 2002). Based on dimerization partners, this result suggests that integrin α6 exists predominantly in the integrin α6β1 dimer at this time (Milner and Campbell, 2002). As development continues, integrin α6 becomes restricted to a subset of developing ganglion cells and the prosensory patches while integrin α3 becomes localized to regions adjacent to those sensory patches. This basic pattern of expression is maintained through at least P0 with integrin α6 expressed predominantly in sensory patches and integrin α3 expressed predominantly in nonsensory regions. There have been a few other reports of integrin expression in the otocyst; however, many of these exist in abstract form only and have not been published as yet.

B. Adhesion Molecules and Differentiation of Sensory Epithelia

In contrast with the development of the otocyst, a number of studies have examined the expression of adhesion molecules in both the chick basilar papilla and in the mammalian organ of Corti. Many adhesion molecules are expressed in consistent patterns in each species; however, there are some intriguing differences. The basilar papilla can be identified by HH stage 30 as an elongated outpocketing from the otocyst that arises from the ventromedial region of otocyst and extends medially toward the midline. The sensory epithelium can be recognized on the dorsal surface as a thickened region that contains both hair cells and supporting cells. The surrounding lateral and ventral regions of the duct will develop as the nonsensory portions of the papilla, including the hyaline cells, cuboidal cells, tegmentum vasculosum, homogene cells, and clear cells (Richardson et al., 1987).

The developing mammalian cochlea first arises at E11.5 also as an outpocketing from the ventromedial region of the otocyst. By E12.5, a thickened region of the epithelium, referred to as Kollicker's organ, can be identified on the dorsal surface of the duct. As development proceeds,

11. Cell Adhesion and Development of the Ear

Kollicker's organ can be subdivided into the greater and lesser epithelial ridges (GER and LER, respectively) (Kelley and Bianchi, 2001). The boundary between these two domains appears to represent the future location of the tunnel of Corti and it is possible that the developing pillar cells play a role in the constriction that is observed at this location. The organ of Corti develops from a population of cells within Kollicker's organ that appears to straddle the division between the GER and LER. Cells within this population, referred to as prosensory cells, will develop as all of the specialized cell types within the organ of Corti, including inner and outer hair cells, Deiter's cells, and pillar cells. Cells located outside of the prosensory population will develop as the inner and outer sulci depending on whether they are located in the GER (inner sulcus) or LER (outer sulcus).

As discussed, expression of *serrate1*, NCAM, N-cadherin, and BEN is maintained in developing sensory patches, including the basilar papilla, from the otocyst stage onward. However, there is heterogeneity in the expression of these molecules between cell types in the sensory patches. BEN is expressed in both hair cells and supporting cells but begins to be downregulated between stage 36 (E9) and stage 38 (E12) (Goodyear *et al.*, 2001). In contrast, expression of N-cadherin and *serrate1* becomes restricted to supporting cells, although N-cadherin may still be expressed in both cell types through HH stage 38 (Raphael *et al.*, 1988; Adam *et al.*, 1998; Cole *et al.*, 2000). NCAM is expressed in both hair cells and supporting cells through HH stage 35 but becomes restricted to hair cells by HH stage 38 (Richardson *et al.*, 1987). Finally, there is a transient expression of the notch ligand *delta1* between HH stage 27 (E5) and HH stage 38 (Adam *et al.*, 1998). *Delta1* is expressed in cells that will develop as hair cells, and this pattern of expression has led to the suggestion that expression of *delta1* could play a role in the specification of cells as hair cells. In contrast with the molecules described previously, expression of E-cadherin and B-cadherin ultimately becomes restricted to nonsensory regions of the duct.

The patterns of expression for the same group of molecules are similar in the developing organ of Corti; however, some interesting differences have also been described. In contrast with the chick, it is not clear whether NCAM expression is maintained in the cells that will develop as the organ of Corti from the otocyst stage or if expression is reinitiated at a later time point (Terkelson *et al.*, 1989; Whitlon and Rutishauser, 1990). Regardless, NCAM is present in the developing sensory epithelium at E16.5 in both mice and rats (Whitlon and Rutishauser, 1990; Simonneau *et al.*, 2003) (Fig. 2). It is somewhat unclear which specific cell types express NCAM at this point. Whitlon *et al.* (1999) demonstrated that NCAM is expressed predominantly on inner and outer hair cells in the mouse, whereas Simonneau *et al.* (2003) indicated a somewhat broader expression on both hair cells and supporting cells of the PSA form of NCAM in the rat at the same time point. At later

336   Matthew W. Kelley

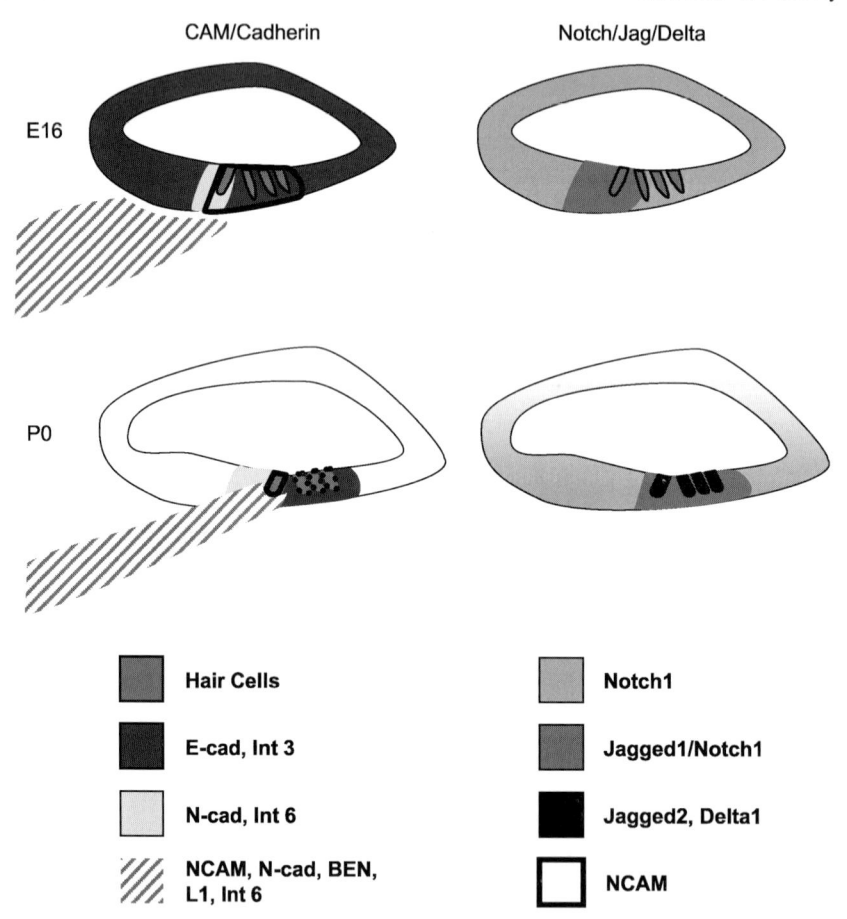

Figure 2 Expression of adhesion molecules in the mammalian cochlear duct during late gestation (E16) and early postnatal (P0) time periods. The developing sensory epithelium is marked by the presence of developing hair cells (white). At E16, E-cadherin and integrin $\alpha3$ are expressed broadly throughout the duct with the exception of a small domain that correlates with the location of the inner hair cell region. At the same time, this region expresses N-cadherin and integrin $\alpha6$. It is not clear whether the domains of N-cadherin and integrin $\alpha6$ exactly overlap. The boundary between domains of N-cadherin and E-cadherin appears to correlate with the boundary between inner and outer hair cell regions. NCAM is expressed throughout the developing sensory epithelium and appears to span the boundary between the N-cadherin and E-cadherin domains. Developing ganglion cells express NCAM, N-cadherin, BEN, L1, and integrin $\alpha6$. At the same time point, *notch1* is broadly expressed throughout the epithelium, while *jagged1* is co-expressed with *notch1* in a subdomain of cells that appears to migrate within the floor of the duct (see text for details). *Jagged2* and *delta1* are expressed in developing hair cells. The degree of overlap between *jagged2/delta1*-positive cells and the *jagged1* domain has not been determined yet. By P0, N-cadherin is still expressed in the developing inner hair cell region. E-cadherin expression has been downregulated in much of the duct but persists in the outer hair cell region. It is not clear whether expression of E-cadherin becomes restricted to hair cells or

developmental time points, in both mice and rats, NCAM is progressively down regulated on all cells except inner hair cells and even in inner hair cells expression becomes restricted to the synaptic zone located at the base of the cell (Fig. 2).

Jagged1 and N-cadherin are both expressed in patterns that suggest a role in boundary formation within the cochlea. At E12.5, *jagged1* is expressed in a broad band of cells located in the neural half of the cochlear duct (Morrison *et al.*, 1999). As development proceeds, the band of *jagged1* expression moves laterally so that by E17, *jagged1* is restricted to the developing organ of Corti (Morrison *et al.*, 1999; Zine *et al.*, 2000). During the same time period, *jagged1* becomes restricted to supporting cells. The movement of the band of expression of *jagged1* from the medial region of the cochlear duct to the more lateral position of the organ of Corti is quite intriguing. This change in position contrasts with results for expression of *serrate1* in the chick basilar papilla, as well as for the expression of *jagged1* in other sensory epithelia within the ear because *jagged1* appears to be initially expressed in a nonsensory region of the duct but then moves to the sensory region (Morrison *et al.*, 1999). Alternatively, it is possible that the cells that will develop as the organ of Corti are initially located in a more medial position within the duct and then move to their final more lateral position. A similar hypothesis has been suggested recently based on the effects of deletion of *fgfr1* (Pirvola *et al.*, 2002); however, the expression of a number of markers of the sensory epithelium, including *math1* and $p27^{kip1}$, suggests that the cells that will develop as the sensory epithelium do not move along the mediolateral axis of the duct (Bermingham *et al.*, 1999; Lanford *et al.*, 2000; Chen and Segil, 1999; Chen *et al.*, 2002).

In addition to *jagged1*, *delta1* is also expressed in the developing cochlea in a pattern that is very similar to the one described in the chick (Morrison *et al.*, 1999). As individual cells within the cochlea begin to develop as hair cells, these cells begin to express *delta1* (Fig. 2). Moreover, in mammals these cells also express a second Notch-ligand, *jagged2* (Lanford *et al.*, 1999). As was discussed for the chick, expression of *delta1* and *jagged2* is believed to play a key role in the determination of cells as hair cells, and in fact,

---

supporting cells within the outer hair cell region. NCAM is still expressed in both inner and outer hair cells; however, expression in outer hair cells is decreasing (light gray color). Expression of *notch1* has been downregulated in some regions of the cochlear duct by P0, but expression persists throughout the floor of the duct, including supporting cells in the organ of Corti. Supporting cells also express *jagged1*. Hair cells have specifically upregulated expression of *jagged2* and *delta1* and downregulated expression of *notch1* and *jagged1*. Expression of NCAM, N-cadherin, BEN, L1, and integrin α6 persists in the developing spiral ganglia. See text and references cited for further details. Note that in order to illustrate as many molecules in a single drawing at single time period, the expression data for some molecules were estimated from results from somewhat different time points.

deletion of *jagged2* results in an overproduction of hair cells in the organ of Corti (Lanford et al., 1999).

N-cadherin is also expressed in the developing cochlea, but unlike the pattern of expression for N-cadherin in the basilar papilla, it is not expressed throughout the developing sensory epithelium. At E16 in rat, N-cadherin expression is restricted to a narrow region of the duct that appears to correlate with the developing inner hair cells and a small number of adjacent cells within the GER (Simonneau et al., 2003) (Fig. 2). At E19.5, N-cadherin expression has been downregulated in developing inner phalangeal cells but is still expressed in inner hair cells and in undifferentiated cells in the GER. At P4, N-cadherin expression has expanded to include approximately two-thirds of the GER and inner hair cells; however, there is still an abrupt boundary of N-cadherin expression that corresponds with the developing pillar cells (Fig. 2). It should be noted that a previous report found ACAM (the chick homolog of N-cadherin) to only be expressed in the developing pillar cells in the gerbil cochlea between P2 and P12 (Nakazawa et al., 1996). It is unclear why these results differ from those reported by Simonneau et al. (2003); however, the antibodies used in the ACAM study was raised against chicken ACAM isolated from cardiac muscle. Given the number of CAMs and the high degree of homology between them, it seems possible that the antibody used might have in fact reacted with a different CAM in the gerbil cochlea.

The expression of N-cadherin is particularly intriguing in light of the embryonic pattern of expression for E-cadherin. At E15 in the mouse and E16 in rat, E-cadherin is expressed in developing outer hair cells, Deiter's cells, Hensen's cells, and possibly pillar cells (Whitlon, 1993; Whitlon et al., 1999; Simonneau et al., 2003). In addition, E-cadherin is expressed broadly within the rest of the cochlear duct with the exception of the region that will develop as inner hair cells (Fig. 2). As development proceeds, E-cadherin expression is maintained in the outer hair cell region of the organ of Corti, as well as in the rest of the cochlear duct. However, there is a specific downregulation in the GER that correlates with the increased expression of N-cadherin in the same region (Simonneau et al., 2003). The differential pattern of expression for N- cadherin and E-cadherin between inner and outer hair cell regions of the developing cochlea suggests that these two regions of the duct may become unique at a fairly early time point during the development of the cochlea (Simonneau et al., 2003). The morphological and physiological characteristics of the inner and outer hair cell domains are distinct; however, the factors that regulate these distinctions are unknown. The expression patterns for N-cadherin and E-cadherin appear to act as early indicators of the generation of those distinctions.

There are some discrepancies in the pattern of expression for E-cadherin during late embryogenesis and the early postnatal period. Whitlon et al.

(1993; 1999) report that in the mouse E-cadherin becomes restricted to developing Deiter's cells and pillar cells by P2. In contrast, Simonneau et al. (2002) illustrate that by E19.5 in the rat expression of E-cadherin is restricted to developing outer hair cells and Hensen's cells and has been downregulated in Deiter's cells and pillar cells. There does not appear to be an obvious explanation for the differences between these two reports. However, it is possible that species differences could exist. Alternatively, the antibodies used could recognize different forms of E-cadherin, suggesting that expression might be maintained in both hair cells and supporting cells in the outer hair cell region, but that the form of E-cadherin expressed could be differentially regulated in different cell types.

The expression of T-cadherin (cadherin 13) has recently been described in the developing organ of Corti. T-cad is a member of the TAG-1 (transiently expressed axonal glycoprotein) family of CAMs and is characterized by the absence of the transmembrane and cytoplasmic domains. Instead, T-cad is linked to the plasma membrane through a glycosylphosphatidylinositol (GPI) anchor. The functions of T-cad are still poorly understood, although the results of several studies have suggested that it may play a greater role in cell signaling than in cell adhesion. In the developing rat cochlea T-cad is only expressed on pillar cells between P0 and at least P12 (Simonneau et al., 2003). Antibody labeling indicated that T-cad is expressed both at the cell surfaces and within the cytoplasm.

Finally, as discussed in the previous section, integrin $\alpha 6$ is expressed in both developing hair cells and supporting cells within the organ of Corti while integrin $\alpha 3$ is expressed in the GER, LER, as well as the developing spiral ligament, stri vascularis, and Reissner's membrane (Davies and Holley, 2002). In addition, a recent study reported that at E16.5 hair cells express mRNA for integrins $\alpha 6$, $\alpha 1$, $\alpha 2$, αv, $\alpha 8$, and $\alpha 3$ (Littlewood Evans and Muller, 2000). The result for integrin $\alpha 3$ appears to conflict with the antibody data from Davies and Holley (2002) (Fig. 2); however, it is possible that the cells express low levels of mRNA that fail to lead to the formation of detectable amounts of protein. The same study also demonstrated that mRNA for intergrin $\beta 1$ is expressed in all hair cells at E16.5.

The results described in the previous two sections are consistent with a potential role for adhesion molecules in multiple aspects of the development of the inner ear. In particular, the observation that many adhesion molecules are expressed in distinct domains within the developing otocyst and cochlea strongly suggests that differential adhesion could play a role in either the regionalization of the epithelium or in the subsequent morphogenesis of different regions of the ear. However, at present extremely limited functional data are available to examine these possibilities. Injection of function-blocking antibodies to NCAM into the mesenchmal region surrounding the chick otic cup at HH stage 11–12 results in defects in the subsequent involution and folding of the cup and in the pinching off from the surface ectoderm to form an encapsulated otocyst (Brown et al., 1998). Considering that NCAM is broadly expressed throughout the otocyst at this stage, these results suggest that NCAM could play a role in morphogenetic changes required for closing of the otocyst. A similar requirement for cellular adhesion has been demonstrated during neural tube folding and closure (Bronner-Fraser et al., 1992), and NCAM is expressed in the neural tube (Tosney et al., 1986; Balak et al., 1997; Zorn and Krieg, 1992; Bally-Cuif et al., 1993). However, no defects in the auditory system have been reported in NCAM knockout mice (Tomasiewicz et al., 1993; Ono et al., 1994), although that does mean that they do not exist. Similarly, inhibition of integrin-ECM interactions using antibodies against laminin, fibronectin, integrin β, or integrin $\alpha 6$ can also perturb the initial invagination of the otocyst (Visconti and Hilfer, 2002).

Targeted mutations have been made in a number of adhesion molecules and at present none of these deletions has been reported to lead to defects in the auditory system. This result does not mean that adhesion molecules are not important for auditory system development. For a number of these deletions, including E-cadherin, N-cadherin, and integrins $\alpha 4$, $\alpha 5$, αv, and $\beta 1$, the animals die prior to formation of the otocyst (Yang et al., 1993, 1995; Larue et al., 1994; Stephens et al., 1995; Radice et al., 1997; McCarty et al., 2002). In addition, functional redundancy may also act to prevent significant defects in animals in which only a single adhesion molecule has been deleted. Moreover, for some of these targeted mutations, defects in the auditory system may be present but an appropriate analysis has not been completed yet.

A recent *in vitro* study has examined the potential role of N-cadherin in the regulation of supporting cell proliferation in the adult chicken. Warchol (2002) isolated epithelial sheets containing hair cells and supporting cells from utricles and established them *in vitro*. Supporting cells located at the outer edges of these explants actively proliferated and, as reported previously, expressed N-cadherin at their cell–cell junctions. To determine the role of N-cadherin, latex beads coated with function-blocking N-cadherin antibodies were applied to the outer edges of these cultures. Neutralization of N-cadherin binding resulted in a decrease in the overall level of proliferation, suggesting that contact inhibition mediated through adherens junctions may play a role in regulating levels of proliferation in the ear. In a second study, Hackett et al. (2002) overexpressed E-cadherin in a cell line derived from a vestibular progenitor cell. Cells expressing E-cadherin were partially inhibited from differentiating as hair cells, suggesting a potential role for E-cadherin in determination of cell fate.

Finally, two spontaneous mutations that affect the Notch signaling pathway have an effect on inner ear development; however, it is difficult to conclude whether these defects are a result of disruptions in adhesion or cell fate. The mouse slalom and head-turner mutants both arise as a result of missense mutations in the extracellular domain of Jagged1 (Kiernan et al., 2001; Tsai et al., 2001). As a result of this mutation, the overall size of the organ of Corti in decreased in these mice. In particular the number of rows of outer hair cells is reduced from three to two. Similarly, preliminary results suggest that targeted deletion of *jagged1* may result in the loss of the posterior semicircular canal cristae. Both of these phenotypes could arise as a result of defects in differential adhesion during the initial specification of sensory versus nonsensory regions of the otocyst. However, defects in specification of cell fate could also give rise to the same phenotypes.

The zebrafish mind bomb mutation also results in a disruption in the Notch signaling pathway as a result of a defect in a ubiquitin ligase that is required for Notch activation (Itoh et al., 2003). Analysis of the inner ears in mind bomb mutants demonstrated that the hair cells are not retained within the epithelium and instead are extruded (Haddon et al., 1999). Obviously, these can be considered to be consistent with a defect in cellular adhesion. However, it is also possible that the defect in adhesion is secondary to the lack of supporting cells within the epithelium, suggesting that the primary affect of Notch signaling is on determination of cell fate.

In summary, at this point, the patterns of expression for many adhesion molecules during the regionalization of the otocyst epithelium certainly suggest a potential role in a number of different cellular events. In particular the mutually exclusive expression patterns for N-cadherin/NCAM/BEN versus E-cadherin/B(P)-cadherin strongly suggest a potential role in the specification of different regions of the otocyst as either sensory or nonsensory. However, from the existing data it is impossible to determine whether any of these molecules actually influence the sensory versus nonsensory decision or if instead the differential expression of these molecules reflects a cell fate decision that has already been made. In that case the role of these adhesion molecules might be to maintain clusters of cells that have become determined to develop with similar fates or to regulate additional aspects of these cells, such as proliferation, that are differentially regulated between sensory and nonsensory regions. Perhaps one of the most intriguing recent results is the demonstration that the expression domains for N-cadherin and E-cadherin mark the developing inner and outer hair cell regions of the mammalian cochlea (Simonneau et al., 2003). Despite overall similarities in gross morphology, inner and outer hair cells are distinct at both a physiological and structural level (Ashmore and Mammano, 2001). Moreover, outer hair cells are fundamentally different from all mammalian hair cells in that they are motile. The observation that these cells initially express E-cadherin, a molecule that is expressed in predominantly nonsensory regions of the ear, raises the question of whether the development of these

C. Adhesion and Neurite Growth in the Inner Ear

As discussed, there are significant data demonstrating that adhesion molecules play a key role in several aspects of neuronal development, including neurite outgrowth and guidance, neuronal fasciculation, and synapse formation. Based on these findings, it seems very likely that adhesion molecules probably play an important role in the development of acoustic and vestibular ganglion cells. As discussed in the section on development of the otocyst, delaminating neuroblasts in the chick express NCAM, BEN, and N-cadherin (Richardson et al., 1987; Raphael et al., 1988; Hrynkow et al., 1998b). In addition, soon after delamination these cells begin to express L1 (approximately stage 19) (Richardson et al., 1987; Mbiene et al., 1989; Goodyear et al., 2001). The timing of L1 expression correlates with the onset of neurite outgrowth. All four of these adhesion molecules are strongly expressed on the processes that developing ganglion cells extend back into the otocyst. Similar patterns of expression have been observed in the mammalian ear; however, the data on expression in recently delaminated neuroblasts are not as complete for mammals (Mothe and Brown, 2001).

The results of several *in vitro* studies have examined the potential role of adhesion molecules in the extension of neurites from chick cochleovestibula ganglion cells and rat spiral ganglion cells. Hrynkow et al. (1998a) established cultures of cochleo vestibular ganglion cells from HH stage 16 to HH stage 21 chicks either in isolation or in co-culture with age-matched otocysts. To examine the effects of NCAM and L1, function-blocking antibodies to both molecules were added to the culture medium. In addition, to examine the effects of PSA modifications of NCAM, PSA was removed from NCAM by treatment with endoneuraminadase-N. As would be expected if cell–cell adhesion is inhibited, treatment with anti-NCAM or anti-L1 or loss of PSA significantly inhibited neurite fasciculation. Moreover, there was an additive effect of treatment with antibodies to both NCAM and L1, suggesting that these molecules may be functionally redundant to some extent. In co-cultures, an increase in neurite ramification was observed after entry into the developing epithelium in the presence of NCAM or L1 antibodies; however, this result may also simply be a result of decreased cell–cell adhesion, leading to extended ramification. In addition to fasciculation, removal of PSA also resulted in an increase in the number of neurites that entered the epithelium. This result was somewhat unexpected since the presence of PSA has been shown to inhibit the effects of NCAM; however, the authors suggest that differential amounts of PSA-NCAM could play a role in some aspects of the tonotopic innervation of the basilar papilla. Further work is clearly needed to address this possibility.

Using a similar experimental design, Aletsee *et al.* (2001) examined the role of integrins, and in particular the $\alpha v \beta 3$ heterodimer, by exposing developing spiral ganglion neurons to culture media containing kistrin, a specific inhibitor of $\alpha v \beta 3$ intergrin heterodimers that is produced in some snake venom. Results indicated that kistrin inhibited the number and length of neurites that were extended from spiral ganglion explants grown on the ECM molecule laminin. While these results are consistent with a role for integrins in neurite outgrowth, it is important to consider that expression of αv and $\beta 3$ integrin in developing spiral ganglion neurons has not been examined yet. That qualification aside, the results are consistent with the hypothesis that interactions between neurites and the ECM regulate outgrowth and extension while subsequent cell–cell interactions regulated through CAMs and cads play a role in the fasciculation of secondary neurites along existing neurite tracks.

D. Adhesion Molecules and Development of Stereocilia Bundles

Perhaps one of the most surprising discoveries related to the role of adhesion molecules and development of the inner ear arose from reverse genetic analysis of deaf mouse mutants. In 2001, the waltzer and Ames waltzer strains of mice, each of which include both hearing and balance disorders, were determined to be caused by mutations in cadherin 23 (Cdh23) and protocadherin 15 (Pcdh15), respectively (DiPalma *et al.*, 2001; Wilson *et al.*, 2001; Alagramam *et al.*, 2001a). More recently, mutations in *CDH23* have been shown to be the cause of Usher's syndrome type 1D in humans (Bork *et al.*, 2001; Bolz *et al.*, 2001; Astuto *et al.*, 2002) and mutations in *PCDH15* have been shown to lead to Usher's syndrome type 1F (Ahmed *et al.*, 2001; Alagramam *et al.*, 2001b; Ben-Yosef *et al.*, 2003).

The basis of the inner ear defects in both of these strains of mice is somewhat suprising. Auditory brainstem responses for Ames waltzer mice were consistent with a peripheral defect; however, overall formation of the organ of Corti was normal at birth, including the presence of a full complement of hair cells and supporting cells in both the cochlea and vestibular sensory epithelia (Alagramam *et al.*, 2000). Spiral ganglion neurons were also present and appeared normal. In contrast, analysis of hair cell stereociliary bundles indicated a significant level of disorganization in the formation of the bundle (Alagramam *et al.*, 2000; Raphael *et al.*, 2001). In particular, the normal shape and cohesion of the bundles was significantly disrupted. In many cases the characteristic staircase arrangement of stereocilia was not present and the overall level of filamentous actin in the hair cells appeared to be significantly decreased. As a result of these defects, hair cells begin to appear pathological within a few weeks after birth, and most hair cells and spiral ganglion neurons die within the first few months of life. Interestingly, expression of *Pcdh15* has been detected in the cochlear and vestibular sensory epithelia as early as E12 (Murcia and Woychik, 2001); however, the effects of mutations in this gene are not observed until the postnatal period.

The basis of the defect in waltzer mice containing a mutation in *Cdh23* appears to be similar to the defect in Ames waltzer. Overall development of the cochlear and vestibular sensory epithelia appeared normal with a full complement of inner and outer hair cells at E18.5; however, defects in stereociliary bundle formation were already evident (DiPalma *et al.*, 2001). Stereocilia were poorly organized, had defects in length, and were decreased in number. The present understanding of the roles of both *Pcdh15* and *Cdh23* is still limited. However, the results of recent studies have suggested that *Chd23* is localized to the developing stereocilia through an interaction with harmonin, a linking protein that interacts with both the actin cytoskeleton and with *Cdh23* (Boeda *et al.*, 2002; Siemens *et al.*, 2002). It is suggested that the extracellular domains of *Cdh23* could then act to link individual developing stereocilia to their neighbors. At this point, it is not clear whether *Pcdh15* might play a similar role during stereociliary bundle development. However, since single mutations in either *Cdh23* or *Pcdh15* are sufficient to cause stereociliary bundle defects, it seems unlikely that these two proteins act redundantly.

The demonstration that mutations in two cadherin molecules leads to specific defects in the formation of the stereociliary bundle demonstrates a new role for adhesion molecules. The unique structure of the hair cell stereociliary bundle apparently requires a unique level of organization. It seems likely that further studies on the biochemical and cellular nature of the effects of *Cdh23* and *Pcdh15* will lead to exciting insights into the unique development of hair cells.

The results of a single study have identified an intriguing role for integrin $\alpha 8$ in stereocilia development. Animals with a targeted deletion of integrin $\alpha 8$ have specific defects in the formation of stereociliary bundles, including immature and fused stereocilia (Littlewood Evans and Muller, 2000). However, although integrin $\alpha 8$ is expressed in all hair cells, stereociliary bundle defects were only observed in the utricle. These results demonstrate a role for integrins in stereocilia formation but suggest that functional redundancy may exist in most hair cell epithelia.

Finally, *Celsr1*, one of the mammalian homologs for *Drosophila* flamingo, has recently been demonstrated to play a role in the establishment of planar polarity in the cochlea (Curtin *et al.*, 2003). The stereocilia on all cochlear hair cells are oriented such that the vertex of the bundle is located on the edge of the cell located closest to the strial edge. Since all stereociliary bundles are directionaly sensitive (Hudspeth and Jacobs, 1979), the development of this orientation is necessary for normal auditory perception (Yoshida and Liberman, 1999). The factors that regulate the formation of this polarization have not been determined; however, similar polarized structures have been identified in *Drosophila*. Analysis of different mutations in flies has identified a group of molecules that regulate polarization in this species (Mlodzik, 2002). These include the cadherin Flamingo, as well as another transmembrane protein (Van Gogh), and the secreted molecule, Wingless. Three recent publications have demonstrated similar roles for mammalian homologs of each of these molecules in the polarization of cochlear hair cell stereociliary bundles (Curtin *et al.*, 2003; Dabdoub *et al.*, 2003; Montcouquiol *et al.*, 2003). In each of these studies, the overall development of stereociliary bundles was normal but the orientation, in particular on outer hair cells, was significantly disrupted. In some regions, stereociliary bundles were randomly oriented around the circumference of the hair cells (Montcouquiol *et al.*, 2003). These results implicate adhesion as an important factor in orienting the stereociliary bundle, but the specific effects of *Celsr1* have not been determined. Both Flamingo and Van Gogh become asymmetrically localized to the distal cell membrane during polarization of the *Drosophila* wing, suggesting that cell–cell adhesion could play a role in polarization (Shimada *et al.*, 2001; Bastock *et al.*, 2003). Similarly, mutations in *vangl2*, a mammalian *van gogh* homolog, lead to defects in the initial movement of the developing kinocilium to the distal region of each developing hair cell (Montcouquiol *et al.*, 2003). Since the movement of the kinocilium represents the earliest sign of stereociliary bundle polarization (Anniko *et al.*, 1979), this result suggests that adhesion could also play a role in directing the kinocilium, and therefore the stereociliary bundle, to the distal region of each hair cell.

IV. Summary

Cellular adhesion plays a key role in a number of unique developmental events, including proliferation, cell fate, morphogenesis, neurite outgrowth, fasciculation, and synaptogensis. The number of families of molecules that can mediate cell adhesion and the number of members of each of those families has continued to increase over time. Moreover, the potential for the formation of different pairs of heterodimers with different binding specificities, and for both homo- and hetero-dimeric interactions suggest that a vast number of specific signaling events can be mediated through the expression of different combinations of adhesion factors at different developmental time points.

By comparison with the number of known adhesion molecules and their potential effects, our understanding of the role of adhesion in ear development is extremely limited. The patterns of expression for some adhesion molecules have been determined for some aspects of inner ear development. Similarly, with a few exceptions, functional data to indicate the roles of these adhesion molecules are also lacking. However, a consideration of even the limited existing data must lead to the conclusion that adhesion molecules play key roles in all aspects of the development of the auditory system. Unique expression domains for different groups of adhesion molecules within the developing otocyst and ear strongly suggest a role in the determination of different cellular domains. Similarly, the specific expression of adhesion molecules on developing neurites and their target hair cells, suggests a key role for adhesion in the establishment of neuronal connections and possible the development of tonotopy. Finally, the recent demonstration that *Cdh23* and *Pcdh15* play specific roles in the formation of the hair cell stereociliary bundle provides compelling evidence for the importance of adhesion molecules in the development of stereocilia. With the imminent completion of the mouse genome, it seems likely that the number of adhesion molecules can soon be fixed and that it will then be possible to generate a more comprehensive map of expression of these molecules within the developing inner ear. At the same time, the generation of new transgenic and molecular technologies promises to provide researchers with new tools to examine the specific effects of different adhesion molecules during inner ear development.

Acknowledgments

The author wishes to thank Dr. Mireille Montcouquiol for reading an earlier version of the manuscript and Dr. Pamela Lanford for helpful conversations on the role of notch and adhesion. Supported by funds from the Intramural Program at NIDCD/NIH.

References

Adam, J., Myat, A., Le Roux, I., Eddison, M., Henrique, D., Ish-Horowicz, D., and Lewis, J. (1998). Cell fate choices and the expression of Notch, Delta and Serrate homologues in the chick inner ear: parallels with Drosophila sense-organ development. *Development* **125,** 4645–4654.
Ahmed, Z. M., Riazuddin, S., Bernstein, S. L., Ahmed, Z., Khan, S., Griffith, A. J., Morell, R. J., Friedman, T. B., and Wilcox, E. R. (2001). Mutations of the protocadherin gene PCDH15 cause Usher syndrome type 1F. *Am. J. Hum. Genet.* **69,** 25–34.

Alagramam, K. N., Zahorsky-Reeves, J., Wright, C. G., Pawlowski, K. S., Erway, L. C., Stubbs, L., and Woychik, R. P. (2000). Neuroepithelial defects of the inner ear in a new allele of the mouse mutation Ames waltzer. *Hear Res.* **148,** 181–191.

Alagramam, K. N., Murcia, C. L., Kwon, H. Y., Pawlowski, K. S., Wright, C. G., and Woychik, R. P. (2001). The mouse Ames waltzer hearing-loss mutant is caused by mutation of Pcdh15, a novel protocadherin gene. *Nat. Genet.* **27,** 99–102.

Alagramam, K. N., Yuan, H., Kuehn, M. H., Murcia, C. L., Wayne, S., Srisailpathy, C. R., Lowry, R. B., Knaus, R., Van Laer, L., Bernier, F. P., *et al.* (2001). Mutations in the novel protocadherin PCDH15 cause Usher syndrome type 1F. *Hum. Mol. Genet.* **10,** 1709–1718.

Aletsee, C., Mullen, L., Kim, D., Pak, K., Brors, D., Dazert, S., and Ryan, A. F. (2001). The disintegrin kistrin inhibits neurite extension from spiral ganglion explants cultured on laminin. *Audiol. Neurootol.* **6,** 57–65.

Anniko, M., Nordemar, H., and Wersall, J. (1997). Genesis and maturation of vestibular hair cells. *Adv. Otorhinolaryngol.* **25,** 7–11.

Appel, F., Holm, J., Conscience, J. F., von Bohlen und Halbach, F., Faissner, A., James, P., and Schachner, M. (1995). Identification of the border between fibronectin type III homologous repeats 2 and 3 of the neural cell adhesion molecule L1 as a neurite outgrowth promoting and signal transducing domain. *J. Neurobiol.* **28,** 297–312.

Ashmore, J. F., and Mammano, F. (2001). Can you still see the cochlea for the molecules? *Curr. Opin. Neurobiol.* **11,** 449–454.

Asou, H. (1992). Monoclonal antibody that recognizes the carbohydrate portion of cell adhesion molecule L1 influences calcium current in cultured neurons. *J. Cell Physiol.* **153,** 313–320.

Astuto, L. M., Bork, J. M., Weston, M. D., Askew, J. W., Fields, R. R., Orten, D. J., Ohliger, S. J., Riazuddin, S., Morell, R. J., Khan, S., *et al.* (2002). CDH23 mutation and phenotype heterogeneity: A profile of 107 diverse families with Usher syndrome and nonsyndromic deafness. *Am. J. Hum. Genet.* **71,** 262–275.

Balak, K., Jacobson, M., Sunshine, J., and Rutishauser, U. (1987). Neural cell adhesion molecule expression in *Xenopus* embryos. *Dev. Biol.* **119,** 540–550.

Bally-Cuif, L., Goridis, C., and Santoni, M. J. (1993). The mouse NCAM gene displays a biphasic expression pattern during neural tube development. *Development* **117,** 543–552.

Barthels, D., Santoni, M. J., Wille, W., Ruppert, C., Chaix, J. C., Hirsch, M. R., Fontecilla-Camps, J. C., and Goridis, C. (1987). Isolation and nucleotide sequence of mouse NCAM cDNA that codes for a Mr 79,000 polypeptide without a membrane-spanning region. *Embo. J.* **6,** 907–914.

Bastock, R., Strutt, H., and Strutt, D. (2003). Strabismus is asymmetrically localised and binds to Prickle and Dishevelled during *Drosophila* planar polarity patterning. *Development* **130,** 3007–3014.

Beggs, H. E., Soriano, P., and Maness, P. F. (1994). NCAM-dependent neurite outgrowth is inhibited in neurons from Fyn-minus mice. *J. Cell Biol.* **127,** 825–833.

Beggs, H. E., Baragona, S. C., Hemperly, J. J., and Maness, P. F. (1997). NCAM140 interacts with the focal adhesion kinase p125(fak) and the SRC-related tyrosine kinase p59(fyn). *J. Biol. Chem.* **272,** 8310–8319.

Ben-Yosef, T., Ness, S. L., Madeo, A. C., Bar-Lev, A., Wolfman, J. H., Ahmed, Z. M., Desnick, R. J., Willner, J. P., Avraham, K. B., Ostrer, H., *et al.* (2003). A mutation of PCDH15 among Ashkenazi Jews with the type 1 Usher syndrome. *N. Engl. J. Med.* **348,** 1664–1670.

Bermingham, N. A., Hassan, B. A., Price, S. D., Vollrath, M. A., Ben-Arie, N., Eatock, R. A., Bellen, H. J., Lysakowski, A., and Zoghbi, H. Y. (1999). Math1: An essential gene for the generation of inner ear hair cells. *Science* **284,** 1837–1841.

Boeda, B., El-Amraoui, A., Bahloul, A., Goodyear, R., Daviet, L., Blanchard, S., Perfettini, I., Fath, K. R., Shorte, S., Reiners, J., et al. (2002). Myosin VIIa, harmonin and cadherin 23, three Usher I gene products that cooperate to shape the sensory hair cell bundle. *Embo. J.* **21**, 6689–6699.

Bokel, C., and Brown, N. H. (2002). Integrins in development: moving on, responding to, and sticking to the extracellular matrix. *Dev. Cell* **3**, 311–321.

Bolz, H., von Brederlow, B., Ramirez, A., Bryda, E. C., Kutsche, K., Nothwang, H. G., Seeliger, M., del, C. S. C. M., Vila, M. C., Molina, O. P., et al. (2001). Mutation of CDH23, encoding a new member of the cadherin gene family, causes Usher syndrome type 1D. *Nat. Genet.* **27**, 108–112.

Bork, J. M., Peters, L. M., Riazuddin, S., Bernstein, S. L., Ahmed, Z. M., Ness, S. L., Polomeno, R., Ramesh, A., Schloss, M., Srisailpathy, C. R., et al. (2001). Usher syndrome 1D and nonsyndromic autosomal recessive deafness DFNB12 are caused by allelic mutations of the novel cadherin-like gene CDH23. *Am. J. Hum. Genet.* **68**, 26–37.

Bowen, M. A., Patel, D. D., Li, X., Modrell, B., Malacko, A. R., Wang, W. C., Marquardt, H., Neubauer, M., Pesando, J. M., Francke, U., et al. (1995). Cloning, mapping, and characterization of activated leukocyte-cell adhesion molecule (ALCAM), a CD6 ligand. *J. Exp. Med.* **181**, 2213–2220.

Brackenbury, R., Thiery, J. P., Rutishauser, U., and Edelman, G. M. (1977). Adhesion among neural cells of the chick embryo. I. An immunological assay for molecules involved in cell-cell binding. *J. Biol. Chem.* **252**, 6835–6840.

Braga, V. M. (2002). Cell-cell adhesion and signalling. *Curr. Opin. Cell Biol.* **14**, 546–556.

Brigande, J. V., Kiernan, A. E., Gao, X., Iten, L. E., and Fekete, D. M. (2000). Molecular genetics of pattern formation in the inner ear. Do compartment boundaries play a role? *Proc. Natl. Acad. Sci. USA* **97**, 11700–11706.

Bronner-Fraser, M., Wolf, J. J., and Murray, B. A. (1992). Effects of antibodies against N-cadherin and N-CAM on the cranial neural crest and neural tube. *Dev. Biol.* **153**, 291–301.

Brown, J. W., Beck-Jefferson, E., and Hilfer, S. R. (1998). A role for neural cell adhesion molecule in the formation of the avian inner ear. *Dev. Dyn.* **213**, 359–369.

Bryant, P. J., Huettner, B., Held, L. I., Jr., Ryerse, J., and Szidonya, J. (1988). Mutations at the fat locus interfere with cell proliferation control and epithelial morphogenesis in *Drosophila*. *Dev. Biol.* **129**, 541–554.

Bryant, J., Goodyear, R. J., and Richardson, G. P. (2002). Sensory organ development in the inner ear: Molecular and cellular mechanisms. *Br. Med. Bull.* **63**, 39–57.

Burns, F. R., von Kannen, S., Guy, L., Raper, J. A., Kamholz, J., and Chang, S. (1991). DM-GRASP, a novel immunoglobulin superfamily axonal surface protein that supports neurite extension. *Neuron* **7**, 209–220.

Campbell, I. G., Foulkes, W. D., Senger, G., Trowsdale, J., Garin-Chesa, P., and Rettig, W. J. (1994). Molecular cloning of the B-CAM cell surface glycoprotein of epithelial cancers: A novel member of the immunoglobulin superfamily. *Cancer Res.* **54**, 5761–5765.

Chen, P., and Segil, N. (1999). p27(Kip1) links cell proliferation to morphogenesis in the developing organ of Corti. *Development* **126**, 1581–1590.

Chen, P., Johnson, J. E., Zoghbi, H. Y., and Segil, N. (2002). The role of Math1 in inner ear development: Uncoupling the establishment of the sensory primordium from hair cell fate determination. *Development* **129**, 2495–2505.

Clark, H. F., Brentrup, D., Schneitz, K., Bieber, A., Goodman, C., and Noll, M. (1995). Dachsous encodes a member of the cadherin superfamily that controls imaginal disc morphogenesis in *Drosophila*. *Genes Dev.* **9**, 1530–1542.

Cole, G. J., Loewy, A., and Glaser, L. (1986). Neuronal cell-cell adhesion depends on interactions of N-CAM with heparin-like molecules. *Nature* **320**, 445–447.

11. Cell Adhesion and Development of the Ear

Cole, G. J., and Burg, M. (1989). Characterization of a heparan sulfate proteoglycan that copurifies with the neural cell adhesion molecule. *Exp. Cell Res.* **182,** 44–60.

Cole, L. K., Le Roux, I., Nunes, F., Laufer, E., Lewis, J., and Wu, D. K. (2000). Sensory organ generation in the chicken inner ear: Contributions of bone morphogenetic protein 4, serrate1, and lunatic fringe. *J. Comp. Neurol.* **424,** 509–520.

Corbel, C., Pourquie, O., Cormier, F., Vaigot, P., and Le Douarin, N. M. (1996). BEN/SC1/DM-GRASP, a homophilic adhesion molecule, is required for in vitro myeloid colony formation by avian hemopoietic progenitors. *Proc. Natl. Acad. Sci. USA* **93,** 2844–2849.

Cremer, H., Chazal, G., Carleton, A., Goridis, C., Vincent, J. D., and Lledo, P. M. (1998). Long-term but not short-term plasticity at mossy fiber synapses is impaired in neural cell adhesion molecule-deficient mice. *Proc. Natl. Acad. Sci. USA* **95,** 13242–13247.

Crossin, K. L., and Krushel, L. A. (2000). Cellular signaling by neural cell adhesion molecules of the immunoglobulin superfamily. *Dev. Dyn.* **218,** 260–279.

Cunningham, B. A., Hemperly, J. J., Murray, B. A., Prediger, E. A., Brackenbury, R., and Edelman, G. M. (1987). Neural cell adhesion molecule: structure, immunoglobulin-like domains, cell surface modulation, and alternative RNA splicing. *Science* **236,** 799–806.

Curtin, J. A., Quint, E., Tsipouri, V., Arkell, R. M., Cattanach, B., Copp, A. J., Henderson, D. J., Spurr, N., Stanier, P., Fisher, E. M., Nolan, P. M., Steel, K. P., Brown, S. D. M., Gray, I. C., and Murdoch, J. N. (2003). Mutation of *Celsr1* disrupts planar polarity of inner ear hair cells and causes severe neural tube defects in the mouse. *Curr. Biol.* Online Publication, May 14, 2003.

D'Eustachio, P., Owens, G. C., Edelman, G. M., and Cunningham, B. A. (1985). Chromosomal location of the gene encoding the neural cell adhesion molecule (N-CAM) in the mouse. *Proc. Natl. Acad. Sci. USA* **82,** 7631–7635.

Dabdoub, A., Donohue, M. J., Brennan, A., Wolf, V., Montcouquiol, M., Sassoon, D. A., Hseih, J. C., Rubin, J. S., Salinas, P. C., and Kelley, M. W. (2003). Wnt signaling mediates reorientation of outer hair cell stereociliary bundles in the mammalian cochlea. *Development* **130,** 2375–2384.

Damsky, C. H., and Ilic, D. (2002). Integrin signaling: It's where the action is. *Curr. Opin. Cell Biol.* **14,** 594–602.

Davies, D., and Holley, M. C. (2002). Differential expression of alpha 3 and alpha 6 integrins in the developing mouse inner ear. *J. Comp. Neurol.* **445,** 122–132.

DeBernardo, A. P., and Chang, S. (1995). Native and recombinant DM-GRASP selectively support neurite extension from neurons that express GRASP. *Dev. Biol.* **169,** 65–75.

DeBernardo, A. P., and Chang, S. (1996). Heterophilic interactions of DM-GRASP: GRASP-NgCAM interactions involved in neurite extension. *J. Cell. Biol.* **133,** 657–666.

Del Pozo, M. A., Kiosses, W. B., Alderson, N. B., Meller, N., Hahn, K. M., and Schwartz, M. A. (2002). Integrins regulate GTP-Rac localized effector interactions through dissociation of Rho-GDI. *Nat. Cell. Biol.* **4,** 232–239.

Di Palma, F., Holme, R. H., Bryda, E. C., Belyantseva, I. A., Pellegrino, R., Kachar, B., Steel, K. P., and Noben-Trauth, K. (2001). Mutations in Cdh23, encoding a new type of cadherin, cause stereocilia disorganization in waltzer, the mouse model for Usher syndrome type 1D. *Nat. Genet.* **27,** 103–107.

Doherty, P., Cohen, J., and Walsh, F. S. (1990). Neurite outgrowth in response to transfected N-CAM changes during development and is modulated by polysialic acid. *Neuron* **5,** 209–219.

Doherty, P., Ashton, S. V., Moore, S. E., and Walsh, F. S. (1991). Morphoregulatory activities of NCAM and N-cadherin can be accounted for by G protein-dependent activation of L- and N-type neuronal Ca2+ channels. *Cell* **67,** 21–33.

Doherty, P., Moolenaar, C. E., Ashton, S. V., Michalides, R. J., and Walsh, F. S. (1992). The VASE exon downregulates the neurite growth-promoting activity of NCAM 140. *Nature* **356,** 791–793.

Doherty, P., Williams, E., and Walsh, F. S. (1995). A soluble chimeric form of the L1 glycoprotein stimulates neurite outgrowth. *Neuron* **14,** 57–66.
Drazba, J., and Lemmon, V. (1990). The role of cell adhesion molecules in neurite outgrowth on Muller cells. *Dev. Biol.* **138,** 82–93.
Edelman, G. M., and Crossin, K. L. (1991). Cell adhesion molecules: Implications for a molecular histology. *Annu. Rev. Biochem.* **60,** 155–190.
Etienne-Manneville, S., and Hall, A. (2001). Integrin-mediated activation of Cdc42 controls cell polarity in migrating astrocytes through PKCzeta. *Cell* **106,** 489–498.
Fehon, R. G., Kooh, P. J., Rebay, I., Regan, C. L., Xu, T., Muskavitch, M. A., and Artavanis-Tsakonas, S. (1990). Molecular interactions between the protein products of the neurogenic loci Notch and Delta, two EGF-homologous genes in Drosophila. *Cell* **61,** 523–534.
Fekete, D. M., and Wu, D. K. (2002). Revisiting cell fate specification in the inner ear. *Curr. Opin. Neurobiol.* **12,** 35–42.
Fischer, G., Kunemund, V., and Schachner, M. (1986). Neurite outgrowth patterns in cerebellar microexplant cultures are affected by antibodies to the cell surface glycoprotein L1. *J. Neurosci.* **6,** 605–612.
Fleming, R. J. (1998). Structural conservation of Notch receptors and ligands. *Semin. Cell. Dev. Biol.* **9,** 599–607.
Formstone, C. J., and Little, P. F. (2001). The flamingo-related mouse Celsr family (Celsr1-3) genes exhibit distinct patterns of expression during embryonic development. *Mech. Dev.* **109,** 91–94.
Frank, M., and Kemler, R. (2002). Protocadherins. *Curr. Opin. Cell. Biol.* **14,** 557–562.
Friedlander, D. R., Milev, P., Karthikeyan, L., Margolis, R. K., Margolis, R. U., and Grumet, M. (1994). The neuronal chondroitin sulfate proteoglycan neurocan binds to the neural cell adhesion molecules Ng-CAM/L1/NILE and N-CAM, and inhibits neuronal adhesion and neurite outgrowth. *J. Cell. Biol.* **125,** 669–680.
Furley, A. J., Morton, S. B., Manalo, D., Karagogeos, D., Dodd, J., and Jessell, T. M. (1990). The axonal glycoprotein TAG-1 is an immunoglobulin superfamily member with neurite outgrowth-promoting activity. *Cell* **61,** 157–170.
Gaiano, N., and Fishell, G. (2002). The role of notch in promoting glial and neural stem cell fates. *Annu. Rev. Neurosci.* **25,** 471–490.
Garrod, D. R., Merritt, A. J., and Nie, Z. (2002). Desmosomal adhesion: Structural basis, molecular mechanism and regulation (Review). *Mol. Membr. Biol.* **19,** 81–94.
Goodyear, R. J., Kwan, T., Oh, S. H., Raphael, Y., and Richardson, G. P. (2001). The cell adhesion molecule BEN defines a prosensory patch in the developing avian otocyst. *J. Comp. Neurol.* **434,** 275–288.
Gottardi, C. J., and Gumbiner, B. M. (2001). Adhesion signaling: How beta-catenin interacts with its partners. *Curr. Biol.* **11,** R792–R794.
Groves, A. K., and Bronner-Fraser, M. (2000). Competence, specification and commitment in otic placode induction. *Development* **127,** 3489–3499.
Haag, T. A., Haag, N. P., Lekven, A. C., and Hartenstein, V. (1999). The role of cell adhesion molecules in *Drosophila* heart morphogenesis: faint sausage, shotgun/DE-cadherin, and laminin A are required for discrete stages in heart development. *Dev. Biol.* **208,** 56–69.
Hackett, L., Davies, D., Helyer, R., Kennedy, H., Kros, C., Lawlor, P., Rivolta, M. N., and Holley, M. (2002). E-cadherin and the differentiation of mammalian vestibular hair cells. *Exp. Cell. Res.* **278,** 19–30.
Haddon, C., Jiang, Y. J., Smithers, L., and Lewis, J. (1998). Delta-Notch signalling and the patterning of sensory cell differentiation in the zebrafish ear: Evidence from the mind bomb mutant. *Development* **125,** 4637–4644.
Haddon, C., Mowbray, C., Whitfield, T., Jones, D., Gschmeissner, S., and Lewis, J. (1999). Hair cells without supporting cells: Further studies in the ear of the zebrafish mind bomb mutant. *J. Neurocytol.* **28,** 837–850.

11. Cell Adhesion and Development of the Ear

Hadjantonakis, A. K., Sheward, W. J., Harmar, A. J., de Galan, L., Hoovers, J. M., and Little, P. F. (1997). Celsr1, a neural-specific gene encoding an unusual seven-pass transmembrane receptor, maps to mouse chromosome 15 and human chromosome 22qter. *Genomics* **45,** 97–104.

Hadjantonakis, A. K., Formstone, C. J., and Little, P. F. (1998). mCelsr1 is an evolutionarily conserved seven-pass transmembrane receptor and is expressed during mouse embryonic development. *Mech. Dev.* **78,** 91–95.

Harrelson, A. L., and Goodman, C. S. (1988). Growth cone guidance in insects: Fasciclin II is a member of the immunoglobulin superfamily. *Science* **242,** 700–708.

Heggem, M. A., and Bradley, R. S. (2003). The cytoplasmic domain of *Xenopus* NF-protocadherin interacts with TAF1/set. *Dev. Cell* **4,** 419–429.

Henry, C. A., Crawford, B. D., Yan, Y. L., Postlethwait, J., Cooper, M. S., and Hille, M. B. (2001). Roles for zebrafish focal adhesion kinase in notochord and somite morphogenesis. *Dev. Biol.* **240,** 474–487.

Hoffman, S., and Edelman, G. M. (1983). Kinetics of homophilic binding by embryonic and adult forms of the neural cell adhesion molecule. *Proc. Natl. Acad. Sci. USA* **80,** 5762–5766.

Holm, J., Appel, F., and Schachner, M. (1995). Several extracellular domains of the neural cell adhesion molecule L1 are involved in homophilic interactions. *J. Neurosci. Res.* **42,** 9–20.

Homayouni, R., Rice, D. S., and Curran, T. (2001). Disabled-1 interacts with a novel developmentally regulated protocadherin. *Biochem. Biophys. Res. Commun.* **289,** 539–547.

Hrynkow, S. H., Morest, D. K., Bilak, M., and Rutishauser, U. (1998). Multiple roles of neural cell adhesion molecule, neural cell adhesion molecule-polysialic acid, and L1 adhesion molecules during sensory innervation of the otic epithelium in vitro. *Neuroscience* **87,** 423–437.

Hrynkow, S. H., Morest, D. K., Brumwell, C, and Rutishauser, U. (1998). Spatio-temporal diversity in the microenvironments for neural cell adhesion molecule, neural cell adhesion molecule-polysialic acid, and L1-cell adhesion molecule expression by sensory neurons and their targets during cochleo-vestibular innervation. *Neuroscience* **87,** 401–422.

Hudspeth, A. J., and Jacobs, R. (1979). Stereocilia mediate transduction in vertebrate hair cells (auditory system/cilium/vestibular system). *Proc. Natl. Acad. Sci. USA* **76,** 1506–1509.

Itoh, M., Kim, C. H., Palardy, G., Oda, T., Jiang, Y. J., Maust, D., Yeo, S. Y., Lorick, K., Wright, G. J., Ariza-McNaughton, L., *et al.* (2003). Mind bomb is a ubiquitin ligase that is essential for efficient activation of Notch signaling by Delta. *Dev. Cell* **4,** 67–82.

Johnston, I. G., Paladino, T., Gurd, J. W., and Brown, I. R. (1990). Molecular cloning of SC1: A putative brain extracellular matrix glycoprotein showing partial similarity to osteonectin/BM40/SPARC. *Neuron* **4,** 165–176.

Kasper, C., Rasmussen, H., Kastrup, J. S., Ikemizu, S., Jones, E. Y., Berezin, V., Bock, E., and Larsen, I. K. (2000). Structural basis of cell–cell adhesion by NCAM. *Nat. Struct. Biol.* **7,** 389–393.

Kelley, M. W., and Bianchi, L. M. (2001). Development and neuronal innervation of the organ of Corti. *In* "Handbook of the Mouse Auditory Research: From Behavior to Molecular Biology" (J. F. Willott, Ed.), pp. 137–156. CRC Press, NY.

Kiernan, A. E., Ahituv, N., Fuchs, H., Balling, R., Avraham, K. B., Steel, K. P., and Hrabe de Angelis, M. (2001). The Notch ligand Jagged1 is required for inner ear sensory development. *Proc. Natl. Acad. Sci. USA* **98,** 3873–3878.

Kimura, Y., Matsunami, H., Inoue, T., Shimamura, K., Uchida, N., Ueno, T., Miyazaki, T., and Takeichi, M. (1995). Cadherin-11 expressed in association with mesenchymal morphogenesis in the head, somite, and limb bud of early mouse embryos. *Dev. Biol.* **169,** 347–358.

Knudsen, K. A., Frankowski, C., Johnson, K. R., and Wheelock, M. J. (1998). A role for cadherins in cellular signaling and differentiation. *J. Cell. Biochem. Suppl.* **30–31,** 168–176.

Kohmura, N., Senzaki, K., Hamada, S., Kai, N., Yasuda, R., Watanabe, M., Ishii, H., Yasuda, M., Mishina, M., and Yagi, T. (1998). Diversity revealed by a novel family of cadherins expressed in neurons at a synaptic complex. *Neuron* **20**, 1137–1151.

Kraynov, V. S., Chamberlain, C., Bokoch, G. M., Schwartz, M. A., Slabaugh, S., and Hahn, K. M. (2000). Localized Rac activation dynamics visualized in living cells. *Science* **290**, 333–337.

Laessing, U., Giordano, S., Stecher, B., Lottspeich, F., and Stuermer, C. A. (1994). Molecular characterization of fish neurolin: A growth-associated cell surface protein and member of the immunoglobulin superfamily in the fish retinotectal system with similarities to chick protein DM-GRASP/SC-1/BEN. *Differentiation* **56**, 21–29.

Lanford, P. J., Lan, Y., Jiang, R., Lindsell, C., Weinmaster, G., Gridley, T., and Kelley, M. W. (1999). Notch signalling pathway mediates hair cell development in mammalian cochlea. *Nat. Genet.* **21**, 289–292.

Lanford, P. J., Shailam, R., Norton, C. R., Gridley, T., and Kelley, M. W. (2000). Expression of Math1 and HES5 in the cochleae of wildtype and Jag2 mutant mice. *J. Assoc. Res. Otolaryngol* **1**, 161–171.

Larue, L., Ohsugi, M., Hirchenhain, J., and Kemler, R. (1994). E-cadherin null mutant embryos fail to form a trophectoderm epithelium. *Proc. Natl. Acad. Sci. USA* **91**, 8263–8267.

Leon, Y., Varela-Nieto, I., and Breen, K. C. (1997). Neural cell adhesion molecule expression in the developing chick otic vesicle. *Biochem. Soc. Trans* **25**, 10S.

Leong, K. G., Hu, X., Li, L., Noseda, M., Larrivee, B., Hull, C., Hood, L., Wong, F., and Karsan, A. (2002). Activated Notch4 inhibits angiogenesis: Role of beta 1-integrin activation. *Mol. Cell Biol.* **22**, 2830–2841.

Lindner, V., Booth, C., Prudovsky, I., Small, D., Maciag, T., and Liaw, L. (2001). Members of the Jagged/Notch gene families are expressed in injured arteries and regulate cell phenotype via alterations in cell matrix and cell-cell interaction. *Am. J. Pathol.* **159**, 875–883.

Linnemann, D., Gaardsvoll, H., Olsen, M., and Bock, E. (1993). Expression of NCAM mRNA and polypeptides in aging rat brain. *Int. J. Dev. Neurosci.* **11**, 71–81.

Littlewood Evans, A., and Muller, U. (2000). Stereocilia defects in the sensory hair cells of the inner ear in mice deficient in integrin alpha8beta1. *Nat. Genet.* **24**, 424–428.

Liu, L., Haines, S., Shew, R., and Akeson, R. A. (1993). Axon growth is enhanced by NCAM lacking the VASE exon when expressed in either the growth substrate or the growing axon. *J. Neurosci. Res.* **35**, 327–345.

Liu, S., Calderwood, D. A., and Ginsberg, M. H. (2000). Integrin cytoplasmic domain-binding proteins. *J. Cell Sci.* **113**, 3563–3571.

Lowell, S., and Watt, F. M. (2001). Delta regulates keratinocyte spreading and motility independently of differentiation. *Mech. Dev.* **107**, 133–140.

Lustig, M., Sakurai, T., and Grumet, M. (1999). Nr-CAM promotes neurite outgrowth from peripheral ganglia by a mechanism involving axonin-1 as a neuronal receptor. *Dev. Biol.* **209**, 340–351.

Mahoney, P. A., Weber, U., Onofrechuk, P., Biessmann, H., Bryant, P. J., and Goodman, C. S. (1991). The fat tumor suppressor gene in *Drosophila* encodes a novel member of the cadherin gene superfamily. *Cell* **67**, 853–868.

Marrs, J. A., and Nelson, W. J. (1996). Cadherin cell adhesion molecules in differentiation and embryogenesis. *Int. Rev. Cytol.* **165**, 159–205.

Mbiene, J. P., Dechesne, C. J., Schachner, M., and Sans, A. (1989). Immunocytological characterization of the expression of cell adhesion molecule L1 during early innervation of mouse otocysts. *Cell Tissue Res.* **255**, 81–88.

McCarty, J. H., Monahan-Earley, R. A., Brown, L. F., Keller, M., Gerhardt, H., Rubin, K., Shani, M., Dvorak, H. F., Wolburg, H., Bader, B. L., *et al.* (2002). Defective associations

11. Cell Adhesion and Development of the Ear

between blood vessels and brain parenchyma lead to cerebral hemorrhage in mice lacking alphav integrins. *Mol. Cell Biol.* **22,** 7667–7677.

Milner, R., and Campbell, I. L. (2002). The integrin family of cell adhesion molecules has multiple functions within the CNS. *J. Neurosci. Res.* **69,** 286–291.

Mlodzik, M. (2002). Planar cell polarization: Do the same mechanisms regulate *Drosophila* tissue polarity and vertebrate gastrulation? *Trends Genet.* **18,** 564–571.

Montcouquiol, M., Rachel, R. A., Lanford, P. J., Copeland, N. G., Jenkins, N. A., and Kelley, M. W. (2003). Identification of Vang12 and Scrb1 as planar polarity genes in mammals. *Nature* **423,** 173–177.

Morrison, A., Hodgetts, C., Gossler, A., Hrabe de Angelis, M., and Lewis, J. (1999). Expression of Delta1 and Serrate1 (Jagged1) in the mouse inner ear. *Mech. Dev.* **84,** 169–172.

Mothe, A. J., and Brown, I. R. (2001). Expression of mRNA encoding extracellular matrix glycoproteins SPARC and SC1 is temporally and spatially regulated in the developing cochlea of the rat inner ear. *Hear Res.* **155,** 161–174.

Muller, D., Stoppini, L., Wang, C., and Kiss, J. Z. (1994). A role for polysialylated neural cell adhesion molecule in lesion-induced sprouting in hippocampal organotypic cultures. *Neuroscience* **61,** 441–445.

Murcia, C. L., and Woychik, R. P. (2001). Expression of Pcdh15 in the inner ear, nervous system and various epithelia of the developing embryo. *Mech. Dev.* **105,** 163–166.

Murphy-Erdosh, C., Napolitano, E. W., and Reichardt, L. F. (1994). The expression of B-cadherin during embryonic chick development. *Dev. Biol.* **161,** 107–125.

Nakazawa, K., Spicer, S. S., and Schulte, B. A. (1996). Focal expression of A-CAM on pillar cells during formation of Corti's tunnel in gerbil cochlea. *Anat. Rec.* **245,** 577–580.

Nose, A., and Takeichi, M. (1986). A novel cadherin cell adhesion molecule: Its expression patterns associated with implantation and organogenesis of mouse embryos. *J. Cell Biol.* **103,** 2649–2658.

Ono, K., Tomasiewicz, H., Magnuson, T., and Rutishauser, U. (1994). N-CAM mutation inhibits tangential neuronal migration and is phenocopied by enzymatic removal of polysialic acid. *Neuron* **13,** 595–609.

Peck, D., and Walsh, F. S. (1993). Differential effects of over-expressed neural cell adhesion molecule isoforms on myoblast fusion. *J. Cell Biol.* **123,** 1587–1595.

Pirvola, U., Ylikoski, J., Trokovic, R., Hebert, J. M., McConnell, S. K., and Partanen, J. (2002). FGFR1 is required for the development of the auditory sensory epithelium. *Neuron* **35,** 671–680.

Pollerberg, G. E., and Mack, T. G. (1994). Cell adhesion molecule SC1/DMGRASP is expressed on growing axons of retina ganglion cells and is involved in mediating their extension on axons. *Dev. Biol.* **165,** 670–687.

Pourquie, O., Coltey, M., Thomas, J. L., and Le Douarin, N. M. (1990). A widely distributed antigen developmentally regulated in the nervous system. *Development* **109,** 743–752.

Pourquie, O., Corbel, C., Le Caer, J. P., Rossier, J., and Le Douarin, N. M. (1992). BEN, a surface glycoprotein of the immunoglobulin superfamily, is expressed in a variety of developing systems. *Proc. Natl. Acad. Sci. USA* **89,** 5261–5265.

Pourquie, O. (2000). Segmentation of the paraxial mesoderm and vertebrate somitogenesis. *Curr. Top. Dev. Biol.* **47,** 81–105.

Rabinowitz, J. E., Rutishauser, U., and Magnuson, T. (1996). Targeted mutation of Ncam to produce a secreted molecule results in a dominant embryonic lethality. *Proc. Natl. Acad. Sci. USA* **93,** 6421–6424.

Radice, G. L., Rayburn, H., Matsunami, H., Knudsen, K. A., Takeichi, M., and Hynes, R. O. (1997). Developmental defects in mouse embryos lacking N-cadherin. *Dev. Biol.* **181,** 64–78.

Raphael, Y., Volk, T., Crossin, K. L., Edelman, G. M., and Geiger, B. (1988). The modulation of cell adhesion molecule expression and intercellular junction formation in the developing avian inner ear. *Dev. Biol.* **128,** 222–235.

Raphael, Y., Kobayashi, K. N., Dootz, G. A., Beyer, L. A., Dolan, D. F., and Burmeister, M. (2001). Severe vestibular and auditory impairment in three alleles of Ames waltzer (av) mice. *Hear Res.* **151,** 237–249.

Rebay, I., Fleming, R. J., Fehon, R. G., Cherbas, L., Cherbas, P., and Artavanis-Tsakonas, S. (1991). Specific EGF repeats of Notch mediate interactions with Delta and Serrate: implications for Notch as a multifunctional receptor. *Cell* **67,** 687–699.

Resh, M. D. (1998). Fyn, a Src family tyrosine kinase. *Int. J. Biochem. Cell Biol.* **30,** 1159–1162.

Reyes, A. A., Schulte, S. V., Small, S., and Akeson, R. (1993). Distinct NCAM splicing events are differentially regulated during rat brain development. *Brain Res. Mol. Brain Res.* **17,** 201–211.

Richardson, G. P., Crossin, K. L., Chuong, C. M., and Edelman, G. M. (1987). Expression of cell adhesion molecules during embryonic induction. III. Development of the otic placode. *Dev. Biol.* **119,** 217–230.

Riley, B. B., Chiang, M., Farmer, L., and Heck, R. (1999). The deltaA gene of zebrafish mediates lateral inhibition of hair cells in the inner ear and is regulated by pax2.1. *Development* **126,** 5669–5678.

Rinkwitz, S., Bober, E., and Baker, R. (2001). Development of the vertebrate inner ear. *Ann. NY Acad. Sci.* **942,** 1–14.

Rutishauser, U., Gall, W. E., and Edelman, G. M. (1978). Adhesion among neural cells of the chick embryo. IV. Role of the cell surface molecule CAM in the formation of neurite bundles in cultures of spinal ganglia. *J. Cell. Biol.* **79,** 382–393.

Sadoul, R., Hirn, M., Deagostini-Bazin, H., Rougon, G., and Goridis, C. (1983). Adult and embryonic mouse neural cell adhesion molecules have different binding properties. *Nature* **304,** 347–349.

Saffell, J. L., Walsh, F. S., and Doherty, P. (1994). Expression of NCAM containing VASE in neurons can account for a developmental loss in their neurite outgrowth response to NCAM in a cellular substratum. *J. Cell Biol.* **125,** 427–436.

Sano, K., Tanihara, H., Heimark, R. L., Obata, S., Davidson, M., St John, T., Taketani, S., and Suzuki, S. (1993). Protocadherins: A large family of cadherin- related molecules in central nervous system. *Embo. J.* **12,** 2249–2256.

Shimada, Y., Usui, T., Yanagawa, S., Takeichi, M., and Uemura, T. (2001). Asymmetric colocalization of Flamingo, a seven-pass transmembrane cadherin, and Dishevelled in planar cell polarization. *Curr. Biol.* **11,** 859–863.

Siemens, J., Kazmierczak, P., Reynolds, A., Sticker, M., Littlewood- Evans, A., and Muller, U. (2002). The Usher syndrome proteins cadherin 23 and harmonin form a complex by means of PDZ-domain interactions. *Proc. Natl. Acad. Sci. USA* **99,** 14946–14951.

Silletti, S., Mei, F., Sheppard, D., and Montgomery, A. M. (2000). Plasmin-sensitive dibasic sequences in the third fibronectin-like domain of L1-cell adhesion molecule (CAM) facilitate homomultimerization and concomitant integrin recruitment. *J. Cell Biol.* **149,** 1485–1502.

Simonneau, L., Gallego, M., and Pujol, R. (2003). Comparative expression patterns of T-, N-, E-cadherins, beta-catenin, and polysialic acid neural cell adhesion molecule in rat cochlea during development: Implications for the nature of Kolliker's organ. *J. Comp. Neurol.* **459,** 113–126.

Small, S. J., and Akeson, R. (1990). Expression of the unique NCAM VASE exon is independently regulated in distinct tissues during development. *J. Cell Biol.* **111,** 2089–2096.

Steinberg, M. S., and McNutt, P. M. (1999). Cadherins and their connections: Adhesion junctions have broader functions. *Curr. Opin. Cell Biol.* **11,** 554–560.

Stephan, J. P., Bald, L., Roberts, P. E., Lee, J., Gu, Q., and Mather, J. P. (1999). Distribution and function of the adhesion molecule BEN during rat development. *Dev. Biol.* **212**, 264–277.

Stephens, L. E., Sutherland, A. E., Klimanskaya, I. V., Andrieux, A., Meneses, J., Pedersen, R. A., and Damsky, C. H. (1995). Deletion of beta 1 integrins in mice results in inner cell mass failure and peri-implantation lethality. *Genes Dev.* **9**, 1883–1895.

Takagi, J., and Springer, T. A. (2002). Integrin activation and structural rearrangement. *Immunol. Rev.* **186**, 141–163.

Takeichi, M., Atsumi, T., Yoshida, C., Uno, K., and Okada, T. S. (1981). Selective adhesion of embryonal carcinoma cells and differentiated cells by Ca2+-dependent sites. *Dev. Biol.* **87**, 340–350.

Tanaka, H., Matsui, T., Agata, A., Tomura, M., Kubota, I., McFarland, K. C., Kohr, B., Lee, A., Phillips, H. S., and Shelton, D. L. (1991). Molecular cloning and expression of a novel adhesion molecule, SC1. *Neuron* **7**, 535–545.

Tang, J., Rutishauser, U., and Landmesser, L. (1994). Polysialic acid regulates growth cone behavior during sorting of motor axons in the plexus region. *Neuron* **13**, 405–414.

Terkelsen, O. B., Bock, E., and Mollgard, K. (1989). NCAM and Thy-1 in special sense organs of the developing mouse. *Anat. Embryol. (Berl.)* **179**, 311–318.

Thanos, S., Bonhoeffer, F., and Rutishauser, U. (1984). Fiber-fiber interaction and tectal cues influence the development of the chicken retinotectal projection. *Proc. Natl. Acad. Sci. USA* **81**, 1906–1910.

Tomasiewicz, H., Ono, K., Yee, D., Thompson, C., Goridis, C., Rutishauser, U., and Magnuson, T. (1993). Genetic deletion of a neural cell adhesion molecule variant (N-CAM-180) produces distinct defects in the central nervous system. *Neuron* **11**, 1163–1174.

Tosney, K. W., Watanabe, M., Landmesser, L., and Rutishauser, U. (1986). The distribution of NCAM in the chick hindlimb during axon outgrowth and synaptogenesis. *Dev. Biol.* **114**, 437–452.

Tsai, H., Hardisty, R. E., Rhodes, C., Kiernan, A. E., Roby, P., Tymowska-Lalanne, Z., Mburu, P., Rastan, S., Hunter, A. J., Brown, S. D., *et al.* (2001). The mouse slalom mutant demonstrates a role for Jagged1 in neuroepithelial patterning in the organ of Corti. *Hum. Mol. Genet.* **10**, 507–512.

Usui, T., Shima, Y., Shimada, Y., Hirano, S., Burgess, R. W., Schwarz, T. L., Takeichi, M., and Uemura, T. (1999). Flamingo, a seven-pass transmembrane cadherin, regulates planar cell polarity under the control of Frizzled. *Cell* **98**, 585–595.

van de Stolpe, A., and van der Saag, P. T. (1996). Intercellular adhesion molecule-1. *J. Mol. Med.* **74**, 13–33.

Visconti, R. P., and Hilfer, S. R. (2002). Perturbation of extracellular matrix prevents association of the otic primordium with the posterior rhombencephalon and inhibits subsequent invagination. *Dev. Dyn.* **223**, 48–58.

Vleminckx, K., and Kemler, R. (1999). Cadherins and tissue formation: Integrating adhesion and signaling. *Bioessays* **21**, 211–220.

Walsh, F. S., Furness, J., Moore, S. E., Ashton, S., and Doherty, P. (1992). Use of the neural cell adhesion molecule VASE exon by neurons is associated with a specific down-regulation of neural cell adhesion molecule-dependent neurite outgrowth in the developing cerebellum and hippocampus. *J. Neurochem.* **59**, 1959–1962.

Walsch, F. S., and Doherty, P. (1997). Neural cell adhesion molecules of the immunoglobulin superfamily: Role in axon growth and guidance. *Annu. Rev. Cell Dev. Biol.* **13**, 425–456.

Warchol, M. E. (2002). Cell density and N-cadherin interactions regulate cell proliferation in the sensory epithelia of the inner ear. *J. Neurosci.* **22**, 2607–2616.

Weinmaster, G., Roberts, V. J., and Lemke, G. (1991). A homolog of *Drosophila* Notch expressed during mammalian development. *Development* **113**, 199–205.

Whitlon, D. S., and Rutishauser, U. S. (1990). NCAM in the organ of Corti of the developing mouse. *J. Neurocytol.* **19,** 970–977.
Whitlon, D. S. (1993). E-cadherin in the mature and developing organ of Corti of the mouse. *J. Neurocytol.* **22,** 1030–1038.
Whitlon, D. S., and Zhang, X. (1997). Polysialic acid in the cochlea of the developing mouse. *Int. J. Dev. Neurosci.* **15,** 657–669.
Whitlon, D. S., Zhang, X., Pecelunas, K., and Greiner, M. A. (1999). A temporospatial map of adhesive molecules in the organ of Corti of the mouse cochlea. *J. Neurocytol.* **28,** 955–968.
Williams, A. F., and Barclay, A. N. (1988). The immunoglobulin superfamily–domains for cell surface recognition. *Annu. Rev. Immunol.* **6,** 381–405.
Williams, E. J., Furness, J., Walsh, F. S., and Doherty, P. (1994). Activation of the FGF receptor underlies neurite outgrowth stimulated by L1, N-CAM, and N-cadherin. *Neuron* **13,** 583–594.
Wilson, S. M., Householder, D. B., Coppola, V., Tessarollo, L., Fritzsch, B., Lee, E. C., Goss, D., Carlson, G. A., Copeland, N. G., and Jenkins, N. A. (2001). Mutations in Cdh23 cause nonsyndromic hearing loss in waltzer mice. *Genomics* **74,** 228–233.
Wu, Q., and Maniatis, T. (1999). A striking organization of a large family of human neural cadherin-like cell adhesion genes. *Cell* **97,** 779–790.
Wu, Q., Zhang, T., Cheng, J. F., Kim, Y., Grimwood, J., Schmutz, J., Dickson, M., Noonan, J. P., Zhang, M. Q., Myers, R. M., *et al.* (2001). Comparative DNA sequence analysis of mouse and human protocadherin gene clusters. *Genome Res.* **11,** 389–404.
Yagi, T., and Takeichi, M. (2000). Cadherin superfamily genes: Functions, genomic organization, and neurologic diversity. *Genes Dev.* **14,** 1169–1180.
Yang, J. T., Rayburn, H., and Hynes, R. O. (1993). Embryonic mesodermal defects in alpha 5 integrin-deficient mice. *Development* **119,** 1093–1105.
Yang, J. T., Rayburn, H., and Hynes, R. O. (1995). Cell adhesion events mediated by alpha 4 integrins are essential in placental and cardiac development. *Development* **121,** 549–560.
Yoshida, K., Watanabe, M., Kato, H., Dutta, A., and Sugano, S. (1999). BH-protocadherin-c, a member of the cadherin superfamily, interacts with protein phosphatase 1 alpha through its intracellular domain. *FEBS Lett.* **460,** 93–98.
Yoshida, N., and Liberman, M. C. (1999). Stereociliary anomaly in the guinea pig: Effects of hair bundle rotation on cochlear sensitivity. *Hear. Res.* **131,** 29–38.
Zine, A., Van De Water, T. R., and de Ribaupierre, F. (2000). Notch signaling regulates the pattern of auditory hair cell differentiation in mammals. *Development* **127,** 3373–3383.
Zorn, A. M., and Krieg, P. A. (1992). Developmental regulation of alternative splicing in the mRNA encoding *Xenopus laevis* neural cell adhesion molecule (NCAM). *Dev. Biol.* **149,** 197–205.
Zuellig, R. A., Rader, C., Schroeder, A., Kalousek, M. B., Von Bohlen und Halbach, F., Osterwalder, T., Inan, C., Stoeckli, E. T., Affolter, H. U., Fritz, A., *et al.* (1992). The axonally secreted cell adhesion molecule, axonin-1. Primary structure, immunoglobulin-like and fibronectin-type-III-like domains and glycosyl-phosphatidylinositol anchorage. *Eur. J. Biochem.* **204,** 453–463.

12
Genes Controlling the Development of the Zebrafish Inner Ear and Hair Cells

Bruce B. Riley
Biology Department, Texas A&M University, College Station, Texas 77843

I. Introduction
   A. Overview
   B. Nomenclature and Staging
II. General Course of Zebrafish Otic Development
III. Otic Induction
   A. Overview
   B. Fgf3 and Fgf8 as Otic Inducers
   C. Wnt8 as an Otic Inducer
   D. *pou2*
   E. *pax2/5/8* Genes
   F. *foxi1*
   G. *dlx3b* and *dlx4b*
   H. Nodal–Fgf Interactions
IV. Patterning of the Placode and Early Vesicle
   A. Overview
   B. Fgf3 Signaling from the Hindbrain
   C. Hedgehog Signaling from the Floorplate and Notochord
   D. *eya1*
   E. *van gogh (vgo)*
   F. *sox10*
   G. *jekyll (jek)*
V. Development of Sensory Epithelia
   A. Overview
   B. *deltaA (dlA)*
   C. E3 Ubiquitin Ligase
   D. *zath1*
   E. *neurogenin1*
   F. *pax2a* and *pax2b*
VI. Auditory and Vestibular Function
   A. Overview
   B. Circler Mutants
   C. Otolith Mutants
   D. A Screen for Hearing Mutants
VII. Conclusions and Prospects
    References

Recent genetic studies in zebrafish have led to significant advances in several areas of inner ear development. Progress has been particularly rapid in elucidating mechanisms of preotic development—processes acting prior to and during induction of the otic placode. Functions required for early patterning of the otic placode and vesicle are in many cases similar in zebrafish and mouse, but there are also some important differences that probably underlie the structural differences between the inner ear in fish vs mammals. Finally, ongoing mutagenesis screens are identifying genes required for auditory and vestibular sensory functions and hair cell differentiation and survival. Analysis of such genes could provide an experimentally amenable model for some kinds of human deafness. © 2003, Elsevier Inc.

I. Introduction

A. Overview

Large-scale mutagenesis screens (Malicki *et al.*, 1996; Whitfield *et al.*, 1996) and smaller focused screens (Appel *et al.*, 1999; Bang *et al.*, 2002; Fritz *et al.*, 1996; Solomon *et al.*, 2003) have identified numerous mutations that perturb the morphology or function of the zebrafish inner ear, and many of the corresponding genes have been identified. These mutations are invaluable resources, yet many additional genes suspected of playing a role in ear development are not yet represented by mutant loci. For such candidate genes, antisense morpholino oligomers can be injected into embryos at the one-cell stage to specifically inhibit translation or splicing of the corresponding transcripts (Draper *et al.*, 2001; Nasevicius and Ekker, 2000). Morpholino-mediated gene knockdown does not totally eliminate gene activity but is often sufficient to produce phenotypes that are indistinguishable from null mutations. Embryos injected with morpholinos are often referred to as morphants. The ability to coinject morpholinos directed against multiple gene products and to use them in conjunction with conventional mutant loci has dramatically increased the range of possible experiments as well as the speed with which genetic interactions can be examined. This chapter reviews how studies of zebrafish mutants and morphants have led to fundamental insights into mechanisms of inner ear development.

B. Nomenclature and Staging

Developmental times are given in hours post fertilization (hpf) at 28.5° C, and where appropriate anatomical staging criteria are also given to facilitate comparison with other vertebrate species. Zebrafish genes are denoted in lowercase italicized text (e.g., *fgf8*) and proteins are denoted in nonitalic text with the first letter capitalized (e.g., Fgf8). In keeping with the standards of other research communities, genes of other vertebrate species are denoted in italics with the first letter capitalized (e.g., *Fgf8*).

II. General Course of Zebrafish Otic Development

Morphological development of the ear has been described in detail by Haddon and Lewis (1996). The otic placode becomes visible by 9–10 somites (13.5–14 hpf) and gives rise to the otic vesicle by 18 somites (18 hpf). Unlike the otic vesicle in amniotes, the otic vesicle in zebrafish forms by cavitation rather than invagination. Otoliths begin to form over the future utricular and saccular maculae as soon as the lumen begins to open. The utricular and saccular maculae mature into pseudostratified epithelia by 25 somites (21.5 hpf) and continue to develop well into adulthood as the body continues to grow (Higgs *et al.*, 2001; Riley *et al.*, 1997). Neuroblasts of the statoacoustical ganglion (SAG) begin to delaminate from the ventral portion of the otic vesicle by 26 somites (22 hpf) and continue through approximately 42 hpf. The nonsensory portions of the semicircular canals and sensory cristae begin to form by 48 hpf. The endolymphatic duct and lagenar macula do not form until 8 days and 10 days postfertilization, respectively (Bever and Fekete, 2002; Riley and Moorman, 2000).

III. Otic Induction

A. Overview

Although the otic placode is not detectable before 9 somites (13.5 hpf), developing preotic tissue can be studied much earlier by visualizing expression of various molecular markers. Preotic markers have been invaluable for following early stages of otic development under various experimental conditions, and genetic studies have shown how they participate in the genetic pathways that regulate otic competence or otic induction. Genetic studies have also identified many of the signaling molecules responsible for inducing expression of preotic markers. In most regards, recent findings in zebrafish are consistent with results of classical embryology studies dealing with otic induction. There have also been a few surprises that have led to a much more detailed understanding of the events controlling otic induction.

B. Fgf3 and Fgf8 as Otic Inducers

One of the clearest and best documented lessons from classical studies in amphibian and avian embryos is that the hindbrain is a rich source of otic-inducing factors (Groves and Bronner-Fraser, 2000; see also Chapter 4 in this volume by Brown *et al.*; Jacobson, 1963a; Stone, 1931; Waddington, 1937; Woo and Fraser, 1998). In zebrafish, grafting hindbrain tissue to the ventral side of the embryo induces ectopic otic vesicles within adjacent (previously ventral) ectoderm (Woo and Fraser, 1998). The best candidates for hindbrain-derived otic-inducing factors are Fgf3 and Fgf8. Both genes are expressed at 50% epiboly in a dorsoventral gradient in the germring (Fürthauer *et al.*, 1997; Phillips *et al.*, 2001). Between 8 hpf and 8.5 hpf (75–85% epiboly), *fgf3* and *fgf8* are expressed in the primordium of rhombomere 4 (r4) in the developing hindbrain. *fgf3* is also expressed in prechordal mesoderm and at low levels in paraxial cephalic mesoderm. The latter expression domains are consistent with experimental embryology studies indicating that mesodermal tissues also emit otic-inducing factors (Harrison, 1945; Ladher *et al.*, 2000; Jacobson, 1963a; Waddington, 1937). Furthermore, zebrafish mutations or morpholinos that block formation of head mesoderm cause a delay in otic induction, despite the fact that hindbrain expression of *fgf3* and *fgf8* occurs on time (Mendonsa and Riley, 1999; Phillips *et al.*, 2001). Disrupting either *fgf3* (by morpholino) or *fgf8* (in *acerebellar* or *ace* mutants [Reifers *et al.*, 1998]) causes a moderate reduction in the size of the otic vesicle and variably impairs formation of sensory patches and placode-derived neurons (Adamska *et al.*, 2000; Leger and Brand, 2002; Phillips *et al.*, 2001). Loss of both *fgf* functions results in near or total ablation of otic tissue. Analysis of early otic markers confirms that otic tissue is never induced properly in the absence of *fgf3* and *fgf8* (Leger and Brand, 2002; Maroon *et al.*, 2002; Phillips *et al.*, 2001). These findings support the notion that Fgf3 and Fgf8 act in a redundant manner to induce otic tissue. It is also likely that there are ligand-specific functions, but there is currently insufficient information about the variety of Fgf receptors involved and their ligand-binding preferences (see Chapter 8 by Wright and Mansour). Nevertheless, it is clear that Fgf3 and Fgf8 play very different roles during later stages of otic development (see later).

Fgf3 appears to play a conserved role in otic induction. In all species examined, Fgf3 is expressed in the hindbrain between the otic anlage (Lambardo *et al.*, 1998; Mahmood *et al.*, 1995, 1996; Wilkinson *et al.*, 1989). Targeted disruption of *Fgf3* in mouse leads to a small malformed otic vesicle (Mansour *et al.*, 1993). The observation that the placode and vesicle still form was initially interpreted to mean that *Fgf3* plays a later role in otic development. However, just as in zebrafish, there is redundancy in the inductive pathway. In mouse, *Fgf8* is not expressed in periotic tissues, but

Fgf10 is expressed in subotic mesoderm. Loss of both *Fgf3* and *Fgf10* specifically ablates otic induction (T. Wright and S. Mansour, personal communication). Misexpression of Fgf3 induces formation of ectopic otic vesicles in chick and *Xenopus* (Lombardo *et al.*, 1998; Vendrell *et al.*, 2000). Chick Fgf19 is expressed in subotic mesoderm and may participate in otic induction (Ladher *et al.*, 2000). Thus, redundancy in Fgf signaling pathways leading to otic induction seems to be a recurrent theme among various vertebrate species, and Fgf3 is always involved.

C. Wnt8 as an Otic Inducer

Wnt8 has also been proposed as a potential otic-inducing factor based on a recent study in chick (Ladher *et al.*, 2000). *Wnt8c* is expressed in the developing chick hindbrain between the prospective otic anlage, and human Wnt8 induces a variety of otic and preotic markers in chick explant cultures. In zebrafish, the *wnt8* locus has an unusual structure, comprising two complete open reading frames (ORFs) encoding distinct but closely related ligands (Lekven *et al.*, 2001). Both ORFs are expressed in the ventrolateral margin of the early gastrula and help regulate AP and DV patterning. In addition, *wnt8*–ORF2 is expressed by 8 hpf (75% epiboly) in the r5/6 anlagen. Loss of *wnt8* function, either mediated by morpholinos or a deletion that removes both ORFs, delays otic induction but does not prevent it because all preotic markers are eventually expressed and small otic vesicles are produced (Riley, unpublished observations). There is a similar delay in expression of *fgf3* and *fgf8* in the hindbrain, suggesting that *wnt8* regulates otic development indirectly by facilitating timely expression of *fgf3* and *fgf8* in the hindbrain. In addition, expression of *wnt8* in the germring may act earlier to set the posterior limit of otic competence (see Section III.F).

D. *pou2*

The POU domain transcription factor gene *pou2* is disrupted in *spiel ohne grenzen* (*spg*) mutants, which develop with small malformed otic vesicles similar to those seen in *ace* (*fgf8*) mutants (Burgess *et al.*, 2002). *pou2* is normally expressed in regions of the neural plate that also express *fgf3* and *fgf8* (Burgess *et al.*, 2002; Hauptmann and Gerster, 1995), and *fgf* gene expression is strongly reduced in *spg* mutants (Belting *et al.*, 2001; Reim and Brand, 2002; Riley, unpublished observations). It has been reported that *pou2* also acts in the midbrain–hindbrain border to make cells competent to respond to Fgf signaling (Reim and Brand, 2002). However, the latter is a cell-autonomous function unlikely to affect preotic cells, which do not detectably express *pou2*. Thus, *pou2* appears to regulate early otic development indirectly by controlling levels of Fgf3 and Fgf8 in the hindbrain.

E. *pax2/5/8* Genes

One of the earliest markers of preotic development is the paired domain transcription factor gene *pax8*, which is first detected in preotic cells at 8.5–9 hpf (85% epiboly) (Pfeffer *et al.*, 1998). It is later expressed throughout the developing placode but is lost at 19–20 h (20–22 somites) just after the otic vesicle forms. Morpholino-mediated knockdown of *pax8* reduces but does not eliminate formation of the otic placode (Riley, unpublished observations). It is possible that the phenotype is ameliorated by two related *pax* genes that are coexpressed at slightly later stages of otic development: *pax2a* (formerly *pax2.1*) is expressed in preotic cells by 11 hpf (3 somites) and *pax2b* (formerly *pax2.2*) is expressed by 13 hpf (8 somites) just prior to morphological development of the placode (Krauss *et al.*, 1991; Pfeffer *et al.*, 1998). Disrupting either or both *pax2* homologues alters development of hair cells in the otic vesicle (see later) but has no effect on placode induction (Whitfield *et al.*, 2002; Riley, unpublished observations). It will be important in future studies to simultaneously disrupt all three *pax* functions to address the possibility that they provide redundant functions in placode induction.

F. *foxi1*

The earliest preotic marker identified to date is the *foxi1*, which encodes a *forkhead* class winged helix transcription factor. *foxi1* is first detected at 50% epiboly in a domain that rapidly expands to include the anteroventral quadrant of the embryo (Solomon *et al.*, 2003; Fig. 1). Fate mapping studies indicate that prospective otic cells originate within this domain (Kozlowski *et al.*, 1997), suggesting that *foxi1* could be a marker of otic competence. By 8 hpf (75% epiboly), *foxi1* expression upregulates at the posterolateral edges of the original domain in a pattern that precedes and fully encompasses the preotic domain of *pax8*, as well as other preotic markers (see Fig. 1). The *foxi1* gene is disrupted in *hearsay* (*hsy*) mutants, which show deficiencies in ear and jaw development. The otic vesicle is either severely reduced in size or is totally ablated. Expression of *pax8* is blocked in all *hsy* mutants, whereas later markers such as *pax2a* are sometimes weakly expressed in erratic patterns during later stages of otic development. Misexpression of *foxi1* induces ectopic expression of *pax8*. It is not yet known whether this is sufficient to induce formation of ectopic otic vesicles because misexpression of *foxi1* leads to embryonic death at later stages (A. Fritz,

12. Zebrafish Ear Development

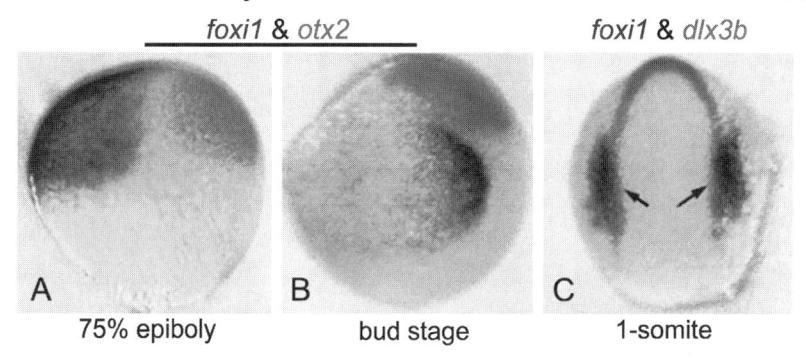

foxi1 & otx2   **foxi1 & dlx3b**

A 75% epiboly   B bud stage   C 1-somite

Figure 1 Expression of *foxi1*. (A, B) Lateral views (anterior to the top, dorsal to the right) showing expression of *foxi1* (black) and the forebrain marker *otx2* (red). At 75% epiboly (A) *foxi1* is expressed uniformly in anteroventral ectoderm. By bud stage (B), *foxi1* has strongly upregulated in preotic placode and downregulated in ventral cells. (C) Dorsal view (anterior to the top) showing *foxi1* (black) and *dlx3* (red) at the 1- somite stage. Expression of *foxi1* fully overlaps the preotic domain of *dlx3*. Reprinted with permission from Solomon, K. S., Kudoh, T., Dawid, I. G., and Fritz, A. (2003). Zebrafish *foxi1* mediates otic placode formation and jaw development. *Development* **130**, 929–940. (See Color Insert.)

personal communication). In any case, these data indicate that *foxi1* regulates an early step in otic induction, and it may also regulate otic competence at earlier stages of gastrulation. The variability of the *hsy* phenotype also indicates that the inductive process is complex and involves multiple pathways that are only partially dependent on *foxi1*.

It is likely that several signaling pathways regulate *foxi1*. Bmp signaling appears to regulate early expression of *foxi1* in the ventral ectoderm. This domain is similar to the expression patterns of *bmp2b*, *bmp4*, and *bmp7* (Dick *et al.*, 2000; Kishimoto *et al.*, 1997), and *foxi1* is not expressed in Bmp pathway mutants (A. Fritz, personal communication). Expression of *foxi1* is restricted from prospective posterior ectoderm by the action of Wnt8, loss of which allows the ventral domain of *foxi1* to expand all the way to the germring (Riley, unpublished data). Later in development, it appears that Fgf and Bmp cooperate to upregulate *foxi1* in the prospective otic region. Localized misexpression of *fgf3* or *fgf8* can induce high-level expression of *foxi1* in the anterior head region, but only in the ectoderm adjacent to the neural plate (Riley, unpublished data).

The *hsy* mutation is remarkable in several regards. First, it is the only known case in which a single gene defect can totally ablate otic development. Second, *hsy* was identified in a simple morphological screen for mutants with inner ear defects and underscores the utility of this time-honored approach in identifying fundamental new functions. Although *foxi1* regulates the earliest stage of preotic development yet identified in zebrafish, it is not yet clear whether it plays a conserved role in other vertebrates. A knockout of mouse *Foxi1* also causes inner ear defects, but the phenotype is considerably milder than in zebrafish and its role in preplacodal development has not yet been examined in detail (Hulander *et al.*, 1998, 2003).

G. *dlx3b* and *dlx4b*

In addition to *foxi1* and *pax8*, there are several genes expressed along the lateral edges of the neural plate—in the "preplacodal domain"—that play a role in otic development. The preplacodal domain appears to represent a distinct ectodermal domain lying between the neural and nonneural (epidermal) ectoderm (Baker and Bronner-Fraser, 2001; Whitfield *et al.*, 2002). Preplacodal markers initially show uniform expression throughout the preplacodal domain and later become restricted to individual placodes. Of the various preplacodal markers identified in zebrafish, *dlx3b* and *dlx4b* (formerly *dlx3* and *dlx7*) are the only two shown to have a clear role in preotic development. These genes are closely linked in a tail-to-tail orientation that has been conserved among most vertebrate *dlx* homologues (Kraus and Lufkin, 1999). *dlx3b* and *dlx4b* are initially detected in the preplacodal domain by 9 hpf, shortly after *pax8* is induced in preotic cells (Akimenko *et al.*, 1994; Ekker *et al.*, 1992; Ellies *et al.*, 1997). By 11 hpf (3 somites), expression of *dlx3b* and *dlx4b* begins to upregulate in the prospective otic and nasal placodes (see Fig. 1C). Expression is subsequently lost in other regions of the preplacodal domain. As the ear develops, *dlx3b* and *dlx4b* continue to be expressed throughout the otic placode but become restricted to dorsal cells after the otic vesicle forms. A deletion that removes both genes (*b380*) causes total ablation of the inner ear (Fritz *et al.*, 1996; Solomon and Fritz, 2002). Preotic expression of *pax8* occurs normally in *b380* mutants, but later otic markers are not detectably expressed. Although the *b380* deletion removes a number of other genes that could potentially contribute to the earless phenotype, coinjection of morpholinos directed against *dlx3b* and *dlx4b* completely blocks expression of *pax2a* in the preplacode. Otic vesicles are eventually produced but are extremely small and lack otoliths. Patterning within these rudimentary otic vesicles has not been examined in detail, but the absence of otoliths suggests that sensory patches fail to form properly. Knocking down *dlx3b* alone results in a more moderate phenotype in which all otic markers are expressed, albeit at reduced levels, and the otic vesicle is of intermediate size and usually contains a single otolith. Knocking down *dlx4b* alone has no effect on ear morphology and most otic markers are expressed normally. It is only when *dlx4b*-MO is coinjected with *dlx3b*-MO that a clear function can be demonstrated. Thus, these two genes are partially redundant, with *dlx3b* contributing disproportionately to the regulation of otic induction.

In addition, analysis of potential autoregulation shows that *dlx3b* and *dlx4b* also have distinct nonoverlapping functions (Solomon and Fritz, 2002). Injection of *dlx4b*-MO reduces *dlx4b* expression but has no effect on *dlx3b*. This could indicate that Dlx4b protein stimulates transcription of its own gene. In contrast, injection of *dlx3b*-MO causes a dramatic upregulation in expression of both *dlx3b* and *dlx4b*, suggesting that *dlx3b* normally inhibits *dlx* gene expression. However, since Dlx proteins are thought to act as transcriptional activators (Kraus and Lufkin, 1999), it seems likely that Dlx3b stimulates expression of some other function that in turn represses *dlx* gene transcription. The significance of this differential regulation is unknown but could reflect a feedback mechanism that serves to maintain an optimal, moderate level of *dlx* gene function.

As with *foxi1*, expression of *dlx3b* and *dlx4b* in the preplacodal domain is probably mediated in part by Bmp signaling. In *swirl* (*bmp2b*) mutants, *dlx3b* is not expressed (Nguyen *et al.*, 1998). This could reflect a direct requirement for transcriptional regulation or, alternatively, it could be an indirect effect stemming from expansion of the neural ectoderm at the expense of nonneural and preplacodal ectoderm. However, studies in *Xenopus* and chick strongly suggest that Bmp signaling directly regulates the expression of *Dlx* genes (Feledy *et al.*, 1999; Pera *et al.*, 1999). In addition, dorsal axial signals also help regulate *Dlx* expression. In chick, for example, it has been shown that grafts of organizer tissue cooperate with Bmp released from beads to induce ectopic *Dlx5*, a gene normally expressed in the preplacodal domain in a manner similar to *dlx3b* and *dlx4b* in zebrafish (Pera *et al.*, 1999). Thus, a balance of ventral Bmp and dorsal axial signals may be required to activate expression of *dlx* genes in a pattern restricted to the preplacodal domain. The identity of the putative dorsal signal is not yet known, but Nodal and Fgf are two candidates.

H. Nodal–Fgf Interactions

There are two Nodal genes in zebrafish, *cyclops* (*cyc*) and *squint* (*sqt*) (Feldman *et al.*, 1998; Rebagliati *et al.*, 1998; Sampath *et al.*, 1998). Loss of both functions leads to complete loss of mesendoderm and narrowing of the neural plate. Otic vesicles are produced, but they are small and malformed. A similar phenotype is caused by the loss of *one-eyed pinhead* (*oep*), which encodes an essential cofactor required for Nodal signaling (Gritsman *et al.*, 1999). Both maternal and zygotic sources of *oep* must be disrupted (MZ-*oep* mutants) to phenocopy the *cyc*–*sqt* double mutant. This can be conveniently achieved by injecting *oep*-MO. Although otic induction is delayed by 1–2 hpf in embryos injected with *oep*-MO, all otic and preotic markers are expressed in an appropriate sequence, albeit in smaller than

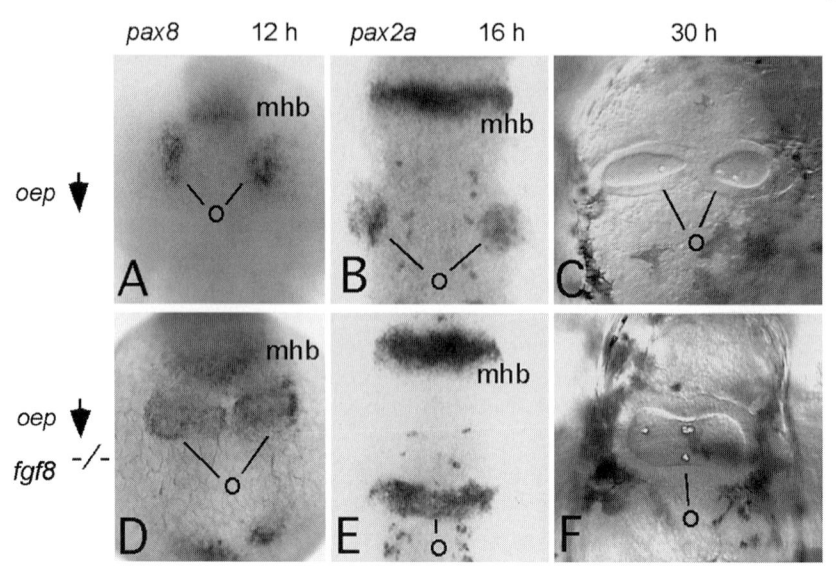

Figure 2 Ear development in embryos deficient in *oep* and *fgf8*. (A–C) Wildtype embryos injected with *oep*-MO. Expression of *pax8* at 12 hpf (A) or *pax2a* at 16 hpf (B) marks the developing otic placodes (o). (C) At 30 hpf, otic vesicles are well formed but often elongate medially and touch at the midline. (D–F) *ace*/+ intercross progeny injected with *oep*-MO. About 25% of injected progeny showed dramatic changes in gene expression and otic vesicle morphology. These are inferred to be *ace* (*fgf8*) mutants. At 12 hpf, the otic domain of *pax8* forms bilateral transverse bands that nearly touch at the midline (D). By 16 hpf, *pax2a* is expressed in a contiguous transverse stripe through the hindbrain (E). At 30 hpf, a single large otic vesicle forms at the midline and fully spans the width of the hindbrain (F). All images are dorsal views with anterior to the top. mhb, midbrain–hindbrain border; o, otic placode or vesicle. (See Color Insert.)

normal domains (Phillips *et al.*, 2001; Riley, unpublished observations). The reduced size and delayed development of otic tissue may reflect loss of mesendodermal signals since *fgf3* and *fgf8* are expressed on time in the hindbrain. In addition, Nodal signaling appears to play a second and previously unanticipated role in otic development as revealed by its interaction with *fgf8*. Since disrupting either Nodal or Fgf8 reduces the size of the otic placode, it was expected that loss of both signals would cause a more severe reduction in otic tissue. Surprisingly, the opposite occurs. Injection of *oep*-MO into *ace* (*fgf8*) mutants results in an enlarged domain of preplacodal markers, including *pax8* (Fig. 2). Moreover, the preplacodal patches of *pax8* expand medially to nearly touch at the midline. By 16 hpf, otic expression of *pax2a* forms a contiguous transverse stripe through the middle of the hindbrain. Later in development, a single large otic vesicle is observed spanning the width of the hindbrain. Otoliths are always produced but their number and distribution are highly variable. This nonadditive phenotype is obtained only when both Fgf8 and Nodal signaling are disrupted, indicating that either pathway is sufficient to restrict otic development from the dorsal midline. As dorsalizing factors, high levels of Fgf8 and Nodal normally help induce and maintain neural ectoderm. Presumably, lower signaling levels are encountered by cells at the edges of the neural plate, i.e., in the preplacodal domain. Nodal may act primarily as an inhibitor of otic fate such that the level of Nodal signaling must drop below a critical threshold to relieve repression. Alternatively, moderate levels of Nodal might act positively to induce preplacodal genes, possibly in cooperation with ventrally expressed Bmp. This might explain the synergism observed between Bmp and node tissue in inducing *Dlx5* in chick embryos (Pera *et al.*, 1999). Similarly, although Fgf signaling is required for otic induction, it appears that only moderate levels are compatible with otic development. Widespread misexpression of *fgf8* results in severe dorsalization of the embryo and impairment of preplacodal development (Fürthauer *et al.*, 1997; Riley, unpublished observations). In the absence of Fgf8 and Nodal signaling, it is possible that cells in the hindbrain are not sufficiently committed to a neural fate and therefore respond to residual Fgf3 by adopting an otic identity. Coinjection of *oep*-MO and *fgf3*-MO is lethal at an early stage (Riley, unpublished observations); therefore the effects of this combination on otic development cannot be assessed.

IV. Patterning of the Placode and Early Vesicle

A. Overview

The first signs of regional specification in the ear can already be detected in the nascent otic placode at 14 hpf (10 somites). Markers of hair cell specification, including various *delta* genes and *zath1* (see later) are expressed in discrete patches in the anterior and posterior ends of the placode (Haddon *et al.*, 1998b; Whitfield *et al.*, 2002; Riley, unpublished observations). These patches, which correspond to the primordia of the utricular and saccular maculae, form in medial placodal cells lying adjacent to rhombomeres 4 and 6. Expression of *nkx5.1* and *pax5* marks the anterior quarter of the otic placode between 16.5 and 17.5 hpf (15–17 somites), reflecting the first detectable asymmetry along the AP axis (Adamska *et al.*, 2000; Pfeffer *et al.*, 1998). Recent findings indicate that signals emitted by the hindbrain, including Fgf3 and Hedgehog, play an important role in specifying these identities. Various other genes have been identified that are necessary for differentiation of dorsal or ventral structures in the otic vesicle. General mechanisms seem to have been highly conserved between zebrafish and mouse, although there are some important differences, too.

B. Fgf3 Signaling from the Hindbrain

The hindbrain continues to express *fgf3* abundantly in r4 until approximately 19 hpf (20 somites) (Kwak *et al.*, 2002; Maroon *et al.*, 2002; Maves *et al.*, 2002; Phillips *et al.*, 2001; Walshe *et al.*, 2002). Gene knockdown experiments suggest that this domain of *fgf3* regulates anterior fates in the medial portion of the otic placode, including specification of the anterior (future utricular) sensory macula (Kwak *et al.*, 2002). Knockdown of *fgf3* suppresses hair cell formation and alters expression of several AP markers in the otic placode and vesicle. Expression of *pax5* and *nkx5.1* is eliminated in *fgf3* morphants. In contrast, the POU-related gene *zp23* (Hauptmann and Gerster, 2000), which is normally limited to the medial wall of the otic vesicle adjacent to r5 and r6, is expanded to include cells adjacent to r4 in *fgf3* morphants. Hindbrain expression of *fgf3* is altered in *valentino* (*val*) mutants, and this causes a corresponding change in ear patterning. The *val* gene encodes a bZip transcription factor orthologous to mouse *MAFb* (*Kreisler*) and is expressed in r5 and r6 (Cordes and Barsh, 1994; Moens *et al.*, 1996, 1998). In *val* mutants, r5 and r6 are not specified properly, and *fgf3* is expressed strongly from r4 through the r5/6 region from 10 hpf (tail bud stage) onward. There are corresponding changes in otic development. For example, *val* mutants express *pax5* and *nkx5.1* throughout the medial wall of the otic vesicle (Fig. 3C, F) and *zp23* is not detected. Furthermore, ectopic hair cells are produced in the medial wall of the otic vesicle adjacent to r5, a region normally devoid of hair cells. All of these ear-patterning defects are suppressed by knocking down *fgf3* in *val* mutants (Kwak *et al.*, 2002).

Unlike *fgf3*, *fgf8* appears to play little or no role in AP patterning in the ear (Kwak *et al.*, 2002). *fgf8* is expressed strongly in r4 through 12 hpf (6 somites) but is no longer detectable by 14 hpf (10 somites). Although *ace* (*fgf8*) mutants often fail to produce a posterior sensory patch, expression of AP markers is not changed. Expression of *fgf8* is not altered in *val* mutants, and loss of *fgf8* function does not reverse the patterning defects seen in *val* mutants. Rather, the mutations interact additively: Double mutants produce otic vesicles that are smaller than in either of the single mutants, but with AP patterning defects and ectopic hair cells adjacent to r5. Thus, while *fgf8* is needed to produce an otic placode of proper size and dimensions, it does not participate directly in the *val–fgf3* pathway that regulates AP patterning.

The *val* mutant phenotype is phenocopied by disruption of *vhnf1*, a homeobox gene normally coexpressed with *val* in the hindbrain (Sun and Hopkins, 2001). In *vhnf1* mutants, *val* expression is severely reduced. Hindbrain expression of *fgf3* has not been reported in this background, but the inner ear shows a similar morphology to *val* mutants, and *pax5* expression is expanded throughout the medial wall of the otic vesicle (Z. Sun, personal

12. Zebrafish Ear Development

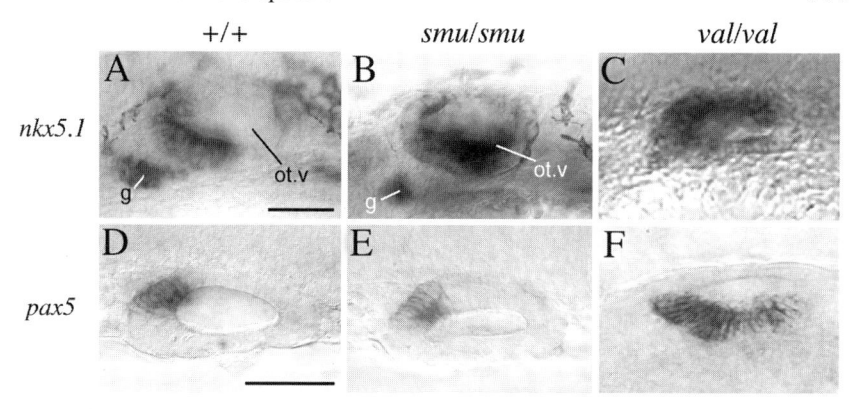

Figure 3 AP patterning defects in *smu* and *val* mutants. (A–C) Lateral views showing expression of *nkx5.1* at 30 hpf in a wildtype embryo (A) or a *smu* mutant (B), or at 24 hpf in a *val* mutant (C). In the *smu* mutant, expression is expanded posteriorly in ventrolateral cells but is normal in medial cells. In contrast, expression in the *val* mutant is expanded posteriorly in medial cells but not in ventrolateral cells. (D–F) Dorsal views showing *pax5* expression in a wildtype (D) or a *smu* mutant (E), or a lateral view of a *val* mutant ear (F). Expression is normal in *smu* mutants but is expanded posteriorly in ventromedial cells in *val* mutants. Scale bars in A–C and D–F: 50 μm. g, Statoacoustical ganglion; ot.v, otic vesicle. Anterior is to the left in all specimens. Dorsal is toward the top in A–C and F. Medial is to the top in D and E. Images in A, B, D, and E were reprinted from Hammond, K., Loynes, H. E., Folarin, A. A., Smith, J., and Whitfield, T. T. (2003). Hedgehog signalling is required for currect anteroposterior patterning of the zebrafish otic vesicle. *Development* **130**, 1403–1417. Images in C and F were reprinted from Kwak, S. J., Phillips, B. T., Heck, R., and Riley, B. B. (2002). An expanded domain of *fgf3* expression in the hindbrain of zebrafish *valentino* mutants results in mis-patterning of the otic vesicle. *Development* **129**, 5279–5287.

communication). Thus, it is likely that *vhnf1* lies upstream of *val*, or in a mutually dependent parallel pathway, such that loss of either function alters ear development in part by causing expansion of the hindbrain domain of *fgf3*.

Contrary to the situation in zebrafish, Fgf3 in mouse is normally upregulated in r5 and r6 soon after the otic placode forms, but upregulation fails to occur in *MAFb* (*kreisler*) mutants (Mahmood *et al.*, 1996; McKay *et al.*, 1996). The otic vesicle is grossly malformed in *MAFb* mutants (Deol, 1964), but most of the described changes reflect relatively late secondary defects caused by hydrops (fluid buildup) resulting from loss of the endolymphatic duct. The endolymphatic duct is one of the first structures formed in the mouse otic vesicle, and its morphogenesis requires Fgf3 (Mansour *et al.*, 1993; McKay *et al.*, 1996). The cochlea, too, is an early structure that requires Fgf3 and is variably deficient in *MAFb* mutants (see Chapter 8 in this volume by Wright and Mansour). However, analysis of early otic markers has not been reported in *MAFb* mutants; therefore a direct comparison with *val* mutants is not yet possible. Such a comparison would be quite interesting given the marked differences in expression of Fgf3 in the hindbrain.

C. Hedgehog Signaling from the Floorplate and Notochord

Three *hedgehog* genes are expressed in midline tissues from the gastrula stage onward: *tiggy-winkle hedgehog* is expressed in the floorplate, *echidna hedgehog* is expressed in the notochord, and *sonic hedgehog* (*shh*) is expressed in both tissues (Currie and Ingham, 1996; Ekker *et al.*, 1995; Krauss *et al.*, 1993). Together these genes regulate AP patterning in the otic placode and vesicle (Hammond *et al.*, 2003), but in a manner that is distinct from regulation by *fgf3*. Due to the multiple redundancies of *hedgehog* genes, loss of any one gene does not strongly affect ear development. However, the zebrafish genome contains only a single homolog of *smoothened* (*smu*), which encodes an essential component of the Hedgehog signal transduction pathway (Chen *et al.*, 2001; Varga *et al.*, 2001). The otic vesicles in *smu* mutants develop with mirror image duplication of anterior structures and loss of posterior structures. A similar phenotype is observed in *chameleon* (*con*) mutants, which are also blocked in Hedgehog signal transduction (Schauerte *et al.*, 1998). Expression of *nkx5.1* is expanded into the posterior end of the ear and the posterior marker follistatin is not expressed. However, this does not constitute a phenocopy of the *val* phenotype. Expansion of *nkx5.1* expression in *val* mutants is limited to the medial wall of the otic vesicle, whereas it is limited to the ventrolateral wall in *con* and *smu* mutants (Hammond *et al.*, 2003; Kwak *et al.*, 2002; see Fig. 3). In addition, expression of the ventrolateral marker *otx1* is not altered in *val* mutants but shows distinctive changes in *con* and *smu* mutants consistent with a mirror image duplication of anterior fates. Finally, *pax5* expression is expanded posteriorly in *val* mutants but is normal in *con* and *smu* mutants (see Fig. 3). These data suggest that Fgf3 and Hedgehog signaling affect different aspects of AP patterning in the ear, with the former predominantly affecting medial cells and the latter more strongly affecting ventrolateral cells. This does not mean, however, that Hedgehog does not affect medial cells. Indeed, misexpression of *shh* causes a mirror image duplication of posterior fates, and *pax5* is no longer expressed (Hammond *et al.*, 2003).

The ear phenotype in Hedgehog pathway mutants is surprising in several regards. First, it is not clear how Hedgehog signaling can affect AP patterning since the relevant ligands appear to be expressed uniformly along this axis. Second, it is also unclear why loss of Hedgehog signaling causes a mirror image duplication of anterior fates (i.e., anterior–middle–anterior) rather than a more uniform anteriorization. There are numerous other examples in which loss or gain of Hedgehog signaling causes mirror image duplications, but in all other reported cases, the axis of duplication is defined by having an asymmetrical source of Hedgehog at one end. The zebrafish ear does not appear to fit this pattern. Presumably, other signaling pathways must act to locally modify or restrict Hedgehog signaling. Furthermore, the *con* and *smu* mutant phenotypes suggest that loss of *hedgehog* activates (or fails to antagonize) a potential anteriorizing factor located in or near the posterior end of the ear. A candidate for such a factor could be the weak Fgf signaling normally seen in r6 (Fürthauer *et al.*, 2001; Maves *et al.*, 2002; Raible and Brand, 2001; Roehl and Nüsslein-Volhard, 2001). Presumably, endogenous levels of Hedgehog signaling cannot normally suppress the much stronger source of Fgf in r4. Third, there are marked differences between zebrafish and mouse with regard to the role of Hedgehog in patterning the otic vesicle. In mouse *Shh* mutants, the ear develops with loss of a subset of ventromedial markers and expansion of dorsolateral markers (Riccomagno *et al.*, 2002). No such defects are observed in zebrafish Hedgehog pathway mutants. Furthermore, AP patterning appears to be normal in mouse *Shh* mutants. The significance of these differences is not yet clear.

D. *eya1*

The dorsoventral axis of the ear is regulated in part by the transcription factor gene *eya1*, a homologue of *eyes-absent* in *Drosophila*. The *eya1* gene is disrupted in *dog-eared* (*dog*) mutants (D. Kozlowski, personal communication), which develop with a complete lack of sensory cristae and poorly organized semicircular canals (Whitfield *et al.*, 1996). The absence of cristae is likely to reflect a direct requirement for *eya1*, which is normally expressed in the ventral epithelium of the otic vesicle (Sahly *et al.*, 1999). Perturbation of semicircular canals could be a secondary consequence of the failure to produce cristae, which are thought to emit signals required for normal morphogenesis of the surrounding nonsensory epithelium. Candidates for such signals are Bmp2b, Bmp4, and Fgf8, all of which are expressed at high levels in the cristae by 48 hpf (Leger and Brand, 2002; Mowbray *et al.*, 2001). *dog* mutants are also deficient in SAG neurons, precursors of which normally begin to delaminate from the ventral epithelium by 22 hpf (26 somites) (Haddon and Lewis, 1996). Newly delaminated neuroblasts transiently express *snail2* and *nkx5.1* (Adamska *et al.*, 2000; Thisse *et al.*, 1996) and only later begin to express neural markers as they differentiate into SAG neurons. In *dog* mutants, the number of *snail2*-expressing cells is strongly reduced (Whitfield *et al.*, 2002), suggesting that too few neuroblasts are specified.

It is interesting that *dog* mutants do not show an earlier and more severe ear phenotype because *eya1* is initially expressed in the preplacodal domain and is later found throughout the otic placode (Sahly *et al.*, 1999). Such an early and widespread pattern might have been expected to regulate early otic or preotic development. In mouse, *Eya1* null mutants produce only a simple epithelial vesicle that fails to undergo further patterning or morphogenesis (Xu *et al.*, 1999). It is likely, however, that zebrafish embryos express additional *eya* homologues not found in mouse since the zebrafish genome often has extra copies of genes reflecting a genome-wide duplication that occurred early in the teleost lineage. If additional *eya* genes are found to be expressed in the ear, it will be important to test their function individually as well as in combination with the *dog* mutation.

E. van gogh (vgo)

Another gene required for development of the DV axis of the ear is *vgo*. In *vgo* mutants, the otic vesicle is small and development of the cristae and semicircular canals is totally blocked at an early stage (Whitfield *et al.*, 1996). Specifically, *otx1* is not expressed in the ventrolateral epithelium at 24 hpf, and *msxC* and *bmp4*, which normally mark the emerging cristae at 48 hpf (Ekker *et al.*, 1992; Mowbray *et al.*, 2001), are not detected in *vgo* mutants. Maculae and otoliths initially form in their normal locations at the anterior and posterior ends of the vesicle. By 4 days, however, the maculae appear to fuse into a small centrally located sensory patch. In addition to the inner ear defects, pharyngeal arches are disorganized and partially fused, and endodermal pouches do not form (Piotrowski and Nüsslein-Volhard, 2000). Arch tissues derived from mesoderm and neural crest are initially normal but become chaotic at later stages without the organizing influence of endodermal pouches. The identity of the *vgo* gene and the nature of its function are not yet known. The ear phenotype could reflect a direct role of *vgo* in the otic vesicle, but loss of signals from pharyngeal endoderm could also impair ear development. Indeed, tissue recombination studies in amphibians indicate that interaction with endoderm is required for proper patterning of the otic vesicle (Jacobson, 1963a,b).

F. sox10

The *sox10* gene is disrupted in *colourless* (*cls*) mutants, which produce very small ears that are superficially similar to those in *vgo* mutants (Dutton *et al.*, 2001a,b; Whitfield *et al.*, 1996). However, the *cls* phenotype is milder since *otx1* is expressed in the correct domain at 24 hpf and the otic vesicle forms two cristae that show the expected expression of *msxC* at 48 hpf. The lateral crista is missing and semicircular canals are severely malformed. *sox10* is strongly expressed throughout the otic vesicle at 18 hpf, indicating that it could directly regulate otic development (Dutton *et al.*, 2001b). In addition, *cls* mutants fail to produce pigment cells due to the death of corresponding neural crest cells (Kelsh and Eisen, 2000). The mutation does not affect mesenchymal populations derived from neural crest so the otic capsule probably develops normally. Whether loss of other neural crest derivatives contributes to the ear phenotype is not known, but neural crest–derived melanocytes regulate the maturation and function of sensory structures in the mammalian cochlea (reviewed by Price and Fisher, 2001).

G. *jekyll (jek)*

The *jek* gene encodes UDP glucose dehydrogenase, a permissive function required for morphogenesis of the semicircular canals (Neuhauss *et al.*, 1996; Walsh and Stainier, 2001). Loss of *jek* function leads to failure of the epithelial protrusions of the semicircular canals to extend beyond their initial outpocketing. UDP glucose dehydrogenase is required for biosynthesis of glycosaminoglycans, essential components of many extracellular matrix molecules, including chondroitin sulfate, heparan sulfate, and hyaluronic acid. Aberrant production of these molecules could cripple multiple developmental processes (reviewed by Lander and Selleck, 2000; Lin and Perrimon, 2000). The hygroscopic activity of chondroitin sulfate and hyaluronic acid is often necessary to expand the extracellular space sufficiently to accommodate cellular rearrangements. Indeed, blocking synthesis of hyaluronic acid in *Xenopus* causes a similar disruption of semicircular canal morphogenesis (Haddon and Lewis, 1991). In addition, heparan sulfate is necessary for signaling interactions via a number of ligands, including Fgf, Wnt, Tgfβ, and Hh. Thus, the ear defects in *jek* mutants could well result from both physical constriction and impaired cell–cell signaling.

V. Development of Sensory Epithelia

A. Overview

In zebrafish, hair cells begin to differentiate at a much earlier stage than in tetrapod vertebrates. The first hair cells are already visible at 18.5 hpf (19 somites) when the lumen of the otic vesicle begins to expand (Riley *et al.*, 1997). These first cells are atypical in several regards and are referred to as tether cells to distinguish them from later forming hair cells. Tether cells initially appear as immature columnar cells whose cell bodies fully span the otic epithelium, yet they possess fully formed kinocilia that serve as tethers for developing otoliths (Fig. 4A). There are invariably two tether cells at anterior and posterior ends of the otic vesicle, marking the future utricular and saccular maculae, respectively. Tether cells acquire the typical morphology of mature hair cells by 21.5 hpf (25 somites) (Fig. 4B) and persist at the macular centers thereafter. Additional hair cells begin to form just after 24 hpf,

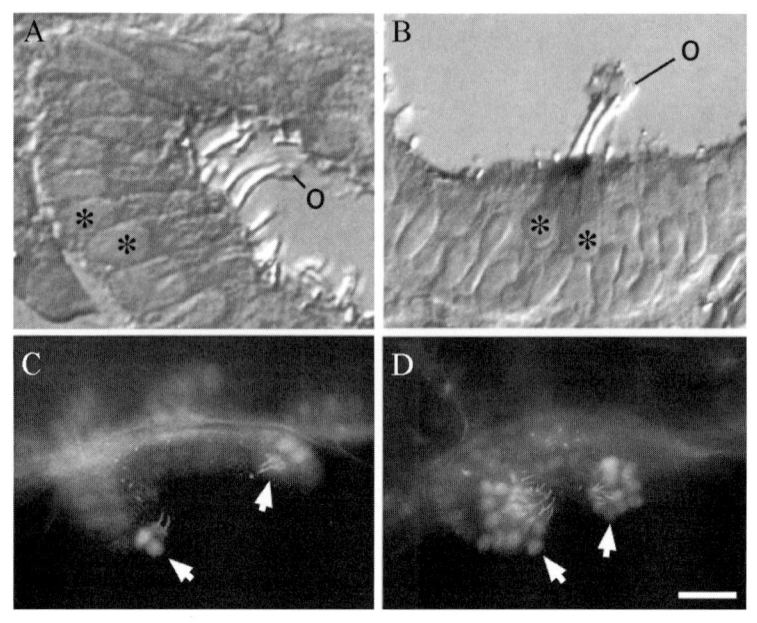

Figure 4 Early hair cell differentiation. (A, B) DIC images of longitudinal sections through anterior sensory epithelia stained with antiacetylated tubulin. Asterisks mark the nuclei of tether cells, the first hair cells to differentiate. (A) Tether cells appear as immature columnar cells at 20 hpf (22 somites), yet they possess kinocilia with attached otoliths (o). Other cells bear short motile cilia, which function to distribute otolith precursor particles. (B) Tether cells take on the morphology of mature hair cells by 21.5 hpf (25 somites). Otolith material (o) is attached to the ends of kinocilia, and the apices of the tether cells stain heavily for acetylated tubulin (brown). (C, D) Dorsolateral views of otic vesicles stained with antimouse Pax2 and anti acetylated tubulin in a wildtype embryo (C) and a dlA^{dx2} mutant (D) at 24 hpf (30 somites). The mouse Pax2 antibody primarily labels zebrafish Pax2a. Hair cells (arrows) are heavily stained for nuclear Pax2 and kinociliary acetylated tubulin. Lower levels of Pax2 staining are visible in intervening cells located within the medial wall of the otic vesicle. The dlA^{dx2} mutant shows a fivefold increase in the number of hair cells. In all panels, anterior is to the left and dorsal is to the top. Scale bar: 5 μm (A, B) or 20 μm.

and these undergo a more typical maturation in which kinocilia form only after the cell body begins to take on the characteristic shape of a mature hair cell. Cristae begin to form by 48 hpf, and the lagenar macula forms around day 10 of larval development (Haddon and Lewis, 1996; Riley and Moorman, 2000). All sensory epithelia continue to grow well into adulthood in a manner commensurate with overall growth of the body (Higgs et al., 2001).

Sensory epithelia are composed of hair cells interspersed with support cells in a regular pattern. It is now generally accepted that both cell types are derived from a common equivalence group in which their respective fates are specified, in part, by Delta–Notch interactions (reviewed by Fekete and

Wu, 2002; Müller and Littlewood-Evans, 2001; Whitfield et al., 2002). Initially, all cells are biased toward the hair cell fate but are prevented from completing differentiation by weak, mutual Delta–Notch signaling. A subset of cells eventually overcome this inhibition and differentiate as hair cells. In the process, they show strong transient upregulation of Delta gene expression. This facilitates "lateral inhibition" by strongly activating Notch receptors on neighboring cells, thereby diverting them onto the support cell pathway. Genetic support for this model has now been obtained in all model vertebrate species, but the strongest support has come from studies in zebrafish.

B. deltaA (dlA)

Multiple delta genes, including *dlA*, are first expressed at low levels in the primordia of the utricular and saccular maculae in the nascent otic placode at 14 hpf (Haddon et al., 1998b). As the otic vesicle forms, delta expression undergoes strong transient upregulation in all newly forming hair cells, beginning with the tether cells. Mutants homozygous for a point mutation in *dlA*, termed dlA^{dx2}, produce a fivefold excess of hair cells and a corresponding reduction in support cells (Riley et al., 1999; Fig. 4C, D). All maculae and cristae are affected similarly. In mouse, disruption of any of the various *Delta* or *Notch* genes expressed in the ear causes considerably milder phenotypes, possibly because of redundancy conferred by coexpression of multiple homologues (Kiernan et al., 2001; Lanford et al., 1999; Zhang et al., 2000; Zine et al., 2000). In zebrafish, too, hair cells coexpress multiple *delta* genes, but the dlA^{dx2} allele encodes a dominant negative protein that appears to antagonize the function of all wildtype Delta homologues. Accordingly, the severity of the dlA^{dx2} phenotype varies with gene dosage: Transheterozygotes carrying one copy of dlA^{dx2} and one copy of a deletion of the *dlA* locus show a 2.7-fold increase in hair cell production, and $dlA^{dx2/+}$ heterozygotes show a 1.7-fold increase (Riley et al., 1999). Thus, the antagonistic activity of the mutant protein is reduced by lowering its concentration and is further reduced by increasing the level of wildtype protein.

C. E3 Ubiquitin Ligase

The *mind bomb* (*mib*) mutation causes the most severe disruption of Delta–Notch signaling yet seen in any vertebrate species. In the inner ear, *mib* mutants produce greatly enlarged maculae containing 10 times too many hair cells and no detectable support cells (Haddon et al., 1998a, 1999; Riley et al., 1999). The maculae expand to cover most of the ventromedial surface of the otic vesicle. No cristae or semicircular canals are produced, possibly reflecting loss of progenitor populations and associated signaling interactions. Posititional cloning of the *mib* gene shows that it encodes an E3 ubiquitin ligase, and biochemical studies show that it functions by ubiquitylating Delta protein, targeting it for destruction via the ubiquitin–proteosome pathway (Itoh *et al.*, 2003). A similar activity was recently shown for *neuralized* (*neur*) in *Drosophila* and *Xenopus* (Deblandre *et al.*, 2001; Lai *et al.*, 2001; Pavlopoulos *et al.*, 2001). There are two nonexclusive models for how *mib* and *neur* facilitate Delta–Notch signaling (reviewed by Krämer, 2001; Lai, 2002). The first model is based on the observation that high-level coexpression of Delta and Notch in the same cell interferes with its ability to receive Delta signals (Jacobsen *et al.*, 1998). In this case, *mib* and *neur* could function cell-autonomously to reduce Delta levels in receiving cells, increasing their sensitivity to Delta signals from neighboring cells. Second, it has been shown in *Drosophila* that, after activating Notch signal transduction, ligand–receptor complexes are endocytosed into the Delta-presenting cell (Parks *et al.*, 2000). Interfering with this process prevents effective Delta–Notch signaling. In this case, *mib* and *neur* could function nonautonomously (in the sending cell) by facilitating Delta turnover, thereby increasing the capacity to send new signals. There is compelling evidence that *neur* facilitates both mechanisms in *Drosophila*, depending on context (Deblandre *et al.*, 2001; Lai *et al.*, 2001; Lai and Rubin, 2001; Parks *et al.*, 2000; Pavlopoulos *et al.*, 2001; Yeh *et al.*, 2000). Similarly, analysis of genetic mosaics in zebrafish suggest that *mib* function is context dependent: It facilitates the ability to send signals during neurogenesis in the brain (Itoh *et al.*, 2003), whereas it facilitates signal reception during hair cell specification in the inner ear (Whitfield *et al.*, 2002).

Despite the similarity in function between *mib* and *neur*, these genes are not orthologous, and zebrafish *neur* and *mib* genes map to different loci (Itoh *et al.*, 2003). The functional relationship between *neur* and *mib* has not been analyzed in zebrafish, nor have *mib* orthologues been functionally analyzed in other species. These remain important goals in future research.

D. *zath1*

Hair cell equivalence groups are initially marked by expression of proneural genes related to *Drosophila atonal*, which encodes a bHLH transcription factor. In rodents, *Math1* (*Murine atonal homologue*) is necessary and sufficient for cells to differentiate as hair cells (Bermingham *et al.*, 1999; Zheng and Gao, 2000; see Chapter 10 by Gao). *Math1* mutant mice produce defective sensory epithelia composed entirely of support cells, whereas misexpression of *Math1* in rat cochlear cultures stimulates production of excess hair cells. The zebrafish orthologue, *zath1*, is first detected in the macular equivalence groups adjacent to r4 and r6 by 14 hpf, and expression strongly upregulates by 19 hpf as hair cells begin to differentiate (Whitfield *et al.*, 2002). The domain of *zath1* expression is greatly expanded in *mib* mutants. Injection of *zath1*-MO impairs hair cell production in wildtype embryos and partially suppresses the ear phenotype in *mib* mutants (Riley, unpublished observations).

In both zebrafish and mouse, loss of *Math1/zath1* function does not block specification of the equivalence group since support cells are still produced despite impairment of hair cell production. Obviously some other gene(s) is (are) sufficient for specification of the equivalence group. Whether *Math1/zath1* shares a role in this process remains to be seen.

E. neurogenin1

Another *atonal* homologue expressed in the zebrafish ear is *neurogenin1* (*ngn1*) (Andermann *et al.*, 2002). Expression begins by 18 hpf in ventromedial cells of the otic vesicle. As the vesicle develops, *ngn1* expression becomes restricted to a small anteroventral patch of cells and is no longer detected by 41 hpf. The majority of *ngn1*-positive cells appear to be newly specified SAG neuroblasts, which lose *ngn1* expression soon after delaminating from the otic vesicle. Injection of *ngn1*-MO completely blocks development of SAG neurons.

Expression of *ngn1* also overlaps with hair cell equivalence groups in the nascent placode, raising the possibility that it could play a role in hair cell specification. Although knockdown of *ngn1* does not appear to affect hair cell production, the possibility remains that *ngn1* and *zath1* are partially redundant in the specification of hair cells, equivalence groups, or both. Disruption of both functions will be required to address this issue.

F. pax2a and pax2b

As the otic vesicle forms, expression of *pax2a* and *pax2b* becomes restricted to the ventromedial epithelium. Expression strongly upregulates in differentiating hair cells by 24 hpf and is subsequently maintained there (Riley *et al.*, 1999; see Fig. 4C, D). Expression begins to downregulate elsewhere in the ear by 30 hpf but continues to accumulate in sensory epithelia as new hair cells form. The *pax2a* gene is disrupted in *no-isthmus* (*noi*) mutants (Brand *et al.*, 1996). Surprisingly, *noi* mutants develop with nearly twice the normal number of hair cells (Riley *et al.*, 1999). This probably reflects the fact that *delta* genes are expressed at reduced levels in *noi* mutants, which presumably weakens lateral inhibition. In contrast to *noi*, knockdown of *pax2b* impairs hair cell formation (Whitfield *et al.*, 2002). Thus, these two genes appear to regulate different stages of hair cell differentiation, with *pax2b* regulating specification or early differentiation and *pax2a* being required for the subsidiary function of lateral inhibition. How *pax2b* function and the hair cell–specifying function of *zath1* relate to one another is not yet known, but preliminary data suggest that they represent parallel pathways that act downstream of Fgf signaling (Riley, unpublished observations).

VI. Auditory and Vestibular Function

A. Overview

Mutant screens have identified a number of genes that control auditory or vestibular function by regulating hair cell physiology or survival, or by regulating development of accessory structures. Several of the identified loci have proven to be homologous to genes associated with human deafness. Others may be fish specific but are still likely to be informative regarding general mechanisms of inner ear function.

B. Circler Mutants

In a screen for mutants with defects in locomotion, eight complementation groups were identified in which the gross morphology of the inner ear is normal but larvae cannot swim in a controlled manner (Granato *et al.*, 1996; Nicolson *et al.*, 1998). Instead they swim in spirals, loops, or upside down, indicating impairment of vestibular function. In addition, all circler mutants fail to demonstrate an acoustic startle response, suggesting loss of auditory function. Several circler loci have been genetically identified but only one, *mariner* (*mar*), has been described in the literature. The *mar* gene encodes Myosin VIIa (Ernest *et al.*, 2000), which in humans is associated with Usher syndrome type 1B (Weil *et al.*, 1995) and in mouse is disrupted in *shaker1* mutants (Gibson *et al.*, 1995). Myosin VIIa is localized within the cross-links between stereocilia in all hair cells. Stereociliary cross-links are believed to facilitate the opening of ion channels during hair cell stimulation. In *mar* mutants, ciliary bundles are splayed and disorganized (Nicolson *et al.*, 1998). In addition, these mutants do not show externally measurable (microphonic) potentials following mechanical stimulation of lateral line neuromasts, suggesting that these hair cells are also severely impaired.

C. Otolith Mutants

Multiple mutations have been identified that lead to loss of otoliths or production of detached otoliths (Malicki *et al.*, 1996; Whitfield *et al.*, 1996). In most cases the developmental basis for the otolith defect is unknown. Although it has been noted that otolith mutants often show severe vestibular deficits, it was not immediately clear whether this resulted from loss of function in the utricle, saccule, or both. However, analysis of one mutant locus, *monolith* (*mnl*), has clarified the nature of otolith formation as well as the relative importance of the utricle and saccule for vestibular function (Riley and Grunwald, 1996; Riley and Moorman, 2000; Riley et al., 1997). A particularly useful feature of *mnl* mutants is that they can be experimentally manipulated to alter the distribution of otoliths within the otic vesicle. Under normal developmental conditions, the majority of *mnl* mutants produce enlarged saccular otoliths but lack utricular otoliths. This phenotype is caused by a differential delay in the ability of anterior and posterior tether cells to bind otolith precursor particles. Binding is initially impaired in all tether cells, but posterior (saccular) tether cells recover faster and often acquire all otolith particles before the anterior (utricular) tether cells recover. The number of particles is limiting and initial phase of "otolith seeding" ceases by 24 hpf, after which otolith deficiencies cannot be corrected. However, briefly immobilizing *mnl* mutants in agarose with the anterior end oriented downward allows the dense otolith particles to settle to the lowest point of the ear. The local increase in concentration facilitates formation of utricular otoliths bilaterally in all mutants. Roughly half are also able to form saccular otoliths. When immobilized in the opposite orientation, 100% lack utricular otoliths and form saccular otoliths only. Analysis of the different phenotypic classes produced in this way showed that it is the utricle alone that mediates vestibular function, at least during early larval development. Mutants lacking utricular otoliths in both ears show loss of balance and motor coordination, they cannot sense gravity, and they invariably die during larval development. Mutants with at least one utricular otolith show nearly normal vestibular function and survive. The presence or absence of saccular otoliths has no measurable effect on vestibular function or survival. Instead, the primary role of the saccular macula seems to be in hearing (Popper and Fay, 1993). Hence, *mnl* mutants with bilateral loss of saccular otoliths are likely to show auditory deficits as adults, although this has yet to be tested. How the utricular and saccular maculae become functionally specialized remains an important unanswered question. Another interesting problem is why loss of vestibular function is lethal. Adult animals readily tolerate vestibular dysfunction due to compensation from visual cues. However, the ability to generate compensatory neural pathways in the brain may initially require an intact vestibular system from which to build. The *mnl* locus has been mapped to a small interval on linkage group 1 but the corresponding gene has not yet been identified (Riley, unpublished data).

D. A Screen for Hearing Mutants

A high-throughput automated screen was recently described for detecting hearing defects in adult zebrafish (Bang *et al.*, 2002). In the screen, groups of 36 fish are tested simultaneously for an auditory startle response following a 400-Hz sound pulse. This frequency stimulates the auditory system but does not appreciably affect the lateral line. Responses are monitored by an automated system that tracks the positions of fish just before and after the sound pulse, and nonresponders are subjected to further testing. In an initial screen of 6596 adults, 72 nonresponders were identified. Most of these were found to have anatomical defects in the swim bladder or Weberian ossicles, accessory structures that make up the sound conduction apparatus in zebrafish (Higgs *et al.*, 2003). Although most of the animals screened were F1 progeny of ENU-mutagenized males, none of the nonresponders transmitted their hearing defects to F2 progeny, suggesting an environmental cause. Nevertheless, the high correlation between nonresponsiveness and mechanical defects in the sound conduction apparatus shows that the screen works as intended. Moreover, fish were screened at 3–4 months of age, and it is possible that rescreening when the fish are older will identify dominant mutations causing late-onset deafness. Although such mutations are expected to be rare, dominant nonsyndromic mutations are associated with perhaps 30 forms of human deafness, most of which show late onset and are progressive (reviewed by Petersen, 2002). Identifying corresponding zebrafish mutations would provide an experimentally tractable model system for studying their functions and etiologies. It is also possible that the animal husbandry could be modified to screen for recessive deafness genes. This approach would be more laborious and space intensive, but potential recessive deafness loci are far more common, making such a screen feasible.

VII. Conclusions and Prospects

Zebrafish is living up to its potential as a useful genetic model system, both in terms of identifying previously unknown functions and also for assessing functions of previously identified candidate genes. Many key regulators of early ear development have been identified. The immediate inducers of otic differentiation appear to be Fgf3 and Fgf8. Prior to this, however, otic competence is influenced by early axial signals as judged by their effects on preotic markers such as (in order of expression) *foxi1*, *pax8*, *dlx3b*, *dlx4b*, *pax2a*, and *pax2b*. Bmp signaling regulates the early domain of *foxi1* in the ventral ectoderm, and upregulation in the preotic domain appears to involve a balance of ventral Bmp and dorsal Fgf signaling. Similarly, *dlx3b* and *dlx4b* are also regulated by Bmp and dorsal axial signals. In the dorsal-most ectoderm, high levels of Fgf and Nodal signaling maintain neural identity and restrict otic development. Wnt8 acts early to set the posterior limit of otic competence and later plays a role in the hindbrain by regulating timely expression of *fgf3* and *fgf8*. Preliminary data suggest that the transcription factors encoded by preotic marker genes represent several distinct pathways, each requiring Fgf signaling, but which are also partially independent. After otic induction, Fgf3 and Hh signaling from the hindbrain specify anterior fates within the developing otic placode. Fgf8 does not contribute to AP patterning but, together with Fgf3, plays an essential role in specification of sensory patches and neuroblasts within the otic epithelium. Hair cells and neuroblasts require expression of proneural transcription factors Zath1 and Ngn1, respectively. Within developing sensory epithelia, Delta–Notch signaling generates the salt-and-pepper pattern of hair cells and support cells by a classical lateral inhibition mechanism. Regulation of DV patterning is less well understood, but *sox10*, *eya1*, *jek*, and *vgo* all play a role.

How such functions are coordinated, and exactly how they regulate the balance of growth, differentiation, and morphogenesis of inner ear structures, are problems that are only beginning to be addressed. With the relative ease of conducting phenotypic analysis, zebrafish will continue to play a prominent role in elucidating the complex genetic interactions regulating otic development. It is also certain that additional essential genes are yet to be identified, a task for which zebrafish is well suited. In many cases, the functions and processes identified in zebrafish are likely to be conserved. Even in cases where gene functions in zebrafish appear to differ from other vertebrate species, understanding how similar structures are produced by different means will lead to a deeper understanding of core regulatory processes. Indeed, the inner ear field has benefited greatly from a long and fruitful history of comparative studies using a variety of vertebrate species. Though a relative newcomer to the field, zebrafish has had a significant impact in many areas of inner ear research. Future zebrafish studies promise to continue in this tradition as ongoing screens come to fruition, additional genes are cloned, and new techniques are developed.

References

Adamska, M., Leger, S., Brand, M., Hadrys, T., Braun, T., and Bober, E. (2000). Inner ear and lateral line expression of a zebrafish *Nkx5-1* gene and its downregulation in the ears of *FGF8* mutant, *ace. Mech. Dev.* **97,** 161–165.

Akimenko, M.-A., Ekker, M., Wegner, J., Lin, W., and Westerfield, M. (1994). Combinatorial expression of three zebrafish genes related to *Distal-Less*: Part of a homeobox gene code for the head. *J. Neurosci.* **14,** 3475–3486.

Andermann, P., Ungos, J., and Raible, D. W. (2002). Neurogenenin1 defines zebrafish cranial sensory ganglia precursors. *Dev. Biol.* **251,** 45–58.

Appel, B., Fritz, A., Westerfield, M., Grunwald, D. J., Eisen, J. S., and Riley, B. B. (1999). Delta-mediated specification of midline cell fates in zebrafish embryos. *Curr. Biol.* **9**, 247–256.

Baker, C. V. H., and Bronner-Fraser, M. (2001). Vertebrate cranial placodes. I. Embryonic induction. *Dev. Biol.* **232**, 1–61.

Bang, P. I., Yelick, P. C., Malicki, J. J., and Sewell, W. F. (2002). High-throughput behavioral screening method for detecting auditory response defects in zebrafish. *J. Neurosci. Methods* **118**, 177–187.

Belting, H. G., Hauptmann, G., Meyer, D., Abdelilah-Seyfried, S., Chitnis, A., et al. (2001). *spiel ohne grenzen/pou2* is required during establishment of the zebrafish midbrain-hindbrain boundary organizer. *Development* **128**, 4165–4176.

Bermingham, N. A., Hassan, B. A., Price, S. D., Vollrath, M. A., Ben-Arie, N., et al. (1999). *Math1*: An essential gene for the generation of inner ear hair cells. *Science* **284**, 1837–1841.

Bever, M. M., and Fekete, D. M. (2002). Atlas of the developing inner ear in zebrafish. *Dev. Dyn.* **223**, 536–543.

Brand, M., Heisenberg, C. P., Jiang, Y. J., Beuchle, D., Lun, K., et al. (1996). Mutations in zebrafish genes affecting the formation of the boundary between midbrain and hindbrain. *Development* **123**, 179–190.

Burgess, S., Reim, G., Chen, W., Hopkins, N., and Brand, M. (2002). The zebrafish *spiel-ohne-grenzen* (*spg*) gene encodes POU domain protein Pou2 related to mammalian *Oct4* and is essential for formation of the midbrain and hindbrain, and for pre-gastrula morphogenesis. *Development* **129**, 905–916.

Chen, W., Burgess, S., and Hopkins, N. (2001). Analysis of the zebrafish *smoothened* mutant reveals conserved and divergent functions of hedgehog activity. *Development* **128**, 2385–2396.

Cordes, S. P., and Barsh, G. S. (1994). The mouse segmentation gene *kr* encodes a novel basic domain-leucine zipper transcription factor. *Cell* **79**, 1025–1034.

Currie, P. D., and Ingham, P. W. (1996). Induction of a specific muscle cell type by a hedgehog-like protein in the zebrafish. *Nature* **382**, 452–455.

Deblandre, G. A., Lai, E. C., and Kintner, C. (2001). *Xenopus* neuralized is a ubiquitin ligase that interacts with Xdelta1 and regulated Notch signaling. *Dev. Cell* **1**, 795–806.

Deol, M. S. (1964). The abnormalities of the inner ear in *kreisler* mice. *J. Embryol. Exp. Morphol.* **12**, 475–490.

Dick, A., Hild, M., Bauer, H., Imain, Y., Maifeld, H., et al. (2000). Essential role of *Bmp7* (*snailhouse*) and its prodomain in dorsoventral patterning of the zebrafish embryo. *Development* **127**, 343–354.

Draper, B. W., Morcos, P. M., and Kimmel, C. B. (2001). Inhibition of zebrafish *fgf8* pre-mRNA splicing with morpholino oligos: A quantifiable method for gene knockdown. *Genesis* **30**, 154–156.

Dutton, K., Dutton, J. R., Pauliny, A., and Kelsh, R. N. (2001a). A morpholino phenocopy of the *colourless* mutant. *Genesis* **30**, 188–189.

Dutton, K. A., Pauliny, A., Lopes, S. S., Elworthy, S., Carney, T., et al. (2001b). Zebrafish *colourless* encodes *sox10* and may specify non-ectomesenchymal neural crest fates. *Development* **128**, 4113–4125.

Ekker, M., Akimenko, M. A., Bremiller, R., and Westerfield, M. (1992). Regional expression of three homeobox transcripts in the inner ear of zebrafish embryos. *Neuron* **9**, 27–35.

Ekker, S. C., Ungar, A. R., Greenstein, P., von Kessler, S. P., Porter, J. A., et al. (1995). Patterning activities of vertebrate *hedgehog* proteins in the developing eye and brain. *Curr. Biol.* **5**, 944–955.

Ellies, D. L., Stock, D. W., Hatch, G., Giroux, G., Weiss, K. M., and Ekker, M. (1997). Relationship between the genomic organization and the overlapping embryonic expression patterns of the zebrafish *dlx* genes. *Genomics* **45**, 580–590.

Ernest, S., Rauch, G. J., Haffter, P., Geisler, R., Petit, C., and Nicolson, T. (2000). *Mariner* is defective in *myosin VIIA*: A zebrafish model for human hereditary deafness. *Hum. Mol. Genet.* **9,** 2189–2196.

Fekete, D. M., and Wu, D. K. (2002). Revisiting cell fate specification in the inner ear. *Curr. Opin. Neurobiol.* **12,** 35–42.

Feldman, B., Gates, M. A., Egan, E. S., Dougan, S. T., Rennebeck, G., *et al.* (1998). Zebrafish organizer development and germ-layer formation require nodal-related signals. *Nature* **395,** 181–185.

Feledy, J. A., Beanan, M. J., Sandoval, J. J., Goodrich, J. S., Lim, J. H., *et al.* (1999). Inhibitory patterning of the anterior neural plate in *Xenopus* by homeodomain factors *Dlx3* and *Msx1. Dev. Biol.* **212,** 455–464.

Fritz, A., Rozowski, M., Walker, C., and Westerfield, M. (1996). Identification of selected gamma ray induced deficiencies in zebrafish using multiplex polymerase chain reaction. *Genetics* **144,** 1735–1745.

Fürthauer, M., Reifers, F., Brand, M., Thisse, B., and Thisse, C. (2001). *sprouty4* acts in vivo as a feedback-induced antagonist of FGF signalling in zebrafish. *Development* **128,** 2175–2186.

Fürthauer, M., Thisse, C., and Thisse, B. (1997). A role for FGF-8 in the dorsoventral patterning of the zebrafish gastrula. *Development* **124,** 4253–4264.

Gibson, F., Walsh, J., Mburu, P., Varela, A., Brown, K. A., *et al.* (1995). A type VII myosin encoded by the mouse deafness gene *shaker-1. Nature* **374,** 62–64.

Granato, M., van Eeden, F. J. M., Schach, U., Trowe, T., Brand, M., *et al.* (1996). Genes controlling and mediating locomotion behavior of the zebrafish embryo and larva. *Development* **126,** 399–413.

Gritsman, K., Zhang, J., Cheng, S., Heckscher, E., Talbot, W. A., and Schier, A. F. (1999). The EGF-CFC protein one-eyed pinhead is essential for nodal signaling. *Cell* **97,** 121–132.

Groves, A. K., and Bronner-Fraser, M. (2000). Competence, specification and commitment in otic placode induction. *Development* **127,** 3489–3499.

Haddon, C., Jiang, Y. L., Smithers, L., and Lewis, J. (1998a). *Delta-Notch* signalling and the patterning of sensory cell differentiation in the zebrafish ear: Evidence from the *mind bomb* mutant. *Development* **125,** 4637–4644.

Haddon, C., and Lewis, J. (1996). Early ear development in the embryo of the zebrafish, *Danio rerio. J. Comp. Neurol.* **365,** 113–128.

Haddon, C., Mowbray, C., Whitfield, T., Jones, D., Gschmeissner, S., and Lewis, J. (1999). Hair cells without supporting cells: Further studies in the ear of the zebrafish *mind bomb* mutant. *J. Neurocytol.* **28,** 837–850.

Haddon, C., Smithers, L., Schneider-Maunoury, S., Coche, T., Henrique, D., and Lewis, J. (1998b). Multiple delta genes and lateral inhibition in zebrafish primary neurogenesis. *Development* **125,** 359–370.

Haddon, C. M., and Lewis, J. H. (1991). Hyaluronan as a propellant for epithelial movement: The development of semicircular canals in the inner ear of *Xenopus. Development* **112,** 541–550.

Hammond, K., Loynes, H. E., Folarin, A. A., Smith, J., and Whitfield, T. T. (2003). Hedgehog signalling is required for correct anteroposterior patterning of the zebrafish otic vesicle. *Development* **130,** 1403–1417.

Harrison, R. G. (1945). Relations in the symmetry in the developing embryo. *Trans. Conn. Acad. Arts. Sci.* **36,** 277–330.

Hauptmann, G., and Gerster, T. (1995). Pou-2a zebrafish gene active during cleavage stages and in the early hindbrain. *Mech. Dev.* **51,** 127–138.

Hauptmann, G., and Gerster, T. (2000). Combined expression of zebrafish *Brn-1-* and *Brn-2-*related POU genes in the embryonic brain, pronephric primordium, and pharyngeal arches. *Dev. Dyn.* **218,** 345–358.

Higgs, D. M., Rollo, A. K., Souza, M. J., and Popper, A. N. (2003). Development of form and function in peripheral auditory structures of the zebrafish (*Danio rerio*). *J. Acoust. Soc. Am.* **113**, 1145–1154.

Higgs, D. M., Souza, M. J., Wilkins, H. R., Presson, J. C., and Popper, A. N. (2001). Age- and size-related changes in the inner ear and hearing ability of the adult zebrafish (*Danio rerio*). *JARO* **3**, 174–184.

Hulander, M., Kiernan, A. E., Blomqvist, S. R., Carlsson, P., Samuelsson, E. J., et al. (2003). Lack of *pendrin* expression leads to deafness and expansion of the endolymphatic compartment in inner ears of *Foxi1* null mutant mice. *Development* **130**, 2013–2025.

Hulander, M., Wurst, W., Carlsson, P., and Ernerbäck, S. (1998). The winged helix transcription factor *Fkh10* is required for normal development of the inner ear. *Nat. Genet.* **20**, 374–376.

Itoh, M., Kim, C. H., Palardy, G., Oda, T., Jiang, Y. J., et al. (2003). Mind Bomb is a ubiquitin ligase that is essential for efficient activation of Notch signaling by Delta. *Dev. Cell* **4**, 67–82.

Jacobsen, T. L., Brennan, K., Martinez Arias, A., and Muskavitch, A. T. (1998). Cis-interactions between Delta and Notch modulate neurogenic signalling in *Drosophila*. *Development* **125**, 4531–4540.

Jacobson, A. G. (1963a). The determination and positioning of the nose, lens, and ear. I. Interactions within the ectoderm, and between the ectoderm and underlying tissues. *J. Exp. Zool.* **154**, 273–284.

Jacobson, A. G. (1963b). The determination and positioning of the nose, lens, and ear. II. The role of the endoderm. *J. Exp. Zool.* **154**, 285–291.

Kelsh, R. N., and Eisen, J. S. (2000). The zebrafish *colourless* gene regulates development of non-ectomesenchymal neural crest derivatives. *Development* **127**, 515–525.

Kiernan, A. E., Ahituv, N., Fuchs, H., Galling, R., Avraham, K. B., et al. (2001). The Notch ligand Jagged1 is required for inner ear sensory development. *Proc. Natl. Acad. Sci. USA* **98**, 3873–3878.

Kishimoto, Y., Lee, K. H., Zon, L., Hammerschmidt, M., and Schulte-Merker, S. (1997). The molecular nature of zebrafish *swirl:* BMP2 function is essential during early dorsoventral patterning. *Development* **124**, 4457–4466.

Kozlowski, D. J., Murakami, T., Ho, R. K., and Weinberg, E. S. (1997). Regional cell movement and tissue patterning in the zebrafish embryo revealed by fate mapping with caged fluorescein. *Biochem. Cell Biol.* **75**, 551–562.

Krämer, H. (2001). Neuralized: Regulating Notch by putting away Delta. *Dev. Cell* **1**, 725–731.

Kraus, P., and Lufkin, T. (1999). Mammalian *Dlx* homeobox gene control of craniofacial and inner ear morphogenesis. *J. Cell. Biochem. Suppl.* **32/33**, 133–140.

Krauss, S., Concordet, J. P., and Ingham, P. W. (1993). A functionally conserved homology of the *Drosophila* segment polarity gene *shh* is expressed in tissues with polarizing activity in zebrafish embryos. *Cell* **75**, 1431–1444.

Krauss, S., Johansen, T., Korzh, V., and Fjose, A. (1991). Expression of the zebrafish paired box gene *pax[zf-b]* during early neurogenesis. *Development* **113**, 1193–1206.

Kwak, S. J., Phillips, B. T., Heck, R., and Riley, B. B. (2002). An expanded domain of *fgf3* expression in the hindbrain of zebrafish *valentino* mutants results in mis-patterning of the otic vesicle. *Development* **129**, 5279–5287.

Ladher, R. K., Anakwe, K. U., Gurney, A. L., Schoenwolf, G. C., and Francis-West, P. H. (2000). Identification of synergistic signals initiating inner ear development. *Science* **290**, 1965–1968.

Lai, E. (2002). Protein degradation: Four E3s for the Notch pathway. *Curr. Biol.* **12**, R74–R78.

Lai, E. C., Deblandre, G. A., Kintner, C., and Rubin, G. M. (2001). *Drosophila* neuralized is a ubiquitin ligase that promotes the internalization and degradation of Delta. *Dev. Cell* **1**, 783–794.

Lai, E. C., and Rubin, G. M. (2001). Neuralized functions cell-autonomously to regulate a subset of Notch-dependent processes during adult *Drosophila* development. *Dev. Biol.* **231**, 217–233.

Lander, A. D., and Selleck, S. B. (2000). The elusive functions of proteoglycans: in vivo veritas. *J. Cell Biol.* **148**, 227–232.

Lanford, P. J., Lan, Y., Jiang, R., Lindsell, C., Weinmaster, G., *et al.* (1999). Notch signalling pathway mediates hair cell development in mammalian cochlea. *Nat. Genet.* **21**, 289–292.

Leger, S., and Brand, M. (2002). Fgf8 and Fgf3 are required for zebrafish ear placode induction, maintenance and inner ear patterning. *Mech. Dev.* **119**, 91–108.

Lekven, A. C., Thorpe, C. J., Waxman, J. S., and Moon, R. T. (2001). Zebrafish *wnt8* encodes two Wnt8 proteins on a bicistronic transcript and is required for mesoderm and neurectoderm patterning. *Dev. Cell* **1**, 103–114.

Lin, X., and Perrimon, N. (2001). Role of heparin sulfate proteoglycans in cell-cell signaling in *Drosophila*. *Matrix Biol.* **19**, 303–307.

Lombardo, A., Isaacs, H. V., and Slack, J. M. W. (1998). Expression and functions of *FGF-3* in *Xenopus* development. *Int. J. Dev. Biol.* **42**, 1101–1107.

Mahmood, R., Kiefer, P., Guthrie, S., Dickson, C., and Mason, I. (1995). Multiple roles for FGF-3 during cranial neural development in the chicken. *Development* **121**, 1399–1410.

Mahmood, R., Mason, I. J., and Morrisskay, G. M. (1996). Expression of *FGF-3* in relation to hindbrain segmentation, otic pit position and pharyngeal arch morphology in normal and retinoic acid exposed mouse embryos. *Anat. Embryol.* **194**, 13–22.

Malicki, J., Schier, A. F., Solnica-Krezel, L., Stemple, D. L., Neuhauss, S. C., *et al.* (1996). Mutations affecting development of the zebrafish ear. *Development* **123**, 275–283.

Mansour, S. L., Goddard, J. M., and Capecchi, M. R. (1993). Mice homozygous for a targeted disruption of the proto-oncogene *int-2* have developmental defects in the tail and inner ear. *Development* **117**, 13–28.

Maroon, H., Walshe, J., Mahmood, R., Keifer, P., Dickson, C., and Mason, I. (2002). Fgf3 and Fgf8 are required together for formation of the otic placode and vesicle. *Development* **129**, 2099–2108.

Maves, L., Jackman, W., and Kimmel, C. B. (2002). FGF3 and FGF8 mediate a rhombomere 4 signaling activity in the zebrafish hindbrain. *Development* **129**, 3825–3837.

McKay, I. J., Lewis, J., and Lumsden, A. (1996). The role of *FGF-3* in early inner ear development: An analysis in normal and *kreisler* mutant mice. *Dev. Biol.* **174**, 370–378.

Mendonsa, E. S., and Riley, B. B. (1999). Genetic analysis of tissue-interactions required for otic placode induction in the zebrafish. *Dev. Biol.* **206**, 100–112.

Moens, C. B., Cordes, S. P., Giorgianni, M. W., Barsh, G. S., and Kimmel, C. B. (1998). Equivalence in the genetic control of hindbrain segmentation in fish and mouse. *Development* **125**, 381–391.

Moens, C. B., Yan, Y. L., Appel, B., Force, A. G., and Kimmel, C. B. (1996). *valentino*: A zebrafish gene required for normal hindbrain segmentation. *Development* **122**, 3981–3990.

Mowbray, C., Hammerschmidt, M., and Whitfield, T. T. (2001). Expression of BMP signalling pathway members in the developing zebrafish inner ear and lateral line. *Mech. Dev.* **108**, 179–184.

Müller, U., and Littlewood-Evans, A. (2001). Mechanisms that regulate mechanosensory hair cell differentiation. *Trends Cell Biol.* **11**, 334–342.

Nasevicius, A., and Ekker, S. C. (2000). Effective targeted gene "knockdown" in zebrafish. *Nat. Genet.* **26**, 216–220.

Neuhauss, S. C. F., Solnica-Krezel, L., Schier, A. F., Zwartkruis, F., Stemple, D. L., *et al.* (1996). Mutations affecting craniofacial development in zebrafish. *Development* **126**, 357–389.

Nguyen, V. H., Schmid, B., Trout, J., Connors, S. A., Ekker, M., and Mullins, M. C. (1998). Ventral and lateral regions of the zebrafish gastrula, including the neural crest progenitors, are established by a *bmp2b/swirl* pathway of genes. *Dev. Biol.* **199**, 93–110.

Nicolson, T., Rüsch, A., Friedrich, R. W., Granato, M., Ruppersberg, J. P., and Nüsslein-Volhard, C. (1998). Genetic analysis of vertebrate sensory hair cell mechanosensation: The zebrafish circular mutants. *Neuron* **20**, 271–283.

Parks, A. L., Klueg, K. M., Stout, J. R., and Muskavitch, M. A. T. (2000). Ligand endocytosis drives receptor dissociation and activation in the Notch pathway. *Development* **127**, 1373–1385.

Pavlopoulos, E., Pitsouli, C., Klueg, K. M., Muskavitch, A. T., Moschonas, N. K., and Delidakis, C. (2001). Neuralized encodes a peripheral membrane protein involved in Delta signaling and endocytosis. *Dev. Cell* **1**, 807–816.

Pera, E., Stein, S., and Kessel, M. (1999). Ectodermal patterning in the avian embryo: Epidermis versus neural plate. *Development* **126**, 63–73.

Petersen, M. B. (2002). Non-syndromic autosomal-dominant deafness. *Clin. Genet.* **62**, 1–13.

Pfeffer, P. L., Gerster, T., Lun, K., Brand, M., and Busslinger, M. (1998). Characterization of three novel members of the zebrafish *Pax2/5/8* family: Dependency of *Pax5* and *Pax8* on the *Pax2.1* (*noi*) function. *Development* **125**, 3063–3074.

Phillips, B. T., Bolding, K., and Riley, B. B. (2001). Zebrafish *fgf3* and *fgf8* encode redundant functions required for otic placode induction. *Dev. Biol.* **235**, 351–365.

Piotrowski, T., and Nüsslein-Volhard, C. (2000). The endoderm plays an important role in patterning the segmented pharyngeal region in zebrafish (*Danio rerio*). *Dev. Biol.* **225**, 339–356.

Popper, A. N., and Fay, R. R. (1993). Sound detection and processing by fish: Critical review and major research questions. *Brain Behav. Evol.* **41**, 14–38.

Price, E. R., and Fisher, D. E. (2001). Sensorineural deafness and pigmentation genes: Melanocytes and the Mitf transcriptional network. *Neuron* **30**, 15–18.

Raible, F., and Brand, M. (2001). Tight transcriptional control of the ETS domain factors Erm and Pea3 by Fgf signaling during early zebrafish development. *Mech. Dev.* **107**, 105–117.

Rebagliati, M. R., Toyama, R., Haffter, P., and Dawid, I. B. (1998). cyclops encodes a nodal-related factor involved in midline signaling. *Proc. Natl. Acad., Sci. USA* **95**, 9932–9937.

Reifers, F., Bohli, H., Walsh, E. C., Crossley, P. H., and Stainier, D. Y. R. (1998). *Fgf8* is mutated in zebrafish *acerebellar* (*ace*) mutants and is required for maintenance of midbrain-hindbrain boundary and somitogenesis. *Development* **125**, 2381–2395.

Reim, G., and Brand, M. (2002). *spiel-ohne-grenzen/pou2* mediates regional competence to respond to Fgf8 during zebrafish early neural development. *Development* **129**, 917–933.

Riccomagno, M. M., Martinu, L., Mulheisen, M., Wu, D. K., and Epstein, D. J. (2002). Specification of the mammalian cochlea is dependent on sonic hedgehog. *Genes Dev.* **16**, 2365–2378.

Riley, B. B., Chiang, M. Y., Farmer, L., and Heck, R. (1999). The *deltaA* gene of zebrafish mediates lateral inhibition of hair cells in the inner ear and is regulated by *pax2.1*. *Development* **126**, 5669–5678.

Riley, B. B., and Grunwald, D. J. (1996). A mutation in zebrafish affecting a localized cellular function required for normal ear development. *Dev. Biol.* **179**, 427–435.

Riley, B. B., and Moorman, S. J. (2000). Development of utricular otoliths, but not saccular otoliths, is necessary for vestibular function and survival in zebrafish. *J. Neurobiol.* **43**, 329–337.

Riley, B. B., Zhu, C., Janetopoulos, C., and Aufderheide, K. J. (1997). A critical period of ear development controlled by distinct populations of ciliated cells in the zebrafish. *Dev. Biol.* **191**, 191–201.

Roehl, H., and Nüsslein-Volhard, C. (2001). Zebrafish *pea3* and *erm* are general targets of FGF8 signalling. *Curr. Biol.* **11**, 503–507.

Sahly, I., Andermann, P., and Petit, C. (1999). The zebrafish *eya1* gene and its expression pattern during embryogenesis. *Dev. Genes Evol.* **209**, 399–410.

Sampath, K., Rubinstein, A. L., Cheng, A. M. S., Liang, J. O., Fekany, K., *et al.* (1998). Induction of the zebrafish ventral brain and floorplate requires cyclopsnodal signalling. *Nature* **395**, 185–189.

Schauerte, H., E., van Eeden, F. J. M., Fricke, C., Odenthal, J., Strähle, U., and Haffter, P. (1998). *Sonic hedgehog* is not required for the induction of medial floorplate cells in the zebrafish. *Development* **125**, 2983–2993.

Solomon, K. S., and Fritz, A. (2002). Concerted action of two *dlx* paralogs in sensory placode formation. *Development* **129**, 3127–3136.

Solomon, K. S., Kudoh, T., Dawid, I. G., and Fritz, A. (2003). Zebrafish *foxi1* mediates otic placode formation and jaw development. *Development* **130**, 929–940.

Stone, L. S. (1931). Induction of the ear by the medulla and its relation to experiments on the lateralis system in amphibia. *Science* **74**, 577.

Sun, Z., and Hopkins, N. (2001). *vhnf1*, the MODY5 and familial GCKD-associated gene, regulates regional specification of the zebrafish gut, pronephros, and hindbrain. *Genes Dev.* **15**, 3217–3229.

Thisse, C., Thisse, B., and Postlethwait, J. H. (1996). Expression of *snail2*, a second member of the zebrafish snail family, in cephalic mesendoderm and presumptive neural crest of wild-type and *spadetail* mutant embryos. *Dev. Biol.* **172**, 86–99.

Varga, Z. M., Amores, A., Lewis, K. E., Yan, Y. L., Postlethwait, J. H., *et al.* (2001). Zebrafish *smoothened* functions in ventral neural tube specification and axon tract formation. *Development* **128**, 3497–3509.

Vendrell, V., Carnicero, E., Giraldez, F., Alonso, M. T., and Schimmang, T. (2000). Induction of inner ear fate by FGF3. *Development* **127**, 155–165.

Waddington, C. H. (1937). The determination of the auditory placode in the chick. *J. Exp. Bio.* **14**, 232–239.

Walsh, E. C., and Stainier, D. Y. R. (2001). UDP-glucose dehydrogenase required for cardiac valve formation in zebrafish. *Science* **293**, 1670–1673.

Walshe, J., Maroon, H., McGonnell, I. M., Dickson, C., and Mason, I. (2002). Establishment of hindbrain segmental identity requires signaling by FGF3 and FGF8. *Curr. Biol.* **12**, 1117–1123.

Weil, D., Blanchard, S., Kaplan, J., Guilford, P., Gibson, F., *et al.* (1995). Defective myosin VIIA gene responsible for Usher syndrome type 1B. *Nature* **374**, 60–61.

Whitfield, T. T., Granato, M., van Eeden, F. J. M., Schach, U., Brand, M., *et al.* (1996). Mutations affecting development of the zebrafish inner ear and lateral line. *Development* **123**, 241–254.

Whitfield, T. T., Riley, B. B., Chiang, M. Y., and Phillips, B. (2002). Development of the zebrafish inner ear. *Dev. Dyn.* **223**, 427–458.

Wilkinson, D. G., Bhatt, S., and McMahon, A. P. (1989). Expression pattern of the FGF-related proto-oncogene *int*-2 suggests multiple roles in fetal development. *Development* **105**, 1331–1136.

Woo, K., and Fraser, S. E. (1998). Specification of the hindbrain fate in the zebrafish. *Dev. Biol.* **197**, 283–296.

Xu, P. X., Adams, J., Peters, H., Brown, M. C., Heaney, S., and Maas, R. (1999). Eya1-deficient mice lack ears and kidneys and show abnormal apoptosis of organ primordia. *Nat. Genet.* **23**, 113–117.

Yeh, E., Zhou, L., Rudzik, N., and Boulianne, G. L. (2000). Neuralized functions cell autonomously to regulate *Drosophila* sense organ development. *EMBO J.* **19**, 4827–4837.

Zhang, N., Martin, G. V., Kelley, M. W., and Gridley, T. (2000). A mutation in the *Lunatic fringe* gene suppresses the effects of a *Jagged2* mutation on inner hair cell development in the cochlea. *Curr. Biol.* **10,** 659–662.

Zheng, J. L., and Gao, W. Q. (2000). Overexpression of *Math1* induces robust production of extra hair cells in postnatal rat inner ears. *Nat. Neurosci.* **3,** 580–586.

Zine, A., Van de Water, T. R., and de Ribaupierre, F. (2000). Notch signaling regulates the pattern of auditory hair cell differentiation in mammals. *Development* **127,** 3373–3383.

13
Functional Development of Hair Cells

Ruth Anne Eatock and Karen M. Hurley
The Bobby R. Alford Department of Otorhinolaryngology and Communicative Sciences, Baylor College of Medicine, Houston, Texas 77030

I. Introduction
II. Maturation of Function
   A. Evoked Activity
   B. Spontaneous Activity
III. Development of Mechanoelectrical Transduction
   A. Hair Bundle Development
   B. Biophysics of Transduction
IV. Developmental Acquisition of Basolateral Ion Channels
   A. Outwardly Rectifying Potassium Conductances: g_{Kv}, g_A, $g_{K,Ca}$, $g_{K,n}$, $g_{K,L}$
   B. Inwardly Rectifying Potassium Conductances: g_{K1}, g_h
   C. Voltage-Gated Sodium Conductances: g_{Na}
   D. Voltage-Gated Calcium Conductances: g_{Ca}
   E. Summary of Developmental Changes in Basolateral Conductances
V. Outer Hair Cell Electromotility
VI. Interactions between Developing Hair Cells and Innervating Nerve Fibers
   A. Hair Cells and Primary Afferents
   B. Efferent Synapses on Hair Cells
VII. Summary and Concluding Remarks
   References

Embryonic hair cells in chicks and mammals have functional transduction channels and voltage-gated outwardly rectifying potassium (K^+) channels, fast inwardly rectifying channels, and voltage-gated sodium (Na^+) and calcium (Ca^{2+}) channels. Together these channels may participate in spiking by the immature hair cells, which may drive rhythmic or bursting activity of neurons at higher levels of the auditory pathway. The electrical activity of immature hair cells may influence afferent synaptogenesis and differentiation indirectly by promoting neurotrophin release or more directly by glutamatergic transmission. With maturation, a number of changes tend to reduce hair cell spiking: Na^+, Ca^{2+}, and fast inwardly rectifying channels may become less numerous, whereas outwardly rectifying K^+ channels become more numerous and diverse. These changes signal the transformation from a developing epithelium with active formation of synaptic contacts to a sensing epithelium where receptor potentials represent the mechanical input in a graded fashion. The composition of the late-arriving outwardly rectifying K^+ channels is specific to the hair cell's type and location in the sensory epithelium and confers specialized properties on the receptor potentials. Fast, Ca^{2+}-gated channels serve high-quality electrical tuning in certain tall hair cells of the chick cochlea. In rodent cochlear hair cells and type I hair cells of chick and rodent vestibular organs, large outwardly rectifying conductances lower the input resistance, enhancing the speed and linearity of voltage responses. © 2003, Elsevier Inc.

I. Introduction

The final cell division that produces a vertebrate hair cell simultaneously produces a sibling that may become a supporting cell or another hair cell (Lang and Fekete, 2001; for reviews of the mechanisms giving rise to one or the other cell type, see Bryant *et al.*, 2002, and Fekete and Wu, 2002). In this chapter we are concerned with the acquisition of features that differentiate hair cells from their sibling supporting cells (Fig. 1): a transducing hair bundle and an abundance of basolateral ion channels, both voltage- and ligand-gated. We also consider how the immature electrical activity of differentiating hair cells before the onset of stimulus-evoked responses may influence development of the rest of the afferent pathway.

In order to place hair cell differentiation in a functional context, we begin by briefly reviewing the time course of the maturation of auditory and vestibular function in rodents and chicks, the subjects of most developmental research on the inner ear. Fig. 2 shows the timing of some key aspects of morphological and functional development for these two model systems. Some hair cell properties are functional before auditory and vestibular function are mature, as assessed by behavioral and electrophysiological assays, whereas others develop with a similar time course. For an excellent comprehensive review of the development of auditory function in chicks and rodents, the reader is referred to Rübsamen and Lippe (1998). The zebrafish is an important emerging model system for development of the inner ear (Whitfield, 2002; see Chapter 11 by Riley), and this chapter presents zebrafish results where those are available.

II. Maturation of Function

A. Evoked Activity

1. Rodent Hearing

In rodents, hearing function begins in the second postnatal week, as evaluated by hearing reflexes and signature potentials generated by various stages in the afferent cascade. Behavioral responses are first evident in mice

13. Functional Development of Hair Cells 391

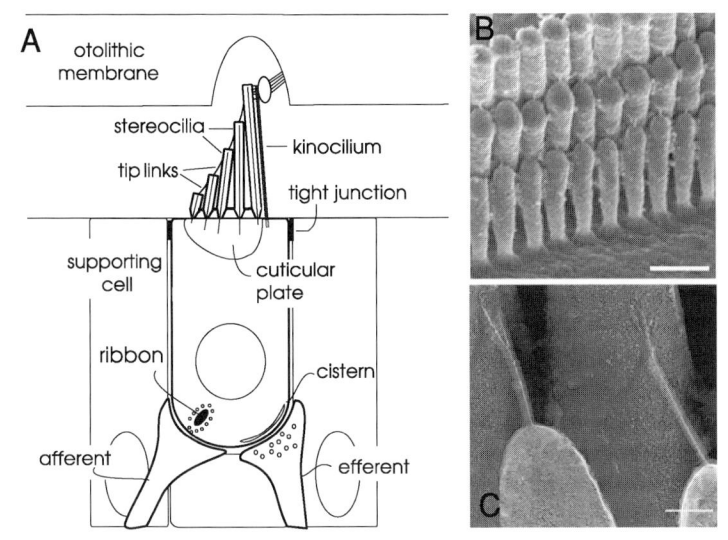

Figure 1 (A) Schematic of a hair cell in the sensory epithelium of the vibration-sensitive frog saccule. Reproduced with permission from Eatock, R. A. (2000). Adaptation in hair cells. *Annu. Rev. Neurosci.* **23,** 285–314. The epithelium comprises hair cells surrounded by supporting cells, plus the terminals of afferent and efferent neurons. The cell bodies of the neurons reside in eighth-nerve ganglia and the brainstem, respectively. *Tight junctions* near the apical (lumenal) surfaces of the cells maintain separate compartments for the extracellular fluids bathing the apical and basolateral cells surfaces (high-K^+, low-Ca^{2+} endolymph, and high-Na^+ perilymph, respectively). The hair bundle is surmounted by an *otolithic membrane*, a gel-like structure that bears calcium carbonate crystals (otoconia; not shown). Linear acceleration of the head causes differential displacement of the epithelium and the denser otoconia, leading to bundle deflection, the effective stimulus to the hair cell. The bundle comprises rows of *stereocilia* that change progressively in height, with the tallest row adjacent to a single *kinocilium*, tethered to the otolithic membrane. Hair cells detect bundle deflection oriented parallel to a line running from the shortest to the tallest row of stereocilia and bisecting the bundle. *Tip links* with the same orientation link stereocilia in adjacent rows and may be the gating springs that apply force to the mechanosensitive transduction channels, and/or a mechanism for adjusting the input to the transduction channels. The stereocilia are anchored in an apical *cuticular plate* via actin rootlets. In the basolateral hair cell compartment, a presynaptic *ribbon* and associated synaptic vesicles (filled with glutamate) face an *afferent* terminal. A postsynaptic *cistern* is apposed to an *efferent* terminal containing vesicles filled with acetylcholine. (B) The staircase of stereocilia in a rat cochlear hair bundle. Scanning electron micrograph. Reproduced with permission from Schneider, M. E., Belyantseva, I. A., Azevedo, R. B., and Kacher, B. (2002). Rapid renewal of auditory hair bundles. *Nature* **418,** 837–838. (C) Tip links connecting two stereociliary pairs in a freeze-etch image of a guinea pig cochlear hair bundle. Reproduced with permission from Kachar, B., Parakkal, M., Kurc, M., Zhao, Y., and Gillespie, P. G. (2000). High-resolution structure of hair-cell tip links. *Proc. Natl. Acad. Sci. USA* **97,** 13336–13341. Scale bars in B and C are 1μm and 100 nm, respectively.

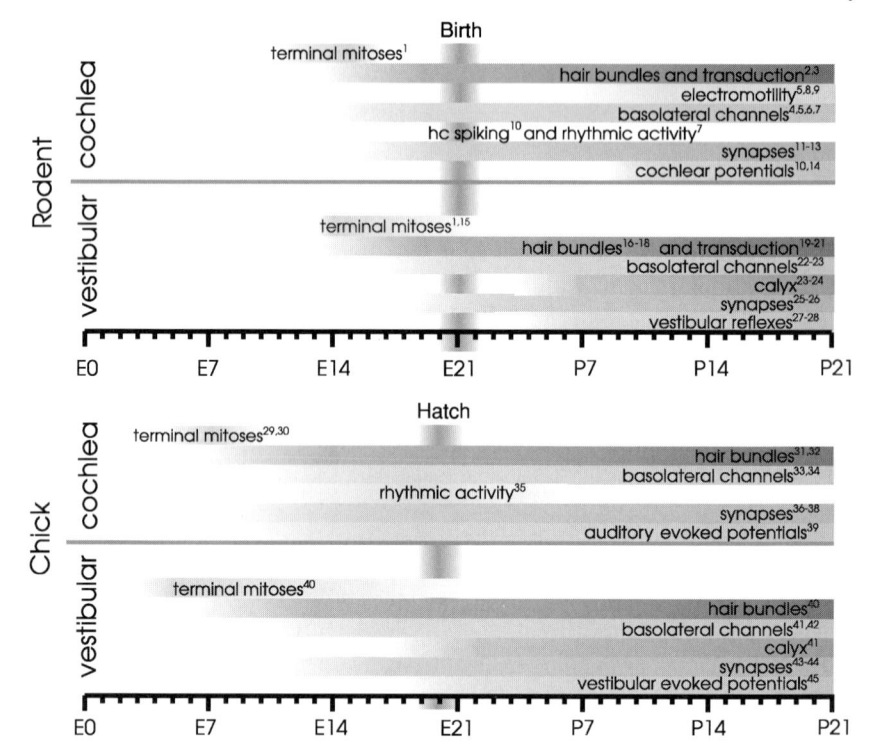

Figure 2 Approximate timing of various developmental changes in cochleas and vestibular organs of rodents and chicks: *terminal mitoses* of hair cells and ganglion cells (primary afferent neurons); *hair bundle* development and mechanoelectrical *transduction*; acquisition of *electromotility* in mammalian outer hair cells; differentiation of *basolateral channels* in hair cells; hair cell *(hc) spiking* and bursting or *rhythmic activity* of primary afferents and higher-order auditory neurons; formation of afferent and efferent *synapses* on hair cells and primary afferent terminals, including the afferent calyx ending on amniote vestibular type I hair cells; and functional measures: *cochlear potentials, vestibular reflexes, auditory evoked potentials* and *vestibular evoked potentials*. References: *Rodent cochlea:* [1]Ruben, 1967; [2]Kros et al., 1992; [3]Holley and Nishida, 1995; [4]Kros et al., 1998; [5]Marcotti and Kros, 1999; [6]Marcotti et al., 1999; [7]Marcotti et al., 2003; [8]He, 1997; [9]Beurg et al., 2001; [10]reviewed in Rübsamen and Lippe, 1998; [11]Echteler et al., 1989; [12]Pujol et al., 1998; [13]Bruce et al., 2000; [14]Mikaelian and Ruben, 1965. *Rodent vestibular organs:* [15]Sans and Chat, 1982; [16]Mbiene et al., 1984; [17]Mbiene and Sans, 1986; [18]Denman-Johnson and Forge, 1999; [19]Géléoc et al., 1997; [20]Holt et al., 1997; [21]G.S.G. Géléoc and J. R. Holt, personal communication; [22]Géléoc et al., 2003; [23]reviewed in Anniko, 1990; [24]Rüsch et al., 1998b; [25]Desmadryl and Sans, 1990; [26]Dechesne et al., 1997; [27]Parrad and Cottereau, 1977; [28]Curthoys, 1979. *Chick cochlea:* [29]Katayama and Corwin, 1989; [30]Adam et al., 1998; [31]Cohen and Fermin, 1978; [32]Tilney et al., 1992b; [33]Griguer and Fuchs, 1996; [34]Fuchs and Sokolowski, 1990; [35]Jones et al., 2001; [36]Sokolowski and Cunningham, 1999; [37,38]Whitehead and Morest, 1985a,b; [39]Saunders et al., 1973. *Chick vestibular:* [40]Goodyear et al., 1999; [41]Masetto et al., 2000; [42]Masetto et al., 2003; [43]Ginzberg and Gilula, 1980; [44]Meza and Hinojosa, 1987; [45]Jones and Jones, 2000a. (See Color Insert.)

at postnatal day (P) 11 (Ehret, 1977). The sensitivity of the behavioral audiogram gradually improves from P11 to P18, and even further by adulthood. This development of the audiogram may reflect development of both sensory and nonsensory components (Werner and Gray, 1998). Sound-evoked responses from the cochlea emerge in the second postnatal week: the cochlear microphonic potential, a measure of outer hair cell transduction, at P8–P12 (in rats and gerbils) and the compound action potential, a measure of cochlear afferent activity, at P11–P12 (in mice and rats) (Rübsamen and Lippe, 1998).

In rodents, the frequencies represented by the base of the cochlea shift from low-to-mid frequencies to high frequencies ("place code" shift; for review, see Romand, 1997). This shift is mirrored at higher stages in the rodent auditory pathway. Whether a similar shift occurs in the basal and middle regions of the chick cochlea is controversial (reviewed in Rübsamen and Lippe, 1998). The term "place code shift" gives an oversimplified impression of the changes in the rodent. Echteler *et al.* (1989) studied developmental changes in the frequency selectivity (tuning) of gerbil cochlear afferents. At a particular basal location in the cochlea, the shift to higher characteristic frequency (CF; sound frequency to which the afferent is most sensitive) occurs in two steps: (1) around hearing onset (P14–P15), a sudden increase in sensitivity at all frequencies; and (2) a subsequent improvement in high-frequency sensitivity (P15–P17). Thus, in the second stage the afferent tuning curve adds a high-frequency-sensitive component (the tuning curve "tip") to a less sensitive, lower-frequency component (the tuning curve "tail"). The addition of the tip is likely to reflect the acquisition of electromotility by cochlear outer hair cells (see Section V) and possibly certain voltage-gated ion channels in both inner and outer hair cells (see Section IV), in addition to maturation of passive material properties of the cochlea (Overstreet *et al.*, 2002).

2. Rodent Vestibular Function

Information about functional development of the vestibular system is less complete, in part because the maturation of vestibular reflexes requires both sensory and motor development, but we do know that some vestibular reflexes are functional before hearing onset. Head rotations compensatory for vestibular input are evident at least as early as P4 in rats (Parrad and Cottereau, 1977). Electrical stimulation of the eighth nerve produces eye movements in newborn rats, showing that the vestibuloocular reflex pathway is functional well before eye opening between P11 and P17 (Curthoys, 1979). The responses of rat semicircular canal afferents to angular velocity increase in signal-to-noise and gain throughout the first postnatal month. The gain increase can be accounted for by an increase in canal size, which is expected to produce more hair bundle deflection per unit of angular head velocity (Curthoys, 1983).

3. Chick Hearing

Development is faster in chicks than in rodents. Thus, although chicks and rodents both have embryonic periods of about 3 weeks, the chick inner ear is morphologically and functionally close to mature at the time of hatching, whereas the rodent inner ear is quite immature at birth. Low-frequency sounds evoke potentials in the chick cochlear nuclei as early as E10, and thresholds decrease thereafter, with particularly marked improvement between E18 and hatching at E20–21, coinciding with the filling of the middle ear with air rather than liquid (Saunders *et al.*, 1973). The hearing of hatchlings is comparable to that of 2-week-old rodents. Behavioral thresholds, particularly at frequencies above 1 kHz, continue to improve postnatally (Gray and Rubel, 1985).

4. Chick Vestibular Function

The chick vestibular system is also functional but immature before hatching. Jones and Jones (1996, 2000a) recorded "linear short-latency vestibular evoked potentials" (VsEPs) in response to linear accelerations. These potentials, like the auditory brainstem response, have both peripheral and central components (Nazareth and Jones, 1998). VsEPs were not detected in embryos before E19 (Jones and Jones, 2000a). From E19 to P22, the properties of the VsEPs (response latency, thresholds, and amplitude/intensity relations) mature.

5. Zebrafish Hearing and Vestibular Function

In the zebrafish, vestibular righting and ocular reflexes (Riley and Moorman, 2000) and acoustical and lateral line startle reflexes (Eaton *et al.*, 1977; Nicolson *et al.*, 1998) are first evident between 3 and 4 days postfertilization (dpf).

Exposing zebrafish larvae to microgravity for various intervals up to 4 dpf produces a deficient vestibuloocular reflex in some animals, consistent with a critical period for vestibular sensory development between 1 and 3 dpf (Moorman *et al.*, 2002). During this interval, vestibular afferents are born and innervate both their peripheral targets (inner ear organs) and central targets in the brainstem. In contrast, Jones *et al.* (2000) found that embryonic exposure to hypergravity or high levels of vibration had no effect on the development of chick vestibular evoked potentials. Chicks were sensitive to these interventions during the second week after hatching, well after the inner ear is mature, suggesting an effect on central development.

B. Spontaneous Activity

Before the onset of hearing, spontaneous activity of the auditory neurons at various levels of the pathway in both chicks and mammals is low, bursting, and sometimes rhythmic (reviewed in Rübsamen and Lippe, 1998). The low spontaneous rates may in part reflect effects of anesthesia in young animals. At the onset of hearing, the rhythmic activity is replaced by steady spontaneous activity. Rhythmic activity is particularly robust in prehatch chicks, where it is present in cochlear ganglion neurons (Jones *et al.*, 2001) and is abolished at higher levels by silencing the cochlea with tetrodotoxin injections or ablation (Lippe, 1994). The activity may therefore arise either in hair cells or in cochlear ganglion cells. Rhythmic activity has been described in other developing systems. In the retina, it appears to have an instructional effect on retinal ganglion projections, mediated at least in part by neurotrophins (Katz and Shatz, 1996). Experiments in which cochleas are ablated before hearing onset show that in both chicks and rodents, cochlear input influences the development of central auditory neurons (Gabriele *et al.*, 2000; Hashisaki and Rubel, 1989).

Rhythmic activity has not been reported in immature vestibular afferents. In rat semicircular canal afferents, the rate of spontaneous activity is low initially, increasing gradually after birth, and the highly regular pattern of firing characteristic of many mature vestibular afferents is first seen at P4–5 in the rat (Curthoys, 1979, 1982). A similar progression was seen in the spontaneous activity of vestibular afferents in inner ear explants made on the day of birth (P0; Desmadryl *et al.*, 1986): Spontaneous rates were initially low and increased as much as 10-fold after 6 days *in vitro* (DIV; equivalent to P6). The percentage of afferents with regular firing increased substantially from almost none at birth to 50% by P6. Regular firing has been recorded as early as E19 in chick (Jones and Jones, 2000b). The change to the highly regular pattern of firing may reflect the acquisition by some afferents of a large after-hyperpolarizing current (I_{AHP}) that determines the interspike interval (Smith and Goldberg, 1986).

In the following sections on development of hair cells and of their interactions with afferent and efferent fibers, the rhythmic activity of immature auditory afferents is a recurring theme. This rhythmic activity may originate in the hair cells, arising through the transient expression of specific voltage-gated and/or transmitter-gated ion channels cochlear.

III. Development of Mechanoelectrical Transduction

In this section we review literature on the morphological development of the hair bundle array, surely one of the most highly ordered and tightly specified structures in the body. We briefly review such developmental information as can be gleaned from studies of hair cell transduction. Despite the fact that most transduction data are obtained during a neonatal period in which significant morphological development occurs, no substantial qualitative changes in the transduction process with development have been described. However, no comprehensive efforts to investigate this process have been published.

A. Hair Bundle Development

1. Bundle Morphology

Hair bundles comprise cross-linked arrays of specialized microvilli, which bear the unfortunately persistent misnomer "stereocilia," arranged in rows of incrementing height to form a staircase profile (see Fig. 1B; see Chapter 2 in this volume by Quint and Steel). At the tall edge of the bundle is a single true microtubule-based cilium, the kinocilium. In mature auditory bundles of mammals and birds, the kinocilium regresses, although its basal body persists. Hair cells detect the component of bundle motion parallel to a line running from the shortest to the tallest rank of stereocilia, which defines the bundle's orientation. The mechanosensitive transduction channels are located near the tips of the stereocilia (Jaramillo and Hudspeth, 1991; Lumpkin and Hudspeth, 1995). Extracellular "tip links" run between rows, angling upward from the tips of shorter stereocilia to the upper flanks of adjacent taller stereocilia, parallel to the bundle's orientation axis (Pickles et al., 1984) (see Fig. 1C). The links have a central zone comprising two filaments wound around each other in a right-handed helix. This central zone is coupled to the stereociliary membrane by 2–3 strands (see Fig. 1C; Kachar et al., 2000; Tsuprun and Santi, 2000). In the "tip link/gating spring" model of transduction, stretch of the links during bundle deflection applies force to the mechanosensitive channels, promoting their opening (Pickles and Corey, 1992). This model has been refined recently to suggest that the distal strands, rather than the helical central portions, are likely to be the elastic elements (Kachar et al., 2000). Alternatively, the tip links may be important for regulating the relative positions of stereocilia but not applying stretch themselves.

The core of each stereocilium is an array of actin filaments (Flock and Cheung, 1977; Flock et al., 1981), cross-linked by fimbrin (Sobin and Flock, 1983) and espin (Zheng et al., 2000b). Many actin filaments terminate at the stereociliary ankle region, where the stereocilium meets the cell body, but a subset of central actin filaments extends as a "rootlet" into the cuticular plate, the hair cell equivalent of the terminal web generally found in epithelial cells. The cuticular plate contains filamentous actin, at cross angles to the rootlets (DeRosier and Tilney, 1989), and other proteins, including myosins

(Hasson et al., 1997; Slepecky and Ulfendahl, 1992). Microtubules penetrate the cuticular plate from below (Furness et al., 1990; Jaeger et al., 1994). A circumferential ring of actin runs around the cuticular plate at the level of the zonula adherens (Hirokawa and Tilney, 1982).

In addition to actin, fimbrin, and espin, stereocilia contain calmodulin and various unconventional myosins (-1c, VI, VIIa, XV) (Hasson et al., 1997; Steel and Kros, 2001). Calmodulin is hypothesized to mediate the Ca^{2+} dependence of an adaptation process by interacting with a myosin motor complex that links the actin core with the transduction channel (Gillespie and Hudspeth, 1994). Calmodulin is also an important regulator of a plasma membrane Ca^{2+} ATPase (PMCA), present in abundance in stereocilia and responsible for the transmembrane Ca^{2+} gradient (Kozel et al., 1988; Yamoah et al., 1998). Information from naturally occurring mutants and transgenic experiments has suggested the following functions for the unconventional myosins that are found in the hair bundle and cuticular plate regions: Myosin-1c is the motor molecule that, together with calmodulin, mediates a major form of Ca^{2+}-dependent transducer adaptation (Holt et al., 2002). Myosin VI, localized to the apical surface of the cuticular plate, may help to anchor the stereociliary membrane to the cuticular plate membrane; without functional myosin VI, the membranes of stereocilia fuse (Self et al., 1999). Myosin XV may be involved in structural development of the actin core of the stereocilium; mutants have short stereocilia and abnormal long actin bundles emanating from the soma (Anderson et al., 2000).

Myosin VIIa is localized in and around the cuticular plate and in the stereocilia (Hasson et al., 1997). In both mice (including shaker-1) and zebrafish with mutations in myosin VIIa, bundles are disorganized (Ernest et al., 2000; Self et al., 1998). Detailed analysis of the bundles in $Myo7a^{6J}$ and $Myo7a^{816SB}$ mouse mutants suggests that the growth of individual stereocilia is normal but that the stereociliary array is fragmented into clumps. This defect may mean that myosin VIIa helps to anchor the stereocilia in the cuticular plate (Self et al., 1998) or maintains the bundle as a unit by tensioning lateral links between stereocilia (Kros et al., 2002). The latter kind of effect was suggested by the effects of the myosin VIIa (*Myo7a*) mutations on hair cell transduction: a negative bias in the resting position of the bundle relative to its operating range (the range of bundle deflections to which it is sensitive) and a hyperadaptation (Kros et al., 2002). Myosin VIIa is homologous to Crinkled, a *Drosophila* protein involved in orienting epithelial cells along the axis orthogonal to the apical–basal axis; by homology, myosin VIIa may participate in orienting hair bundles (see Section III.A.5).

The stereociliary plasma membrane houses mechanoelectrical transduction channels of unknown molecular identity at just one-to-several per stereocilium (see discussion in Ricci and Fettiplace, 1997); PMCA molecules;

and, at least in cochlear hair cells, purinergic receptor channels (P2X receptors; Housley et al., 1998) of uncertain function.

2. Bundle Growth

a. Chick Bundles. The morphological stages in hair bundle development have been thoroughly examined in the chick cochlea by the Tilneys and DeRosier and their colleagues (Tilney and DeRosier, 1986; Tilney and Saunders, 1983; Tilney et al., 1983, 1986, 1988; reviewed in Tilney et al., 1992a). In this organ, hair bundle formation takes place in clearly defined stages that occur simultaneously throughout the organ (Fig. 3). Hair bundles begin (E8) with a central kinocilium and small microvilli protruding from the apical surface. Thus, bundles are not oriented at the outset. Between E8 and E10, the microvilli grow to a uniform height and the kinocilium moves to the edge of the cell's apical surface, defining bundle orientation. The staircase then begins to form: stereocilia elongate in successive rows, beginning with the row adjacent to the kinocilium (which will be the tallest row), widen, acquire tapered ankles, and extend rootlets into the cuticular plate. In the chick cochlea, staircase formation takes place in two stages (E10.5–E13 and E16–P1); in the intervening pause, excess microvilli are resorbed.

b. Rodent Bundles. In rodent inner ear organs, hair cell differentiation is staggered, i.e., at any one time point, hair cells are at multiple stages of differentiation, from E13 through to the second or third postnatal week (e.g., Denman-Johnson and Forge, 1999; Mbiene and Sans, 1986; Mbiene et al., 1984; Nishida et al., 1998; Rüsch et al., 1998b; Zine and Romand, 1996). In the cochlea, most differentiation markers begin basally and spread apically, and this is true for bundle differentiation as well (Kaltenbach et al., 1994; Zine and Romand, 1996). The apical turn is delayed by about 2 days relative to the basal turn (Zine and Romand, 1996). The initial stage of bundle formation, characterized by a central kinocilium and multiple microvilli (similar to E8 bundles in the chick cochlea, Fig. 3), is evident at E13.5 in mouse vestibular organs (Denman-Johnson and Forge, 1999; Mbiene and Sans, 1986; Mbiene et al., 1984) and at E17 in the basal turn of the rat cochlea (Zine and Romand, 1996).

By birth, cochlear hair bundles in the hamster (Fig. 4) have an eccentric kinocilium and elongate microvilli (immature stereocilia), corresponding to bundle development at about E10 in the chick cochlea (compare Figs. 3 and 4A). The neonatal rat cochlea appears to be more advanced, with bundles closer to the E13 stage in the chick cochlea (see Fig. 3; compare Fig. 4C [hamster] with Fig. 2 [rat] in Zine and Romand, 1996). The stages in bundle development in the rodent cochlea resemble those in the chick

13. Functional Development of Hair Cells

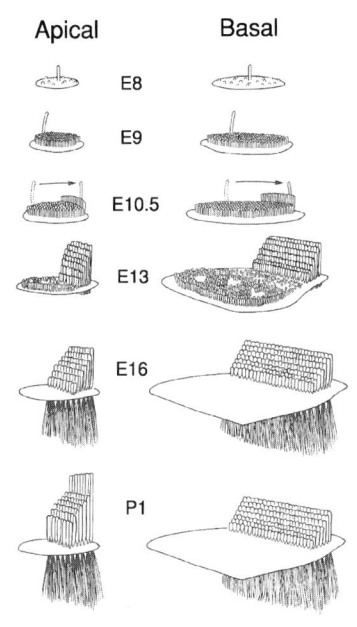

Figure 3 Development of hair bundles in the chick cochlea. Modified and reproduced with permission from Tilney, L. G., Tilney, M. S., and DeRosier, D. J. (1992). Actin filaments, stereocilia, and hair cells: How cells count and measure. *Annu. Rev. Cell Biol.* **8**, 257–274. *Top to bottom*: Bundles from the apical (distal), low-CF end of the chick cochlea (*left*) and from the basal (proximal), ultimately high-CF end (*right*), from E8 to posthatching (P1). (E8) Tiny stereocilia are randomly distributed on the apical surface of the hair cell. A single, taller kinocilium is apparent in the middle of the apical surface. (E9) The kinocilium has moved to the periphery of the cell and the apical surface of the hair cell is covered with a dense carpet of immature stereocilia of uniform length. (E10.5) The tallest row of the stereocilia has begun to form behind the kinocilium. The apical surface has continued to expand and fill with immature stereocilia. (E13) The staircase profile of the stereocilia has formed. The number of stereocilia still greatly exceeds the mature number. Stereocilia have begun to taper into ankles at their bases and to send rootlets into the cuticular plate. (E16) The stereociliary rootlets reach mature length. Extraneous stereocilia have been resorbed. (P1) The stereocilia in most cochlear bundles have reached their mature height and thickness.

cochlea, except that there is no evidence for a pause in stereociliary elongation (Kaltenbach *et al.*, 1994). An excess of stereocilia are made and then lost. In the middle part of the cochlea, the number per bundle rises from about 200 at P0 to 300 at P2–3, falling precipitously to 150 at P5 and to 100 by P20. By P14, almost all of the kinocilia have disappeared, leaving the notch that produces the W shape of mature outer hair cell bundles.

The orientation of actin filaments in the stereocilium predicts that actin monomers are added at the distal tip. This was confirmed recently by Schneider *et al.* (2002), who transfected neonatal cultures of the rat organ of

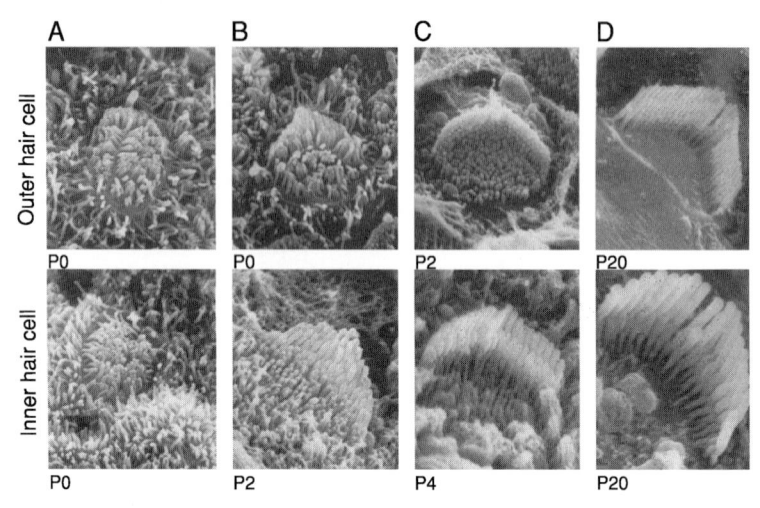

Figure 4 Development of hair bundles in the hamster cochlea. Reproduced from Kaltenbach, J. A., Falzarano, P. R., and Simpson, T. H. (1994). Postnatal development of the hamster cochlea. II. Growth and differentiation of stereocilia bundles, *J. Comp. Neurol.* **350,** 181–198. Reproduced with permission of Wiley-Liss, Inc., a subsidiary of John Wiley & Sons, Inc. *Top and bottom rows*: Outer hair cells and inner hair cells, respectively. Ages are indicated in the lower left hand corner of each electron micrograph. (A) Newly differentiated hair cells have clusters of stereocilia that can be differentiated from apical microvilli on surrounding cells (stage I, Kaltenbach *et al.*, 1994). (B) Stage 2 bundles have a clearly established tall edge and smaller, shorter, and more loosely organized stereocilia elsewhere. (C) In stage 3 bundles, the staircase becomes clear. The apical surface outside the staircase is still covered with short stereocilia. (D) In stage 4, extraneous stereocilia are resorbed.

Corti with plasmids containing β-actin linked to green fluorescent protein. Actin monomers are added to filaments at the tips of the stereocilia at a high rate. Even in 10- to 15-day-old cultures, the actin filaments "treadmill" toward the base by incorporating new actin monomers at their tips.

A cuticular plate antibody (CP1) stains nascent apical structures beginning at E14 in the mouse cochlea (Nishida *et al.*, 1998). Based on the staining patterns obtained with this monoclonal antibody, Nishida and colleagues propose that the cuticular plate originates as a peripheral ring, then a thin central plate forms. The mature form of the cuticular plate is present by P4–P6, at the same time as the V-shaped bundle arrays.

c. Zebrafish Bundles. Relatively little is known about the timing of bundle development in the zebrafish inner ear. Haddon and Lewis (1996) saw bundle-like structures in an otolith organ as early as 1 dpf, when the otoliths are small irregular granules. The long bundles of semicircular canal hair cells are visible *in vivo* at 2.5 dpf (Nicolson *et al.*, 1998), before the onset of

13. Functional Development of Hair Cells

vestibular, acoustical, and lateral line reflexes. By 5 dpf there are 20 properly oriented hair cells per crista (Haddon and Lewis, 1996).

3. Staircase Formation

The formation of the staircase is particularly interesting because it is both peculiar to vertebrate hair bundles and implicated in the tip link/gating spring hypothesis. The heights of the ranks of stereocilia within a bundle are presumably set by the molecules that regulate the length of the actin bundles that form each stereociliary core. This in turn may be determined by molecules that affect polymerization at the growing tip and/or depolymerization at the base. Candidate molecules include Rho GTPases, well-known regulators of actin polymerization in other systems (Kollmar, 1999). Another likely candidate is the stereociliary actin-bundling protein, espin. In jerker mice, which have deficient levels of espin, hair bundles are short (Zheng et al., 2000b). Furthermore, excess production of espin stimulates the growth of extra-long microvilli in transfected cells of a kidney cell line (Loomis et al., 2003).

How do stereociliary ranks rise to (or stop at) specific heights? One idea is that there is feedback from the transduction process on this process: As the stereocilia grow, the tip link becomes stretched and applies force to the transduction channels, opening channels. The resulting Ca^{2+} influx through the open channels is then envisioned as signaling an end to actin filament elongation (Pickles et al., 1991a; Tilney et al., 1988). In support of this model, Pickles and Rouse (1991) found that chick cochlear bundles grown in the presence of a transduction channel blocker are very disorganized. In contrast, in a preliminary report, Bryant et al. (2003) found that the transduction channel blocker neomycin had no obvious effect on staircase development in chick cochlear cultures made on E6.5.

4. Tip Link Formation

Tip links are seen as each successive row of the staircase develops (Kaltenbach et al., 1994; Pickles et al., 1991b). Clues about how tip links grow are provided by a study of their regeneration in cultured chick cochleas (Zhao et al., 1996). Tip links were broken with extracellular application of a calcium chelator, BAPTA. Over the next 24 h, multiple linkages sprouted from the tips of the stereocilia, with a relatively high percentage off axis and/or below the normal tip link positions on the stereocilia. With time, these inappropriate links were cleaned up, so that by 24 h the treated bundles were close to normal in appearance. Similar changes are described in developing bundles (Kaltenbach et al., 1994; Pickles et al., 1991b). Over the same period, transduction recovered. The tip link regeneration did not depend on protein synthesis and was inhibited by increasing intracellular Ca^{2+} with an ionophore. Zhao et al. (1996) suggest that tip links are moved into position by motors within the stereocilium, possibly myosin molecules moving along the actin core.

5. Bundle Orientation

Bundle orientation must be organized both within a hair cell (to create a bundle's characteristic polarity) and within the epithelium (to produce the stereotypic arrays of bundle orientations on the epithelial surface). These developments are of utmost functional importance because the bundle's orientation defines its axis of functional sensitivity (Flock, 1965; Shotwell et al., 1981).

The early formation of the kinocilium in developing and regenerating hair bundles, and its stereotypic and singular position at the center of the bundle's tall edge, have suggested that it plays a role in establishing bundle orientation (e.g., Sobkowicz et al., 1995). Other possibilities include the striated rootlet that emanates from the kinociliary basal body, as it does from centrioles in other ciliated cells, and assemblies of microtubules or microfilaments (see discussion in Denman-Johnson and Forge, 1999). A very different hypothesis arose through observations in the chick cochlea, where bundles reorient during development as they come into contact with the growing tectorial membrane. Cotanche and Corwin (1991) suggested that the reorientation results from traction exerted by the growing tectorial membrane. Whether this mechanism works in the chick cochlea has been debated (see Lewis and Davies, 2002), but it cannot work in other hair cell organs in which the development of the overlying tectorial/otolithic structure lacks the right timing or pattern of formation to mediate bundle orientation. For example, in parts of some lizard cochleas (e.g., the alligator lizard; Mulroy, 1974), there is no overlying tectorial structure, yet the bundles form a precise array. In the mammalian cochlea, bundle orientation is approximately established before the tectorial membrane is formed (Coleman et al., 1995), although there is gradual refinement of the orientation in the first 2 postnatal weeks (Dabdoub et al., 2003). Finally, the traction model seems implausible in otolith organs, where bundle orientations form complex arrays and where the overlying structures exert shearing forces in different directions, specified by the sum of linear forces applied to the head at any instant in time.

In the saccule and utricle, bundles reverse orientation at a line running through a specialized swath of macula called the striola. On either side of the line, bundle orientation changes very gradually with position along the reversal line, such that neighboring bundles have similar orientations. The otolithic gel ("otolithic membrane") is acquired simultaneously across the width of the organ, beginning at very early stages. In mouse otolith organs (Denman-Johnson and Forge, 1999), otolithic gel substance is apparent as early as E12.5 in dilated rough endoplasmic reticulum of supporting cells and overlying their apical surfaces. Small otoconia appear by E16.5, and an organized gel with otoconia is present by E18.5. At E15.5, the line of bundle reversal becomes evident and a second wave of bundle formation begins (see also Mbiene *et al.*, 1984). The addition of new bundles does not disrupt the orientation map and the line of reversal, indicating that those properties are set for the population rather than on a bundle-by-bundle basis.

That mechanical feedback of some kind, either direct or through an intermediary such as Ca^{2+}, affects the orientation of tip links is suggested by the observations cited earlier on developing (Kaltenbach *et al.*, 1994; Pickles *et al.*, 1991b) and regenerating (Zhao *et al.*, 1996) tip links. Tip links that form at odd angles and odd places appear to be pruned, leaving only those at appropriate orientations. Furthermore, in chick cochleas cultured for a day (with tectorial membranes detached), the number of inappropriate links increases (Zhao *et al.*, 1996), suggesting that normal mechanical input may be required to continuously prune inappropriate tip links. Finally, many hair bundles in otolith organs have tip links that are 30° off the bundle's axis of bilateral symmetry, reflecting the tight packing of the stereocilia (Takumida *et al.*, 1989). One might speculate that this off-axis orientation is tolerated because mechanical stimuli arrive from all angles in otolith organs. In contrast, mechanical stimuli to cochleas have a uniform orientation normal to the long axis of the epithelium so that it is clearly advantageous to align all bundles with that stimulus angle. Indeed, the misalignment of some hair bundles that occurs naturally in some guinea pig cochleas decreases hearing sensitivity (Yoshida and Liberman, 1999).

Bundle orientation is an example of planar cell polarity (PCP) in epithelia: polarization of the epithelial cell along an axis orthogonal to its apical–basal axis. The signaling mechanisms that control hair bundle orientations in inner ear epithelia may be homologous to those that set up the regular array of hairs in *Drosophila* wings (Eaton, 1997; Lewis and Davies, 2002). Orientation of the fly-wing hairs is controlled by a two-tier PCP signaling pathway involving (1) communication of a directional signal to individual cells by a cadherin protein (Fat), and (2) alignment of neighbor–neighbor orientations via the transmembrane Wnt receptor, Frizzled (Ma *et al.*, 2003). Frizzled on the distal side of a cell recruits the cytoplasmic protein, Prickle, to the proximal side of the adjacent cell (Tree *et al.*, 2002), setting up similar orientations among neighbors. A homologous pathway may operate in hair cell epithelia. Mutations in protocadherin 23, homologous to *Drosophila* Fat, and myosin VIIa, homologous to *Drosophila* Crinkled (another PCP protein), produce misoriented bundles (Di Palma et al., 2001; Self et al., 1998). A Wnt protein (Wnt7a) is expressed in developing pillar cells in the organ of Corti, and *in vitro* application of Wnt7a or of Frizzled-related Wnt inhibitors to embryonic cochlear cultures causes bundle misalignment (Dabdoub et al., 2003).

B. Biophysics of Transduction

Much of what we know about hair cell mechanotransduction in mammals comes from excised sensory epithelia of the neonatal mouse cochlea and utricle. These epithelia are morphologically immature in many ways. In the P0 mouse organ of Corti, certain areas (tunnel of Corti, inner sulcus) have not yet been sculpted, and, relevant to mechanoelectrical transduction, the tectorial membrane is not yet fully formed and hair bundles have many small microvilli/stereocilia that are later resorbed (Coleman et al., 1995; Kaltenbach and Falzarano, 1994; Kaltenbach et al., 1994; Zine and Romand, 1996). Nevertheless, these hair cells generate large transduction currents in excised preparations in response to bundle deflections affected by a fluid jet stimulus (Kros et al., 1992). From a preliminary report on transduction in mouse organ of Corti cultures made from the basal cochlea at E16.5, transduction currents can be recorded from inner hair cells about 3 days later (E19–E20) and outer hair cells 1 day later still (Bryant et al., 2003).

The P0 mouse utricle is also immature, with some hair cells not yet born and a number of cells apparently transitional between supporting cells and hair cells with immature hair bundles (Rüsch et al., 1998b). In hair cells chosen for their relatively mature-looking bundles, bundle deflection evokes robust transduction currents with properties that do not appear to change dramatically over the first 10 postnatal days (Fig. 5; M. A. Vollrath and R. A. Eatock, unpublished results). Even as early as E16.5, transduction currents are robust and show prominent adaptation (G. S. G. Géléoc and J. R. Holt, personal communication).

Changes with transduction during development may become apparent as we look harder. One improvement would be to use stimuli that better reveal the functional consequences of developmental changes in bundle morphology and stiffness. In the frog saccule, stiffness of the otolithic gel and bundles combined is dominated by bundle stiffness (Benser et al., 1993), suggesting that the natural stimulus to the bundle may be better approximated by a flexible probe than by the rigid probe stimulus used in the experiments of Fig. 5. When bundles are deflected by a flexible probe (a quasi-force step) rather than a stiff probe (a displacement step), interbundle differences in instantaneous stiffness will affect the initial angular deflection and the stimulus to the gating springs. Furthermore, with a flexible probe, Ca^{2+}-mediated adaptation and amplification processes can feed back on the

13. Functional Development of Hair Cells

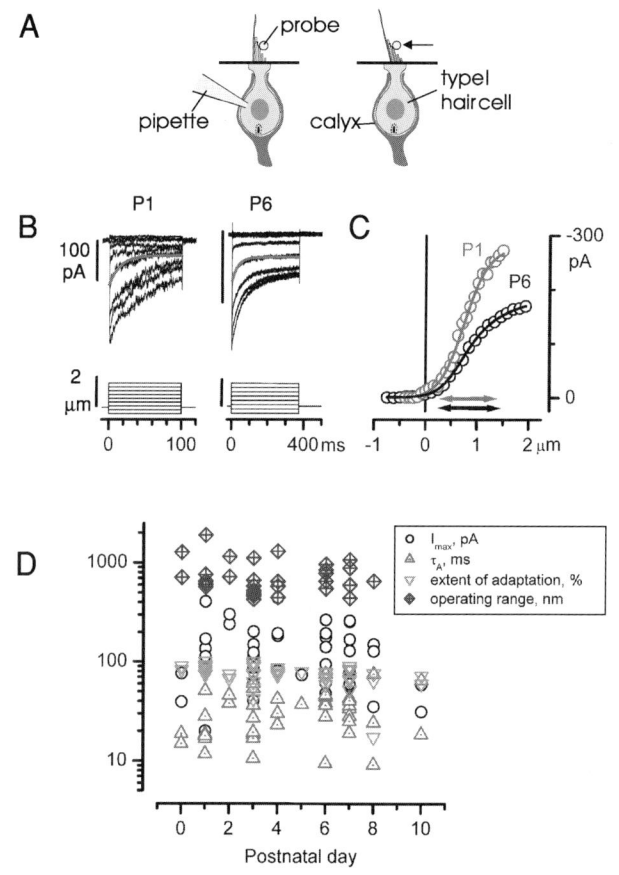

Figure 5 Transduction properties in mouse utricular hair cells do not appear to change systematically during the first postnatal week (M. A. Vollrath and R. A. Eatock, unpublished results). (A) Bundles were deflected along their sensitive axes by moving a rigid glass probe against the back of the bundle (the probe is shown end on); step deflections had a rise time of 1 ms. Transduction currents were recorded in whole-cell voltage clamp mode with a glass pipette. A deflection toward the tall edge, as shown, produces an inward transduction current. The cells were bathed in a standard high-Na^+ saline and the pipette contained a standard high-K^+ solution, with 10 mM EGTA. (B) Transduction currents evoked by families of step bundle deflections, in hair cells in a P1 and a P6 utricle, respectively. Stimulus amplitudes (*lower panels*) are given as displacements of the probe (and bundle) parallel to the plane of the epithelium. Scale bars apply to both P1 and P6 data. The half-maximal responses were fit by a single-exponential function (*red lines*), with time constants (τ_A) of 15 ms (P1) and 34 ms (P6). This difference in the time constants is not reflective of a trend with age, as shown in D. The currents from the younger cell look noisier because they were not averaged. (C) Current-displacement relations, obtained by plotting peak currents in B as functions of bundle displacement from the resting position. The data are fit by second-order Boltzmann functions. The lines at the bottom show the *operating ranges* for each cell: displacements corresponding to 10–90% of the maximal response. (D) The *maximum current* (I_{max}); *adaptation time constant* at half-maximal stimulation

bundle mechanics, causing dynamic changes in the stimulus to the gating springs (Hudspeth et al., 2000).

IV. Developmental Acquisition of Basolateral Ion Channels

Mature hair cells express a rich variety and abundance of ion channels that serve to modulate the receptor potential. A mature hair cell of a particular morphology and location in the sensory epithelium is likely to express a set of conductances that distinguishes it from its neighbors and from hair cells in other kinds of sensory epithelia. Patch clamp experiments on hair cells at different ages have shown that the mature repertoire is attained through both expansion, acquisition of new channel types, and pruning, elimination of channel types that represent an immature phase. In this section we consider the channels that are either Ca^{2+}- or voltage-gated (Table I). A recurring theme is that immature hair cells have electrophysiological properties—especially certain voltage-gated conductances—that favor spiking. Hair cell spiking may underlie the spontaneous rhythmic activity noted at various stages of the auditory pathway in immature animals (see Section II.B) and may promote synaptogenesis, as envisioned in other systems (Katz and Shatz, 1996).

A. Outwardly Rectifying K$^+$Conductances: g_{Kv}, g_A, $g_{K,Ca}$, $g_{K,n}$, $g_{K,L}$

In most hair cells, the most abundant ion channels are outwardly rectifying K$^+$ channels. With a couple of notable exceptions ($g_{K,n}$ and $g_{K,L}$; see Sections IV.A.3 and IV.A.4), these channels have a low open probability at the resting potential, which increases with depolarization. Thus, they are activated by the depolarization initiated by inward transduction currents during positive bundle deflections (see Fig. 5B and C). Because the membrane potential is positive to the K$^+$ equilibrium potential, K$^+$ flows outward through the open channels (hence the term "outwardly rectifying"), tending to repolarize the membrane potential toward its resting value. In mature organs, the nature of these ion channels varies with hair cell type and location in the sensory epithelium. In cochleas, the best known variations are longitudinal and therefore correlate with the best frequency. In vestibular organs, variations with longitudinal position and relative to

---

(τ_A, obtained as shown in A) *extent of adaptation* expressed as a percentage *[100 × (I_{max} − $I_{steady-state}$)/I_{max}]*; and *operating range*, as functions of postnatal age. Variability does not appear to be systematic with age in this period, but recordings were not taken from immature-looking bundles (e.g., bundles in the first two columns in Fig. 4). (See Color Insert.)

Na$^+$ channels		Negatively inactivating	Alligator cochlea[23], fish saccular[3]; rodent cochlear[24]; amniote vestibular[25]	Spiking
		Positively inactivating	Immature rat utricle[26]	
Ca^{2+} channels	L	Negatively activating, sustained	Frog saccular,[6,27] crista[28] and amphibian papilla[7]; turtle,[10] chick,[29] and rodent cochlear[30]; rodent vestibular[31]	Transmitter release, activation of BK channels (electrical tuning)	Ca$_V$1.3 (α_{1D}) (CACNA1D)[41]
	N		Frog saccular[27]		
	R		Frog crista[28]		
ACh receptor channels		Neuronal nicotinic	Weakly-cation selective, Ca^{2+}-permeable	Chick short hair cells[31]; mammalian outer hair cells[32–34]	Efferent-evoked ipsp: inhibition, de-tuning	$\alpha9/\alpha10$ heteromers[34,35]

Hair cell types: *KV, K, DR conductances*: [1]Armstrong and Roberts, 1998. [2]Housley et al., 1989; Masetto et al., 1994. [3]Sugihara and Furukawa, 1995; Steinacker and Romero, 1991. [4]Goodman and Art, 1996. [5]Lang and Correia, 1989; Masetto and Correia, 1997a,b; Eatock et al., 1998; Masetto et al., 2000; Brichta et al., 2002.

A conductances: [6]Hudspeth and Lewis, 1988a,b. [7]Smotherman and Narins, 1999. [8]Sugihara and Furukawa, 1999. [9]Murrow and Fuchs, 1990. [5]Lang and Correia, 1989; Masetto and Correia, 1997a,b; Eatock et al., 1998; Masetto et al., 2000; Brichta et al., 2002.

K,Ca-BK conductances: [6]Hudspeth and Lewis, 1988a,b.; [7]Smotherman and Narins, 1999. [10]Art and Fettiplace, 1987; Eatock et al., 1993. [11]Fuchs et al., 1988. [12]Marcotti et al., 2003

K,Ca-SK conductances: [13]Yuhas and Fuchs, 1999. [14]Dulon et al., 1998; Oliver et al., 2000.

K,n conductance: [15]Housley and Ashmore, 1992; Marcotti and Kros, 1999; Oliver et al., 2003.

K,L (also KI) conductance: [16]Correia and Lang, 1990; Rennie and Correia, 1994; Rüsch and Eatock, 1996a; Ricci et al., 1996; Masetto et al., 2000; Brichta et al., 2002.

Table I *(Continued)*

K1 conductances: [17]Holt and Eatock, 1995. [18]Sugihara and Furukawa, 1996. [19]Goodman and Art, 1996. [20]Marcotti et al., 1999. [21]Ohmori, 1985; Rüsch et al., 1998b.
h conductances: [17]Holt and Eatock, 1995. [18]Sugihara and Furukawa, 1996. [22]Rüsch and Eatock, 1996b; Rüsch et al., 1998b.
Voltage-gated Na⁺ conductances: [23]Evans and Fuchs, 1987. [8]Sugihara and Furukawa, 1989. [24]Witt et al., 1994; Oliver et al., 1997. [25]Rüsch and Eatock, 1997; Masetto et al., 2003. [26]Lenman et al., 1999.
Voltage-gated Ca⁺ conductances: [6]Hudspeth and Lewis 1988a,b. [27]Rodriguez-Contreras and Yamoah, 2001. [28]Martini et al., 2000. Perin et al., 2001.
[7]Smotherman and Narins, 1999. [10]Art and Fettiplace, 1987. [29]Zidanic and Fuchs, 1995. [30]Platzer et al., 2000. [31]Bao et al., 2003.
ACh receptor channels: [32]Fuchs and Murrow, 1992a. [33]Dulon and Lenoir, 1996. [34]Elgoyhen et al., 1994, 2001. [35]Weisstaub et al., 2002.
Molecules: *KV K, DR conductances:* [36]Navaratnam et al., 1997.
A conductances: [37]Rajeevan et al., 1999.
K, Ca-BK conductances: [38]Jiang et al., 1997; Rosenblatt et al., 1997; Navaratnam et al., 1997; Jones et al., 1998; Ramanathan et al., 2000; Langer et al., 2003.
K, Ca-SK conductances: [13]Yuhas and Fuchs, 1999. [14]Dulon et al., 1998; Oliver et al., 2000.
K,n conductance: [39]Kharkovets et al., 2000; Oliver et al., 2003.
K1 conductances: [40]Navaratnam et al., 1997.
Voltage-gated Ca⁺ conductances: [41]Kollmar et al., 1997a,b; Platzer et al., 2000; Bao et al., 2003.
ACh receptor channels: [34]Elgoyhen et al., 1994, 2001.

chick cochlea is tonotopically organized, with high characteristic frequencies represented basally and a smooth decrease in characteristic frequency with distance from the basal end. In the mature animal, short and tall hair cells are both found both apically and basally. Short hair cells express g_A (Murrow and Fuchs, 1990) and $g_{K,Ca}$ carried by SK channels ($g_{K,Ca-SK}$) (Yuhas and Fuchs, 1999); the latter generates the inhibitory postsynaptic potential in response to release of acetylcholine (ACh) from efferent terminals. Tall hair cells express a delayed rectifier, g_K, and $g_{K,Ca}$ carried by BK channels ($g_{K,Ca-BK}$) (Fuchs and Evans, 1990); the latter participates in the electrical resonance that contributes to acoustical tuning in these hair cells (Fuchs *et al.*, 1988).

Griguer and Fuchs (1996) were able to distinguish some of these differences in outwardly rectifying conductances very early in development, at E10. The most basal embryonic hair cells, which look morphologically like the short hair cells of mature chick cochleas, have a relatively fast g_A, whereas the most apical hair cells resemble tall hair cells in having a slower g_K with more positive voltage ranges of activation and inactivation. The lack of short hair cell–like currents in the apical zone at E10–E14 suggests that differentiation of short hair cells proceeds from base to apex. The slow delayed rectifier of apical hair cells may be encoded by $cK_V3.1$ (KCNC or *Shaw* family), a putative delayed rectifier subunit that is preferentially expressed in the apical half of the cochlea (Navaratnam *et al.*, 1997). Sokolowski and colleagues have shown that a number of Kv1 (KCNA, Shaker) subunits are expressed beginning at different times in the chick inner ear, from as early as E3 (Duzhyy *et al.*, 2003; Rajeevan *et al.*, 1999).

Despite the similarities, the embryonic K^+ currents are distinguishable from their mature counterparts by their sensitivity to K^+ channel blockers (Griguer and Fuchs, 1996) and most strikingly by the relatively late acquisition of $g_{K(Ca)-BK}$ by tall hair cells in the basal half of the cochlea (CFs > 100 Hz). $g_{K(Ca)-BK}$ turns on rather abruptly at E18 (Fuchs and Sokolowski, 1990; Griguer and Fuchs, 1996), near hatching (E19–E21, depending on incubation temperature) and concomitant with a dramatic increase in hearing sensitivity (Saunders *et al.*, 1973). The acquisition of $g_{K(Ca)-BK}$ has a major impact on the voltage response of hair cells to depolarizing currents (Fuchs and Sokolowski, 1990). Before $g_{K(Ca)-BK}$ acquisition, tall hair cells generate large, slow, repetitive Ca^{2+} spikes in response to sustained depolarizing current. After $g_{K(Ca)-BK}$ acquisition, the cells produce higher frequency damped oscillations (electrical resonances). These resonances may serve acoustic tuning (Fuchs *et al.*, 1988).

The properties of $g_{K(Ca)-BK}$ vary along the length of the chick and turtle cochleas. These differences help to establish the tonotopic gradient of electrical tuning of hair cells with CFs above 100 Hz. Within this range, lower-CF hair cells have slower $g_{K(Ca)-BK}$ channels than do higher-CF hair cells (Duncan and Fuchs, 2003). The underlying channels comprise pore-forming α subunits (KCNMA subunits) and accessory β subunits (KCNMB subunits) that modulate the channel's Ca^{2+} sensitivity. The tonotopic gradient in the kinetic properties of the channels (Duncan and Fuchs, 2003; Wu et al., 1995) may reflect changes with cochlear position in both β subunits and splice variants of the α-subunit products of a single gene (Jiang et al., 1997; Jones et al., 1998; Navaratnam et al., 1997; Rosenblatt et al., 1997; reviewed in Fettiplace and Fuchs, 1999). Below 100 Hz, electrical tuning is achieved with delayed rectifiers and inward rectifiers (Fettiplace and Fuchs, 1999; Goodman and Art, 1996). It is not clear how these gradients in ion channel expression are set up in development.

2. Outward Rectifiers in Rodent Cochlear Hair Cells

Mature mammalian inner hair cells express several outwardly rectifying currents: a large, fast, Ca^{2+}-dependent K^+ current ($g_{K,f}$); a slower delayed rectifier with a component ($g_{k,4AP}$) that is sensitive to block by 4-aminopyridine ($g_{K,4AP}$); and a linopirdine-sensitive current ($g_{K,n}$). $g_{K,n}$ is a relatively negatively activating current that was originally described in outer hair cells (Housley and Ashmore, 1992) and that is likely to comprise KCNQ4 subunits (Kharkovets et al., 2000; Marcotti and Kros, 1999; Oliver et al., 2003), possibly in combination with other KCNQ subunits. Subunits of the same K^+ channel family (KCNQ2 and KCNQ3) make up the M current of neurons (Wang et al., 1998). In one form of DFNA2, a dominant human deafness, the KCNQ4 subunit has a point mutation in its pore region (Kubisch et al., 1999). In inner hair cells, Marcotti et al., (2003) have shown that $g_{K,4AP}$ and $g_{K,n}$ combine to form the slow outwardly rectifying current that has been called $I_{K,s}$ (Kros and Crawford, 1990).

The mature complement of inner hair cell outward rectifiers is acquired over a period of several weeks following terminal mitosis (see Fig. 2 and a more detailed schematic in Fig. 6A). The earliest recordings, at E14.5 in the mouse, show a small delayed rectifier (Marcotti et al., 2003). The total outwardly rectifying current gradually becomes larger and slower up to about P10. Several changes occur at about the same time as the onset of hearing: From P8 to P20 the voltage activation range of the total current becomes more negative, reflecting the acquisition of two negatively activating K^+ conductances, $g_{K,n}$ and $g_{k,4AP}$ the 4-AP–sensitive component of $I_{K,s}$. It is not clear whether $g_{K,4AP}$ is a new conductance or a modified form of the neonatal 4-AP–sensitive conductance (see Fig. 6A). In the rat cochlea, in situ hybridization with probes for KCNQ4 reveals mRNA expression in inner hair cells of the basal turn as early as P0 (Beisel et al., 2000), a week earlier than the onset of $g_{K,n}$. If, as is widely assumed, $g_{K,n}$ channels do include KCNQ4 subunits (rather than other types of KCNQ subunits), then somehow there is a delay between the expression of mRNA and the functional expression of the channels, reflecting perhaps other subunits or posttranslational changes in KCNQ4.

The total outward current acquires a fast component when a fast Ca^{2+}-dependent K^+ conductance ($g_{K,f}$) appears, at P10 in basal hair cells (Marcotti et al., 2003) and 2 days later in apical hair cells (Kros et al., 1998). Messenger RNA for the subunits responsible for $I_{K,f}$ (KCNMA and KCNMB1) is first detected in the rat cochlea about 1–3 days earlier (Langer et al., 2003). Note that the timing of this acquisition—at the onset of hearing—is similar to that for $g_{K,Ca-BK}$ in the chick cochlea. The timing of $I_{K,f}$ expression may be at least partly controlled by thyroid hormone: The appearance of $I_{K,f}$ is delayed in mice null for thyroid hormone receptor β (TRβ) (Rüsch et al., 1998a). This receptor is expressed in the greater epithelial ridge of the immature cochlea, an appropriate site for guiding hair cell development (Bradley et al., 1994).

In rodent cochlea, unlike in chick and turtle cochlea, there is little expression of splice variants for BK channel subunits and no tonotopic variation in their expression (Langer et al., 2003). This suggests that differential expression of BK transcripts does not contribute to tonotopic variation in frequency tuning in mammalian cochlea, unlike the results in chick and turtle cochlea (see Section IV.A.1). This is not unexpected. Because of constraints on the speed of ion channel activation, electrical tuning via ion channels may not work above several kHz, thereby missing much of the mammalian range. Rather than contributing high-quality electrical tuning, it is argued that $I_{K,f}$ both speeds up the receptor potential of inner hair cells by reducing the membrane time constant and opposes spiking (Kros et al., 1998; Marcotti et al., 2003). Figure 7 shows that neonatal mouse inner hair cells respond to small current injections with spikes (see Fig. 7B) and that the spikes disappear when $I_{K,f}$ is acquired (see Fig. 7E). As we see, acquisition of $g_{K,f}$ is just one of several developmental changes that reduces the tendency of hair cells to spike as they mature.

Mouse *outer* hair cells also show significant electrophysiological changes with the onset of hearing. These include the abrupt appearance of $g_{K,n}$ (Marcotti and Kros, 1999), which is a more prominent fraction of the total outwardly rectifying conductance in outer hair cells (particularly basal hair cells) than in inner hair cells. Again, $g_{K,n}$ reduces the input resistance and therefore the membrane time constants of cochlear hair cells, expanding the frequency range of their voltage responses. Outer hair cells also show weak expression of BK channel transcripts (KCNMA and KCNMB), beginning later (after P16) than expression in inner hair cells (Langer et al., 2003).

A) Mouse inner hair cells

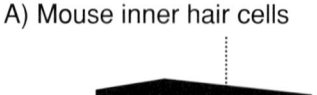

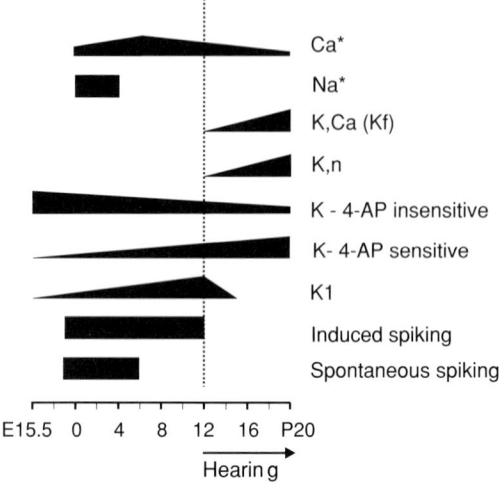

Conductance
Ca*
Na*
K,Ca (Kf)
K,n
K - 4-AP insensitive
K- 4-AP sensitive
K1
Induced spiking
Spontaneous spiking

E15.5 0 4 8 12 16 P20 → Hearing

B) Chick crista hair cells

Bouton afferents
Efferents
Type I
Calyces

Conductance
Ca
Kv
KA
Na*
K,Ca
h
K,L
K1

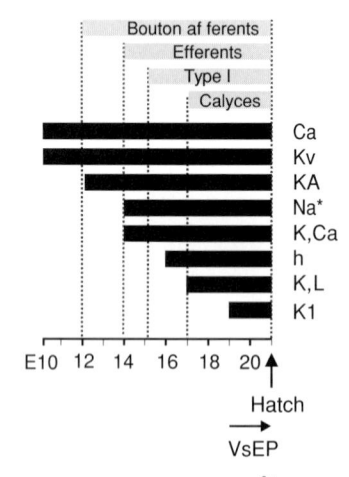

E10 12 14 16 18 20 ↑ Hatch → VsEP

Figure 6 Timing of acquisition of voltage- and Ca^{2+}-gated conductances. (A) In mouse inner hair cells. Modified with permission from Marcotti, W., Johnson, S. L., Holley, M. C., and Kros, C. J. (2003). Developmental changes in the expression of potassium currents of embryonic, and mature mouse inner hair cells. *J. Physiol.* **548**, 383–400. Ca^{2+} and Na^+ conductance data are derived from Beutner and Moser, 2001. (B) In chick crista vestibular hair cells. Modified with permission from Masetto, S., Perin, P., Malusa, H., Zucca, G., and Valli, P. (2000). Membrane properties of chick semicircular canal hair cells in situ during embryonic development. *J. Neurophysiol.* **83**, 2740–2756. Na^+ conductance data are derived from Masetto *et al.*, 2003. Conductances are identified by their subscripts; note that "K,Ca" represents $g_{K,Ca-BK}$. In the mouse data, changes in conductance amplitudes are shown where known. Note that the development for the chick is accelerated relative to the mouse, with a mature complement of conductances present by hatching.

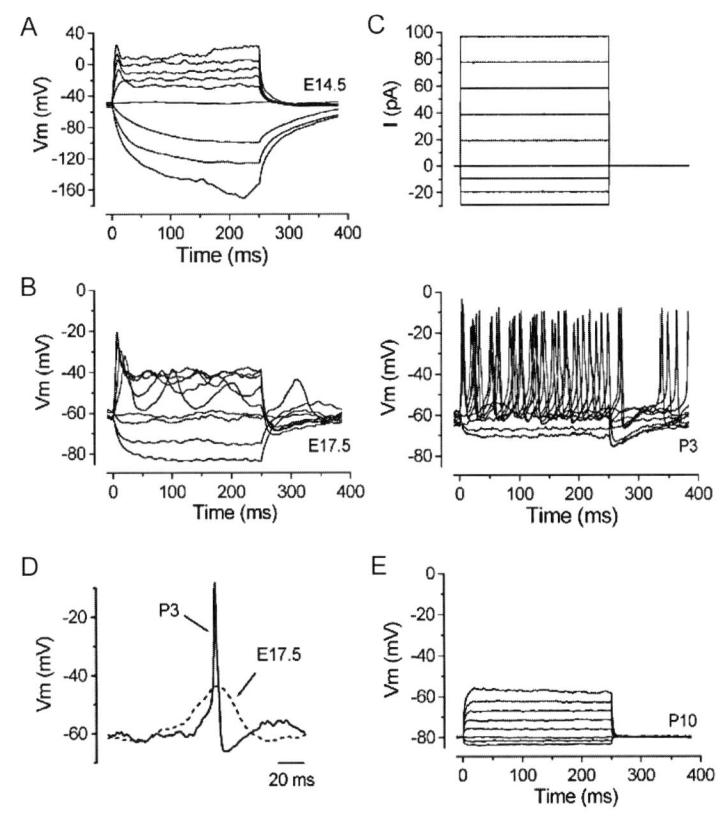

Figure 7 Developmental changes in the excitability of mouse inner hair cells. Reproduced with permission from Marcotti, W., Johnson, S. L., Holley, M. C., and Kros, C. J. (2003). Developmental changes in the expression of potassium currents of embryonic, neonatal and mature mouse inner hair cells. *J. Physiol.* **548**, 383–400. Hair cell voltage was recorded with the patch clamp method in current clamp mode. The current protocol is shown in C. (A) At E14.5, just after terminal mitoses of cochlear hair cells, hyperpolarizing currents evoked large, slow voltage changes because there is very little voltage-activated current in this range (i.e., the cell had a high input resistance). Depolarizing currents activated an outward rectifier that tended to repolarize the cell, producing an onset peak in the voltage response. (B) At E17.5 (*left*), spontaneous oscillations in membrane potential occurred (see zero-current trace in B; a single oscillation is shown at higher gain in D). By P3 (*right*), these oscillations have transformed to large, fast action potentials (spikes; see waveform of one spike in D). (C) Current protocol. Hyperpolarizing and depolarizing current steps were delivered; there was no holding current. (D) Spontaneous regenerative behavior. A single voltage oscillation at E17.5 and a single spike at P3. (E) By P10, just before the onset of hearing, the acquisition of negatively activating outward rectifiers has made the resting potential more negative and the response to injected currents small and relatively linear.

3. Outward Rectifiers in Chick Vestibular Hair Cells

Chick vestibular hair cells have mature complements of ion channels by the time of hatching (E21), 2 days after vestibular evoked potentials are detected (see Fig. 2). These ion channels vary with location in the epithelium and with hair cell type. The vestibular epithelia of birds, mammals, and reptiles have two morphological hair cell types, I and II, defined according to the form of the afferent synaptic contact. Primary vestibular afferents make conventional bouton endings on type II hair cells, each bouton opposite a single presynaptic ribbon structure in the hair cell. On type I hair cells, in contrast, the afferent forms a large cup-shaped calyx ending that engulfs much of the basolateral membrane (Wersäll and Bagger-Sjöbäck, 1974) (see schematic in Fig. 5A). Mature type I and II cells are also distinguished by their outwardly rectifying K^+ conductances. Type II cells express a variety of voltage- and Ca^{2+}-gated conductances that are activated by depolarization from the resting potential, whereas type I cells express an unusually large and negatively activating conductance, called $g_{K,L}$ or g_{KI}, that is substantially activated at the cells' resting potentials (Chen and Eatock, 2000; Correia and Lang, 1990; Rennie and Correia, 1994; Ricci et al., 1996; Rüsch and Eatock, 1996a). As described in Section IV.A.4, this conductance is acquired rather late in hair cell differentiation.

Sokolowski et al. (1993) documented changes in voltage- and Ca^{2+}-gated currents in hair cells and precursor cells in chick otocysts, organ cultured for varying periods from E3. At E3 there were only small round cells, about half of which expressed a small outwardly rectifying K^+ current that activated very positively (positive to 0 mV). By 9 days in vitro, hair cells could be recognized, and these expressed a much larger delayed rectifier K^+ current that activated positive to −30 mV. By 12 DIV, a number of cells expressed an additional A-type current, with fast activation and inactivation. By 17 DIV, some cells expressed a small Ca^{2+}-dependent K^+ current as well (consistent with the onset of $g_{K,Ca\text{-}BK}$ in chick cochlear hair cells at E18; see Section IV.A.1). By this stage the currents of hair cells from the cultured otocyst resembled those of type II hair cells dissociated from semicircular canal sensory epithelia (cristas) of 2- to 3-week-old posthatch chicks. No hair cells expressed $g_{K,L}$, the negatively activating conductance that is typical of type I cells.

Using the chick otocyst culture system, Sokolowski and Cunningham (1999) found that different neurotrophic factors—retinoic acid, brain-derived neurotrophic factor (BDNF), and neurotrophin-3 (NT-3)—enhance or inhibit expression of basolateral conductances. Such factors are present in the developing inner ear in various amounts, depending on the organ or even location in the organ (Rubel and Fritzsch, 2002). Neurotrophins within the inner ear epithelia are known to influence growth and differentiation of

13. Functional Development of Hair Cells 417

inner ear afferents (see Section VI.A.1). The results of Sokolowski and Cunningham (1999) suggest that neurotrophins may also help to set up gradients in expression of hair cell basolateral channels.

Masetto et al. (2000) surveyed the time course of acquisition of voltage-gated currents in slices of semicircular canal cristas, made acutely between E10 and E21. Their results are summarized in Fig. 7B. At E10, just as synapses are beginning to form (Ginzberg and Gilula, 1980), embryonic hair cells already express a delayed rectifier conductance (g_{Kv}). By E12, when afferent synapses are evident, some cells acquire an inactivating K$^+$ conductance, g_A. By E14, when efferent synapses are first reported, $g_{K,Ca}$ is evident. Thus, the progression in these acute crista preparations is similar in important ways to that described by Sokolowski et al. (1993) for their chick otocyst cultures; the earlier reported onset of some conductances in the acute slices may reflect the shorter sampling interval in that study. By using a slice preparation rather than dissociated hair cells, Masetto et al. (2000) were able to show regional variation in K$^+$ channel acquisition. For example, the cells that acquire g_A are from the edge of the crista (the planum). In the acute crista slices, $g_{K,L}$ first appeared at E17, 2 days after type I hair cells begin to acquire their distinctive shape and partial calyces and simultaneously with the appearance of full calyx endings (see Fig. 6B). The coincidence of the appearance of $g_{K,L}$ and calyces *in vivo* raises the possibility that calyx formation induces $g_{K,L}$. Data from the developing rodent utricle (see Section IV.A.4) and the regenerating pigeon utricle, however, show that $g_{K,L}$ can occur without calyces and calyces can occur without $g_{K,L}$, respectively.

Avian auditory and vestibular organs are capable of substantial regeneration following damage to hair cells (for a recent review of hair cell regeneration, see Bermingham-McDonogh and Rubel, 2003). Correia and colleagues have investigated the recovery of ion channels in regenerating hair cells in the mature vestibular organs of pigeons (Correia et al., 2001; Masetto and Correia, 1997a,b). Supporting cells, from which the regenerating hair cells derive, have virtually no voltage-gated K$^+$ channels. The ionic currents return, with appropriate regional variation, in about the same order as they are acquired in development. Thus, the developmental signals that set up the regional and cell type variations in ion channels are also available in regenerating epithelia. The signal that drives the appearance of $g_{K,L}$, however, is delayed or attenuated, because it takes months for $g_{K,L}$ to reappear.

4. Outward Rectifiers in Rodent Vestibular Hair Cells

The earliest recordings from a rodent utricle are from a preliminary report of experiments in the mouse (Géléoc et al., 2003). At E16.5, there is a small outwardly rectifying K$^+$ current that activates positive to resting potential

($V_{1/2} \sim -20$ mV). Over the next 2 days the conductance doubles in size. The *incidence* of $g_{K,L}$ (percentage of hair cells expressing $g_{K,L}$) increases rapidly in the first postnatal week to mature levels by P7–P8 ($\sim 60\%$, corresponding to the percentage of type I cells in the mature rodent utricle; Rüsch et al., 1998b). The time course of acquisition of $g_{K,L}$ is not affected by culturing the denervated organs at P0, before calyx formation takes place (Rüsch et al., 1998b). Thus, calyces do not induce $g_{K,L}$. (The cultured type I hair cells also acquire their distinctive morphological features, including the flask shape, despite the absence of calyces.)

In hair cells from *rat* utricles, $g_{K,L}$ appears with a slightly delayed schedule relative to the mouse data. Figure 8A shows voltage-gated currents from rat utricular type I cells at P4, P13, and P21 and from a type II cell at P21. Until P7, all outward K^+ currents activate positive to resting potential ($V_{1/2}$ values positive to -50 mV; Fig. 8B and C). Then the negatively activating current, $g_{K,L}$, appears and for the second postnatal week the $V_{1/2}$ values of type I cells range from -25 mV to -90 mV. By P18, all $V_{1/2}$ values in type I cells are negative to -60 mV and we also start to see very large maximal conductances (see arrows, Fig. 8C). It is not yet clear whether this change represents acquisition of new K^+ channel subunits or modification of previous channels. It has been suggested that $g_{K,L}$ is a KCNQ4 channel, like $g_{K,n}$ in rodent outer hair cells. One reason to suggest this is that KCNQ4 is first detected in vestibular epithelia at about the same time as $g_{K,L}$ appears (Kharkovets et al., 2000). Arguing against this are the strong biophysical and pharmacological differences between $g_{K,L}$ and $g_{K,n}$, the KCNQ4 conductance of outer hair cells (Hurley et al., 2002; Rennie et al., 2001), and observations that most KCNQ4 immunoreactivity is in the calyx rather than the hair cell (Lysakowski and Price, 2003). Another possibility is that the change in $V_{1/2}$ with age (see Fig. 8C) reflects modulation of the voltage range of activation through a process other than new subunit expression. The activation range of $g_{K,L}$ is known to be sensitive to phosphorylation (Eatock et al., 2002).

Although $g_{K,L}$ is not induced by calyx formation, *in vivo* the two features appear simultaneously and may be functionally linked. The unusual shape of the calyx ending has long fed speculation that it sustains unconventional transmission: retrograde (Scarfone et al., 1988), electrical (Yamashita and Ohmori, 1990), ephaptic (Gulley and Bagger-Sjoback, 1979; Hamilton, 1968), or via K^+ accumulation in the synaptic cleft (Goldberg, 1996). There is immunocytochemical evidence that KCNQ4 subunits are expressed postsynaptically in the calyx ending (Kharkovets et al., 2000; Lysakowski and Price, 2003). The presence of negatively activating conductances both pre- and postsynaptically may serve ephaptic transmission via the direct flow of K^+ current out of the hair cell conductance ($g_{K,L}$) across the cleft and into the calyx ending via KCNQ4 channels (Trussell, 2000).

13. Functional Development of Hair Cells 419

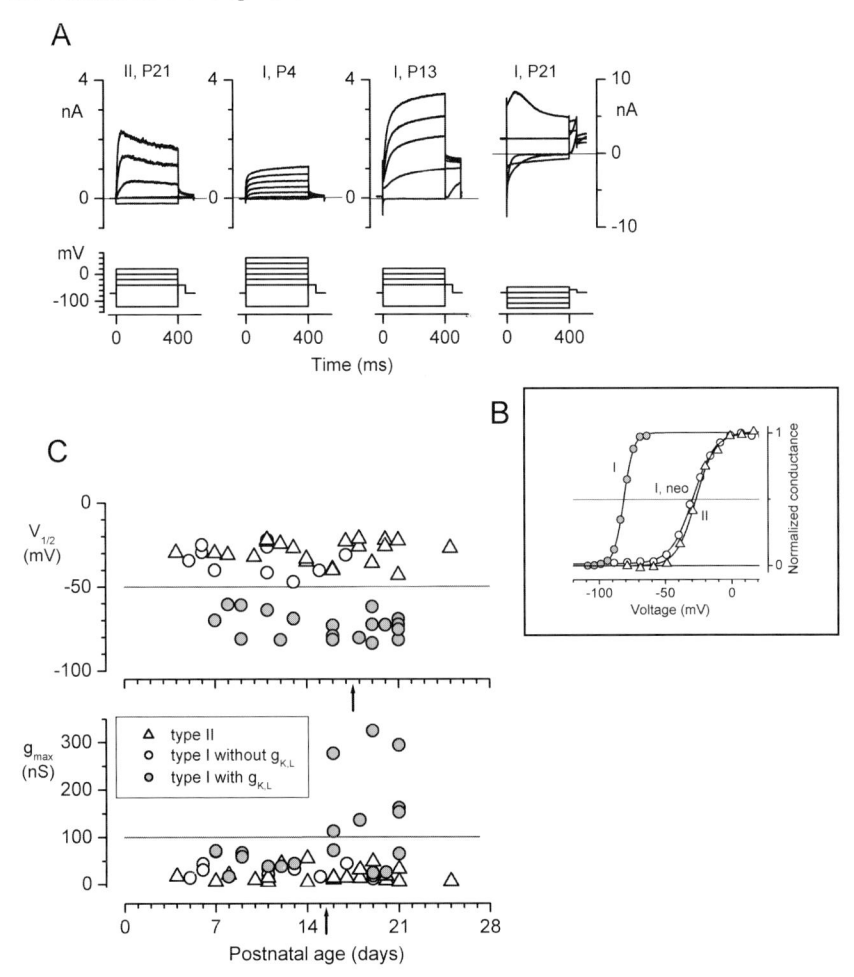

Figure 8 Postnatal changes in the outwardly rectifying current of rat utricular hair cells (K. M. Hurley and R. A. Eatock, unpublished results). $g_{K,L}$, a negatively activating delayed rectifier specific to type I cells, appears after the first postnatal week and becomes large during the third postnatal week. Hair cells were dissociated enzymatically and mechanically. Whole-cell currents were recorded in voltage clamp mode with the perforated patch method, which preserves internal second messengers. (A) Voltage-activated currents from a type II cell at P21 and from type I hair cells at P4, P13, and P21. The voltage protocols used are shown in B. (B) Activation curves for a neonatal type I hair cell ($V_{1/2} = -32$ mV); a type II cell at P21 ($V_{1/2} = -27$ mV); and a type I cell at P21 (I, $V_{1/2} = -85$ mV). Activation curves were generated from tail currents at the offset of iterated voltage steps similar to those shown in A (but for cells with $g_{K,L}$, longer voltage steps [≥ 1 s] were needed because of the slow kinetics of $g_{K,L}$). The curves are normalized to the maximum conductance in order to emphasize differences in $V_{1/2}$. (C) Parameters ($V_{1/2}$ and g_{max}) from fits of activation curves like those in B, plotted against postnatal age. In the first postnatal week, $V_{1/2}$ and g_{max} values were similar for all cells. $g_{K,L}$ first appeared at P7 (filled circles, $V_{1/2}$ plot). During the second postnatal week (arrows), type I

B. Inwardly Rectifying K⁺ Conductances (g_{K1}, g_h)

In hair cells, inwardly rectifying currents activate with increasing hyperpolarization negative to about -50 mV (Goodman and Art, 1996; Holt and Eatock, 1995). Two kinds have been described. One (g_{K1}) has very fast kinetics, is K⁺-selective, and is blocked by Ba^{2+}. The other (g_h) is much slower and is a mixed K⁺/Na⁺ conductance; in standard solutions, its reversal potential is ~ -40 mV. Much of the activation range of g_{K1} is negative to its reversal potential (near the equilibrium potential for K⁺, E_K, ~ -85 mV in standard solutions) and all of the activation range of g_h is negative to its reversal potential (~ -40 mV). Negative to the reversal potential, positive ions flow inward—hence the term "inwardly rectifying." It should be noted, however, that current through g_{K1} is outward over the physiological range of voltages, which are positive to E_K. Inward rectifiers contribute to the resting conductance and resting potential of the hair cell (reviewed in Holt and Eatock, 1995). Furthermore, g_{K1} can contribute to electrical tuning: as the hair cell depolarizes, rapid deactivation of g_{K1} can boost the depolarization (Goodman and Art, 1996).

The fast inward rectifier, g_{K1}, tends to appear early in development, though a bit later than the earliest outward rectifier. In chick otocysts that were organ cultured on E3, g_{K1} first appeared after 9 days in culture (Sokolowski *et al.*, 1993). Although after many days in culture the hair cells came to resemble chick crista cells, Masetto *et al.* (2000) found no evidence for g_{K1} in embryonic chick crista slices until E19. Thus, the otocyst-derived hair cells may have been more heterogeneous at early stages in the *in vitro* differentiation of the hair cells than at later stages. The inward rectifier in chick cochlear hair cells has been identified as cIRK1 (Kir2.1; Navaratnam *et al.*, 1997). In the slices, a small g_h was recorded in the same regions beginning at E14.

In the mouse cochlea, both inner and outer hair cells express g_{K1} at birth (Marcotti *et al.*, 1999). In outer hair cells, g_{K1} declines to a negligible amplitude by P6. For inner hair cells, we know how g_{K1} changes between E14.5 and P20 (Marcotti *et al.*, 1999, 2003). In g_{K1} is not evident at E14.5, when a small outward rectifier is found, first appearing at E15.5. g_{k1} increases in size until the onset of hearing (P12), after which it rapidly decreases to a small current (see Fig. 6A). The significance of the transient increase in expression is not known. The action of g_{K1} to boost regenerative behavior in turtle cochlear hair cells (Goodman and Art, 1996) suggests that transient

---

cells with positively activating outward rectifiers (open circles) were no longer seen, and the amplitude of the negatively activating conductance became very large in many type I cells (note absence of open circles after P17 in $V_{1/2}$ plot and note g_{max} values >100 nS after P15 in g_{max} plot).

expression of g_{K1} in immature hair cells may represent an adaptation to boost spiking during synaptogenesis (see Section IV.C). Indeed, blocking g_{K1} in a P4 inner hair cell with external Cs^+ eliminated spiking in response to depolarizing currents (Marcotti *et al.*, 1999). Thus, the disappearance of g_{K1} at the onset of hearing may be one of several coordinated changes that reduce the tendency of the hair cells to spike once sound detection is paramount.

In the mouse utricle, g_{K1} is present in most hair cells at P0 and persists through maturity (Rüsch *et al.*, 1998b). In contrast, g_h is acquired by both type I and type II cells in the first postnatal week, over the same period that $g_{K,L}$ is acquired by type I hair cells. Thus, inward rectifiers may appear with similar timing in mammalian and chick vestibular organs. The data from rodent utricles argue that type I and type II hair cells undergo electrophysiological differentiation concurrently, rather than the type I hair cell being a "super-differentiated" type II hair cell (Eatock and Rüsch, 1997), but the data from regenerating pigeon utricles are consistent with the latter (Masetto and Correia, 1997b; see Section IV.A.3). These kinds of relations will be better understood when the molecular constituents of the ion channels have been resolved.

C. Voltage-Gated Na^+ Conductances: g_{Na}

Voltage-gated Na^+ currents are expressed by some mature hair cells in diverse organs, including the alligator cochlea (Evans and Fuchs, 1987), goldfish saccule (Sugihara and Furukawa, 1989), chick crista (Masetto *et al.*, 2003; Sokolowski *et al.*, 1993), and guinea pig cochlea (Witt *et al.*, 1994). The voltage range of activation is positive to resting potential, with reported $V_{1/2}$ values between -35 and -45 mV. As in other cell types, Na^+ currents in hair cells are transient because the channels inactivate very rapidly. Hair cells are unusual, however, in that the reported voltage range of inactivation for g_{Na} in hair cells is often very negative, with voltages of half-maximal inactivation, $V_{1/2,in}$, between -90 and -110 mV. The somewhat puzzling consequence is that the channels fully inactivate at the usual resting potential (-50 to -70 mV). To be activated, therefore, such channels must first experience strong hyperpolarizations to relieve the inactivation. Inhibitory mechanical stimuli and efferent-activated K^+ currents can provide hyperpolarizing inputs. It is also possible that *in vitro* recording conditions artificially shift the inactivation range to more negative potentials. For example, reports are generally at room temperature. For rat cochlear outer hair cells, raising the temperature from $23°$ to $37°$ C shifts the inactivation range by $+8$ mV (Oliver *et al.*, 1997). In other experiments, the use of enzymatic dissociation protocols or washout of internal second messengers during recording may have altered the Na$^+$ channel properties.

In the chick crista, hair cells begin to express g_{Na} at E14 and continue to express it through to maturity (Masetto et al., 2003). It is expressed by most hair cells, except for type II cells at one end of the crista (near the *planum semilunatum*). To examine the influence of g_{Na} on the voltage behavior of the hair cells, Masetto et al. (2003) compared the hair cells' voltage responses to current steps in control conditions and with tetrodotoxin present to block g_{Na}. As expected, the Na$^+$ current augments spikes at the onset of the current step—but only if the cell is prehyperpolarized negative to -80 mV to relieve inactivation and then is suddenly depolarized. It is not clear how often such conditions would be met *in vivo*; not only must there be a substantial hyperpolarizing input, but also the depolarization must be fast enough to permit current activation before inactivation takes over. This may be a problem for vestibular organs, for which the mass of the head attenuates stimulus energy at high frequencies.

In rodents, g_{Na} is more prominent in immature cochleas and vestibular organs than at maturity. Between P0 and P9, virtually all outer hair cells in the rat cochlea express a negatively inactivating g_{Na} (Oliver et al., 1997). The size of this conductance increases two- to threefold from P0 to P3 in basal OHCs and from P0 to P4–P8 in apical OHCs, after which it rapidly returns to the P0 level, by P6 in basal OHCs and by P9 in apical OHCs. At P18 it is present only in about half of OHCs, and in those it is about half its P0 value. In a preliminary report Chabbert et al. (2003) observed that g_{Na} in rat utricular hair cells decreases in incidence from P1 to P8, becoming negligible by P21. Reverse transcription polymerase chain reaction (RT-PCR) experiments suggested that the channels comprise Na$_v$ 1.2 and/or Na$_v$ 1.6 subunits.

There is one report of a hair cell Na$^+$ conductance with a more conventional inactivation range ($V_{1/2,in} = -25$ mV), which was transiently expressed in the immature rat utricle (P3; Lennan et al., 1999). These recordings were made with 0 ATP in the pipette, which may explain the discrepancy between these results and other reports of g_{Na} in hair cells. In neonatal mouse utricles, $V_{1/2,in}$ values of -100 mV and -70 mV have both been seen (Rüsch and Eatock, 1997; J. R. A. Wooltorton and R. A. Eatock, unpublished observations).

The postnatal reduction of Na$^+$ currents in rat outer hair cells and rat utricular hair cells suggests that g_{Na} may contribute to spiking in neonatal hair cells and, possibly, to spontaneous rhythmic activity at other levels. In their preliminary report, Chabbert et al. (2003) provide an attractive possible link between Na$^+$ spikes and the formation or stabilization of hair cell–afferent synapses. Brain-derived neurotrophin factor (BDNF) is made by vestibular sensory epithelia and is required for successful innervation of those epithelia

13. Functional Development of Hair Cells

by vestibular nerve fibers (Bianchi et al., 1996). Chabbert and colleagues report that BDNF release from immature rat utricles has a tetrodotoxin-sensitive component that is not seen in mature rat utricles. Since tetrodotoxin specifically blocks voltage-gated Na^+ channels, this result suggest that Na^+ channel activity, presumably in the immature hair cells, promotes BDNF release.

D. Voltage-Gated Ca^{2+} Conductances: g_{Ca}

Ca^{2+} influx through voltage-gated Ca^{2+} channels in hair cells mediates afferent transmitter release, may contribute in some cases to slow Ca^{2+} spikes, and activates Ca^{2+}-activated K^+ channels. Ca^{2+} channels are localized at presynaptic active zones next to synaptic dense bodies or ribbons (Roberts et al., 1990). Hair cells express an L-type Ca^{2+} channel, $Ca_v1.3$ (also called α_{1D}) (Bao et al., 2003; Kollmar et al., 1997a,b; Ramakrishnan et al., 2002). In some cases they may express an additional channel type, either N-type (Rodriguez-Contreras and Yamoah, 2001) or R-type (Martini et al., 2000). In a preliminary report, Brandt et al. (2003) describe how neonatal hair cells in $Ca_v1.3$ null mutants express a residual Ca^{2+} current carried by other L-type channels ($Ca_v1.2$, α_{1C}) and R-type channels. Although the $Ca_v1.2$ Ca^{2+} channels do permit a small amount of exocytosis by the null mutant hair cells, they cannot sustain cochlear function in that the null mutants are deaf (Platzer et al., 2000).

g_{Ca} is evident early in embryonic development: at E10 in chick crista slices (Masetto et al., 2000) and E16.5 in mouse cochlear inner hair cells (Marcotti et al., 2003). In mouse inner hair cells, Beutner and Moser (2001) showed that depolarizing voltage steps evoke Ca^{2+} currents and synaptic vesicle exocytosis (measured by increments in hair cell capacitance) that increase in parallel from P0 to peak values at P6, then fall back to levels below the P0 level by P14–P25 (Fig. 9). The decrease after the first postnatal week may reflect developmental pruning of active zones and/or presynaptic ribbons (Sobkowicz et al., 1982, 1986) and the loss of associated Ca^{2+} channels.

At P6, spontaneous Ca^{2+} action potentials were recorded from some hair cells and shown to provoke exocytosis. Beutner and Moser (2001) suggest that these Ca^{2+} spikes may promote neurotrophin release, underlie the bursting activity that characterizes the immature auditory nervous system, and set up Ca^{2+} oscillations within the hair cells that may have important developmental functions, such as regulation of gene expression. The frequency of Ca^{2+} spikes influences differentiation in cultured developing *Xenopus* neurons, specifically the maturation of the time course of a delayed rectifier and the appearance of neurotransmitter (Gu and Spitzer, 1995).

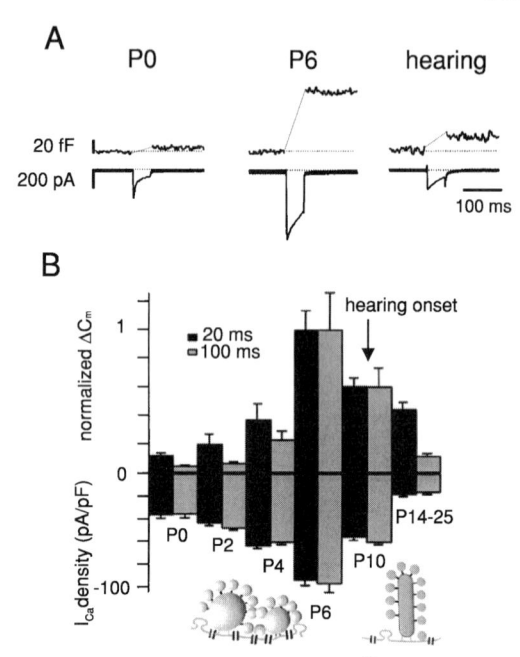

Figure 9 Developmental changes in voltage-gated Ca^{2+} current and voltage-dependent capacitance increments, reflecting synaptic vesicle exocytosis. Cell capacitance is proportional to cell surface area and increases as vesicles fuse with the plasmal membrane. Mouse inner hair cells. Reproduced with permission from Beutner D., and Moser, T. (2001). The presynaptic function of mouse cochlear inner hair cells during development of hearing. *J. Neuro. Sci.* **21,** 4593–4599. (A) Depolarizing steps (not shown; 50 ms duration, to -5 mV from the holding potential of -80 mV) evoke increments in cell capacitance (C_m, *upper traces*) and inward Ca^{2+} currents (*lower traces*). The C_m trace is blanked out during the voltage step, when C_m cannot be accurately read. Both the capacitance change and the inward Ca^{2+} current are largest at P6 (*middle panel*). This is quantified for a number of hair cells in B, which shows the increment in C_m (*top histogram*) and the Ca^{2+} current density (current divided by total cell capacitance; *bottom histogram*), in response to depolarizing steps to -5 mV of 20 ms or 100 ms duration (*black and gray bars, respectively*). The relatively large currents and exocytosis at P6 may reflect the presence of multiribbon active zones at this stage (*left schematic*), which become pruned to a single ribbon in mature hair cells (*right schematic*).

E. Summary of Developmental Changes in Basolateral Conductances

The earliest available recordings from immature hair cells show them expressing small outwardly rectifying K^+ conductances. These are quickly joined by voltage-gated Ca^{2+} channels, voltage-gated Na^+ channels, and fast inwardly rectifying channels (g_{K1}), all of which act to boost the effects of depolarizing inputs—the Ca^{2+} and Na^+ channels by rapidly activating and feeding back positively with more inward current and the inwardly rectifying

K⁺ channels by rapidly deactivating, feeding back positively by reducing outward current. These channels may contribute to spontaneous spiking behavior in the hair cells, which may help to establish connections both in the periphery and, as the spiking propagates through the system, more centrally. At about the onset of hearing and vestibular reflexes, the hair cells undergo multiple changes that reduce their excitability: decreases in g_{Na}, g_{Ca}, and g_{K1} and increases in outwardly rectifying K⁺ conductances. In the chick cochlea, the acquisition of $g_{K(Ca)-BK}$ just around hearing onset enhances electrical tuning at acoustic frequencies. In the mammalian cochlea and hair cells of amniote vestibular organs, late arriving outward rectifiers include K⁺ conductances that activate negative to the resting potential. Such conductances reduce the size and speed up voltage responses to transduction currents. In the mammalian auditory system, the enhanced speed may be important for signaling and transmitting frequencies in the thousands of Hertz. In the vestibular system the acquisition of large outward rectifiers may have less import with respect to the speed of signaling given the relatively low frequencies of head movements, but it may help to linearize the receptor potentials of type II hair cells (Holt *et al.*, 1999). In type I hair cells, the acquisition of $g_{K,L}$ attenuates the voltage response of the hair cell but may contribute to unconventional forms of transmission across the unusual calyx synapse.

V. Outer Hair Cell Electromotility

Isolated outer hair cells expand and contract lengthwise when hyperpolarized and depolarized, respectively (Brownell *et al.*, 1985). This fast "electromotility" may be the "cochlear amplifier" that imparts high sensitivity at high frequencies to mammalian cochleas (but see Gale and Ashmore, 1997), possibly in combination with a stereociliary mechanism (Hudspeth *et al.*, 2000). Electromotility is associated with a nonlinear membrane capacitive current, which may reflect the charge movement that drives the motility (Santos-Sacchi, 1991). Both depend critically on the protein prestin, a member of a large anion transporter family that is present at high density in the outer hair cell membrane (Liberman *et al.*, 2002; Zheng *et al.*, 2000a). It is thought that the transmembrane voltage affects the conformation of the protein, which somehow—possibly through changes in protein area—changes the length of the outer hair cell. Other properties peculiar to the outer hair cell may be important in amplifying the basic electromotility of this membrane protein—for example, the specialized cytoskeleton (Holley and Ashmore, 1988), the subsurface cisternae (Dieler *et al.*, 1991), and an unusual membrane lipid composition (Oghalai *et al.*, 2000). As described later, prestin and other proteins and structures that may play supporting or amplifying roles in electromotility are acquired at about the same time that gerbil afferents at the basal end of the cochlea add a high-frequency tip to the tuning curve, i.e., acquire the cochlear amplifier (Echteler et al., 1989; see Section II.A.1).

In gerbil outer hair cells, isolated acutely or from cultures made on P0, the incidence of the electromotile response (percentage of cells showing electromotility) rises as a sigmoidal function of time, from negligible at P5 to 100% by the onset of hearing at P12 (He, 1997). In outer hair cells isolated from rats, the density ($\mu F/cm^2$) of the nonlinear capacitance rises with a similar time course, from negligible at <P5 to nearly mature levels by P15 (Belyantseva et al., 2000; Beurg et al., 2001). Similarly, the amplitude of the electromotile length change increases (Fig. 10A; He, 1997), doubling between P6–P8 and P12–P14 (Beurg et al., 2001). These physiological changes correlate in time with increased expression of prestin, as assayed with fluorescent immunocytochemistry (Fig. 10B; Belyantseva et al., 2000). At P6, prestin-like immunoreactivity is dispersed over the entire basolateral membrane of outer hair cells. By P8 it is strongly localized to the "lateral wall," the specialized electromotile membrane that runs from the apical cuticular plate to the basolateral nucleus. Interestingly, in thyroid hormone–deficient animals, this redistribution never takes place (Weber et al., 2002). This abnormality is therefore another candidate for the hearing defect in hypothyroidism, along with the delayed expression of $I_{K,f}$ by inner hair cells (Rüsch et al., 1998a; see Section IV.A.3).

The sugar transporter, GLUT-5, is also expressed at high levels in the mature outer hair cell membrane (Nakazawa et al., 1995) and mediates an effect of sugar transport on the voltage sensitivity of electromotility (Géléoc et al., 1999). Presumably, the membrane particles evident at high density in freeze-fracture images of the outer hair cell's lateral membrane are both prestin and GLUT-5 proteins (Kalinec et al., 1992). GLUT-5 expression, as indicated by fluorescence immunocytochemistry, rises immediately following the onset of hearing, from negligible at P12 to mature levels by P18 (Belyantseva et al., 2000). Activity of this transporter may be important not just for energy metabolism but also as a regulator of the voltage dependence of electromotility, possibly through direct mechanical interactions with prestin in the membrane.

The subsurface cisternae, a collection of cisterns that ring the elongate soma of the outer hair cell, begin to appear at the same time as both electromotility (Souter et al., 1995; Weaver and Schweitzer, 1994) and the cisterns that are postsynaptic to efferent boutons (Pujol et al., 1998). The function of the subsurface cisternae is not known, although they have been implicated in electromotility by their localization along the electromotile lateral membrane of outer hair cells, the timing of their appearance in development, and their sensitivity to agents that disrupt electromotility (Dieler et al., 1991).

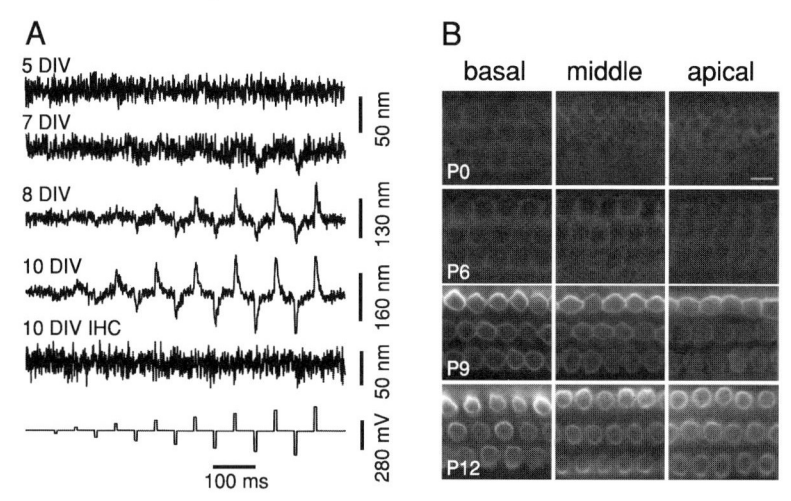

Figure 10 Developmental acquisition of electromotility and of the electromotile protein, prestin. (A) Longitudinal movements of gerbil outer hair cells, cultured on P0 and isolated after 5–10 days *in vitro* (DIV). Reproduced with permission from He, D. Z. (1997). Relationship between the development of outer hair cell electromotility and efferent innervation: A study in cultured organ of Corti of neonatal gerbils. *J. Neurosci.* **17,** 3634–3643. The hair cell was drawn into a "microchamber," a polished glass pipette, through which incrementing steps of voltage were applied (*bottom trace*). Electromotile responses can be seen for large steps at 7 DIV, increasing in amplitude with further time *in vitro*. An inner hair cell (IHC, 10 DIV) did not show electromotility. (B) Prestin immunoreactivity appears in the lateral membranes of rat outer hair cells after P6. Reproduced with permission from Belyantseva, I. A., Adler, H. J., Curi, R., Frolenkov, E. I., Kachar, B. (2000). Expression and localization of prestin and the sugar transporter GLUT-5 during development of electromotility incochlear outer hair cells. *J. Neurosci.* **20,** RC116. Organs of Corti from P0–P12 rats were divided into basal, middle, and apical segments and stained with fluorescent antibody to prestin protein. The focal plane is about midway along the cell bodies of the outer hair cells. Scale bar:10 μm.

VI. Interactions between Developing Hair Cells and Innervating Nerve Fibers

In the mature mammalian cochlea, radial afferents (type I neurons) innervate inner hair cells in a 10:1 to 20:1 pattern, spiral afferents (type II neurons) innervate outer hair cells, lateral olivocochlear efferents innervate the terminals of radial afferents below the inner hair cells, and medial olivocochlear efferents innervate outer hair cells. Afferent neurons undergo terminal mitoses 2–3 days in advance of hair cells: in the mouse, terminal mitoses occur at E12–13 and E15–16 for spiral ganglion neurons and cochlear hair cells, respectively (Ruben, 1967). Afferent fibers are seen in the greater epithelial ridge of the developing organ of Corti before morphological differentiation of hair cells is evident (Pujol et al., 1998). Radial afferents, which innervate inner hair cells in the mature cochlea, show considerable sprouting early in development to multiple inner hair cells and to outer hair cells. Pruning before hearing onset produces the mature pattern, with most radial afferents innervating just one inner hair cell (with exceptions at the cochlear apex; Liberman et al., 1990) and with spiral afferents innervating outer hair cells.

Efferent nerve fibers to the inner ear emanate from cell groups in the brainstem and project onto hair cells, principally short hair cells in the bird cochlea, outer hair cells in the mammalian cochlea, and type II hair cells in amniote vestibular organs. Efferents also terminate on primary afferent endings that contact both cochlear as vestibular hair cells. Several lines of reasoning suggest that inner ear efferent neurons arose from brainstem motorneurons (Barritt et al., 1999; Simmons, 2002).

Efferent fibers enter the developing inner ear very early. In rodents, they penetrate the sensory epithelia at around E14. Early in development, efferents make functional synapses on the cell bodies of inner hair cells, as discussed in Section VI.B. These efferent terminals may be from lateral olivocochlear efferents (Pujol et al., 1998), which ultimately project mostly onto primary afferent endings below the hair cells (but possibly also onto synapses on spiny projections from the inner hair cell; Sobkowicz et al., 2002). Additionally or alternatively, the efferent terminals on inner hair cell bodies may belong to medial olivocochlear efferents (Simmons, 2002). Medial olivocochlear efferents reach their mature targets, outer hair cells, between P6 and P12 (Cole and Robertson, 1992; Lenoir et al., 1980; Robertson et al., 1989; Simmons et al., 1990). Thus, during synaptogenesis, rodent cochlear hair cells are exuberantly innervated by multiple afferents and efferents. The connections are pruned and retargeted postnatally.

In the developing chick cochlea, afferent fibers penetrate the presumptive cochlear epithelium on E6–E7 and form synapses beginning on E8–E9 (Sokolowski and Cunningham, 1999; Whitehead and Morest, 1985a,b). Efferents may arrive at about the same time (Fritzsch et al., 1998; Sokolowski and Cunningham, 1999) but efferent synapses form over a long period, continuing for weeks past hatching (Sokolowski and Cunningham, 1999).

A. Hair Cells and Primary Afferents

1. Hair Cells Differentiate Independently of Innervation

The early arrival of afferent fibers beneath hair cell epithelia has long suggested the possibility that afferents direct hair cell differentiation. The preponderance of evidence does not support this idea. Diverse experiments, from *in vitro* culture of early otocysts (Raymond, 1987; Van de Water,

13. Functional Development of Hair Cells 429

1976) to gene knockout experiments (Ma *et al.*, 2000), have indicated that differentiated morphological attributes of hair cells can be acquired without innervation by afferents or efferents.

Consistent with these morphological studies, many of the differentiating electrophysiological changes described previously have been observed in cultured inner ear explants. Usually rodent inner ear cultures are made shortly after birth (P0–P3). The distal endings of the primary afferents are severed, leading to their rapid degeneration. Denervation of the hair cells at this time and subsequent short-term culture (<2 weeks) have no strong qualitative effect on the differentiation of mechanosensitive or voltage-gated ion channels or of electromotility, as assessed by comparing results from cultured and acutely excised preparations (Beurg *et al.*, 2001; He, 1997; Kros *et al.*, 1998; Rüsch *et al.*, 1998b). This is true even for the acquisition of conductances that normally appear coincidentally with synaptogenesis (e.g., $g_{K,L}$ in type I hair cells; see Section IV.A.4). This not to say that there are no effects of culturing, which may include the effects of denervation. For example, Beurg *et al.* (2001) saw a reduction in the nonlinear charge density of the membrane in cultured outer hair cells relative to acutely dissociated outer hair cells, presumably reflecting decreased density of prestin proteins.

2. Hair Cells Influence Afferent Innervation Via Neurotrophins

In contrast, afferent nerve fibers clearly need hair cells in order to reach their targets and/or to maintain their presence in the developing epithelium. One critical mechanism for this effect is the neurotrophin–Trk receptor cascade (for a review of its impact in the inner ear, see Rubel and Fritzsch, 2002). During development, hair cells in vestibular epithelia and the apical cochlea release BDNF, which binds to TrkB receptors on eighth-nerve afferents. Hair cells in the basal cochlea also release neurotrophin-3 (NT-3), which binds to TrkC receptors on basal cochlear afferents. Null mutants for neurotrophins or their receptors make morphologically differentiated hair cell epithelia but lack appropriate innervation.

Neurotrophins may be needed not just for appropriate targeting and maintenance of synapses but also for differentiation and/or maintenance of the appropriate electrophysiological profile of the afferent neurons. Recent results from Davis and collaborators (Adamson *et al.*, 2002) show how neurotrophins can influence the expression of voltage-gated ion channels in postnatal eighth-nerve afferent neurons. Exposing cultured spiral ganglion neurons from any cochlear region to BDNF or NT-3 promotes the expression of ion channels characteristic of neurons from the basal or apical regions of the cochlea, respectively. (Although both neurotrophins are present in the cochlea, their distribution along its length is somewhat controversial; see discussion in Adamson et al., 2002.) It is not clear how neurotrophins change the expression of ion channels in the afferents. Acute effects of neurotrophins on afferent activity have been observed. BDNF and NT-3 cause transient increases in intracellular Ca^{2+} in early postnatal rat vestibular afferents (Montcouquiol et al., 1997), and iontophoresis of NT-3 into the synaptic region of the adult guinea pig cochlea *in vivo* potentiates afferent activity (Oestreicher et al., 2000).

In addition to neurotrophin-mediated effects, transmission at the hair cell–afferent synapse might influence targeting by and differentiation of afferent neurons through glutamatergic mechanisms, e.g., gene expression activated by Ca^{2+} entry through NMDA receptors (Takasu et al., 2002). It is very likely that such mechanisms are active before the inner ear is fully functional, given the evidence we have already described for functionality at preceding stages in the hair cell. Because the transduction apparatus functions at an early stage, hair cells will produce voltage changes in response to bundle movements evoked by Brownian motions or vibrations. As discussed in Section IV.E, the presence of Na^+, Ca^{2+}, and fast inwardly rectifying channels in the basolateral hair cell membrane may amplify these potentials into spikes. The hair cells have voltage-gated Ca^{2+} channels and exocytose in response to both exogenously applied depolarizations and Ca^{2+} spikes at P0 and probably earlier (Beutner and Moser, 2001). Finally, glutamate receptors are present on afferent terminals at early stages. In the mouse (Puyal et al., 2002), NMDA receptors (NR1) are first seen in statoacoustical ganglion neurons on E11, when neurons are growing into the otocyst, followed shortly by AMPA receptor subunits of the GluR3 variety. GluR2 AMPA subunits appear during the period of synaptogenesis, beginning at E15 in the vestibular ganglion and E17 in the spiral ganglion (Puyal et al., 2002).

B. Efferent Synapses on Hair Cells

In finding their targets in the inner ear, efferents seem to depend more on afferents than on neurotrophins. Efferents may find their way by piggybacking on growing afferents in a process that depends on ephrins (Cowan et al., 2000; reviewed in Simmons, 2002).

All efferent terminals release acetylcholine (ACh). In birds, cochlear efferents are exclusively cholinergic, but in mammals there is evidence for GABA, dopamine, and CGRP, especially in the lateral olivocochlear efferents, which synapse principally on primary afferent endings below inner hair cells (see review by Simmons, 2002). Medial olivocochlear efferents and chick cochlear efferents appear to release just ACh onto outer and short hair cells, respectively. In both hair cell types, the receptor is the α9/α10 neuronal nicotinic ACh receptor (nAChR), an ionotropic receptor that upon binding ACh opens and fluxes cations, including Ca^{2+} (Elgoyhen et al., 1994, 2001; Weisstaub et al., 2002). $\alpha 9$ but not $\alpha 10$ subunits can form functional channels by themselves (Elgoyhen et al., 2001). A preliminary report suggests that similar receptors occur on type II hair cells in amniote vestibular organs (Holt et al., 2003). The $\alpha 9/\alpha 10$ receptors are functionally coupled to small-conductance Ca^{2+}-activated K channels (SK channels), as elucidated by Fuchs and colleagues in the chick short hair cell (Fuchs and Murrow, 1992b) and by Dulon and colleagues in the mammalian outer hair cell (Blanchet et al., 1996; Dulon et al., 1998). ACh opens the channel of the $\alpha 9/\alpha 10$ receptor, through which Ca^{2+} enters and activates SK channels. Efflux of K^+ through SK channels mediates a hyperpolarizing inhibitory postsynaptic potential (ipsp), which is much larger than the excitatory potential resulting from inward current through the $\alpha 9/\alpha 10$ receptor channel. Thus, the dominant effect of efferent transmission is inhibitory. In turtle cochlear hair cells, which receive both afferent and efferent terminals, the ipsps reduce tuning directly by shunting electrical resonance (Art et al., 1985). In chick and mammalian cochleas, the efferent inhibition acts principally on one class of hair cells, while a different class are the main source of afferent information. In the mammalian cochlea, efferent actions on *outer* hair cells reduce the gain of the *inner* hair cells at CF (i.e., the sharpness of tuning; Brown and Nuttall, 1984; Wiederhold and Kiang, 1970), presumably through the mechanical effects of outer hair cells on the cochlear partition (Brownell et al., 1985; Mountain, 1980).

An interesting aspect of efferent development in mammals is the evidence that medial olivocochlear efferents form functional synapses on young inner hair cells before or at the same time as they form synapses on their mature targets, principally the first row of outer hair cells (summarized by Simmons, 2002). Inner hair cells express mRNA for $\alpha 9$ subunits from E18 through maturity (Simmons and Morley, 1998) but express mRNA for $\alpha 10$, the silent partner in the receptor, only from E21 to P15, during the period of transient innervation by medial olivocochlear efferents (Morley and Simmons, 2002). With the onset of hearing, the bulk of the efferent innervation on hair cells shifts to outer hair cells, although some efferent endings on inner hair cell somata persist (Vetter et al., 1999).

Is there a developmental function to this transient efferent innervation of inner hair cells? One possibility is that it plays a role in the rhythmic firing of immature auditory afferents (see Section II.B). Walsh et al. (1998) found that cutting the efferent input to the cochlea in neonatal cats transformed the rhythmic firing of immature auditory afferents into continuous firing. Glowatzki and Fuchs (2000) suggest that this effect may be mediated through the inner hair cells. In an *in vitro* preparation of rat organ of Corti (P7–P13), they showed that the efferent synapses on the immature inner hair cells are functional by recording inhibitory postsynaptic currents (ipscs) from the hair cells. The ipscs were either endogenous in origin, from efferent synapses that persist despite being cut off from their parent cell bodies in the brainstem, or evoked by exogenously applied ACh. Ca^{2+} spiking, readily induced in the inner hair cells by depolarizing current injections (as Beutner and Moser [2001] found in the neonatal mouse organ of Corti), was inhibited by ipsps. Thus, intermittent inhibitory input from efferents in the immature cochlea may help set up afferent rhythmicity or bursting by interrupting spontaneous spiking in the hair cells. If so, eliminating the rhythmicity by cutting the efferents or knocking out the $\alpha 9$ subunit has relatively subtle effects (see later), whereas eliminating all output from immature cochleas has dramatic effects on the development of brainstem auditory structures (Gabriele *et al.*, 2000; Hashisaki and Rubel, 1989).

In outer hair cells ACh responses, subsurface cisternae, and efferent innervation are acquired concurrently with electromotility in the week leading up to hearing onset, (see Section V). In gerbil outer hair cells, He and colleagues first detected ACh-evoked outward currents (presumably carried by SK channels activated by $\alpha 9/\alpha 10$ receptors) at P5–P6 in the basal turn of the cochlea (He and Dallos, 1999; He *et al.*, 2001). The incidence and size of the ACh response gradually increased until it reached mature levels at P11 (He and Dallos, 1999), just before hearing onset.

Although the overall picture of how efferent innervation changes during cochlear development suggests a possible instructional role for efferents, in fact the impact of eliminating efferent input on cochlear development is relatively modest, with the caveat that complete and selective elimination of efferent innervation has yet to be done. (A number of experiments, including overexpression of the $\alpha 9$ subunit [Maison *et al.*, 2002], suggest that medial olivocochlear efferents protect the cochlea from acoustical injury.) The developmental increase in ACh responsiveness in outer hair cells has been seen in cultured, denervated cochleas that were made on P0 (He *et al.*, 2001), just as it has been seen in acutely excised cochleas (He and Dallos, 1999). Recall also that outer hair cells acquire electromotility in denervated cultured cochleas (see Section IV.A.1).

Instead of denervating surgically, Vetter *et al.* (1999) silenced a major component of the efferent pathway by knocking out the $\alpha 9$ subunit of the $\alpha 9/\alpha 10$ ACh receptor. The results were subtle. In wildtype animals, outer hair cells are initially contacted by single relatively large efferent endings, but multiple efferent terminals are more typical in mature cochleas. In the null mutants, large single terminals were present even in mature animals. Vetter *et al.* (1999) suggest that in wildtype animals, activity at the efferent synapses leads to competition between terminals to produce the mature configuration.

VII. Summary and Concluding Remarks

The descriptions by different groups of developmental changes in the basolateral conductances of hair cells, though not complete, have suggested a framework for thinking about the functional impact of these changes. The evidence to date shows no influence of innervation on hair cell differentiation. In contrast, hair cell epithelia control afferent innervation through neurotrophin release. The electrical activity of immature hair cells, arising from complements of ion channels that are distinct from those in mature hair cells, may influence afferent synaptogenesis and differentiation indirectly by promoting neurotrophin release, or more directly by glutamatergic transmission. Embryonic hair cells in both chicks and mammals have functional transduction channels and voltage-gated outwardly rectifying K^+ channels, fast inwardly rectifying channels, and voltage-gated Na^+ and Ca^{2+} channels. Together these channels may participate in spiking, which may drive the rhythmic or bursting activity of immature auditory neurons at higher levels of the auditory pathway. Given the capability for transduction and exocytosis in neonatal mouse cochleas, spontanous activity in hair cells is likely to drive afferent activity well before the onset of sound-evoked activity. A transient inhibitory input from efferent terminals onto inner hair cells may also contribute to the bursting, rhythmic, spontaneous activity of immature auditory neurons.

With maturation, a number of changes tend to reduce hair cell spiking. In some hair cells, the numbers of voltage-gated Na^+, voltage-gated Ca^{2+} channels, and/or fast inward rectifier channels diminish, whereas the numbers and types of outwardly rectifying K^+ channels ($g_{K,Ca}$, $g_{K,f}$, $g_{K,n}$, or $g_{K,L}$) increase. This parallels the transformation from a developing epithelium, with active formation of new and in some cases transient synaptic contacts, to a sensing epithelium, where receptor potentials represent the mechanical input in a graded, nonspiking fashion that varies with location in the epithelium and/or hair cell type. The composition of the underlying conductances confers these variations. In the chick cochlea, the acquisition of the fast $g_{K,Ca-BK}$ confers high-quality electrical tuning on tall hair cells. Mammalian cochlear hair cells and type I vestibular hair cells, in contrast, acquire outwardly rectifying conductances that are activated at resting potential, reducing input resistance and time constants. Functionally, these changes both linearize and speed up the voltage response to input currents. Enhanced linearity may be an important effect for vestibular signals that drive motor reflexes and for the electromotile response of outer hair cells, whereas enhanced speed is important in enabling timing information at acoustical frequencies. The changes in outer hair cells that accompany electromotility occur in the same time frame as the maturation of basolateral conductances, significantly later than the onset of transduction capability and in time to account for the acquisition of sharp tuning by mammalian cochlear afferents.

The notion that hair cell spiking is an important driving force in the development of central pathways is attractive but lacking specific evidence. More supporting data on hair cell spiking are required, ideally from hair cells in quasi-intact, transducing epithelia (as in Fig. 7). In addition, the hypothesis should be tested by selectively eliminating such spiking without interfering with hair cell transduction and transmission. The ability to manipulate ion channel expression selectively with either null mutations or hair cell–specific promoters (Maison *et al.*, 2002) may allow us to dissect the relative importance of individual ion channels in hair cell spiking and, ultimately, the differentiation of the auditory and vestibular systems. The null mutation for the principal hair cell Ca^{2+} channel, $Ca_v1.3$, does not test the hypothesis because although it presumably knocks out Ca^{2+} spiking, it also knocks out most hair cell afferent transmission (Brandt *et al.*, 2003; Platzer *et al.*, 2000). Eliminating the hair cells' Na^+ channels or expressing outward rectifiers early in development might more selectively affect spiking by immature hair cells.

Acknowledgments

Work in our laboratory is supported by grants from the National Institute of Deafness and Communication Disorders (to RAE) and the Deafness Research Foundation (to KMH) and by Karim Al-Fayed Neurobiology of Hearing funds. We thank Dr. Julian Wooltorton and Jasmine Garcia for their comments on the manuscript.

References

Adam, J., Myat, A., Le, R. I., Eddison, M., Henrique, D., *et al.* (1998). Cell fate choices and the expression of Notch, Delta and Serrate homologues in the chick inner ear: Parallels with *Drosophila* sense-organ development. *Development* **125**, 4645–4654.

Adamson, C. L., Reid, M. A., and Davis, R. L. (2002). Opposite actions of brain-derived neurotrophic factor and neurotrophin-3 on firing features and ion channel composition of murine spiral ganglion neurons. *J. Neurosci.* **22**, 1385–1396.

Anderson, D. W., Probst, F. J., Belyantseva, I. A., Fridell, R. A., Beyer, L., *et al.* (2000). The motor and tail regions of myosin XV are critical for normal structure and function of auditory and vestibular hair cells. *Hum. Mol. Genet.* **9**, 1729–1738.

Anniko, M. (1990). Development of the vestibular system. *In* "Development of Sensory Systems in Mammals" (J. R. Coleman, ed.), pp. 341–400. Wiley, New York.

Armstrong, C. E., and Roberts, W. M. (1998). Electrical properties of frog saccular hair cells: Distortion by enzymatic dissociation. *J. Neurosci.* **18**, 2962–2973.

Art, J. J., Crawford, A. C., Fettiplace, R., and Fuchs, P. A. (1985). Efferent modulation of hair cell tuning in the cochlea of the turtle. *J. Physiol.* **360**, 397–421.

13. Functional Development of Hair Cells

Art, J. J., and Fettiplace, R. (1987). Variation of membrane properties in hair cells isolated from the turtle cochlea. *J. Physiol.* **385**, 207–242.

Bao, H., Wong, W. H., Goldberg, J. M., and Eatock, R. A. (2003). Voltage-gated calcium channel currents in type I and type II hair cells isolated from the rat crista. *J. Neurophysiol.* March 20 (Epub ahead of print).

Barritt, L. C., Fritzsch, B., and Beisel, K. W. (1999). Characterization of G-protein betagamma expression in inner ear. *Brain Res. Mol. Brain Res.* **68**, 42–54.

Beisel, K. W., Nelson, N. C., Delimont, D. C., and Fritzsch, B. (2000). Longitudinal gradients of KCNQ4 expression in spiral ganglion and cochlear hair cells correlate with progressive hearing loss in DFNA2. *Brain Res. Mol. Brain Res.* **82**, 137–149.

Belyantseva, I. A., Adler, H. J., Curi, R., Frolenkov, G. I., and Kachar, B. (2000). Expression and localization of prestin and the sugar transporter GLUT-5 during development of electromotility in cochlear outer hair cells. *J. Neurosci.* **20**, RC116.

Benser, M. E., Issa, N. P., and Hudspeth, A. J. (1993). Hair-bundle stiffness dominates the elastic reactance to otolithic membrane shear. *Hear. Res.* **68**, 243–252.

Bermingham-McDonogh, O., and Rubel, E. W. (2003). Hair cell regeneration: Winging our way towards a sound future. *Curr. Opin. Neurobiol.* **13**, 119–126.

Beurg, M., Bouleau, Y., and Dulon, D. (2001). The voltage-sensitive motor protein and the Ca2+-sensitive cytoskeleton in developing rat cochlear outer hair cells. *Eur. J. Neurosci.* **14**, 1947–1952.

Beutner, D., and Moser, T. (2001). The presynaptic function of mouse cochlear inner hair cells during development of hearing. *J. Neurosci.* **21**, 4593–4599.

Bianchi, L. M., Conover, J. C., Fritzsch, B., DeChiara, T., Lindsay, R. M., and Yancopoulos, G. D. (1996). Degeneration of vestibular neurons in late embryogenesis of both heterozygous and homozygous BDNF null mutant mice. *Development* **122**, 1965–1973.

Blanchet, C., Erostegui, C., Sugasawa, M., and Dulon, D. (1996). Acetylcholine-induced potassium current of guinea pig outer hair cells: Its dependence on a calcium influx through nicotinic-like receptors. *J. Neurosci.* **16**, 2574–2584.

Bradley, D. J., Towle, H. C., and Young, W. S., III (1994). Alpha and beta thyroid hormone receptor (TR) gene expression during auditory neurogenesis: Evidence for TR isoform-specific transcriptional regulation in vivo. *Proc. Natl. Acad. Sci. USA* **91**, 439–443.

Brandt, A., Striessnig, J., and Moser, T. (2003). Impact of Ca^{2+} channels on the development of cochlear inner hair cells. *Asso. Res. Otolaryngol. Abstracts* **26**, 110.

Brichta, A. M., Aubert, A., Eatock, R. A., and Goldberg, J. M. (2002). Regional analysis of whole cell currents from hair cells of the turtle posterior crista. *J. Neurophysiol.* **88**, 3259–3278.

Brown, M. C., and Nuttall, A. L. (1984). Efferent control of cochlear inner hair cell responses in the guinea-pig. *J. Physiol.* **354**, 625–646.

Brownell, W. E., Bader, C. R., Bertrand, D., and de Ribaupierre, Y. (1985). Evoked mechanical responses of isolated cochlear outer hair cells. *Science* **227**, 194–196.

Bruce, L. L., Christensen, M. A., and Warr, W. B. (2000). Postnatal development of efferent synapses in the rat cochlea. *J. Comp. Neurol.* **423**, 532–548.

Bryant, J. E., Marcotti, W., Kros, C. J., and Richardson, G. P., (2003). FM1-43 enters hair cells from the onset of mechano-electrical transduction in both mouse and chick cochlea. *Assoc. Res. Otolaryngol. Abstracts* **26**,107.

Chabbert, C., Mechaly, I., Sieso, V., Travot, C., Couraud, F., *et al.* (2003). Transient expression of voltage-gated Na^+ channels and activity-dependent BDNF release during late synaptogenesis in rat utricle. *Asso. Res. Otolaryngol. Abstracts* **26**, 35.

Chen, J. W. Y., and Eatock, R. A. (2000). Major potassium conductance in type I hair cells from rat semicircular canals: Characterization and modulation by nitric oxide. *J. Neurophysiol.* **84**, 139–151.

Cohen, G. M., and Fermin, C. D. (1978). The development of hair cells in the embryonic chick's basilar papilla. *Acta Otolaryngol.* **86**, 342–358.

Cole, K. S., and Robertson, D. (1992). Early efferent innervation of the developing rat cochlea studied with a carbocyanine dye. *Brain Res.* **575**, 223–230.

Coleman, G. B., Kaltenbach, J. A., and Falzarano, P. R. (1995). Postnatal development of the mammalian tectorial membrane. *Am. J. Otol.* **16**, 620–627.

Correia, M. J., and Lang, D. G. (1990). An electrophysiological comparison of solitary type I and type II vestibular hair cells. *Neurosci. Lett.* **116**, 106–111.

Correia, M. J., Rennie, K. J., and Koo, P. (2001). Return of potassium ion channels in regenerated hair cells: Possible pathways and the role of intracellular calcium signaling. *Ann. N. Y. Acad. Sci.* **942**, 228–240.

Cotanche, D. A., and Corwin, J. T. (1991). Stereociliary bundles reorient during hair cell development and regeneration in the chick cochlea. *Hear. Res.* **52**, 379–402.

Cowan, C. A., Yokoyama, N., Bianchi, L. M., Henkemeyer, M., and Fritzsch, B. (2000). EphB2 guides axons at the midline and is necessary for normal vestibular function. *Neuron* **26**, 417–430.

Curthoys, I. S. (1979). The development of function of horizontal semicircular canal primary neurons in the rat. *Brain Res.* **167**, 41–52.

Curthoys, I. S. (1982). Postnatal development changes in the response of rat primary horizontal semicircular canal neurons to sinusoidal angular accelerations. *Exp. Brain Res.* **47**, 295–300.

Curthoys, I. S. (1983). The development of function of primary vestibular neurons. *In* "Development of Auditory and Vestibular Systems" (R. Romand, ed.), pp. 425–461. Academic Press, New York.

Dabdoub, A., Donohue, M. J., Brennan, A., Wolf, V., Montcouquiol, M., *et al.* (2003). Wnt signaling mediates reorientation of outer hair cell stereociliary bundles in the mammalian cochlea. *Development* **130**, 2375–2384.

Dechesne, C. J., Kauff, C., Stettler, O., and Tavitian, B. (1997). Rab3A immunolocalization in the mammalian vestibular end-organs during development and comparison with synaptophysin expression. *Brain Res. Dev. Brain Res.* **99**, 103–111.

Denman-Johnson, K., and Forge, A. (1999). Establishment of hair bundle polarity and orientation in the developing vestibular system of the mouse. *J. Neurocytol.* **28**, 821–835.

DeRosier, D. J., and Tilney, L. G. (1989). The structure of the cuticular plate, an in vivo actin gel. *J. Cell Biol.* **109**, 2853–2867.

Desmadryl, G., Raymond, J., and Sans, A. (1986). *In vitro* electrophysiological study of spontaneous activity in neonatal mouse vestibular ganglion neurons during development. *Brain Res.* **390**, 133–136.

Desmadryl, G., and Sans, A. (1990). Afferent innervation patterns in crista ampullaris of the mouse during ontogenesis. *Brain Res. Dev. Brain Res.* **52**, 183–189.

Dieler, R., Shehata-Dieler, W. E., and Brownell, W. E. (1991). Concomitant salicylate-induced alterations of outer hair cell subsurface cisternae and electromotility. *J. Neurocytol.* **20**, 637–653.

Di Palma, F., Holme, R. H., Bryda, E. C., Belyantseva, I. A., Pellegrino, R., *et al.* (2001). Mutations in Cdh23, encoding a new type of cadherin, cause stereocilia disorganization in waltzer, the mouse model for Usher syndrome type 1D. *Nat. Genet.* **27**, 103–107.

Dulon, D., and Lenoir, M. (1996). Cholinergic responses in developing outer hair cells of the rat cochlea. *Eur. J. Neurosci.* **8**, 1945–1952.

Dulon, D., Luo, L., Zhang, C., and Ryan, A. F. (1998). Expression of small-conductance calcium-activated potassium channels (SK) in outer hair cells of the rat cochlea. *Eur. J. Neurosci.* **10**, 907–915.

Duncan, R. K., and Fuchs, P. A. (2003). Variation in large-conductance, calcium-activated potassium channels from hair cells along the chicken basilar papilla. *J. Physiol.* **547**, 357–371.

13. Functional Development of Hair Cells

Duzhyy, D. E., Harvey, M. C., and Sokolowski, B. H. A. (2003). Cloning, expression and localization of ion channel genes in the cochlea. *Asso. Res. Otolaryngol. Abstracts* **26**, 107.

Eatock, R. A. (2000). Adaptation in hair cells. *Annu. Rev. Neurosci.* **23**, 285–314.

Eatock, R. A., Hurley, K. M., and Vollrath, M. A. (2002). Mechanoelectrical and voltage-gated ion channels in mammalian vestibular hair cells. *Audiol. Neurootol.* **7**, 31–35.

Eatock, R. A., and Rüsch, A. (1997). Developmental changes in the physiology of hair cells. *Semin. Cell Dev. Biol.* **8**, 265–275.

Eatock, R. A., Rüsch, A., Lysakowski, A., and Saeki, M. (1998). Hair cells in mammalian utricles. *Otolaryngol. Head Neck Surg.* **119**, 172–181.

Eatock, R. A., Saeki, M., and Hutzler, M. J. (1993). Electrical resonance of isolated hair cells does not account for acoustic tuning in the free-standing region of the alligator lizard's cochlea. *J. Neurosci.* **13**, 1767–1783.

Eaton, R. C., Farley, R. D., Kimmel, C. B., and Schabtach, E. (1977). Functional development in the Mauthner cell system of embryos and larvae of the zebra fish. *J. Neurobiol.* **8**, 151–172.

Eaton, S. (1997). Planar polarization of *Drosophila* and vertebrate epithelia. *Curr. Opin. Cell Biol.* **9**, 860–866.

Echteler, S. M., Arjmand, E., and Dallos, P. (1989). Developmental alterations in the frequency map of the mammalian cochlea. *Nature* **341**, 147–149.

Ehret, G. (1977). Postnatal development in the acoustic system of the house mouse in the light of developing masked thresholds. *J. Acoust. Soc. Am.* **62**, 143–148.

Elgoyhen, A. B., Johnson, D. S., Boulter, J., Vetter, D. E., and Heinemann, S. (1994). Alpha 9: An acetylcholine receptor with novel pharmacological properties expressed in rat cochlear hair cells. *Cell* **79**, 705–715.

Elgoyhen, A. B., Vetter, D. E., Katz, E., Rothlin, C. V., Heinemann, S. F., and Boulter, J. (2001). alpha10: A determinant of nicotinic cholinergic receptor function in mammalian vestibular and cochlear mechanosensory hair cells. *Proc. Natl. Acad. Sci. USA* **98**, 3501–3506.

Ernest, S., Rauch, G. J., Haffter, P., Geisler, R., Petit, C., and Nicolson, T. (2000). Mariner is defective in myosin VIIA: A zebrafish model for human hereditary deafness. *Hum. Mol. Genet.* **9**, 2189–2196.

Evans, M. G., and Fuchs, P. A. (1987). Tetrodotoxin-sensitive, voltage-dependent sodium currents in hair cells for the alligator cochlea. *Biophys. J.* **52**, 649–652.

Fekete, D. M., and Wu, D. K. (2002). Revisiting cell fate specification in the inner ear. *Curr. Opin. Neurobiol.* **12**, 35–42.

Fettiplace, R., and Fuchs, P. A. (1999). Mechanisms of hair cell tuning. *Annu. Rev. Physiol.* **61**, 809–834.

Flock, A. (1965). Transducing mechanisms in the lateral line canal organ receptors. *Cold Spring Harb. Symp. Quant. Biol.* **30**, 133–145.

Flock, A., and Cheung, H. C. (1977). Actin filaments in sensory hairs of inner ear receptor cells. *J. Cell Biol.* **75**, 339–343.

Flock, A., Cheung, H. C., Flock, B., and Utter, G. (1981). Three sets of actin filaments in sensory cells of the inner ear. Identification and functional orientation determined by gel electophoresis, immunofluorescence, and electron microscopy. *J. Neurocytol.* **10**, 133–147.

Fritzsch, B., Barald, K. F., and Lomax, M. I. (1998). Early embryology of the vertebrate ear. In "Development of the Auditory System" (E. W. Rubel, A. N. Popper, and R. R. Fay, eds.), pp. 80–145. Springer-Verlag, New York.

Fuchs, P. A., and Evans, M. G. (1990). Potassium currents in hair cells isolated from the cochlea of the chick. *J. Physiol.* **429**, 529–551.

Fuchs, P. A., and Murrow, B. W. (1992a). A novel cholinergic receptor mediates inhibition of chick cochlear hair cells. *Proc. R. Soc. Lond B Biol. Sci.* **248**, 35–40.

Fuchs, P. A., and Murrow, B. W. (1992b). Cholinergic inhibition of short (outer) hair cells of the chick's cochlea. *J. Neurosci.* **12**, 800–809.

Fuchs, P. A., Nagai, T., and Evans, M. G. (1988). Electrical tuning in hair cells isolated from the chick cochlea. *J. Neurosci.* **8**, 2460–2467.

Fuchs, P. A., and Sokolowski, B. H. A. (1990). The acquisition during development of Ca-activated potassium currents by cochlear hair cells of the chick. *Proc. R. Soc. Lond. B. Biol. Sci.* **241**, 122–126.

Furness, D. N., Hackney, C. M., and Steyger, P. S. (1990). Organization of microtubules in cochlear hair cells. *J. Electron Microsc. Tech.* **15**, 261–279.

Gabriele, M. L., Brunso-Bechtold, J. K., and Henkel, C. K. (2000). Plasticity in the development of afferent patterns in the inferior colliculus of the rat after unilateral cochlear ablation. *J. Neurosci.* **20**, 6939–6949.

Gale, J. E., and Ashmore, J. F. (1997). An intrinsic frequency limit to the cochlear amplifier. *Nature* **389**, 63–66.

Géléoc, G., Lennan, G. W. T., Richardson, G. P., and Kros, C. J. (1997). A quantitative comparison of mechanoelectrical transduction in vestibular and auditory hair cells of neonatal mice. *Proc. R. Soc. Lond. B Biol. Sci.* **264**, 611–621.

Géléoc, G. S., Casalotti, S. O., Forge, A., and Ashmore, J. F. (1999). A sugar transporter as a candidate for the outer hair cell motor. *Nat. Neurosci.* **2**, 713–719.

Géléoc, G. S. G., Risner, J. R., and Holt, J. R. (2003). Prenatal acquisition of voltage-gated conductances in vestibular hair cells of the developing mouse embryo. *Asso. Res. Otolaryngol. Abstracts* **26**, 271.

Gillespie, P. G., and Hudspeth, A. J. (1994). Pulling springs to tune transduction: Adaptation by hair cells. *Neuron* **12**, 1–9.

Ginzberg, R. D., and Gilula, N. B. (1980). Synaptogenesis in the vestibular sensory epithelium of the chick embryo. *J. Neurocytol.* **9**, 405–424.

Glowatzki, E., and Fuchs, P. A. (2000). Cholinergic synaptic inhibition of inner hair cells in the neonatal mammalian cochlea. *Science* **288**, 2366–2368.

Goldberg, J. M. (1996). A theoretical analysis of intercellular communication between the vestibular type I hair cell and its calyx ending. *J. Neurophysiol.* **76**, 1942–1957.

Goodman, M. B., and Art, J. J. (1996). Positive feedback by a potassium-selective inward rectifier enhances tuning in vertebrate hair cells. *Biophys. J.* **71**, 430–442.

Goodyear, R. J., Gates, R., Lukashkin, A. N., and Richardson, G. P. (1999). Hair-cell numbers continue to increase in the utricular macula of the early posthatch chick. *J. Neurocytol.* **28**, 851–861.

Gray, L., and Rubel, E. W. (1985). Development of absolute thresholds in chickens. *J. Acoust. Soc. Am.* **77**, 1162–1172.

Griguer, C., and Fuchs, P. A. (1996). Voltage-dependent potassium currents in cochlear hair cells of the embryonic chick. *J. Neurophysiol.* **75**, 508–513.

Gu, X., and Spitzer, N. C. (1995). Distinct aspects of neuronal differentiation encoded by frequency of spontaneous Ca^{2+} transients. *Nature* **375**, 784–787.

Gulley, R. L., and Bagger-Sjöbäck, D. (1979). Freeze-fracture studies on the synapse between the type I hair cell and the calyceal terminal in the guinea-pig vestibular system. *J. Neurocytol.* **8**, 591–603.

Haddon, C., and Lewis, J. (1996). Early ear development in the embryo of the zebrafish, *Danio rerio*. *J. Comp. Neurol.* **365**, *113–128*.

Hamilton, D. W. (1968). The calyceal synapse of type I vestibular hair cells. *J. Ultrastruct. Res.* **23**, 98–114.

Hashisaki, G. T., and Rubel, E. W. (1989). Effects of unilateral cochlea removal on anteroventral cochlear nucleus neurons in developing gerbils. *J. Comp. Neurol.* **283**, 5–73.

13. Functional Development of Hair Cells

Hasson, T., Gillespie, P. G., Garcia, J. A., Macdonald, R. B., Zhao, Y. D., et al. (1997). Unconventional myosins in inner-ear sensory epithelia. *J. Cell Biol.* **137**, 1287–1307.

He, D. Z. (1997). Relationship between the development of outer hair cell electromotility and efferent innervation: A study in cultured organ of Corti of neonatal gerbils. *J. Neurosci.* **17**, 3634–3643.

He, D. Z., and Dallos, P. (1999). Development of acetylcholine-induced responses in neonatal gerbil outer hair cells. *J. Neurophysiol.* **81**, 1162–1170.

He, D. Z., Zheng, J., and Dallos, P. (2001). Development of acetylcholine receptors in cultured outer hair cells. *Hear. Res.* **162**, 113–125.

Hirokawa, N., and Tilney, L. G. (1982). Interactions between actin filaments and between actin filaments and membranes in quick-frozen and deeply etched hair cells of the chick ear. *J. Cell Biol.* **95**, 249–261.

Holley, M. C., and Ashmore, J. F. (1988). A cytoskeletal spring in cochlear outer hair cells. *Nature* **335**, 635–637.

Holley, M. C., and Nishida, Y. (1995). Monoclonal antibody markers for early development of the stereociliary bundles of mammalian hair cells. *J. Neurocytol.* **24**, 853–864.

Holt, J. C., Xue, J. T., and Goldberg, J. M. (2003). Synaptic mechanisms underlying afferent responses to efferent activation in the turtle posterior canal. *Assoc. Res. Otolaryngol. Abstracts* **26**, 33.

Holt, J. R., Corey, D. P., and Eatock, R. A. (1997). Mechanoelectrical transduction and adaptation in hair cells of the mouse utricle, a low-frequency vestibular organ. *J. Neurosci.* **17**, 8739–8748.

Holt, J. R., and Eatock, R. A. (1995). The inwardly rectifying currents of saccular hair cells from the leopard frog. *J. Neurophysiol.* **73**, 1484–1502.

Holt, J. R., Gillespie, S. K., Provance, D. W., Shah, K., Shokat, K. M., et al. (2002). A chemical-genetic strategy implicates myosin-1c in adaptation by hair cells. *Cell* **108**, 371–381.

Holt, J. R., Vollrath, M. A., and Eatock, R. A. (1999). Stimulus processing by type II hair cells in the mouse utricle. *Ann. N. Y. Acad. Sci.* **871**, 15–26.

Housley, G. D., and Ashmore, J. F. (1992). Ionic currents of outer hair cells isolated from the guinea-pig cochlea. *J. Physiol.* **448**, 73–98.

Housley, G. D., Norris, C. H., and Guth, P. S. (1989). Electrophysiological properties and morphology of hair cells isolated from the semicircular canal of the frog. *Hear. Res.* **38**, 259–276.

Housley, G. D., Raybould, N. P., and Thorne, P. R. (1998). Fluorescence imaging of Na^+ influx via P2X receptors in cochlear hair cells. *Hear. Res.* **119**, 1–13.

Hudspeth, A. J., Choe, Y., Mehta, A. D., and Martin, P. (2000). Putting ion channels to work: Mechanoelectrical transduction, adaptation and amplification by hair cells. *Proc. Natl. Acad. Sci. USA* **97**, 11765–11772.

Hudspeth, A. J., and Lewis, R. S. (1988a). A model for electrical resonance and frequency tuning in saccular hair cells of the bull-frog. *Rana catesbeiana. J. Physiol.* **400**, 275–297.

Hudspeth, A. J., and Lewis, R. S. (1988b). Kinetic analysis of voltage- and ion-dependent conductances in saccular hair cells of the bull-frog. *Rana catesbeiana. J. Physiol.* **400**, 237–274.

Hurley, K. M., and Eatock, R. A. (2002). Heterogeneous expression of K^+ conductances in vestibular type I hair cells. *Assoc. Res. Otolaryngol. Abstracts* **25**, 61.

Jaeger, R. G., Fex, J., and Kachar, B. (1994). Structural basis for mechanical transduction in the frog vestibular sensory apparatus: II. The role of microtubules in the organization of the cuticular plate. *Hear. Res.* **77**, 207–215.

Jaramillo, F., and Hudspeth, A. J. (1991). Localization of the hair cell's transduction channels at the hair bundle's top by iontophoretic application of a channel blocker. *Neuron* **7**, 409–420.

Jiang, G. J., Zidanic, M., Michaels, R. L., Michael, T. H., Griguer, C., and Fuchs, P. A. (1997). CSlo encodes calcium-activated potassium channels in the chick's cochlea. *Proc. R. Soc. Lond B Biol. Sci.* **264**, 731–737.

Jones, E. M., Laus, C., and Fettiplace, R. (1998). Identification of Ca(2+)-activated K+ channel splice variants and their distribution in the turtle cochlea. *Proc. R. Soc. Lond B Biol. Sci.* **265**, 685–692.

Jones, S. M., and Jones, T. A. (1996). Short latency vestibular evoked potentials in the chicken embryo. *J. Vestib. Res.* **6**, 71–83.

Jones, S. M., and Jones, T. A. (2000a). Ontogeny of vestibular compound action potentials in the domestic chicken. *J. Assoc. Res. Otolaryngol.* **1**, 232–242.

Jones, S. M., Warren, L. E., Shukla, R., Browning, A., Fuller, C. A., and Jones, T. A. (2000). The effects of hypergravity and substrate vibration on vestibular function in developing chickens. *J. Gravit. Physiol.* **7**, 31–44.

Jones, T. A., and Jones, S. M. (2000b). Spontaneous activity in the statoacoustic ganglion of the chicken embryo. *J. Neurophysiol.* **83**, 1452–1468.

Jones, T. A., Jones, S. M., and Paggett, K. C. (2001). Primordial rhythmic bursting in embryonic cochlear ganglion cells. *J. Neurosci.* **21**, 8129–8135.

Kachar, B., Parakkal, M., Kurc, M., Zhao, Y., and Gillespie, P. G. (2000). High-resolution structure of hair-cell tip links. *Proc. Natl. Acad. Sci. USA* **97**, 13336–13341.

Kalinec, F., Holley, M. C., Iwasa, K. H., Lim, D. J., and Kachar, B. (1992). A membrane-based force generation mechanism in auditory sensory cells. *Proc. Natl. Acad. Sci. USA* **89**, 8671–8675.

Kaltenbach, J. A., and Falzarano, P. R. (1994). Postnatal development of the hamster cochlea. I. Growth of hair cells and the organ of Corti. *J. Comp. Neurol.* **340**, 87–97.

Kaltenbach, J. A., Falzarano, P. R., and Simpson, T. H. (1994). Postnatal development of the hamster cochlea. II. Growth and differentiation of stereocilia bundles. *J. Comp. Neurol.* **350**, 187–198.

Katayama, A., and Corwin, J. T. (1989). Cell production in the chicken cochlea. *J. Comp. Neurol.* **281**, 129–135.

Katz, L. C., and Shatz, C. J. (1996). Synaptic activity and the construction of cortical circuits. *Science* **274**, 1133–1138.

Kharkovets, T., Hardelin, J. P., Safieddine, S., Schweizer, M., El-Amraoui, A., et al. (2000). KCNQ4, a K$^+$ channel mutated in a form of dominant deafness, is expressed in the inner ear and the central auditory pathway. *Proc. Natl. Acad. Sci. USA* **97**, 4333–4338.

Kollmar, R. (1999). Who does the hair cell's 'do? Rho GTPases and hair-bundle morphogenesis *Curr. Opin. Neurobiol.* **9**, 394–398.

Kollmar, R., Fak, J., Montgomery, L. G., and Hudspeth, A. J. (1997). Hair cell-specific splicing of mRNA for the α_{1D} subunit of voltage-gated Ca^{2+} channels in the chicken's cochlea. *Proc. Natl. Acad. Sci. USA* **94**, 14889–14893.

Kollmar, R., Montgomery, L. G., Fak, J., Henry, L. J., and Hudspeth, A. J. (1997). Predominance of the α_{1D} subunit in L-type voltage-gated Ca^{2+} channels of hair cells in the chicken's cochlea. *Proc. Natl. Acad. Sci. USA* **94**, 14883–14888.

Kozel, P. J., Friedman, R. A., Erway, L. C., Yamoah, E. N., Liu, L. H., et al. (1998). Balance and hearing deficits in mice with a null mutation in the gene encoding plasma membrane Ca2+-ATPase isoform 2. *J. Biol. Chem.* **273**, 18693–18696.

Kros, C. J., and Crawford, A. C. (1990). Potassium currents in inner hair cells isolated from the guinea-pig cochlea. *J. Physiol.* **421**, 263–291.

Kros, C. J., Marcotti, W., van Netten, S. M., Self, T. J., Libby, R. T., *et al.* (2002). Reduced climbing and increased slipping adaptation in cochlear hair cells of mice with Myo7a mutations. *Nat. Neurosci.* **5**, 41–47.

Kros, C. J., Ruppersberg, J. P., and Rüsch, A. (1998). Expression of a potassium current in inner hair cells during development of hearing in mice. *Nature* **394**, 281–284.

Kros, C. J., Rüsch, A., and Richardson, G. P. (1992). Mechano-electrical transducer currents in hair cells of the cultured neonatal mouse cochlea. *Proc. R. Soc. Lond. B Biol. Sci.* **249**, 185–193.

Kubisch, C., Schroeder, B. C., Friedrich, T., Lütjohann, B., El-Amraoui, A., *et al.* (1999). KCNQ4, a novel potassium channel expressed in sensory outer hair cells, is mutated in dominant deafness. *Cell* **96**, 437–446.

Lang, D. G., and Correia, M. J. (1989). Studies of solitary semicircular canal hair cells in the adult pigeon: II. Voltage-dependent ionic conductances. *J. Neurophysiol.* **62**, 935–945.

Lang, H., and Fekete, D. M. (2001). Lineage analysis in the chicken inner ear shows differences in clonal dispersion for epithelial, neuronal, and mesenchymal cells. *Dev. Biol.* **234**, 120–137.

Langer, P., Grunder, S., and Rüsch, A. (2003). Expression of Ca2+-activated BK channel mRNA and its splice variants in the rat cochlea. *J. Comp. Neurol.* **455**, 198–209.

Lennan, G. W. T., Steinacker, A., and Lehouelleur, J. (1999). Ionic currents and current-clamp depolarisations of type I and type II hair cells from the developing rat utricle. *Pflugers Arch.* **438**, 40–46.

Lenoir, M., Shnerson, A., and Pujol, R. (1980). Cochlear receptor development in the rat with emphasis on synaptogenesis. *Anat. Embryol. (Berl)* **160**, 253–262.

Lewis, J., and Davies, A. (2002). Planar cell polarity in the inner ear: How do hair cells acquire their oriented structure? *J. Neurobiol.* **53**, 190–201.

Liberman, M. C., Dodds, L. W., and Pierce, S. (1990). Afferent and efferent innervation of the cat cochlea: Quantitative analysis with light and electron microscopy. *J. Comp. Neurol.* **301**, 443–460.

Liberman, M. C., Gao, J., He, D. Z., Wu, X., Jia, S., and Zuo, J. (2002). Prestin is required for electromotility of the outer hair cell and for the cochlear amplifier. *Nature* **419**, 300–304.

Lippe, W. R. (1994). Rhythmic spontaneous activity in the developing avian auditory system. *J. Neurosci.* **14**, 1486–1495.

Loomis, P. A., Sekerkova, G., Zheng, L., Mugnaini, E., and Bartles, J. R. (2003). Does hair cell espin determine the steady-state length of the stereocilium? *Asso. Res. Otolaryngol. Abstracts* **26**, 10.

Lumpkin, E. A., and Hudspeth, A. J. (1995). Detection of Ca^{2+} entry through mechanosensitive channels localizes the site of mechanoelectrical transduction in hair cells. *Proc. Natl. Acad. Sci. USA* **92**, 10297–10301.

Lysakowski, A., and Price, S. D. (2003). Potassium channel localization in sensory epithelia of the rat inner ear. *Assoc. Res. Otolaryngol. Abstracts* **26**, 107.

Ma, D., Yang, C. H., McNeill, H., Simon, M. A., and Axelrod, J. D. (2003). Fidelity in planar cell polarity signalling. *Nature* **421**, 543–547.

Ma, Q., Anderson, D. J., and Fritzsch, B. (2000). Neurogenin 1 null mutant ears develop fewer, morphologically normal hair cells in smaller sensory epithelia devoid of innervation. *J. Assoc. Res. Otolaryngol.* **1**, 129–143.

Maison, S. F., Luebke, A. E., Liberman, M. C., and Zuo, J. (2002). Efferent protection from acoustic injury is mediated via alpha9 nicotinic acetylcholine receptors on outer hair cells. *J. Neurosci.* **22**, 10838–10846.

Marcotti, W., Géléoc, G. S., Lennan, G. W., and Kros, C. J. (1999). Transient expression of an inwardly rectifying potassium conductance in developing inner and outer hair cells along the mouse cochlea. *Pflugers Arch.* **439**, 113–122.

Marcotti, W., Johnson, S. L., Holley, M. C., and Kros, C. J. (2003). Developmental changes in the expression of potassium currents of embryonic, neonatal and mature mouse inner hair cells. *J. Physiol.* **548**, 383–400.

Marcotti, W., and Kros, C. J. (1999). Developmental expression of the potassium current $I_{K,n}$ contributes to maturation of mouse outer hair cells. *J. Physiol.* **520**, 653–660.

Martini, M., Rossi, M. L., Rubbini, G., and Rispoli, G. (2000). Calcium currents in hair cells isolated from semicircular canals of the frog. *Biophys. J.* **78**, 1240–1254.

Masetto, S., Bosica, M., Correia, M. J., Ottersen, O. P., Zucca, G., et al. (2003). Na^+ currents in vestibular type I and type II hair cells of the embryo and adult chicken. *J. Neurophysiol.*, April 17 (Epub ahead of print).

Masetto, S., and Correia, M. J. (1997a). Ionic currents in regenerating avian vestibular hair cells. *Int. J. Dev. Neurosci.* **15**, 387–399.

Masetto, S., and Correia, M. J. (1997b). Electrophysiological properties of vestibular sensory and supporting cells in the labyrinth slice before and during regeneration. *J. Neurophysiol.* **78**, 1913–1927.

Masetto, S., Perin, P., Malusa, A., Zucca, G., and Valli, P. (2000). Membrane properties of chick semicircular canal hair cells in situ during embryonic development. *J. Neurophysiol.* **83**, 2740–2756.

Masetto, S., Russo, G., and Prigioni, I. (1994). Differential expression of potassium currents by hair cells in thin slices of frog crista ampullaris. *J. Neurophysiol.* **72**, 443–455.

Mbiene, J. P., Favre, D., and Sans, A. (1984). The pattern of ciliary development in fetal mouse vestibular receptors. *Anat. Embryol.* **170**, 229–238.

Mbiene, J. P., and Sans, A. (1986). Differentiation and maturation of the sensory hair bundles in the fetal and postnatal vestibular receptors of the mouse: A scanning electron microscopy study. *J. Comp. Neurol.* **254**, 271–278.

Meza, G., and Hinojosa, R. (1987). Ontogenetic approach to cellular localization of neurotransmitters in the chick vestibule. *Hear. Res.* **28**, 73–85.

Mikaelian, D., and Ruben, R. J. (1965). Development of hearing in the normal CBA-J mouse. *Acta Otolaryngol.* **59**, 451–461.

Montcouquiol, M., Valat, J., Travo, C., and Sans, A. (1997). Short-term response of postnatal rat vestibular neurons following brain-derived neurotrophic factor or neurotrophin-3 application. *J. Neurosci. Res.* **50**, 443–449.

Moorman, S. J., Cordova, R., and Davies, S. A. (2002). A critical period for functional vestibular development in zebrafish. *Dev. Dyn.* **223**, 285–291.

Morley, B. J., and Simmons, D. D. (2002). Developmental mRNA expression of the alpha10 nicotinic acetylcholine receptor subunit in the rat cochlea. *Brain Res. Dev. Brain Res.* **139**, 87–96.

Mountain, D. C. (1980). Changes in endolymphatic potential and crossed olivocochlear bundle stimulation alter cochlear mechanics. *Science* **210**, 71–72.

Mulroy, M. J. (1974). Cochlear anatomy of the alligator lizard. *Brain Behav. Evol.* **10**, 69–87.

Murrow, B. W., and Fuchs, P. A. (1990). Preferential expression of transient potassium current (I_A) by "short" hair cells of the chick's cochlea. *Proc. R. Soc. Lond. B Biol. Sci.* **242**, 189–195.

Nakazawa, K., Spicer, S. S., and Schulte, B. A. (1995). Postnatal expression of the facilitated glucose transporter, GLUT 5, in gerbil outer hair cells. *Hear. Res.* **82**, 93–99.

Navaratnam, D. S., Bell, T. J., Tu, T. D., Cohen, E. L., and Oberholtzer, J. C. (1997). Differential distribution of Ca2+-activated K+ channel splice variants among hair cells along the tonotopic axis of the chick cochlea. *Neuron* **19**, 1077–1085.

Nazareth, A. M., and Jones, T. A. (1998). Central and peripheral components of short latency vestibular responses in the chicken. *J. Vestib. Res.* **8**, 233–252.

Nicolson, T., Rüsch, A., Friedrich, R. W., Granato, M., Ruppersberg, J. P., and Nusslein-Volhard, C. (1998). Genetic analysis of vertebrate sensory hair cell mechanosensation: The zebrafish circler mutants. *Neuron* **20**, 271–283.

Nishida, Y., Rivolta, M. N., and Holley, M. C. (1998). Timed markers for the differentiation of the cuticular plate and stereocilia in hair cells from the mouse inner ear. *J. Comp. Neurol.* **395**, 18–28.

Oestreicher, E., Knipper, M., Arnold, A., Zenner, H. P., and Felix, D. (2000). Neurotrophin 3 potentiates glutamatergic responses of IHC afferents in the cochlea in vivo. *Eur. J. Neurosci.* **12**, 1584–1590.

Oghalai, J. S., Zhao, H. B., Kutz, J. W., and Brownell, W. E. (2000). Voltage- and tension-dependent lipid mobility in the outer hair cell plasma membrane. *Science* **287**, 658–661.

Ohmori, H. (1985). Mechano-electrical transduction currents in isolated vestibular hair cells of the chick. *J. Physiol.* **359**, 189–217.

Oliver, D., Klocker, N., Schuck, J., Baukrowitz, T., Ruppersberg, J. P., and Fakler, B. (2000). Gating of Ca2+-activated K+ channels controls fast inhibitory synaptic transmission at auditory outer hair cells. *Neuron* **26**, 595–601.

Oliver, D., Knipper, M., Derst, C., and Fakler, B. (2003). Resting potential and submembrane calcium concentration of inner hair cells in the isolated mouse cochlea are set by KCNQ-type potassium channels. *J. Neurosci.* **23**, 2141–2149.

Oliver, D., Plinkert, P., Zenner, H. P., and Ruppersberg, J. P. (1997). Sodium current expression during postnatal development of rat outer hair cells. *Pflugers Arch.* **434**, 772–778.

Overstreet III, E. H., Temchin, A. N., and Ruggero, M. A. (2002). Passive basilar membrane vibrations in gerbil neonates: Mechanical bases of cochlear maturation. *J. Physiol.* **545**, 279–288.

Parrad, J., and Cottereau, P. (1977). [Appearance of rotatory reactions in the newborn rat (author's transl)]. *Physiol. Behav.* **18**, 1017–1020.

Perin, P., Masetto, S., Martini, M., Rossi, M. L., Rubbini, G., *et al.* (2001). Regional distribution of calcium currents in frog semicircular canal hair cells. *Hear. Res.* **152**, 67–76.

Pickles, J. O., Comis, S. D., and Osborne, M. P. (1984). Cross-links between stereocilia in the guinea pig organ of Corti, and their possible relation to sensory transduction. *Hear. Res.* **15**, 103–112.

Pickles, J. O., and Corey, D. P. (1992). Mechanoelectrical transduction by hair cells. *Trends Neurosci.* **15**, 254–259.

Pickles, J. O., and Rouse, G. W. (1991). Effects of streptomycin on development of the apical structures of hair cells in the chick basilar papilla. *Hear. Res.* **55**, 244–254.

Pickles, J. O., Rouse, G. W., and von Perger, M. (1991a). Morphological correlates of mechanotransduction in acousticolateral hair cells. *Scanning Microsc.* **5**, 1115–1124.

Pickles, J. O., von Perger, M., Rouse, G. W., and Brix, J. (1991b). The development of links between stereocilia in hair cells of the chick basilar papilla. *Hear. Res.* **54**, 153–163.

Platzer, J., Engel, J., Schrott-Fischer, A., Stephan, K., Bova, S., *et al.* (2000). Congenital deafness and sinoatrial node dysfunction in mice lacking class D L-type Ca2+ channels. *Cell* **102**, 89–97.

Pujol, R., Lavigne-Rebillard, M., and Lenoir, M. (1998). Development of sensory and neural structures in the mammalian cochlea. *In* "Development of the Auditory System" (E. W. Rubel, A. N. Popper, and R. R. Fay, eds.), pp. 146–192. Springer-Verlag, New York.

Puyal, J., Sage, C., Dememes, D., and Dechesne, C. J. (2002). Distribution of alpha-amino-3-hydroxy-5-methyl-4 isoazolepropionic acid and N-methyl-D-aspartate receptor subunits in the vestibular and spiral ganglia of the mouse during early development. *Brain Res. Dev. Brain Res.* **139**, 51–57.

Rajeevan, M. S., Hu, S., Sakai, Y., and Sokolowski, B. H. (1999). Cloning and expression of Shaker alpha- and beta-subunits during inner ear development. *Brain Res. Mol. Brain Res.* **66**, 83–93.

Ramakrishnan, N. A., Green, G. E., Pasha, R., Drescher, M. J., Swanson, G. S., Perin, P. C., Lakhani, R. S., Ahsan, S. F., Hatfield, J. S., Khan, K. M., and Drescher, D. G. (2002). Voltage-gated Ca^{2+} channel $Ca_v1.3$ subunit expressed in the haircell epithelium of the trout *Oncorhynchus mykiss*: Cloning and comparison across vertebrate classes. *Brain Res. mol. Brain Res.* **109**, 49–83.

Ramanathan, K., Michael, T. H., and Fuchs, P. A. (2000). Beta subunits modulate alternatively spliced, large conductance, calcium-activated potassium channels of avian hair cells. *J. Neurosci.* **20**, 1675–1684.

Raymond, J. (1987). In vitro differentiation of mouse embryo statoacoustic ganglion and sensory epithelium. *Hear. Res.* **28**, 45–56.

Rennie, K. J., and Correia, M. J. (1994). Potassium currents in mammalian and avian isolated type 1 semicircular canal hair cells. *J. Neurophysiol.* **71**, 317–329.

Rennie, K. J., Weng, T., and Correia, M. J. (2001). Effects of KCNQ channel blockers on K^+ currents in vestibular hair cells. *Am. J. Physiol. Cell Physiol.* **280**, C473–C480.

Ricci, A. J., and Fettiplace, R. (1997). The effects of calcium buffers on mechanoelectrical transduction in turtle hair cells. *Biophys. J.* **72**, A266.

Ricci, A. J., Rennie, K. J., and Correia, M. J. (1996). The delayed rectifier, I_{K1} is the major conductance in type 1 vestibular hair cells across vestibular end organs. *Pflugers Arch.* **432**, 34–42.

Riley, B. B., and Moorman, S. J. (2000). Development of utricular otoliths, but not saccular otoliths, is necessary for vestibular function and survival in zebrafish. *J. Neurobiol.* **43**, 329–337.

Roberts, W. M., Jacobs, R. A., and Hudspeth, A. J. (1990). Colocalization of ion channels involved in frequency selectivity and synaptic transmission at presynaptic active zones of hair cells. *J. Neurosci.* **10**, 3664–3684.

Robertson, D., Harvey, A. R., and Cole, K. S. (1989). Postnatal development of the efferent innervation of the rat cochlea. *Brain Res. Dev. Brain Res.* **47**, 197–207.

Rodriguez-Contreras, A., and Yamoah, E. N. (2001). Direct measurement of single-channel Ca^{2+} currents in bullfrog hair cells reveals two distinct channel subtypes. *J. Physiol.* **534**, 669–689.

Romand, R. (1997). Modification of tonotopic representation in the auditory system during development. *Prog. Neurobiol.* **51**, 1–17.

Rosenblatt, K. P., Sun, Z. P., Heller, S., and Hudspeth, A. J. (1997). Distribution of Ca2+-activated K+ channel isoforms along the tonotopic gradient of the chicken's cochlea. *Neuron* **19**, 1061–1075.

Rubel, E. W., and Fritzsch, B. (2002). Auditory system development: Primary auditory neurons and their targets. *Annu. Rev. Neurosci.* **25**, 51–101.

Ruben, R. J. (1967). Development of the inner ear of the mouse: A radioautographic study of terminal mitoses. *Acta Otolaryngol. (Stockh)* Suppl. **220**, 1–31.

Rübsamen, R., and Lippe, W. R. (1998). The development of cochlear function. In "Development of the Auditory System" (E. W. Rubel, A. N. Popper, and R. R. Fay, eds.), pp. 193–270. Springer-Verlag, New York.

Rüsch, A., and Eatock, R. A. (1996a). A delayed rectifier conductance in type I hair cells of the mouse utricle. *J. Neurophysiol.* **76**, 995–1004.

Rüsch, A., and Eatock, R. A. (1996b). Voltage responses of mouse utricular hair cells to injected currents. *Ann. N. Y. Acad. Sci.* **781**, 71–84.

13. Functional Development of Hair Cells

Rüsch, A., and Eatock, R. A. (1997). Sodium currents in hair cells of the mouse utricle. In "Diversity in Auditory Mechanics" (E. R. Lewis, G. R. Long, R. F. Lyon, C. R. Steele, P. M. Narins, and E. Hecht-Poinar, eds.), pp. 549–555. World Scientific Press, Singapore.

Rüsch, A., Erway, L. C., Oliver, D., Vennstrom, B., and Forrest, D. (1998a). Thyroid hormone receptor beta-dependent expression of a potassium conductance in inner hair cells at the onset of hearing. *Proc. Natl. Acad. Sci. USA* **95**, 15758–15762.

Rüsch, A., Lysakowski, A., and Eatock, R. A. (1998b). Postnatal development of type I and type II hair cells in the mouse utricle: Acquisition of voltage-gated conductances and differentiated morphology. *J. Neurosci.* **18**, 7487–7501.

Sans, A., and Chat, M. (1982). Analysis of temporal and spatial patterns of rat vestibular hair cell differentiation by tritiated thymidine radioautography. *J. Comp. Neurol.* **206**, 1–8.

Santos-Sacchi, J. (1991). Reversible inhibition of voltage-dependent outer hair cell motility and capacitance. *J. Neurosci.* **11**, 3096–3110.

Saunders, J. C., Coles, R. B., and Gates, G. R. (1973). The development of auditory evoked responses in the cochlea and cochlear nuclei of the chick. *Brain Res.* **63**, 59–74.

Scarfone, E., Demêmes, D., Jahn, R., De Camilli, P., and Sans, A. (1988). Secretory function of the vestibular nerve calyx suggested by the presence of vesicles, synapsin I, and synaptophysin. *J. Neurosci.* **8**, 4640–4645.

Schneider, M. E., Belyantseva, I. A., Azevedo, R. B., and Kachar, B. (2002). Rapid renewal of auditory hair bundles. *Nature* **418**, 837–838.

Self, T., Mahoney, M., Fleming, J., Walsh, J., Brown, S. D. M., and Steel, K. P. (1998). Shaker-1 mutations reveal roles for myosin VIIA in both development and function of cochlear hair cells. *Development* **125**, 557–566.

Self, T., Sobe, T., Copeland, N. G., Jenkins, N. A., Avraham, K. B., and Steel, K. P. (1999). Role of myosin VI in the differentiation of cochlear hair cells. *Dev. Biol.* **214**, 331–341.

Shotwell, S. L., Jacobs, R., and Hudspeth, A. J. (1981). Directional sensitivity of individual vertebrate hair cells to controlled deflection of their hair bundles. *Ann. N.Y. Acad. Sci.* **374**, 1–10.

Simmons, D. D. (2002). Development of the inner ear efferent system across vertebrate species. *J. Neurobiol.* **53**, 228–250.

Simmons, D. D., Manson-Gieseke, L., Hendrix, T. W., and McCarter, S. (1990). Reconstructions of efferent fibers in the postnatal hamster cochlea. *Hear. Res.* **49**, 127–139.

Simmons, D. D., and Morley, B. J. (1998). Differential expression of the alpha 9 nicotinic acetylcholine receptor subunit in neonatal and adult cochlear hair cells. *Brain Res. Mol. Brain Res.* **56**, 287–292.

Slepecky, N. B., and Ulfendahl, M. (1992). Actin-binding and microtubule-associated proteins in the organ of Corti. *Hear. Res.* **57**, 201–215.

Smith, C. E., and Goldberg, J. M. (1986). A stochastic after hyperpolarization model of repetitive activity in vestibular afferents. *Biol. Cyber.* **54**, 41–51.

Smotherman, M. S., and Narins, P. M. (1999). The electrical properties of auditory hair cells in the frog amphibian papilla. *J. Neurosci.* **19**, 5275–5292.

Sobin, A., and Flock, A. (1983). Immunohistochemical identification and localization of actin and fimbrin in vestibular hair cells in the normal guinea pig and in a strain of the waltzing guinea pig. *Acta Otolaryngol.* **96**, 407–412.

Sobkowicz, H. M., Rose, J. E., Scott, G. E., and Slapnick, S. M. (1982). Ribbon synapses in the developing intact and cultured organ of Corti in the mouse. *J. Neurosci.* **2**, 942–957.

Sobkowicz, H. M., Rose, J. E., Scott, G. L., and Levenick, C. V. (1986). Distribution of synaptic ribbons in the developing organ of Corti. *J. Neurocytol.* **15**, 693–714.

Sobkowicz, H. M., Slapnick, S. M., and August, B. K. (1995). The kinocilium of auditory hair cells and evidence for its morphogenetic role during the regeneration of stereocilia and cuticular plates. *J. Neurocytol.* **24**, 633–653.

Sobkowicz, H. M., Slapnick, S. M., and August, B. K. (2002). Differentiation of spinous synapses in the mouse organ of corti. *Synapse* **45**, 10–24.

Sokolowski, B. H., and Cunningham, A. M. (1999). Patterns of synaptophysin expression during development of the inner ear in the chick. *J. Neurobiol.* **38**, 46–64.

Sokolowski, B. H. A., Stahl, L. M., and Fuchs, P. A. (1993). Morphological and physiological development of vestibular hair cells in the organ-cultured otocyst of the chick. *Dev. Biol.* **155**, 134–146.

Souter, M., Nevill, G., and Forge, A. (1995). Postnatal development of membrane specialisations of gerbil outer hair cells. *Hear. Res.* **91**, 43–62.

Steinacker, A., and Romero, A. (1991). Characterization of voltage-gated and calcium-activated potassium currents in toadfish saccular hair cells. *Brain Res.* **556**, 22–32.

Steel, K. P., and Kros, C. J. (2001). A genetic approach to understanding auditory function. *Nat. Genet.* **27**, 143–149.

Sugihara, I., and Furukawa, T. (1989). Morphological and functional aspects of two different types of hair cells in the goldfish sacculus. *J. Neurophysiol.* **62**, 1330–1343.

Sugihara, I., and Furukawa, T. (1995). Potassium currents underlying the oscillatory response in hair cells of the goldfish sacculus. *J. Physiol.* **489**, 443–453.

Sugihara, I., and Furukawa, T. (1996). Inwardly rectifying currents in hair cells and supporting cells in the goldfish sacculus. *J. Physiol.* **495**, 665–679.

Takasu, M. A., Dalva, M. B., Zigmond, R. E., and Greenberg, M. E. (2002). Modulation of NMDA receptor-dependent calcium influx and gene expression through EphB receptors. *Science* **295**, 491–495.

Takumida, M., Bagger-Sjöbäck, D., Wersäll, J., and Harada, Y. (1989). The effect of gentamicin on the glycocalyx and the ciliary interconnectons in vestibular sensory cells: A high-resolution scanning electron microscopic investigation. *Hear. Res.* **37**, 163–170.

Tilney, L. G., and DeRosier, D. J. (1986). Actin filaments, stereocilia, and hair cells of the bird cochlea. IV. How the actin filaments become organized in developing stereocilia and in the cuticular plate. *Dev. Biol.* **116**, 119–129.

Tilney, L. G., Egelman, E. H., DeRosier, D. J., and Saunder, J. C. (1983). Actin filaments, stereocilia, and hair cells of the bird cochlea. II. Packing of actin filaments in the stereocilia and in the cuticular plate and what happens to the organization when the stereocilia are bent. *J. Cell Biol.* **96**, 822–834.

Tilney, L. G., and Saunders, J. C. (1983). Actin filaments, stereocilia, and hair cells of the bird cochlea. I. Length, number, width, and distribution of stereocilia of each hair cell are related to the position of the hair cell on the cochlea. *J. Cell Biol.* **96**, 807–821.

Tilney, L. G., Tilney, M. S., and Cotanche, D. A. (1988). Actin filaments, stereocilia, and hair cells of the bird cochlea. V. How the staircase pattern of stereociliary lengths is generated. *J. Cell Biol.* **106**, 355–365.

Tilney, L. G., Tilney, M. S., and DeRosier, D. J. (1992). Actin filaments, stereocilia, and hair cells: How cells count and measure. *Annu. Rev. Cell Biol.* **8**, 257–274.

Tilney, L. G., Tilney, M. S., Saunders, J. S., and DeRosier, D. J. (1986). Actin filaments, stereocilia, and hair cells of the bird cochlea. III. The development and differentiation of hair cells and stereocilia. *Dev. Biol.* **116**, 100–118.

Tree, D. R., Shulman, J. M., Rousset, R., Scott, M. P., Gubb, D., and Axelrod, J. D. (2002). Prickle mediates feedback amplification to generate asymmetric planar cell polarity signaling. *Cell* **109**, 371–381.

Trussell, L. (2000). Mutant ion channel in cochlear hair cells causes deafness. *Proc. Natl. Acad. Sci. USA* **97**, 3786–3788.

Tsuprun, V., and Santi, P. (2000). Helical structure of hair cell stereocilia tip links in the chinchilla cochlea. *J. Assoc. Res. Otolaryngol.* **1**, 224–231.

Van de Water, T. R. (1976). Effects of removal of the statoacoustic ganglion complex upon the growing otocyst. *Ann. Otol. Rhino. Laryngol.* **85**, 1–32.

Vetter, D. E., Liberman, M. C., Mann, J., Barhanin, J., Boulter, J., et al. (1999). Role of alpha9 nicotinic ACh receptor subunits in the development and function of cochlear efferent innervation. *Neuron* **23**, 93–103.

Walsh, E. J., McGee, J., McFadden, S. L., and Liberman, M. C. (1998). Long-term effects of sectioning the olivocochlear bundle in neonatal cats. *J. Neurosci.* **18**, 3859–3869.

Wang, H. S., Pan, Z., Shi, W., Brown, B. S., Wymore, R. S., et al. (1998). KCNQ2 and KCNQ3 potassium channel subunits: Molecular correlates of the M-channel. *Science* **282**, 1890–1893.

Weaver, S. P., and Schweitzer, L. (1994). Development of gerbil outer hair cells after the onset of cochlear function: An ultrastructural study. *Hear. Res.* **72**, 44–52.

Weber, T., Zimmermann, U., Winter, H., Mack, A., Kopschall, I., et al. (2002). Thyroid hormone is a critical determinant for the regulation of the cochlear motor protein prestin. *Proc. Natl. Acad. Sci. USA* **99**, 2901–2906.

Weisstaub, N., Vetter, D. E., Elgoyhen, A. B., and Katz, E. (2002). The alpha9alpha10 nicotinic acetylcholine receptor is permeable to and is modulated by divalent cations. *Hear. Res.* **167**, 122–135.

Werner, L. A., and Gray, L. (1998). Behavioral studies of hearing development. In "Development of the Auditory System" (E. W. Rubel, A. N. Popper, and R. R. Fay, eds.), pp. 12–79. Springer-Verlag, New York.

Wersäll, J., and Bagger-Sjöbäck, D. (1974). Morphology of the vestibular sense organ. In "Handbook of Sensory Physiology. Vestibular System. Basic Mechanisms" (H. H. Kornhuber, ed.), pp. 123–170. Springer-Verlag, New York.

Whitehead, M. C., and Morest, D. K. (1985a). The development of innervation patterns in the avian cochlea. *Neuroscience* **14**, 255–276.

Whitehead, M. C., and Morest, D. K. (1985b). The growth of cochlear fibers and the formation of their synaptic endings in the avian inner ear: A study with the electron microscope. *Neuroscience* **14**, 277–300.

Whitfield, T. T. (2002). Zebrafish as a model for hearing and deafness. *J. Neurobiol.* **53**, 157–171.

Wiederhold, M. L., and Kiang, N. Y. S. (1970). Effects of electric stimulation of the crossed olivocochlear bundle on single auditory-nerve fibers in the cat. *J. Acoust. Soc. Am.* **48**, 950–965.

Witt, C. M., Hu, H. Y., Brownell, W. E., and Bertrand, D. (1994). Physiologically silent sodium channels in mammalian outer hair cells. *J. Neurophysiol.* **72**, 1037–1040.

Wu, Y. C., Art, J. J., Goodman, M. B., and Fettiplace, R. (1995). A kinetic description of the calcium-activated potassium channel and its application to electrical tuning of hair cells. *Prog. Biophys. Mol. Biol.* **63**, 131–158.

Yamashita, M., and Ohmori, H. (1990). Synaptic responses to mechanical stimulation in calyceal and bouton type vestibular afferents studied in an isolated preparation of semicircular canal ampullae of chicken. *Exp. Brain Res.* **80**, 475–488.

Yamoah, E. N., Lumpkin, E. A., Dumont, R. A., Smith, P. J. S., Hudspeth, A. J., and Gillespie, P. G. (1998). Plasma membrane Ca^{2+}-ATPase extrudes Ca^{2+} from hair cell stereocilia. *J. Neurosci.* **18**, 610–624.

Yoshida, N., and Liberman, M. C. (1999). Stereociliary anomaly in the guinea pig: Effects of hair bundle rotation on cochlear sensitivity. *Hear. Res.* **131**, 29–38.

Yuhas, W. A., and Fuchs, P. A. (1999). Apamin-sensitive, small-conductance, calcium-activated potassium channels mediate cholinergic inhibition of chick auditory hair cells. *J. Comp. Physiol [A]* **185**, 455–462.

Zhao, Y., Yamoah, E. N., and Gillespie, P. G. (1996). Regeneration of broken tip links and restoration of mechanical transduction in hair cells. *Proc. Natl. Acad. Sci. USA* **93**, 15469–15474.

Zheng, J., Shen, W., He, D. Z., Long, K. B., Madison, L. D., and Dallos, P. (2000a). Prestin is the motor protein of cochlear outer hair cells. *Nature* **405**, 149–155.

Zheng, L., Sekerkova, G., Vranich, K., Tilney, L. G., Mugnaini, E., and Bartles, J. R. (2000b). The deaf jerker mouse has a mutation in the gene encoding the espin actin-bundling proteins of hair cell stereocilia and lacks espins. *Cell* **102**, 377–385.

Zidanic, M., and Fuchs, P. A. (1995). Kinetic analysis of barium currents in chick cochlear hair cells. *Biophys. J.* **68**, 1323–1336.

Zine, A., and Romand, R. (1996). Development of the auditory receptors of the rat: A SEM study. *Brain Res.* **721**, 49–58.

14

The Cell Cycle and the Development and Regeneration of Hair Cells

Allen F. Ryan
Departments of Surgery/Otolaryngology and Neurosciences
University of California San Diego School of Medicine and San Diego Veterans Administration Medical Center
La Jolla, California 92093

I. Introduction
II. The Cell Cycle and the Regulation of Cell Proliferation
III. Hair Cell Development and Regeneration
IV. Cell Cycle Events during Hair Cell Development
V. The Cell Cycle during Hair Cell Regeneration
VI. Conclusions
    References

The replication of cells occurs by a process known as the cell cycle, which is under multiple forms of regulation. Cell proliferation is an integral part of the development of the inner ear. Generation of the otic sensory epithelia involves coordinated periods of cell division. However, commitment of cells in the developing sensory epithelia to a hair cell fate and the differentiation of hair cells from committed precursors are postmitotic processes, which occur after the establishment of a zone of nonproliferating cells within the epithelium. The suppression of cell division by regulators of the cell cycle is both temporally and functionally linked to hair cell development. Animals lacking cell cycle regulatory proteins can exhibit supernumerary hair cells and exhibit hearing loss. The absence of hair cell regeneration in mammals appears to be related to both mitotic and postmitotic factors. Unlike in fish and birds, cell proliferation is rare in mammalian sensory epithelia following hair cell damage. Cell division can be dramatically increased in mammalian vestibular sensory epithelia by a variety of factors. However, the resulting cells rarely adopt a hair cell fate. As with the cell cycle, our increased understanding of the factors that determine hair cell fate selection may allow us to overcome barriers to hair cell regeneration in mammals. © 2003, Elsevier Inc.

I. Introduction

As is true for the ontogeny of all tissues, the development of the inner ear is intimately tied to the process of cellular replication. The great majority of the cells of the inner ear appear to originate from the otic placode and nearby mesenchymal tissue (e.g., Lang *et al.*, 2000; Morsli *et al.*, 1998). Thus a small mass of placodal and adjacent mesenchymal cells contains the precursors for most of the many different populations and the vastly larger numbers of mature cochlear cells. These original precursors must go through many cycles of cell division and progressive specialization to produce the different cell types of the mature inner ear. An exception is the small number of cells that originate elsewhere in the organism and enter the inner ear later during development. An example is the neural crest cells that migrate into the inner ear after the placodal stage and become the intermediate cells of the stria vascularis and the dark cells of the vestibular system (Steel and Barkway, 1989). However, most cochlear cells arise locally through cell division.

As a general rule, cells become more restricted in their potential to assume different cell phenotypes as they undergo successive replications and are exposed to various cellular environments. Also as a general rule, the more specialized cells become, the more likely they are to be incapable of replication. The highly specialized cells of the adult mammalian cochlea, including its hair cells, many types of supporting cells, and neurons, are examples of this. In particular, in the adult mammalian inner ear all cells that are capable of assuming a hair cell phenotype are postmitotic. Thus when the cochlea of a mammal is damaged, the organ of Corti is unable to produce replacement hair cells. In contrast, the inner ear sensory epithelia of fish grow continuously and add new hair cells throughout life from active stem cell populations (Corwin, 1981). Although the cochlear epithelia of mature birds do not add additional hair cells (Katayama and Corwin, 1989), there is evidence of at least a degree of ongoing proliferation balanced by the extrusion of cells from the utricular epithelium of the chick (Wilkins *et al.*, 1999). Fish, amphibians, and birds display vigorous hair cell regeneration after hair cell loss.

The fact that cell proliferation occurs in the damaged and even the normal inner ear of lower organisms suggests that if the regulation of the cell cycle in cochlear cells were understood, it might be possible to induce hair cell regeneration in mammals as well. This possibility has led to increased research on the cell cycle in the sensory epithelia of the developing and damaged inner ear.

II. The Cell Cycle and the Regulation of Cell Proliferation

The growth and replication of multicellular organisms from a fertilized ovum requires extensive cellular replication, including duplication of the genetic material within each cell. The problem of regulating cell division during embryonic development requires that the final number of cells and the specific cell types in the various tissues be carefully determined. To accomplish this, developmental interactions in the form of cell–cell interactions, growth factor signaling and hormones, and poorly understood mechanisms of size control and morphogenesis all converge to influence the cell cycle machinery, which responds to these various inputs either by cell division or by blocking the progress of the cell cycle. Similarly, cell division also plays an essential, ongoing role in homeostasis of the adult organism, replacing damaged and dying cells in many organ systems (Nabel, 2002). Control of the process of cell division is thus under strict control, designed to limit replication to appropriate settings. Breakdown in this control can lead to malignancy or cell death.

The cell cycle is a highly ordered set of events, involving cellular growth and chromosome duplication and culminating in the division of a single cell into two daughter cells. This process of cellular replication has been divided into intervals, defined by the events that occur in each stage of the cycle. The phases of the cell cycle are termed G_1, S, G_2, and M. G_1 stands for GAP 1 and identifies the gap between mitosis and the onset of DNA synthesis. G_1 and G_2 are considered growth phases during which cells increase in size and accumulate intracellular resources preparatory to the start of cell division and mitosis, respectively. The S, or synthesis, stage is when DNA replication occurs. In the G_2 (GAP 2) stage, during the interval between the end of DNA replication and actual cell division, the cell prepares for mitosis. The M, or mitosis, stage is when the cell divides and segregates the duplicated chromosomes into the two daughter cells. The progression through mitosis is divided into a series of morphologically observable events. During prophase the nuclear envelope dissolves and chromosomes begin to condense; during metaphase the condensed chromosomes align along the axis of the cell in preparation for segregation to the daughter cells; during anaphase the aligned chromosomes begin to separate; they continue separating during telophase, completing this process and moving to the opposite ends of the cell. At the end of telophase the daughter nuclei reassemble and the cytokinesis occurs, partitioning the cytoplasm between the daughters. After M phase, the daughter cells re-enter G_1, where they can either initiate another round of cell division or withdraw from the cell cycle into the resting or G_0 state (Nurse, 2000).

Cell cycle checkpoints monitor progress through the cycle, particularly at three important transitions: from G_1 to S, from G_2 to M, and from metaphase to anaphase. These checkpoints block entry to the next stage if the previous step is incomplete or damage to the cell has occurred (Hartwell and Weinert, 1989). For instance, DNA damage checkpoints can delay the cell cycle, allowing for DNA to be repaired, or if repair is unsuccessful, directing the destruction of the cell by the process of apoptosis (Murakami and Nurse, 2000).

The cell cycle is thus subject to complex regulation by the developmental signals required for morphogenesis, as well as by the surveillance systems

Table I Cyclin/Cdk Regulation of the Mammalian Cell Cycle

Cyclin	Protein kinase	Process regulated
Cyclin D1–3	CDK4, 6	G_1-phase progression
Cyclin E	CDK2	G_1 to S-phase
Cyclin A	CDK2	S-phase progression
Cyclin A	CDK1	S through G_2
Cyclin B	CDK1	M-phase

that monitor and regulate cellular homeostasis and regeneration. This developmental and homeostatic information is integrated by a family of enzymes known as cyclin-dependent kinases (CKDs) (e.g., Nabel, 2002). Specific members of the CDK family are responsible for coordinating cell cycle progression at each stage of the cell cycle mentioned previously (Table I) and also are crucial for arresting the cell cycle at the various checkpoints that monitor cellular integrity (Hartwell and Weinert, 1989). In mammalian development, the G_1 phase of the cell cycle appears to be the time during which many of the decisions related to development and cellular differentiation are made (Elledge et al., 1996), and thus the activity of the G_1 CDKs, CDK4/6, and CDK2 (see Table 1) are the crucial regulatory event for cells exiting the cell cycle and becoming postmitotic.

As is true for all the members of the CDK family, the G_1 CDKs are regulated both positively and negatively by phosphorylation of CDK itself; by the binding of required, positively acting cofactors, the G_1 cyclins (cyclin D1–3); as well as by their interaction with two families of small protein inhibitors, the cyclin-dependent kinase inhibitors (CKIs) (Nabel, 2002; Sherr and Roberts, 1999). Following mitosis, Cyclin D is first upregulated in early G_1 and it interacts with CDK4/6. Through this interaction CDK4/6 activity is increased. Cyclin D accumulation is itself under complex developmental control involving both transcriptional and posttranscriptional mechanisms. Once it becomes active, the most important known substrate for CDK4/6 is the retinoblastoma protein (RB), which upon phosphorylation releases its inhibitory effects on the transcription factor E2F, thus allowing the transcription of genes necessary for the continued transit of G_1. These include genes necessary for DNA synthesis, as well as another late stage G_1 cyclin, cyclin E. As levels of cyclin E rise, CDK2 activity increases, leading to higher levels of pRB phosphorylation and ultimately to a commitment by the cell to enter another round of DNA synthesis and cell division (Sherr and Roberts, 1999).

A further level of regulation of CDK activity and thus cell cycle progression is achieved through the activity of the CKIs, including the Cip/Kip family and the Ink4 family. The Ink4 family is capable of binding to the

CDK alone and blocking cyclin binding, whereas the Cip/Kip family members bind to the CDK/cyclin complex and inhibit CDK activity (Sherr and Roberts, 1999). In dividing cells, these proteins are regulated in a cell cycle–dependent manner and are involved in the orderly transition from one stage of the cell cycle to the next. However, the regulated appearance of one CKI during inner ear development, p27^{Kip1}, plays a crucial role in timing the exit from the cell cycle for the progenitors that will form the normally postmitotic organ of Corti.

A common feature of the action of cyclins, as well as the CKIs, is that although their appearance enables entry to key stages of the cell cycle such as mitosis or G_1, their continued activity will often block passage to the next stage. Therefore complete passage through each stage requires an increase in cyclin–Cdk activity, followed by removal of the activity. Removal is achieved by a ubiquitin-mediated destruction of the cyclin protein, dependent upon successful completion of that stage of the cell cycle and induction of the next (Koepp et al., 1999), allowing for the continued progression through another round of cell division. Similarly, cell cycle exit is thought to be largely controlled by the developmentally coordinated downregulation of cyclin synthesis and, importantly, by the appearance of CKIs. Thus, both cyclins and CKIs are known to be regulated by a large variety of growth stimulatory and inhibitory pathways (Elledge et al. 1996).

During embryonic development, the cell cycle machinery just outlined is regulated to control organ and tissue morphogenesis through controlled cell proliferation within specific cell lineages. However, some cell types, upon exiting the cell cycle and entering a quiescent state, can upon the correct stimulation emerge from G_0 to re-enter G_1 and begin the process of cell division again in order to renew or regenerate specific tissues. In contrast, many cells, such as neurons and hair cells, become terminally postmitotic, in which case they never re-enter the cell cycle and if stimulated to do so will frequently undergo cell death (Liu and Greene, 2001). The difference between these two types of cell cycle exits represents the major problem for regeneration research and in the case of sensory hair cell regeneration begs the question of how many cold-blooded vertebrates and birds are able to overcome the obstacle of the postmitotic state, whereas mammals are not.

III. Hair Cell Development and Regeneration

The development of hair cells has been extensively treated in previous chapters and so is not described here in detail. However, the mammalian species in which the most information is available regarding the relationship between cell cycle and hair cell development is the mouse, and events in this species are briefly recapped here.

The cochlear duct of the mouse originates late in the 10th day postconception (dpc) as an outpouching of the otocyst (e.g., Morsli et al., 1998). The duct elongates to form approximately one half turn by 12.5 dpc. The epithelium of the duct also beings to differ between its dorsal and ventral walls at about 12.5 dpc (Retzius, 1884; Sher, 1971). The ventral wall, which will form the organ of Corti, inner sulcus, and spiral limbus, is thicker than the dorsal wall, which will develop into the stria vascularis and Reissner's membrane. By 14.5 dpc the ventral epithelium has developed a greater and lesser epithelial ridge in the basal turn, but the component cells remain morphologically undifferentiated. The first cells to be recognized morphologically are the inner hair cells of the midbasal region of the cochlear duct, which separate from the basement membrane and become rounded at the border between the lesser and greater epithelial ridges around 15.5 dpc in the base. The basal turn outer hair cells can subsequently be recognized morphologically at 16.5 dpc by their lighter cytoplasm and position. Morphological differentiation of cochlear cells then continues in a wave that progresses along the basal-to-apical axis of the cochlea. By 18.5 dpc, a single row of recognizable inner hair cells and three rows of outer hair cells are present along most of the cochlear duct.

Molecular markers of hair cell differentiation appear prior to morphological differentiation. Morsli et al. (1998) found that lunatic fringe is an early marker for the sensory cell region of the cochlear epithelium of the mouse, appearing from 12 to 13 dpc. In the mouse, Math1 is the earliest marker of hair cell differentiation (Bermingham et al., 1999), beginning sometime between 13.5 and 14.5 dpc in the midbasal turn of the cochlea. Expression then spreads in basal to apical wave that briefly precedes the appearance of another molecular marker of hair cell differentiation, myosin VIIa (Chen et al., 2002), and is followed closely by morphological differentiation. In addition, at the leading edge of the basal to apical gradient of differentiation it can be seen that inner hair cells begin to differentiate earlier than outer hair cells, evidence of a second medial to lateral gradient of hair cell differentiation that occurs orthogonally across the sensory epithelium (Li and Ruben, 1979; Lim and Anniko, 1985; Sher, 1971). This can also be observed with molecular markers by the sequential appearance in hair cells of first Math1 and subsequently myosin VIIa (Chen and Segil, 1999; Chen et al., 2002; Sahly et al., 1997). Patterning of the sensory region of the cochlear duct is complete by ~17.5 dpc, yielding the characteristic pattern of one row of inner hair cells and three rows of outer hair cells. Several additional markers of hair cell differentiation appear in the hair cells of the basal turn during the 14th dpc. These include Brn3.1 (Brn3c, POU4f3; Erkman et al., 1996; Ryan, 1997) and the Delta ligand Jagged2 (Lanford et al., 1999). Genes associated with hair cell function, such as ion transport molecules (e.g., Furuta et al., 1998), neuronal receptors (e.g., Luo et al., 1995, 1998),

14. Hair Cells and the Cell Cycle   455

and the motor protein prestin (Zheng *et al.*, 2003), appear later in development, usually several days prior to the onset of function at around postnatal day 10.

Once formed, mature hair cells in mammals are stable unless they are damaged. That is, new cells are not added and the existing cells do not recycle. Thus in humans the hair cells that we are born with typically survive throughout many decades of life. Moreover, data from many studies, including the behavior of hearing thresholds and the characteristics of temporal bone histopathology in humans (Schuknecht, 1993), suggest that mammalian cochlear hair cells, once lost, are not regained. However, vigorous regeneration of hair cells after their loss due to noise or drug-induced injury was first noted in birds and fish in the late 1980s (e.g., Corwin and Cotanche, 1988; Ryals and Rubel, 1988). Hair cell regeneration in these lower vertebrate species typically involves the cell cycle. Cells within the epithelium divide, giving rise to new cells, some of which differentiate into hair cells. The process of proliferative regeneration has been well documented and is illustrated schematically in Fig. 1. Severe injury to the sensory epithelium results in the loss of hair cells, often through extrusion of the damaged cell from the sensory epithelium. Hair cell damage stimulates nearby supporting cells to leave their resting state and enter the cell cycle. The nuclei of these cells appear to enter the cell cycle at the bottom of the epithelium near the basement membrane (Raphael *et al.*, 1994). As the cell cycle progresses the nuclei rise and the cells enter M phase with the nuclei near the apical surface. After telophase the cycle may be repeated, or the daughter cells may adopt either the hair cell or the supporting cell fate, presumably depending on signals from adjacent cells, and differentiate appropriately.

It should be noted that not all hair cell regeneration requires cell division since regeneration can be observed in mitotically blocked sensory epithelial cultures (Baird *et al.*, 1996, 2000). Transformation of differentiated supporting cells into hair cells has been proposed as a mechanism for this regeneration (Adler and Raphael, 1996; Adler *et al.*, 1997). In addition, damaged hair cells that lose most of the characteristics of hair cells, including stereocilia and even apical contact with the surface of the sensory epithelium, can reestablish surface contact and regenerate stereociliary arrays (Sobkowicz *et al.*, 1996).

IV. Cell Cycle Events during Hair Cell Development

The developing sensory epithelia of the inner ear are characterized by cell proliferation during their early expansion. Up to approximately 12.5 dpc in the basal mouse cochlea, cells in various stages of M phase can be observed morphologically in the sensory epithelium. At this stage the

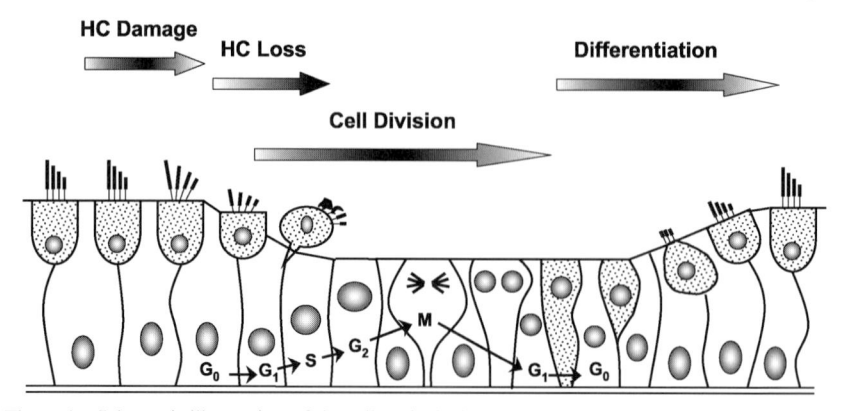

Figure 1 Schematic illustration of the cell cycle during hair cell damage and regeneration in the chick. Modified from Bermingham-McDonogh, O., and Rubel, E. W. (2003). Hair cell regeneration: Winging our way towards a sound future. *Curr. Opin. Neurobiol.* **13**, 119–126.

epithelium consists of several pseudo-stratified layers of cells. Dividing cells are typically observed in the upper region of the epithelium, immediately below the apical epithelial surface. Cell division in the region destined to become the sensory epithelium is not observed after about 12.5 dpc. Whereas M phase occurs near the surface of the epithelium, there is evidence that other events in the cell cycle do not take place at this location. In the developing chick cochlea, staining with markers of the cell cycle has suggested that cells enter G_1 at the bottom of the epithelium (Tsue *et al.*, 1994). As they progress through S and G_2, the nuclei of the cells ascend to the upper epithelium for M phase. After cell division, the nuclei of cells that continue to proliferate return to the bottom of the epithelium to re-enter G_1. Thus proliferation in inner ear sensory epithelium involves a pattern of intrakinetic nuclear migration similar to that described in early neurogenesis (Sauer, 1935; Sidman *et al.*, 1959; Takahashi *et al.*, 1993). The progeny of inner ear epithelial proliferative events eventually differentiate to form the various cell types of the organ of Corti, including hair cells.

The relationship between hair cell development and the cell cycle has been studied using a variety of techniques, including thymidine and BrdU incorporation, expression of proteins related to the cell cycle, and mutation of genes controlling the cell cycle. These studies have not always produced consistent results. Some of this disagreement may result from the processes of cochlear morphogenesis. There is evidence that elongation of the cochlear duct occurs from the addition of cells from the basal end of the epithelium, resulting in early birth dates for apical hair cells. In contrast, the differentiation of hair cells clearly progresses from the base to the apex, resulting in earlier molecular and morphological maturation of basal turn hair cells.

14. Hair Cells and the Cell Cycle

As noted earlier, an exception appears to be the hook region in the extreme base. Maturation appears to begin in the midbasal turn and to spread in both the apical and basal directions from this point.

Ruben (1967) studied the terminal mitoses of cochlear cells by injecting tritiated thymidine into pregnant mice of different gestational ages and quantifying the labeling of cochlear cell nuclei in adulthood. Animals for which injection occurred on the 12th dpc showed labeling in all cochlear turns, but with a greater number in the apex than the base. Injection on the 13th dpc produced the greatest degree of labeling, and the upper basal and lower middle turns showed the most labeled cells. Injection on the 14th dpc resulted in labeling primarily in the basal turn. Injection on the 15th dpc produced labeling of only a few cells, all in the extreme basal turn. These data suggested that the addition of cells to the basal turn of the cochlea by cell division may be responsible for elongation of the organ of Corti.

The careful longitudinal studies of Ruben (1967) have been the basis for our understanding of cell division during organ of Corti development in the mouse since they were conducted more than 30 years ago. To date no similarly complete study has been published. Recent pulse labeling studies with BrdU have confirmed Ruben's observation that the vast majority of cells in the organ of Corti exit the cell cycle in a relatively narrow window of time, around 13.5 dpc (Chen *et al.*, 2003). Injection of BrdU at 12.5 dpc followed by evaluation of incorporation 8 h later, indicated that cells throughout the cochlear epithelium, including in the region of the greater epithelial ridge where the organ of Corti will develop, were labeled. Injections 1 day later, on 13.5 dpc, indicated the presence of a zone of non-proliferating cells (ZNPC) running the length of the cochlear duct (Fig 2A, C, E). This ZNPC represents the postmitotic progenitors that will soon initiate the wave of basal to apical differentiation described earlier.

The basis for the formation of the ZNPC appears to be the tightly regulated expression of the cell cycle inhibitor p27^{Kip1} (Chen and Segil, 1999). p27^{Kip1} is a member of the Cip/Kip family of CKIs and interacts with both CDK4/6 and CDK2 to arrest the cell cycle (Sherr and Roberts, 1999). Like the formation of the ZNPC, p27^{Kip1} expression appears between 12.5 and 13.5 dpc (Fig. 2B, D, F) and correlates spatially with the observed region of cell cycle exit (Chen *et al.*, 2003). Examination of *p27^{Kip1}* null mice indicated that in the absence of p27^{Kip1} the ZNPC did not form (Chen and Segil, 1999) and supernumerary hair cells were present in the adult organ of Corti, suggesting that an overreplication of sensory precursors had occurred (Chen and Segil, 1999; Lowenheim *et al.*, 1999). However, although the cells of the nascent organ of Corti no longer exit the cell cycle in a synchronous wave in the absence of p27^{Kip1}, they do eventually exit the cell cycle and become patterned into irregular rows of inner and outer hair cells, indicating that in the absence of p27^{Kip1} other mechanisms are able to bring about cell cycle

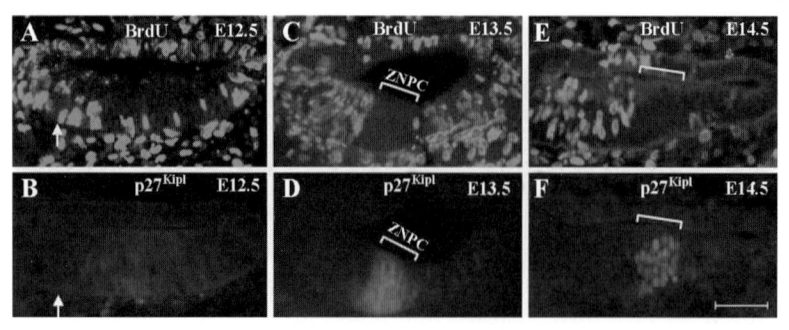

Figure 2 Expression of the CKI p27Kip1 defines a zone of nonproliferation in the developing inner ear, just prior to the specification of hair cell fate. A comparison of BrdU and p27^{Kip1} labeling in the developing cochlea. Expression of p27^{Kip1} begins on E13.5 and corresponds to cessation of mitosis as detected by BrdU labeling. From Chen, P., Zindy, F., Abdala, C., Liu, F., Li, X., *et al.* (2003). Progressive hearing loss in mice lacking the cyclin-dependent kinase inhibitor Ink4d. *Nat. Cell Biol.* **5**, 422–426. (See Color Insert.)

exit and that patterning is relatively unaffected. A possible candidate for this role is p19^{Ink4d}, a member of the Ink4 family of CKIs. In p19^{Ink4d} null mice, exit from the cell cycle and differentiation and patterning in the organ of Corti appear normal (Chen *et al.*, 2003). However, *in situ* hybridization indicates that p19^{Ink4d} expression overlaps that of p27^{Kip1} within the ZNPC at 14.5 dpc, suggesting that in the absence of p27^{Kip1}, Ink4d may be able to partially substitute to achieve cell cycle exit.

As the hair cells differentiate, expression of p27^{Kip1} is downregulated in mammals, although it continues to be produced in supporting cells (Chen *et al.*, 2003). The reasons for this downregulation are not clear. Interestingly, p27^{Kip1} continues to be expressed in the adult cochlear hair cells of birds (Torchinsky *et al.*, 1999), which are somewhat less specialized than those of mammals.

Lineage studies also support a postmitotic origin of hair cells. Fekete *et al.* (1998) studied retrovirally labeled lineages within the chick cochlear epithelium. Those lineages consisting of only two cells, and which thus incorporated the retrovirus during the final cell division, consisted of either two hair cells, two supporting cells, or one of each cell type. Thus the cells of the sensory epithelium retained flexibility until after the final cell division.

Flexibility in cell fate determination is also retained for some time following cell cycle exit. For example, treatment of the mouse or rat organ of Corti with retinoic acid has been shown to induce extra hair cells from 16 dpc until several days after birth, without the production of additional cells in the epithelium (Chardin and Romand, 1997; Kelley *et al.*, 1993). Moreover, Abdouh *et al.* (1994) found that merely culturing the postmitotic organ of Corti in media without retinoic acid resulted in numerous extra hair

14. Hair Cells and the Cell Cycle 459

cells. In addition, destruction of a developing hair cell can result in an adjacent supporting cell adopting the hair cell fate (Kelley *et al.*, 1995). Thus cells that normally chose a supporting cell fate can be induced to adopt a hair cell fate, long after cell division has ceased in the sensory region of the epithelium, under the influence of exogenous factors and signals from adjacent cells.

Between 14.5 and 17.5 dpc, the cochlear duct lengthens from approximately $\frac{3}{4}$ of a turn to its final $1\frac{3}{4}$ turns (Morsli *et al.*, 1998); however, the vast majority of cells of the organ of Corti have already exited the cell cycle prior to this period of growth (Ruben, 1967). Early in this process, the region of the cochlear duct that gives rise to the sensory epithelium is 5–6 cells thick. Math1 first appears in columns of cells that run through the 5–6 cells thick cochlear epithelium at the medial edge of the $p27^{Kip1}$-positive ZNPC (Chen *et al.*, 2003). As the wave of differentiation patterns the organ of Corti from base to apex between 14.5 and 17.5 dpc, the epithelium thins to the characteristic two-cell thickness made up of differentiated hair cells and supporting cells. Since expression of $p27^{Kip1}$ and cell cycle exit occur prior to the onset of hair cell differentiation as assayed by the expression of Math1 and consequently no, or very few, new cells are added during this phase of the elongation of the cochlear duct, it was suggested that the elongation of the organ of Corti, within the still dividing cells of the cochlear duct, results from the radial intercalation of the cells of the thickened 14.5 dpc cochlear epithelium (Chen *et al.*, 2003).

Finally, although *$p27^{Kip1}$* null mice are viable, albeit larger (Fero *et al.*, 1996), they also exhibit severe hearing loss (Chen and Segil, 1999; Lowenheim *et al.*, 1999). Although the reasons for this are not clear, it seems possible that the supernumerary cells (both hair cells and supporting cells) compromise the mechanics of cochlear function. If so, this provides an example of the importance of coordinating the cell cycle with cell differentiation during embryonic development. It also provides a caution for those seeking to restore hearing through gene therapy aimed at stimulating the regeneration of hair cells.

V. The Cell Cycle during Hair Cell Regeneration

The relationship between hair cell regeneration and the cell cycle has been well studied in birds, amphibians, and fish, in which robust, proliferative hair cell regeneration is observed. For example, Stone and Cotanche (1994) observed that after a brief noise exposure that results in hair cell damage and loss some cells in the affected region entered S phase, as determined by BrdU incorporation, within 18 h. S phase entry peaked at 42–72 h, but additional cells continued to enter the cell cycle for several days. Labeled cells at longer survival times tended to occur in clusters, suggesting that several rounds of replication were stimulated by hair cell damage. As expected, entry of cells into S phase was restricted to cochlear regions exhibiting hair cell loss. This is consistent with a role of undamaged hair cells in suppressing entry of adjacent supporting cells into the cell cycle. Similarly, Hashino et al. (1995) found that a wave of cell proliferation followed the death of hair cells without a latency of less than 48 h in a base-to-apex direction during and after kanamycin administration in chicks. Warchol and Corwin (1996) studied supporting cells after restricted laser ablation in hair cells in the chicken cochlea. They found that supporting cells entered the cell cycle beginning 16 h after hair cell death and that only cells within 200 μm of the lesion were stimulated to proliferate, suggesting a local and perhaps diffusible signal.

The identity of cells that enter the cell cycle was also examined by Corwin and Cotanche (1988), who found that acoustical overstimulation in chicks completely eliminated hair cells from the region that would later regenerate. Since this left only supporting cells in this region, they concluded that supporting cells and not hair cells were the cells that proliferate. This was confirmed by Tucci and Rubel (1990) using gentamicin. Presson and Popper (1990) compared the ultrastructure of proliferating cells and supporting cells in the sensory epithelium of the fish and found them to be indistinguishable. Finally, Balak et al. (1990) performed laser ablation of hair cells in the amphibian lateral line and directly observed the division of adjacent supporting cells.

The signals that contribute to proliferation may not be limited to regions of obvious hair cell damage. Bhave et al. (1995) used phase-specific markers to examine entry into G_1, S, and G_2 after a single high dose of gentamicin. They compared responses in the region of the chick cochlea exhibiting hair cell loss to an apical region that did not. They found that a marked decrease in statin, a marker for cells not in the cell cycle (G_0), occurred in supporting cells 3 days after gentamicin in both the region of hair cell loss and in the region that did not lose hair cells. This suggests that cells in both regions enter G_1. Moreover, proliferating cell nuclear antigen (PNCA), a subunit of DNA polymerase associated with G_1, S, and G_2 phases, was also observed in both regions, also consistent with entry into the cell cycle. However, BrdU incorporation, indicating S phase entry, was again restricted to the region of hair cell damage. These data suggest that entry into the cell cycle is broadly stimulated by hair cell damage, perhaps via a diffusible factor such as a growth factor. Moreover, the results indicate that undamaged hair cells do not suppress entry into the cell cycle but prevent cells from progressing from G_1 to S phase via factors that act locally. Signaling mediated by cell–cell contact, as with membrane-bound ligand–receptor systems, is one possibility.

14. Hair Cells and the Cell Cycle

It is clear that during regeneration in lower vertebrates, signals related to hair cell damage induce supporting cells to enter the cell cycle. A number of efforts have been made to identify such signals. Several laboratories have demonstrated that various growth factors that are active on other epithelial systems can increase cell proliferation in chick inner ear epithelia, including insulin-like growth factor and insulin (Oesterle et al. 2000) as well as transforming growth factor-α and tumor necrosis factor-α (Warchol, 1999). Cell contacts also appear to regulate proliferation. Warchol (2002) found that the extracellular matrix molecule fibronectin was a powerful stimulant of cell proliferation in chicken utricular epithelium *in vitro*. He also found that cell density was inversely related to cell proliferation, consistent with cell–cell interactions. Proliferation was blocked by anti-cadherin 23 antibodies, suggesting that this cadherin provides a substrate for the interaction of adjacent cells.

Intracellular signal transduction pathways involved in regulating proliferation of avian hair cells were examined by Navaratnam et al. (1996), who found that increasing cAMP levels increased proliferation in the chick cochlear sensory epithelium. Witte et al. (2001) used inhibitors of several signal transduction pathways to evaluate the regulation of proliferation in chick vestibular sensory epithelium. They found that S phase entry was almost completely blocked by inhibitors of phosphatidyl inositol 3 kinase, the related kinase target of rapamycin (TOR), or the Erk mitogen activated protein kinase (MAPK), but was only partially blocked by inhibition of protein kinase C. The signaling cascades in which these molecules participate are known to be activated by growth factor and matrix molecule receptors (e.g., Aletsee et al., 2001, 2003) and all have been linked to the stimulation of proliferation in other epithelia and so are consistent with the stimulation of sensory epithelial proliferation as mentioned earlier. TOR in particular is a central controller of eukaryotic growth and proliferation (Jacinto and Hall, 2003), and its involvement is not surprising.

In the mammalian cochlea, neither cell proliferation nor hair cell regeneration are observed after hair cell damage. However, limited proliferative responses are observed in the supporting cells of damaged vestibular sensory epithelia. Progression of the proliferated cells to hair cells occurs, but it appears to be rare (Malgrange et al., 2002, Warchol et al., 1993). The normally limited proliferation of mammalian vestibular epithelia can be dramatically increased by treatment with growth factors such as tumor necrosis factor-α, epithelial growth factor and insulin (Kuntz and Oesterle, 1998; Oesterle and Hume, 1999; Yamashita and Oesterle, 1995), glial growth factor (Montcouquiol and Corwin, 2001), and heregulin (Zheng et al., 1999). Proliferation can be enhanced by the cAMP donor forskolin and blocked by inhibitors of a downstream protein kinase (Montcouquiol and Corwin, 2001), suggesting the involvement of cAMP signaling. To the extent that it has been studied, the regulation of proliferation in mammalian vestibular epithelium appears to be similar to that observed in the chick (Montcouquiol and Corwin, 2001). This suggests that the factors restricting vestibular hair cell regeneration in mammals may occur at points after the regulation of cell proliferation.

VI. Conclusions

Regulation of the cell cycle plays an important role in the development of hair cells. The suppression of cell division early in the development of the organ of Corti helps to determine the placement of hair cells in the mature organ of Corti and to some extent regulates the number of hair cells. However, the determination of hair cell fate appears to be a postmitotic process. Cell proliferation plays a central role in hair cell regeneration in lower vertebrates, with proliferation of supporting cells providing a pool of new cells, some of which adopt the hair cell fate to replace lost cells. However, proliferation is less prevalent in the damaged sensory epithelia of normal mammals and is restricted to the vestibular sensory epithelia. Inducing cells within the mammalian sensory epithelium to replicate is a major goal of hair cell regeneration research. However, although it has proven possible to dramatically increase cell division within the mammalian vestibular sensory epithelium *in vitro*, progression of these cells into the hair cell phenotype is limited. In the organ of Corti, comparable stimulation of cell division has not proved possible, much less induction of hair cells.

For hair cell regeneration to occur in mammals, the inhibition of cell proliferation that occurs in the organ of Corti must be overcome. Increased understanding of factors that contribute to cell cycle inhibition, including the role of cyclin kinase inhibitors such as $p27^{Kip1}$, may help to increase cell proliferation in the damaged mammalian organ of Corti. However, experience in mammalian vestibular epithelium indicates that significant hurdles also exist at the stage of cell fate determination. As with the cell cycle, our increasing understanding of the factors that determine the hair cell fate may provide tools with which to overcome barriers to hair cell fate determination in the adult mammalian inner ear.

Acknowledgments

Neil Segil of the House Ear Institute read the manuscript and provided extensive critical commentary and suggestions. His contributions are gratefully acknowledged.

References

Abdouh, A., Despres, G., and Romand, R. (1994). Histochemical and scanning electron microscopic studies of supernumerary hair cells in embryonic rat cochlea in vitro. *Brain Res.* **660**, 181–191.

Adler, H. J., Komeda, M., and Raphael, Y. (1997). Further evidence for supporting cell conversion in the damaged avian basilar papilla. *Int. J. Dev. Neurosci.* **15**, 375–385.

Adler, H. J., and Raphael, Y. (1996). New hair cells arise from supporting cell conversion in the acoustically damaged chick inner ear. *Neurosci. Lett.* **205**, 17–20.

Aletsee, C., Beros, A., Mullen, L., Palacios, C., Pak, K., *et al.* (2001). Ras/MEK but not p38 signaling mediates neurite extension from spiral ganglion neurons. *JARO* **2**, 377–387.

Aletsee C., Brors, D., Mlynski, R., Ryan, A. F., and Dazert, S. (2003). Wortmannin, a specific inhibitor of phosphatidylinositol-3-kinase, influences neurotrophin-induced spiral ganglion growth. *Laryngo rhino otologie* **81**, 189–195, German.

Baird, R. A., Steyger, P. S., and Schuff, N. R. (1996). Mitotic and nonmitotic hair cell regeneration in the bullfrog vestibular otolith organs. *Ann. N. Y. Acad. Sci.* **781**, 59–70.

Baird, R. A., Burton, M. D., Fashena, D. S., and Naeger, R. A. (2000). Hair cell recovery in mitotically blocked cultures of the bullfrog saccule. *Proc. Natl. Acad. Sci. USA* **9**, 11722–11729.

Balak, K. J., Corwin, J. T., and Jones, J. E. (1990). Regenerated hair cells can originate from supporting cell progeny: Evidence from phototoxicity and laser ablation experiments in the lateral line system. *J. Neurosci.* **10**, 2502–2512.

Bermingham, N., Hassan, B., Price, S., Vollrath, M., Ben-Arie, N., *et al.* (1999). Math1: An essential gene for the generation of inner ear hair cells. *Science* **284**, 1837–1841.

Bhave, S. A., Stone, J. S., Rubel, E. W., and Coltrera, M. D. (1995). Cell cycle progression in gentamicin-damaged avian cochleas. *J. Neurosci.* **15**, 4618–4628.

Chardin, S., and Romand, R. (1997). Factors modulating supernumerary hair cell production in the postnatal rat cochlea in vitro. *Int. J. Dev. Neurosci.* **15**, 497–507.

Chen, P., Zindy, F., Abdala, C., Liu, F., Li, X., *et al.* (2003). Progressive hearing loss in mice lacking the cyclin-dependent kinase inhibitor Ink4d. *Nat. Cell Biol.* **5**, 422–426.

Chen, P., Johnson, J. E., Zoghbi, H. Y., and Segil, N. (2002). The role of Math1 in inner ear development: Uncoupling the establishment of the sensory primordium from hair cell fate determination. *Development* **129**, 2495–2505.

Chen, P., and Segil, N. (1999). p27(Kip1) links cell proliferation to morphogenesis in the developing organ of Corti. *Development* **126**, 1581–1590.

Corwin, J. T., and Cotanche, D. A. (1988). Regeneration of sensory hair cells after acoustic trauma. *Science* **240**, 1774–1776.

Corwin, J. T. (1981). Postembryonic production and aging in inner ear hair cells in sharks. *J. Comp. Neurol.* **201**, 541–553.

Elledge, S. J., Winston, J., and Harper, J. (1996). A question of balance: The role of cyclin-kinase inhibitors in development and tumorigenesis. *Trends Cell Biol.* **6**, 388–397.

Erkman, L., McEvilly, R. J., Luo, L., Ryan, A. E., Hoosmand, F., *et al.* (1996). Role of transcription factors Brn-3.1 and Brn-3.2 in auditory and visual system development. *Nature* **381**, 603–606.

Fekete, D. M., Muthukumar, S., and Karagogeos, D. (1998). Hair cells and supporting cells share a common progenitor in the avian inner ear. *J. Neurosci.* **18**, 7811–7821.

Fero, M. L., Rivkin, M., Tasch, M., Porter, P., Carow, C. E., *et al.* (1996). A syndrome of multiorgan hyperplasia with features of gigantism, tumorigenesis, and female sterility in p27(Kip1)-deficient mice. *Cell* **85**, 733–744.

Furuta, H., Luo, L., Hepler, K., and Ryan, A. F. (1998). Evidence for differential regulation of calcium by outer versus inner hair cells: Plasma membrane Ca-ATPase gene expression. *Hear. Res.* **123**, 10–26.

Hartwell, L. H., and Weinert, T. (1989). Checkpoints: Controls that ensure the order of cell cycle events. *Science* **246**, 629–634.

Hashino, E., TinHan, E. K., and Salvi, R. J. (1995). Base-to-apex gradient of cell proliferation in the chick cochlea following kanamycin-induced hair cell loss. *Hear. Res.* **88**, 156–168.

Jacinto, E., and Hall, M. N. (2003). TOR signaling in bugs, brain and brawn. *Nat. Rev. Molec. Cell Biol.* **4**, 117–126.

Katayama, A., and Corwin, J. T. (1989). Cell production in the chicken cochlea. *J. Comp. Neurol.* **281**, 129–135.

Kelley, M. W., Xu, X. M., Wagner, M. A., Warchol, M. E., and Corwin, J. T. (1993). The developing organ of Corti contains retinoic acid and forms supernumerary hair cells in response to exogenous retinoic acid in culture. *Development* **119**, 1041–1055.

Kelley, M. W., Talreja, D. R., and Corwin, J. T. (1995). Replacement of hair cells after laser microbeam irradiation in cultured organs of Corti from embryonic and neonatal mice. *J. Neurosci.* **15**, 3013–3026.

Koepp, D. M., Harper, J. W., and Elledge, S. J. (1999). How the cyclin became a cyclin: Regulated proteolysis in the cell cycle. *Cell* **97**, 431–434.

Kuntz, A. L., and Oesterle, E. C. (1998). Transforming growth factor alpha with insulin stimulates cell proliferation in vivo in adult rat vestibular sensory epithelium. *J. Comp. Neurol.* **399**, 413–423.

Lanford, P. J., Lan, Y., Jiang, R., Lindsell, C., Weinmaster, G., et al. (1999). Notch signaling pathway mediates hair cell development in mammalian cochlea. *Nat. Genet* **21**, 289–292.

Lang, H., Bever, M. M., and Fekete, D. M. (2000). Cell proliferation and cell death in the developing chick inner ear: Spatial and temporal patterns. *J. Comp. Neurol.* **417**, 205–220.

Li, C. W., and Ruben, R. J. (1979). Further study of the surface morphology of the embryonic mouse cochlear sensory epithelia. *Otolaryngol. Head Neck Surg.* **87**, 479–485.

Lim, D. J., and Anniko, M. (1985). Developmental morphology of the mouse inner ear. A scanning electron microscopic observation. *Acta Otolaryngol. Suppl.* **422**, 1–69.

Liu, D. X., and Greene, L. A. (2001). Neuronal apoptosis at the G1/S cell cycle checkpoint. *Cell Tissue Res.* **305**, 217–228.

Lowenheim, H., Furness, D. N., Kill, J., Zinn, C., Gultig, K., et al. (1999). Gene disruption of p27(Kip1) allows cell proliferation in the postnatal and adult organ of Corti. *Proc. Natl. Acad. Sci. USA* **96**, 4084–4088.

Luo, L., Brumm, D., and Ryan, A. F. (1995). Distribution of non-NMDA glutamate receptor mRNAs in the developing rat cochlea. *J. Comp. Neurol.* **361**, 372–382.

Luo, L., Bennett, T., Jung, H. H., and Ryan, A. F. (1998). Developmental expression of alpha 9 acetylcholine receptor mRNA in the rat cochlea and vestibular inner ear. *J. Comp. Neurol.* **393**, 320–331.

Malgrange, B., Belachew, S., Thiry, M., Nguyen, L., Rogister, B., et al. (2002). Proliferative generation of mammalian auditory hair cells in culture. *Mech. Dev.* **112**, 79–88.

Montcouquiol, M., and Corwin, J. T. (2001). Brief treatments with forskolin enhance S-phase entry in balance epithelia from the ears of rats. *J. Neurosci.* **21**, 974–982.

Morsli, H., Choo, D., Ryan, A. F., Johnson, R., and Wu, D. K. (1998). Development of the mouse inner ear and origin of its sensory organs. *J. Neurosci.* **18**, 3327–3335.

Murakami, H., and Nurse, P. (2000). DNA replication and damage checkpoints and meiotic cell cycle controls in the fission and budding yeasts. *Biochem. J.* **349**, 1–12.

Nabel, E. G. (2002). CDKs and CKIs: Molecular targets for tissue remodeling. *Nat. Rev. Drug Discov.* **1**, 587–598.

14. Hair Cells and the Cell Cycle

Navaratnam, D. S., Su, H. S., Scott, S. P., and Oberholtzer, J. C. (1996). Proliferation in the auditory receptor epithelium mediated by a cyclic AMP-dependent signaling pathway. *Nat. Med.* **2**, 1136–1139.

Nurse, P. (2000). A long twentieth century of the cell cycle and beyond. *Cell* **100**, 71–78.

Oesterle, E. C., Bhave, S. A., and Coltrera, M. D. (2000). Basic fibroblast growth factor inhibits cell proliferation in cultured avian inner ear sensory epithelia. *J. Comp. Neurol.* **424**, 307–326.

Oesterle, E. C., and Hume, C. R. (1999). Growth factor regulation of the cell cycle in developing and mature inner ear sensory epithelia. *J. Neurocytol.* **28**, 877–887.

Presson, J. C., and Popper, A. N. (1990). Possible precursors to new hair cells, support cells, and Schwann cells in the ear of a post-embryonic fish. *Hear. Res.* **46**, 9–21.

Raphael, Y., Adler, H. J., Wang, Y., and Finger, P. A. (1994). Cell cycle of transdifferentiating supporting cells in the basilar papilla. *Hear. Res.* **80**, 53–63.

Retzius, G. (1884). "Das Gehörorgan der Wirbeltiere. II. Das Gehörorgan der Reptilien, der Vögel und Sägetiere." Samson and Wallin, Stockholm, Sweden.

Ruben, R. J. (1967). Development of the inner ear of the mouse: A radioautographic study of terminal mitoses. *Acta Otolaryngol. Suppl.,* **220**, 1–44.

Ryals, B. M., and Rubel, E. W. (1988). Hair cell regeneration after acoustic trauma in adult Coturnix quail. *Science* **240**, 1772–1774.

Ryan, A. F. (1997). Transcription factors and the control of inner ear development. *Stem Cell Dev. Biol.* **8**, 249–256.

Sahly, I., El-Amraoui, A., Abitbol, M., Petit, C., and Dufier, J. L. (1997). Expression of myosin VIIA during mouse embryogenesis. *Anat. Embryol.* **196**, 159–170.

Sauer, F. C. (1935). Mitosis in the neural tube. *J. Comp. Neurol.* **62**, 377–405.

Schuknecht, H. F. (1993). "Pathology of the Ear," ed 2, Lea & Febiger, Philadelphia.

Sher, A. E. (1971). The embryonic and postnatal development of the inner ear of the mouse. *Acta Otolaryngol. Suppl.* **285**, 1–77.

Sherr, C. J., and Roberts, J. M. (1999). CDK inhibitors: Positive and negative regulators of G1-phase progression. *Gene Dev.* **13**, 1501–1512.

Sidman, R. L., Miale, I. L., and Feder, N. (1959). Cell proliferation and migration in the primitive ependymal zone: An autoradiographic study of histogenesis in the nervous system. *Exp. Neurol.* **1**, 322–333.

Sobkowicz, H. M., August, B. K., and Slapnick, S. M. (1996). Post-traumatic survival and recovery of the auditory sensory cells in culture. *Acta Otolaryngol.* **116**, 257–262.

Steel, K. P., and Barkway, C. (1989). Another role for melanocytes: Their importance for normal stria vascularis development in the mammalian inner ear. *Development* **107**, 453–463.

Stone, J. S., and Cotanche, D. A. (1994). Identification of the timing of S phase and the patterns of cell proliferation during hair cell regeneration in the chick cochlea. *J. Comp. Neurol.* **341**, 50–67.

Takahashi, T., Nowakowski, R. S., and Caviness, V. S. (1993). Cell cycle parameters and patterns of nuclear movement in the neocortical proliferative zone of the fetal mouse. *J. Neurosci.* **13**, 820–833.

Torchinsky, C., Messana, E. P., Arsura, M., and Cotanche, D. A. (1999). Regulation of p27Kip1 during gentamicin mediated hair cell death. *J. Neurocytol.* **28**, 913–924.

Tsue, T. T., Watling, D. L., Weisleder, P., Coltrera, M. D., and Rubel, E. W. (1994). Identification of hair cell progenitors and intermitotic migration of their nuclei in the normal and regenerating avian inner ear. *J. Neurosci.* **14**, 140–152.

Tucci, D. L., and Rubel, E. W. (1990). Physiologic status of regenerated hair cells in the avian inner ear following aminoglycoside ototoxicity. *Otolaryngol. Head Neck Surg.* **103**, 443–450.

Warchol, M. E., and Corwin, J. T. (1996). Regenerative proliferation in organ cultures of the avian cochlea: Identification of the initial progenitors and determination of the latency of the proliferative response. *J. Neurosci.* **16**, 5466–5477.

Warchol, M. E., Lambert, P. R., Goldstein, B. J., Forge, A., and Corwin, J. T. (1993). Regenerative proliferation in inner ear sensory epithelia from adult guinea pigs and humans. *Science* **259**, 1619–1622.

Warchol, M. E. (2002). Cell density and N-cadherin interactions regulate cell proliferation in the sensory epithelia of the inner ear. *J. Neurosci.* **22**, 2607–2616.

Warchol, M. E. (1999). Immune cytokines and dexamethasone influence sensory regeneration in the avian vestibular periphery. *J. Neurocytol.* **28**, 889–900.

Wilkins, H. R., Presson, J. C., and Popper, A. N. (1999). Proliferation of vertebrate inner ear supporting cells. *J. Neurobiol.* **39**, 527–535.

Witte, M. C., Montcouquiol, M., and Corwin, J. T. (2001). Regeneration in avian hair cell epithelia: Identification of intracellular signals required for S-phase entry. *Eur. J. Neurosci.* **14**, 829–838.

Yamashita, H., and Oesterle, E. C. (1995). Induction of cell proliferation in mammalian inner-ear sensory epithelia by transforming growth factor alpha and epidermal growth factor. *Proc. Natl. Acad. Sci. USA* **92**, 3152–3155.

Zheng, J. L., Frantz, G., Lewis, A. K., Sliwkowski, M., and Gao, W. Q. (1999). Heregulin enhances regenerative proliferation in postnatal rat utricular sensory epithelium after ototoxic damage. *J. Neurocytol.* **28**, 901–912.

Zheng, J., Long, K. B., Matsuda, K. B., Madison, L. D., Ryan, A. F., and Dallos, P. D. (2003). Genomic characterization and expression of mouse prestin, the motor protein of outer hair cells. *Mamm. Genome* **14**, 87–96.

Index

A

Acousticolateral placode, 4
  patterning defects in, 369f
Adhesion molecules
  BEN as, 329
  cadherins as, 326–27
  cell adhesion molecules and, 323–30
  DM-GRASP as, 329
  domain structure of, 324f
  inner ear and, 342–43
  inner ear development and, 330–43
  integrins as, 326, 328
  in mammalian cochlear duct, 336f
  neurite growth and, 342–43
  Notch as, 329–30
  otocyst expression of, 330–34
  SC-1 as, 329
  sensory epithelia and, 334–42
  stereocilia bundles and, 343–45
Afferent fibers
  exuberant/unconnected, 30
  guidance of, 25–29
  survival of, 29–32
Afferent innervation
  hair cells and, 429–30
  NTs and, 429–30
Amphibian
  inner ear induction, 126–28
Amphioxus, 6
Atrial chambers
  of salps, 5
Auditory function
  in chicks, 390
  in rodents, 390
  in zebrafish, 378–81

B

Basic helix-loop-helix (BHLH) genes, 14, 15f, 21, 22, 23, 28
  heterodimerization of, 14
  of NeuroD family, 19

Basilar papilla, 33
Basolateral conductances
  developmental changes in, 424–25
Basolateral ion channels
  development acquisition of, 405–25
BDNF. *See* Brain-derived neurotrophic factor
BEN. *See* Bursal epithelium & neurons
BHLH genes. *See* Basic helix-loop-helix genes
Birds
  inner ear induction in, 128–31
BMP signals. *See* Bone morphogenetic protein signals
BMP2 gene, 95
BMP4 genes, 24
Bone morphogenetic protein (BMP) signals
  FGF and, 165f
  gene expression patterns and, 164–65
  in semicircular canal formation, 164–65
BOR syndrome. *See* Branchio-oto-renal syndrome
Brain-derived neurotrophic factor (BDNF), 16, 27, 29
Branchial arch skeletal development
  secreted factors/receptors in, 96–98
  transcription factors in, 98–102
Branchial arch skeleton
  ossicle development and, 94
Branchial arches
  of ear, 85
Branchio-oto-renal (BOR) syndrome, 87
Brn3a, 26
Brn3c, 28
Bursal epithelium & neurons (BEN)
  as adhesion molecules, 329

C

Cadherins
  as adhesion molecules, 326–27
Caenorhabditis
  touch receptor of, 11
CAMs. *See* Cell adhesion molecules

468 Index

CDK regulation. *See* Cyclin-dependent kinase regulation
Cell adhesion molecules (CAMs)
  adhesion molecules and, 323–30
  inner ear/hair cell development and, 321–46
Cell cycle
  cell proliferation and, 450–53
  definition of, 449
  during hair cell damage, 456f
  during hair cell development, 455–59
  during hair cell regeneration, 459–62
  hair cells and, 449–62
Cell cycle regulators
  hair cell progenitors and, 309–10
  specific growth factors and, 309–10
Cell delamination, 25
Cell differentiation
  from otic vesicle, 178
Cell differentiation/myelination
  IGF$_{-1}$ and, 195
Cell fate commitment, 26
Cell growth/proliferation/survival
  IGF1 and, 193–95
Cell neuroprotection/regeneration
  IGF$_{-1}$ and, 195–96
Cell proliferation
  cell cycle and, 450–53
Cell states
  during otic neurogenesis, 181t
Cell survival loop
  neurotrophin and, 29
Cellular function conservation
  of Math1 gene, 21
Cellular/molecular information
  neural determination/differentiation and, 179
Chick cochlea
  hair cells of, 407–12
  outward rectifiers in, 407–12
Chick otic ectoderm
  molecular markers of, 130t
Chick otic placodes
  commitment steps of, 154t
Chick vestibular hair cells
  outward rectifiers in, 415–16
Chicks
  auditory function in, 390
  hair bundles of, 398
  hearing function in, 394
  vestibular function in, 394
Ciliated sensory neurons, 24

Circler mutants
  vestibular function and, 378
Cloning technology, 46
Cochlea, 4, 178
  hair cell stereociliary bundles of, 61f
  vestibular system innervation and, 20f
Cochlear hair cells
  development model of, 313f
  in rodents, 412–14
Cochlear neurons
  early development of, 179–87
Cochleas/vestibular organs
  developmental changes in, 392f
Cochleovestibular ganglion (CVG), 177, 178, 179
Commitment states
  of chick otic placodes, 154t
  of inner ear anlagen, 152–56
Connexins, 69
Conservation. *See also* Molecular conservation
  of genetic networks, 138–39
Craniate ear
  morphological evolution of, 6, 6f
Craniate vertebrates, 5
Cupula, 4
CVG. *See* Cochleovestibular ganglion
Cyclin-dependent kinase (CDK) regulation
  of mammalian cell cycle, 452t
Cyclin regulation
  of mammalian cell cycle, 452t
Cytoskeleton
  of hair cells, 46
  specialization of, 59–75

D

DACH. *See* Dachshund genes
Dachshund (DACH) genes, 8, 9
Deafness, hereditary
  therapies for, 47
Delaminating cells, 25
Delta/Notch gene expression, 21
  distribution of, 22
DeltaA genes
  sensory epithelia and, 375
Diffusible factors
  in ear neurogenesis, 179
  of inner ear neurogenesis, 184t
Diffusible signals
  transcription factors and, 162–65

Index

Diploblasts
  sensory cells of, 12f
Dlx homeobox gene family, 48, 99, 100, 133
  of Drosophila, 159
Dlx3b gene, 120
  otic induction and, 364-65
Dlx4b gene
  otic induction and, 364-65
Dlx5 gene, 166-68
Domains/borders
  of inner ear development, 160-62
Dorsal funiculus & ventral midline-immunoglobulin-like restricted axonal surface protein (DM-GRASP)
  as adhesion molecules, 329
Drosophila, 57, 58
  Dlx homeobox gene family, 159
  eye development of, 8
  homeobox genes in, 156f
  mechanosensory bristle of, 11
  PCP of, 62
  sensory organ development of, 20-21, 181
  transcription regulators in, 157

E

E3 ubiquitin ligase
  sensory epithelia and, 375-76
Ear, early developing
  NT in, 214-15
  Trk induction in, 214-15
Ear development
  FGF signaling in, 232-51
  OEP/FGF8 deficiency and, 366f
Ear evolution
  overview of, 4-6
Ear lineages, 4
Ear morphogenesis
  implementing/expanding genes for, 6-11
Ear neurogenesis
  diffusible factors in, 179
Ear pinna, 84, 85
  outer ear and, 87-92
Ectodermal cells, 13
Ectopic expression
  of foxil, 124
Ectopic fibroblast growth factor (FGF) expression
  otic placode formation and, 238-39
  preplacodal ectoderm and, 238-39

Ectopic hair cells
  from GER cells, 297-98
Edn1 gene, 96
Efferent synapses
  on hair cells, 430-32
Embryonic cochlea
  NT-3 in, 211f
Embryonic dysmorphogenesis, 7
Embryonic ear
  in retinoid receptor transcripts, 266f
  synthetic/catabolic enzymes in, 266f
Embryonic induction, 116
Embryonic inner ear
  Trk receptors in, 209-12
Endochondral ossification
  ossicle development and, 94
Endolymph
  anatomy of, 83
  formation of, 4
Endolymph homeostasis, 69-71
  disrupting mutations of, 53-54
ENU mutagenesis. *See* Ethylnitrosourea mutagenesis
Ephrin ligands/receptors, 27
Epithelial-mesenchymal interactions, 97
Epithelial neuroblasts, 177
ErbB2 genes
  null mutation of, 27
Ethylnitrosourea (ENU) mutagenesis, 58
Evoked activity
  hearing function and, 390-94, 395
Evolution
  functionality of, 1
  selective advantage in, 1
  vertebrate ear and, 1-35
  of vestibular organ, 4
External acoustic meatus (EAM), 84, 84f
  development mechanism of, 92-94
  outer ear and, 92-94
Extra cochlear inner hair cells
  with paraffin sectioning, 304f
Extracellular membranes, 74
Eya. *See* Eye absent gene
Eye
  development of, 2
  evolution, 3, 3f
Eye absent 1 (Eya1) gene, 48
  placode patterning and, 371-72
  vesicle patterning and, 371-72
Eye absent (Eya) gene, 7, 8, 9, 23, 105, 126

F

FGFRs. *See* Fibroblast growth factor receptors
FGFs. *See* Fibroblast growth factors
Fiber guidance
  hair cells and, 27–29
Fibroblast growth factor 3 (FGF3)
  inner ear morphogenesis and, 243–45
  otic ganglion formation and, 243–45
  otic induction and, 360–61
Fibroblast growth factor 8 (FGF8), 96
  inner ear morphogenesis and, 246–47
  otic ganglion formation and, 246–47
  otic induction and, 360–61
Fibroblast growth factor 10 (FGF10)
  inner ear morphogenesis and, 245–46
  otic ganglion formation and, 245–46
Fibroblast growth factor (FGF) control induction
  pattern formation and, 162–64
Fibroblast growth factor (FGF) family, 121, 123, 128–29
  fish inner ear induction and, 118–22
  of growth factors, 134–35
  as inner ear inducers, 131–32
  signaling, 123, 132
  targeting of, 119
Fibroblast growth factor receptor 2b (FGFR2b)
  inner ear morphogenesis and, 247
  otic placode formation and, 247
Fibroblast growth factor receptors (FGFRs)
  FGF signaling and, 227–29
  inner ear neurogenesis and, 188–90
  in mammals, 132–33, 133t
  in mouse ear development, 234f
  phylogenetic tree of, 230f
  schematic diagram of, 228f
Fibroblast growth factor (FGF) signal blocking
  otic placode formation and, 239–40
Fibroblast growth factor (FGF) signal transduction, 229–32
Fibroblast growth factor (FGF) signaling
  downstream consequences of, 240–42
  in ear development, 232–51
  FGFRs and, 227–29
  FGFs and, 226–27
  hair cell precursors and, 248–49
  from hindbrain, 368–69
  in inner ear development, 237–51
  inner ear induction and, 115
  inner ear morphogenesis and, 243–47
  innervation and, 225–52
  in middle ear development, 235–37
  in otic ganglion formation, 243–47
  in outer ear development, 232–34
  outer hair cells and, 248–49
  pillar cells and, 249–51
  schematic diagram of, 228f
  sensory epithelia development in, 247–51
Fibroblast growth factors (FGFs), 7, 10–11, 177, 179
  BMP signals and, 165f
  encoding of, 227
  FGF signaling and, 226–27
  inner ear neurogenesis and, 188–90
  in mouse ear development, 234f
  schematic diagram of, 228f
First pharyngeal pouch
  and middle ear, 102
Fish inner ear induction, 118–26
  FGF gene family and, 118–22
Fly scolopidial attachment
  myosin VIIa in, 21
Force generators
  outer hair cells as, 68–69
Forkhead genes, 7, 23, 124
Foxg1 gene, 7
Foxi1
  ectopic expression of, 124
  expression of, 363f
  null mutant analyses of, 7
  otic induction and, 362–64

G

Ganglionar neuroblast (Nb$_g$), 177, 187
GATA3 gene, 7, 8, 19
GDNF. *See* Glial cell line-derived neurotrophic factor
Gene control
  of hair cell differentiation, 294–310
  of zebrafish hair cells, 357–81
  of zebrafish inner ear, 357–81
Gene expression
  BMP signals and, 164–65
  in otocyst, 161f
  in presumptive otic tissue, 124–26, 130–31, 133–34

Index

Gene identification
  otic vesicle specification and, 154
  zebrafish of, 358
Gene mutation
  hearing/balance defects and, 50
General placode induction
  steps of, 139–41
Genetic determinants
  of middle ear, 94–105
  of outer ear, 87–105
Genetic networks
  conservation of, 138–39
Genetic revolution, 46–50
Genome sequence
  of human, 46
  of mice, 46
Genome websites, 50t
Gentamicin-damaged inner ears
  hair cells in, 312
GER cells. *See* Greater epithelial ridge cells
GFP. *See* Green fluorescent protein gene, 48
Glia-axon interactions, 27
Glial cell line-derived neurotrophic factor (GDNF), 207
  family, 217–19
  inner ear receptors and, 218–19
Goosecoid (Gsc) gene
  tympanic ring and, 92, 93
Greater epithelial ridge (GER) cells
  ectopic hair cells from, 297–98
  morphological/immunocytochemical conversion of, 299f
Green fluorescent protein (GFP) gene, 48
Growth factors
  FGF family of, 134–35
Gsc gene. *See* Goosecoid gene

H

Hagfish ear, 33
Hair bundle
  of chick cochlea, 399f
  in chicks, 398
  development, 396–404
  morphology, 396–98
  orientation of, 402–4
  of rodents, 398–400
  staircase formation of, 401
  tip link formation of, 401–2
  of zebrafish, 400–401
Hair bundle polarity
  establishment of, 60–63
  maintenance of, 63–65
  stereociliary bundle and, 60–65
Hair cell adaptation
  stereociliary bundle and, 66–68
Hair cell damage/regeneration
  during cell cycle, 456f
Hair cell development
  cell adhesion molecules and, 321–46
  cell cycle during, 455–59
  in higher vertebrates, 293–314
  model of, 313f
  regeneration and, 453–55
  of zebrafish, 357–81
Hair cell differentiation
  cell cycle regulators and, 309–10
  gene control of, 294–310
  helix-loop-helix transcription factors and, 296–308
  Hes1 gene and, 302–8, 308f
  innervation and, 428–29
  Math1 gene and, 297–302, 305–8
  notch signaling and, 295–96
  outer hair cells and, 68–69
  RA and, 280–82
  stereociliary bundle and, 59–68
  in utricles, 301–2
  in zebrafish, 374f
Hair cell fate determination
  Math1 gene and, 302
  sensory primordium and, 302
Hair cell patterning
  in sensory patches, 57–59
Hair cell precursors
  FGF signaling and, 248–49
  sensory epithelia development and, 248–49
Hair cell production
  Hes1 gene and, 303–4
  in utricle, 303–4
Hair cell progenitors
  cell cycle regulators and, 309–10
  specific growth factors and, 309–10
Hair cell regeneration
  during cell cycle, 459–62
  development and, 453–55
Hair cell stereociliary bundles
  of cochlea, 61f
Hair cells, 11, 15
  afferent innervation and, 429–30
  brain and, 23

Hair cells (*cont.*)
  Ca^{2+}-gated conductances in, 408f–410f
  cell cycle and, 449–62
  of chick cochlea, 407–12
  cytoskeletal components of, 46
  developmental molecular biology of, 13
  ectopically induced, 299–301, 300f
  efferent synapses on, 430–32
  fiber guidance and, 27–29
  functional development of, 389–434
  in Gentamicin-damaged inner ears, 312
  immature, stereociliary bundle of, 59, 60f
  innervating nerve fibers and, 427–33
  maturation/maintenance of, 310
  in mature inner ears, 311–12
  neuron populations and, 33–34
  Ngn1 null mutation and, 16f
  outward rectifiers in, 407–12, 412–14
  primary afferents and, 428–30
  production/regeneration of, 310–12
  in rodents, 412–14
  role of, 25–29
  in sensory epithelium, 391f
  survival of, 71–73
  transcription factors and, 310
  types of, 408f–410f
  ultrastructural features of, 299–301, 300f
  voltage in, 408f–410f
Haploid insufficiency, 9
Hearing
  mechanics of, 178
Hearing/balance defects
  gene mutation and, 50
Hearing function
  in chicks, 394
  evoked activity and, 390–94
  maturation of, 390–95
  spontaneous activity and, 395
  in zebrafish, 394
Hearing impairment
  social impact of, 46
Hearing loss, 46
Hearing mutant screen
  for zebrafish, 380
Hedgehog signaling
  from floorplate, 370–71
  from notochord, 370–71
Helix-loop-helix (HLH) gene regulation
  heterochronic alteration of, 23–24
Helix-loop-helix (HLH) transcription factors
  hair cell differentiation and, 296–308
Hes1 gene
  differential expression of, 304–5
  extra inner hair cells and, 302–3, 303f
  hair cell differentiation and, 302–8, 308f
  hair cell production and, 303–4
  Math1 and, 305–8
  as negative regulator, 302–8
  targeted disruption of, 302–3
Hes5 gene
  differential expression of, 304–5
Hes5 RNA, 306f
Heterochronic alteration
  of HLH gene regulation, 23–24
Higher vertebrates
  hair cell development, 293–314
Hindbrain
  FGF signaling from, 368–69
  neuronal connections to, 1
Hindbrain patterning
  homeobox genes and, 165–66
  inner ear development and, 165–66
HLH. *See* Helix-loop-helix
Hmx3 genes, 45
Homeobox-containing genes
  pattern formation and, 157–59
Homeobox genes
  in Drosophila, 156f
  hindbrain patterning and, 165–66
  homeotic-type phenotypes and, 157
  in mice, 156f
  Prx1/Prx2 as, 53
  of sonic hedgehog, 51
Homeodomain-containing transcription factors, 151–71
Homeotic-type phenotypes
  homeobox genes and, 157
Hox genes, 158
Hoxa1 mouse mutant
  inner ear morphogenesis of, 276f
Hoxa2 gene, 90, 91, 98, 99, 104
  tympanic ring and, 93
Human genetics
  deafness of, 47
  mouse genetics and, 47
Human genome sequence, 46

Index

Hydrodynamic sensors
  tunicate atrial chambers as, 5

I

Id genes. *See* Inhibitors of differentiation genes
Immature hair cell
  stereociliary bundle of, 59, 60f
Immature neuronal precursors (INPs), 186
Implementing/expanding genes
  for ear morphogenesis, 6–11
Induction process
  necessity/sufficiency in, 135
Inhibitors of differentiation (Id) genes, 14
Inner ear
  adhesion molecules in, 342–43
  complexity of, 116
  early development stages of, 152
  early morphogenesis of, 50–55
  fate maps, 160
  formation of, 116
  molecular inducers of, 118–24, 128–30
  mutant phenotypes of, 88t–89t
  neurite growth in, 342–43
  neuronal losses in, 212t
  nonproliferation zone in, 458f
  ontogeny of, 449–50
  patterning process, 151–71
  sensory epithelia development in, 247–51
  studies review on, 116–17
Inner ear anlagen
  commitment states of, 152–56
  patterning of, 152–62
Inner ear development, 50
  adhesion molecules and, 330–43
  cell adhesion molecules and, 321–46
  domains/borders, 160–62
  FGF signaling in, 237–51
  gross malformations of, 51
  hindbrain patterning and, 165–66
  mouse genetics and, 45–75
  neurotrophic factors in, 207–19
  patterning genes during, 278–80
  RA and, 261–83
  stages of, 262–63
  truncated Trk receptors during, 216
  of zebrafish, 357–81
Inner ear inducing molecules
  temporal expression of, 120f
  in zebrafish/chick/mouse, 119

Inner ear induction
  among species, 141–42
  in amphibians, 126–28
  in birds, 128–31
  FGF signaling and, 115
  in fish, 118–26
  genes involved in, 118
  molecular basis of, 115–42
  perspectives on, 134–41
Inner ear morphogenesis
  FGF signaling in, 243–47
  FGF3 and, 243–45
  FGF8 and, 246–47
  FGF10 and, 245–46
  FGFR2b and, 247
  of Hoxa1 mouse mutant, 276f
  mammalian, 293–94
  otic vesicle morphogenesis and, 242–43
Inner ear neurogenesis, 184–87
  diffusible factors of, 184t, 196
  extrinsic factors in, 187–97
  FGFs/FGFRs role in, 188–90
  IGFs and, 192–93
  molecular basis of, 184–85
  NGF and, 190–92
  proneural gene Ngn1 in, 185
Inner ear phenotype
  of NT, 212–14
  of NT-3, 212–14
  of Trk null mutant mice, 212–14
Inner ear placode
  molecular mechanisms of, 154
  specification/determination events in, 153
Inner ear receptors
  GDNF and, 218–19
  RA metabolic enzymes and, 265–69
Inner ear sensory epithelia
  NT-3 in, 210f
Inner hair cell bundles
  from tasmanian devil, 64f
Innervating nerve fibers
  hair cells and, 427–33
Innervation
  FGF signaling and, 225–52
  hair cell differentiation and, 428–29
INPs. *See* Immature neuronal precursors
Insect sensilla development, 24
Insulin-like growth factor-1 (IGF$_{-1}$), 177, 179
  cell differentiation/myelination and, 195

Insulin-like growth factor-1 (IGF$_{-1}$) (cont.)
    cell growth/proliferation/survival and, 193–95
    cell neuroprotection/regeneration and, 195–96
Insulin-like growth factors (IGFs)
    inner ear neurogenesis and, 192–93
Integrins
    as adhesion molecules, 328
Internal proprioreceptors, 22
Intracellular signaling pathways, 182f, 183
Intrinsic/extrinsic signals
    interactions between, 177–97
Invagination, 9
Invertebrate sensory cell
    developmental factors of, 20–22

J

Jackson Laboratory, 46
Jag1 gene expression, 55
Jekyll gene
    placode patterning and, 373
    vesicle patterning and, 373

K

KCNQ4 ion channel, 69
Knockins, 47–48
Knockout ears
    otocyst patterning genes and, 167f
Knockout studies, 134
Knockouts, 45, 48
Kreisler gene product, 131

L

LacZ, 48
Lateral inhibition, 14
Loss-of-function experiments, 14

M

Mammalian atonal homologue 1 (Math1) gene, 5, 15, 23, 28, 58
    cellular function conservation of, 21
    extra cochlear hair and, 297
    hair cell differentiation and, 297–302, 305–8
    hair cell fate determination and, 302
    Hes1 gene and, 305–8
    loss of, 18f
    misexpression of, 297
    of mouse, 22
    neurogenins/ role and, 23
    null mutation of, 16, 17f
    overexpression of, 298f
    sensory primordium and, 302
    targeted gene deletion of, 297
Mammalian cell cycle
    CDK regulation, 452t
    cyclin regulation of, 452t
Mammalian cochlear duct
    adhesion molecules in, 336f
Mammalian fibroblast growth factor receptors (FGFRs), 132–33
    interactions of, 133t
Mammalian hair cell
    development of, 293–314
Mammalian inner ear
    induction of, 131–34
    morphogenesis of, 293–94, 294
Mammalian peripheral auditory system
    components of, 225, 226
Math1 gene. *See* Mammalian atonal homologue 1 gene
Mechanoelectrical transduction, 4, 69
    biophysics of, 404–5
    development of, 395–405
    vertebrate ear and, 4
Mechanosensory cells, 13
Mechanosensory transducers, 5, 13
Merkel cells, 14
Mesenchymal epithelial interactions
    disruptive mutations in, 51–53
Mesendodermal tissue
    of developing ear, 123
    otic placode induction and, 122–23
    in zebrafish otic placode induction, 122–23
Middle ear
    anatomy of, 83–87, 84f
    development of, 235–37
    embryological development of, 84–87, 86f
    FGF signaling in, 235–37
    first pharyngeal pouch and, 102
    function of, 103–5
    genetic determinants of, 94–105
    muscles/cavity of, 102–3
    ossicle development of, 94–102
    ossicles of, 87
    tympanic membrane and, 103
Molecular conservation
    in vertebrate ear, 1–35

Index

Molecular inducers
  of inner ear, 118–24, 128–30
  tissue sources and, 134–36
Molecular markers
  during otic neurogenesis, 181t
  otic placode of, 117
Molecular mechanisms
  of inner ear placode, 154
  of outer/middle ear formation, 83–106
Morphodynamic mechanisms, 2
  evolutionarily conserved, 5
Morphogenesis, 2
  neurosensory processes and, 4
  pattern formation and, 165–70
Morphogens
  retinoids as, 271–78
Morphological conservation
  of hearing organs, 5
Morphological evolution, 5
Morphological/immunocytochemical conversion
  of GER cells, 299f
Morphostatic cascade, 8
Morphostatic mechanisms, 2
Mouse, tasmanian devil
  inner hair cell bundles from, 64f
Mouse ear development
  FGFs/FGFRs in, 234f
Mouse genetics
  definition of, 47–48
  genome sequence and, 46
  homeobox genes in, 156f
  human genetics and, 47
  inner ear development and, 45–75
  otic ectoderm molecular markers of, 135t
Mouse hair cell bundles
  at P4, 63f
Mouse inner ear
  components of, 52f
Mouse inner hair cells
  developmental changes in, 418f
Mouse utricular hair cells
  transduction properties in, 406f
MP_e. *See* Multipotent precursor
Msx1, 99
Multipotent precursor (MP_e), 187
Mutagenesis screens
  chemical/radiation-induced, 45
Mutant phenotypes
  of inner ear, 88t–89t
  of outer ear, 88t–89t
Myosin
  stereociliary bundles of, 62
Myosin VIIa
  in fly scolopidial attachment, 21

N

Nb_g. *See* Ganglionar neuroblast
Negative regulator
  Hes1 gene as, 302–8
Nerve growth factor (NGF), 177, 179
  inner ear neurogenesis and, 190–92
Neural cell adhesion molecule (NCAM), 323–26
Neural crest development, 87
  controlling factors of, 95–96
  ossicle development and, 95–96
Neural defects
  neurogenesis and, 73–74
Neural determination/differentiation
  cellular/molecular information and, 179
Neurite growth
  adhesion molecules and, 342–43
  inner ear in, 342–43
NeuroD family, 26
  of bHLH genes, 19
  null mutation of, 19, 20f, 25
Neuroectoderm
  otocyst and, 274–78
  RA and, 274–78
Neuroepithelium
  development of, 55–75
Neurogenesis
  neural defects and, 73–74
Neurogenic placodes, 179
Neurogenin-like genes, 22
Neurogenin1 (NGN1)
  sensory epithelia and, 377
Neurogenins
  Math1 gene and, 23
Neuron populations
  hair cells and, 33–34
Neuronal losses
  in inner ear, 212t
Neuronal precursor cells, 13
Neurosensory evolution
  of ear, 33
Neurosensory processes
  morphogenesis and, 4
Neurotrophic factors
  definition of, 208

Neurotrophic factors (*cont.*)
  in inner ear development, 207–19
  of NT-3, 29
Neurotrophic hypothesis
  neurotrophin system and, 208–9
Neurotrophin-3 (NT-3), 207
  in embryonic cochlea, 211f
  inner ear phenotype of, 212–14
  in inner ear sensory epithelia, 210f
  neurotrophic factors of, 29
Neurotrophin brain-derived neurotrophic factor (BDNF), 15f
Neurotrophin system, 208–17
  neurotrophic hypothesis and, 208–9
  p75 receptor and, 217
  Trk receptors and, 209
Neurotrophins (NTs), 27, 30, 177, 207
  afferent innervation and, 429–30
  cell survival loop and, 29
  deficiency of, 27
  in early developing ear, 214–15
  neurotrophin receptor null mutations and, 31, 32f
  during postnatal inner ear development, 215–16
NGF. *See* Nerve growth factor
Ngn1 gene, 18, 22, 23, 24
  loss of, 18f
  null mutation of, 16f
Nkx5-1 gene, 166–68
Nkx5-1 gene expression
  otic vesicle formation and, 155f
  patterns of, 160
Nodal-fibroblast growth factor (FGF) interactions
  in zebrafish, 365–67
Nodal genes
  in zebrafish, 365–67
Non-Hox genes, 158
Notch
  as adhesion molecules, 329–30
  gene expression, 55
Notch ligand Jagged1 gene, 51
Notch signaling
  hair cell differentiation and, 295–96
  pathway, 45
NT-3. *See* Neurotrophin-3
NTs. *See* Neurotrophins
Null mutant analyses
  of Foxi1 gene, 7
Null mutation
  of Math1, 16, 17f
  of Ngn1 gene, 16f
Numb & Prospero factors, 21

O

Octavolateralis hypothesis, 4
Oep gene function. *See* One-eyed pinhead gene function
One-eyed pinhead (Oep) gene function, 122
ONs. *See* Otic neurons
Organ of Corti, 55, 60
  of adult mouse, 56f
Ossicle development
  branchial arch skeleton and, 94
  endochondral ossification and, 94
  of middle ear, 94–102
  neural crest development and, 95–96
Otic capsule
  morphogenesis of, 271–74
  retinoids and, 271–74
Otic commitment
  molecular basis of, 151–71
Otic development
  in zebrafish, 359
Otic ectoderm, 125
Otic ectoderm molecular markers
  of chicks, 130t
  of mice, 135t
  of Xenopus, 128t
  of zebrafish, 127t
Otic ganglion formation
  FGF signaling in, 243–47
  FGF3 and, 243–45
Otic induction
  dlx3b and, 364–65
  dlx4b and, 364–65
  FGF3 and, 360–61
  FGF8 and, 360–61
  foxi1 and, 362–64
  pax2/5/8 genes and, 362
  pou2 and, 361–62
  wnt8 and, 361
  in zebrafish, 359–67
Otic neurogenesis, 180f
  cell states during, 181t
  molecular markers during, 181t

Index

Otic neuronal precursors
  delamination of, 185
Otic neurons (ONs), 187
  early development of, 177–97
  generation of, 178
  growth factors of, 177–97
  sequential formation of, 178
Otic placode, 1
  development of, 1
  formation of, 237–47
  induction, 122–23, 129, 136, 237–42, 269–71
  molecular markers of, 117
  otic vesicle and, 118
Otic placode formation
  ectopic FGF expression on, 238–39
  FGF signal blocking and, 239–40
  FGFR2b and, 247
  preplacodal ectoderm and, 238–39
Otic placode induction
  mesendodermal tissue and, 122–23
  RA and, 269–71
Otic tissue, presumptive
  gene expression in, 124–26, 130–31, 133–34
  in receptor targets, 136–38
Otic vesicle
  cell differentiation from, 178
  formation of, 242
  gene identification and, 154
  morphogenesis and, 242–43
  Nkx5-1 gene expression and, 155f
  otic ganglion formation and, 242–43
  otic placode and, 118
  specification of, 154
Otoconia, 4
Otocyst
  adhesion molecules and, 330–34
  descent of, 24, 25
  development of, 277f
  early compartmentalization of, 45
  expressed gene mutations in, 51
  gene expression domains in, 161f
  knockout ears and, 167f
  neuroectoderm and, 274–78
  patterning genes of, 167f
  retinoid pathways and, 277f
Otolith mutants
  vestibular function and, 378–79
Outer ear
  anatomy of, 83–87, 84f
  ear pinna development and, 87–92
  embryological development of, 84–87, 86f
  external acoustic meatus and, 92–94
  FGF signaling in, 232–34
  genetic determinants of, 87–105
  molecular mechanisms of, 83–106
  mutant phenotypes of, 88t–89t
Outer hair cells
  FGF signaling and, 248–49
  as force generators, 68–69
  hair cell differentiation and, 68–69
  sensory epithelia development and, 248–49
Outward rectifiers
  in chick cochlear hair cells, 407–12
  in chick vestibular hair cells, 415–16
  in rodent cochlear hair cells, 412–14
  in rodent vestibular hair cells, 416–18
Oval window formation, 104–5

P

P75 receptor
  neurotrophin system and, 217
Paraffin sectioning
  extra cochlear inner hair cells with, 304f
Pattern formation
  FGF control induction and, 162–64
  homeobox-containing genes and, 157–59
  morphogenesis and, 165–70
Patterning defects
  of acousticolateral placode, 369f
Patterning genes
  inner ear development and, 78–80
  RA and, 278–80
Pax genes, 19, 124
Pax2/5/8 genes
  otic induction and, 362
Pax2 genes, 45, 166
Pax2a
  sensory epithelia and, 377–78
Pax2b
  sensory epithelia and, 377–78
Pax8 gene, 121
PCP. *See* Planar cell polarity
Pendred syndrome, 45
Peripheral fiber projections, 25
Pillar cells
  FGF signaling and, 249–51
Placodal ectoderm
  thickening of, 117
Placodal origin
  of sensory neurons, 179

Placode patterning
  Eya1 gene and, 371–72
  jekyll gene and, 373
  sox10 gene and, 372–73
  van gogh gene and, 372
  in zebrafish, 367–73
Planar cell polarity (PCP), 60
  of *Drosophila,* 62
Postnatal inner ear development
  NTs during, 215–16
Potassium conductances
  inwardly rectifying, 419–21
  outwardly rectifying, 405–18
Pou-domain-containing transcription factor, 28
Pou2
  otic induction and, 361–62
Pou4f3 gene, 71–72
Prechordal mesendoderm, 123
Preneural genes, 13, 14
Preplacodal ectoderm, 140
  ectopic FGF expression and, 238–39
  ectopic fibroblast growth factor (FGF) expression and, 238–39
  otic placode formation and, 238–39
Prestin gene, 75
Presumptive otic tissue
  gene expression in, 124–26, 130–31, 133–34
  in receptor targets, 136–38
Preyer/s reflex test, 49–50
Primary afferents
  hair cells and, 428–30
Primary neuron primordia, 25
Primary sensory neurons
  with acquired identities, 26
Progenitor cells
  hair cells and, 57–59
Proliferation/apoptosis
  semicircular canal formation and, 168–70
Proneural gene Ngn1
  in inner ear neurogenesis, 185
Prx1 gene, 101
  tympanic ring and, 92, 93
Prx1/Prx2
  as homeobox genes, 53
Ptx1 gene, 101

R

RA. *See* Retinoic acid
Random mutagenesis approach, 48
Receptor targets
  in presumptive otic tissue, 136–38
Reissner/s membrane, 9
Retinoic acid (RA), 53
  action mechanism of, 263f
  embryopathy of, 265
  hair cell differentiation and, 280–82
  during inner ear development, 278–80
  inner ear development and, 261–83
  neuroectoderm and, 274–78
  otic placode induction and, 269–71
  patterning genes and, 278–80
  pleiotropic functions of, 261
  signaling, 97
  synthesis pathways of, 263f
Retinoic acid (RA) homeostasis, 268–69
Retinoic acid (RA) metabolic enzymes
  inner ear receptors and, 265–69
Retinoic acid (RA) synthesis
  synthetic/catabolic enzymes and, 267
Retinoid pathways
  otocyst development and, 277f
Retinoid receptor transcripts
  in embryonic ear, 266f
Retinoids
  binding of, 264
  metabolism of, 263–65
  as morphogens, 271–78
  otic capsule and, 271–74
  receptors of, 263–65
Retinol, 263
  conversion of, 264
Rodent cochlear hair cells
  outward rectifiers in, 412–14
Rodent hearing
  maturation of, 390–94
Rodent utricular hair cells
  outwardly rectifying current changes in, 420f
Rodent vestibular function, 393–94
Rodent vestibular hair cells
  outward rectifiers in, 416–18
Rodents
  auditory function in, 390
  cochlear hair cells in, 412–14
  hair bundles of, 398–400

Index

S

Saccule, 4
Salps
  atrial chambers of, 5
  development of, 23
SC-1
  as adhesion molecules, 329
SE gene. *See* Short ear gene
Sea squirt. *See* Trunicate Ciona
Secreted factors/receptors
  in branchial arch skeletal development, 96–98
Segregated sensory patches
  incomplete segregation of, 34
Selective advantage, 1
Semaphorins, 27
Semicircular canal formation
  BMP signals and, 164–65
  morphogenesis of, 169f
  proliferation/apoptosis and, 168–70
Sensory cells
  in atrium, 22
  of diploblasts, 12f
  formation of, 5
  in hemichordates, 11
  in invertebrates, 11
  of Mollusca, 29
Sensory epithelia
  adhesion molecules and, 334–42
  deltaA genes and, 375
  development of, 247–51, 373–78
  e3 ubiquitin ligase and, 375–76
  of ear, 33
  hair cells in, 391f
  morphogenesis of, 373–74
  NGN1 and, 377
  pax2a/pax2b and, 377–78
  sensory neurons and, 24
  zath1 and, 376–77
  in zebrafish, 373–78
Sensory epithelia development
  in FGF signaling, 247–51
  hair cell precursors and, 248–49
  in inner ear, 247–51
  pillar cells and, 249–51
Sensory evolution, 5
Sensory neurons
  evolution of, 23–24
  formation of, 13
  placodal origin of, 179
  precursor cells of, 24
  sensory epithelia and, 24
Sensory patch primordia, 55
  hair cell fate determination and, 302
  Math1 gene and, 302
Sensory patches
  affecting mutations of, 54–55
  formation of, 56–57
  hair cell patterning in, 57–59
  specification of, 54–55
Shh gene. *See* Sonic hedgehog gene
Short ear (SE) gene, 90
Signaling molecules, 151–71
Sine oculis (SIX) genes, 8, 9
SIX. *See* Sine oculis genes
Sodium conductances
  voltage-gated, 421–23, 424f
SOHO gene expression
  patterns of, 160
Sonic hedgehog (Shh) gene
  homeobox genes and, 51
  null mutations in, 53
Sox9a gene, 125
Sox10 gene
  placode patterning and, 372–73
  vesicle patterning and, 372–73
Specific growth factors
  cell cycle regulators and, 309–10
  hair cell differentiation and, 309–10
  hair cell progenitors and, 309–10
Spontaneous activity
  hearing function and, 395
Sprouty genes, 241
Staircase formation
  of hair bundle, 401
Stapes, 104
Statocyst hypothesis, 5
Stereocilia bundles, 59–68
  adhesion molecules and, 343–45
  development of, 343–45
  hair bundle polarity and, 60–65
  hair cell adaptation and, 66–68
  hair cell differentiation and, 59–68
  of immature hair cell, 59, 60f
  of myosin, 62
  transduction channel and, 65–66
Stria vascularis, 9
Styloid process, 86–87
Synthetic/catabolic enzymes
  in embryonic ear, 266f
  in RA synthesis, 267

T

Targeted gene deletion
 of Math1 gene, 297
Targeted mutagenesis, 48, 75
Tasmanian devil mice
 inner hair cell bundles from, 64f
Tectorial membranes, 4
TGFβ-related genes, 20–21
Tip link formation
 of hair bundle, 401–2
Tissue sources
 molecular inducers and,
  134–36
Transcription factor(s)
 in branchial arch skeletal development,
  98–102
 diffusible signals and,
  162–65
 genes, 45
 hair cells and, 310
 link for, 5
 Pax6 as, 2
Transcription regulators
 in Drosophila, 157
Transduction channel
 stereociliary bundle and,
  65–66
Transduction properties
 in mouse utricular hair cells, 406f
Transforming proneuronal clusters, 24
Transgenetic mouse technology, 48
Transit-amplifying cell population, 186
Trk. *See* Tyrosine kinase
Truncated tyrosine kinase
  Trk) receptors
 during inner ear development, 216
Trunicate Ciona, 22
Tunicate atrial chambers
 as hydrodynamic sensors, 5
Tympanic membrane
 construction of, 103
 middle ear and, 103
Tympanic ring
 Gsc gene and, 92, 93
 Hoxa2 gene and, 93
 Prx1 gene and, 92, 93
Tyrosine kinase (Trk) induction
 in early developing ear, 214 15
Tyrosine kinase (Trk) null
  mutant mice
 inner ear phenotype of, 212–14

Tyrosine kinase (Trk) receptors
 in embryonic inner ear, 209–12
 neurotrophin system and, 209
 truncated, 216

U

Utricles
 hair cell differentiation in, 301–2
 hair cell production in, 303–4

V

Van gogh gene
 placode patterning and, 372
 vesicle patterning and, 372
Vertebrate ear
 evolution of, 1–35
 mechanoelectrical transduction and, 4
 molecular conservation in, 1–35
Vertebrate hair cell development,
  22–23
Vertebrate inner ear
 structure of, 321–23
Vertebrate neurogenesis
 scheme for, 181–84
Vertebrates, higher
 hair cell development,
  293–314
Vesicle patterning
 eya1 gene and, 371–72
 jekyll gene and, 373
 sox10 gene and, 372–73
 van gogh gene and, 372
Vestibular fibers
 misrouting of, 30
Vestibular function
 in chicks, 394
 circler mutants and, 378
 otolith mutants and, 378–79
 in rodents, 393–94
 in zebrafish, 378–81, 394
Vestibular neurons
 early development of,
  179–87
Vestibular organ
 evolution of, 4
Vestibular system innervation
 cochlea and, 20f
Vitamin A, 263, 265
 reproduction and, 265

Index

W

Winged helix genes. *See* Forkhead genes
Wnt1/Wnt3a genes, 95, 96
Wnt8
   otic induction and, 361

X

Xenopus
   otic ectoderm molecular markers of, 128t
Xenopus laevis, 126, 127

Z

Zath1
   sensory epithelia and, 376–77
Zebrafish, 7
   auditory function in, 378–81
   gene expression in, 124–25
   gene identification of, 358
   hair bundle in, 400–401
   hair cell differentiation in, 374f
   hearing function in, 394
   hearing mutant screen for, 380
   inner ear development of, 357–81
   nodal-FGF interactions in, 365–67
   nodal genes in, 365–67
   otic development in, 359
   otic ectoderm molecular markers of, 127t
   otic induction in, 359–67
   otic placode induction and, 122–23
   placode patterning in, 367–73
   sensory epithelia in, 373–78
   vesicle patterning in, 367–73
   vestibular function in, 378–81, 394
Zebrafish ear development
   nomenclature/staging of, 358
Zic-genes, 14

Contents of Previous Volumes

Volume 47

1. **Early Events of Somitogenesis in Higher Vertebrates: Allocation of Precursor Cells during Gastrulation and the Organization of a Moristic Pattern in the Paraxial Mesoderm**
   Patrick P. L. Tam, Devorah Goldman, Anne Camus, and Gary C. Shoenwolf

2. **Retrospective Tracing of the Developmental Lineage of the Mouse Myotome**
   Sophie Eloy-Trinquet, Luc Mathis, and Jean-François Nicolas

3. **Segmentation of the Paraxial Mesoderm and Vertebrate Somitogenesis**
   Olivier Pourqulé

4. **Segmentation: A View from the Border**
   Claudio D. Stern and Daniel Vasiliauskas

5. **Genetic Regulation of Somite Formation**
   Alan Rawls, Jeanne Wilson-Rawls, and Eric N. Olsen

6. **Hox Genes and the Global Patterning of the Somitic Mesoderm**
   Ann Campbell Burke

7. **The Origin and Morphogenesis of Amphibian Somites**
   Ray Keller

8. **Somitogenesis in Zebrafish**
   Scott A. Halley and Christiana Nüsslain-Volhard

9. **Rostrocaudal Differences within the Somites Confer Segmental Pattern to Trunk Neural Crest Migration**
   Marianne Bronner-Fraser

Volume 48

1 Evolution and Development of Distinct Cell Lineages Derived from Somites
Beate Brand-Saberi and Bodo Christ

2 Duality of Molecular Signaling Involved in Vertebral Chondrogenesis
Anne-Hélène Monsoro-Burq and Nicole Le Douarin

3 Sclerotome Induction and Differentiation
Jennifer L. Docker

4 Genetics of Muscle Determination and Development
Hans-Henning Arnold and Thomas Braun

5 Multiple Tissue Interactions and Signal Transduction Pathways Control Somite Myogenesis
Anne-Gaëlle Borycki and Charles P. Emerson, Jr.

6 The Birth of Muscle Progenitor Cells in the Mouse: Spatiotemporal Considerations
Shahragim Tajbakhsh and Margaret Buckingham

7 Mouse–Chick Chimera: An Experimental System for Study of Somite Development
Josiane Fontaine-Pérus

8 Transcriptional Regulation during Somitogenesis
Dennis Summerbell and Peter W. J. Rigby

9 Determination and Morphogenesis in Myogenic Progenitor Cells: An Experimental Embryological Approach
Charles P. Ordahl, Brian A. Williams, and Wilfred Denetclaw

Volume 49

1 The Centrosome and Parthenogenesis
Thomas Küntziger and Michel Bornens

2 γ-Tubulin
Berl R. Oakley

3 γ-Tubulin Complexes and Their Role in Microtubule Nucleation
Ruwanthi N. Gunawardane, Sofia B. Lizarraga, Christiane Wiese, Andrew Wilde, and Yixian Zheng

4 γ-Tubulin of Budding Yeast
Jackie Vogel and Michael Snyder

5 The Spindle Pole Body of *Saccharomyces cerevisiae:* Architecture and Assembly of the Core Components
Susan E. Francis and Trisha N. Davis

6 The Microtubule Organizing Centers of *Schizosaccharomyces pombe*
Iain M. Hagan and Janni Petersen

7 Comparative Structural, Molecular, and Functional Aspects of the *Dictyostelium discoideum* Centrosome
Ralph Gräf, Nicole Brusis, Christine Daunderer, Ursula Euteneuer, Andrea Hestermann, Manfred Schliwa, and Masahiro Ueda

8 Are There Nucleic Acids in the Centrosome?
Wallace F. Marshall and Joel L. Rosenbaum

9 Basal Bodies and Centrioles: Their Function and Structure
Andrea M. Preble, Thomas M. Giddings, Jr., and Susan K. Dutcher

10 Centriole Duplication and Maturation in Animal Cells
B. M. H. Lange, A. J. Faragher, P. March, and K. Gull

11 Centrosome Replication in Somatic Cells: The Significance of the G_1 Phase
Ron Balczon

12 The Coordination of Centrosome Reproduction with Nuclear Events during the Cell Cycle
Greenfield Sluder and Edward H. Hinchcliffe

13 Regulating Centrosomes by Protein Phosphorylation
Andrew M. Fry, Thibault Mayor, and Erich A. Nigg

14 The Role of the Centrosome in the Development of Malignant Tumors
Wilma L. Lingle and Jeffrey L. Salisbury

15 The Centrosome-Associated Aurora/Ipl-like Kinase Family
T. M. Goepfert and B. R. Brinkley

16 Centrosome Reduction during Mammalian Spermiogenesis
   G. Manandhar, C. Simerly, and G. Schatten

17 The Centrosome of the Early *C. elegans* Embryo: Inheritance, Assembly, Replication, and Developmental Roles
   Kevin F. O'Connell

18 The Centrosome in *Drosophila* Oocyte Development
   Timothy L. Megraw and Thomas C. Kaufman

19 The Centrosome in Early *Drosophila* Embryogenesis
   W. F. Rothwell and W. Sullivan

20 Centrosome Maturation
   Robert E. Palazzo, Jacalyn M. Vogel, Bradley J. Schnackenberg, Dawn R. Hull, and Xingyong Wu

Volume 50

1 Patterning the Early Sea Urchin Embryo
   Charles A. Ettensohn and Hyla C. Sweet

2 Turning Mesoderm into Blood: The Formation of Hematopoietic Stem Cells during Embryogenesis
   Alan J. Davidson and Leonard I. Zon

3 Mechanisms of Plant Embryo Development
   Shunong Bai, Lingjing Chen, Mary Alice Yund, and Zinmay Rence Sung

4 Sperm-Mediated Gene Transfer
   Anthony W. S. Chan, C. Marc Luetjens, and Gerald P. Schatten

5 Gonocyte–Sertoli Cell Interactions during Development of the Neonatal Rodent Testis
   Joanne M. Orth, William F. Jester, Ling-Hong Li, and Andrew L. Laslett

6 Attributes and Dynamics of the Endoplasmic Reticulum in Mammalian Eggs
   Douglas Kline

7 Germ Plasm and Molecular Determinants of Germ Cell Fate
   Douglas W. Houston and Mary Lou King

Volume 51

1. **Patterning and Lineage Specification in the Amphibian Embryo**
   Agnes P. Chan and Laurence D. Etkin

2. **Transcriptional Programs Regulating Vascular Smooth Muscle Cell Development and Differentiation**
   Michael S. Parmacek

3. **Myofibroblasts: Molecular Crossdressers**
   Gennyne A. Walker, Ivan A. Guerrero, and Leslie A. Leinwand

4. **Checkpoint and DNA-Repair Proteins Are Associated with the Cores of Mammalian Meiotic Chromosomes**
   Madalena Tarsounas and Peter B. Moens

5. **Cytoskeletal and Ca^{2+} Regulation of Hyphal Tip Growth and Initiation**
   Sara Torralba and I. Brent Heath

6. **Pattern Formation during *C. elegans* Vulval Induction**
   Minqin Wang and Paul W. Sternberg

7. **A Molecular Clock Involved in Somite Segmentation**
   Miguel Maroto and Olivier Pourquié

Volume 52

1. **Mechanism and Control of Meiotic Recombination Initiation**
   Scott Keeney

2. **Osmoregulation and Cell Volume Regulation in the Preimplantation Embryo**
   Jay M. Baltz

3. **Cell–Cell Interactions in Vascular Development**
   Diane C. Darland and Patricia A. D'Amore

4. **Genetic Regulation of Preimplantation Embryo Survival**
   Carol M. Warner and Carol A. Brenner

Volume 53

1. **Developmental Roles and Clinical Significance of Hedgehog Signaling**
   Andrew P. McMahon, Philip W. Ingham, and Clifford J. Tabin

2. **Genomic Imprinting: Could the Chromatin Structure Be the Driving Force?**
   Andras Paldi

3. **Ontogeny of Hematopoiesis: Examining the Emergence of Hematopoietic Cells in the Vertebrate Embryo**
   Jenna L. Galloway and Leonard I. Zon

4. **Patterning the Sea Urchin Embryo: Gene Regulatory Networks, Signaling Pathways, and Cellular Interactions**
   Lynne M. Angerer and Robert C. Angerer

Volume 54

1. **Membrane Type-Matrix Metalloproteinases (MT-MMP)**
   Stanley Zucker, Duanqing Pei, Jian Cao, and Carlos Lopez-Otin

2. **Surface Association of Secreted Matrix Metalloproteinases**
   Rafael Fridman

3. **Biochemical Properties and Functions of Membrane-Anchored Metalloprotease-Disintegrin Proteins (ADAMs)**
   J. David Becherer and Carl P. Blobel

4. **Shedding of Plasma Membrane Proteins**
   Joaquín Arribas and Anna Merlos-Suárez

5. **Expression of Meprins in Health and Disease**
   Lourdes P. Norman, Gail L. Matters, Jacqueline M. Crisman, and Judith S. Bond

6. **Type II Transmembrane Serine Proteases**
   Qingyu Wu

7. **DPPIV, Seprase, and Related Serine Peptidases in Multiple Cellular Functions**
   Wen-Tien Chen, Thomas Kelly, and Giulio Ghersi

8 The Secretases of Alzheimer's Disease
  Michael S. Wolfe

9 Plasminogen Activation at the Cell Surface
  Vincent Ellis

10 Cell-Surface Cathepsin B: Understanding Its Functional Significance
  Dora Cavallo-Medved and Bonnie F. Sloane

11 Protease-Activated Receptors
  Wadie F. Bahou

12 Emmprin (CD147), a Cell Surface Regulator of Matrix Metalloproteinase Production and Function
  Bryan P. Toole

13 The Evolving Roles of Cell Surface Proteases in Health and Disease: Implications for Developmental, Adaptive, Inflammatory, and Neoplastic Processes
  Joseph A. Madri

14 Shed Membrane Vesicles and Clustering of Membrane-Bound Proteolytic Enzymes
  M. Letizia Vittorelli

Volume 55

1 The Dynamics of Chromosome Replication in Yeast
  Isabelle A. Lucas and M. K. Raghuraman

2 Micromechanical Studies of Mitotic Chromosomes
  M. G. Poirier and John F. Marko

3 Patterning of the Zebrafish Embyro by Nodal Signals
  Jennifer O. Liang and Amy L. Rubinstein

4 Folding Chromosomes in Bacteria: Examining the Role of Csp Proteins and Other Small Nucleic Acid-Binding Proteins
  Nancy Trun and Danielle Johnston

Volume 56

1. **Selfishness in Moderation: Evolutionary Success of the Yeast Plasmid**
   Soundarapandian Velmurugan, Shwetal Mehta, and Makkuni Jayaram

2. **Nongenomic Actions of Androgen in Sertoli Cells**
   William H. Walker

3. **Regulation of Chromatin Structure and Gene Activity by Poly(ADP-Ribose) Polymerases**
   Alexei Tulin, Yurli Chinenov, and Allan Spradling

4. **Centrosomes and Kinetochores, Who needs 'Em? The Role of Noncentromeric Chromatin in Spindle Assembly**
   Priya Prakash Budde and Rebecca Heald

5. **Modeling Cardiogenesis: The Challenges and Promises of 3D Reconstruction**
   Jeffrey O. Penetcost, Claudio Silva, Maurice Pesticelli, Jr., and Kent L. Thornburg

6. **Plasmid and Chromosome Traffic Control: How ParA and ParB Drive Partition**
   Jennifer A. Surtees and Barbara E. Funnell

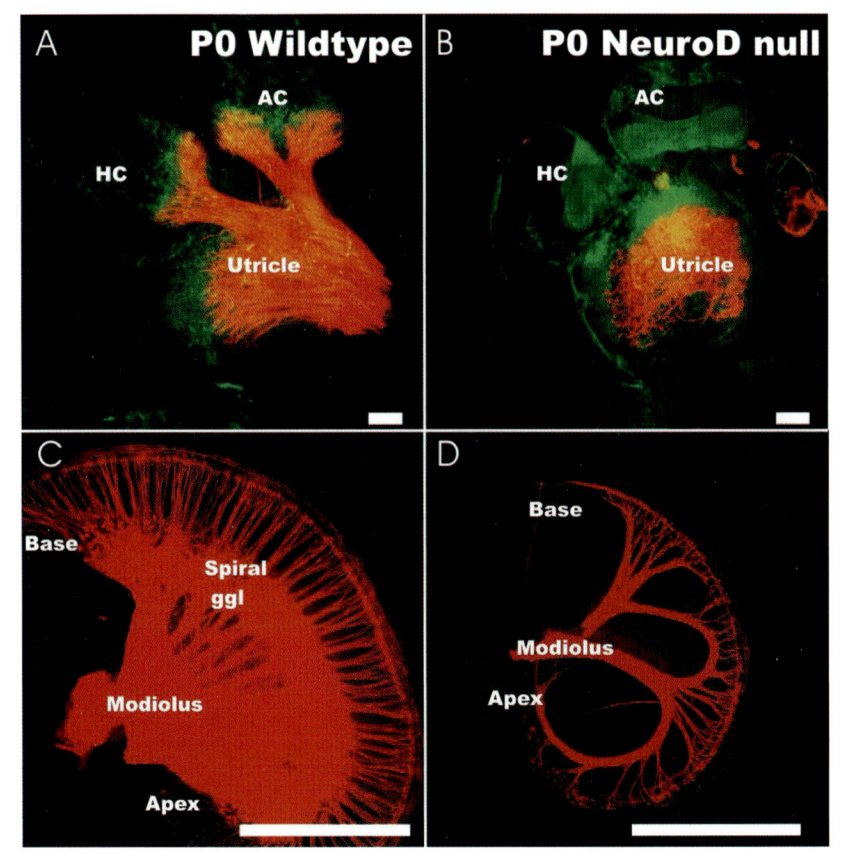

Chapter 1, Figure 7 The effect of a NeuroD null mutation on the pattern of innervation of the cochlea and vestibular system is revealed with DiI tracing and confocal microscopy. All vestibular and cochlear epithelia show hair cells, but some hair cells have no innervation. In the vestibular system, the utricle shows a somewhat disorganized innervation in the NeuroD mutant (B) but may lack all innervation of canal cristae (A, B). The cochlea shows a reduction in size and a loss of most spiral sensory neurons (C, D). The remaining neurons expand their peripheral processes to innervate most inner hair cells and a few outer hair cells. Bar indicates 1 mm (C, D) and 100 μm (A, B).

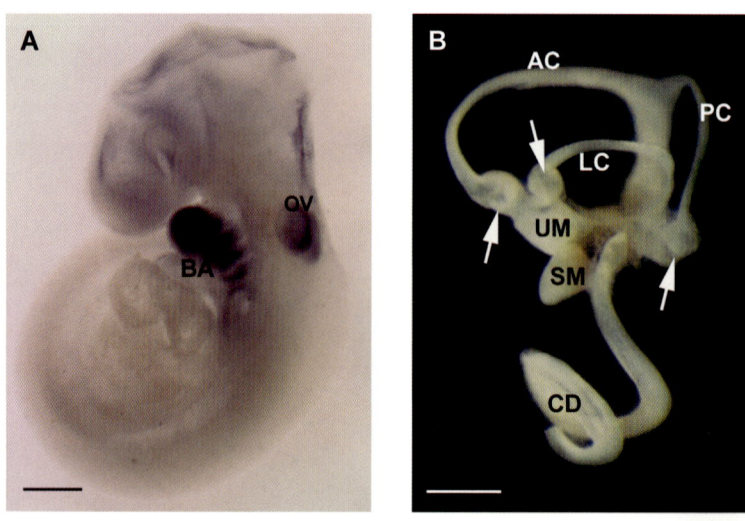

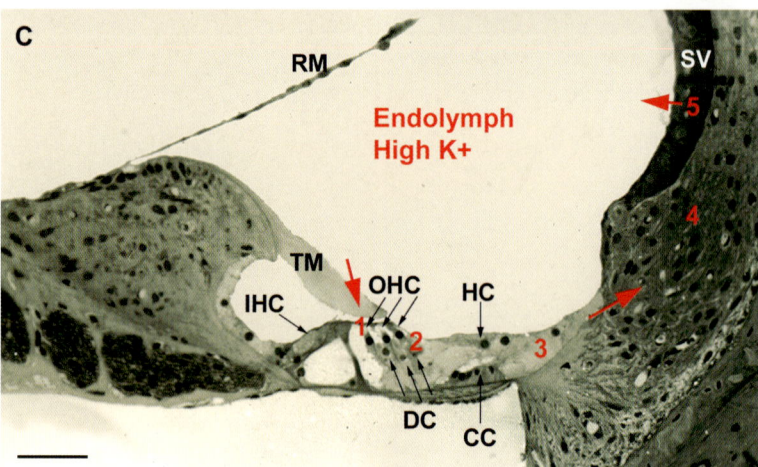

Chapter 2, Figure 1 Components of the mouse inner ear. (A) RNA *in situ* hybridization of a mouse at E9.5 showing *Dlx5* expression in the otic vesicle (OV) and branchial arches (BA). Anterior to the left: (B) Paint-filled inner ear from a P0 mouse. The cochlear portion contains the cochlear duct (CD), whereas the vestibular portion contains the saccular macula (SM); utricular macula (UM); and lateral, posterior, and anterior semicircular canals (LC, PC, AC). Arrows point to the ampullae, containing the cristae, which are also part of the vestibular system. Anterior to the left, ventral down: (C) Semithin section of the cochlear duct from an adult mouse, stained with toluidine blue. RM, Reissner's membrane; SV, stria vascularis; TM, tectorial membrane; IHC, inner hair cell; OHC, outer hair cell; DC, Deiter's cells; HC, Hensens's cells; CC, Claudius' cells. Medial to the left. Numbering illustrates the sequence of steps in the hypothesized potassium recycling pathway, referred to in the text in detail (see Section III.D). Scale bars = 500 μm (A and B), 25 μm, (C).

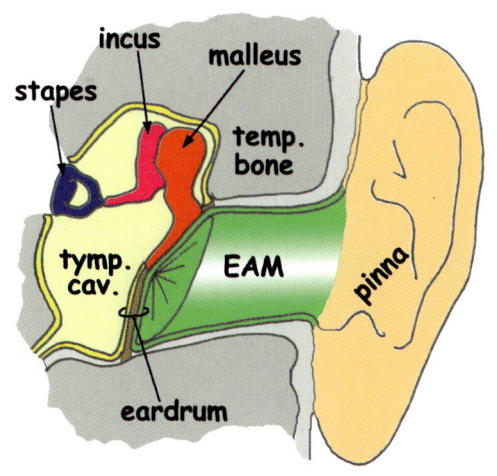

Chapter 3, Figure 1 Anatomy of the outer and middle ears. The outer ear is composed of the pinna and the external acoustic meatus (EAM). The eardrum (or tympanic membrane) is located at the end of the EAM. It is composed of three layers provided by the epithelia of the EAM and middle ear cavity and fibrous tissue between the two epithelia. The middle ear is located on the tympanic cavity (within the temporal bone). The main components of the middle ear are the ossicles: malleus, incus, and stapes. Other components, such as muscles, nerves, and arteries, are not shown for the sake of simplicity.

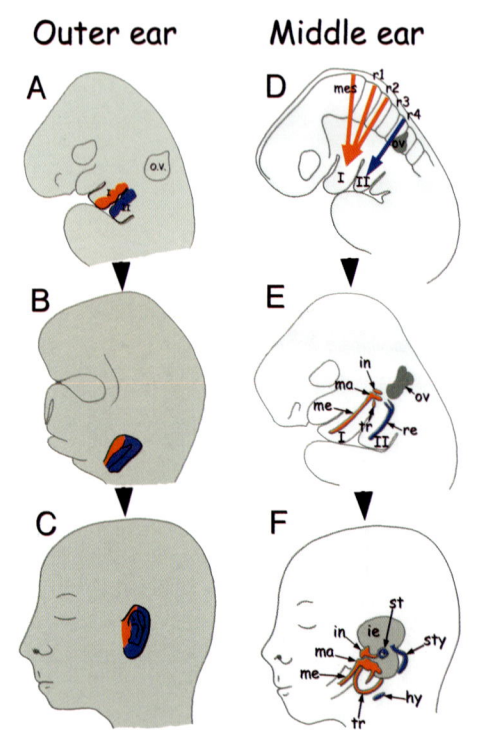

Chapter 3, Figure 2 Embryological development of the outer and middle ears. (A) The outer ear develops from hillocks on the first (I) and second (II) branchial arches (represented in red and blue, respectively). (B, C) As development proceeds, they give rise to specific areas of the pinna. (D) The middle ear ossicles develop from the cranial neural crest. Crest cells from the caudal mesencephalon (mes) and rhombomere 1 and 2 (r1 and r2), shown in red, populate the first branchial arch (I). r4 neural crest (blue) populates the second arch (II). (E, F) From the first arch develop the Meckel's cartilage (me), malleus (ma), incus (in), and tympanic ring (tr). Skeletal development in the second arch starts with Reichert's cartilage (re), which will eventually form the stapes (st), styloid process (sty), and lesser horn of the hyoid bone (hy). The otic vesicle (ov) will give rise to the inner ear (ie).

A   **FISH**

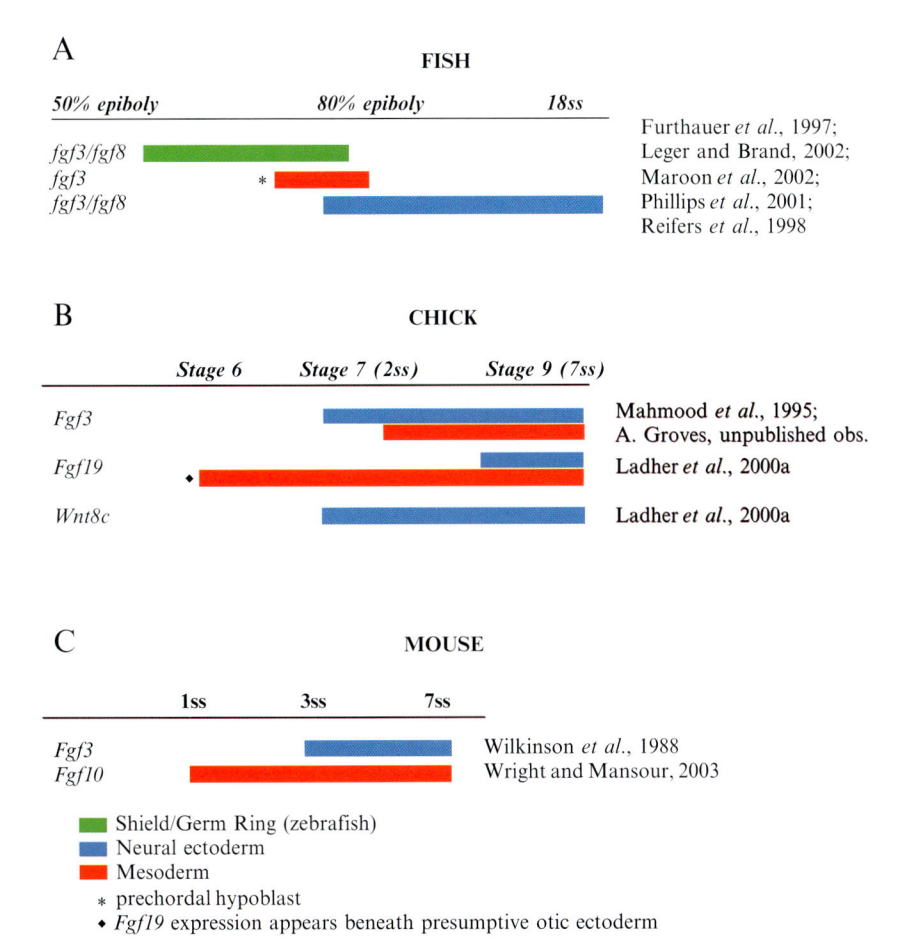

| 50% epiboly | 80% epiboly | 18ss |

fgf3/fgf8
fgf3
fgf3/fgf8

Furthauer et al., 1997;
Leger and Brand, 2002;
Maroon et al., 2002;
Phillips et al., 2001;
Reifers et al., 1998

B   **CHICK**

| Stage 6 | Stage 7 (2ss) | Stage 9 (7ss) |

Fgf3

Fgf19

Wnt8c

Mahmood et al., 1995;
A. Groves, unpublished obs.
Ladher et al., 2000a
Ladher et al., 2000a

C   **MOUSE**

| 1ss | 3ss | 7ss |

Fgf3
Fgf10

Wilkinson et al., 1988
Wright and Mansour, 2003

■ Shield/Germ Ring (zebrafish)
■ Neural ectoderm
■ Mesoderm
 * prechordal hypoblast
 ◆ *Fgf19* expression appears beneath presumptive otic ectoderm

Chapter 4, Figure 1 Summary of the temporal expression of candidate inner ear inducing molecules in zebrafish, chick, and mouse. The approximate duration of expression for each factor in potential ear-inducing tissues is shown relative to the indicated developmental stages. Factors of neural origin are represented by blue bars, factors of mesodermal origin are represented by red bars, and factors expressed in the zebrafish shield and germring are represented by a green bar. The location of *Fgf19* expression in chick changes progressively from mesoderm beneath the neural plate to mesoderm beneath presumptive otic ectoderm to the ventral neural tube. The approximate time of the initial appearance of *Fgf19* expression in mesoderm beneath presumptive otic ectoderm is indicated by a diamond(◆).

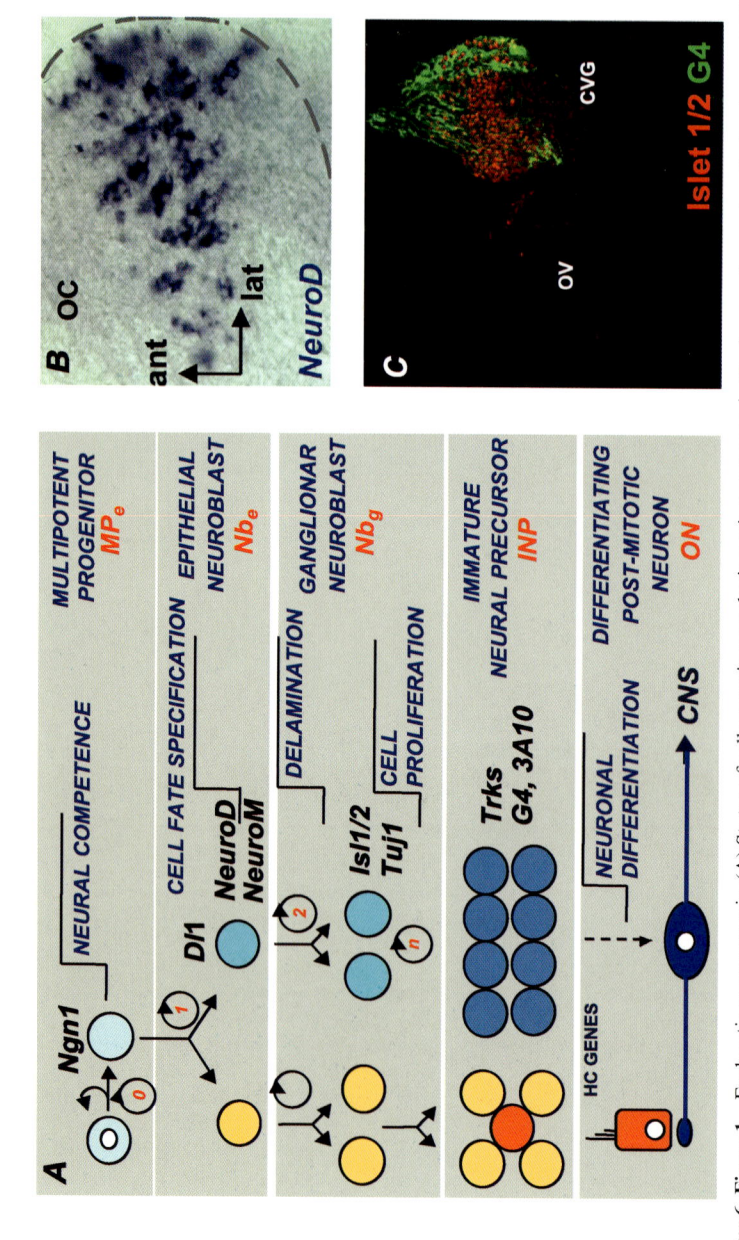

Chapter 6, Figure 1 Early otic neurogenesis. (A) States of cell commitment during otic neurogenesis. Early neural genes in otic development. The neural lineage is labeled in different tones of blue and the sensory lineage in different tones of yellow. (B) Epithelial neuroblasts singled out by NeuroD expression. *In situ* hybridization with a NeuroD probe in HH stage 14 chick otic cup. (C) Islet-1 and G4 expression in ganglionar neuroblasts. Double immunocytological recognition with an anti-Islet-1/2 antibody and an anti-G4 antibody.

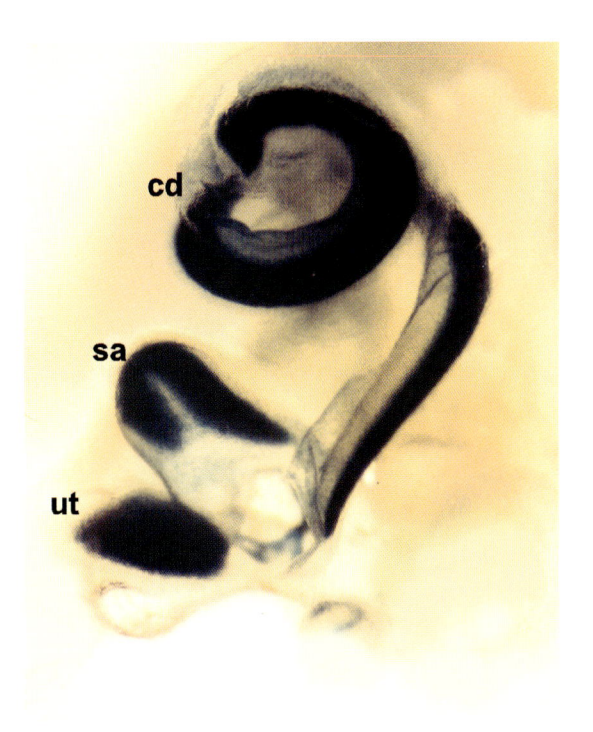

Chapter 7, Figure 1 *Nt-3* expression in the inner ear sensory epithelia of an *Nt-3* knockout mouse at birth as revealed by *lacZ* staining. We are grateful to Dr. L. F. Reichardt for donating the *Nt-3* mutant mouse line to us. Cd, cochlear duct; sa, saccular macula; ut, utricular macula.

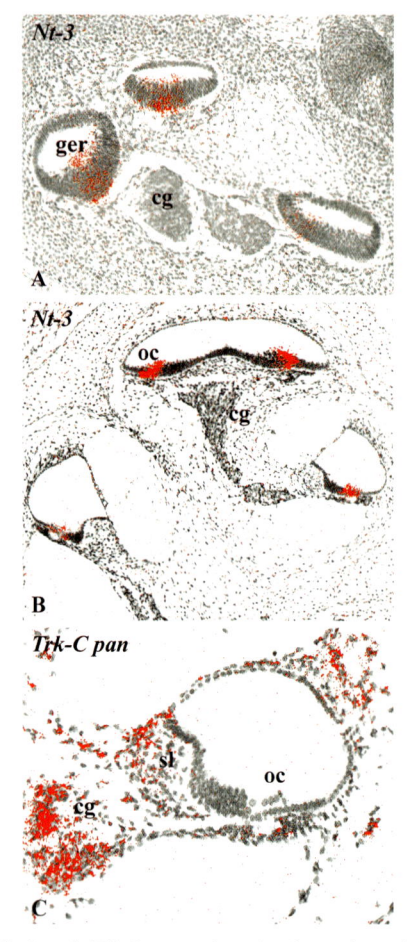

Chapter 7, Figure 2 *Nt-3* and *TrkC* expressions in the embryonic cochlea of the mouse as revealed by *in situ* hybridization. (A) At embryonic day 13.5, the greater epithelial ridge (ger) of the nascent cochlear duct shows *Nt-3* expression. (B) At birth, *Nt-3* is expressed in the organ of Corti (oc) of the cochlear duct. (C) The *pan-TrkC* probe detects both the catalytic and truncated forms of *TrkC*. Based on the use of isoform-specific probes, the full-length (catalytic) isoform is expressed exclusively in the neurons of the cochlear ganglion (cg). Truncated *TrkC* is expressed, in addition to the ganglion, in mesenchymal regions such as the spiral limbus (sl) and the domain next to the lateral epithelial wall.

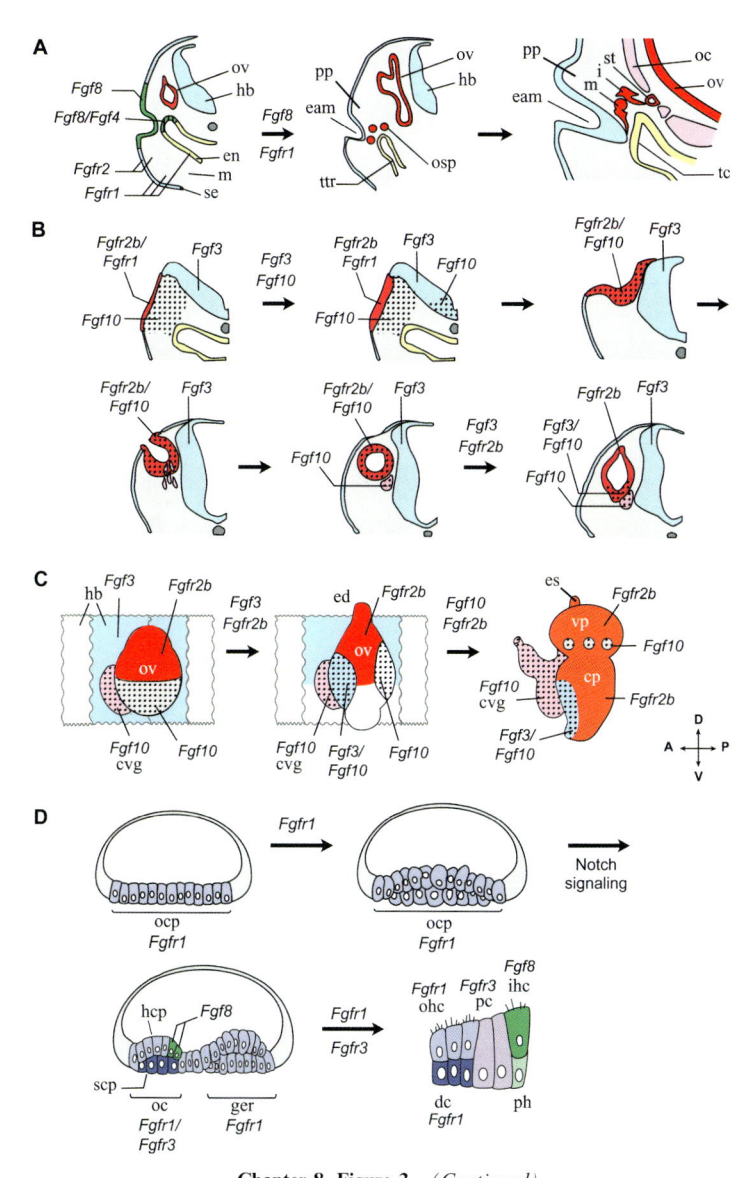

Chapter 8, Figure 3 *(Continued).*

Chapter 8, Figure 3 Summary of FGFs and FGFRs involved in mouse ear development. (A) Outer and middle ear development. The left diagram depicts the patterns of gene expression in the otic region immediately prior to the initiation of outer and middle ear development. *Fgfr1*, expressed throughout the surface ectoderm (se), mesenchyme (m), and endoderm (en), is required both for pinna and ossicle development. *Fgf8* (green highlight), expressed in the first branchial cleft of the se and in the pharyngeal pouch en, is also required for both pinna and ossicle development. *Fgf4*, expressed in the pharyngeal pouch en, is required for pinna development, but potential roles in middle ear development have not been examined. A role for *Fgfr2* in se and m has not been established but is suggested because FGF8 is not a good ligand for FGFR1 isoforms but does activate FGFR2c. The middle and right diagrams show the progressive FGF signal-dependent development of the outer and middle ears. Additional abbreviations: eam, external auditory meatus; hb, hindbrain; i, incus; m, malleus; oc, otic capsule; osp, ossicle precursors; ov, otic vesicle; p, pinna; pp, pinna primordium; st, stapes; tc, tympanic cavity; ttr, tubotympanic recess. (B) The earliest stages of inner ear development showing the expression patterns of *Fgf* and *Fgfr* genes required for otic vesicle formation. The otic placode, which thickens, invaginates and forms a closed vesicle, is shown in red. The neurons of the eighth ganglion, which derive from the otic epithelium, are shown in pink. The hindbrain and non-otic ectoderm are blue; the endoderm is yellow; and the mesenchyme is not colored, except in the region of *Fgf10* expression (dotted). Other regions of *Fgf10* expression are also dotted and retain their assigned colors. *Fgf3* and *Fgfr2b* are generally expressed throughout the tissues indicated, except at the last stage depicted, in which *Fgf3* is confined with *Fgf10* to the most ventral region of the vesicle, the rest of which expresses *Fgfr2b*. *Fgf3* expressed in the hindbrain and *Fgf10* expressed in the mesenchyme, acting through *Fgfr2b* (and potentially *Fgfr1*) expressed in the otic placode, are required redundantly for induction of the placode. All components of this signaling system continue to be expressed in changing patterns after the placode is induced, but specific roles for FGF signaling in invagination and vesicle formation have not been established. See text for a description of the roles of FGF signaling in the early stages of zebrafish development. (C) FGF signaling in the early morphogenesis of the otic vesicle. Lateral views of the otic vesicle (ov) and cochleovestibular ganglion (cvg, pink) from shortly after vesicle closure through endolymphatic duct (ed) induction and the earliest stages of semicircular canal formation. *Fgf3* (blue), expressed in the hindbrain (hb), acting through *Fgfr2b* (red) expressed in the dorsal half of the ov is required for ed induction. Later functions for *Fgf3* expressed in the anterior ov are possible but not yet explored. *Fgf10* (dotted) is expressed throughout this period in the cvg but does not play a unique role in its formation. *Fgf10*, expressed initially in the ventral ov, then in both the anterior and posterior ov, and eventually in the developing sensory patches of the cochlear primordium (cp) and vestibular primordia (vp) acts through *Fgfr2b*, expressed in nonsensory tissue, and may be required for the initial (dorsal) outpouching of the semicircular canal plates but is not required for (ventral) cochlear development. (D) FGF signaling in the proliferation and differentiation of the organ of Corti (oc). Sequential diagrams depict sections taken through the developing cochlear duct in which signaling through FGFR1, activated by an unknown ligand, stimulates proliferation of undifferentiated organ of Corti precursors (ocp). Notch signaling is then required for the cell fate decision separating hair cell precursors (hcp) from supporting cell precursors (scp). Cells in both the primitive oc and greater epithelial ridge (ger) express *Fgfr1*, whereas only cells in the primitive oc coexpress *Fgfr3*. Signaling through FGFR1, stimulated by an unknown ligand, is required for differentiation of the hcps to outer hair cells (ohc). Signaling through FGFR3, possibly stimulated by FGF8, is required for the differentiation of scps to pillar cells (pc). Other abbreviations: dc, Deiter's cell; ph; phalangeal cell.

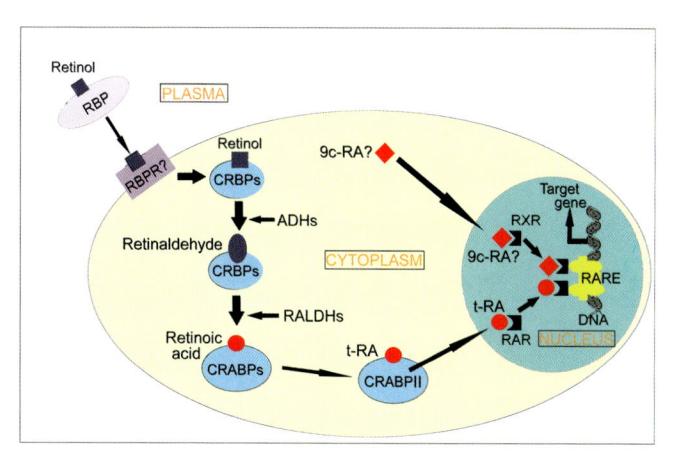

Chapter 9, Figure 1 Pathways for the synthesis and mechanism of action of retinoic acid. Retinol in plasma is bound to a retinol binding protein (RBP) and transported toward target cells. This complex might interact with a retinol binding protein receptor (RBPR?), facilitating the transfer of retinol into the cell. In the cytoplasm, retinol is bound to two molecules known as cellular retinol binding proteins (CRBPs). Retinol is converted into retinoic acid (RA) in two successive steps by two different sets of enzymes: aldehyde dehydrogenases (ADHs) and retinaldehyde dehydrogenases (RALDHs). Then, RA interacts with two cellular RA binding proteins (CRABPs) where the complex CRABP2-RA could be internalized into the nucleus. In the nucleus, RA effects are mediated by specific receptors that in turn modulate gene expression. Two families of nuclear receptors are involved: the retinoid acid receptors (RARs), activated both by *all-trans* RA (t-RA) and *9-cis* RA (9c-RA), and the retinoid X receptors (RXRs), activated exclusively by 9c-RA. RARs and RXRs act on RA response genes by heterodimerization in binding to specific sequences known as RA response elements (RAREs).

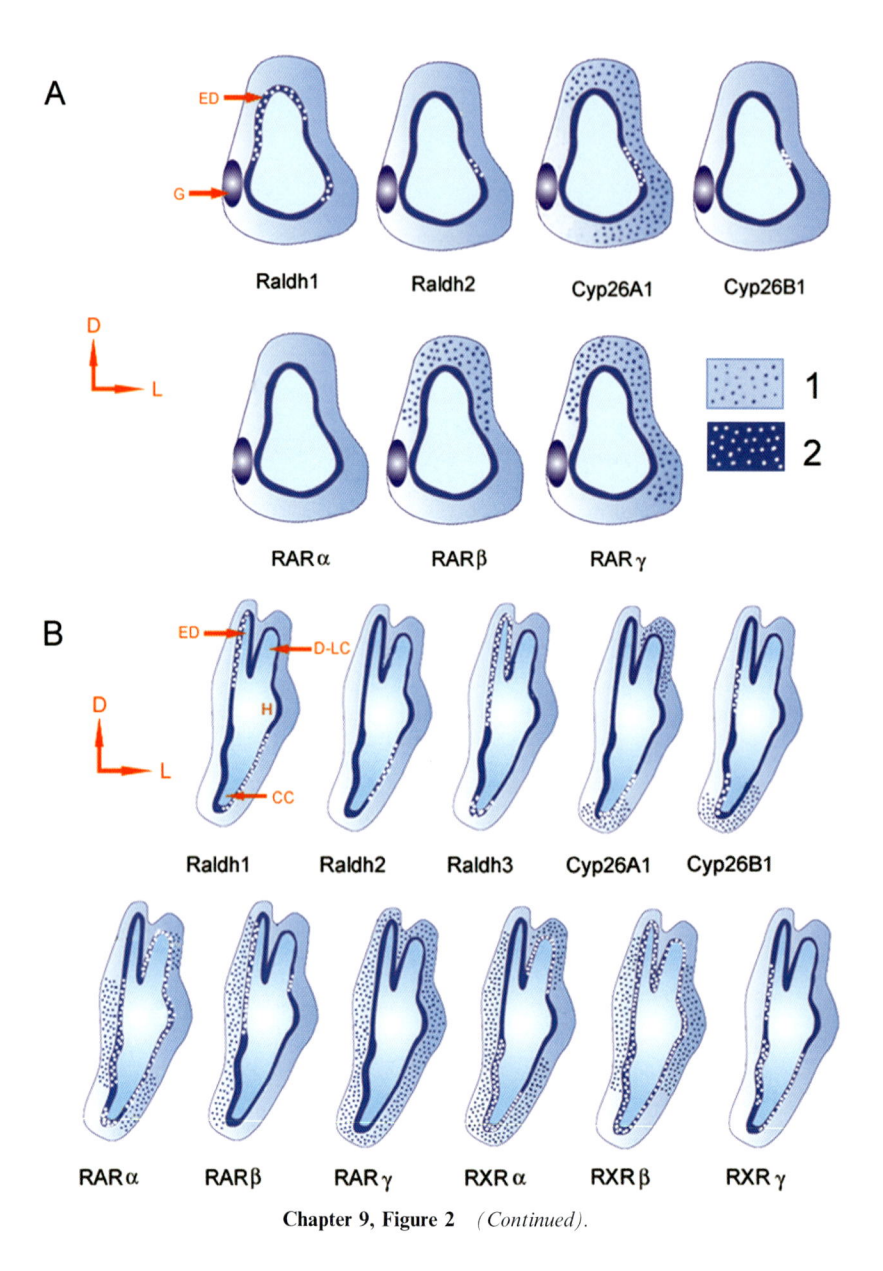

Chapter 9, Figure 2 *(Continued).*

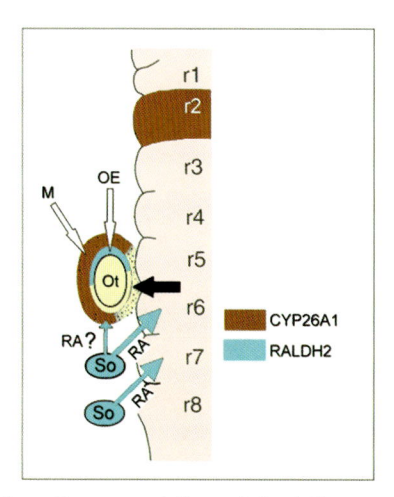

Chapter 9, Figure 4 Schematic representation of the influence of retinoid pathways on otocyst development. Two complementary pathways may be related to inner ear ontogenesis. One from outside the otocyst with retinoic acid (RA) synthesized by the retinaldehyde dehydrogenase 2 (RALDH2) from the somite mesoderme (So), which influences the posterior region of the hindbrain such as rhombomeres 5 and 6 (r5–r6) and also may be directly the otocyst (RA?). The catabolic enzyme CYP26A1 presents a complementary distribution in r2 that may contribute to the generation of a precise balance of RA into the posterior hindbrain. RA is known to be crucial for r5 and r6 specification through *hoxa1*, which in turn seems to be required indirectly for the normal development of the otocyst (black arrow) through a cascade of factors in which FGF3 might be involved. The second pathway is inside the presumptive inner ear where the synthetic and catabolic enzymes are present in a complementary fashion, respectively in the otic epithelium (OE) and the mesenchyme (M). It is possible that the retinoid pathway in the prospective inner ear is independent from outside otocyst RA and functions without the influence of neighboring regions.

Chapter 9, Figure 2 Distribution of synthetic and catabolic enzymes and retinoid receptor transcripts in the embryonic ear. (A) Transcript distribution of retinaldehyde dehydrogenase enzymes (*Raldh1*, *Raldh2*), catabolic enzymes (*Cyp26A1*, *Cyp26B1*), and RA receptors (RARα, RARβ, and RARγ) in E10.5 mouse. Synthetic enzyme transcripts (*Raldh1* and *2*) are mainly restricted to some regions of the otic epithelium whereas the catabolic enzyme (*Cyp26A1*) is largely present in the mesenchyme. Only two RA receptors (RARβ and RARγ) are detected at this stage of development and restricted to mesenchyme tissues mainly around the presumptive endocochlear duct (ED). VII–VIII ganglion complex (G). (B) E12.5 stage. *Raldh1* and *2* transcripts are present in the epithelium of the endolymphatic duct (ED) and the presumptive cochlear canal (CC). *Cyp26* gene expression is mainly observed in restricted regions of the ventral otic epithelium and mesenchyme tissues around the presumptive cochlear canal and restricted in the dorsoanterior semicircular canals for Cyp26A1. All retinoid receptors are abundant in the mesenchyme except RXRγ, which is restricted to the otic epithelium. RARs and RXRs can be observed in some regions of the otic epithelium, except RARγ, which is restricted to the surrounding mesenchyme. 1, Distribution of transcripts in mesenchyme and surrounding tissues of the otocyst; 2, distribution of transcripts in the otic epithelium.

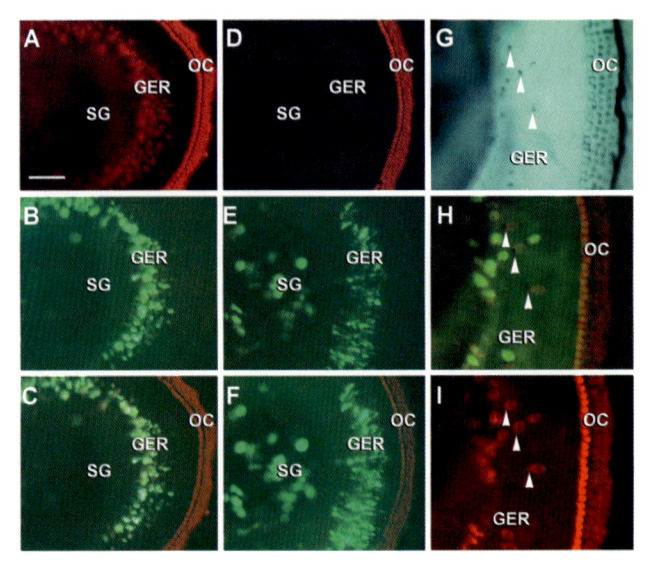

Chapter 10, Figure 1 Overexpression of *Math1* leads to robust production of extra hair cells in the GER of P0 rat cochlear explant cultures. (A–C) Myosin VIIa (red, A) and EGFP (green, B) double immunocytochemistry (double exposure, C) of a culture transfected with the pRK5–Math1–EGFP plasmid. (D–F) Myosin VIIa (red, D) and EGFP (green, E) double immunocytochemistry (double exposure, F) of a control culture transfected with the pRK5–EGFP plasmid. (G–I) Triple labeling (lectin labeling in G, myosin and EGFP double labeling in H, and myosin VIIa labeling in I) of a culture transfected with the pRK5–Math1–EGFP plasmid shows the presence of stereociliary bundles (arrowheads) in the hair cells produced in the GER region. GER, Greater epithelial ridge; OC, organ of Corti; SG, spiral ganglion. Bar: 60 μm for A–F; 30 μm for G–I. Modified with permission from *Nature* from Zheng, J., and Gao, W. Q. (2000). Overexpression of Math1 induces robust production of extra hair cells in postnatal rat inner ears. *Nat. Neurosci.* **3**, 580–586.

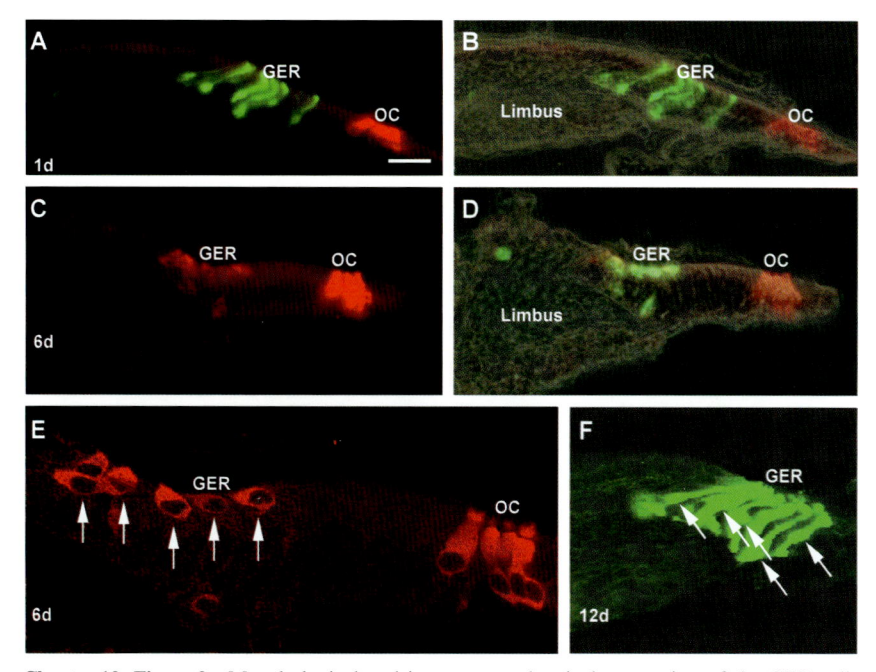

Chapter 10, Figure 2 Morphological and immunocytochemical conversion of the GER cells into hair cells. Cross sections of cultures at 1 day (A, B) and 6 days (C, D) after transfection with the pRK5–Math1–EGFP plasmid, respectively. (A, B) Myosin VIIa (red) and EGFP (green) double labeling (A) and superimposed image of A and phase contrast image from the same area, respectively, verify that the transfected cells are GER epithelial cells that display a radial pattern from the base to the roof and show that these cells are myosin VIIa negative at 1 day following transfection. The transfected cells are not located in the limbus. (C, D) Myosin VIIa labeling (C) and myosin VIIa, EGFP, and phase contrast superimposed image (D) show that the transfected GER cells have converted into hair cells at 6 days after transfection. (E) High magnification confocal image of anti-myosin VIIa antibody labeling shows the EHC in the GER and normal hair cells in the OC region in C. The arrows in E indicate the big nuclei in the bottom of the EHC. (F) High magnification confocal image of the EGFP-positive cells in the GER region of a culture transfected with the pRK5–EGFP plasmid at 12 days following transfection. The radial morphology of the transfected GER cells is retained throughout the 12-day culture period. The arrows in F indicate the small nuclei located in the middle of the radial GER cells. Bar: 40 μm for A–D, 20 μm for E–F. Modified with permission from *Nature* from Zheng, J., and Gao, W. Q. (2000). Overexpression of Math1 induces robust production of extra hair cells in postnatal rat inner ears. *Nat. Neurosci.* **3,** 580–586.

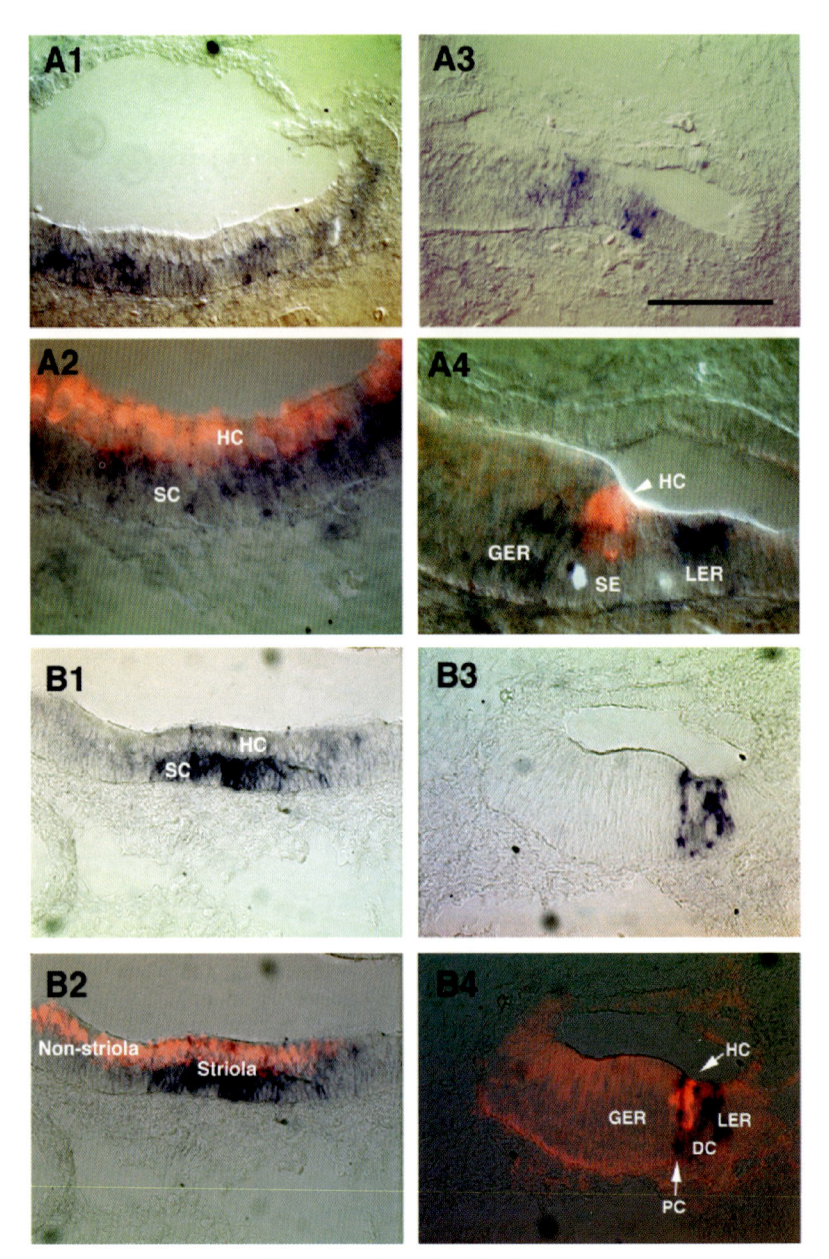

Chapter 10, Figure 6 *(Continued)*.

Chapter 10, Figure 6 Nonradioactive *Hes1* and *Hes5* RNA *in situ* hybridization labeling in the inner ear. (A) Low and high magnification images of *Hes1* expression in E17.5 rat utricular (A1, A2) and cochlear (A3, A4) sections, respectively. (A2, A4) Myosin VIIa immunocytochemical labeling (mediated by Texas-red conjugated secondary antibody) of the sections shown in A1, A3, respectively. Note that specific *Hes1* signals are seen in supporting cell (SC) layer but not hair cell (HC) layer of the utricular sensory epithelium. In the cochlea, *Hes1* signal is seen in the GER and LER cells but is minimal in the sensory epithelium (SE). Hair cells (red labeling in A2, A4) are devoid of *Hes1* signal. Hybridizing the sections with sense control probes under the same experimental conditions does not show any staining (data not shown). (B) *Hes5* expression in E17.5 rat utricular (B1, B2) and cochlear (B3, B4) sections, respectively. (B2, B4) Myosin VIIa immunocytochemical labeling (mediated by Texas-red secondary antibody) of the sections shown in B1 and B3, respectively. Note that specific *Hes5* signals are seen in Deiter's cells (DC), pillar cells (PC), and the LER cells in the cochlea (B4) and in striola supporting cells (SC) in the utricle (B2). Hair cells (red labeling in B2, B4), the connective tissue, and nonsensory epithelial cells (not shown) are devoid of *Hes5* signal. Bar: 50 µm for A1, A3, B1–B4; 25 µm for A2, A4. Modified with permission from Zheng, J., Shou, J., Guillemot, F., Kageyama, R., and Gao, W. Q. (2000). Hes1 is a negative regulator of inner ear hair cell differentiation. *Development* **127,** 4551–4560.

Chapter 10, Figure 7 (*Continued*).

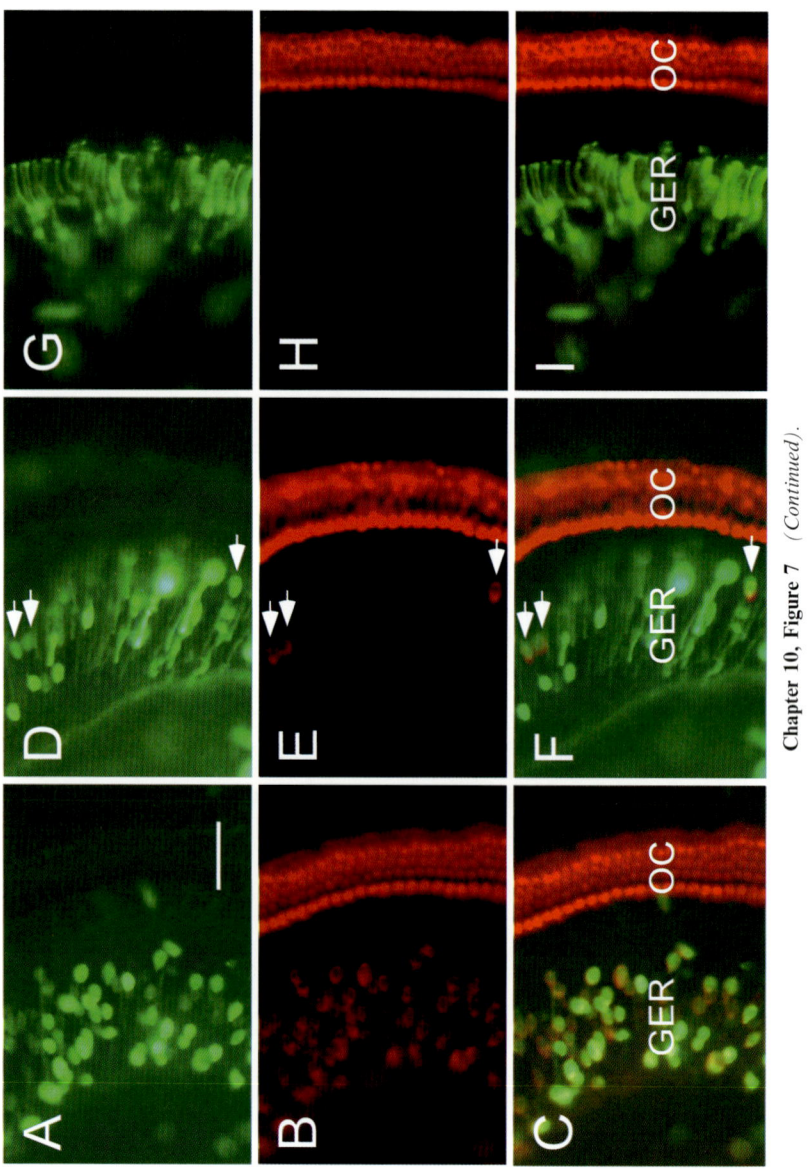

Chapter 10, Figure 7 *Hes1* blocks hair cell differentiation induced by *Math1*. EGFP (green, A, D, G) and myosin VIIa (red, B, E, H) double immunocytochemistry (double exposure, C, F, I) of the cultures transfected with the pRK5–Math1–EGFP plasmid (A–C), a mixture of equal amounts of pSVCMV–Hes1 and pRK5–Math1–EGFP plasmid (D–F) and a mixture of pSVCMV–Hes1 and pRK5–EGFP plasmid at a ratio of 5:1 (G–I), respectively. Note that whereas virtually all *Math1* transfected GER cells become hair cells (A–C), only a very small number of GER cells in the cultures cotransfected with *Hes1* and *Math1* (arrows in D–F) are able to differentiate into hair cells. All *Hes1* transfected GER cells remain myosin VIIa negative (G–I). Bar: 25 μm. Modified with permission from Zheng, J., Shou, J., Guillemot, F., Kageyama, R., and Gao, W. Q. (2000). Hes1 is a negative regulator of inner ear hair cell differentiation. *Development* **127,** 4551–4560.

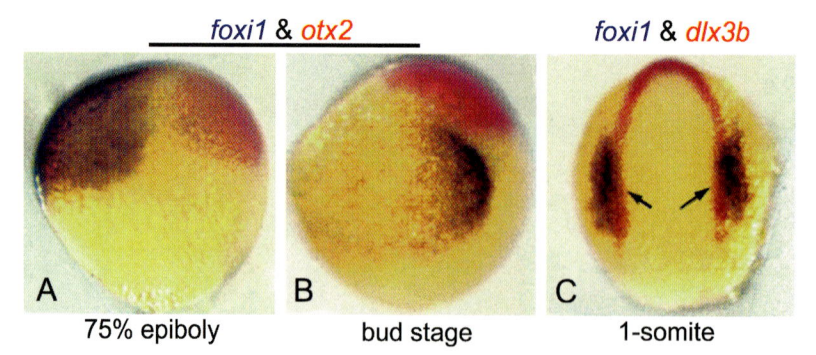

Chapter 12, Figure 1 Expression of *foxi1*. (A, B) Lateral views (anterior to the top, dorsal to the right) showing expression of *foxi1* (black) and the forebrain marker *otx2* (red). At 75% epiboly (A) *foxi1* is expressed uniformly in anteroventral ectoderm. By bud stage (B), *foxi1* has strongly upregulated in preotic placode and downregulated in ventral cells. (C) Dorsal view (anterior to the top) showing *foxi1* (black) and *dlx3* (red) at the 1- somite stage. Expression of *foxi1* fully overlaps the preotic domain of *dlx3*. Reprinted with permission from Solomon, K. S., Kudoh, T., Dawid, I. G., and Fritz, A. (2003). Zebrafish *foxi1* mediates otic placode formation and jaw development. *Development* **130**, 929–940.

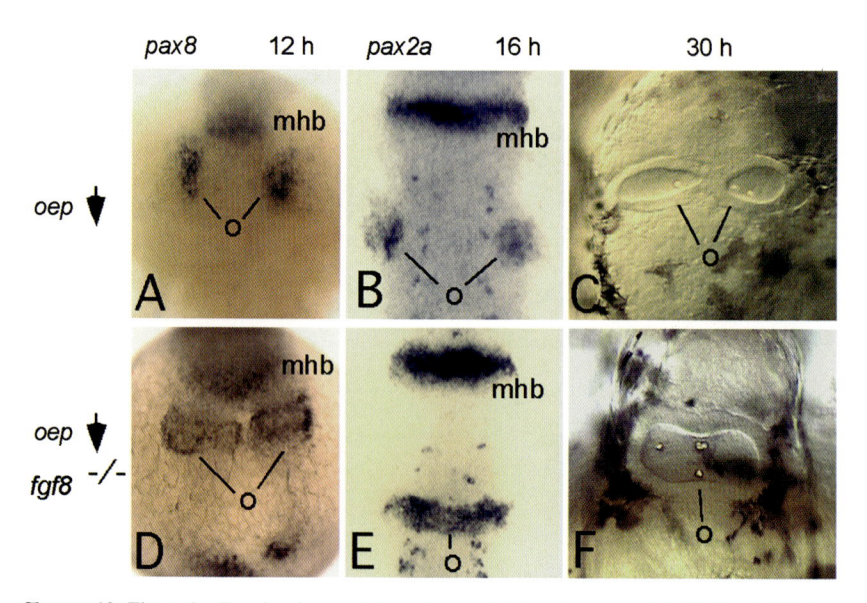

Chapter 12, Figure 2 Ear development in embryos deficient in *oep* and *fgf8*. (A–C) Wildtype embryos injected with *oep*-MO. Expression of *pax8* at 12 hpf (A) or *pax2a* at 16 hpf (B) marks the developing otic placodes (o). (C) At 30 hpf, otic vesicles are well formed but often elongate medially and touch at the midline. (D–F) *ace*/+ intercross progeny injected with *oep*-MO. About 25% of injected progeny showed dramatic changes in gene expression and otic vesicle morphology. These are inferred to be *ace* (*fgf8*) mutants. At 12 h, the otic domain of *pax8* forms bilateral transverse bands that nearly touch at the midline (D). By 16 h, *pax2a* is expressed in a contiguous transverse stripe through the hindbrain (E). At 30 h, a single large otic vesicle forms at the midline and fully spans the width of the hindbrain (F). All images are dorsal views with anterior to the top. mhb, midbrain–hindbrain border; o, otic placode or vesicle.

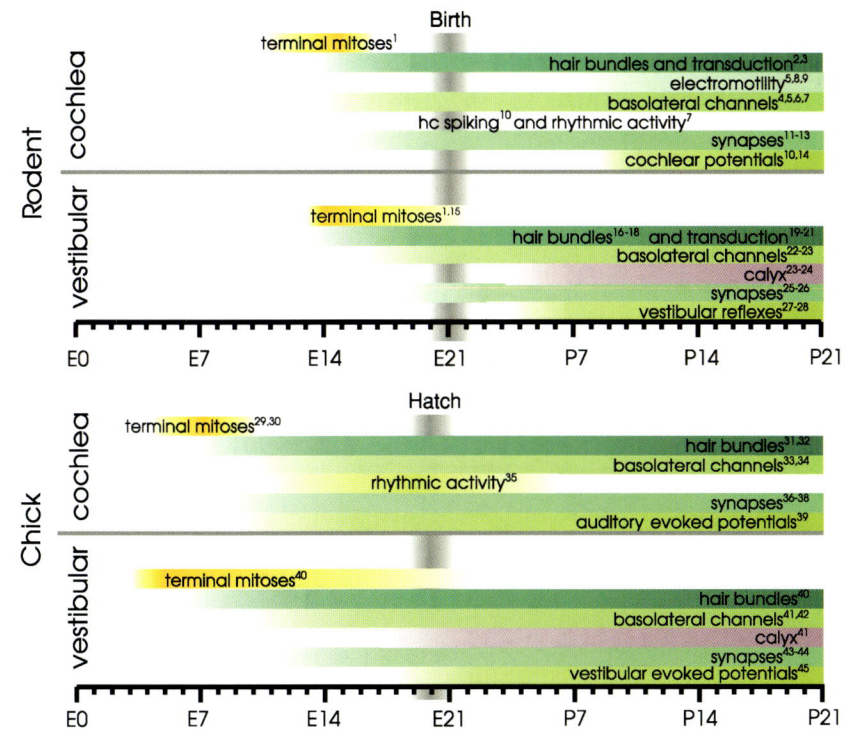

Chapter 13, Figure 2 Approximate timing of various developmental changes in cochleas and vestibular organs of rodents and chicks: *hair bundle* development; *terminal mitoses* of hair cells and ganglion cells (primary afferent neurons); differentiation of *basolateral channels* in hair cells; acquisition of *electromotility* in mammalian outer hair cells; bursting or *rhythmic activity* of primary afferents and/or higher-order auditory neurons; hair cell *(hc) spiking*; formation of afferent and efferent *synapses* on hair cells and primary afferent terminals; and functional measures: *cochlear potentials*, *vestibular reflexes*, *auditory evoked potentials*, and *vestibular evoked potentials*. References: *Rodent cochlea*: [1]Ruben, 1967; [2]Kros et al., 1992; [3]Holley and Nishida, 1995; [4]Kros et al., 1998; [5]Marcotti and Kros, 1999; [6]Marcotti et al., 1999; [7]Marcotti et al., 2003; [8]He, 1997; [9]Beurg et al., 2001; [10]reviewed in Rübsamen and Lippe, 1998; [11]Echteler et al., 1989; [12]Pujol et al., 1998; [13]Bruce et al., 2000; [14]Mikaelian and Ruben, 1965. *Rodent vestibular organs*: [15]Sans and Chat, 1982; [16]Mbiene et al., 1984; [17]Mbiene and Sans, 1986; [18]Denman-Johnson and Forge, 1999; [19]Géléoc et al., 1997; [20]Holt et al., 1997; [21]G. S. G. Géléoc and J. R. Holt, personal communication; [22]Géléoc et al., 2003; [23]reviewed in Anniko, 1990; [24]Rüsch et al., 1998b; [25]Desmadryl and Sans, 1990; [26]Dechesne et al., 1997; [27]Parrad and Cottereau, 1977; [28]Curthoys, 1979. *Chick cochlea*: [29]Katayama and Corwin, 1989; [30]Adam et al., 1998; [31]Cohen and Fermin, 1978; [32]Tilney et al., 1992; [33]Griguer and Fuchs, 1996; [34]Fuchs and Sokolowski, 1990; [35]Jones et al., 2001; [36]Sokolowski and Cunningham, 1999; [37,38]Whitehead and Morest, 1985a,b; [39]Saunders et al., 1973. *Chick vestibular*: [40]Goodyear et al., 1999; [41]Masetto et al., 2000; [42]Masetto et al., 2003; [43]Ginzberg and Gilula, 1980; [44]Meza and Hinojosa, 1987; [45]Jones and Jones, 2000a.

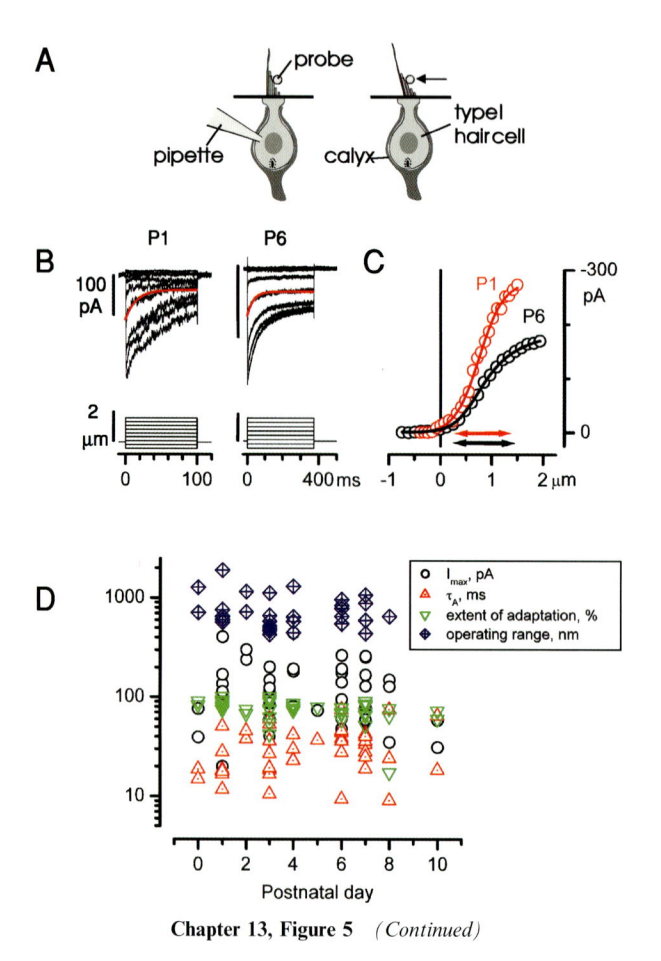

Chapter 13, Figure 5 *(Continued)*

Chapter 13, Figure 5 Transduction properties in mouse utricular hair cells do not appear to change systematically during the first postnatal week (M. A. Vollrath and R. A. Eatock, unpublished results). (A) Bundles were deflected along their sensitive axes by moving a rigid glass probe against the back of the bundle (the probe is shown end on); step deflections had a rise time of 1 ms. Transduction currents were recorded in whole-cell voltage clamp mode with a glass pipette. A deflection toward the tall edge, as shown, produces an inward transduction current. The cells were bathed in a standard high-Na^+ saline and the pipette contained a standard high-K^+ solution, with 10 mM EGTA. (B) Transduction currents evoked by families of step bundle deflections, in hair cells in a P1 and a P6 utricle, respectively. Stimulus amplitudes (*lower panels*) are given as displacements of the probe (and bundle) parallel to the plane of the epithelium. Scale bars apply to both P1 and P6 data. The half-maximal responses were fit by a single-exponental function (*red lines*), with time constants (τ_A) of 15 ms (P1) and 34 ms (P6). This difference in the time constants is not reflective of a trend with age, as shown in D. The currents from the younger cell look noisier because they were not averaged. (C) Current-displacement relations, obtained by plotting peak currents in B as functions of bundle displacement from the resting position. The data are fit by second-order Boltzmann functions. The lines at the bottom show the *operating ranges* for each cell: displacements corresponding to 10–90% of the maximal response. (D) The *maximum current* (I_{max}); *adaptation time constant* at half-maximal stimulation (τ_A, obtained as shown in A; *extent of adaptation* expressed as a percentage $[100 \times (I_{max} - I_{steady-state})/I_{max}]$; and *operating range*, as functions of postnatal age. Variability does not appear to be systematic with age in this period, but recordings were not taken from immature-looking bundles (e.g., bundles in the first two columns in Fig. 4).

Science
QL 951
.C8
vol.57

QL951 .C8
vol.57
Current topics in
developmental biology

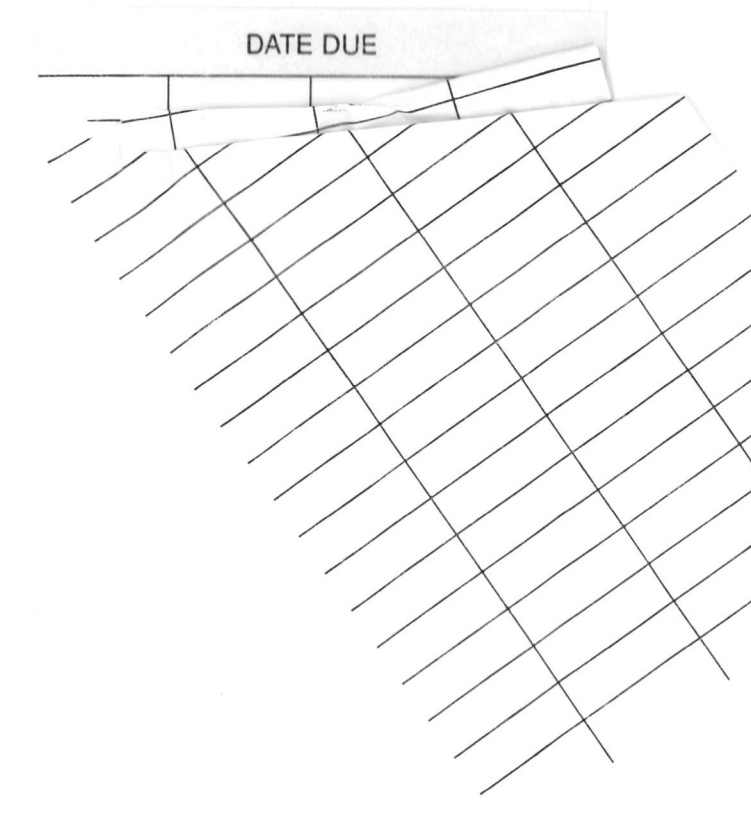

DATE DUE